Topological Methods for Set-Valued Nonlinear Analysis

Topological Methods for Set-Valued Nonlinear Analysis

ENAYET U TARAFDAR
The University of Queensland, Australia

&

MOHAMMAD S R CHOWDHURY
Lahore University of Management Sciences, Pakistan

World Scientific

NEW JERSEY • LONDON • SINGAPORE • BEIJING • SHANGHAI • HONG KONG • TAIPEI • CHENNAI

Published by

World Scientific Publishing Co. Pte. Ltd.

5 Toh Tuck Link, Singapore 596224

USA office: 27 Warren Street, Suite 401-402, Hackensack, NJ 07601

UK office: 57 Shelton Street, Covent Garden, London WC2H 9HE

British Library Cataloguing-in-Publication Data
A catalogue record for this book is available from the British Library.

TOPOLOGICAL METHODS FOR SET-VALUED NONLINEAR ANALYSIS

ISBN-13 978-981-270-467-2
ISBN-10 981-270-467-1

Printed in Singapore by World Scientific Printers

To

Our beloved professor Dr. Eneyet Ullah Tarafdar, his wife — Mrs. Sangha Mitra, his daughter — Hashina Tarafdar and his two sons — Abir Tarafdar and Ashique Tarafdar.

Preface

This book is a monograph of a significant and recent publications in non-linear analysis involving set-valued mappings. A map $T : X \to 2^Y$ is said to be a set-valued mapping if for each $x \in X$, $T(x) \subset Y$.

We need analysis, topology and geometry, i.e., a mixture of these three fields, in studying the theory of set-valued mappings. There have been a significant number of publiactions in this area of research over the last 40 years. These have become possible because there are huge applications in the fileds of Physics, Biology, Control Theory, Optimization, Economics and Game Theory.

We shall cover the following topics in this book: contraction mappings, fixed point theorems, minimax inequalities, end points, variational inequalities, generalized variational inequalities, and generalized quasi-variational inequalities, equilibrium analysis in economics, best approximation and fixed point theorems, topological degree theory,and non-expansive types of mappings and fixed point theorems.

In Chapter 5, we shall present variational inequalities, quasi-variational equalities and equilibrium analysis in economics. We have applied the topological methods to study the equilibrium analysis in economics. We shall discuss them in more details in the Introduction Chapter. In Chapter 6, we shall discuss best approximation and fixed point theorems for set-valued mappings in topological vector spaces. Finally, in Chapters 7 and 8 we shall present some aspects of degree theories for set-valued mappings and non-expansive types of mappings and fixed point theorems in locally convex topological vector spaces.

We are very much grateful to Professor Dr. Ken Smith at the Dept. of Mathematics of the University of Queensland for his tremendous help in making this publication possible by compiling the manuscript into Latex format. We are also thankful to Dr. Bevan Thompson of the same department for all his administrative help and encouragement in completing this project.

After the sudden and unfortunate death of Dr. Enayet Tarafdar in November, 2002, I continued with the project and tried to finish it with the help of my other friends and well wishers who were working in this area of research. In this direction, I would like to mention the names of Dr. George Yuan and Dr. Peter Watson who tried to help me in finishing this project with their valuable suggestons and inputs.

Thanks goes to Dr. Yuan for partially completing the Promotional Questionnaire of this book before its publication.

Finally, my heartfelt gratitude and thanks goes to Mrs. Sangha Mitra — wife of Dr. Tarafdar, Hashina Tarafdar — daughter of Dr. Tarafdar and his two sons — Abir Tarafdar and Ashique Tarafdar and also to my wife — Mrs. Fowzia Akhter. Abir Tarafdar helped in getting the back up of the original soft copy of this project of Dr. Tarafdar. Mrs. Sangha Mitra, Hashina Tarafdar and Ashique Tarafdar helped in searching and collecting some of the documents left at their home which were very helpful for the completion of Dr. Tarafdar's unfinished project. My wife, Fowzia Akhter, helped me invaluably in completing the typing of some chapters of this project.

Lastly, I am very happy to acknowledge that Professor Dr. Zakaullah Khan, the Previous Head of the Dept. of Mathematics of the International Islamic University in Islamabad, Pakistan — my previous work place, helped me so generously that it created an environment for me to complete the proof-reading and final revision of the manuscript of this project for submission to the World Scientific U.K. within the specific deadline of August, 2007.

Mohammad Showkat Rahim Chowdhury
Associate Professor
Centre for Advanced Studies in Mathematics
Lahore University of Management Sciences (LUMS)
Lahore-54792, Pakistan
October 10, 2007

Contents

Chapter 1

Introduction

Our main objective in this book is to study some aspects of non-linear analysis which involve set-valued mappings. However, a single valued mapping $T : X \to Y$ of a non-empty set X into a non-empty set Y can be regarded as a set-valued mapping by considering one point $\{T(x)\}$ for each each $x \in X$.

The various aspects of fixed points, minimax inequalities, end points, variational inequalities, generalized variational inequalities, and generalized quasi-variational inequalities, equilibrium analysis in economics, best approximation and fixed point theorems, topological degree theory,and non-expansive types of mappings and fixed point theorems, and related topics are considered in this book.

It is well known that fixed point theory is very important in mathematics. The close relationship between fixed point theory and mathematical economics can be illustrated in many ways. The usefullness of Brouwer's fixed point theorem was recognized by John Von Neumann when he developed the foundations of game theory in 1928.

Fixed point and coincedence theorems for set-valued mappings and their applications to minimax theorems and economics originated from the works of John Von Neumann (Neumann (1937)) (see also Neumann (1928b), Neumann (1928a), Neumann and Morgenstern (1944) and Neumann and Morgenstern (1947)). Then the theory was advanced by Kakutani (1941), Fan (1952) and others (see Zeidler's book (Zeidler (1985))).

In most of the economic papers appearing in any journals of economics, one can find the terms economic equilibria, Pareto optimum in abundance. Pareto talked about the optimum which has come to be known popularly as Pareto Optimum (Pareto allocation). In the last century, new discipline called mathematical economics — has evolved into a highly developed and fast growing branch of mathematics blended with the components of economy, games, econometrics, psychology and many related areas.

In fact, in a fascinating article Franklin (1983) (incidentally has a book, see Franklin (1980)) wrote: In 1969 a spokesman for the Nobel foundation welcomed the new prize subject, economics, as "the oldest of the arts, the youngest of the sciences". It might be fair to say that economics became a science when it started making significant use of mathematics.

In this book, we have applied the topological methods to study the equilibrium analysis in economics, i.e., to prove the existence of equilibrium of social economics. It seems that in this area nothing dominates more significantly than fixed point theory of set-valued mappings. In fact, Nobel laureate Debreu (1959) proved two fundamental theorems of mathematical economics by using Kakutani's fixed point theorem.

Let E be a topological vector space and A a non-empty subset of E. If $S, T : A \to 2^E$ are correspondences, then $T \cap S : A \to 2^E$ is a correspondences defined by $(T \cap S)(x) = T(x) \cap S(x)$ for each $x \in A$.

Ding, Kim and Tan introduced the notions of correspondences of class $\mathcal{L}_\theta^*$, $\mathcal{L}_\theta^*$-majorant of ϕ at x and $\mathcal{L}_\theta^*$- majorized correspondences in Ding, Kim, and Tan (1992) as follows:

Let X be a topological space, Y be a non-empty subset of a vector space $E, \theta : X \to E$ be a map and $\phi : X \to 2^Y$ be a correspondence. Then (1) ϕ is said to be of class $\mathcal{L}_\theta^*$ if for every $x \in X$, $con\phi(x) \subset Y$ and $\theta(x) \notin con\phi(x)$ and for each $y \in Y$, $\phi_{-1}(y) = \{x \in X : y \in \phi(x)\}$ is open in X; (2) a correspondence $\phi_x : X \to 2^Y$ is said to be an $\mathcal{L}_\theta^*$ majorant of ϕ at x if there exists an open neighborhood N_x, of x in X such that (a) for each $z \in N_x$, $\phi(z) \subset \phi_x(z))$ and $\theta(z) \notin con\phi_x(z)$ (b) for each $z \in X$, $con\phi_x, (z) \subset Y$ and (c) for each $y \in Y, \phi_1(y)$ is open in X; (3) ϕ is $\mathcal{L}_\theta^*$-majorized if for each $x \in X$ with $\phi(x) \neq \emptyset$, there exists an $\mathcal{L}_\theta^*$-majorant of ϕ at x.

In view of Yannelis and Prabhakar (1983, p. 239, Lemma 5.1), Ding, Kim and Tan's notions of the correspondence ϕ being of class $\mathcal{L}_\theta^*$ or $\mathcal{L}_\theta^*$-majorized generalize the notions $\phi \in \mathcal{C}(\mathcal{X}, \mathcal{Y}, \theta)$ or $\mathcal{C}$-majorized respectively which were introduced by Tulcea (1986, p. 2). Ding, Kim and Tan pointed out that their map $\theta : X \to E$ is less restrictive than that of [Tulcea (1986)], where $\theta : X \to Y$. In most applications, either (I) X and Y are non-empty subsets of the same topological vector space E and $\theta(x) = x$ for all $x \in X$, or (II) $X = \Pi_{i=I} X_I$ and $\theta(x) = \pi_j(x)$ for all $x \in X$, where $\pi_j : X \to X_j$ is the projection of X onto X_j and X_j and Y are non-empty subsets of the same topological vector space E.

Ding, Kim and Tan observed that when $X = Y$ and is convex (and $\theta(x) = x$ for all $x \in X$), the notion of correspondence of class $\mathcal{L}_\theta^*$ coincides with the notion of correspondence of class $\mathcal{L}$ introduced by [Yannelis and Prabhakar (1983)] and the notions of $\mathcal{L}_\theta^*$-majorant of ϕ at x and $\mathcal{L}_\theta^*$-majorized correspondence generalize the notions of $\mathcal{L}$-majorant of ϕ at x and $\mathcal{L}$-majorized correspondence respectively also introduced by [Yannelis and Prabhakar (1983)]. In the special case (I), where $\theta = 1_x$, the identity map on X or (II), where $\theta = \pi_j$, $\mathcal{L}^*$ is written in place of $\mathcal{L}_\theta^*$ if there is no ambiguity.

It should be noted that if ϕ is $\mathcal{L}_\theta{}^*$-majorized, then for $x \in X, \theta(x) \notin con\phi(x)$ and $con\phi(x) \subset Y$.

Let I be a (possibly infinite) set of agents. For each agent $i \in I$, let its choice set or strategy set X_i be a non-empty set in a topological vector space. Let $X = \Pi_{i\in I}X_I$. If $i \in I$, let $\pi_i : X \to X_i$ be the projection of X onto X_I and for $x \in X$, let x_i denote the projection $\pi_i(x)$ of x on X_i. Let $P_i : X \to 2^{x_i}$ be an *irreflexive* preference correspondence, i.e., $x_i \notin P_i(x)$ for all $x \in X$. Following [Gale and Mas-Colell (1978)], the collection $(X_i, P_i)_{i\in I}$ will be called a *qualitative game*. A point $\hat{x} \in X$ is said to be an *equilibrium* of that game if $P_i(\hat{x}) = \emptyset$ for all $i \in I$. For each $i \in I$, let A_i be a non-empty subset of X; if $i \in I$ is arbitrarily fixed, we define

$$\Pi_{j\neq I, j\in i}A_I \oplus A_I = \{x = (x_k)_{k\in I} \in X : x_k \in A_k \text{ for each } k \in I\}$$

Let I be a (finite or an infinite) set of agents. An *abstract economy* $\Gamma = (X_i, A_i, B_i, P_i)_{i\in I}$ is defined as a family of ordered quadruples (X_i, A_i, B_i, P_i), where X_i is a topological space, $A_i : \Pi_{j\in I}X_j \to 2^{X_i}$ and $B_i\Pi_{j\in I}X_j \to 2^{X_i}$ are constraint correspondences and $P_i\Pi_{j\in I}X_j \to 2^{X_i}$ is a preference correspondence. An *equilibrium* for Γ is a point $\hat{x} \in X = \Pi_{i\in I}X_i$ such that for each $i \in I, \hat{x}_i \in clB_i(\hat{x})$ and $A_i(\hat{x}) \cap P_i(\hat{x}) = \emptyset$. When $A_i = B_i$ for each $i \in I$, our definitions of an abstract economy and an equilibrium coincide with the standard definitions; e.g., in Borglin and Keiding (1976, p. 315), or in Yannelis and Prabhakar (1983, p. 242).

In the following chapters, if E is a topological vector space, we shall denote the dual space of E, i.e. the vector space of all continuous linear functionals on E, by E^* and the pairing between E^* and E by $\langle w, x\rangle$ for each $w \in E^*$ and $x \in E$, and by $Re\langle w, x\rangle$ the real part of the pairing between $w \in E^*$ and $x \in E$. Unless otherwise stated, if A is a subset of E, we shall denote by 2^A the family of all non-empty subsets A and by clA the closure in E, and by coA, the convex hull of A. Also, we shall denote by $\mathcal{F}(\mathcal{A})$ the family of all non-empty finite subsets of A, by $\mathbb{R}$ the set of all real numbers and $\mathbb{R}^+ = \{r \in \mathbb{R} : r > 0\}$.

Let E be a topological vector space. For each $x_0 \in E$, each non-empty subset A of E and each $\epsilon > 0$, let $W(x_0; \epsilon) := \{y \in E^* : |\langle y, x_0\rangle| < \epsilon\}$ and $U(A; \epsilon) := \{y \in E^* : \sup_{x\in A} |\langle y, x\rangle| < \epsilon\}$. Let $\sigma\langle E^*, E\rangle$ be the topology on E^* generated by the family $\{W(x; \epsilon) : x \in E \text{ and } \epsilon > 0\}$ as a sub-base for the neighbourhood system at 0 and $\delta\langle E^*, E\rangle$ be the topology on E^* generated by the family $\{U(A; \epsilon) : A$ is a non-empty bounded subset of E and $\epsilon > 0\}$ as a base for the neighbourhood system at 0. We note that E^*, when equipped with the topology $\sigma\langle E^*, E\rangle$ or the topology $\delta\langle E^*, E\rangle$, becomes a locally convex Hausdorff topological vector space. Furthermore, for a net $\{y_\alpha\}_{\alpha\in\Gamma}$ in E^* and for $y \in E^*$, (i) $y_\alpha \to y$ in $\sigma\langle E^*, E\rangle$ if and only if $\langle y_\alpha, x\rangle \to \langle y, x\rangle$ for each $x \in E$ and (ii) $y_\alpha \to y$ in $\delta\langle E^*, E\rangle$ if and only if $\langle y_\alpha, x\rangle \to \langle y, x\rangle$ uniformly for $x \in A$ for each non-empty bounded subset A of E. The topology $\sigma\langle E^*, E\rangle$ (respectively, $\delta\langle E^*, E\rangle$) is called the *weak* topology* (respectively, the *strong topology*) on E^*. If $p \in E$, $\hat{p}$ is the linear functional on E^* defined by $\hat{p}(f) = f(p)$ for each $f \in E^*$.

Let X be a non-empty subset of E. Then X is a *cone* in E if X is convex and $\lambda X \subset X$ for all $\lambda \geq 0$. If X is a cone in E, then $\widehat{X} = \{w \in E^* : Re\langle w, x\rangle \geq 0 \text{ for all } x \in X\}$ is also a cone in E^*, called the *dual cone* of X.

We shall now state a result of S. C. Fang (e.g. see [Chan and Pang (1982)] and [Shih and Tan (1986), p. 59]) with a little modification, as follows, made in Lemma 2.4.2 in Tan (1994):

Lemma 1.1 *Let X be a cone in a Hausdorff topological vector space E and $T : X \to 2^{E^*}$ be a map. Then the following statements are equivalent:*
(a) There exist $\hat{y} \in X$ and $\hat{w} \in T(\hat{y})$ such that $Re\langle \hat{w}, \hat{y} - x\rangle \leq 0$ for all $x \in X$.
(b) There exist $\hat{y} \in X$ and $\hat{w} \in T(\hat{y})$ such that $Re\langle \hat{w}, \hat{y}\rangle = 0$ and $\hat{w} \in \widehat{X}$.

Let $y \in E$. Then the *inward set* of y with respect to X is the set $I_X(y) = \{x \in E : x = y + r(u - y)$ for some $u \in X$ and $r > 0\}$. We shall denote by $\overline{I_X(y)}$ the closure of $I_X(y)$ in E.

Let X and Y be topological spaces and $T : X \to 2^Y$. Then T is said to be:

upper (respectively, *lower*) *semicontinuous* at $x_0 \in X$ [Berge (1963), p. 109] if for each open set G in Y with $T(x_0) \subset G$ (respectively, $T(x_0) \cap G \neq \emptyset$), there exists an open neighbourhood U of x_0 in X such that $T(x) \subset G$ (respectively, $T(x) \cap G \neq \emptyset$) for all $x \in U$;

upper (respectively, *lower*) *semicontinuous* on X if T is upper (respectively, lower) semicontinuous at each point of X.

Moreover, T is said to be continuous on X if it is both upper semi-continuous and lower semi-continuous on X.

Let X be a non-empty subset of E and $T : X \to 2^{E^*}$. Then T is said to be:

(*i*) *monotone* (on X)[Browder (1976), p. 79] if for each $x, y \in X$, each $u \in T(x)$ and each $w \in T(y)$, $Re\langle w - u, y - x\rangle \geq 0$;

(*ii*) *semi-monotone* [Bae, Kim, and Tan (1993), pp. 236–237] (on X) if for each $x, y \in X$, $\inf_{u\in T(x)} Re\langle u, y - x\rangle \leq \inf_{w\in T(y)} Re\langle w, y - x\rangle$.

It is clear that if T is monotone, then T is semi-monotone. The converse is in general false, see Example 2 in [Bae et al. (1993)].

A real-valued function $\psi : X \to \mathbb{R}$ defined on a convex subset X of E is said to be *quasi-concave* if for every real number α the set $\{x \in X : \psi(x) > \alpha\}$ is convex.

If X is a topological space and $\{U_\alpha : \alpha \in \mathcal{A}\}$ is an open cover for X, then a partition of unity subordinated to the open cover $\{U_\alpha : \alpha \in \mathcal{A}\}$ is a family $\beta_\alpha : \alpha \in \mathcal{A}\}$ of continuous real-valued functions $\beta_\alpha : X \to [0, 1]$ such that
(1) $\beta_\alpha(y) = 0$ for all $y \in X\backslash U_\alpha$,
(2) {support $\beta_\alpha : \alpha \in \mathcal{A}$} is locally finite and
(3) $\Sigma_{\alpha\in\mathcal{A}}\beta_\alpha(y) = 1$ for each $y \in X$.

Let X be a non-empty subset of a topological vector space E and $T : X \to 2^{E^*}$ be a map. Then the generalized variational inequality problem associated with X and T is to find $\hat{y} \in X$ such that generalized variational inequality $\sup_{w\in T(\hat{y})} Re\langle w, \hat{y} - x\rangle \leq 0$ for all $x \in X$ holds, or to find $\hat{y} \in X$ and $\hat{w} \in T(\hat{y})$ such that $Re\langle \hat{w}, \hat{y} - x\rangle \leq 0$ for all $x \in X$ holds. When T is single-valued, a generalized variational inequality is called a variational inequality. Browder (1965b) and Hartman and Stampacchia

(1966) first introduced variational inequalities. Since then, there have been many generalizations, e.g., see Allen (1977), Bae et al. (1993), Baiocchi and Capelo (1984), H. Brézis and Stampacchia (1972), Browder (1968), Dugundji and Granas (1978), Kinderlehrer and Stampacchia (1980), Shih and Tan (1984), Shih and Tan (1988a), Shih and Tan (1988c), Tan (1983) and Yen (1981), etc.

Let X and Y be subsets of a vector space E such that $co(X) \subset Y$. Then $F : X \longrightarrow 2^Y$ is called a KKM-map if for each $A \in \mathcal{F}(X)$, $co(A) \subset \cup_{x\in A}F(x)$. Note that if F is a KKM-map, then $x \in F(x)$ for all $x \in X$.

In this research monograph, we shall use H to denote a Hilbert space with inner product $\langle\ ,\ \rangle$ and its corresponding induced norm $\|\cdot\|$.

If X is a non-empty subset of H, we shall denote by $\partial_H(X)$ the boundary of X in H. We shall denote by $bc(H)$ the family of all non-empty bounded closed subsets of H. If $x \in H$ and $r > 0$, let $B_r(x) = \{y \in H : \|x - y\| < r\}$.

Let K be a non-empty closed convex subset of H. For each $x \in H$, there is a unique point $\pi_K(x)$ in K such that

$$\|x - \pi_K(x)\| = \inf_{z\in K} \|x - z\|.$$

$\pi_K(x)$ is called the projection of x on K and is characterized as follows [Kinderlehrer and Stampacchia (1980), Theorem 1.2.3, p. 9]:

Proposition 1.1 *Let K be a non-empty closed convex subset of H. Then for each $x \in H$ and $y \in K$, $y = \pi_K(x)$ if and only if $Re\langle x - y, z - y\rangle \leq 0$ for all $z \in K$.*

Suppose that E and F denote two vector spaces over a scalar field Φ (either the real field or the complex field).

Now, suppose that X is a nonempty subset of E and that $\langle\ ,\ \rangle : F \times E \to \Phi$ is a bilinear functional.

Suppose E, F and X are as above. Let $T : X \to 2^F$ be a set-valued mapping, $f : X \to F$ and $\eta : X \times X \to E$ be two single-valued mappings, and $h : X \times X \to \mathbb{R}$ be a real-valued function.

The generalized variational-like inequality (in short, $GVLI(T, \eta, h, X, F)$) is: find $\hat{y} \in X$ and $\hat{w} \in T(\hat{y})$ such that

$$\langle \hat{w}, \eta(\hat{y}, x)\rangle + h(\hat{y}, x) \leq 0, \quad \forall x \in X. \tag{1.1}$$

The generalized variational-like inequalities were used by Chowdhury and Tan in the simplest form in Chowdhury and Tan (1996) using the name of the operators as generalized variational inequalities (GVI). Later, Tarafdar and Ding gave a generalization of these GVI and called these operators as generalized variational-like inequalities (GVLI) in their paper in Ding and Tarafdar (2000).

The $GVLI(T, \eta, h, X, F)$ given in (1.1) includes various variational inequalities studied in Chan and Pang (1982), Chowdhury and Tan (1996), Ding and Tarafdar (1996), Ding and Tarafdar (1994), Ding and Tarafdar (1995), Fang and

Peterson (1982), Harker and Pang (1990), Jou and Yao (1993), Noor (1988), Noor (1992), Noor (1995), J. Parida and Kumar (1989),and Rockafeller (1970), as special cases. In particular, when $\eta(\hat{y}, x) = \hat{y} - x$ and there exists $h' : X \to \mathbb{R}$ such that $h(\hat{y}, x) = h'(\hat{y}) - h'(x)$ then the GVLI (1.1) reduces to Chowdhury and Tan's GVLI in Chowdhury and Tan (1996).

Suppose that X is a nonempty subset of a topological vector space E. A function $\phi : X \times X \to \mathbb{R} \cup \{\pm\infty\}$ is said to be 0-diagonally concave (in short, 0-DCV) in the second argument Zhou and Chen (1988), if for any finite set $\{x_1, \cdots, x_n\} \subset X$ and any $\lambda_i \geq 0$, $\sum\limits_{i=1}^{n} \lambda_i = 1$, we have $\sum_{i=1}^{n} \lambda_i \phi(y, x_i) \leq 0$ where $y = \sum\limits_{i=1}^{n} \lambda_i x_i$.

An *ordered set* is a non-empty set X with a binary relation $\leq$ defined on it that is reflexive, transitive and anti-symmetric.

A *lattice* is an ordered set such that $\sup\{x, y\}$ and $\inf\{x, y\}$ exist for each pair, x, y in X.

An *ordered vector space* $(L, \leq)$ is a vector space L over the reals such that L is an ordered set and $f \leq g$implies $f + h \leq g + h$ for all h in L and $\alpha f \leq g$ for all $\alpha \geq 0$.

An ordered vector space L which is also a lattice is said to be a *Riesz space*. The set $L^+ = \{f \in L / f \geq 0\}$ is called the *positive cone* of L.

Let L be a Riesz space. Then for $f \in L$ we put $f^+ = f \vee 0, f_- = (-f) \vee 0$ and $|f| = f \vee (-f)$, where $x \vee y$ is the supremum of the two elements x and y.

A linear functional $f : L \to \mathbb{R}$ is said to be *order-bounded* whenever f maps order-intervals of the form $[-u, u] = \{a \in L | -u \leq a \leq u\}$, where $u \in L^+$, into bounded subsets of the real line. The vector space of all order-bounded linear functionals on L is called the *order-dual* of L and is denoted by $L^\sim$. In $L^\sim$, an ordering $\geq$ is introduced by saying $f \geq g$ whenever $f(u) \geq g(u)$ for all $u \in L^+$.

Variational inequalities are offshoot of fixed point theorems. Many problems, e.g., Euler-Lagrange equations – which could be dealt with direct method of calculus of variations, can now be solved by variational inequalities. One such has been done by Browder (1970).

The topic of variational inequalities has gained importance in analysis in the last forty five years both theoretically and practically. The variational inequality theory has many dversified applications. The important developments in variational inequality theory are formulations that variational inequalities can be used to study problems of fluid flow through porous media, contact problems in elasticity, transposition problems and economic equilibria. Moreover, there are applications in control problems with a quadratric objective functional where the control equations are partial differential equations. There are intimate interconnections between variational inequalities, stochastic differential equations, and stochastic optimization.

An important area is Mathematical Programming and is known as Complementarity Theory. It was proved that if the set involved in a variatioal inequality and a complementarity problem is a convex cone, then both problems are equivalent.

We remark here that the development of variatioal inequalities can be viewed as the simultaneous pursuit of two different lines of research: On the one side, it reveals the fundamental facts on the qualitative behaviour of solutions (such as its existence, uniqueness and regularity) to important classes of problems. On the other side, it enables us to develop highly efficient and powerful new numerical methods to solve, for example, free and moving boundary value problems and the general equilibrium problems. Among the most effective numerical techniques are projection methods and its variant forms, linear approximation method, relaxation method, auxiliary principle and penalty function techniques. In addition to these methods, the finite element technique which is also being applied for the approximate solution of variational inequalities, have been obtained by many mathematicians.

In recent years, various extensions and generalizations of variational inequalities have been considered and studied. It is clear that in variational inequalities formulation, the convex set involved does not depend on solutions. If the convex set does depend on solutions, then the variational inequalities are called quasi-variational inequalities.

Now, suppose X, E and E^* are same as defined above. Then, given a (point-to-set) map $S : X \to 2^X$ and and a (point-to-point) map $T : X \to E^*$, the quasi-variational inequality (QVI) problem is to find a point $\hat{y} \in S(\hat{y})$ such that $Re\langle T(\hat{y}), \hat{y} - x\rangle \leq 0$ for all $x \in S(\hat{y})$. The QVI was introduced by Bensousson and Lions in 1973 (see, e.g., [Bensousson and Lions (1973)]) in connection with *impulse control*. A recent work concerning the QVI may be found in Mosco [Mosco (1976)].

Again, if we consider a point-to-set map $T : X \to 2^{E^*}$, then the generalized quasi-variational inequality (GQVI) problem is to find a point $\hat{y} \in S(\hat{y})$ and a point $\hat{w} \in T(\hat{y})$ such that $Re\langle \hat{w}, \hat{y} - x\rangle \leq 0$ for all $x \in S(\hat{y})$.

In 1982, for the study of operations research, mathematical programming and optimization theory, Chan and Pang (Chan and Pang (1982)) first introduced the so-called generalized quasi-variational inequalities in finite dimensional spaces. In 1985, Shih and Tan (Shih and Tan (1985)) were the first to study the GQVI in infinite dimensional locally convex Hausdorff topological vector spaces. Since then, there have been a numerous generalizations of the existence theorems on generalized quasi-variational inequalities. We shall present different types of variational inequalities, quasi-variational equalities and the equilibria of generalized games in Chapter 5.

In Chapter 6, we shall discuss best approximation and fixed point theorems for set-valued mappings in topological vector spaces. In Chapter 7, we give a presentation of the classical theory of the degree of a mapping as given by Kronecker and Brouwer. We also discuss its extension by Leray and Schauder to mappings in infinite dimensional Banach spaces of the form $I - Q$, with Q compact. The details of the existence and uniqueness of these degrees as defined by the additional properties of additivity, homotopy invariance, and normalization will be elaborated. Further, we shall discuss a self-contained exposition of the recent extension of these existence and uniqueness results for the degree functions for nonlinear mappings of monotone type from a reflexive Banach space X to its conjugate space X^*. We shall find out how such mappings arise from the combination of the ideas of fixed point theory and the somewhat different circle of ideas associated with the direct method of the calculus of variations. The concept of degree of mapping in all these forms is one of the most effective tools for studying the properties of existence and multiplicity of solutions of nonlinear equations.

Finally, in Chapter 8 we shall present some aspects of non-expansive types of mappings and fixed point theorems in locally convex topological vector spaces.

Chapter 2

Contraction Mappings

2.1 Contraction Mapping Principle in Uniform Topological Spaces and Applications

Let (X, ρ) be a metric space. Then a mapping f of X into itself is called a contraction on X if there exists a real number r with $0 < r < 1$ such that $\rho(f(x), f(y)) \leq r\rho(x, y)$ for all points x and y in X. Banach contraction mapping principle states that if (X, ρ) is a complete metric space and f is a contraction on X, then f has a unique fixed point $a \in X$ such that $f^n(x) \to a$ for each $x \in X$. This principle is well known for its wide scope of applications in analysis. It is, therefore, of some interest to extend this principle in complete Hausdroff uniform spaces which are generalizations of complete metric spaces.

Let (X, h) be a uniform space, h being the uniformity, i.e., the family of entourages. Taylor (Taylor (1972)) has introduced the following definitions:

Let $\mathcal{B}$ be a base of h. If f maps X into itself; then

(a) f is said to be $\mathcal{B}$-nonexpansion on X if $(x, y) \in H$ implies $(f(x), f(y)) \in H$ for each $H \in \mathcal{B}$.

(b) f is said to be $\mathcal{B}$-contraction on X if, for each $H \in \mathcal{B}$, there is a $K \in \mathcal{B}$ such that $(x, y) \in H \circ K$ implies $(f(x), f(y)) \in H$.

(c) f is said to be asymptotically regular if for each $x \in X$ and entourage $H \in h$ there is a positive integer n_0 such that $(f^n(x), f^{n+1}(x)) \in H$ for $n \geq n_0$.

The following result is obtained in (Taylor (1972)(see Tarafdar (1974), Lemma 1.5)).

Let (X, h) be a complete well-chained Hausdorff uniform space and $\mathcal{B}$ a base for h. If f is a $\mathcal{B}$-contraction on X, then f has a unique fixed point $a \in X$ such that $f^n(x) \to a$ for each $x \in X$. (For definition of well-chained uniform space see (Tarafdar (1974), p. 166).)

This result is not an exact analogue of Banach contraction mapping principle in the sense that an additional condition of the space being well chained is imposed on X.

In Section 2.2 of this chapter we shall obtain an exact analogue of Banach contraction mapping principle on a complete Hausdorff uniform space by giving a suitable definition of contraction mapping on a uniform space (which will reduce to the well-known definition of contraction mapping stated in the beginning when the uniform space is a metric space).

In (J. B. Diaz and Metcalf (1969)) Diaz and Metcalf obtained a series of results on the cluster set of successive approximations in a metric space by using mainly the non-expansion and contraction of a mapping with respect to the set of fixed points of the mapping. The main source of this work of Diaz and Metcalf was a paper of Tricomi (Tricomi (1916)) which is concerned with iteration of a real function. In Section 2.3 we shall present most of the results of (J. B. Diaz and Metcalf (1969)) extended to a uniform space. In that section we shall present some applications of the results of the Sections 2.2 and 2.3 in locally convex linear topological spaces. We should point out that the theory of non-expansive mappings has been growing very rapidly and a good account of the existing literature can be obtained in (Defigueiredo (1967)).

2.2 Banach Contraction Mapping Principle in Uniform Spaces

Let (x, h) be a uniform space, h being the uniformity. The uniform topology induced by h will be denoted by τ_h. A family $[\rho_\alpha : \alpha \in I]$ of pseudometrics on X is called an associated family for the uniformity h on X if the family $[H(\alpha, \epsilon) : \alpha \in I, \epsilon > 0]$, where $H(\alpha, \epsilon) = [(x, y) : \rho_\alpha(x, y) < \epsilon]$, is a subbase for h. (For definition of subbase and base for h, see Kelley (Kelley (1955)).) A family $[\rho_\alpha : \alpha \in I]$ of pseudometrics on X is called an augmented associated family for h if $[\rho_\alpha : \alpha \in I]$ is an associated family for h and has the additional property that, given $\alpha, \beta \in I$, there is a $\nu \in I$ such that $\rho_\nu(x, y) \geq \max(\rho_\alpha(x, y), \rho_\beta(x, y))$ for all $(x, y) \in X \times X$. An associated family and an augmented associated family for h will be respectively denoted by $A(h)$ and $A^*(h)$.

It is well known that if (X, h) is a uniform space and $[\rho_\alpha : \alpha \in I] = A^*(h)$, then the family $[H(\alpha, \epsilon) : \alpha \in I, \epsilon > 0]$ is a base for h (see Thron (Thron (1966), p. 179) or (Kelley (1955), pp. 188–189)). It is also well known that for each uniformity h on X, there exists a family $[\rho_\alpha : \alpha \in J]$, of pseudometrics on X that determines a unique uniformity h on X such that $A(h) = [\rho_\alpha : \alpha \in J]$ and $A(h)$ can be enlarged to $A^*(h)$ by adjoining to $A(h)$ all the pseudometrics of the form $\max[\rho_{\alpha_k} : k = 1, 2, \ldots, n]$, where $[\alpha_1, \alpha_2, \ldots, \alpha_n]$ is an arbitrary finite subset of the index set J (for details see (Thron (1966), p. 177)).

We now present the following definitions in (Tarafdar (1974), pp. 210–211). Note that all the results of this section are taken from (Tarafdar (1974)).

Let (X, h) be a uniform space and let $[\rho_\alpha : \alpha \in I] = A^*(h)$. If f maps X into itself, then

(i) f is said to be $A^*(h)$-non-expansion on X, or simply non-expansion on X, if for each $\alpha \in I$, $\rho_\alpha(f(x), f(y)) \leq \rho_\alpha(x, y)$ for all $(x, y) \in X \times X$;

(*ii*) f is said to be $A^*(h)$-contracton on X if for each $\alpha \in I$, there exists a real number $r(\alpha)$ with $0 < r(\alpha) < 1$ such that for all $(x, y) \in X \times X$ we have $\rho_\alpha(f(x), f(y)) \leq r(\alpha)\rho_\alpha(x, y)$ (we note that the above inequality implies $\rho_\alpha(f(x), f(y)) = 0$ if $\rho_\alpha(x, y) = 0$);

(*iii*) f is said to be $A^*(h)$-asymptotically regular on X, or simply asymptotically regular on X, if for each $x \in X$ and $\alpha \in I$,

$$\lim_{n\to\infty} \rho_\alpha(f^n(x), f^{n+1}(x)) = 0.$$

Remark 2.1 (Tarafdar (1974)) If f is $A^*(h)$-non-expansion, -contraction, or -asymptotically regular on X, then it is trivial to see that f is also $A(h)$-non-expansion, -contraction, or -asymptotically regular, respectively, on X. The converse of this is also true, i.e., if f is $A(h)$-non-expansion, -contraction, or -asymptotically regular on X, then f is $A^*(h)$-non-expansion, -contraction, or -asymptotically regular, respectively, on X. We prove it for the case of contraction. The case of non-expansion, and asymptotically regularity follows similarly.

Let f be $A(h)$-contraction. Let $A(h) = [\rho_\alpha : \alpha \in J]$. Let $\rho \in A^*(h)$ be arbitrary. If $\rho \in A(h)$, then $\rho = \rho_\alpha$ for some $\alpha \in J$. Hence there will exist a real number $\tau(\alpha)$ with $0 < \tau(\alpha) < 1$ satisfying the condition of Definition (ii) as f is $A(h)$-contraction. If $\rho \notin A(h)$, then $\rho = \max[\rho_{\alpha_k} : k = 1, 2, 3, \ldots, n]$ for some finite subset $[\alpha_1, \alpha_2, \ldots, \alpha_n]$ of J. Let $\tau = \max[\tau(\alpha_k) : k = 1, 2, 3, \ldots, n]$, where $\tau(\alpha_k)$'s are obtained from the definition of $A(h)$-contraction of f. We will assert that τ which clearly satisfies the relation $0 < \tau < 1$ is the required number for ρ. Let $(x, y) \in X \times X$. Then $\rho(x, y) = \rho_{\alpha_m}(x, y)$ for some $m = 1, 2, 3, \ldots, n$ and $\rho(f(x), f(y)) = \rho_{\alpha_j}(f(x), f(y))$ for some $j = 1, 2, 3, \ldots, n$.

Now noting that f is $A(h)$-contraction, and $\rho_{\alpha_j} \in A(h)$ and that $\rho_{\alpha_j}(x, y) \leq \rho_{\alpha_m}(x, y)$, we have

$$\begin{aligned}\rho(f(x), f(y)) &= \rho_{\alpha_j}(f(x), f(y)) \leq \tau(\alpha_j)\rho_{\alpha_j}(x, y) \leq \tau(\alpha_j)\rho_{\alpha_m}(x, y)\\ &= \tau(\alpha_j)\rho(x, y) \leq \tau\rho(x, y) \quad \text{as} \quad \tau(\alpha_j) \leq \tau.\end{aligned}$$

Clearly, τ depends on $\alpha_1, \alpha_2, \ldots, \alpha_n$ and hence on β for which $\rho = \rho_\beta \in A^*(h) = [\rho_\alpha : \alpha \in I]$.

Thus we see that it does not matter whether we use $A(h)$ or $A^*(h)$ in the above definitions.

Remark 2.2 (Tarafdar (1974)) It is easy to see that the Definition (c) of Taylor (1972) for asymptotic regularity stated in the beginning of Section 2 is equivalent to the Definition (iii) (Tarafdar (1974), p. 211) above. Also, if we take $\mathcal{B} = [H(\alpha, \epsilon) : \alpha \in I, \epsilon > 0]$, where $[\rho_\alpha : \alpha \in I] = A^*(h)$, then we see that Definition (a) of Taylor (1972) for $\mathcal{B}$-non-expansion coincides with Definition (i) (Tarafdar (1974), p. 211) above for non-expansion. However, Definition (b) of Taylor for $\mathcal{B}$-contraction is not, in general, equivalent to the above Definition (ii) (Tarafdar (1974), p. 211) for contraction. We can see this by comparing Theorem 2.1 below, the first lemma in Taylor (1972) and the discussion following this first lemma in Taylor (1972).

Remark 2.3 (Tarafdar (1974)) If (X, h) is replaced by a metric space (X, d), then the above Definitions (i), (ii) and (iii) (Tarafdar (1974), p. 211) reduce respectively to the well known definitions for non-expansion, contraction and asymptotic regularity on the metric space (X, d).

The following result is Theorem 1.1 in Tarafdar (1974):

Theorem 2.1 (Banach Contraction Mapping Principle) *Let (X, h) be a Hausdorff complete uniform space and let $[\rho_\alpha : \alpha \in I] = A^*(h)$. Let f be a contraction on X.*

Then f has a unique fixed point $a \in X$ such that $f^n(x) \to a$ in τ_h-topology for each $x \in X$.

Proof. Let $x_0 \in X$. Let $x_n = f(x_{n-1}) = f^n(x_0), n = 1, 2, \ldots$. Let $\alpha \in I$ be arbitrary. If m and n are positive integers with $m < n$, then

$$\begin{aligned}
\rho_\alpha(x_m, x_n) &= \rho_\alpha(f^m(x_0), f^n(x_0)) = \rho_\alpha(f^m(x_0), f^m f^{n-m}(x_0)) \\
&\leq \{\tau(\alpha)\}^m \rho_\alpha(x_0, f^{n-m}(x_0)) = \{\tau(\alpha)\}^m \rho_\alpha(x_0, x_{n-m}) \\
&\leq \{\tau(\alpha)\}^m [\rho_\alpha(x_0, x_1) + \rho_\alpha(x_1, x_2) + \cdots + \rho_\alpha(x_{n-m-1}, x_{n-m})] \\
&\leq \{\tau(\alpha)\}^m \rho_\alpha(x_0, x_1)[1 + \tau(\alpha) + \cdots + \{\tau(\alpha)\}^{n-m-1} \\
&< \{\tau(\alpha)\}^m \frac{\rho_\alpha(x_0, x_1)}{(1 - \tau(\alpha))} \to 0 \quad \text{as } m, n \to \infty.
\end{aligned}$$

Hence $\{x_n\}_{n=1}^\infty$ is a ρ_α-Cauchy sequence (i.e., a Cauchy sequence in ρ_α-topology). Since $\alpha \in I$ is arbitrary, $\{x_n\}_{n=1}^\infty$ is a ρ_α-Cauchy sequence for each $\alpha \in I$. Let $S_p = \{x_n : n \geq p\}$ for all positive integers p and let $\mathcal{B}$ be the filter basis $\{S_p : p = 1, 2, \ldots\}$. Then, since $\{x_n\}_{n=1}^\infty$ is a ρ_α-Cauchy sequence for each $\alpha \in I$, it is easy to see that the filter basis $\mathcal{B}$ is Cauchy in the uniform space (X, h). To see this we first note that the family $[H(\alpha, \epsilon) : \alpha \in I, \epsilon > 0]$ is a base for h as $A^*(h) = [\rho_\alpha : \alpha \in I]$. Now let $H \in h$ be an entourage. Then there exist a $\nu \in I$ and $\epsilon > 0$ such that $H(\nu, \epsilon) \subset H$. Now since $\{x_n\}_{n=1}^\infty$ is a ρ_ν-Cauchy sequence in X, there exists a positive integer p such that $\rho_\nu(x_m, x_n) < \epsilon$ for $m \geq p$, $n \geq p$. This implies that $S_p \times S_p \subset H(\nu, \epsilon)$. Thus given any $H \in h$, we can find a $S_p \in \mathcal{B}$ such that $S_p \times S_p \subset H$. Hence $\mathcal{B}$ is a Cauchy filter in (X, h). Since (X, h) is complete and Hausdorff, the Cauchy filter $\mathcal{B} = \{S_p\}$ converges to a unique point $a \in X$ in the τ_h-topology. Thus $\tau_h\text{-}\lim S_p = a$. Now since f is a ρ_α-continuous for each $\alpha \in I$, it follows that f is τ_h-continuous. Hence $f(a) = f(\tau_h\text{-}\lim S_p) = \tau_h\text{-}\lim f(S_p) = \tau_h\text{-}\lim S_{p+1} = a$. Thus a is a fixed point of f.

We now complete the proof of our theorem by showing that a is the only fixed point of f. We assume that f has another fixed point b such that $a \neq b$ and deduce an absurdity from this assumption. Since (X, h) is a Hausdorff space and $a \neq b$, there is an index $\beta \in I$ such that $\rho_\beta(a, b) \neq 0$. Since f is a contraction on X, $\rho_\beta(a, b) = \rho_\beta(f(a), f(b)) \leq \tau(\beta)\rho_\beta(a, b)$ which is absurd as $0 < \tau(\beta) < 1$ and $\rho_\beta(a, b) \neq 0$. This together with the fact that $\tau_h\text{-}\lim S_p = a$ implies $\tau_h\text{-}\lim x_n = a$ completes the proof. □

Note: When $I = \{1\}$, we obtain the famous Banach contraction mapping principle in a complete metric space (Banach (1922), pp. 133–181).

For more on contraction mapping principle we refer to Lee (Lee (1977)) and Morales (Morales (1980) (1980)).

Remark 2.4 The following facts are noted in Chu and Diaz (Chu and Diaz (1965)) in metric space situation.

1. If $f : (X, \tau_h) \to (X, \tau_h)$ of a complete Hausdorff uniform topological space (X, τ_h) into itself such that f^N is a contraction mapping for some positive integer N, then f has a unique point $x_0 \in X$.

By our Theorem 2.1, f^N has a unique point $x_0 \in X$. Now for each $\alpha \in I$, $\rho_\alpha(f(x_0), x_0) = \rho_\alpha(ff^N(x_0), f^N(x_0)) = \rho_\alpha(f^N(f(x_0), f^N(x_0)) \leq \tau_\alpha \rho_\alpha(f(x_0), x_0)$ which implies $\rho_\alpha(f(x_0), x_0) = 0$ as $0 < \tau_\alpha < 1$. Since X is Hausdorff, $f(x_0) = x_0$. Since a fixed point of f is necessarily a fixed point of f^N, x_0 is the unique point of f.

In fact the following is true.

2. If $f : X \to X$ is a mapping of a non-empty set X into itself such that f^N has a unique fixed point $x_0 \in X$ for some positive integer N, then x_0 is also the unique fixed point of f. Indeed, $f(x_0) = f(f^N(x_0)) = f^N(f(x_0))$. Thus by the uniqueness of fixed point of f^N, $f(x_0) = x_0$.

3. If $f, g : X \to X$ is a mapping of X into itself such that f commutes with g, i.e., $fg = gf$ and f has a unique fixed point $x_0 \in X$, then x_0 is also a fixed point of g.

Evidently, $g(x_0) = gf(x_0) = fg(x_0)$. Thus $g(x_0)$ is a fixed point of f. Hence by the uniqueness of fixed point of f, $g(x_0) = x_0$.

Example 2.1 (The example of I.I. Glick cf. Chu and Diaz (Chu and Diaz (1965)) of discontinuous mapping T with T^2 is contracting.) Let $X = C([0,1])$ be the complete metric space with usual supremum metric ρ, i.e., $\rho(f,g) = \sup_{a \leq t \leq 1} |(f - g)(t)|$. Let H be Hamel basis of X containing the set $A = \{e^x, 1, x, x^2, \ldots\}$ of linearly independent vectors.

Now we define a mapping T on H by

$$T(e^x) = \frac{1}{2}.1,\ T(1) = \frac{1}{2}.e^x, \quad \text{and } T(h) = \frac{1}{2}.h \ \text{ if } h \neq e^x \ \text{ and } 1.$$

T can be extended to $C[0,1]$ by $T(x) = \sum_{i=1}^n \alpha_i T(h_i)$ when $x = \sum_{i=1}^n \alpha_i h_i$, n is a positive integer, $\alpha_i \neq 0$ for $i = 1, 2, \ldots, n$, and $h_i \in H$ for $i = 1, 2, \ldots, n$; also let $T(0) = 0$. Then it can be easily checked that $T^2 = I$ and, theerefore, T^2 is contracting. Now we see that T is not continuous at e^x, that is

$$\lim_{n \to \infty} T(\sum_{k=0}^{n} \frac{1}{k!} x^k) \neq T(e^x) = \frac{1}{2}$$

as $\lim_{n\to\infty}[T(1 + \sum_{k=1}^\infty \frac{1}{k!}x^k)] = \lim_{n\to\infty}[\frac{1}{2}e^x + \frac{1}{2}\sum_{k=1}^n \frac{1}{k!}x^k] = e^x - \frac{1}{2}$.

2.2.1 *Successive Approximation*

In this section We shall present extensions of some of the results of Diaz and Metcalf (J. B. Diaz and Metcalf (1969)) to Hausdorff uniform spaces.

We shall begin with the following lemmas:

Lemma 2.1 (Tarafdar (1974), p. 213) *Let (X, h) be a uniform space and let $[\rho_\alpha : \alpha \in I] = A^*(h)$. If X is τ_h-compact, then X is ρ_α-compact for each $\alpha \in I$. (A topological space (X, τ) is τ-compact if every τ-open covering of X has a finite subcovering.)*

Lemma 2.2 (Tarafdar (1974), p. 213) *Let (X, h) be a uniform space and let $[\rho_\alpha : \alpha \in I] = A^*(h)$. Let $\{x_n\}_{n\in J}$ be a net in X.*

(a) If $\{x_n\}_{n\in J}$ is τ_h-convergent and converges to x, then for each $\alpha \in I$, it is ρ_α-convergent and converges to x. Conversely, if $\{x_n\}_{n\in J}$ is ρ_α-convergent and converges to x for each $\alpha \in I$, then it is τ_h-convergent and converges to x.

(b) If y is τ_h-cluster point of the net $\{x_n\}_{n\in J}$, then y is a ρ_α-cluster point of the net $\{x_n\}_{n\in J}$ for each $\alpha \in I$.

The proofs of the above two lemmas are trivial and hence omitted.

Lemma 2.3 (Tarafdar (1974), p. 213) *Let (X, h) be a Hausdorff uniform space and let $[\rho_\alpha : \alpha \in I] = A^*(h)$. If A and B are a disjoint pair of τ_h-compact subsets of X, then there exists at least one $\beta \in I$ such that $\rho_\beta(A, B) > 0$, that is $\rho_\beta(x, y) > 0$ for all $x \in A$ and all $y \in B$.*

Proof. Since X is τ_h-Hausdorff, it follows that A and B are disjoint τ_h-closed and τ_h-compact subsets of X. Hence we can find a symmetric entourage $W \in h$ such that $W(A) \cap W(B) = \emptyset$ (e.g., see (Schubert (1964), Theorem 5, p. 117) where for any subset C of X, $W(C) = \{y \in X : (x, y) \in W, x \in C\}$; i.e., $W(C) = \cup_{x\in C} W(x)$ where $W(x) = \{y \in X : (x, y) \in W\}$.

With this symmetric entourage W we can construct a sequence $\{W_n\}$ of symmetric entourages such that $W_n \circ W_n \circ W_n \subset W_{n-1}$, $W_1 = W \cap W^{-1} = W$ and $W_0 = X$ and we can show that there exists a pseudometric $\rho \in A^*(h)$ such that for each positive integer n, $W_n \subset [(x, y) : \rho(x, y) \leq 2^{-n}] \subset W_{n-1}$ (for details see Thron (Thron (1966), pp. 178–179)). We first assert that for no pair (x, y) of points with $x \in A$, $y \in B$ we have $\rho(x, y) = 0$. We suppose contrary to what we wish to prove that for some pair (p, q) of points with $p \in A$ and $q \in B$, we have $\rho(p, q) = 0$. Then since $\rho(p, q) = 0$, $(p, q) \in W_n$ for each n and hence $(p, q) \in W_1 = W$ in particular. Then clearly, $p \in W(A)$ and $p \in W(B)$. This contradicts $W(A) \cap W(B) = \emptyset$. Thus we have proved our assertion. Next we prove that $\rho(A, B) > 0$. Since $\rho \in A^*(h)$ and A and B are τ_h-compact, we have, by Lemma 2.1, that A and B are both ρ-compact.

Hence, if $\rho(A, B) = 0$, then there would exist a pair (x, y) of points with $x \in A$ and $y \in B$ such that $\rho(x, y) = 0$ which would contradict our established assertion that $\rho(x, y) \neq 0$ for each pair (x, y) of points with $x \in A$ abd $y \in B$. Hence

$\rho(A, B) > 0$. As $\rho \in A^*(h)$, $\rho = \rho_\beta$ for some $\beta \in I$. Thus we have proved the lemma. □

Notation: Let (X, h) be a uniform space and let $f : X \to X$ be a mapping of X into X. Then $F(f)$ will be the set of fixed points of f, i.e., $F(f) = \{x \in X : f(x) = x\}$. For any $x \in X$, $L(x)$ will denote the set of all cluster points of the net (sequence) of iterates $\{f^n(x)\}_{n=1}^\infty$, i.e., $L(x)$ is the set of all $y \in X$ such that $\{f^{n_i}(x)\}_{i\in J} \to y$ in τ_h-topology for some subnet $\{f^{n_i}(x)\}_{i\in J}$ of the net $\{f^n(x)\}_{n=1}^\infty$.

The following is an extension of Theorem 6 in (J. B. Diaz and Metcalf (1969)) to the uniform space.

Theorem 2.2 (Tarafdar (1974), p. 214) *Let (X, h) be a non-empty Hausdorff uniform space and let $[\rho_\alpha : \alpha \in I] = A^*(h)$. Let $f : X \to X$ be τ_h-continuous. Also let*

(a) $f(X)$ be τ_h-compact; and

(b) f be asymptotically regular on X.

Then for each $x \in X$, the τ_h-cluster set $L(x)$ is a non-empty τ_h-closed and τ_h-connected subset of $F(f)$. In case $L(x)$ is just one point, then τ_h-$\lim f^m(x)$ exists and belongs to $F(f)$. In case $L(x)$ contains more than one point, then it is contained in the τ_h-boundary of $F(f)$. [The τ_h-boundary of a subset K of X = τ_h-closure of $K - \tau_h$-Int K where Int K stands for the interior of K.]

Proof. (Tarafdar (1974), pp. 214–217) The sequence $\{f^n(x)\}_{n=1}^\infty$ being also a net in the compact set $f(X)$ has a cluster point. Hence $L(x)$ is non-empty. We prove the rest of the theorem in few steps.

(i) $L(x)$ is a subset of $F(f)$. Let $y \in L(x)$. Then there is a subnet $\{f^{n_j}(x)\}_{j\in J}$ of the net $\{f^n(x)\}_{n=1}^\infty$ such that $f^{n_j}(x) \to y$ in τ_h-topology. Also since f is τ_h-continuous, the net $f^{n_j+1}(x) \to f(y)$ in τ_h-topology. Hence by Lemma 2.2, for each $\alpha \in I$, the net $f^{n_j}(x) \to y$ and the net $f^{n_j+1}(x) \to f(y)$ in the ρ_α-topology of X. Let $\alpha \in I$ be arbitrary. Then for each $j \in J$, we have

$$\rho_\alpha(f(y), y) \leq \rho_\alpha(f(y), f^{n_j+1}(x)), +\rho_\alpha(f^{n_j+1}(x), f^{n_j}(x)) + \rho_\alpha(f^{n_j}(x), y). \tag{2.1}$$

Let $\emptyset > 0$ be arbitrarily chosen. Then since in ρ_α-topology of X, $f^{n_j+1}(x) \to f(y)$ and $f^{n_j}(x) \to y$ and since by asymptotic regularity we have $\rho_\alpha(f^{n_j+1}(x), f^{n_j}(x)) \to 0$, we can find a $p \in J$ such that for all $n_j \geq n_p$ we have simultaneously

$$\rho_\alpha(f^{n_j+1}(x), f(y)) < \frac{\epsilon}{3}, \quad \rho_\alpha(f^{n_j+1}(x), f^{n_j}(x)) < \frac{\epsilon}{3}, \quad \text{and} \quad \rho_\alpha(f^{n_j}(x), y) < \frac{\epsilon}{3}. \tag{2.2}$$

Now from (2.1), and (2.2) we have $\rho_\alpha(f(y), y) < \epsilon$. Sincce ϵ is arbitrary, $\rho_\alpha(f(y), y) = 0$. Again since α is arbitrary, $\rho_\alpha(f(y), y) = 0$ for each $\alpha \in I$. Finally since X is Hausdorff, it follows that $f(y) = y$. (The uniform space X is Hausdorff iff, given two distinct points x and y, there is a $\beta \in I$ such that $\rho_\beta(x, y) \neq 0$.)

(ii) $L(x)$ is a τ_h-closed subset of $F(f)$. It is well known that the cluster set of any net is always closed. Thus $L(x)$ is a τ_h-closed subset of $F(f)$ as we have proved in (i) that $L(x) \subset F(f)$.

(iii) We now prove that $L(x)$ is a τ_h-connected subset of $F(f)$. Although our proof is similar to the proof of the corresponding part of Theorem 6 in J. B. Diaz and Metcalf (1969), our Lemma 2.3 will play the crucial role. If $L(x)$ consists of a single point, then there is nothing to prove. So, we may suppose that $L(x)$ consists of more than one point. We assume that $L(x)$ is not a τ_h-connected subset of $F(f)$ and deduce a contradiction from this assumption. Since $L(x)$ is not a τ_h-connected subset of $F(f)$, $L(x) = S_1 \cup S_2$ where S_1 and S_2 are both non-empty and τ_h-closed and $S_1 \cap S_2 = \emptyset$. Also since $f(X)$ is compact, it follows that S_1 and S_2 are both τ_h-compact. Thus S_1 and S_2 are disjoint τ_h-closed and τ_h-compact subsets of X.

Now for each $\alpha \in I$, let

$$S_1^\alpha = \{y \in F(f) : \rho_\alpha(y, S_1) \leq \frac{1}{4}\rho_\alpha(S_1, S_2)$$

and

$$S_2^\alpha = \{y \in F(f) : \rho_\alpha(y, S_2) \leq \frac{1}{4}\rho_\alpha(S_1, S_2).$$

$F(f)$, being a τ_h-closed subset of τ_h-compact set $f(X)$, is τ_h-compact. Hence $F(f)$ is, by Lemma 2.1, ρ_α-compact for each $\alpha \in I$. Hence S_1^α and S_2^α being ρ_α-closed in $F(f)$, are both ρ_α-compact subsets of $F(f)$ for each $\alpha \in I$.

We first prove that $\lim_{m\to\infty} \rho_\alpha(f^m(x), S_1^\alpha \cup S_2^\alpha) = 0$ for each $\alpha \in I$. We suppose that this is not true for some $\beta \in I$ and obtain a contradiction. Then there would exist $\epsilon > 0$ and a subsequence $\{f^{n_i}(x)\}_{i=1}^\infty$ of the sequence $\{f^m(x)\}_{m=1}^\infty$ such that

$$\rho_\beta(f^{n_i}(x), S_1^\beta \cup S_2^\beta) \geq \epsilon > 0, \quad \text{for each } i = 1, 2, \ldots. \tag{2.3}$$

Now the subsequence $\{f^{n_i}(x)\}_{i=1}^\infty$, being a net in the τ_h-compact set $f(X)$, has a cluster point, say, z. Then obviously $z \in L(x)$ and z is a ρ_β-cluster point of the sequence $\{f^{n_i}(x)\}_{i=1}^\infty$ by Lemma 2.2. Hence there is a subsequence $\{f^{n_{p_j}}(x)\}_{j=1}^\infty$ of the sequence $\{f^{n_i}(x)\}_{i=1}^\infty$ such that $\lim_{j\to\infty} \rho_\beta(f^{n_{p_j}}(x), z) = 0$ because ρ_β-topology of X satisfies the first axiom of countability. Now since $z \in L(x) = S_1 \cup S_2 \subset S_1^\beta \cup S_2^\beta$, we have

$$\rho_\beta(f^{n_{p_j}}(x), S_1^\beta \cup S_2^\beta) \leq \rho_\beta(f^{n_{p_j}}(x), z), \;\; j = 1, 2, \ldots.$$

Hence

$$\lim_{j\to\infty} \rho_\beta(f^{n_{p_j}}(x), S_1^\beta \cup S_2^\beta) = 0$$

which contradicts (2.3).

Thus we have proved that

$$\lim_{m\to\infty} \rho_\alpha(f^m(x), S_1^\alpha \cup S_2^\alpha) = 0 \quad \text{for each} \quad \alpha \in I.$$

Since S_1 and S_2 are disjoint, τ_h-closed and τ_h-compact subsets of X, by Lemma 2.3, there exists at least one $\nu \in I$ such that $\rho_\nu(S_1, S_2) > 0$. We now prove that

S_1^ν and S_2^ν are disjoint. If $p \in S_1^\nu \cap S_2^\nu$ then, since S_1 and S_2 are both ρ_ν-compact subsets (by Lemma 2.1) of $F(f)$, there would exist points $a \in S_1$ and $b \in S_2$ such that $\rho_\nu(p, S_1) = \rho_\nu(a, p)$ and $\rho_\nu(p, S_2) = \rho_\nu(p, b)$.

Hence $0 < \rho_\nu(S_1, S_2) \leq \rho_\nu(a, b) \leq \rho_\nu(a, p) + \rho_\nu(p, b) = \rho_\nu(p, S_1) + \rho_\nu(p, S_2) \leq \frac{1}{2}\rho_\nu(S_1, S_2)$ which is absurd. Thus S_1^ν and S_2^ν are disjoint. Also we have noted earlier that both S_1^ν and S_2^ν are ρ_ν-closed ρ_ν-compact subsets of $F(f)$. Hence $\rho_\nu(S_1^\nu, S_2^\nu) > 0$. In summary we have

$$\rho_\nu(S_1^\nu, S_2^\nu) > 0; \ \lim_{m\to\infty} \rho_\nu(f^m(x), f^{m+1}(x)) = 0 \text{ and } \lim_{m\to\infty} \rho_\nu(f^m(x), S_1^\nu \cup S_2^\nu) = 0.$$

In view of the last two limits we can find a positive integer M such that, for all $m \geq M$,

$$\rho_\nu(f^m(x), f^{m+1}(x)) < \frac{\rho_\nu(S_1^\nu, S_2^\nu)}{3} \quad \text{and} \quad \rho_\nu(f^m(x), S_1^\nu \cup S_2^\nu) < \frac{\rho_\nu(S_1^\nu, S_2^\nu)}{3}.$$

The rest of the proof is similar to that in (J. B. Diaz and Metcalf (1969)) and we repeat this for the sake of clarity and completeness. It follows from the last inequality that for any $m \geq M$, we have either

$$\rho_\nu(f^m(x), S_1^\nu) < \frac{\rho_\nu(S_1^\nu, S_2^\nu)}{3} \tag{2.4}$$

or,

$$\rho_\nu(f^m(x), S_2^\nu) < \frac{\rho_\nu(S_1^\nu, S_2^\nu)}{3}. \tag{2.5}$$

The set of integers $m \geq M$ satisfying (2.4) is not empty as S_1 is not empty and the set of integers $m \leq M$ satisfying (2.5) is not empty as S_2 is not empty. Hence we can find a positive integer $n \geq M$ such that we have both

$$\rho_\nu(f^n(x), S_1^\nu) < \frac{\rho_\nu(S_1^\nu, S_2^\nu)}{3} \quad \text{and} \quad \rho_\nu(f^{n+1}(x), S_2^\nu) < \frac{\rho_\nu(S_1^\nu, S_2^\nu)}{3}.$$

[For any $m_1 \geq M$ such that $\rho_\nu(f^{m_1}(x), S_1^\nu) < \frac{\rho_\nu(S_1^\nu, S_2^\nu)}{3}$ there always exists a positive integer $m_2 > m_1$ such that $\rho_\nu(f^{m_2}(x), S_2^\nu) < \frac{\rho_\nu(S_1^\nu, S_2^\nu)}{3}$. n can be chosen to be one less than smallest such m_2.]

By using ρ_ν-compactness of S_1^ν and S_2^ν we have

$$\rho_\nu(S_1^\nu, S_2^\nu) \leq \rho_\nu(f^n(x), S_1^\nu) + \rho_\nu(f^n(x), f^{n+1}(x)) + \rho_\nu(f^{n+1}(x), S_2^\nu).$$

But then by what we have proved above

$$\rho_\nu(S_1^\nu, S_2^\nu) < \frac{\rho_\nu(S_1^\nu, S_2^\nu)}{3} + \frac{\rho_\nu(S_1^\nu, S_2^\nu)}{3} + \frac{\rho_\nu(S_1^\nu, S_2^\nu)}{3} = \rho_\nu(S_1^\nu, S_2^\nu)$$

which is impossible. Hence our original assumption that $L(x)$ is not a τ_h-connected subset of $F(f)$ is wrong. Thus we have proved that $L(x)$ is a τ_h-connected subset of $F(f)$.

(iv) If $L(x)$ consists of a single point, then we prove that τ_h-$\lim f^m(x)$ exists. Let us denote the only cluster point of $\{f^m(x)\}_{m=1}^{\infty}$ by p. We prove that τ_h-$\lim f^m(x) = p$. Let us assume that τ_h-$\lim f^m(x) \neq p$ and deduce a contradiction from this assumption. Then there is a subnet $\{f^{m_k}(x)\}_{k\in K}$ of the net $\{f^m(x)\}_{m=1}^{\infty}$ such that $\{f^{m_k}(x)\}_{k\in K}$ has no subnet converging to p in the τ_h-topology. But since $\{f^{m_k}(x)\}_{k\in K}$ is a net in the τ_h-compact set $f(X)$, it has a subnet converging to a point, say, q in the τ_h-topology, i.e., it has a τ_h-cluster point q. Clearly, $p \neq q$ and q is also a τ_h-cluster point of the net $\{f^m(x)\}_{m=1}^{\infty}$. Hence $L(x)$ consists of at least two distinct points p and q. This contradiction proves that τ_h-$\lim f^m(x) = p$.

(v) Finally, we prove that if $L(x)$ consists of more than one point, then it is in the τ_h-boundary of $F(f)$. Since $L(x)$ consists of more than one point, it is clear that $f^m(x) \notin F(f)$ for any $m = 0, 1, 2, \ldots,$ where $f^0(x) = x$. Let $y \in L(x)$ be arbitrary. If y belonged to $Int\ F(f)$, then it would follow that $Int\ F(f)$ would contain $f^k(x)$ for some positive integer k which would contradict the assumption that $L(x)$ contains more than one point. Hence $y \in \tau_h$-boundary of $F(f)$. Thus $L(x) \subset \tau_h$-boundary of $F(f)$. □

The following theorem is the generalization of some parts of the main result (Theorem 2) of (J. B. Diaz and Metcalf (1969)) to uniform space.

Theorem 2.3 (Tarafdar (1974), p. 214) *Let $X, h)$ be a Hausdorff uniform space and let $[\rho_\alpha : \alpha \in I] = A^*(h)$. Let $fX \to X$ be τ_h-continuous. Also suppose that*

(i) $F(f)$ is non-empty and compact;

(ii) for each $x \in X$, with $x \notin F(f)$ we have for each $\alpha \in I$, $\rho_\alpha(f(x), F(f)) < \rho_\alpha(x, F(f))$ if $\rho_\alpha(x, F(f)) \neq 0$ and $\rho_\alpha(f(x), F(f)) = 0$ if $\rho_\alpha(x, F(f)) = 0$.

Then for each $x \in X$ the set $L(x)$ is a closed subset of $F(f)$. If $L(x)$ consists of more than one point, then $L(x)$ is contained in the τ_h-boundary of $F(f)$.

Proof. We have nothing to prove if $L(x)$ is empty. So we may suppose that $L(x)$ is non-empty. Again, if $x \in F(f)$ or $f^k(x) \in F(f)$ for some integer $k \geq 1$, then obviously, τ_h-$\lim f^m(x)$ exists and belongs to $F(f)$ and thus the theorem is proved in this case, i.e., we assume that $f^m(x) \notin F(f)$ for each $m = 0, 1, 2, \ldots$.

(1) First we prove that for each $\alpha \in I$, $\lim_{m\to\infty} \rho_\alpha(f^m(x), F(f))$ exists and non-negative. Let α be arbitrary. Now if $\rho_\alpha(f^k(x), F(f)) = 0$ for some integer $k = 0, 1, 2, \ldots,$ then by the second part of condition (ii) of the theorem every element of the sequence $\{\rho_\alpha(f^m(x), F(f))\}_{m=1}^{\infty}$ starting from the kth element (from the first element if $k = 0$) is zero. Consequently, $\lim_{m\to\infty} \rho_\alpha(f^m(x), F(f)) = 0$. In the remaining case, i.e., when $\rho_\alpha(f^m(x), F(f)) \neq 0$ for each $m = 0, 1, 2, \ldots,$ the sequence of positive numbers $\{\rho_\alpha(f^m(x), F(f))\}_{m=1}^{\infty}$ is a decreasing sequence by virtue of the first part of condition (ii) of the theorem and, therefore, $\lim_{m\to\infty} \rho_\alpha(f^m(x), F(f))$ exists and is non-negative.

(2) Next we prove that $L(x)$ is a subset of $F(f)$. Let $y \in L(x)$. It suffices to prove that $y \in F(f)$. We assume that $y \notin F(f)$ and arrive at a contradiction from this

assumption. Since $y \in L(x)$, there is a subnet $\{f^{n_j}(x)\}_{j\in J}$ of the net $\{f^m(x)\}_{m=1}^{\infty}$ such that $f^{n_j}(x) \to y$ in the τ_h-topology of X. Also since f is continuous, the net $f^{n_j+1}(x) \to f(y)$ in the τ_h-topology of X. Now by our Lemma 2.2, for each $\alpha \in I$, $f^{n_j}(x) \to y$ and $f^{n_j+1}(x) \to f(y)$ in the ρ_α-topology of X. The rest of the proof depends on the simple fact that if $\{f^{n_j}(x)\}_{j\in J}$ is a subnet of $\{f^m(x)\}_{m=1}^{\infty}$ then $\{f^{n_j+1}(x)\}_{j\in J}$ is also a subnet of $\{f^m(x)\}_{m=1}^{\infty}$.(1)

Let $\alpha \in I$ be arbitrary. Then noting that in ρ_α-topology of X, $f^{n_j}(x) \to y$ and $f^{n_j+1}(x) \to f(y)$ and using the continuity of the real valued function $\rho_\alpha(x, F(f))$, $x \in X$, we have

$$\rho_\alpha(f^{n_j}(x), F(f)) \to \rho_\alpha(y, F(f)) \quad \text{and} \quad \rho_\alpha(f^{n_j+1}(x), F(f)) \to \rho_\alpha(f(y), F(f)).$$

Hence in view of the fact that $\lim_{m\to\infty} \rho_\alpha(f^m(x), F(f))$ exists and that $\{\rho_\alpha(f^{n_j}(x), F(f))\}_{j\in J}$ and $\{\rho_\alpha(f^{n_j+1}(x), F(f))\}_{j\in J}$ are both subnets of the net $\{\rho_\alpha(f^m(x), F(f))\}_{j\in J}$ of real numbers, we have $\{\rho_\alpha(f(y), F(f)) = \rho_\alpha(y, F(f))$. Now since by assumptions $y \notin F(f)$, the above equality together with condition (ii) of the theorem implies that $\rho_\alpha(y, F(f)) = 0$. Since α is arbitrary, $\rho_\alpha(y, F(f)) = 0$ for each $\alpha \in I$. But this contradicts our Lemma 2.3 as $\{y\}$ and $F(f)$ are disjoint pair of τ_h-compact subsets of Hausdorff uniform space X. Hence $y \in F(f)$.

(3) That $L(x)$ is closed is well known and the proof that $L(x)$ is in the τ_h-boundary of $F(f)$ when it consists of more than one point is exactly the same as given in part (v) of the proof of Theorem 2.2. □

The next theorem is an extension of Theorem 3 in (J. B. Diaz and Metcalf (1969)) to uniform space.

Theorem 2.4 (Tarafdar (1974), p. 218) *Let (X, h) be a Hausdorff uniform space and let $[\rho_\alpha : \alpha \in I] = A^*(h)$. Let $f : X \to X$ be τ_h-continuous. Further suppose that*

(i) $F(f)$ is non empty;

(ii) for each $x \in X$, with $x \notin F(f)$ and each $p \in F(f)$, we have for each $\alpha \in I$, $\rho_\alpha(f(x), p) < \rho_\alpha(x, p)$ if $\rho_\alpha(x, p) \neq 0$, and $\rho_\alpha(f(x), p) = 0$ if $\rho_\alpha(x, p) = 0$.

Then for each $x \in X$, either $\{f^m(x)\}_{m=1}^{\infty}$ has no τ_h-convergent subnet, or τ_h- $\lim f^m(x)$ *exists and belongs to $F(f)$.*

Proof. We have nothing to prove if $L(x)$ is empty. So, we may assume that $L(x)$ is non-empty. If $x \in F(f)$, or $f^k(x) \in F(f)$ for some integer $k \geq 1$, then obviously τ_h-$\lim f^m(x)$ exists and belongs to $F(f)$ and, therefore, the theorem is proved in this case. Hence we assume that $f^m(x) \notin F(f)$ for each $m = 0, 1, 2, \ldots$ and prove our theorem in the following steps.

(a) In this step we prove that for each $\alpha \in I$, $\lim_{m\to\infty} \rho_\alpha(f^m(x), p)$ exists and non-negative, where p is any point belonging to $F(f)$. Such a p exists by condition (i) of the theorem. The proof is similar to the proof of the part 1 of our previous theorem. Let $\alpha \in I$ be arbitrary. Now if $\rho_\alpha(f^k(x), p) = 0$ for some integer $k = 0, 1, 2, \ldots$, then by the second part of condition (ii) of our theorem each element of the sequence $\{\rho_\alpha(f^m(x), p)\}_{m=1}^\infty$ starting from the kth element (from the first element if $k = 0$) is zero and hence $\lim_{m\to\infty} \rho_\alpha(f^m(x), p) = 0$ in this case.

If $\rho_\alpha(f^m(x), p) \neq 0$ for each $m = 0, 1, 2, \ldots$, then by the first part of condition (ii) of the theorem the sequence of positive numbers $\{\rho_\alpha(f^m(x), p)\}_{m=1}^\infty$ is a decreasing and, therefore, $\lim_{m\to\infty} \rho_\alpha(f^m(x), p)$ exists and is non-negative.

(b) We next prove that $L(x)$ is a subset of $F(f)$. Here again the proof is very much similar to the proof of part 2 of the previous theorem. Let $y \in L(x)$. Then there is a subnet $\{f^{n_j}(x)\}_{j\in J}$ of the net $\{f^m(x)\}_{m=1}^\infty$ such that $f^{n_j}(x) \to y$ in the τ_h-topology of X. Also by the τ_h-continuity of f, $f^{n_j+1}(x) \to f(y)$. Hence by our Lemma 2.2, for each $\alpha \in I$ we have that $f^{n_j}(x) \to y$ and $f^{n_j+1}(x) \to f(y)$ in the ρ_α-topology of X. Then by using the above two limits and the continuity of the real valued function $\rho_\alpha(x, p)$, $x \in X$, we have

$$\rho_\alpha(f^{n_j}(x), p) \to \rho_\alpha(y, p) \quad \text{and} \quad \rho_\alpha(f^{n_j+1}(x), p) \to \rho_\alpha(f(y), p).$$

Now, since we have proved in part (a) that $\lim_{m\to\infty} \rho_\alpha(f^m(x), p)$ exists, we have $\rho_\alpha(f(y), p) = \rho_\alpha(y, p)$ because of the same reason given in part 2 of the previous theorem, i.e. because $\{f^{n_j}(x)\}_{j\in J}$ and $\{f^{n_j+1}(x)\}_{j\in J}$ are both subnets of the net $\{f^m(x)\}_{m=1}^\infty$. We now assume that $y \notin F(f)$ and readily deduce a contradiction from this assumption. Since $y \notin F(f)$ and $\rho_\alpha(f(y), p) = \rho_\alpha(y, p)$, it follows from condition (ii) of our theorem that $\rho_\alpha(y, p) = 0$. Since α is arbitrary, we have $\rho_\alpha(y, p) = 0$ for each $\alpha \in I$. This implies that $y = p$ as X is τ_h-Hausdorff. This is a condition because of the fact that $p \in F(f)$. Thus we have proved that $y \in L(x)$.

(c) We now prove that $L(x)$ contains at most one point. We suppose that $L(x)$ contains two points, p and q. By Lemma 2.2, p and q are also ρ_α-cluster points of the net $\{f^m(x)\}_{m=1}^\infty$ for each $\alpha \in I$. Let α be arbitrary. Then there are two subsequences $\{f^{m_i}(x)\}_{i=1}^\infty$ and $\{f^{n_i}(x)\}_{i=1}^\infty$ of the sequence $\{f^m(x)\}_{m=1}^\infty$ such that $f^{m_i}(x) \to p$ and $f^{n_i}(x) \to q$ in the ρ_α-topology of X (as the ρ_α-topology of X satisfies the first axiom of countability) i.e.,

$$\lim_{i\to\infty} \rho_\alpha(f^{m_i}(x), p) = 0 \quad \text{and} \quad \lim_{i\to\infty} \rho_\alpha(f^{n_i}(x), q) = 0.$$

We can select a subsequence $\{m'_i\}_{i=1}^\infty$ of $\{m_i\}_{i=1}^\infty$ such that $m'_i > n_i$ for $i = 1, 2, \ldots$. Then $f^{m'_i}(x) \to p$ in the ρ_α-topology of X, i.e., $\lim_{i\to\infty} \rho_\alpha(f^{m'_i}(x), p) = 0$. Also we have $f^m(x) \notin F(f)$ for each $m = 0, 1, 2, \ldots$ and $q \in F(f)$ by part (b). We show that $f^{m'_i}(x) \to q$ in the ρ_α-topology of X, i.e., $\lim_{i\to\infty} \rho_\alpha(f^{m'_i}(x), q) = 0$. We prove this by considering two cases (A) and (B).

(A) $\rho_\alpha(f^{m'_k}(x), q) = 0$ for some $k = 1, 2, \ldots$.

Then $\rho_\alpha(f^{m'_{k+\tau}}(x), q) = 0$ for each $\tau = 1, 2, \ldots$. For, let $m'_{k+\tau} - m'_k = t$. Then by the second part of condition (ii) of the theorem

$$0 = \rho_\alpha(f^{m'_k}(x), q) = \rho_\alpha(f^{m'_k+1}(x), q) = \rho_\alpha(f^{m'_k+t}(x), q) = \rho_\alpha(f^{m'_k+\tau}(x), q).$$

Hence it follows that in this case $\lim_{i\to\infty} \rho_\alpha(f^{m'_i}(x), q) = 0$.

(B) $\rho_\alpha(f^{m'_i}(x), q) = 0$.

Then by the first part of condition (ii) of the theorem

$$\rho_\alpha(f^{m'_i}(x), q) < \rho_\alpha(f^{m'_i - 1}(x), q) < \cdots < \rho_\alpha(f^{n_i}(x), q)$$

Hence $\lim_{i\to\infty} \rho\rho_\alpha(f^{m'_i}(x), q) = 0$ in this case either as $\lim_{i\to\infty} \rho_\alpha(f^{n_i}(x), q) = 0$. Thus we have $\lim_{i\to\infty} \rho_\alpha(f^{m'_i}(x), q) = 0$. Now by the triangle inequality,

$$\rho_\alpha(p, q) \le \rho_\alpha(p, f^{m'_i}(x)) + \rho_\alpha(f^{m'_i}(x), q) \quad \text{for} \quad i = 1, 2, \ldots.$$

Hence $\rho_\alpha(p, q) = 0$ as $\lim_{i\to\infty} \rho_\alpha(p, f^{m'_i}(x)) = 0$ and $\rho_\alpha(f^{m'_i}(x), q) = 0$. Since α is arbitrary, $\rho_\alpha(p, q) = 0$ for each $\alpha \in I$. Since X is τ_h-Hausdorff, this implies that $p = q$.

(d) Finally, we prove that if $L(x)$ consists of just one point, then τ_h-$\lim f^m(x)$ exists. Let $\{y\} = L(x)$. Then there is a subnet $\{f^{n_j}(x)\}_{j\in J}$ of the net $\{f^m(x)\}_{m=1}^\infty$ such that τ_h-$\lim f^{n_j}(x) = y$. Now by Lemma 2.2, $f^{n_j}(x) \to y$ in the ρ_α-topology of X for each $\alpha \in I$. Let $\alpha \in I$ be arbitrarily chosen. Let $\epsilon > 0$. Then since $f^{n_j}(x) \to y$ in the ρ_α-topology of X, there is a $s \in J$ such that $\rho_\alpha(f^{n_j}(x), y) < \epsilon$ for all $j \succ s$ and hence for all $n_j \ge n_s$ where $\succ$ is the relation in J. We now show that for all positive integers $m \ge n_s$, $\rho_\alpha(f^m(x), y) < \epsilon$. We have at most two cases:

Case 1. $0 = \rho_\alpha(f^{n_s}(x), y) < \epsilon$.

Case 2. $0 < \rho_\alpha(f^{n_s}(x), y) < \epsilon$.

In Case 1, $0 = \rho_\alpha(f^{n_s}(x), y) = \rho_\alpha(f^{n_s+1}(x), y) = \cdots = \rho_\alpha(f^m(x), y) < \epsilon$.

In Case 2, $\rho_\alpha(f^m(x), y) < \rho_\alpha(f^{m-1}(x), y) < \cdots < \rho_\alpha(f^{n_s}(x), y) < \epsilon$.

Thus in all cases, $\rho_\alpha(f^m(x), y) < \epsilon$ whenever $m \ge n_s$. Hence $f^m(x) \to y$ in the ρ_α-topology of X. Now since α is arbitrary, $f^m(x) \to y$ in the ρ_α-topology of X for each $\alpha \in I$. Hence by Lemma 2.2, τ_h-$\lim f^m(x) = y$. □

Remark 2.5 The Remark 10 in (J. B. Diaz and Metcalf (1969)) concerning the work of Edelstein [(Edelstein (1962)), Theorem 1 and 3.2] applies equally here.

Corollary 2.4.1 *Suppose in addition to the hypotheses of Theorem 2.4 that for each $x \in X$, $L(x) \ne \emptyset$, then for each $x \in X$, $\{f^n(x)\}_{n=1}^\infty$ converges in τ_h-topology to a fixed point of f.*

Proof. This follows immediately from the above Theorem 2.4. □

Remark 2.6 If we assume in the above Theorem 2.4 that $f(X)$ is compact, then this will insure the additional condition assumed in the above corollary, i.e., for each $x \in X$, $L(x) \ne \emptyset$.

The next theorem is patterned after the Theorem 3.1 of (J. B. Diaz and Metcalf (1969)).

Theorem 2.5 (Tarafdar (1974), p. 221) *Let (X, h) be a Hausdorff uniform space and let $[\rho_\alpha : \alpha \in I] = A^*(h)$. Let $f : x \to X$ be τ_h-continuous. Also let*

(*i*) *$F(f)$ be non-empty;*

(*ii*) *for each $x \in X$ with $x \notin F(f)$ and each $p \in F(f)$, we have for each $\alpha \in I$, $\rho_\alpha(f(x), p) \leq \rho_\alpha(x, p)$;*

(*iii*) *f be asymptotically regular on X.*

Then for each $x \in X$, either $\{f^m(x)\}_{m=1}^{\infty}$ contains no τ_h-convergent subnet, or τ_h-$\lim f^m(x)$ exists and belongs to $F(f)$.

Proof. The proof that $L(x) \subset F(f)$ is exactly the same as given in the part (i) of the proof of Theorem 2.2. The proof that $L(x)$ contains at most one point can be obtained from the part (c) of the proof of the above Theorem 2.4 by ignoring the case A and replacing all the strict inequality signs appearing in case B by $\leq$ signs. The rest of the proof of this theorem can be obtained from part (d) of the Theorem 2.4 by ignoring Case 1 and replacing all the strict inequality signs by $\leq$ in the proof of Case 2. □

Corollary 2.5.1 *Let (X, h) be a Hausdorff uniform space and let $[\rho_\alpha : \alpha \in I] = A^*(h)$. Let f be an asymptotically regular τ_h-continuous mapping of a subset $Y \subset X$ into Y. Also suppose that*

(*i*) *$f(Y)$ is τ_h-compact;*

(*ii*) *for each $x \in Y$ with $x \notin F(f)$, we have for each $\alpha \in I$, $\rho_\alpha(f(x), p) \leq \rho_\alpha(x, p)$.*

Then for each $x \in Y$, the sequence $\{f^m(x)\}_{m=1}^{\infty}$ converges in the τ_h-topology to a fixed point of f.

Proof. Since f is asymptotically regular on Y on $f(Y)$ is compact, we have by Theorem 2.2 that $L(x) \neq \emptyset$ for each $x \in Y$ and $F(f) \neq \emptyset$. Hence the corollary follows from the above Theorem 2.5. □

An Extended Remark. Since our results will concern only with locally convex linear Hausdorff topological spaces, (E, τ) will denote a locally convex linear Hausdorff topological space throughout the rest of this section.

It is well known that given (E, τ), there always exists a family $[\rho_a l : \alpha \in I]$ of seminorms on E which generates the topology τ in E. More specifically, there always exists a family $[\rho_a l : \alpha \in I]$ of seminorms on E such that the family of scalar multiple rU, $r > 0$, of finite intersections $U = \bigcap_{k=1}^{n} U_{\alpha_k}$, where $U_{\alpha_k} = \{x : p_{\alpha_k}(x) \leq 1\}$, forms a neighborhood base at 0 for the topology τ (see (Schaefer (1966), p. 48), or (Kothe (1960), p. 203)).

In the sequel this zero neighborhood base will be denoted by $\mathcal{B}$.

Now for each $\alpha \in I$, the function $\rho_\alpha : E \times E \to \mathbb{R}$ defined by $\rho_\alpha(x, y) = p_\alpha(x - y)$ for each pair $(x, y) \in E \times E$ is a pseudometric on E. Thus by what we noted in the beginning of Section 2.1 the family $[\rho_\alpha : \alpha \in I$ of pseudometrics on E (obtained from the family $[p_\alpha : \alpha \in I$ as above) determines a unique uniformity h on E such that $A^*(h) = [\rho_\alpha : \alpha \in I$. It is well known that $\tau_h = \tau$ (e.g. see (Schaefer (1966), p. 16)).

Now it is straightforward to see that the following definitions (i'), (ii') and (iii') are equivalent respectively to the definitions (i), (ii) and (iii) given in Section 2.1.

Let $\mathcal{U}$ be the family of zero neighborhoods in E. Then we have seen that $\mathcal{B}$ defined above is a base for $\mathcal{U}$. Let f maps $X \subset E$ into X, then

(i') f is said to be nonexpansion on X if $x - y \in U$ implies $f(x) - f(y) \in U$ for each $U \in \mathcal{B}$ and $(x, y) \in X \times X$.

(ii') f is said to be a contraction on X if for each $U \in \mathcal{B}$, there is a real number τ_U, $0 < \tau_U < 1$, such that $x - y \in U$ implies $f(x) - f(y) \in \tau_U U$ for each $U \in \mathcal{B}$ and $(x, y) \in X \times X$.

(iii') f is asymptotically regular on X if for each $x \in X$ and $U \in \mathcal{U}$, there is a positive integer n_0 such that $f^n(x) - f^{n+1}(x) \in U$ for $n \geq n_0$.

[For $(i) \iff (i')$ and $(iii) \iff (iii')$ see Remark 2.2 and (Taylor (1972)). We prove that $(ii') \iff (ii)$. To prove this we first show that f is a contraction with respect to $A^*(h)$. Let $\rho_\alpha \in A(h)$ and $(x, y) \in X \times X$. Also let $\rho_\alpha(x, y) = \tau$. Then $x - y \in \tau U_\alpha = U \in \mathcal{B}$ where $U_\alpha = \{x : p_\alpha(x) \leq 1\}$. Hence by (ii') there is a real number τ_U, $0 < \tau_U < 1$,such that $f(x) - f(y) \in \tau_U U$. This implies that $\rho_\alpha(f(x), f(y)) \leq \tau_U \rho_\alpha(x, y)$. Clearly τ_U depends on α. Hence we can write $\tau_U = \tau(\alpha)$.

Similarly we prove that $(ii) \Rightarrow (ii')$. This can be done as follows. Let $U \in \mathcal{B}$. Then $U = \tau \bigcap_{k=1}^n U_{\alpha_k}$, $\tau > 0$ and $U_{\alpha_k} = \{x : p_{\alpha_k}(x) \leq 1\}$. Let ρ_{α_k} be the corresponding pseudometrics. Choose $\tau_U = \max(\tau(\alpha_k) : k = 1, 2, \ldots, n)$ where α_k's are obtained from the definitions (ii).]

Since now we will be concerned with only the locally convex topology τ in E, unless otherwise stated, all topological concepts such as continuity, convergence, closedness, etc. will hereafter be meant with respect to the topology τ in E.

A subset X of E is called starshaped if there is a point $p \in X$ such that for each $x \in X$ and real t with $0 < t < 1$, $tx + (1 - t)p \in X$. p is called the star centre of X. Every convex subset of E is thus starshaped. □

Lemma 2.4 (Tarafdar (1974), p. 223) *Let X be a complete bounded starshaped subset of E and let f be nonexpansion on X. Then 0 lies in the closure of $(I - f)X$, i.e., in $\overline{(I - f)X}$, where I is the identity map on X.*

Proof. For each t, $0 < t < 1$, we define $f_t(x) = tf(x) + (1 - t)p$, $x \in X$ p being the star centre of X, f_t is a self mapping on X as X is starshaped. Let $U \in \mathcal{B}$ be arbitrary. Let $x - y \in U$. Then $f_t(x) - f_t(y) = t(f(x) - f(y)) \in tU$ as f is nonexpansion on X. Thus f_t is a contraction on X. Now since X is complete and by our Theorem 2.1 f_t has a unique fixed point x_t, say, in X. Now

$$\begin{aligned}(I - f)(x_t) &= x_t - f(x_t)\\ &= x_t - (f_t(x_t) - (1 - t)p)/t, \text{ from the definition of } f_t,\\ &= (1 - 1/t)(x_t - p) \to 0 \text{ as } t \to 1, \text{ because } X \text{ is bounded.}\end{aligned}$$ □

Remark 2.7 Note that in the proof of the above lemma we have not used the fact that X is connected.

Let us express the condition (ii) of Theorem 2.5 and condition (ii) of Theorem 2.4 by saying respectively that f is nonexpansion on X with respect to $F(f)$ and f is contractive (in the terminology of M. Edelstein (Edelstein (1962))) on X with respect to $F(f)$. Thus in (E, τ) f is nonexpansion on $X \subset E$ with respect to $F(f)$ if $x - p \in U$ implies $f(x) - p \in U$ for each $U \in \mathcal{B}$ and $x \in X$ with $x \notin F(f)$ and $p \in F(f)$. f is contractive on X with respect to $F(f)$ if the condition (ii) of Theorem 2.4 holds where ρ_α's are obtained from the corresponding p_α's.

Theorem 2.6 (Tarafdar (1974), p. 223) *Let f be a nonexpansion on a bounded complete starshaped subset X of E. Also let $(I - f)X$ be closed. Then f has a fixed point.*

Proof. By Lemma 2.4, $0 \in \overline{(I - f)X} = (I - f)X$. Hence there is a point $p \in X$ such that $(I - f)(p) = 0$, i.e., $f(p) = p$. □

Corollary 2.6.1 (Tarafdar (1974), p. 223) *Let f be a nonexpansion on a compact starshaped subset X of E. Also suppose that f is contractive on X with respect to $F(f)$. Then for each $x \in X$, the sequence $\{f^m(x)\}_{m=1}^{\infty}$ converges to a fixed point of f.*

Proof. By continuity of $(I - f)$, $(I - f)X$ is compact and hence closed. Now by Theorem 2.3 $F(f)$ is non-empty. Also since $f(X)$ is compact, $L(x) \neq 0$ for each $x \in X$. Hence the corollary follows from Theorem 2.4. □

Theorem 2.7 (Tarafdar (1974), p. 224) *Let f be a continuous asymptotically regular mapping on a closed bounded subset X of E. Also suppose that $(I - f)$ maps closed and bounded subsets of X into closed subset of E. Then for each $x \in X$, $L(x)$ is non-empty and a closed subset of $F(f)$. If in addition f is nonexpansion on X with respect to $F(f)$, then for each $x \in X$, the sequence $\{f^m(x)\}_{m=1}^{\infty}$ converges to a fixed point of f.*

Proof. The proof is identical to the proof of Theorem 3.3 in (Taylor (1972)) because the nonexpansion of f with respect to $F(f)$ is only used there. Alternatively, we prove in the same way as in Theorem 3.3 of (Taylor (1972)) that $F(f) \neq \emptyset$ and $L(x) \neq \emptyset$ for each $x \in X$ and then we refer to our Theorem 2.5. □

Remark 2.8 Clearly this theorem includes the Theorem 3.3 of (Taylor (1972)) which includes the Theorem 6 of Browder and Petryshyn (Browder and Petryshyn (1966)). Also we note that the present theorems weaken the conditions of Theorem 3.3 of (Taylor (1972)) in exactly the same way as the Theorem 3.4 of Diaz and Metcalf (J. B. Diaz and Metcalf (1969)) does to the conditions of Theorem 6 of (Browder and Petryshyn (1966)).

([1]) Note that if n is the corresponding function for the first subnet, i.e., $n(j) = n_j$ for each $j \in J$, then the function $n^1(j) = n(j) + 1 = n_j + 1$ is the required function for the latter subnet.

Diaz and Metcalf have obtained the following result (J. B. Diaz and Metcalf (1969), Theorem 4):

Lemma 2.5 (Tarafdar (1976), p. 341) *Let K be a closed convex subset of a strictly convex Banach space X; $f : K \to K$ a self mapping on K satisfying*

$$(A)\ \|f(x) - f(y)\| \leq \|x - y\| \ \forall x, y \in K;$$

and suppose $f(K)$ is contained in a compact set $K_1 \subset K$. Then for each $x \in K$, the sequence $\{U_n^\lambda(x)\}_{n=1}^\infty$ of iterates converges to a fixed point of f where $U_\lambda : K \to K$ is the mapping defined by

$$U_\lambda(y) = \lambda f(y) + (1 - \lambda)y, y \in K,\ \ 0 < \lambda < 1.$$

With $\lambda = 1/2$ this result reduces to a result of Edelstein (Edelstein (1966)), which, in turn, is an extension of a result of Krasonsel'skii (Krasonsel'skii (1955)).

Barbuti and Guerra (Barbuti and Guerra (1971), pp. 29–31) have obtained the same result by replacing the non-expansive condition (A) by a weaker condition, namely

$$(B)\ \|f(x) - f(y)\| \leq a\|x - y\| + b(\|f(x) - x\| + \|f(y) - y\|),\ \forall x, y \in K,$$

where $a \geq 0, b \geq 0$ and $a + 2b \leq 1$.

In here we obtain this result of Tarafdar (Tarafdar (1976)) on a locally convex linear topological space with strict convexity suitably defined on it and with a contraction condition weaker than (B), so that our result even in Banach space setting will include the result of Barbuti and Guerra as a special case.

To prove our result we will apply our Theorem 2.4 on uniform space by the author (Tarafdar (1974)). (In fact, this result can be viewed as an application of this theorem.)

Definition 2.1 (Tarafdar (1976), p. 342) A seminorm p on a linear space is said to be strictly convex if for each pair x, y of points of E with $p(x - y) \neq 0$, $p(x) = p(y) = 1$, we have always $p(\lambda x + \mu y) < 1$ whenever $\lambda, \mu > 0$ and $\lambda + \mu = 1$.

A locally convex linear topological space (or a locally convex topological vector space) hereafter abbreviated by l.c.l.t. space E, whose topology is generated by a family $[p_\alpha : \alpha \in I]$ of seminorms on E is said to be strictly convex if p_α is strictly convex for each $\alpha \in I$.

Example 2.2 (Strictly convex l.c.l.t. space) Let A be any non-empty set and X_α a strictly convex normed linear space with norm $\|\cdot\|_\alpha$ for each $\alpha \in A$. We consider the cartesian product $\prod_{\alpha \in A} X_\alpha$ and for each $\alpha \in A$, we define

$$p_\alpha(x) = \|x_\alpha\|_\alpha \quad \text{for each} \quad x = \{x_\alpha\} \in \prod_{\alpha \in A} X_\alpha.$$

Then $[p_\alpha : \alpha \in A]$ is a family of seminorms on $\prod_{\alpha \in A} X_\alpha$ and, therefore, generates locally convex topology (which is also the product topology) on $\prod_{\alpha\in A} X_\alpha$. It is now easy to see that $\prod_{\alpha\in A} X_\alpha$ is strictly convex.

Theorem 2.8 (Tarafdar (1976), p. 343) *Let E be a quasicomplete separated and strictly convex l.c.l.t. space; Y a closed convex subset of E and $f : Y \to Y$ be a continuous self-mapping on Y such that $f(Y)$ is contained in a compact set $Y_1 \subset Y$. Further, suppose that*

(C) for each $\alpha \in I$, there exist functions $d_1(\alpha, \cdots)$ and $d_2(\alpha, \cdots)$ of $Y \times Y$ into $[0, \infty)$ such that $d_1(\alpha, x, y) + 2d_2(\alpha, x, y) \le 1$ for all $x, y \in Y$ and for any $x, y \in Y$,

$$p_\alpha(f(x)-f(y)) \le \alpha_1(\alpha)p_\alpha(x-y)+\alpha_2(\alpha)p_\alpha(f(x)-x)+p_\alpha(f(y)-y), \quad \text{if } p_\alpha(x-y) \neq 0$$

and $p_\alpha(f(x) - f(y)) = 0$ if $p_\alpha(x - y) = 0$ where $\alpha_i(\alpha) = d_i(\alpha, x, y)$, $i = 1, 2$.

Then for each $x \in Y$, the sequence $\{U_\lambda^n(x)\}_{n=1}^\infty$ of iterates where $U_\lambda : Y \to Y$ is the mapping defined by

$$U_\lambda(y) = \lambda f(y) + (1 - \lambda)y, \ \ y \in Y, \ \ 0 < \lambda < 1$$

converges to a fixed point of f.

Proof. $U_\lambda(Y) \subset Y$ as Y is convex. Also by definition of U_λ and by virtue of the Tychonoff fixed point theorem (see Chapter 4) we have

$$F(f) = F(U_\lambda) \neq \emptyset \tag{2.6}$$

where $F(f)$ and $F(U_\lambda)$ denote the set of fixed points of f and U_λ respectively. Since E is quasi-complete, it follows (see (Kothe (1960), p. 241)) that the closed convex hull of $f(Y) \cup \{x\}$ is compact for each $x \in Y$.

$$\text{Hence, for each } x \in Y, \text{ the sequence } \{U_\lambda^n(x)\}_{n=1}^\infty \text{ being in this compact set} \tag{2.7}$$

has a convergent subnet.

Now for each $u \in F(f)$ and $x \in Y$ with $x \notin F(f)$ we have by (C) for each $\alpha \in I$, if $p_\alpha(x - u) \neq 0$,

$$\begin{aligned} p_\alpha(f(x) - u) &= p_\alpha(f(x) - f(u)) \\ &\le a(\alpha)p_\alpha(x - u) + b(\alpha)[p_\alpha(f(x) - x) + p_\alpha(f(u) - u)] \\ &\le a(\alpha)p_\alpha(x - u) + b(\alpha)[p_\alpha(f(x) - u) + p_\alpha(x - u), \end{aligned}$$

where $a(\alpha) = d_1(\alpha, x, u)$ and $b(\alpha) = d_2(\alpha, x, u)$. It follows that

$$p_\alpha(f(x) - u) = 0 \quad \text{if} \quad p_\alpha(x - u) = 0 \tag{2.8}$$

and since $\frac{(a(\alpha)+b(\alpha))}{1-b(\alpha)} \le 1$, it follows that

$$p_\alpha(f(x) - u) \le p_\alpha(x - u). \tag{2.9}$$

Now

$$p_\alpha(U_\lambda(x) - u) = p_\alpha(U_\lambda(x) - U_\lambda(u)) \tag{2.10}$$
$$= p_\alpha(\lambda(f(x) - u) + (1-\lambda)(x-u)). \tag{2.11}$$

From (2.8), and (2.10) we have

$$p_\alpha(U_\lambda(x) - u) = 0 \quad \text{if} \quad p_\alpha(x-u) = 0. \tag{2.12}$$

If $p_\alpha(x-u) \neq 0$, then from (2.10) we have

$$p_\alpha(U_\lambda(x) - u) = p_\alpha(x-u)\left[p_\alpha(\lambda\frac{f(x)-u}{p_\alpha(x-u)} + (1-\lambda)\frac{(x-u)}{(p_\alpha(x-u))}\right]. \tag{2.13}$$

If strictly inequality holds in (2.9), then from (2.13) we obtain

$$p_\alpha(U_\lambda(x) - u) < p_\alpha(x-u). \tag{2.14}$$

If equality holds in (2.9), then strict convexity reduces (2.13) to (2.14).

Thus (2.6), (2.7), (2.12), and (2.14) ensure that all the conditions required in Theorem 2.4 are fulfilled for the mapping U_λ. Hence the conclusion of our theorem follows from Theorem 2.4. □

Remark 2.9 In Banach space situation contractions of type (C) have been considered by Hardy and Rogers (Hardy and Rogers (1973)), Chi Song Wong (Wong (1974)) and others (see references in (Wong (1974))).

2.3 Further Generalization of Banach Contraction Mapping Principle

A nonempty set P with a partial order relation $\leq$ (reflexive, antisymmetric and transitive) is called a partially ordered set and is denoted by $(P, \leq)$. A subset C of $(P, \leq)$ is called a chain if, given any two elements x, y of C, either $x \leq y$ or $y \leq x$. An element $x \in P$ is called a maximal element if $x \leq y \Rightarrow x = y$.

Let A be a nonempty subset of $(P, \leq)$. An element $x \in P$ is called a lower (resp. an upper) bound of A if $x \leq a$ (resp. $a \leq x$) for all $a \in A$. A lower (resp. an upper) bound of A is called infimum (resp. supremum) of A if it is greater than or equal to (resp. less than or equal to) each lower (resp. each upper bound of A). If the supremum or infimum of A exists, it is unique (see (Shafer and Sonnenschein (1975), pp. 43–44)) and is, respectively denoted $\sup A$ and $inf A$.

In this section we have proved several fixed point theorems which are generalization of the fixed point Theorem 2.1 and are generalization of the corresponding fixed point theorems of Caristi and Kirk (Caristi and Kirk (1975)) on metric spaces to uniform topological spaces. Later in this section we have shown the interplay of order and pseudometrics.

2.3.1 *Fixed Point Theorems for Some Extension of Contraction Mappings on Uniform Spaces*

Lemma 2.6 (Cantor Intersection Theorem in Uniform Space) *Let (X, h) be complete Hausdorff uniform topological space with uniform topology τ_h and $[\rho_\alpha : \alpha \in I] = A^*(h)$. Let $\{X_n : n = 1, 2, \dots\}$ be a sequence of nonempty closed set with finite intersection property such that for each $\alpha \in I$,*

$$\delta_\alpha(X_n) = \sup\{\rho_\alpha(x, y) : x, y \in X_n\} \to 0 \ \textit{as} \ \ n \to \infty.$$

Then $A = \bigcap_{n=1}^{\infty} X_n$ consists of a single point x_0, i.e. $\bigcap_{n=1}^{\infty} X_n = \{x_0\}$.

Proof. First we prove that A cannot contain two distinct points. Let $x, y \in A$. Suppose $x \neq y$. Then since X is Hausdorff, $\rho_\alpha(x, y) = \epsilon > 0$ for some $\alpha \in I$. Now since $\delta_\alpha(X_n) \to 0$ as $n \to \infty$, there exists a positive integer N_0 such that $\delta_\alpha(X_n) < \epsilon$ for all $n \geq N_0$. It follows that $\rho_\alpha(x, y) \leq \delta_\alpha(X_{N_0}) < \epsilon$ which is a contradiction. Thus $x = y$.

Now we prove that $A \neq \emptyset$. We choose a point x_n from each X_n for each n. Then the sequence $\{x_n\}$ is a ρ_α-Cauchy sequence for each $\alpha \in I$. To see this, let $\epsilon > 0$ be given. Let $\alpha \in I$ be fixed but arbitrary. Then since $\delta_\alpha(X_n) \to 0$, there exists a positive integer N such that $\delta_\alpha(X_n) < \frac{\epsilon}{2}$ for all $n \geq N$. Now for all $m, n \geq N$, we have $x_m \in X_m, x_n \in X_n$ and $z \in X_m \cap X_n$, $\rho_\alpha(x_m, x_n) \leq \rho_\alpha(x_m, z) + \rho_\alpha(z, x_n) \leq \delta_\alpha(X_m) + \delta_\alpha(X_n) < \frac{\epsilon}{2} + \frac{\epsilon}{2} = \epsilon$. Next, let for each positive integer p, $S_p = \{x_n : n \geq p\}$ and let β be the filter basis $\{S_p : p = 1, 2, \dots\}$. Then since $\{x_n\}_{n=1}^{\infty}$ is ρ_α-Cauchy for each $\alpha \in I$, it is easy to see that the filter basis β is Cauchy in the uniform space (X, h). Since (X, h) is complete, the filter basis $\beta = \{S_p\}$ converges to a unique point $x_0 \in X$ in the uniform topology τ_h. It follows that $\tau_h\text{-}\lim_{n\to\infty} x_n = \tau_h\text{-}\lim S_p = x_0$ and, therefore, $\rho_\alpha(x_n, x_0) \to 0$ for each $\alpha \in I$. Now we prove $x_0 \in X_n$ for each n. Let n_0 be fixed but arbitrary. Let G be an τ_h-open neighborhood of x_0. Then there exists $r > 0$ and $\rho_\alpha \in A^*(h)$ such that $S_r(\alpha, x_0) \subset G$. Now since $\rho_\alpha(x_n, x_0) \to 0$, there exists a positive integer N_1 such that $\rho_\alpha(x_n, x_0) < \frac{r}{2}$ for all $n \geq N_1$ and since $\delta_\alpha(X_n) \to 0$, there exists a positive integer N_2 such that $\delta_\alpha(X_n) < \frac{r}{2}$ for all $n \geq N_2$. Let $N = \max(N_1, N_2)$. By finite intersection property, there is point $u \in X_N \cap X_{n_0}$. Now $\rho_\alpha(u, x_0) \leq \rho_\alpha(u, x_N) + \rho_\alpha(x_N, x_0) \leq \delta_\alpha(X_N) + \rho_\alpha(x_N, x_0) < \frac{r}{2} + \frac{r}{2} = r$. Thus $u \in X_{n_0} \cap G$. This implies that x_0 is a limit point of x_{n_0} and $x_0 \in \overline{X}_{n_0} = X_{n_0}$. Since n_0 is arbitrary, $x_0 \in A$. □

Now we include some extensions of Banach contraction principle which are generalization of the corresponding results of Caristi and Kirk from metric (Caristi and Kirk (1975)), see also Ekland (Ekland (1972)) spaces to uniform spaces.

Theorem 2.9 *Let (X, h) be a complete uniform space as in Lemma 2.6 and $T : X \to X$ a mapping, not necessarily continuous. Assume that for each $\alpha \in I$ and each $\epsilon > 0$, there exists $\delta(\alpha, \epsilon) > 0$ such that $T[S_\epsilon(\alpha, x)] \subset S_\epsilon(\alpha, x)$, whenever $0 \neq \rho_\alpha(x, T(x)) < \delta(\alpha, \epsilon)$.*

Then, if for each $\alpha \in I$ $\rho_\alpha(T(u), T^{n+1}(u)) \to 0$ for some $u \in X$, the sequence $\{T^n(u)\}$ converges to a fixed point of T.

Proof. We let $T(u) = u_1, u_2 = T(u_1) = T^2(u), \ldots, T^n(u) = u^n, \ldots$. We show that $\{u_n\}$ is ρ_α-Cauchy for each $\alpha \in I$. Let $\alpha \in I$ be arbitrary. Given $\frac{\epsilon}{2} > 0$ we can choose a sufficiently large positive integer N such that $\rho_\alpha(u_n, u_{n+1}) < \delta(\alpha, \frac{\epsilon}{2})$ for all $n \geq N$. Since $\rho_\alpha(u_N, u_{N+1}) < \delta(\alpha, \frac{\epsilon}{2})$ we have $T[S_{\frac{\epsilon}{2}}(\alpha, u_N)] \subset S_{\frac{\epsilon}{2}}(\alpha, u_N)$ which implies that $u_{N+1} = T(u_N) \in S_{\frac{\epsilon}{2}}(\alpha, u_N)$. Thus by induction $T^k(u_N) = u_{N+k} \subset S_{\frac{\epsilon}{2}}(\alpha, u_N)$ for all $k \geq 0$ and hence $\rho_\alpha(u_m, u_n) \leq \rho_\alpha(u_m, u_N) + \rho_\alpha(u_N, u_m) < \frac{\epsilon}{2} + \frac{\epsilon}{2} = \epsilon$ for all $m, n \geq N$. Hence $\{u_N\}$ is ρ_α-Cauchy for all $\alpha \in I$. As noted in the lemma $\beta = \{S_p\}$, where $S_p = \{u_n : n \geq p\}$ is a Cauchy filter which converges to a point $x_0 \in X$ as (X, h) is complete. It follows that $\rho_\alpha(u_n, x_0) \to 0$ for each $\alpha \in I$. Now we prove that $Tx_0 = x_0$. If possible, suppose $Tx_0 \neq x_0$. Since (X, h) is Hausdorff, there must exist one $\alpha \in I$ such that $\rho_\alpha(x_0, Tx_0) = r > 0$. Then we can choose an $u_n \in S_{\frac{r}{3}}(\alpha, x_0)$ such that $\rho_\alpha(u_n, u_{n+1}) < \delta(\alpha, \frac{r}{3})$. We would then have $T[S_{\frac{r}{3}}(\alpha, u_n)] \subset S_{\frac{r}{3}}(\alpha, u_n)$ by the hypothesis. But this would imply $Tx_0 \in S_{\frac{r}{3}}(\alpha, u_n)$ and $\rho_\alpha(Tx_0, u_n) \geq \rho_\alpha(Tx_0, x_0) - \rho_\alpha(x_0, u_n) = \frac{2}{3}r$, i.e. $Tx_0 \notin S_{\frac{r}{3}}(\alpha, u_n)$. Thus we have contradiction. It follows that $\rho_\alpha(x_0, Tx_0) \neq 0$ for all $\alpha \in I$. Hence $Tx_0 = x_0$. □

Corollary 2.9.1 *Let (X, ρ) be a complete metric space and $T : X \to X$ a mapping, not necessarily continuous. Assume that for each $\epsilon > 0$, there is a $\delta(\epsilon) > 0$ such that $T[S_\epsilon(x)] \subset S_\epsilon(x)$ whenever $\rho(x, T(x)) < \delta(\epsilon)$. Then, if $\rho(T^n(u), T^{n+1}(u)) \to 0$ for some u, the sequence $\{T^n(u)\}$ converges to a fixed point of T.*

Proof. Take $I = \{1\}$ in Theorem 2.9. □

Theorem 2.10 *Let (x, h) be a complete Hausdorff uniform space as in Lemma 2.6 and T a mapping, not necessarily continuous. Let for each $\alpha \in I$, there is a monotone non-decreasing (not necessarily continuous) function $\phi_\alpha : \mathbb{R}^+ \to \mathbb{R}^+$ with $\phi_\alpha^n(t) \to 0$ for each fixed $t > 0$ such that $\rho_\alpha(Tx, Ty) \leq \phi_\alpha[\rho_\alpha(x, y)]$.*

Then T has a unique fixed point u, and $T^n(x) \to u$ for each $x \in X$.

Proof. First we begin by noting that for each $\alpha, \phi_\alpha(t) < t$ with each $t > 0$. For if $t \leq \phi_\alpha(t)$, then by monotonicity $\phi_\alpha(t) \leq \phi_\alpha^2(t)$ and by induction $t \leq \phi_\alpha(t) \leq \phi_\alpha^2(t) \leq \cdots \leq \phi_\alpha^n(t) \leq \cdots$. This would imply $t \leq 0$ which would be contradiction. Now let $\alpha \in I$ be arbitrary but fixed. By hypothesis we have $\rho_\alpha(T^n(x), T^{n+1}(x)) \leq \phi_\alpha^n[\rho_\alpha(x, T(x)]$. Assume $\rho_\alpha(x, T(x) \neq 0$.

Thus $\rho_\alpha(T^n(x), T^{n+1}(x)) \to 0$ for each $x \in X$. Now given a real number $\epsilon > 0$, we choose $\delta(\alpha, \epsilon) = [\epsilon - \phi_\alpha(\epsilon)] > 0$. Then if $\rho_\alpha(x, Tx) < \delta(\alpha, \epsilon)$ and $y \in S_{\delta(\alpha,\epsilon)}(\alpha, x)$, we have $\rho_\alpha(Ty, x) \leq \rho_\alpha(Ty, Tx) + \rho_\alpha(Tx, x) \leq \phi_\alpha[\rho_\alpha(y, x)] + \delta(\alpha, \epsilon) < \phi_\alpha(\epsilon) + \epsilon - \phi_\alpha(\epsilon) = \epsilon$. Thus $T(y) \in S_\epsilon(\alpha, x)$, whenever $T(x) \in S_{\delta(\alpha,\epsilon)}(\alpha, x)$, i.e. $T[S_\epsilon(\alpha, x)] \subset S_\epsilon(\alpha, x)$.

Hence all the conditions of the Theorem 2.9 hold. The conclusion of Theorem 2.10 follows from Theorem 2.9. □

Corollary 2.10.1 *Let (X,ρ) be a complete metric space and $T : X \to X$ be a mapping, not necessarily continuous. Assume that there is a monotone non-decreasing function $\phi : \mathbb{R}^+ \to \mathbb{R}^+$ with $\phi^n(t) \to 0$ as $n \to \infty$ for each $t > 0$ such that $\rho(T(x), Ty) \leq \phi[\rho(x,y)]$ for all $x, y \in X$. Then there is a unique fixed u of T and $T^n(x) \to u$ for each $x \in X$.*

Proof. Let $I = \{1\}$ and apply Theorem 2.9. □

Corollary 2.10.2 (Boyd and Wong (1969)) *Let (X,ρ) be a complete metric space and $T : X \to X$ a mapping satisfying $\rho(T(x), T(y)) \leq \phi[\rho(x,y)]$ for all $x, y \in X$ where $\phi : \mathbb{R}^+ \to \mathbb{R}^+$ ia any function such that (i) ϕ is non-decreasing, (ii) $\phi(t) < t$ for each $t > 0$ and (iii) ϕ is right continuous. Then the conclusion of Corollary 2.10.1 holds.*

Proof. We can easily verify that $\phi^n(t) \to 0$ for each $t > 0$. The conclusion follows from Corollary 2.10.1. □

Remark 2.10 Theorem 2.9 and hence Theorem 2.10 is a generalization of Banach Contraction Mapping Principle on uniform topological space. To see this let $\epsilon > 0$ be given. Let $\delta(\alpha,\epsilon) = \epsilon - \lambda_\alpha\epsilon$ where λ_α is the contraction constant for $\alpha \in I$. Now let $y \in T(S_\epsilon(\alpha,x))$. Then $y = T(u)$ for some $u \in S_\epsilon(\alpha,x)$. Then, if $\rho_\alpha(x,Tx) < \delta(\alpha,\epsilon), \rho_\alpha(y,x) = \rho_\alpha(T(u),x) \leq \rho_\alpha(T(u),T(x)) + \rho_\alpha(T(x),x) \leq \lambda_\alpha\epsilon + \epsilon - \lambda_\alpha\epsilon = \epsilon$. This proves $T[S(\alpha,x)] \subset S_\epsilon(\alpha,x)$, whenever $\rho_\alpha(x,Tx) < \delta(\alpha,\epsilon)$.
In our next theorem we allow the contraction constants to depend on $\rho_\alpha(x,y)$.

Theorem 2.11 *Let (X,h) be a complete uniform Hausdorff space and $T : X \to X$ a mapping, not necessarily continuous. Let for each $\alpha \in I$, $\rho_\alpha(T(x),T(y)) \leq \psi_\alpha(x,y)\rho_\alpha(x,y)$ for all $x, y \in X$, where $\psi_\alpha : X \times X \to \mathbb{R}^+$ satisfies the condition: for any closed interval $[a,b] \subset \mathbb{R}^+ - \{0\}$,*

$$\sup[\psi_\alpha(x,y) : a \leq \rho_\alpha(x,y) \leq b] = \lambda_\alpha(a,b) < 1$$

Then T has a unique fixed point u and $T^n(x) \to u$ for each $x \in X$.

Proof. We note that $\rho_\alpha(x,y) = 0$ implies that $\rho_\alpha(Tx,Ty) = 0$ by the given condition. Let α be arbitrary but fixed. For each $x \in X$, the sequence $\{\rho_\alpha(T^n(x),T^{n+1}(x))\}$ being nonincreasing converges to some real number $t_\alpha \geq 0$. We will prove that $t_\alpha = 0$. If possible, $t_\alpha > 0$. Then $\rho_\alpha(T^n(x),T^{n+1}(x)) \in [t_\alpha, t_\alpha + 1]$ for all $n \geq N$ for some positive integer N. We set $c_\alpha = \lambda_\alpha(t_\alpha, t_\alpha+1)$. Then by induction $t_\alpha \leq \rho_\alpha(T^{N+k}(x),T^{N+k+1}(x)) \leq c_\alpha^k\rho_\alpha(T^N(x),T^{N+1}(x)) \leq c_\alpha^k(t_\alpha+1)$ for all integers $k > 0$. This leads to a contradiction as $c_\alpha < 1$ (choose sufficiently large k so as to make $c^k(t_\alpha+1) \leq t_\alpha$). Now let $\epsilon > 0$ be given. Set $\lambda_\alpha = \lambda_\alpha[\frac{\epsilon}{2},\epsilon]$. Choose $\delta(\alpha,\epsilon) = \min[\frac{\epsilon}{2}, \epsilon(1-\lambda_\alpha)]$. Let $0 \neq \rho_\alpha(x,T(x) < \delta(\alpha,\epsilon)$ and $y \in S_\epsilon(\alpha,x)$. Then $\rho_\alpha(T(y),x) \leq \rho_\alpha(T(y),T(x) + \rho_\alpha(T(x),x)$.

We now consider the following two cases:

(*i*) $\rho_\alpha(y,x) < \frac{\epsilon}{2}$. Then $\rho_\alpha(T(y),x) \le \rho_\alpha(y,x) + \rho_\alpha(T(x),x) < \frac{\epsilon}{2} + \frac{\epsilon}{2} = \epsilon$;

(*ii*) $\frac{\epsilon}{2} \le \rho_\alpha(y,x) \le \epsilon$. Then $\rho_\alpha(T(y),x) \le \psi_\alpha(y,x)\rho_\alpha(y,x) + \rho_\alpha(T(x),x) < \lambda_\alpha\epsilon + \epsilon(1-\lambda_\alpha) = \epsilon$.

Thus $T[S_\epsilon(\alpha,x)] \subset S_\epsilon(\alpha,x)$ whenever $\rho_\alpha(x,T(x)) < \delta(\alpha,\epsilon)$. Now the conclusion of the theorem follows from Theorem 2.9. □

Corollary 2.11.1 *Let (X,ρ) be a complete metric spaces and $T: X \to X$ be a mapping, not necessarily continuous. Let $\rho(T(x),T(y)) \le \psi(x,y)\rho(x,y)$ for all $x,y \in X$, where $\psi: X \times X \to \mathbb{R}^+$ satisfies the condition: for any closed interval $[a,b] \subset \mathbb{R}^+ - \{0\}$,*

$$\sup\{\psi(x,y) : a \le \rho(x,y) \le b\} = \lambda(a,b) < 1.$$

Then T has a unique fixed point $u \in X$ and $T^n(x) \to 0$ for each $x \in X$.

Proof. Let $I = \{1\}$ and apply Theorem 2.11. □

Now we give another kind of generalization of Banach Contraction Theorem on uniform spaces.

Theorem 2.12 *Let (X,h) be a complete uniform Hausdorff space. For each $\alpha \in I$, let $\phi_\alpha : X \to \mathbb{R}^+$ be a non-negative function (not necessarily continuous) satisfying:*

$$(*) \qquad \inf\{\phi_\alpha(x) + \phi_\alpha(y) : \rho_\alpha(x,y) \ge a\} = \mu_\alpha(a) > 0 \quad \textit{for all} \quad a > 0.$$

Then each sequence $\{x_n\}$ with $\phi_\alpha(x_n) \to 0$ for each $\alpha \in I$ converges to one and the same point $u \in X$.

Proof. For each $\alpha \in I$, let $A_\alpha(n) = \{x \in X : \phi_\alpha(x) \le \phi_\alpha(x_n)\}$ for each positive integer n. $A_\alpha(n) \ne \phi$ as $x_n \epsilon A_\alpha(n)$. By using $\phi_\alpha(x_n) \to 0$, we can easily see that the sequence $[A_\alpha(n) : n = 1,2,\ldots]$ has finite intersection property. We now prove that $\delta_\alpha(A_\alpha(n)) \to 0$. Let $\epsilon > 0$ be given. Since $\mu_\alpha(\epsilon) > 0$, there exists a positive integer N such that $\phi_\alpha(x_n) < \frac{1}{2}\mu_\alpha(\epsilon)$ for all $n \ge N$. Then for all $n \ge N$, and $x,y \in A_\alpha(n), \phi_\alpha(x) + \phi_\alpha(y) < \mu_\alpha(\epsilon)$. It then follows from $(*)$, $\rho_\alpha(x,y) < \epsilon$. Thus $\delta_\alpha(A_\alpha(n)) < \epsilon$ for all $n \ge N$. Hence $\delta_\alpha(A_\alpha(n)) \to 0$. Now let $A_n = \{x \in X : \phi_\alpha(x) \le \phi_\alpha(x_n)$ for all $\alpha \in I\} = \cap_{\alpha\in I} A_\alpha(n)$. Since for each $n, A_n \subseteq A_\alpha(n)$. We have $\delta_\alpha(A_n) = \sup\{\rho_\alpha(x,y) : x,y \in A_n\} \le \delta_\alpha(A_\alpha(n))$. Hence $\delta_\alpha(A_n) \to 0$. Now by Lemma 2.6, $\cap_{n=1}^\infty \overline{A}_n = \{u\}$ for some $u \in X$. Now since $x_n \in A_n$ for each n, it is easy to prove $\tau_h - \lim_{n\to\infty} x_n = u$. Hence $\rho_\alpha(x_n,u) \to 0$ for each $\alpha \in I$.

Let $\{y_n\}$ be another sequence with $\phi_\alpha(y_n) \to 0$ for each $\alpha \in I$ and $y_n \to y$ as proved for sequence $\{x_n\}$. Then $\phi_\alpha(x_n) + \phi_\alpha(y_n) \to 0$ for each $\alpha \in I$. Let α be fixed but arbitrary and $\epsilon > 0$ be given. Then there exists a positive integer N such that $\phi_\alpha(x_n) + \phi_\alpha(y_n) < \mu_\alpha(\epsilon)$ for all $n \ge N$. Now by virtue of $(*)$, $\rho_\alpha(x_n,y_n) < \epsilon$ for all $n \ge N$. Thus $\rho_\alpha(x_n,y_n) \to 0$. Since α is arbitrary, $\rho_\alpha(x_n,y_n) \to 0$ for each $\alpha \in I$. Hence $\rho_\alpha(x,y) = 0$ for each $\alpha \in I$. Since X is Hausdorff, $x = y$. □

Now taking $I = \{1\}$ in the above theorem we have the following corollary, (see Dugundji and Granas (Dugundji and Granas (1982)).

Corollary 2.12.1 *Let (X, ρ) be a complete metric space and $\phi : X \to \mathbb{R}^+$ be a non-negative function satisfying:*

$$\inf\{\phi(x) + \phi(y) : \rho(x, y) \geq a\} = \mu(a) > 0 \quad \textit{for all} \quad a > 0.$$

Then for each sequence $\{x_n\}$ in X with $\phi(x_n) \to 0$ converges to one and the same point $u \in X$.

Theorem 2.13 *Let (X, h) be a complete uniform Hauddorff space and $T : X \to X$ a continuous mapping. For each $\alpha \in I$, assume that $\phi_\alpha(x) = \rho_\alpha(x, Tx)$ satisfies the condition $(*)$ above and $\inf_{x \in X} \rho_\alpha(x, T(x)) = 0$. Then there is a unique fixed point of T.*

Proof. Theorem follows immediately from Theorem 2.12. □

Remark 2.11 Theorem 2.13 is a generalization of Banach contraction mapping principle. To see this let $\rho_\alpha(T(x), T(y)) \leq \lambda(\alpha)\rho_\alpha(x, y) x, y \in X$. Then $\phi_\alpha(x) = \rho_\alpha(x, T(x))$ satisfies $(*)$ of Theorem 2.12. Indeed $[1 - \lambda(\alpha)]\rho_\alpha(x, y) \leq \rho_\alpha(x, y) - \rho_\alpha(T(x), T(y)) \leq \rho_\alpha(x, T(x)) + \rho_\alpha(y, T(y))$. Also it is easy to see that $\inf_{x \to X} \rho_\alpha(x, T(x)) = 0$ from the fact that $\rho_\alpha(T^n(x), T^{n+1}(x)) \to 0$ from each $x \in X$.

As before we have the following corollary.

Corollary 2.13.1 *Let (X, ρ) be a complete metric space and T a continuous mapping. Assume $\phi(x) = \rho(x, T(x))$ has the property:*

$$\inf\{\phi(x) + \phi(y) : \rho(x, y) \geq a\} = \mu(a) > 0 \quad \textit{for all} \quad a > 0$$

and $\inf_{x \in X} \rho(x, T(x)) = 0$. Then T has a unique fixed point.

2.3.2 *An Interplay Between the Order and Pseudometric Partial Ordering in Complete Uniform Topological Space*

In this section we display the interconnection between the order and pseudometric partial ordering in a complete uniform space. Let (X, h) be a uniform space and $A^*(h) = [\rho_\alpha : \alpha \in I]$.

For each $\alpha \in I$, let $\phi_\alpha : X \to \mathbb{R}$ be a real valued function. Then for each $\alpha \in I$, we can define a preorder relation (reflexive and transitive) $(\leq_\alpha$ on X by $x \leq_\alpha y$ if and only if $\rho_\alpha(x, y) \leq \phi_\alpha(x) - \phi_\alpha(y)$. Observe that transitivity of $\leq_\alpha$ follows from the triangle inequality of ρ_α. Thus we have a family $\{\leq_\alpha : \alpha \in I\}$ of preorder relations on X. From these relations we now define another partial order relation $\leq$ on X by $x \leq y$ if and only if $x \leq_\alpha y$ (as X is Hausdorff, it is to see that $\leq$ is antisymmetric).

Theorem 2.14 *Let (X, h) be a complete Hausdorff uniform space and let for each $\alpha \in I$, $\phi_\alpha : X \to \mathbb{R}$ a lower semi-continuous function with a common finite lower bound. Then given $x_0 \in X$, there is at least one maximal element $\overline{x}$ with respect to $\leq$ (as defined above) with $x_0 \leq \overline{x}$.*

Proof. For any $z \in X$ and each $\alpha \in I$, denote the set $\{x \in X : z \leq_\alpha x\}$ by $T_\alpha(z)$. For each $\alpha \in I, T_\alpha(z) = \{x \in X : \phi_\alpha(x) + \rho_\alpha(z,x) \leq \phi_\alpha(z)\}$ is a closed set by virtue of the lower semicontinuity of the map $x \to \phi_\alpha(x) + \rho_\alpha(z,x)$.
Hence $T(z) = \{x \in X : z \leq x\} = \{x \in X : z \leq_\alpha x$ for each $\alpha \in I\} = \cap_{\alpha \in I} T_\alpha(z)$ is a closed set. Note $T(z)$ is nonempty for all $z \in X$ as $z \in T(z)$.
Now let $x_0 \in X$ be given. We construct an ascending sequence $x_0 \leq x_1 \leq x_2 \leq \cdots$ inductively. We choose $x_1 \in T(x_0)$ so that $\phi_\alpha(x_1) \leq 1 + \inf[\phi_\alpha \backslash T(x_0)]$ for all $\alpha \in I$. (Note this could be done as ϕ_α has common lower bound independent of α.) Next we choose $x_2 \in T(x_1)$ so that $\phi_\alpha(x_2) \leq \frac{1}{2} + \inf[\phi_\alpha \backslash T(x_1)]$. We continue this process to obtain the sequence $\{x_n\}$ inductively with $x_n \in T(x_{n-1})$ and $\phi_\alpha(x_n) \leq \frac{1}{n} + \inf[\phi_\alpha \backslash T(x_{n-1})]$. Clearly $T(x_0) \supset T(x_1) \supset \cdots$ is a descending sequence of nonempty closed sets. Next we show that $\delta_\alpha[T(x_n)] \to 0$ for each $\alpha \in I$. Let $u \in T(x_n) \subset T(x_{n-1})$. Then $\phi_\alpha(u) \geq \inf[\phi_\alpha \backslash T(x_{n-1})] \geq \phi_\alpha(x_n) - \frac{1}{n}$ for each $\alpha \in I$. Now since $x_n \leq u$, we have $x_n \leq_\alpha u$ for each $\alpha \in I$ and hence $\rho_\alpha(x_n, u) \leq \phi_\alpha(x_n) - \phi_\alpha(u) < \frac{1}{n}$ for each $\alpha \in I$. Thus $\delta_\alpha[T(x_n)] \leq \frac{2}{n}$. Hence by the Cantor Intersection Theorem in Uniform Space $\cap_{n=0}^{\infty} T(x_n) = \{\overline{x}\}$. Since $\overline{x} \in T(x_n)$ for each n, we have $x_n \leq \overline{x}$. Also $\overline{x}$ is a maximal element. Let, if possible, $\overline{x} \leq v$. Then $x_n \leq \overline{x} \leq v$ and $v \in \cap_{n=0}^{\infty} T(x_n) = \{\overline{x}\}$. Thus $\overline{x} = v$. □

The above theorem extends the result of Bronsted (Brondsted (1974)).

Corollary 2.14.1 *Let (X, ρ) be a complete metric space and let $\phi : X \to \mathbb{R}$ be a lower semicontinuous function with finite lower bound. Then given $x_0 \in X$, there is maximal element $\overline{x} \leq x_0$ with respect to $\leq$, where $\leq$ is defined on X by: $x \leq y$ if and only if $\rho(x,y) \leq \phi(x) - \phi(y)$.*

Proof. Take $I = \{1\}$ in the previous theorem. □

Theorem 2.15 *Let (X, h) be a complete Hausdorff uniform space and let for each $\alpha \in I$, $\phi_\alpha : X \to \mathbb{R}$ a lower semicontinuous function with a commom finite lower bound. Let $T : X \to X$ be a mapping (not necessarily continuous) such that for each $\alpha \in I$, $\rho_\alpha(x, T(x)) \leq \phi_\alpha(x) - \phi_\alpha(Tx), x \in X$. Then T has a fixed point.*

Proof. By the preceeding theorem there is a maximal element $\overline{x}$. Since $\rho_\alpha(\overline{x}, T(\overline{x})) \leq \phi_\alpha(\overline{x}) - \phi_\alpha(T(\overline{x}))$, we have $\overline{x} \leq_\alpha T(\overline{x})$ for each $\alpha \in I$. Thus $\overline{x} \leq T(\overline{x})$. Hence $\overline{x} = T(\overline{x})$ by the maximality of $\overline{x}$. □

Corollary 2.15.1 *Let (X, ρ) be a complete metric space and $\phi : X \to \mathbb{R}$ a lower semicontinuous function with a finite lower bound. Let $T : X \to X$ be a mapping such that $\rho(x, T(x)) \leq \phi(x) - \phi(T(x))$ for each $x \in X$. Then T has a fixed point.*

Proof. Take $I = \{1\}$ and apply Theorem 2.15. □

2.4 Changing Norm

At the outset we note that if we have two equivalent norms $\|\cdot\|$ and $\|\cdot\|_1$ in a Banach space E (i.e., if we have $m\|x\|_1 \leq \|x\| \leq M\|x\|_1$ for all $x \in E$ with some constants m and M), then a mapping which is Lipschitzian in one norm remains Lipschitzian in other. Thus for a mapping $T : E \to E$ which is Lipschitzian in one norm with Lipschitz constant ≥ 1, it is of paramount interest to seek an equivalent norm which makes T Lipschitzian with the corresponding Lipschitz constant < 1. To the best of the knowledge of the author this idea was first applied by Bielecki (Bielecki (1956), pp. 265–268) and Chu and Diaz (Chu and Diaz (1964/65, pp. 351–363)).

In this section $E = \mathcal{C}[0,T]$ is the Banach space of all continuous real valued functions defined on $[0,T]$ with the norm $\|u\| = \sup_{0\leq t\leq T} |u(t)|, \;\; u \in \mathcal{C}[0,T]$.

Now for any $\lambda > 0$, the norm $\|\cdot\|_\lambda$ in E defined by

$$\|u\|_\lambda = \sup_{0\leq t\leq T} e^{-\lambda t}|u(t)|, \;\; u \in E = \mathcal{C}[0,T].$$

As $e^{-\lambda t}\|u\| \leq \|u\|_\lambda \leq \|u\|$ two norms are equivalent and E is also complete with respect to the later norm.

Now let us first consider the following Volterra integral equation of the second kind

$$u(t) = f(t) + \int_0^t F(t,s)u(s)ds, \;\; 0 \leq t \leq T \tag{2.15}$$

where the given function f and kernel F are assumed to be continuous in $[0,T]$ and $[0,T] \times [0,T]$ respectively.

Let $E = (\mathcal{C}[0,T], \|\cdot\|)$, where $\|\cdot\|$ is the Sup norm as defined above. If we define the mapping $T : E \to E$ by

$$T(u)(t) = f(t) + \int_0^t F(t,s)u(s)ds, \;\; 0 \leq t \leq T,$$

then for any $v_1, v_2 \in \mathcal{C}[0,T]$,

$$\|T(v_1) - T(v_2)\| \leq \int_0^t \sup_{00\leq s,t\leq T} |f(t,s)| \sup_{0\leq s\leq T} |v_1(s) - v_2(s)|ds \leq aN\|v_1 - v_2\|,$$

where $N = \sup_{0\leq s,t\leq T}|F(t,s)|$.

Thus T is not, in general, a contraction mapping unless the product $aN < 1$. Hence in order to apply the contraction mappings principle one is forced to restrict either the size of the interval $0 \leq t \leq T$ where the solution is defined or the "size" of the kernel $F(x,y)$.

On the other hand, if we take $E = (\mathcal{C}[0,T], \|\cdot\|_\lambda)$ and define the mapping $T : E \to E$ by

$$T(u)(t) = f(t) + \int_0^t F(t,s)u(s)ds, \ \ 0 \le t \le T,$$

then for any $v_1, v_2 \in E$,

$$\begin{aligned}\|T(v_1) - T(v_2)\|_\lambda &\le N \sup_{0\le t\le T} e^{-\lambda t} \int_0^t |v_1(s) - v_2(s)|ds \\ &= N \sup_{0\le t\le T} e^{-\lambda t} \int_0^t e^{\lambda s} e^{-\lambda s} |v_1(s) - v_2(s)|ds \\ &\le N\|v_1 - v_2\|_\lambda \sup_{0\le t\le T} e^{-\lambda t} \int_0^t e^{\lambda s}.ds \\ &= N\|v_1 - v_2\|_\lambda \sup\nolimits_{0\le t\le T} \frac{1 - e^{-\lambda t}}{\lambda} \\ &\le \frac{N}{\lambda}\|v_1 - v_2\|_\lambda.\end{aligned}$$

Now it is clear that with sufficiently large λ, T is a contraction on $(\mathcal{C}[0,T], \|\cdot\|_\lambda)$. Hence by the contraction mapping principle the integral equation (2.15) has a unique solution $u \in \mathcal{C}[0,T]$. □

Next, we consider another Volterra integral equation of second kind

$$u(t) = f(t) + \int_0^t F(t,s,u(s))ds, \ \ 0 \le t \le T \tag{2.16}$$

where the given function f and the kernel F are assumed to be continuous on the finite interval $[0,T]$ and $[0,T] \times [0,T] \times \mathbb{R}$ respectively.

It is known (e.g. Tricomi (Tricomi (1957))) that by employing the usual method of successive approximations each of the integral equations (2.15) and has a unique solution $u \in \mathcal{C}[0,T]$ for any continuous kernel. However, our interest in here is limited to the application of contraction mapping principle.

Theorem 2.16 *Let the kernel function* $F : [0,T] \times [0,T] \times \mathbb{R} \to \mathbb{R}$ *be continuous and satisfy a Lipschitz condition*

$$|F(t,s,x) - F(t,s,y)| \le \lambda|x - y|$$

for all $s,t \in [0,T]$ *and* $x,y \in \mathbb{R}$. *Then for any* $f \in \mathcal{C}[0,1]$, *the integral equation* (2.16)

$$u(t) = f(t) + \int_0^t F(t,s,u(s))ds, \ \ 0 \le t \le T$$

has a unique solution $u \in \mathcal{C}[0,T]$. *Moreover, for any arbitrary* $u_0 \in \mathcal{C}[0,T]$, *the sequence* $\{u_n\}$ *of iterates defined by* $u_{n+1}(t) = f(t) + \int_0^t F(t,s,u_n(s))ds$ *converges uniformly to the unique solution* u.

Proof. Let $E = (\mathcal{C}[0,T], \|\cdot\|_\lambda)$. We define the mapping $T : E \to E$ by

$$T(v)(t) = f(t) + \int_0^t F(t,s,v(s))ds, \quad v \in \mathcal{C}[0,T].$$

Then it is easy to see that $u \in \mathcal{C}[0,T]$ is a solution of the integral equation (2.16) if and only if u is a fixed point of T.

Now for any $v_1, v_2 \in E$,

$$\begin{aligned}
\|T(v_1) - T(v_2)\|_\lambda &\leq \sup_{0\leq t\leq T} e^{-\lambda t} \int_0^t |F(t,s,v_1(s)) - F(t,s,v_2(s))|ds \\
&\leq \lambda \sup_{0\leq t\leq T} e^{-\lambda t} \int_0^t |v_1(s) - v_2(s)|ds \\
&= \lambda \sup_{0\leq t\leq T} e^{-\lambda t} \int_0^t e^{\lambda s} e^{-\lambda s} |v_1(s) - v_2(s)|ds \\
&\leq \lambda \|v_1 - v_2\|_\lambda \sup_{0\leq t\leq T} e^{-\lambda t} \int_0^t e^{\lambda s} ds \\
&= \lambda \|v_1 - v_2\|_\lambda \sup_{0\leq t\leq T} \frac{[e^{-\lambda t}(e^{\lambda t} - 1)]}{\lambda} \\
&= \|v_1 - v_2\|_\lambda \sup_{0\leq t\leq T} [1 - e^{-\lambda t}] \\
&\leq (1 - e^{-\lambda t}) \|v_1 - v_2\|_\lambda.
\end{aligned}$$

Thus T is a contraction mapping with contraction constant $r = (1 - e^{-\lambda t}) < 1$. Hence by contraction mapping principle, T has a unique $u \in \mathcal{C}[0,T]$ which is the unique solution of (2.16).

The rest of the theorem is obvious from the contraction mapping principle when the sequence $\{T^n(u_0)\}$ of iterates starting from any $u_0 \in \mathcal{C}[0,T]$ is considered. □

Corollary 2.16.1 *Let $G : [0,T] \times \mathbb{R} \to \mathbb{R}$ satisfy a Lipschitz condition*

$$|G(s,x) - G(s,y)| \leq \lambda |x - y|$$

for all $s \in [0,T]$ and $x, y \in \mathbb{R}$. Then the initial value problem

$$\frac{du}{dp} = G(s,u), \quad u(0) = 0 \tag{2.17}$$

has a unique solution $u \in \mathcal{C}[0,T]$.

Proof. If we take $f(t) \equiv 0$ and $F(t,s,u) = G(s,u)$ in (2.15), then we get the integral equation

$$u(t) = \int_0^t G(s,u(s))ds. \tag{2.18}$$

Now it is trivial to see that $u \in \mathcal{C}[0,T]$ is a solution of (2.18) if and only if u is a solution of (2.17). The corollary follows from the observation that (2.18) has by virtue of Theorem 2.16, a unique solution $u \in \mathcal{C}[0,T]$. □

Next we consider a simple but useful theorem due to (Chu and Diaz (1964/65, pp. 351–363)) with some applications to some functional equations and Goursat problem for the wave equations, also due to (Chu and Diaz (1964/65, pp. 351–363)).

Theorem 2.17 *Let $T : X \to X$ be a mapping of a non-empty set X into itself and $K : X \to X$ another self mapping possessing a right inverse K^{-1}, i.e., $KK^{-1} = I_X$, the identity mapping on X. Then the mapping T has a fixed point if and only if $K^{-1}TK$ has a fixed point.*

Proof. First, let T have a fixed point $x_0 \in X$. Then $K^{-1}x_0 = K^{-1}Tx_0 = K^{-1}T(KK^{-1})x_0 = K^{-1}TK(K^{-1}x_0)$. Thus $K^{-1}x_0$ is a fixed point of $K^{-1}TK$.

Next, let $\bar{x} \in X$ is a fixed point of $K^{-1}TK$. Then $K\bar{x} = KK^{-1}TK\bar{x} = T(K\bar{x})$. Hence $K\bar{x}$ is a fixed point of T. □

Corollary 2.17.1 *Let (X,h) be a complete Hausdorff uniform space and $T, K : X \to X$ be two self mappings such that K has a right inverse K^{-1} and $K^{-1}TK$ is a contraction mapping. Then T has a unique fixed point.*

Proof. By contraction mapping principle $K^{-1}TK$ has a unique fixed point. Hence Theorem 2.17 implies that T has a unique fixed point. □

Corollary 2.17.2 *Let (X,h) be a complete Hausdorff uniform spacc and $T, K : X \to X$ be two self mappings such that K has a right inverse K^{-1} and $K^{-1}T^nK$ is a contraction mapping for some integer n. Then T has a unique fixed point.*

Proof. By Corollary 2.17.1 T^n has a unique fixed point $x_0 \in X$. Now $Tx_0 = T(T^nx_0) = T^n(Tx_0)$. By the uniqueness of fixed point of T^n, $Tx_0 = x_0$. It follows that x_0 is a unique fixed point of T (see also Remark 2.4). □

As applications of the above corollaries we consider the functional equation:

$$u(t) = f(t) + g(t)u(a(t)), \tag{2.19}$$

where the functions f, g, and a are given and continuous on the finite closed interval $[0,T]$ and $0 \leq a(t) \leq T$. Picard (Picard (1927), pp. 158–161) considered the equation (2.19) in the theory of boundary value problems of hyperbolic problems. He showed the existence of a unique solution of (2.19) by the successive substitution of the function $u(a(t))$, i.e.,

$$\begin{aligned} u(t) &= f(t) + g(t)u(a(t)) \\ &= f(t) + g(t)[f(a(t)) + g(a(t))u(a^2(t)] \\ &= \cdots \\ &= f(t) + g(t)f(a(t)) + g(t)g(a(t))f(a^2(t)) \\ &\quad + g(t)g(a(t)) \cdots + g(a^n(t))u(a^{n+1}(t)). \end{aligned}$$

Picard proved that under certain conditions, which are detailed in the remark following the next theorem, the process converges to a unique solution. Much simpler proof (even with conditions weaker than those of Picard) is obtained in the following theorem.

Theorem 2.18 *Let the functions f, g and a be given, continuous in $[0,T]$ and satisfy the followings:*

(a) $a(0) = 0$ *and* $a(t) < t$ *whenever* $0 < t \leq T$; *and*

(b) *for some* ρ *with* $0 < \rho < 1$, *there exists* $\delta = \delta(\rho)$ *with* $0 < \delta \leq T$ *such that* $|g(t)| < \rho$ *for* $0 \leq t \leq \delta$.

Then there exists a unique solution of the functional equation (2.19).

Proof. Since by assumption $t - a(t) \geq 0$ for $0 \leq t \leq T$, we have $0 < e^{-\lambda(t-a(t))} < 1$ for all $\lambda > 0$.

Hence by assumption (b),

$$|g(t)e^{-\lambda(t-a(t))}| < \rho \quad \text{for} \quad 0 \leq t \leq \delta \quad \text{and all} \quad \lambda > 0.$$

Now suppose that $\delta \leq t \leq T$. Then as $t > a(t)$ for all $t > 0$, it follows that $m = m(\rho) = \inf_{\delta \leq t \leq T}[t - a(t)] > 0$.

Thus $\delta \leq t \leq T$ and for $\lambda > 0$,

$$|g(t)e^{-\lambda(t-a(t))}| \leq e^{-\lambda m} \sup_{0 \leq t \leq T} |g(t)|.$$

Now choosing sufficiently large λ, we can make the right side smaller than ρ. Hence for such λ, say $\bar{\lambda}$, we have

$$|g(t)e^{-\bar{\lambda}(t-a(t))}| < \rho \quad \text{whenever} \quad 0 \leq t \leq T. \qquad \square$$

We give two proofs, the first one is based on our previous consideration, i.e., changing the norm and the application of Corollary 2.17.1.

2.4.1 *Changing the Norm*

Let $T : (\mathcal{C}[0,T], \|\cdot\|_{\bar{\lambda}}) \to (\mathcal{C}[0,T], \|\cdot\|_{\bar{\lambda}})$ defined by $T(u)(t) = f(t) + g(t)u(a(t)),\ 0 \leq t \leq T$. where $\|u\|_{\bar{\lambda}} = \sup_{0 \leq t \leq T} e^{-\bar{\lambda}(t-a(t))}|u(a(t))|$. Then for any $v_1, v_2 \in \mathcal{C}[0,T]$, by what has been done already,

$$\begin{aligned}\|T(v_1) - T(v_2)\|_{\bar{\lambda}} &\leq \sup_{0 \leq t \leq T}|g(t)| \sup_{0 \leq t \leq T} e^{-\bar{\lambda}(t-a(t))}|v_1(t) - v_2(t)| \\ &\leq \rho\|v_1 - v_2\|_{\bar{\lambda}}.\end{aligned}$$

As $0 < \rho < 1$, T is a contraction on $\mathcal{C}[0,T]$. Hence T has a unique fixed point $u(t) \in \mathcal{C}[0,T]$, i.e., there is a unique solution of (2.19). $\square$

Using Corollary 2.17.1 Let $K(u(t)) = e^{\bar{\lambda}t}u(t)$, $u \in \mathcal{C}[0,T]$ and T is as defined above. Then $K^{-1}(u(t)) = e^{-\bar{\lambda}t}u(t)$ and

$$\begin{aligned}K^{-1}TK(u(t)) &= K^{-1}(T(e^{\bar{\lambda}t}u(t)) \\ &= K^{-1}[f(t) + g(t)e^{\bar{\lambda}a(t)}u(a(t))] \\ &= e^{-t}f(t) + g(t)e^{-\bar{\lambda}(t-a(t))}u(a(t)).\end{aligned}$$

Hence considering $K^{-1}TK$ as a mapping of $(\mathcal{C}[0,T], \|\cdot\|)$ into itself, where $\|\cdot\|)$ is the usual sup norm, we have for any $v_1, v_2 \in \mathcal{C}[0,T]$,

$$\begin{aligned}\|K^{-1}TK(v_1) - K^{-1}TK(v_2)\| &\leq [\sup_{0\leq t\leq T}|g(t)e^{-\bar{\lambda}(t-a(t))}|]\|v_1 - v_2\| \\ &\leq \rho\|v_1 - v_2\|.\end{aligned}$$

Thus $K^{-1}TK$ is a contraction. Hence in this case the result follows from Corollary 2.17.1. □

Remark 2.12 The condition (b) of Theorem 2.18 will be guranteed if we assume that $|g(0)| < 1$. For then for any ρ with $|g(0)| < \rho < 1$, we will have by the continuity of g at 0, a real number $\delta = \delta(p)$ such that $0 < \delta \leq T$ and $|g(t) - g(0)| < \rho - |g(0)|$ for $0 \leq t \leq \delta$, i.e., $|g(t)| < \rho$ for $0 \leq t \leq \delta$.

In addition to continuity of f, g and a, Picard assumed that (i) $a(x)$ is a linear function, $\alpha(x) = \beta(x)$ where β is a constant satisfying $0 < \beta < 1$ (the condition (b) is evidently stronger than the condition (a)), (ii) the function $f(x)$ satisfies an inequality of the form $|f(x)| \leq cx\mu$ with positive constant c and μ on $0 \leq t \leq T$ (which is not assumed here) and (iii) the function g satisfies an inequality of the form $|g(t)| < e^{At}$ with constant $A > 0$ on $0 \leq t \leq T$ (which implies $|g(0)| < 1$).

As application of Corollary 2.17.2, next we consider the systems of functional equations of the form:

$$u(x) = f(x) + k_1(x)v(\alpha(x)) \tag{2.20}$$
$$v(y) = g(y) + k_2(y)u(\beta(y)). \tag{2.21}$$

We assume that

(*)

- (i) $f(x)$, $k_1(x)$, $\alpha(x)$ and $g(y)$, $k_2(y)$, $\beta(y)$ are continuous on $0 \leq x \leq a$ and $0 \leq y \leq b$ respectively;
- (ii) $0 \leq \alpha(x) \leq b$ and $0 \leq \beta(y) \leq a$;
- (iii) $|k_1(0)k_2(0)| < 1$ and
- (iv) $\alpha(0) = \beta(0) = 0$, and $\beta(\alpha(x)) < x$; $\alpha(\beta(y)) < y$ on $0 < x \leq a$ and $0 < y \leq b$ respectively.

In this case we take the Banach space $(E, \|\cdot\|)$, where $E = \mathcal{C}[0,a] \times \mathcal{C}[0,b]$ and

$$\|(u,v)\| = \max[\sup_{0\leq x\leq a}|u(x)|, \sup_{0\leq y\leq b}|v(y)|],$$

the mapping $T : E \to E$ defined by

$$T(u,v) = (f(x) + k_1(x)v(\alpha(x)), g(y) + k_2(y)u(\beta(y))$$

and the mapping $K : E \to E$ defined by

$$K(u,v) = (e^{\lambda x}u(x), e^{\lambda y}v(y)).$$

Under the asssumption $(*)$ we can show that T has a unique fixed point $(u,v) \in E$ which will be the unique solution of (2.20).

Now $K^{-1}(u,v) = (e^{-\lambda x}u, e^{-\lambda}v)$, and

$$TK(u,v) = (f(x) + k_1(x)e^{\lambda\alpha(x)}v(\alpha(x)), g(y) + k_2(y)e^{\lambda\beta(y)}u(\beta(y)).$$

It can be easily seen that

$$T^2K = TTK(u,v) = (\bar{u}, \bar{v})$$

where

$\bar{u} = f(x) + k_1(x)g(\alpha(x)) + k_1(x)k_2(\alpha x)e^{\lambda\beta(\alpha(x))}u(\beta(\alpha(x)))$ and

$\bar{v} = g(y) + k_2(y)f(\beta(y)) + k_1(\beta(y))k_2(y)e^{\lambda\alpha(\beta(y))}v(\alpha(\beta(y)))$.

Hence $K^{-1}T^2K = K^{-1}(\bar{u}, \bar{v}) = (\hat{u}, \hat{v})$, where

$\hat{u} = [f(x) + k_1(x)g(\alpha(x))]e^{-\lambda x} + k_1(x)k_2(\alpha(x))e^{-\lambda(x-\beta(\alpha(x)))}u(\beta(\alpha(x)))$ and

$\hat{v} = [g(y) + k_2(y)f(\beta(y))]e^{-\lambda x} + k_1(\beta(y))k_2(y)e^{-\lambda(y-\alpha(\beta(y)))}v(\alpha(\beta(y)))$.

Now we can use an argument similar to the one used in the proof of Theorem 2.18 to show that under the condition $(*)$, $K^{-1}T^2K$ is a contraction mapping. Hence by Corollary 2.17.2, T has a unique fixed point. Thus we have proved the following theorem:

Theorem 2.19 *Under the condition $(*)$, the functional equation (2.20) has a unique solution.*

These functional equations have applications in boundary value problems for partial differential equations of hyperbolic type. Here we consider an example, the Goursat problem for the wave equations:

$$\begin{cases} u_{xy}(x,y) = f(x,y) & \text{in } R = \{(x,y) : 0 \le x \le a,\ \ 0 \le y \le b\}, \\ u(x,\alpha(x)) = \sigma(x), & \text{if } 0 \le x \le a, \\ u(\beta(y),y) = \tau(y), & \text{if } 0 \le y \le b, \\ \sigma(0) = \tau(0) = 0 \end{cases} \tag{2.22}$$

With $0 \le \alpha(x) \le b$ and $0 \le \beta(y) \le a$ for $0 \le x \le a$ and $0 \le y \le b$ respectively.

Assuming u, u_x, u_y and u_{xy} to be continuous, any solution of $u_{xy} = f(x,y)$ (in the classical sense) must be of the form

$$u(x,y) = \theta(x) + \phi(y) + \int_0^x \int_0^y f(\zeta,\eta)d\eta, d\zeta, \tag{2.23}$$

where θ and ϕ are C^1-function in $0 \leq x \leq a$ and $0 \leq y \leq b$ respectively.

Substituting the boundary conditions in (2.23) we obtain

$$\sigma(x) = u(x,\alpha(x)) = \theta(x) + \phi(\alpha(x)) + \int_0^x \int_0^{\alpha(x)} f(\zeta,\eta)d\eta, d\zeta$$

$$\tau(y) = u(\beta(y),y) = \theta(\beta(y)) + \phi(y) + \int_0^{\beta(y)} \int_0^y f(\zeta,\eta)d\eta, d\zeta.$$

Differentiating the last two equations we obtain a system of functional equations for $\theta'(x)$ and $\phi'(y)$:

$$\theta'(x) = [\sigma'(x) - \int_0^{\alpha(x)} f(x,\eta)d\eta - \alpha'(x)\int_0^x f(\zeta,\alpha(x))d\zeta] - \alpha'(x)\phi'(\alpha(x)) \tag{2.24}$$

$$\phi'(y) = [\tau'(y) - \int_0^{\beta(y)} f(\zeta,y)d\zeta - \beta'(y)\int_0^y f(\beta(y),\eta)d\eta] - \beta'(y)\theta'(\beta(y)). \tag{2.25}$$

Under the assumption that $\alpha(x), \sigma(x)$ and $\beta(y), \tau(y)$ are of C^1 on their respective class, f is continuous in R and $\alpha(0) = \beta(0) = 0$, it can be easily seen that (2.22) is equivalent to the system of functional equations (2.24) which is a special case of (2.20).

Let h and h' be two Hausdorff uniform topologies, on a non-empty set X, which are generated by the families of pseudometrics $\{\rho_\alpha : \alpha \in I\}$ and $\{\rho'_\alpha : \alpha \in I\}$ respectively. h and h' are said to be equivalent if there exist families $\{a_\alpha : \alpha \in I\}$ and $\{b_\alpha : \alpha \in I\}$ of positive constants such that for each $\alpha \in I$,

$$a_\alpha \rho_\alpha(x,y) \leq \rho'_\alpha(x,y) \leq b_\alpha \rho_\alpha(x,y) \quad \text{for all} \quad x,y \in X.$$

Throughout the rest of the section all uniform spaces are assumed to be Hausdorff. Let T be a mapping of a uniform space $(X, \{\rho_\alpha : \alpha \in I\})$ into itself satisfying a Lipschitz condition

$$\text{for each} \quad \alpha \in I, \rho_\alpha(T(x),T(y)) \leq a_\alpha \rho_\alpha(x,y) \quad \text{for} \quad x,y \in X. \tag{2.26}$$

(Note that $\rho_\alpha(x,y) = 0$ implies $\rho_\alpha(T(x),T(y)) = 0$.)

It follows that for each positive integer n we have

$$\text{for each} \quad \alpha \in I, \rho_\alpha(T^n(x),T^n(y)) \leq a_\alpha^n \rho(x,y) \quad \text{for} \quad x,y \in X. \tag{2.27}$$

A mapping satisfying (2.26) is said to be Lipschitz continuous. If T^n is a contraction mapping for some positive integer n, then by Remark 2.2 (1) and also by Corollary 2.17.2 (taking $K = I_X$ = identity mapping on X), T has a unique fixed point. The following theorem will imply that if T^n is a contraction mapping for a fixed positive integer n, then T is also a contraction mapping with some equivalent uniformity.

Proposition 2.1 *Let $T : (X,h) \to (X,h)$ be a mapping of a uniform space $(X,h) = (X, \{\rho_\alpha : \alpha \in I\})$ into itself such that, for a fixed positive integer n, T^n satisfies (2.27). Then for each $\alpha \in I$, the function ${\rho'}_\alpha : X \times X \to \mathbb{R}$ defined by*

$$\rho'_\alpha(x,y) = \rho_\alpha(x,y) + \frac{1}{a_\alpha}\rho_\alpha(T(x),T(y)) + \cdots + \frac{1}{a_\alpha^{n-1}}\rho_\alpha(T^{n-1}(x),T^{n-1}(y)) \quad (2.28)$$

is a pseudometric on X and T satisfies (2.26) with respect to ρ'_α, i.e., for each $\alpha \in I$

$$\rho'_\alpha(T(x),T(y)) \leq a_\alpha \rho(x,y) \quad \text{for} \quad x,y \in X. \quad (2.29)$$

Two uniformities $\{\rho_\alpha : \alpha \in I\}$ and $\{{\rho'}_\alpha : \alpha \in I\}$ are equivalent if and only if T is Lipschitz continuous.

Proof. That for each $\alpha \in I$, ρ'_α is a pseudometric is obvious. Now,

$$\begin{aligned}
\rho'_\alpha(T(x),T(y)) &= \rho_\alpha(T(x),T(y)) + \cdots + \frac{1}{a_\alpha^{n-1}}\rho_\alpha(T^n(x),T^n(y)) \\
&\leq \rho_\alpha(T(x),T(y)) + \cdots + \frac{1}{a_\alpha^{n-2}}\rho_\alpha(T^{n-1}(x),T^{n-1}(y)) \\
&\quad + \frac{1}{a_\alpha^{n-1}}\rho_\alpha(T^n(x),T^n(y)) \\
&\leq \rho_\alpha(T(x),T(y)) + \cdots + \frac{1}{a_\alpha^{n-2}}\rho_\alpha(T^{n-1}(x),T^{n-1}(y)) + a_\alpha\rho_\alpha(x,y) \\
&= a_\alpha \rho'_\alpha(x,y).
\end{aligned}$$

Hence (2.29) is proved. Next, it is clear that $\rho_\alpha(x,y) \leq \rho'_\alpha(x,y)$ for each $\alpha \in I$. Hence if T is Lipschitz continuous with Lipschitz constant b_α with respect to ρ_α, then it follows that T is Lipschitz continuous with respect to ρ'_α with Lipschitz constant $a_\alpha b_\alpha$. Conversely, if for each $\alpha \in I$, T is ρ_α-Lipschitz, then powers of T is ρ_α-Lipschitz continuous. Assuming that for each $\alpha \in I$,

$$\rho_\alpha(T^k(x),T^k(y)) \leq a_{k(\alpha)}\rho_\alpha(x,y) \quad \text{for} \quad k = 1,2\ldots,n-1,$$

we have

$$\rho_\alpha(x,y) \leq \rho'_\alpha(x,y) \leq b_\alpha \rho_\alpha(x,y) \quad (2.30)$$

where $b_\alpha = 1 + a_{\alpha(1)}{a_\alpha}^{-1} + \cdots + a_{\alpha(n-1)}{a_\alpha}^{1-n}$. □

Corollary 2.19.1 *Let $T : (X,h) \to (X,h)$ be a mapping of a uniform space $(x,h) = (X, \{\rho_\alpha : \alpha \in I\})$ into itself such that T^n is a contraction mapping with contraction constants $\{r_\alpha : \alpha \in I\}$. Then T is ρ'_α-contraction with contraction constants $(r_\alpha)^{1/n}$ for each $\alpha \in I$.* □

Proof. It can be easily seen that the corollary follows from the Proposition 2.1. □

2.4.2 *On the Approximate Iteration*

For $T : X \to X$ and $x_0 \in X$, define the sequence of iterates $\{x_n\}$ by $x_k = T^k(x_0)$. Let x^* be a unique fixed point of T and $\{y_k\}$ be a sequence of points in X as will be defined in the following theorem. It is a question of some theoretical and practical to get the information of $\rho_\alpha(x_k, y_k)$ and $\rho_\alpha(y_k, x^*)$ for each $\alpha \in I$.

The following theorem follows from Proposition 2.1.

Theorem 2.20 *Let $T : (X, h) \to (X, h)$ be a Lipschitz continuous mapping of a complete uniform space $(X, h) = (X, \{\rho_\alpha : \alpha \in I\})$ into itself with Lipschitz constants $\{a_\alpha : \alpha \in I\}$ and T^n be contraction with some positive integer n, say, for each $\alpha \in I$, $\rho_\alpha(T^k(x), T^k(y)) \le a_{\alpha(k)}\rho(x, y)$ for $k = 1, 2, \ldots, n$, where $a_{\alpha(n)} = a_\alpha^n < 1$. Let x^k be the unique fixed point (obtained by Corollary 2.19.1) and let $\{y_k\}$ be the sequence in X such that for each $\alpha \in I$, $\epsilon_{k(\alpha)} = \rho_\alpha(y_{k+1}, T(y_k))$. Then for each $\alpha \in I$,*

$$\begin{aligned}\rho_\alpha(x^*, y_{k+1}) &\le \rho'_\alpha(x^*, y_{k+1}) \le b_\alpha A_{k(\alpha)}\\ \rho_\alpha(x^*, y_{k+1}) &\le \rho'_\alpha(x^*, y_{k+1}) \le b_\alpha B_{k(\alpha)},\end{aligned}$$

where b_α is as defined in (2.30),

$$\begin{aligned}A_{k(\alpha)} &= (1 - a_\alpha{}^{-1})[\epsilon_{k(\alpha)} + a_\alpha \rho_\alpha(y_{k+1}, y_k)\\ B_{k(\alpha)} &= \rho_\alpha(x_{k+1}, y_{k+1}) + a_\alpha^{k+1}\rho_\alpha(x_0, y_0) + \textstyle\sum_{i=0}^{k} a_\alpha{}^{k-i}\epsilon_i.\end{aligned}$$

The next proposition follows immediately from Proposition 2.1 taking $I = \{1\}$ and $\|x\| = \rho(x, 0)$.

Proposition 2.2 *Let $(X, \|\cdot\|)$ be a Banach space and let A be a bounded linear operator of X into itself such that $\|A^n\| = \alpha^n$ ($\|A\|$ denotes the operator norm). Then the function $\|\cdot\|_1 : X \to \mathbb{R}$ defined by*

$$\|x\|_1 = \|x\| + \frac{1}{\alpha}\|A(x)\| + \cdots + \frac{1}{\alpha^{n-1}}\|A^{n-1}(x)\|$$

defines a norm equivalent to the original norm $\|\cdot\|$ and for the corresponding operator norm of A, we have $\|A\|_1 \le \alpha$.

As an immediate consequence we obtain $\|A\|_1 \le \|A^n\|^{1/n}$, which is one half of Gelfand's theorem (see Riesz and Sz-Nagy (1955, p. 425)) on spectral radius δ of A, namely $\delta \le \inf \|A^n\|^{1/n}$.

The results begining from Proposition 2.1 to the end of this section are due to Walter (Walter (1976)) for the metric space and Theorem 2.20 in metric space case with $n = 1$ is due to Ostrowski (e.g. see Ortega and Rheinboldt (1970), Theorem 12.2.1). The authors are not aware if these have been done in uniform space by anyone.

2.5 The Contraction Mapping Principle Applied to the Cauchy-Kowalevsky Theorem

In this section we present a result of Walter (Walter (1976)). The Cauchy-Kowalevsky Theorem asserts that under certain conditions the following partial differential equation has a unique solution satisfying the initial data in some neighborhood of the initial data.

$$\frac{\partial^p u(t,x)}{\partial t^p} = F\left(t, x_1, \ldots, x_n, u, \ldots, \frac{\partial^k}{\partial t^k}\frac{\partial^{|\alpha|} u}{\partial x^\alpha}, \ldots\right) \tag{2.31}$$

with initial data given by

$$\frac{\partial^j u(0,x)}{\partial t^j} = \phi_j(x), \qquad j = 0, \ldots, p-1 \tag{2.32}$$

where the order of the differential operators satisfies $k + |\alpha| \leq p$ and $k < p$.

The specific conditions employed by Cauchy and Kowalevsky are that the ϕ_k's are analytic in a neighborhood of $x_0 = (x_1^0, \ldots, x_n^0)$ and the right hand side of equation (2.31) is analytic in a neighborhood of the point

$$\left(0, x_0, \phi_0(x_0), \ldots, \frac{\partial^{|\alpha|}\phi_j(x_0)}{\partial x^\alpha}, \ldots\right).$$

Furthermore, the initial data is not defined on a characteristic surface.

This problem is called Cauchy's problem and the initial conditions are called Cauchy data. It was first solved by Cauchy and, in a more general and simplified way, by Sophie v. Kowalevsky. In this section existence and uniqueness of a solution to this problem will be illustrated via an application of the Contraction Mapping Principle.

The first task is to rewrite equation (2.31) in a simpler form: a quasilinear first order system

$$\frac{\partial u(t,x)}{\partial t} = \sum_{j=1}^{n} B_j(t,x,u)\frac{\partial u}{\partial x_j} + c(t,x,u). \tag{2.33}$$

We illustrate this with a second order equation, but the same method holds for the more general equation (2.31). Consider

$$\frac{\partial^2 u}{\partial t^2} = f(t, x, u, u_t, u_x, u_{xt}, u_{xx}) \tag{2.34}$$

with initial data

$$u(0,x) = \phi_0(x) \tag{2.35}$$

$$u_t(0,x) = \phi_1(x). \tag{2.36}$$

Here u_t stands for $\frac{\partial u}{\partial t}$ and similarly for the other subscripts. The variable x is an element of the reals.

Rewrite equation (2.34) as a first order system by defining new variables

$$v = u_t \quad \text{and} \quad w = u_x$$

which leads to the system

$$u_t(t,x) = v(t,x) \tag{2.37}$$

$$v_t(t,x) = f(t,x,u,v,w,v_x,w_x) \tag{2.38}$$

$$w_t(t,x) = v_x(t,x) \tag{2.39}$$

with initial data

$$u(0,x) = \phi_0(x) \tag{2.40}$$

$$v(0,x) = \phi_1(x) \tag{2.41}$$

$$w(0,x) = \phi_0'(x). \tag{2.42}$$

Now any first order system of the form

$$\frac{\partial \mathbf{u}(t,x)}{\partial t} = \mathbf{g}(t,x,\mathbf{u},\mathbf{u}_x) \tag{2.43}$$

$$\mathbf{u}(0,x) = \phi(x) \tag{2.44}$$

(where $\mathbf{u}$, $\mathbf{g}$ and ϕ are vectors) can be transformed to a quasilinear system by the substitution $\mathbf{v} = \mathbf{u}_x$. This gives

$$\frac{\partial \mathbf{v}}{\partial t} = \frac{\partial \mathbf{u}_x}{\partial t} = \frac{\partial \mathbf{u}_t}{\partial x} = \frac{\partial \mathbf{g}}{\partial x}(t,x,\mathbf{u},\mathbf{v}) \tag{2.45}$$

$$= \mathbf{g}_x + \mathbf{g}_{\mathbf{u}}\mathbf{v} + \mathbf{g}_{\mathbf{v}}\mathbf{v}_x. \tag{2.46}$$

Thus we arrive at a quasilinear first order system in the variables $(\mathbf{u}, \mathbf{v})$:

$$\mathbf{u}_t = \mathbf{g}(t,x,\mathbf{u},\mathbf{v}) \tag{2.47}$$

$$\mathbf{v}_t = \mathbf{g}_x + \mathbf{g}_{\mathbf{u}}\mathbf{v} + \mathbf{g}_{\mathbf{v}}\mathbf{v}_x. \tag{2.48}$$

2.5.1 *Geometric Preliminaries*

Since we will make use of an important result of Nagumo (Nagumo (1951)), the setting of the problem will be in the complex plane $\mathbb{C}$, or more generally $\mathbb{C}^n$. Discussions about the real-valuedness of solutions will be deferred until later.

Let Ω be an open set in $\mathbb{C}^n$ with nonempty boundary $\Gamma = \partial\Omega$, so Ω is not all of $\mathbb{C}^n$ but it may be unbounded. Let $d(z) = \text{dist}(z, \Gamma)$ be the distance from $z \in \Omega$ to Γ measured in the maximum norm $|z| = \max_{i=1,\dots,n} |z_i|$.

The set $G \subset \mathbb{R} \times \mathbb{C}^n$ (in the real case) or $G \subset \mathbb{C}^{n+1}$ (complex case) is defined as those points (t, z) with $z \in \Omega$ and $|t| < \eta d(z)$ for some $\eta > 0$ to be specified below.

In the real case the set G is a double cone with base Ω and pitch η.

Let Ω_t be those $z \in \Omega$ with $(t, z) \in G$, which is equivalent to $d(z) > |t|/\eta$. Then Ω_t is just the t-level sections of G. Define $d(t, z) = d(z) - |t|/\eta$ which is the distance from $z \in \Omega_t$ to the boundary of Ω_t.

As a straightforward application of the triangle inequality ($d(z) \leq |z-z'|+d(z')$) one can show that

$$z \in \Omega_t, |z - z'| = r < d(t, z) \Rightarrow z' \in \Omega_t \text{ and } d(t, z') \geq d(t, z) - r. \tag{2.49}$$

2.5.2 *The Linear Problem*

To illustrate the methods used we first prove existence and uniqueness for a special case of the quasilinear problem, the *linear* problem:

$$\frac{\partial u}{\partial t} = A(t, z)u + \sum_{j=1}^{n} B_j(t, z)\frac{\partial u}{\partial z_j} + c(t, z) \quad \text{in } G \tag{2.50}$$

$$u(0, z) = \phi(z) \quad \text{in } \Omega \tag{2.51}$$

where u, c, and ϕ are m-vectors and B_j are $m \times m$ matrices.

Some authors of textbooks consider only the linear problem, but in this case, the linear problem is useful in illustrating the methods to be employed in the quasilinear problem.

Now integration in t leads to an integral equation of Volterra type:

$$u(t, z) = g(t, z) + \int_0^t \left[A(\tau, z)u(\tau, z) + \sum_{j=1}^{n} B_j(\tau, z)\frac{\partial u(\tau, z)}{\partial z_j} \right] d\tau \tag{2.52}$$

where

$$g(t, z) = \phi(z) + \int_0^t c(\tau, z)d\tau.$$

A solution of (2.52) is by definition continuous in G and holomorphic in z for fixed t (in the real case) or holomorphic in z and t in the complex case. It is clear that a solution of equation (2.52) is a solution of the differential equation (2.50) satisfying the initial conditions (2.51) and vice versa.

Throughout we use the maximum norm for vectors $|u| = \max_{i=1,\dots,m} |u_i|$ and the corresponding operator norm for matrices: $|A| = \max_k \sum_j |a_{jk}|$.

The following lemma is due to Nagumo (Nagumo (1951)) and it gives a bound on $\frac{\partial f}{\partial z_j}$ in terms of f (Nagumo (1951)).

Lemma 2.7 (Nagumo (1951)) *Let $f : \Omega \to \mathbb{C}^m$ be holomorphic and $p \geq 0$. Then*

$$|f(z)| \leq \frac{c}{d^p(z)} \quad \textit{for all } z \in \Omega \tag{2.53}$$

$$\Longrightarrow \left| \frac{\partial f(z)}{\partial z_j} \right| \leq C_p \frac{c}{d^{p+1}(z)} \quad \textit{for all } z \in \Omega \tag{2.54}$$

where

$$C_p = (1+p)\left(1+\frac{1}{p}\right)^p < e(p+1)$$

and $C_0 = 1$.

Proof. For a function ψ of a complex variable $z \in \mathbb{C}$, it is straightforward to show from the Cauchy integral formula

$$\psi'(z) = \frac{1}{2\pi i}\int_{|z-z'|=r} \frac{\psi(z')}{(z-z')^2}dz' \tag{2.55}$$

that the estimate for $|\psi'(z)|$ holds:

$$|\psi'(z)| \leq \frac{1}{r}\max_{|z-z'|=r} |\psi(z')|.$$

Applying this in the z_j direction to each f_i gives

$$\left|\frac{\partial f_i(z)}{\partial z_j}\right| \leq \frac{1}{r}\max_{|z_j-z_j'|=r} |f_i(z_1,\ldots,z_j',\ldots,z_n)| \tag{2.56}$$

$$\leq \frac{1}{r}\max_{|z_j-z_j'|=r} \frac{c}{d^p(z_1,\ldots,z_j',\ldots,z_n)} \tag{2.57}$$

$$\leq \frac{1}{r}\frac{c}{(d(z)-r)^p} \tag{2.58}$$

because $d(z) \leq d(z') + r$. Choosing $r = d(z)/(p+1)$ gives the desired result.

Now the Cauchy-Kowalevsky theorem is formulated and proved for the linear problem in the case with real 'time' t.

Theorem 2.21 *Assume that*

(*i*) *the functions* $A(t,z)$, $B_j(t,z)$ *and* $c(t,z)$ *are continuous in* G, *holomorphic in* z *for fixed* t, *and the initial data* $\phi(z)$ *is holomorphic in* Ω.

(*ii*) *there exists positive constants* α, β_j, γ, δ *and* p *such that*

$$|A(t,z)| \leq \frac{\alpha}{d(t,z)}, \qquad |B_t(t,z)| \leq \beta_j$$

$$|c(t,z)| \leq \frac{\gamma}{d^{p+1}(t,z)}, \; |\phi(z)| \leq \frac{\delta}{d^p(z)}.$$

(*iii*) $\alpha/p + (1+1/p)^p\sum_j \beta_j < 1/\eta$ *(which can be achieved by choosing* η *sufficiently small).*

Then the problem (2.52) *has a unique solution* u *in* G *and it satisfies*

$$|u(t,z)| \leq \frac{c}{d^p(t,z)}$$

in G.

Proof. The first step is to find an appropriate metric space and metric to pose the fixed point problem in. Define E to be the set of continuous functions $u \in C(G, \mathbb{C}^m)$ with u holomorphic in z and $||u|| < \infty$ where

$$||u|| = \sup_{(t,z)\in G} |u(t,z)|d^p(t,z).$$

Since convergence in this norm implies uniform convergence on compact subsets of G, the limit is holomorphic in z and E is complete.

The integral equation (2.52) can be written

$$u(t,z) = g(t,z) + (Tu)(t,z) \tag{2.59}$$

where

$$g(t,z) = \phi(z) + \int_0^t c(\tau,z)d\tau$$

and

$$(Tu)(t,z) = \int_0^t \left[A(\tau,z)u(\tau,z) + \sum_{j=1}^n B_j(\tau,z)\frac{\partial u(\tau,z)}{\partial z_j}\right]d\tau$$

is a linear operator.

To apply the contraction principle, we need to verify that $g \in E$ and $T : E \to E$ is well defined and a contraction.

It is clear that g is holomorphic in z. To show $||g(t,z)|| < \infty$ the following estimate will be useful, and is not difficult to verify:

$$\left|\int_0^t \frac{d\tau}{d^{p+1}(\tau,z)}\right| = \int_0^{|t|} \frac{d\tau}{(d(z)-\tau/\eta)^{p+1}} < \frac{\eta}{p\, d^p(t,z)}. \tag{2.60}$$

It follows from assumption (ii) and the definition of g that $g \in E$.

Now to estimate $||Tu||$ we have from the definition of the norm that

$$|u(t,z)| \le \frac{||u||}{d^p(t,z)}. \tag{2.61}$$

Applying Nagumo's lemma (Nagumo (1951)) to the region Ω_t with distance function $d(t,z)$ we have

$$\left|\frac{\partial u(t,z)}{\partial z_j}\right| \le C_p \frac{||u||}{d^{p+1}(t,z)}. \tag{2.62}$$

From assumption (ii) and the last two inequalities we find

$$|Au| \le ||A||.||u|| \tag{2.63}$$

$$\le \frac{\alpha||u||}{d^{p+1}(t,z)}, \tag{2.64}$$

$$\left| B_j \frac{\partial u}{\partial z_j} \right| \leq C_p \beta_j \frac{||u||}{d^{p+1}(t,z)}.$$

It follows that

$$|Tu(t,z)| \leq \int_0^t \left[\frac{\alpha ||u||}{d^{p+1}(\tau,z)} + \sum_j C_p \beta_j \frac{||u||}{d^{p+1}(\tau,z)} \right] d\tau \tag{2.65}$$

$$= \left(\alpha ||u|| + C_p \sum_j \beta_j ||u|| \right) \int_0^t \frac{d\tau}{d^{p+1}(\tau,z)} \tag{2.66}$$

$$\leq ||u|| \frac{\eta}{p} \left(\alpha + C_p \sum_j \beta_j \right) \frac{1}{d^p(t,z)} \tag{2.67}$$

where the last line is calculated from the estimate (2.60).

After multiplying both sides by $d^p(t,z)$ and using the definition of the norm, it follows that

$$||Tu|| \leq q||u|| < \infty$$

where $q = \eta/p(\alpha + C_p \sum_j \beta_j)$, and so $Tu \in E$ for $u \in E$. As T is a linear operator it need only be shown that $q < 1$ to deduce that T is a contraction. Using the expression for C_p then

$$q = \eta/p(\alpha + (1+p)(1+1/p)^p \Sigma_j \beta_j) \tag{2.68}$$

$$= \eta(\alpha/p + (1+1/p)^{p+1} \Sigma_j \beta_j) \tag{2.69}$$

$$< 1 \tag{2.70}$$

from assumption (iii). Therefore T is a contraction and equation (2.59) has a unique fixed point in E which corresponds to a solution of the differential equation (2.50).

To complete the proof one must show that there is no possibility of there existing a solution outside E. To this end, let u^* be a solution defined on some open subset D of G with D containing the point $(0, z_0)$. Choose r small enough so that

$$G^* := \{(t,z) : |t| < \eta d^*(z), z \in B_r(z_0)\} \subset D$$

and G^* is some positive distance from the boundary of D. Here $d^*(z) = r - |z - z_0|$.

Let E^* be the corresponding Banach space as defined above. Applying the theorem to $B_r(z_0)$ it follows that there uniquely exists a solution u defined on G^*. If it can be shown that $u^* \in E^*$ then by uniqueness $u = u^*$ on G^*. But $u^* \in E^*$ because it is bounded on G^* so its norm is finite.

In the case of complex time, we can directly conclude that $u = u^*$ in D because equality of holomorphic functions in the small implies the same in the large. The proof would then be complete.

In the case of real time, we repeat the above arguments at a new point $(0, z_1)$, eventually establishing equality of u and u^* in a small strip: $(t,z) \in D$ and $|t| < \alpha$.

By repeating the arguments at an new time $t = \pm\alpha$, we eventually obtain $u = u^*$ in all of D.

As for the solution having real values, one needs the assumption that A, B_j, c and ϕ are real valued for real values of z. Then $g(t,z)$ is real valued for real z and the sequence of functions $u_0 = g$, $u_{k+1} = g + Tu_k$ have this property as well. As is known from the contraction mapping principle, this sequence converges to the unique fixed point, and this limit is real valued for real values of z. □

Another consequence of the contraction mapping principle is that the fixed point varies continuously with the right hand side. Thus for suitable variations in the functions A, B_j, c and ϕ, the solution varies continuously with respect to the metric. Once again see Walter (Walter (1976)) for the details.

2.5.3 *The Quasilinear Problem*

For simplicity we look for a solution to the quasilinear problem

$$\frac{\partial u}{\partial t} = \sum_{j=1}^{n} B_j(t,z,u)\frac{\partial u}{\partial z_j} + c(t,z,u) \tag{2.71}$$

which takes values in the set $\Omega = B_R(0)$ instead of a more general, perhaps unbounded domain. Furthermore the initial data will be

$$u(0,z) = 0$$

as a simple transformation will reduce the general problem to this case. As before G is the set (t,z) with $|t| < \eta d(z)$ and $z \in B_R(0)$ where $d(z)$ is the distance from z to $\partial B_R(0)$.

Theorem 2.22 *Assume that either (Real Case) the functions $B_j(t,z,u)$ and $c(t,z,u)$ are continuous in $G \times B_R(0)$ and holomorphic in (z,u) for fixed t, or (Complex Case) $B_j(t,z,u)$ and $c(t,z,u)$ are holomorphic in all variables.*

Further assume the following estimates hold:

$$|c| \leq \frac{\gamma}{\sqrt{d(t,z)}}\,, \quad d(t,z)|c(t,z,u) - c(t,z,v)| \leq \gamma'|u-v|$$

$$|B_j| \leq \beta_j\,, \qquad \sqrt{d(t,z)}|B_j(t,z,u) - B_j(t,z,v)| \leq \beta_j'|u-v|.$$

If $\eta > 0$ is such that

$$2\eta\sqrt{R}(\beta+\gamma) < R, \tag{2.72}$$

$$\frac{\eta}{p}(\beta' + \beta C_p + \gamma') < 1 \tag{2.73}$$

$$\eta(3\sqrt{3}(\beta+\gamma) + 2\beta) \leq 1 \tag{2.74}$$

where $\beta = \sum_j \beta_j$, $\beta' = \sum_j \beta_j'$, then the quasilinear system has a unique solution u satisfying the initial conditions $u(0,z) = 0$ and existing in G.

Proof. We show the nonlinear integral operator

$$Su(t,z) = \int_0^t \left[\sum_j B_j(s,z,u) \frac{\partial u(s,z)}{\partial z_j} + c(s,z,u) \right] ds \tag{2.75}$$

has a fixed point in the Banach space E defined as before with the same norm

$$||u|| = \sup_{(t,z)\in G} |u(t,z)| d^p(t,z).$$

Note that unlike the linear case, the operator S is not defined on all of E because the functions B_j and c are holomorphic for $u \in B_R(0)$ only. So we need a proper subset of E which S maps to itself. Let F be the subset of E with the properties that each $u \in F$ satisfies

$$|u(t,z)| \leq \rho \quad \text{and} \quad \left| \frac{\partial u(t,z)}{\partial z_j} \right| \leq \frac{1}{\sqrt{d(t,z)}}$$

where $\rho = 2\eta\sqrt{R}(\beta+\gamma) < R$. For $u \in F$ let $v = Su$. We show $v \in F$. Now

$$|v_t(t,z)| \leq \sum_j |B_j|.|u_{z_j}| + |c| \leq \frac{\beta+\gamma}{\sqrt{d(t,z)}}.$$

Using the estimate

$$\int_0^{|t|} \frac{ds}{\sqrt{d(s,z)}} \leq 2\eta\sqrt{R}$$

integration in t yields

$$|v(t,z)| \leq 2\eta\sqrt{R}(\beta+\gamma) < R$$

so the first defining condition for membership in F is satisfied. Now to estimate the derivative of v, Nagumo's Lemma (Nagumo (1951)) implies

$$\left| \frac{\partial c(u)}{\partial z_k} \right| \leq \frac{\gamma C}{d^{3/2}(t,z)}, \quad |u_{z_j z_k}| \leq \frac{C}{d^{3/2}(t,z)}$$

and

$$\left| \frac{\partial B_j(u)}{\partial z_k} \right| \leq \frac{\beta_j}{d}$$

where $C = C_{1/2} = 3/2\sqrt{3}$. Thus we obtain

$$|v_{t,z_k}| \leq \frac{1}{d^{3/2}(t,z)}(\beta C + \beta + \gamma C)$$

and integration gives

$$|v_{z_k}| \leq \frac{2\eta}{\sqrt{d(t,z)}}(C(\beta+\gamma)+\beta) \leq \frac{1}{\sqrt{d(t,z)}}$$

hence $v \in F$, so $S : F \to F$. Now to show S is a contraction on F we estimate the difference $Su - Sv$ for $u, v \in F$.

We have

$$(Su - Sv)_t = \sum_j (B_j(u) - B_j(v))u_{z_j} + \sum_j B_j(v)(u_{z_j} - v_{z_j}) + c(u) - c(v)$$

so by the assumptions

$$|(Su - Sv)_t| \leq \sum_j \beta'_j \frac{|u - v|}{d} + \sum_j \beta_j |u_{z_j} - v_{z_j}| + \frac{\gamma'}{d}|u - v|.$$

Since $|u - v| \leq ||u - v||/d^p$ and $|u_{z_j} - v_{z_j}| \leq C_p||u - v||/d^{p+1}$ then

$$|(Su - Sv)_t| \leq \frac{||u - v||}{d^{p+1}}(\beta' + \beta C_p + \gamma')$$

and after integration

$$|Su - Sv| \leq \frac{||u - v||}{d^p}\frac{\eta}{p}(\beta' + \beta C_p + \gamma').$$

Thus by the definition of the norm we find

$$||Su - Sv|| \leq \frac{\eta}{p}(\beta' + \beta C_p + \gamma')||u - v||$$

hence S is a contraction and the proof is complete. □

To show that solutions are real valued for real values of z one uses successive approximations starting at $u_0 = 0$ as in the linear case. Furthermore, the solution depends continuously on the initial condition and the functions B_j and c.

Note finally that the set F constructed in the proof is in fact compact (a straightforward application of the Ascoli Arzela theorem) and convex. Thus existence follows from the Schauder fixed point theorem.

Remark 2.13 Walter's paper (Walter (1976)) contains many facts which are of historical and practical value: M. Nagumo (Nagumo (1951)) solves the quasilinear system (2.33) with zero initial values by transforming it into an operator equation in an appropriate Banach space. Then he shows that the solution is a fixed point of the operator equation, the existence of which is obtained by the Schauder fixed point theorem (see Chapter 4).

Keller and Schneider gave a proof of the classical Cauchy-Kowalevsky (C-K) theorem by reducing the problem to an application of the Schauder fixed point theorem which resembles Nagumo's proof (Nagumo (1951)). Another approach which was developed in sixties and gave rise to various generalized versions of C-K theorem is the notion of a scale of Banach spaces, i.e. a collection $\{B_\rho\}_{\rho>0}$ of Banach spaces with property that $0 < \rho < \sigma$ implies that $B_\sigma \subset B_\rho$ and $\|u\|_\rho \leq \|u\|_\sigma$ for any $u \in B_\sigma$.

These notions are used in Vol.III of Gelfand and Silov (1967, A.2.1–2.3), and Ovsjannikov. Also the works of Tarafdar, Nirenberg, Ovsjannikov, and others contributed to linear and nonlinear abstract C–K Theorems.

2.6 An Implicit Function Theorem for a Set of Mappings and Its Application to Nonlinear Hyperbolic Boundary Value Problem as Application of Contraction Mapping Principle

As an application of Banach contraction mapping principle in a complete metric space we first prove the implicit theorem for a set of mappings, which in turn, will be used to prove the existence of solution of nonlinear hyperbolic boundary value problem [Tarafdar and Husain (1998)]. This implicit theorem contains the well-known implicit theorem for a single mapping in Banach space. Given a Banach space E, $x_0 \in E$ and $\delta > 0$, we will use the notation $S_\delta(x_0, E) = \{x \in E : \|x - x_0\| < \delta\}$. Where $\|\cdot\|$ is the norm in E. We will need the following notions. Let A be any nonempty set and F a normed space, then a mapping $f : A \to F$ is called bounded if $f(A)$ is bounded, or equivalently if $\sup \{\|f(t)\| : t \in A\}$ is finite. The set $B_F(A)$ of all bounded mappings of A into F is a real (resp complex) vector space if F is real (vesp. complex), as $\|f(t) + g(t)| \leq \|f(t)\| + \|g(t)\|$ for any $f, g, \in B_F(A)$. Furthermore we can easily check that $\|f\| = \sup\{\|f(t)\| : f \in A\}$ is a norm on $B_F(A)$. We need the following lemmas due to Dieudonné (Diedudonné (1969)). The results of this section are mainly from (Tarafdar and Husain (1998)).

Lemma 2.8 *If F is a Banach space, $B_F(A)$ is a Banach space.*

Proof. Let $\{f_n\}$ be a cauchy sequence in $B_F(A)$. Then given a real number $\varepsilon < 0$, there exists a positive integer N such that $\|f_m - f_n\| \leq \varepsilon$ for all $m, n \geq N$. Hence for each $t \in A$, $\|f_m(t) - f_n(t)\| \leq \varepsilon$ for $m, n \geq N$, i.e. for each $t \in A$, $\{f_n(t)\}$ is a cauchy sequence in F. Since F is complete, $\{f_n(t)\}$ converges to an element $f(t) \in F$ for each $t \in A$. Hence it follows that $\|f_m(t) - f(t)\| \leq \varepsilon$ for any $t \in A$ and all $m \geq N$. Thus $\|f(t)\| \leq \|f_N(t)\| + \varepsilon$ for all $t \in A$. Hence f is bounded, i.e. $f \in B_F(A)$. Furthermore, we have $\|f_m - f\| \leq \varepsilon$ for all $m \geq N$. Thus $\{f_n\}$ converges to f in $B_F(A)$. □

Let U be an open subsset of a metric space E and $C_F(U)$ the vector space of all continuous mappings of U into the normed space F. Let $C_F^\infty(U)$ denote the set of all bounded continuous mappings of U into F, i.e. $C_F^\infty(U) = C_F(U) \cap B_F(U)$.

Lemma 2.9 *The subspace $C_F^\infty(U)$ is closed in $B_F(U)$.*

Proof. Let $\{f_n\}$ be a sequence of $C_F^\infty(U)$ converging to $f \in B_F(U)$. Given real number $\varepsilon > 0$, there exists a positive integer N such that $\|f_n - f\| \leq \varepsilon/3$ for all $n \geq N$. Let t_0 be any arbitary but fixed point of U. Then by the continuity of f_N, there exists an open neighborhood V of t_0 in E such that $\|f_N(t) - f_N(t_0)\| \leq \varepsilon/3$ for any $t \in V \cap U$. Since $\|f_N(t) - f(t)\| \leq \|f_N - f\| \leq \varepsilon/3$ for all $t \in U$. Thus

$\|f(t) - f(t_0)\| \le \|f(t) - f_N(t)\| + \|f_N(t) - f_N(to)\| + \|f_N(t_0) - f(t_0)\| \le \varepsilon$ for all $t \in V \cap V$. Hence f is continuous. □

For a normed linear space $(X, \|\cdot\|)$ with $0 \in X$ and a real number $r > 0$, we use the notation $S_r(0, X) = \{x \in X : \|x\| < r\}$.

Lemma 2.10 *Let E and F be Banach spaces, $U = S_\alpha(O, E)$ and $V = S_\beta(O, F)$, where $\alpha > 0$ and $\beta > 0$ are real numbers. Let f be a continuous mapping of $U \times V$ into F such that $\|f(x, y_1) - f(x, y_2)\| \le k \|y_1 - y_2\|$ for all $x \in U, y_1, y_2 \in V$, where k is a constant with $0 < k < 1$. Also let $\|f(x, 0)\| < \beta(1 - k)$ for all $x \in U$. Then there is a unique continuous mapping Φ of U into V such that $\Phi(x) = f(x, \Phi(x))$ for all $x \in U$.*

Proof. We give a variant of the proof given by Dieudonné (Diedudonné (1969)). Let $T: C_F^\infty(U) \to C_F^\infty(U)$ be a mapping defined by $(T\,\Phi)(x) = f(x, \Phi(x)), \Phi \in C_F^\infty(U)$. Now it is easy to see that $T(\Phi)$ is continuous. Also for each $\Phi \in C_F^\infty(U)$,

$$\|(T\Phi)(x)\| \le \|f(x, \Phi(x)) - f(x, 0)\| + \|f(x, 0)\| \le k \|\Phi(x) - 0\| + \beta(1 - k) < \infty$$

Thus $T\Phi \in C_F^\infty(U)$ wherever $\Phi \in C_F^\infty$. Now for all Φ_1, Φ_2, $\in C_F^\infty(U)$.

$$\begin{aligned} \|T(\Phi_1) - T(\Phi_2)\| &= \|f(x, \Phi_1, (x)) - f(x, \Phi_2(x))\| \\ &\le k \|\Phi_1(x) - \Phi_2(x)\| \le k \|\Phi_1 - \Phi_2\|. \end{aligned}$$

Hence T is a contraction mapping and there is a unique fixed point $\Phi \in C_F^\infty(U)$.

Now for any $x \in U$, we define inductively a sequence of points Φ_n of V such that $\Phi_0 = 0$, $\Phi_n = f(x, \Phi_{n-1})$, $n \ge 1$. Then $\|\Phi_n - \Phi_{n-1}\| \le k^{n-1} \|\Phi_1\|$, $\Phi_n = T(\Phi_{n-1})$.

Hence $(*)$ $\|\Phi_n\| \le (1 + k + k^2 + \cdots + k^{n-1}) \|\Phi\| \le \|\Phi_1\| / (1 - k) < \beta$ and $\Phi_n \in C_F^\infty(U)$. Thus by contraction mapping principle $\Phi_n \to \Phi$. Hence by $(*)$ $\|\Phi(x)\| \le \|f(x, 0)\| / (1 - k) < \beta$ for any $x \in U$.

Hence Φ is a mapping of U into V.

Smiley (1985) has obtained an existence theorem for the nonlinear abstract hyperbolic boundary value problem in which the equation is of the form

$$\frac{\partial^2 u}{\partial t^2} + Au = \varepsilon g[u] \tag{2.76}$$

where A is a strongly elliptic operator, uniformly in an open bounded set $\Omega \subset \mathbb{R}^n$ and g is a nonlinear Nemytsky operator generated by a real valued function. The equation (2.76) is to be satisfied weakly in the cylindrical domain $(0, T) \times \Omega$. The main technique used by Smiley (1985) is blending together the abstract techniques of Lions and Magenes [LM1] with the alternative methods of Cesari (1976) and Hale (1967). After reducing the problem (2.76) into a Cesari's alternative form, he was then able to use the implicit function theorem in proving the existence of solutions of (2.76) for sufficiently small ε. The boundary conditions used for the equation (2.76) will be made explicit in Subsection 2.6.3.

As an application to his abstract existence theory, Smiley (1985) derived the following theorem first established by Rabinowitz (1967).

Theorem 2.23 *Let $g \in C^3(-\infty, \infty)$, $g'(u) \geqslant \beta > 0$, $g(0) = 0$ and $f \in H^2(G)$ be 2π-periodic in t where $G = (0, 2\pi) \times (0, \pi)$, then there exists $\varepsilon_0 > 0$ such that for each $\varepsilon \in (-\varepsilon_0, \varepsilon_0)$ the problem*

$$u_{tt} - u_{xx} = \varepsilon[g(u) + f], \ -\infty < t < \infty, \ 0 < x < \pi\,;$$
$$u(t, 0) = u(t, \pi) = 0, \ -\infty < t < \infty\,;$$
$$u(t + 2\pi, x) = u(t, x), \ -\infty < t < \infty, \ 0 < x < \pi$$

has a nontrivial solution.

The above problem and similar problems have also been considered by many authors, e.g. H. and L. (1978), Cesari (1965), Lovicarova (1972), Nakamo (1976), Rabinowitz (1967), Vejvoda (1964).

In this section we have obtained a more general abstract existence theorem for the equation (2.76) and as an application of our abstract theorem we have also derived the above theorem. Our technique follows an approximating procedure in the line of Tarafdar (1980) and an application of an implicit function theorem involving a set of mappings which is proved in Subsection 2.6.1. This type of implicit function theorem has its own interest and hopefully will find application elsewhere.

Finally we should mention that in this section the symbols and notations of Smiley (1985) have been maintained and free use of preliminary results obtained thereof has been made.

2.6.1 *An Implicit Function Theorem for a Set of Mappings*

In this subsection we prove an implicit function theorem for a set of mappings and use this theorem in the latter part of our paper to obtain an abstract existence theorem for nonlinear hyperbolic boundary value problem. All Banach spaces in this section are assumed to be real.

Definition 2.2 Let E and F be Banach spaces and U an open subset of E. Let I be an index set and $\{u_\alpha : \alpha \in I\}$ be a set of points in U. A family $\{f_\alpha : \alpha \in I\}$ of mappings each defined on U and taking values in F is said to be equicontinuously Fréchet differentiable on U with respect to the set $\{u_\alpha : \alpha \in I\}$ if f_α is continuously differentiable on U for each $\alpha \in I$ and given $\varepsilon > 0$, there exists $\delta > 0$ (independent of α) such that for each $\alpha \in I$, $\|Df_\alpha(u) - Df_\alpha(u_\alpha)\| < \varepsilon$ whenever $u \in S_\delta(u_\alpha, E)$, where $Df(v)$ stands for the derivative of f at $u \in U$.

Remark 2.14 If $f_\alpha : U \subset E \to F$ is differentiable on U for each $\alpha \in I$ and $\|Df_\alpha(u) - Df_\alpha(v)\| \leq C_0 \|u - v\|$ for all $\alpha \in I$, $u, v \in U$ and for some constant C_0, then clearly $\{f_\alpha : \alpha \in I\}$ is equicontinuously differentiable on U with respect to each set $\{u_\alpha : \alpha \in I\}$ of points in U.

Theorem 2.24 (Implicit Function Theorem) *Let E, F, G be Banach spaces $\{f_\alpha : \alpha \in I\}$ a family of mappings of an open subset A of $E \times F$ into G. Let $\{(x_0, y_\alpha) : \alpha \in I\}$ be a set of points in A such that $f_\alpha(x_0, y_\alpha) = 0$ for all $\alpha \in I$. Assume further:*

(*a*) *$\{f_\alpha : \alpha \in I\}$ is equicontinuously differentiable on A with respect to the set $\{(x_0, y_\alpha) : \alpha \in I\}$;*

(*b*) *(1) For each $\alpha \in I$, the partial derivative $D_2 f_\alpha(x_0, y_\alpha)$ (i.e. the partial derivative with respect to variables in F) is a linear homeomorphism of F onto G and*
(2) the set $\{(D_2 f_\alpha(x_0, y_\alpha))^{-1} : \alpha \in I\}$ is uniformly bounded in the operator norm, i.e. there exists a constant C such that

$$\left\|(D_2 f_\alpha(x_0, y_\alpha))^{-1} z\right\| \le C \|z\| \quad \textit{for all} \quad z \in G \quad \textit{and all} \quad \alpha \in I;$$

(*c*) *given $\varepsilon > 0$, there exists a $\delta > 0$ such that for all $\alpha \in I$,*

$$\|f_\alpha(x, y_\alpha) - f_\alpha(x_0, y_\alpha)\| < \varepsilon, \quad \textit{whenever} \ \|x - x_0\| < \delta.$$

Then there exist an open neighborhood U_0 of x_0 and a number $\beta_0 > 0$ such that for each open connected neighborhood U of x_0 with $U \subset U_0$ there exists for each $\alpha \in I$ a unique continuous mapping $\Phi_\alpha : U \to S_{\beta_0}(y_\alpha, F)$ with the property that

$$\Phi_\alpha(x_0) = y_\alpha, \ (x, \Phi_\alpha(x)) \in A \quad \textit{and} \quad f_\alpha(x, \Phi_\alpha(x)) = 0 \quad \textit{for all} \quad x \in U.$$

Moreover, for each $\alpha \in I$, Φ_α is continuously differentiable on U and the derivatives are given by

$$\Phi'_\alpha(x) = -(D_2 f_\alpha(x, \Phi_\alpha(x)))^{-1} \cdot D_1 f_\alpha(x, \Phi_\alpha(x)). \tag{2.77}$$

Proof. We denote by T_α the linear homeomorphism $D_2 f_\alpha(x_0, y_\alpha)$ of F onto G. For each $\alpha \in I$, let us consider the mapping $g_\alpha : A \to F$ defined by $g_\alpha(x, y) = y - T_\alpha^{-1} \cdot f_\alpha(x, y)$. Now since $T_\alpha^{-1} T_\alpha = I$, the identity on F, we have for $(x, y_1) \in A$ and $(x, y_2) \in A$, $\alpha \in I$,

$$g_\alpha(x, y_1) - g_\alpha(x, y_2) = T_\alpha^{-1} \left(D_2 f_\alpha(x_0, y_\alpha) \cdot (y_1 - y_2) - (f_\alpha(x, y_1) - f_\alpha(x, y_2))\right). \tag{2.78}$$

By condition [(b)](2) we can choose sufficiently small $\varepsilon > 0$ such that $\varepsilon \left\|T_\alpha^{-1}\right\| < 1/2$ (by choosing $\varepsilon > 0$ small enough to make $\varepsilon C < 1/2$). Also by condition (a) there exist $\alpha_0 > 0$ and $\beta_0 > 0$ such that for all $\alpha \in I$, $\|Df_\alpha(x, y) - Df_\alpha(x_0, y_\alpha)\| < \varepsilon$ whenever $\|x - x_0\| < \alpha_0$ and $\|y - y_\alpha\| < \beta_0$, i.e. whenever $x \in S_{\alpha_0}(x_0, E)$ and $y \in S_{\beta_0}(y_\alpha, F)$. Now using the continuous differentiability of each f_α and mean value theorem (see (Diedudonné (1969)) we have for each $\alpha \in I$,

$$\begin{aligned}
&\|f_\alpha(x, y_1) - f_\alpha(x, y_2) - D_2 f_\alpha(x_0, y_\alpha)\cdot(y_1 - y_2)\| \\
&\quad = \|f_\alpha(x, y_1) - f_\alpha(x, y_2) - D f_\alpha(x_0, y_\alpha)\cdot((x, y_1) - (x, y_2))\| \\
&\quad \le \|y_1 - y_2\| \sup_{(x,y)\in S} \|D f_\alpha(x, y) - D f_\alpha(x_0, y_\alpha)\| \\
&\quad \text{(where } S \subset A \text{ is a segment joining } (x, y_1) \text{ and } (x, y_2).) \\
&\quad \le \varepsilon \|y_1 - y_2\|, \qquad (2.79)
\end{aligned}$$

whenever $x \in S_{\alpha_0}(x_0, E) = U_1$ and $y_1, y_2 \in S_{\beta_0}(y_\alpha, F)$. Now from (2.78) and (2.79) we have for each $\alpha \in I$,

$$\|g_\alpha(x, y_1) - g_\alpha(x, y_2)\| \le \varepsilon \left\|T_\alpha^{-1}\right\| \|y_1 - y_2\| \le \frac{1}{2}\|y_1 - y_2\| \qquad (2.80)$$

whenever $x \in U_1$ and $y_1, y_2 \in S_{\beta_0}(y_\alpha, F)$.

Also by condition (c) there exists $\delta > 0$ such that

$$\begin{aligned}
\|g_\alpha(x, y_\alpha) - y_\alpha\| &= \left\|-T_\alpha^{-1}\cdot f_\alpha(x, y_\alpha)\right\| \le \left\|T_\alpha^{-1}\right\| \|f_\alpha(x, y_\alpha)\| \\
&= \left\|T_\alpha^{-1}\right\| \|f_\alpha(x, y_\alpha) - f_\alpha(x_0, y_\alpha)\| < \left\|T_\alpha^{-1}\right\| \varepsilon\beta_0 < \frac{\beta_0}{2} \qquad (2.81)
\end{aligned}$$

whenever $x \in S_\delta(x_0, E) = U_2$.

Now taking $U_0 = U_1 \cap U_2$ we see that (2.80) and (2.81) both hold whenever $x \in U_0$ and $y_1, y_2 \in S_{\beta_0}(y_\alpha, F)$.

Thus we can apply Lemma 2.10 to the mapping $\tilde{g}_\alpha$ for each $\alpha \in I$ where $\tilde{g}_\alpha$ is defined by

$$\tilde{g}_\alpha(x', y') = g_\alpha(x_0 + x', y_\alpha + y') - y_\alpha$$

in a small neighborhood of $(0, 0)$.

When we do this, we obtain for each $\alpha \in I$ a unique continuous mapping $\Phi_\alpha : U_0 \to V_\alpha = S_{\beta_0}(y_\alpha, F)$ having the property that $f_\alpha(x, \Phi_\alpha(x)) = 0$ for all $x \in U_0$ and also $\Phi_\alpha(x_0) = y_\alpha$, since $f_\alpha(x_0, y_\alpha) = 0$.

For the rest of the proof we repeat the argument of Diedudonné (1969) for each fixed $\alpha \in I$. Let $\alpha \in I$ be arbitrary but fixed. Next we prove that if $U \subset U_0$ is an open connected neighborhood of x_0, Φ_α is the unique continuous mapping of U into F such that $\Phi_\alpha(x_0) = y_\alpha$, $(x, \Phi_\alpha(x)) \in A$ and $f_\alpha(x, \Phi_\alpha(x)) = 0$ for all $x \in U$ and if ψ_α be a second mapping satisfying these conditions, then $\Phi_\alpha = \psi_\alpha$ on U.

Let $M = \{x \in U : \Phi_\alpha(x) = \psi_\alpha(x)\}$. Then $x_0 \in M$ and M is closed as Φ_α and ψ_α are continuous. It will thus suffice to prove that M is open, for then it will imply that $M = U$ as U is connected. Now since $x \to D_2 f_\alpha(x, \Phi_\alpha(x))$ is continuous in U_0, by virtue of Lemma 2.10 (replacing if necessary U_0 by a smaller neighborhood), we can assume that $D_2 f_\alpha(x, \Phi_\alpha(x))$ is a linear homeomorphism of F onto G for

$x \in U_0$. Let $a \in M$. Then by the first part of theorem already proved, there exists an open neighborhood $U_a \subset U$ and open neighborhood $V_a \subset V_\alpha$ of $b_\alpha = \Phi_\alpha(a)$ such that, for any $x \in U_\alpha$, $\Phi_\alpha(x)$ is the only solution y_α of the equation $f_\alpha(x, y_a) = 0$ such that $y_\alpha \in V_\alpha$.

Now since ψ_α is continuous at a and $\Psi_\alpha(\alpha) = \Phi_a(a)$, there is a neighborhood $W \subset U_a$ such that $\psi_a(x) \in V_a$ for $x \in W$. Hence it follows that $\Phi_\alpha(x) = \psi_\alpha(x)$ for $x \in W$, both being solutions of $f_\alpha(x, y_\alpha) = 0$. Thus M is open.

Next, we show that Φ_α is continuously differentiable in U_0.

For x and $x+s$ in U_0, put $t = \Phi_\alpha(x+s) - \Phi_\alpha(x)$. Then by what we have already proved, $f_\alpha(x+s, \Phi_\alpha(x)+t) = f_\alpha(x+s, \Phi_\alpha(x+s)) = 0$ and $t \to 0$ as $s \to 0$. Hence by (2.81), for a given $x \in U_0$ and for any $\delta > 0$, there is $r > 0$ such that $\|s\| \le r$ implies

$$\|f_\alpha(x+s,\, \Phi_\alpha(x)+t) - f_\alpha(x, \Phi_\alpha(x)) - S_\alpha(x)\cdot s - T_\alpha(x)\cdot t\| \le \delta(\|s\| + \|t\|),$$

where $S_\alpha(x) = D_1 f_\alpha(x, \Phi_\alpha(x))$ and $T_\alpha(x) = D_2 f_\alpha(x, \Phi_\alpha(x))$, which reduces to

$$\|S_\alpha(x)\cdot s + T_\alpha(x)\cdot t\| \le \delta(\|s\| + \|t\|)$$

Now as $T_\alpha(x)$ is a linear homeomorphism,

$$\left\|(T_\alpha^{-1}(x)_0 S_\alpha(x))\cdot s + t\right\| = \left\|T_\alpha^{-1}(x)_0 S_\alpha(x)\cdot s + T_\alpha^{-1}(x)_0 T_\alpha(x)\cdot t\right\| \tag{2.82}$$

$$\le \delta \left\|T_\alpha^{-1}(x)\right\| (\|s\| + \|t\|) \tag{2.83}$$

Let δ be so chosen that $\delta \left\|T^{-1}(x)\right\| \le 1$. Then substituting $a = 2\left\|T_\alpha^{-1}(x)_0 S_\alpha(x)\right\| + 1$, we have from (2.82)

$$\begin{aligned}\|t\| &= \left\|T_\alpha^{-1}(x)_0 S_\alpha(x)\cdot s + t - T_\alpha^{-1}(x)_0 S_\alpha(x)\cdot s\right\| \\ &\le \delta\left\|T_\alpha^{-1}(x)\right\|(\|s\| + \|t\|) + \left\|T_\alpha^{-1}(x)_0 S_\alpha(x)\right\| \|s\| \\ &= \frac{1}{2}(\|s\| + \|t\|) + \frac{a-1}{2}\|s\|.\end{aligned}$$

Hence

$$\|t\| \le \alpha \|s\|. \tag{2.84}$$

Finally from (2.82) and (2.84) we have

$$\left\|t + (T_\alpha^{-1}(x)_0 S_\alpha(x))\cdot s\right\| \le \delta(a+1)\left\|T_\alpha^{-1}(x)\right\| \|s\|$$

whenever $\|s\| < r$. Hence it follows from the definition of t, Φ_α is differentiable at the point x and has the derivative $T_\alpha^{-1}(x)_0 S_\alpha(x)$ which is the right hand side of (2.77). Now from (2.82), (2.84) and (2.77), it follows that Φ_α is continuously differentiable in U_0. □

Remark 2.15 An important fact which will be crucial for our application of the above theorem is that if the set $\{y_\alpha : \alpha \in I\}$ is bounded, then for each $x \in U$ the set $\{\Phi_\alpha(x) : \alpha \in I\}$ is also bounded. This can be seen easily. Indeed, if $\|y_\alpha\| \le C_1$ for all $\alpha \in I$, then

$$\|\Phi_\alpha(x)\| \le \|\Phi_\alpha(x) - y_\alpha\| + \|y_\alpha\| \le \beta_0 + C_1 \quad \text{for all } \alpha \in I.$$

We now consider the special case when $y_\alpha = y_0$ for all $\alpha \in I$.

Corollary 2.24.1 *Let all the conditions of Theorem* 1.1 *hold with* $y_\alpha = y_0$ *for all* $\alpha \in I$. *Then there exist an open neighborhood* U_0 *of* x_0 *in* E *and an open neighborhood* V_0 *of* y_0 *such that for each open connected neighborhood* U *of* x_0 *with* $U \subset U_0$, *there exists for each* $\alpha \in I$ *a unique continuous mapping* $\Phi_\alpha : U \to V_0$ *with the property that* $\Phi_\alpha(x_0) = y_0$, $(x, \Phi(x)) \in A$ *and* $f_\alpha(x, \Phi_\alpha(x)) = 0$ *for all* $x \in U$. *Moreover, for all* $\alpha \in I$, Φ_α *is continuously differentiable on* U *and the derivatives are given by* (2.77).

Taking $I = \{1\}$ in Theorem 2.24, we obtain the following well-known Implicit Function in Banach spaces.

Corollary 2.24.2 *Let* E, F, G *be Banach spaces and* f *a continuously differentiable mapping of an open subset* A *of* $E \times F$ *into* G. *Let* (x_0, y_0) *be a point of* A *such that* $f(x_0, y_0) = 0$ *and that the partial derivative* $D_2 f(x_0, y_0)$ *be a linear homeomorphism of* F *onto* G. *Then there exist an open neighborhood* U_0 *of* x_0 *and a number* $\beta_0 > 0$ *such that for each connected* U *of* x_0 *with* $U \subset U_0$, *there exists a unique continuous mapping* $\Phi : U \to S_{\beta_0}(y_0, F)$ *with the property* $\Phi(x_0) = y_0$, $(x, \Phi(x)) \in A$ *and* $f(x, \Phi(x)) = 0$ *for all* $x \in U$. *Furthermore,* Φ *is continuously differentiable in* U *and its derivative is given by*

$$\Phi'(x) = -(D_2 f(x, \Phi(x))^{-1} D_1(f(x, \Phi(x)).$$

In the next corollary we deduce the well-known inverse theorem in Banach spaces.

Corollary 2.24.3 *Let* E *and* F *be two Banach spaces and* f *a continuously differentiable mapping of an open neighborhood* U *of* $x_0 \in E$ *into* F. *If* $f'(x_0)$ *is a linear homeomorphism of* E *onto* F, *then there is an open neighborhood* U_0 *of* x_0 *with* $U_0 \subset U$ *such that the restriction of* f *to* U_0 *is homeomorphism of* U_0 *onto an open neighborhood of* $y_0 = f(x_0)$ *in* F. *Furthermore, the inverse mapping* f/U_0 *is continuously differentiable and* $[(f/U_0)^{-1}]'(y_0) = [f'(x_0)]^{-1}$.

Proof. We will apply Corollary 2.24.2 by interchanging the role of x and y. Set $y_0 = f(x_0)$. Let $A = F \times U$ and $g : A \to F$ defined by $g(y, x) = -y + f(x)$, $y \in F$ and $x \in U$. Then $D_2 g(y_0, x_0) = f'(x_0)$ which is a linear homeomorphism of E onto F, $D_1 g(y, x) = I$, the identity on F and $g(y_0, x_0) = 0$. Hence by Corollary 2.24.2 we obtain an open neighborhood V of y_0 and a continuously differentiable mapping $\Phi : V \to E$ such that $\Phi(y_0) = x_0$; $(y, \Phi(y)) \in A$ and $g(y, \Phi(y)) = 0$ for all $y \in V$, i.e. $\Phi(V) \subset U$ and $f(\Phi(y)) = y$ for all $y = V$. The last equality implies that Φ is one-to-one. Hence it follows that $f^{-1} = \Phi$ is a homeomorphism of V onto

$\Phi(V) \subset U$. Thus $f^{-1}(V) = \Phi(V)$ is an open set in E and f is a homeomorphism of $U_0 = \Phi(V)$ onto V. Furthermore,

$$[(f/U_0)^{-1}]'(y_0) = \Phi'(y_0) = -(D_2g(x_0, \Phi(x_0))^{-1}D_1(g(x_0, \Phi(x_0)) = [f'(x_0)]^{-1}.$$

□

2.6.2 *Notations and Preliminaries*

We use the following symbols and notations in the rest of this section. Let $(V, \|\cdot\|, ((\cdot, \cdot)))$ and $(H, |\cdot| (\cdot, \cdot))$ be two separable Hilbert spaces with $V \subset H$ and V dense in H where $\|\cdot\|$, $((\cdot, \cdot))$ and $|\cdot|$, $(\cdot, \cdot)$ are norms and inner products of V and H respectively.

Let $A : V \to V^*$ be a continuous linear mapping of V into its dual V^*. We define a bilinear form on V by $a(u, v) = (Au, v)$ and assume that $a(\cdot, \cdot)$ is symmetric and coercive, i.e. there exist constants $\alpha_0 \geq 0$ and $\alpha_1 > 0$ such that

$$a(u, u) + \alpha_0 |u|^2 \geq \alpha_1 \|u\|^2. \tag{2.85}$$

Then

$$\|u\|_A = (a(u, u) + \alpha_0 |u|^2)^{1/2} \tag{2.86}$$

is a norm on V and is equivalent to the norm $\|\cdot\|$ on V. Henceforth $\|u\|_A$ will be referred to as $\|\cdot\|$, the norm on V.

Let $\{w_1, w_2, \dots\}$ be a complete orthogonal basis for V, consisting of eigenvectors for the operator A. Hence the set of finite linear combinations of w's is dense in V. Let $\{\lambda_1, \lambda_2, \dots\}$ be the corresponding eigenvalues, i.e. $Aw_i = \lambda_i w_i$ for $i = 1, 2, \dots$. We do not rule out the possibility that $\lambda_i = \lambda_j$ for some $i \neq j$.

Let $L^2(0, T; V)$ and $L^2(0, T; H)$ denote the Hilbert spaces (cf. Lasota and Myjak (1996a)) of measurable functions from the interval $(0, T)$ into V and H respectively satisfying

$$\|u\|^2_{L^2(0,T;V)} = \int_0^T \|u(t)\|^2 \, dt < \infty$$

and

$$\|u\|^2_{L^2(0,T;H)} = \int_0^T |u(t)|^2 \, dt < \infty.$$

We also define the space

$$W(0, T) = \left\{u \in L^2(0, T; V) : u' = \frac{du}{dt} \in L^2(0, T; H)\right\}$$

where the derivative u' is the weak derivative[1] in the sense

$$\int_0^T u'(t)\Phi(t)dt = -\int_0^T u(t)\Phi'(t)dt \qquad \text{(Bochner integral)}$$

for all $\Phi \in C_0^\infty(0, T)$, the set of infinity differentiable functions with compact support in $(0, T)$. Analogously we define weak periodic derivative by replacing $C_0^\infty(0, T)$ by $C_{\text{per}}^\infty(0, T)$, the set of functions $\Phi \in C^\infty(R)$ which are T-periodic.

The inner product in $W(0, T)$ is defined by

$$(u, w)_W = \int_0^T [a(u, w) + \alpha_0(u, w) + (u', w')]dt. \tag{2.87}$$

In this setting as in Smiley (1985) we consider the problem

$$\frac{d^2u}{dt^2} + Au = \varepsilon g[u],\ 0 < t < T \tag{2.88}$$

$$B_1 u = B_2 u = 0, \tag{2.89}$$

where $B_1 u = B_2 u = 0$ denote either Dirichlet, Neumann or periodic boundary conditions.

2.6.3 *Results of Smiley on Linear Problem*

The boundary condition subspace W_B of $W(0, T)$ corresponding to three boundary conditions are defined as follows (cf. Smiley (1982) and Smiley (1985))

$$W_B = \begin{cases} W_0(0, T) = \text{closure of } C_0^\infty(0, T;\, V), & B_1 u = u(0),\ B_2 u = u(T); \\ W(0, T) & B_1 u = u'(0),\ B_2 u = u'(T); \\ W_{\text{per}}(0, T) = \text{closure of } C_{\text{per}}^\infty(0, T, V), & B_1 u = u(0) - u(T), \\ & B_2 u = u'(0) - u'(T); \end{cases}$$

where the closures are taken with respect to the norm topology of $W(0, T)$, induced by the inner product (2.87), $C_0^\infty(0, T;\, V)$ and $C_{\text{per}}^\infty(0, T;\, V)$ denote the sets of C^∞ functions from $\mathbb{R}$ to V which have compact support in $(0, T)$ and are T-periodic respectively.

Let $\mathfrak{B}$ be a subset of W_B and $g : \mathfrak{B} \to L^2(0, T;\, H)$ be a mapping (possibly nonlinear). Then $u \in \mathfrak{B}$ is said to be a weak solution of the problem (2.88)–(2.89) if

$$\int_0^T [-(u', w') + a(u, w)]dt = \varepsilon \int_0^T (g[u], w)dt \quad \text{for all } w \in W_B. \tag{2.90}$$

[1] See Schwartz (1957)

We define a symmetric form $B(\cdot, \cdot)$ by

$$B(u, w) = \int_0^T [-(u', w') + a(u, w)]dt. \tag{2.91}$$

Then $u \in \mathfrak{B}$ is a weak solution of the problem (2.88)–(2.89) if

$$B(u, w) = \varepsilon(g[u], w)_{L^2(0,T;H)} \quad \text{for all } w \in W_B. \tag{2.92}$$

$B(\cdot, \cdot)$ is obviously continuous on $W_B \times W_B$.

With $\varepsilon = 0$, the equation (2.92) reduces to the homogeneous problem

$$B(u, w) = 0, \qquad \text{for all } w \in W_B. \tag{2.93}$$

Let $X_0 = \{u \in W_B : u \text{ satisfies (2.93)}\}$ which is a closed subspace of W_B. If $\dim X_0 \geq 1$, the problem is said to be at resonance. Such problems are not well-posed in the sense of Hadamard.

Let Σ denote the set of eigenvalues for the corresponding scalar problem:

$$\Phi'' + \mu\Phi = 0, \ 0 < t < T,$$
$$B_1 u = B_2 u = 0.$$

Here we quote the result of Smiley (1982).

Theorem 2.25 (Fredholm Alternative) *If $\lambda_i \in \Sigma$ for all but a finite number of indices $i = 1, 2, \ldots,$ then for $f \in L^2(0, T; H)$ the weak solutions of*

$$\frac{d^2u}{dt^2} + Au = f, \ 0 < t < T, \tag{2.94}$$

exist if and only if $(f, u)_{L^2(0,T;H)} = 0$ for all $u \in X_0$. If solutions exist, then there is a unique solution $u_0 \in X_0^{\perp}$, the orthogonal complement of X_0 in W_B.

Let I be the set of indices for which $\lambda_i \in \Sigma$. We reorder the set I by $\{1, 2, 3, \ldots\}$ so that $\lambda_1 \lambda_2, \ldots, \lambda_n, \cdots \in \Sigma$ (relabelling if necessary). Let us denote the real valued eigen function corresponding to the eigenvalue $\mu_i = \lambda_i \in \Sigma$ of the scalar problem by $e_i(t)$. Then clearly $e_i(t)w_i \in X_0$ where w_i is the eigen vector of A corresponding to the eigen value λ_i of A. We now denote the linear span of $\{e_1(t)w_1, e_2(t)w_2, \ldots, e_n(t)w_n\}$ by X_n. Thus we have a sequence $\{X_n\}$ of finite dimensional subspaces of X_0 under the hypothesis of Theorem 2.25.

Let $Y_0, Y_1 \subset L^2(0, T; H)$ be defined as follows:

$$Y_1 = \{u : (u, u_0)_{L^2(0,T;H)} = 0, \text{ for all } u_0 \in X_0\}$$
$$Y_0 = \{u : (u, v)_{L^2(0,T;H)} = 0, \text{ for all } v \in Y_1\}.$$

Then we have $L^2(0, T; H) = Y_0 \oplus Y_1$ and $W_B = X_0 \oplus X_0^{\perp}$ where $X_0^{\perp}$ is the orthogonal complement of X_0 in W_B. Obviously $X_0 \subset Y_0$. Let $P : W_B \to X_0$ and $Q : L^2(0, T; H) \to Y_0$ be respectively the orthogonal projections of W_B onto X_0 and

$L^2(0, T; H)$ onto Y_0. Then $I - P : W_B \to X_1 = X_0^{\perp}$ and $I - Q = L^2(0, T; H) \to Y_1$ are orthogonal projections onto X_1 and Y_1 respectively.

For each positive integer n, let $P_n : X_0 \to X_n$ be the orthogonal projection of X_0 onto X_n and $Q_n : Y_0 \to X_n$ be that of Y_0 onto X_n so that $P_n P : W_B \to X_n$ and $Q_n Q : L^2(0, T : H) \to X_n$ are orthogonal projections.

We also define

$$\|u\|_{W^\infty} = \|u\|_{L^\infty(0,T;V)} + \|u'\|_{L^\infty(0,T;H)}$$

and

$$X_0^\infty = \{u \in X_0 : \|u\|_{W^\infty} < \infty\} \tag{2.95}$$

$$X_1^\infty = \{u \in X_1 : \|u\|_{W^\infty} < \infty\} \tag{2.96}$$

and

$$W_B^\infty = X_0^\infty \oplus X_1^\infty.$$

It is easily seen that W_B^∞, X_0^∞ and X_1^∞ are Banach spaces and the restriction of P to W_B^∞ is a continuous projection of W_B^∞ onto X_0^∞.

Remark 2.16 It can be seen that the norm defined by $\|u\| = \|Pu\|_{W^\infty} + \|(I - P)u\|_{W^\infty}$ is equivalent to $\|u\|_{W^\infty}$ defined above.

We will need the following results which we write as lemmas (for proof see Smiley (1982) and Smiley (1985)).

Lemma 2.11 *Under the hypothesis of Theorem 2.25 there is a positive constant C such that for each $f \in Y_1$ and the corresponding unique solution $u_1 \in X_1 \subset W_B$,*

$$\|u_1\|_{W^\infty} \le C \|f\|_{L^2(0,T;H)} .$$

That is, $\|K_B\|_{W^\infty} \le C \|f\|_{L^2(0,T;H)}$ where K_B is the inverse of the restriction of L_B to X_1^∞ and L_B is the linear mapping defined by $L_B u = \dfrac{d^2u}{dt^2} + Au$ and $D(L_B) = \left\{u \in W_B : \dfrac{d^2u}{dt^2} + Au = f \text{ for some } f \in Y_1\right\}$. Thus $K_B : Y_1 \to X_1^\infty$ is continuous from $L^2(0, T; H)$-topology to W^∞-topology.

Lemma 2.12 *For each $u \in X_0$ there exists a positive constant C_1 such that*

$$\|u\|_{W^\infty} \le C_1 \|u\|_W .$$

Also for all $u \in W(0, T)$ we have

$$\|u\|_W \le \sqrt{2T} \|u\|_{W^\infty} .$$

Proof. For the sake of convenience we assume the boundary conditions are $u(0) = u(T) = 0$, the proof in other cases will be similar. Thus $\Sigma = \{(n\pi/T)^2 : n = 1, 2, \dots\}$. Now since $u \in X_0$, $u(t) \sim \sum_{i\in I} u_i \sin(\sqrt{\lambda}\, t) w_i$, and since $B(u, w) = 0$ for all $w \in X_0$ by (2.93),

$$\|u\|_W^2 = \|u\|_W^2 + B(u,u) = \int_0^T (2|u'|^2 + \alpha_0 |u|^2) dt \tag{2.97}$$

$$\geq \int_0^T |u'|^2 dt. \tag{2.98}$$

We note that $\delta_{i,j} = ((w_i, w_j)) = (\lambda_i + \alpha_0)(w_i, w_j)$ and hence $|u'(t)|^2 = (u'(t), u'(t)) = \sum_{i\in I} u_i^2 \lambda_i \cos^2(\sqrt{\lambda_i} t) \cdot (\lambda_i + \alpha_0)^{-1}$.

Integrating we obtain

$$2\int_0^T |u'(t)|^2 dt = \frac{T}{2} \sum_{i\in I} u_i^2 \frac{\lambda i}{\lambda_i + \alpha_0} \leq \|u\|_W^2 .$$

From the same equality we have

$$|u'(t)|^2 \leq \sum_{i\in I} u_i^2 \frac{\lambda_i}{\lambda_i + \alpha_0} \leq \frac{2}{T} \|u\|_W^2 .$$

Also we have

$$\|u(t)\|^2 = ((u(t), u(t)) = \sum_{i\in I} u_i^2 \sin^2(\sqrt{\lambda} t) \leq \sum_{i\in I} u_i^2.$$

Now since $\lambda_i \geq (\pi/T)^2$ for all $i \in I$, it follows that

$$\|u(t)\|^2 \leq \sum_{i\in I} u_i^2 \leq \frac{(\pi/T)^2 + \alpha_0}{(\pi/T)^2} \sum_{i\in I} u_i^2 \frac{\lambda_i}{\lambda_i + \alpha_0} \leq C \|u\|_W^2 .$$

Combining these estimates we can show that there is a constant $C > 0$ such that $\|u\|_{W\infty} \leq C \|u\|_W$. We can conclude the proof by observing

$$\|u\|_W \leq \sqrt{2T} \|u\|_{W\infty} \quad \text{for all } u \in W(0,T).$$ □

Remark 2.17 By virtue of the last inequality above we obtain that K_B is also continuous from $L^2(0, T : H)$-topology to W_B-topology.

Lemma 2.13 *For all $u \in W_B$, we have*

$$Pu = Qu \quad \textit{and } P_n Pu = Q_n Qu.$$

Proof. The proof of the first part is similar to that of Lemma 3.3 in Smiley (1985). Let $u = u_0 + u_1$, $u_0 \in X_0$ and $u_1 \in X_1$. Then $Pu = u_0$. Also since $X_0 \subset Y_0$, $Qu_0 = u_0$. It will thus suffice to show that $Qu_1 = 0$. To this end we show that $X_1 \subset Y_1$. Let $u_1 \in X_1$ be arbitrary and $w = \Phi w_i \in X_0$ where Φ is a real valued function and w_i is an eigenvector of A. Now $B(w, u_1) = 0$ as $w \in X_0$ and $(w, u_1)_W = 0$ as $u_1 \in X_1 = X_0^{\perp}$. Thus,

$$\begin{aligned} 0 &= B(w, u_1) + (w, u_1)_W \\ &= \int_0^T [-(w', u_1') + a(w, u_1) + a(w, u_1) \\ &\quad + \alpha_0(w, u_1) + (w', u_1')]dt \\ &= \int_0^T [2a(w, u_1) + \alpha_0(w, u_1)]dt = \int_0^T (2\lambda_i + \alpha_0)(w, u_1)dt. \end{aligned} \tag{2.99}$$

Thus u_1 is $\perp$ to every element $w \in X_0$ of the form Φw_i with respect to $L^2(0, T; H)$ inner product and is therefore, $\perp$ to every finite linear combination of the elements of this form. But since the set of such finite linear combinations are dense in X_0 endowed with W_B-topology (see Smiley (1982)), it follows from this $u_1 \in Y_1$. Thus we have proved that $X_1 \subset Y_1$.

Now let n be a positive integer and let

$$X_0 = X_n \oplus U_n \quad \text{where} \quad P_n X_0 = X_n \quad \text{and} \quad (P - P_n P)X_0 = U_n$$

and

$$Y_0 = X_n \oplus V_n \quad \text{where} \quad Q_n Y_0 = X_n \quad \text{and} \quad (Q - Q_n Q)Y_0 = V_n.$$

Let $u \in W_B$. Then by first part $Pu = Qu$.

Let $Pu = Qu = x + y$ where $x \in X_n$ and $y \in U_n$.

Then $P_n Pu = x$ and $Q_n Qx = x$ as $x \in X_n$. If we prove that $Q_n Qy = 0$, then it will follow that $P_n Pu = Q_n Qu$.

Thus it will suffice as above to prove that $U_n \subset V_n$.

First we recall that X_n = linear span of $\{e_1(t)w_1, e_2(t)w_2, \ldots, e_n(t)w_n\}$.

Let $v \in U_n$ be arbitrary. Then for $k = 1, 2, \ldots, n$, $B(e_k(t)w_k, v) = 0$ as $e_k(t)w_k \in X_n \subset X_0$ and $v \in U_n \subset W_B$. Thus as above we obtain

$$\int_0^T (e_k(t)w_k, v)dt = 0 \quad \text{for } k = 1, 2, \ldots, n.$$

This implies that $v \perp X_n$ with respect to $L^2(0, T; H)$ inner product. Thus $v \in V_n$. This completes the proof. □

2.6.4 *Alternative Problem and Approximate Equations*

In this section we will reduce our problem to the Cesari's alternative type of problems. Let, as before, $L_B u = \left(\frac{d^2}{dt^2} + A\right) u$ where domain $D(L_B) = \left\{ w \in W_B : \frac{d^2 u}{dt^2} + Au = f \text{ for some } f \in Y_1 \right\}$. Equivalently $D(L_B) = X_0^\infty \oplus K_B(Y_1)$. Since what follows in this section there is nothing to distinguish between three boundary conditions we will denote L_B and K_B simply by L and K.

Lemma 2.14 *Under the hypothesis of Theorem* 2.25, *the operators* P, Q, L *and* K *introduced in the previous section and above satisfy:*

(*i*) $K(I - Q)Lu = (I - P)u$ *for all* $u \in D(L)$;
(*ii*) $LPu = QLu$ *for all* $u \in D(L)$;
(*iii*) $LK(I - Q)u = (I - Q)u$ *for all* $u \in L^2(0, T; H)$.

The proof is simple and is therefore omitted.

We now consider the following approximate equations.

For each positive integer n we consider the equation

$$L_n u = \varepsilon g[u] \tag{2.100}$$

where $L_n u = Lu + (Q - Q_n Q)u$, $u \in D(L)$.

It is easy to see that $X_n = \ker L_n$.

Lemma 2.15 (1) *u is a weak solution of the problem* (2.88)–(2.89), *i.e.* $Lu = \varepsilon g[u]$ *if and only if*

$$u = Pu + \varepsilon K(I - Q)g[u] \tag{2.101}$$

and

$$Qg[u] = 0 \qquad (\textit{bifurcation equation}). \tag{2.102}$$

(2) *u is a solution of the approximate equation* (2.100), *i.e.* $L_n u = \varepsilon g[u]$ *if and only if*

$$u = Pu + \varepsilon K(I - Q)g[u], \tag{2.103}$$

$$(Q - Q_n Q)u = \varepsilon Q g[u], \tag{2.104}$$

and

$$Q_n Q g[u] = 0. \tag{2.105}$$

Proof. The proof of (1) follows immediately from (i), (ii) and (iii) of Lemma 2.14. To prove (2), let $L_n u = \varepsilon g[u]$, i.e. $Lu + (Q - Q_n Q)u = \varepsilon g[u]$. Applying $K(I - Q)$

both sides we obtain (2.103) by virtue of (i). Applying Q_nQ to both sides we obtain (2.104) and and applying $(Q - Q_nQ)$ to both sides we have

$$(Q - Q_nQ)u = \varepsilon(Q - Q_nQ)g[u] = \varepsilon Qg[u] \quad \text{by (2.104).}$$

Next let (2.103), (2.104) and (2.105) hold. Then

$$\begin{aligned} Lu &= \varepsilon(I - Q)g[u] \quad \text{by applying } L \text{ to (2.103)} \\ &= \varepsilon g[u] - (Q - Q_nQ)u \quad \text{by (2.78).} \end{aligned}$$

Thus $L_nu = \varepsilon g[u]$. □

Lemma 2.16 *Let $g : \mathfrak{B} \subset W_B^\infty \to W_B^\infty$ be a mapping. We consider the mapping*

$$\begin{aligned} &F : \mathbb{R} \times \mathfrak{B} \to W_B^\infty \quad \textit{defined by} \\ &F(\varepsilon, u) = Pg(u) + (I - P)u - \varepsilon K(I - Q)g(u), \ u \in \mathfrak{B} \end{aligned}$$

and for each positive integer n,

$$F_n : \mathbb{R} \times \mathfrak{B} \to W_B^\infty \quad \textit{defined by}$$
$$F_n(\varepsilon, u) = P_nPg(u) + (I - P_nP)u - \varepsilon K(I - Q)g(u) - \varepsilon(Q - Q_nQ)g(u),$$

$u \in \mathfrak{B}$. *Then (ε, u) is a weak solution of the problem* (2.88)–(2.89) *if and only if $F(\varepsilon, u) = 0$ and (ε, u) is a solution of the approximate equation* (2.100) *if and only if $F_n(\varepsilon, u) = 0$.*

Proof. It is clear that if (2.101) and (2.102) hold, then $F(\varepsilon, u) = 0$. Suppose now that $F(\varepsilon, u) = 0$. Then $0 = PF(\varepsilon, u) = Pg(u) = Qg(u)$ by Lemma 2.12 as $g(u) \in W_B$. Also $0 = (I - P)F(\varepsilon, u) = (I - P)u - \varepsilon K(I - Q)g(u)$ which gives (2.101). Thus the conclusion follows from Lemma 2.15. Similarly if (2.103), (2.104) and (6.23) hold, then clearly $F_n(\varepsilon, u) = 0$. Conversely suppose that $F_n(\varepsilon, u) = 0$. Then using Lemma 2.12 we have $0 = P_nPF_n(\varepsilon, u) = P_nPg(u)$ which gives (2.105). $0 = (P - P_nP)F_n(\varepsilon, u) = (P - P_nP)u - \varepsilon(Q - Q_nQ)g(u) = (P - P_nP)u - \varepsilon Qg(u)$. Thus the lemma follows from Lemma 2.15. □

Remark 2.18 The mapping $g : \mathfrak{B} \to W_B^\infty$ in the above lemma is not necessarily of the form $g[u]$. In fact we will be interested in g of the form $g[u] + f$ where $f \in W_B^\infty$.

An Abstract Existence Theorem

In the rest of this subsection we deal with the existence of solutions of the approximate equations $L_nu = \varepsilon g[u]$. We show by applying the implicit function theorem established in Subsection 2.6.1 that for sufficiently small ε there exists $u_n(\varepsilon)$ such that, $L_nu_n(\varepsilon) = \varepsilon g(u_n(\varepsilon))$ for each positive integer n, i.e. given $\varepsilon > 0$ sufficiently small each approximate equation has a solution.

Throughout the rest of this subsection we will assume that the hypothesis of Theorem 2.25 holds.

Lemma 2.17 *Let $g : \mathfrak{B} \to W_B^\infty$ be a continuously (Fréchet) differentiable mapping defined on an open subset $\mathfrak{B}$ of W_B^∞. Then for each positive integer n, F_n is continuously differentiable mapping from $\mathbb{R} \times \mathfrak{B}$ into W_B^∞. The derivative, i.e. the linear action of the differential $DF_n(\varepsilon_0, u) \in L(\mathbb{R} \times W_B^\infty, W_B^\infty)$, the space of bounded linear mappings of $\mathbb{R} \times W_B^\infty$ into W_B^∞ is given by*

$$DF_n(\varepsilon_0, u_0) \cdot (\delta, h) = -\delta K(I - Q)g(u_0) - \delta(Q - Q_nQ)g(u_0) + P_nPg'(u_0) \cdot h + (I - P_nP)h - \varepsilon_0 K(I - Q)g'(u_0) \cdot h - \varepsilon_0(Q - Q_nQ)g'(u_0) \cdot h. \tag{2.106}$$

Also the partial derivative of F_n with respect to $u \in \mathfrak{B}$ evaluated at $(0, u_0) \in \mathbb{R} \times \mathfrak{B}$ is given by

$$D_uF_n(0, u_0) = P_nPg'(u_0) + (I - P_nP). \tag{2.107}$$

Furthermore if there exists points $\{u_n : n = 1, 2, \dots\} \subset \mathfrak{B}$ such that:

(0) *given $\varepsilon > 0$, there exists $\delta > 0$ such that for all n*

$$\|g(u) - g(u_n)\|_{W^\infty} < \varepsilon \quad \textit{whenever } \|u - u_n\|_{W^\infty} < \delta;$$

(00) *given $\varepsilon > 0$, there exists $\delta > 0$ such that for all n*

$\|Dg(u) - Dg(u_n)\| = \|g'(u) - g'(u_n)\| < \varepsilon$ whenever $\|u - u_n\|_{W^\infty} < \delta$ (the norm on the left-hand side is the operator norm). Then the family $\{F_n : n = 1, 2, \dots\}$ is equicontinuously differentiable on $\mathfrak{B}$ with respect to each set $\{(t_0, u_n) : n = 1, 2, \dots\}$.

Proof. P, P_nP, Q, Q_nQ, K being continuous linear mappings are continuously differentiable and are their own derivatives. Thus by chain rule (see Dieudonné [6]) F_n is continuously differentiable for each n. The partial derivatives are given by

$$D_\varepsilon F_n(\varepsilon_0, u_0) \cdot \delta = -\delta K(I - Q)g(u_0) - \delta(Q - Q_nQ)g(u_0)$$

and

$$D_uF_n(\varepsilon_0, u_0) \cdot h = P_nPg'(u_0) \cdot h + (I - P_nP) \cdot h - \varepsilon_0 K(I - Q)g'(u_0) \cdot h - \varepsilon_0(Q - Q_nQ)g'(u_0) \cdot h. \tag{2.108}$$

Now since $DF_n(\varepsilon_0, u_0) \cdot (\delta, h) = D_\varepsilon F_n(\varepsilon_0, u_0) \cdot \delta + D_uF_n(\varepsilon_0, u_0) \cdot h$ (see Dieudonné Diedudonné (1969)) we have proved (2.17).

From (2.108) we have

$$D_uF_n(0, u_0) \cdot h = P_nPg'(u_0) \cdot h + (I - P_nP)h = [P_nPg'(u_0) + (I - P_nP)] \cdot h \quad \text{for all } h \in W_B^\infty.$$

Hence we have proved (2.107).

Using (2.17) we have

$$[DF_n(t, u) - DF_n(t_0, u_n)] \cdot (\delta, h) = DF_n(t, u) \cdot (\delta, h) - DF_n(t_0, u_n) \cdot (\delta, h)$$
$$= -\delta K(I - Q)[g(u) - g(u_n)] - \delta(Q - Q_nQ)[g(u) - g(u_n)]$$
$$+ P_nP[g'(u) - g'(u_n)] \cdot h - (t - t_0)K(I - Q)[g'(u) - g'(u_n)] \cdot h$$
$$- (t - t_0)(Q - Q_nQ)[g'(u) - g'(u_n)] \cdot h. \tag{2.109}$$

By our Lemma 2.13, $P = Q$, $P_nP = Q_nQ$ and by using our Lemma 2.12 we can show that $\|P_n\|_{W_B^\infty} \leq \overline{C}$, a constant for all n, where $\|P_n\|_{W_B^\infty}$ is the operator norm in $L(W_B^\infty, W_B^\infty)$ (see Remark 2.19 below). Noting this we obtain from (2.109) that for all n,

$$\|[DF_n(t, u) - DF_n(t_0, u_n)] \cdot (\delta, h)\|_{W^\infty}$$
$$\leq |\delta| \, [\|K\| \, \|(I - Q)\| + (\|Q\| + \overline{C}\,\|Q\|)] \, \|g(u) - g(u_n)\|_{W^\infty}$$
$$+ \|P\| \, \|g'(u) - g'(u_n)\| \, \|h\|_{W^\infty} + |t - t_0| \, [\|K\| \, \|(I - Q)\|$$
$$+ (\|Q\| + \overline{C}\,\|Q\|)] \, \|g'(u) - g'(u_n)\| \, \|h\|_{W^\infty}$$
$$\leq r\,|\delta| \, \|g(u) - g(u_n)\|_{W^\infty}$$
$$+ [\|P\| \, \|g'(u) - g'(u_n)\| + r\,|t - t_0| \, \|g'(u) - g'(u_n)\|] \, \|h\|_{W^\infty} \tag{2.110}$$

where $r = \|K\| \, \|(I - Q\| + (\|Q\| + \overline{C}\,\|Q\|)$.

Now $\varepsilon > 0$ be given. Then by condition (0) there exists a $\delta_1 > 0$ such that for all n,

$$\|g(u) - g(u_n\|_{W^\infty} < \frac{\varepsilon}{r} \quad \text{whenever} \quad \|u - u_n\|_{W^\infty} < \delta_1,$$

and by condition (00) there exist $\delta_2 > 0$ and $\delta_3 > 0$ such that for all n,

$$\|g'(u) - g'(u_n)\| < \frac{\varepsilon}{2\,\|P\|} \quad \text{whenever} \quad \|u - u_n\|_{W^\infty} < \delta_2,$$

and

$$\|g'(u) - g'(u_n)\| < \frac{\sqrt{\varepsilon}}{\sqrt{2r}} \quad \text{whenever} \quad \|u - u_n\|_{W^\infty} < \delta_3.$$

Taking $\delta = \min(\delta_1, \delta_2\delta_3)$ we see from (2.110) that for all n and all $(\delta, h) \in \mathbb{R} \times W^\infty$.

$$\|[DF_n(t, u) - DF_n(t_0, u_n)] \cdot (\delta, h)\|_{W^\infty}$$
$$\leq |\delta|\,\varepsilon + \left[\frac{\varepsilon}{2} + r\frac{\sqrt{\varepsilon}}{\sqrt{2r}}\frac{\sqrt{\varepsilon}}{\sqrt{2r}}\right] \|h\|_{W^\infty} = \varepsilon[|\delta| + \|h\|_{W^\infty}] = \varepsilon\,\|(\delta, h)\|$$

whenever

$$\|u - u_n\|_{W^\infty} < \delta \quad \text{and} \quad |t - t_0| < \frac{\sqrt{\varepsilon}}{\sqrt{2r}},$$

i.e. whenever

$$\|(t,\, u) - (t_0,\, u_n)\| < \delta + \frac{\sqrt{\varepsilon}}{\sqrt{2r}}.$$

Thus we have proved that for all n,

$$\|DF_n(t,\, u) - DF_n(t_0,\, u_n)\| < \varepsilon \quad \text{whenever} \quad \|(t,\, u) - (t_0,\, u_n)\| < \overline{\delta}$$

where $\overline{\delta} = \delta + \sqrt{\varepsilon}/\sqrt{2r}$. Hence $\{F_n\}$ is equicontinuously differentiable on $\mathfrak{B}$ with respect to each set of the form $\{(t_0,\, u_n) : n = 1,\, 2, \ldots\}$. □

Remark 2.19 In the above lemma we have used the fact that $\|P_n\| \le \overline{C}$, a constant for all n. This can be seen easily by using our Lemma 2.11 as follows: $P_n : X_0 \to X_n$. Hence for all $0 \ne x \in X_0$,

$$\frac{\|P_n x\|_{W^\infty}}{\|x\|_{W^\infty}} \le \frac{C_1 \|P_n x\|_W}{(\sqrt{2T})^{-1} \|x\|_W} \le \frac{C_1}{(\sqrt{2T})^{-1}} \sup_{0 \ne x \in W_B^\infty \cap X_0} \frac{\|P_n x\|_W}{\|x\|_W} \le \overline{C} \|P_n\|_W \le \overline{C}$$

where

$$\overline{C} = \frac{C_1}{(\sqrt{2T})^{-1}}$$

as P_n being an orthogonal projection in W_B has norm $= 1$.

Theorem 2.26 *Let $\mathfrak{B}$ be an open subset of W_B^∞ and $g : \mathfrak{B} \to W_B^\infty$ a continuously differentiable mapping. Assume that for $n = 1,\, 2, \ldots,$ there exist points $u_n \in X_n \cap \mathfrak{B}$ satisfying the following:*

- (a)′ *g satisfies (0) and (00) of Lemma 2.17 with respect to the set $\{u_n : n = 1,\, 2, \ldots,\}$;*
- (b)′ *$P_n P g(u_n) = 0$;*
- (c)′ *for each n the restriction $\hat{T}_n$ of $P_n P g'(u_n)$ to X_n is a linear homeomorphism;*
- (d)′ *the sequence $\{\|\hat{T}_n^{-1}\|\}$ is uniformly bounded, i.e. there exists a constant C such that $\left\|\hat{T}_n^{-1} u\right\|_{W^\infty} \le C \|u\|_{W^\infty}$ for all $u \in X_n$ and all n;*
- (e)′ *the sequence of bounded linear mappings $g'(u_n)$ is uniformly bounded in operator norm, i.e. there exists a constant C_1 such that $\|g'(u_n) \cdot u\|_{W^\infty} \le C_1 \|u\|_{W^\infty}$ for all $u \in W^\infty$ and all n;*
- (f)′ *the sequence $\{\|g(u_n)\|_{W^\infty}\}$ is bounded.*

Then there exists an $\varepsilon_0 > 0$ and $\beta > 0$ such that for each $n = 1,\, 2, \ldots,$ there exists a continuous mapping $\Phi_n : (-\varepsilon_0,\, \varepsilon_0) \to V_n = S_\beta(u_n,\, W_B^\infty)$ (open sphere with centre at u_n and radius $= \beta$) with property that $\Phi_n(0) = u_n$ and $F_n(\varepsilon,\, \Phi_n(\varepsilon)) = 0$ for all $\varepsilon \in (-\varepsilon_0,\, \varepsilon_0)$, i.e. $\Phi_n(\varepsilon)$ is a solution of the approximate equation $L_n u = \varepsilon g(u)$.

Moreover Φ_n is continuously differentiable for each n and the derivatives are given by

$$\Phi_n'(\varepsilon) = -[D_u F_n(\varepsilon,\, \Phi_n(\varepsilon))]^{-1} \cdot D_\varepsilon F_n(\varepsilon,\, \Phi_n(\varepsilon)).$$

Proof. We are going to show that the family $\{F_n : n = 1, 2, \dots\}$ satisfies all the conditions of our implict function Theorem 2.24 so that the conclusions of our theorem will follow from those of Theorem 2.24. By virtue of our condition (a)$'$ it follows from Lemma 2.17 that the family $\{F_n : n = 1, 2, \dots\}$ is equicontinuously differentiable on $\mathbb{R}\times\mathfrak{B}$ with respect to the set $\{(0,\, u_n)\}$ of points. Thus the condition (a) of Theorem 2.24 is fulfilled. Also since $u_n \in X_n$ for all n and by (b)$'$ $P_n P g(u_n) = 0$ for all n, we obtain

$$F_n(0,\, u_n) = P_n P g(u_n) + (I - P_n P) u_n = 0 \quad \text{for all } n.$$

Next we prove that for each n, $D_u F_n(0,\, u_n)$ is a linear homeomorphism of W_B^∞ onto W_B^∞. We have already seen that our Lemma 2.17 is valid. Hence from (2.107) we have

$$D_u F_n(0,\, u_n) = P_n P g'(u_n) + (I - P_n P).$$

Let us denote the linear mapping $P_n P g'(u_n)$ by T_n.

Let $h = h_0 + h_1 \in W_B^\infty$ where $h_0 = P_n P h$ and $h_1 = (I - P_n P)h$. Then $D_u F_n(0,\, u_n) \cdot h = T_n h + h_1$ where $T_n h = P_n P g'(u_n) \cdot h \in X_n$. Now if $T_n h + h_1 = 0$, then $T_n h = 0$ and $h_1 = 0$ because of direct sum decomposition. But then $h = h_0$ and $T_n h_0 = 0$. However by assumption (c)$'$ $T_n h_0 = \hat{T}_n h_0$ implies $h_0 = 0$. Thus $D_u F_n(0,\, u_n) \cdot h = 0$ implies $h = 0$. Hence $D_u F_n(0,\, u_n)$ is an injection of W_B^∞ into itself. We now prove that $D_u F_n(0,\, u_n)$ is onto. Let $h = P_n P h + (I - P_n P)h = h_0 + h_1 \in W_B^\infty$. Then $h_0 - T_n h_1 \in X_n$. Thus $v_0 = \hat{T}_n^{-1}(h_0 - T_n h_1) \in X_n$ is well defined. Let $v_1 = h_1$ and $v = v_0 + v_1$. Then

$$\begin{aligned}
D_u F_n(0, u_n) \cdot v &= P_n P g'(u_n) \cdot v + (I - P_n P) v \\
&= T_n v + (I - P_n P) v \\
&= T_n v_0 + T_n v_1 + v_1 \\
&= \hat{T}_n(\hat{T}_n^{-1}(h_0 - T_n h_1)) + T_n h_1 + h_1 \\
&= h_0 - T_n h_1 + T_n h_1 + h_1 = h. \qquad (2.111)
\end{aligned}$$

Thus we have proved that $D_u F_n(0,\, u_n)$ is onto. Therefore, by the open mapping theorem $D_u F_n(0,\, u_n)$ is an open mapping. Hence $D_u F_n(0,\, u_n)$ is a linear homeomorphism of W_B^∞ onto W_B^∞. Now from (2.109) and using (d)′ and (e)′ we have

$$\begin{aligned}
\left\|[D_uF_n(0,\,u_n)]^{-1}\cdot h\right\|_{W_B^\infty} &= \left\|\hat{T}_n^{-1}(h_0 - T_n h_1) + h_1\right\|_{W^\infty} \\
&\le \left\|\hat{T}_n^{-1}\right\| [\|h_0\|_{W^\infty} + \|T_n\|\,\|h_1\|_{W^\infty}] + \|h_1\|_{W^\infty} \\
&\le C[\|h_0\|_{W^\infty} + \|P_n\|\,\|P\|\,\|g'(u_n)\|\,\|h_1\|_{W^\infty}] + \|h_1\|_{W^\infty} \\
&\le C[\|h_0\|_{W^\infty} + \|P\|\,\overline{C}C_1\,\|h_1\|_{W^\infty}] + \|h_1\|_{W^\infty} \\
&\quad \text{(where } \overline{C} \text{ is the constant obtained from Remark 2.19)} \\
&\le C_0[\|h_0\|_{W^\infty} + \|h_1\|_{W^\infty}] \\
&\quad \text{where } C_0 = \max(C, CC_1\overline{C}\,\|P\| + 1) \le \overline{C}_0\,\|h\|_{W^\infty} \text{ by Remark 2.16.}
\end{aligned}$$

Clearly the constant $\overline{C}_0$ is independent of n. Thus we have proved that $\left\|[D_uF_n(0,\,u_n)]^{-1}\right\| \le \overline{C}_0$ for all n and hence the condition (b) of Theorem 2.24 is fulfilled.

Finally from the definition of F_n we obtain

$$F(t,\,u_n) - F(0,\,u_n) = t[(I-Q)g(u_n) + (Q - Q_nQ)g(u_n)].$$

Hence by using (f)′ and noting Remark 2.19 we have

$$\begin{aligned}
\|F(t,\,u_n) - F(0,\,u_n)\|_{W^\infty} &\le |t\|\,[\|(I-Q)\| + C_3\|Q\|]\|g(u_n)\|_{W^\infty} \\
&\le C_2[\|(I-Q)\| + C_3\,\|Q\|]\,|t|
\end{aligned}$$

(because $\|g(u_n)\|_{W^\infty} \le C_2$ for all n by (f)′ where $C_3 = (1+\overline{C})$, $\overline{C}$ of Remark 2.19).

$$\le C_4\,|t|\,, \quad \text{where } C_4 = C_2[\|(I-Q)\| + C_3\,\|Q\|]$$

and is independent of n. The condition (c) of Theorem 2.24 clearly follows from this. Thus we have verified all the conditions of Theorem 2.24 and therefore, the proof of our Theorem 2.26 is complete. □

Corollary 2.26.1 *If in addition to the conditions of Theorem 2.24, $\{\|u_n\|_{W^\infty}\}$ is also bounded, then for each $\varepsilon \in (-\varepsilon_0,\,\varepsilon_0)$ the sequence $\{\Phi_n(\varepsilon)\}_{n=1}^\infty$ is also bounded in W^∞-norm.*

Proof. This follows from Remark 2.15. □

We now suppose that $\{\|u_n\|_{W^\infty}\}$ is bounded. Then by Lemma 2.11 and Remark 2.18, for each $\varepsilon \in (-\varepsilon_0,\,\varepsilon_0)$ $\{\|\Phi_n(\varepsilon)\|_{W_B}\}$ is bounded. Thus for each $\varepsilon \in (-\varepsilon_0,\,\varepsilon_0)$ $\Phi_n(\varepsilon) \to \Phi(\varepsilon) \in W_B$ weakly (passing to a subsequence, if necessary) in W_B. Noting this we can now state the following.

Corollary 2.26.2 *In addition to the conditions of Theorem 2.24 assume that $\{\|u_n\|_{W^\infty}\}$ is bounded and for each $\varepsilon \in (-\varepsilon_0,\,\varepsilon_0)$, $g[\Phi_n(\varepsilon)] \to g[\Phi(\varepsilon)]$ weakly in W_B. Then for each $\varepsilon \in (-\varepsilon_0,\,\varepsilon_0)$, $\Phi(\varepsilon)$ is a weak solution of problem (2.88)–(2.89).*

Proof. We first prove that for each $\varepsilon \in (-\varepsilon_0, \varepsilon_0)$, $Qg[\Phi(\varepsilon)] = 0$. By (2.104) $Q_n Q\Phi(\varepsilon) = 0$ for all n. Let $\varepsilon \in (-\varepsilon_0, \varepsilon_0)$ be arbitrary. Let n_0 be arbitrary but fixed, then clearly $Q_{n_0} Qg[\Phi_m(\varepsilon)] = 0$ for all $m \geq n_0$. By Lemma 2.13, $P_{n_0} Pg[\Phi_m(\varepsilon)] = 0$ for $m \geq n_0$. Since $P_{n_0} P$ is weakly continuous in W_B and $g[\Phi_m(\varepsilon)] \to g[\Phi(\varepsilon)]$ weakly in W_B, we have $P_{n_0} Pg[\Phi_m(\varepsilon)] = Q_{n_0} \Phi g[\Phi_m(\varepsilon)] \to P_{n_0} Pg[\Phi(\varepsilon)]$ weakly in W_B. Hence $P_{n_0} Pg[\Phi(\varepsilon)] = 0$. Since n_0 is arbitrary $P_n Pg[\Phi(\varepsilon)] = 0$ for all n. This implies that $Pg[\Phi(\varepsilon)] = Qg[\Phi(\varepsilon)] = 0$. Thus the condition (2.102) holds. On the other hand from (2.103) we have

$$\Phi_n(\varepsilon) = P(\Phi_n(\varepsilon)) + \varepsilon K(I - Q)g[\Phi_n(\varepsilon)],$$

taking weak limit in W_B we obtain

$$\Phi(\varepsilon) = P(\Phi(\varepsilon)) + \varepsilon K(I - Q)g[\Phi(\varepsilon)] \quad \text{which is (2.101).}$$

Thus $\Phi(\varepsilon)$ is a weak solution of problem (2.88)–(2.89). □

Corollary 2.26.3 (Smiley (1985)) *Let $\mathcal{B}$ be an open subset of W_B^∞ and let $g : \mathcal{B} \to W_B^\infty$ be a continuously differentiable map. If there exists $u_0 \in X_0$ such that*

$$Pg[u_0] = 0$$

and the restriction of the map $Pg'[u_0]$ to the subspace X_0^∞ is a linear homeomorphism onto X_0^∞, then there exists an $\varepsilon_0 > 0$ and a unique continuous map:, $u : (-\varepsilon_0, \varepsilon_0) \to \mathcal{B}$ such that $u(0) = u_0$ and $F(\varepsilon, u(\varepsilon)) = 0$ for all $|\varepsilon| < \varepsilon_0$.

Furthermore, u is continuously differentiable and

$$u'(\varepsilon) = -[D_u F(\varepsilon, u(\varepsilon))]_0 D_\varepsilon F(\varepsilon, u(\varepsilon)).$$

Proof. Define $u_n = P_n u_0$ in Lemma 2.17, Theorem 2.26 and Corollaries 2.26.1 and 2.26.2. Then Corollary 2.26.3 follows from these results. □

2.6.5 *Application to Nonlinear Wave Equations — A Theorem of Paul Rabinowitz*

We consider the problem

$$u_{tt} - u_{xx} = \varepsilon[g(u) + f], \qquad -\infty < t < \infty,\ 0 < x < \pi \tag{2.112}$$

$$u(t, 0) = u(t, \pi) = 0, \qquad -\infty < t < \infty \tag{2.113}$$

$$u(t + 2\pi, x) = u(t, x), \qquad -\infty < t < \infty,\ 0 < x < \pi \tag{2.114}$$

where $g : \mathbb{R} \to \mathbb{R}$ and $g[u] = g(u) + f$ for convenience of presentation. Here $g(u)$ is the Nemytsky operator generated by g. The argument for the more general case $g[u] = g(t, x, u) + f(t, x)$ can be modified without difficulty.

Theorem 2.27 *If $g \in C^3(-\infty, \infty)$, $g'(u) \geq \beta > 0$, $g(0) = 0$ and $f \in W^{\infty}_{\text{per}}H(G)$ is 2π-periodic in t where $G = (0, 2\pi) \times (0, \pi)$, then there exists $\varepsilon_0 > 0$ such that for each $\varepsilon \in (-\varepsilon_0, \varepsilon_0)$ the problem (2.112)–(2.114) has a solution.*

Before we prove our theorem, we need some preliminaries. For details we refer to Smiley (1985).

In order to make our preceding results applicable we take $T = 2\pi$, $V = H^1_0(0, \pi)$, $H = L^2(0, \pi)$ and $A = -(\partial^2/\partial x^2)$ with the following norms in V and H respectively

$$\|u\|^2_V = \frac{2}{\pi}\int_0^\pi |u_x|^2\, dx \quad \text{and} \quad \|u\|^2_H = \frac{2}{\pi}\int_0^\pi |u|^2\, dx.$$

We also take $G = (0, 2\pi) \times (0, \pi)$ so that $L^2(0, T; H) = L^2(G)$ and $W_B = W_{\text{per}}(0, 2\pi) = \{u \in H^1(G) : u(t, \cdot) \in H^1(0, \pi)$ for a.e. $t \in (0, 2\pi)$ and $u(t + 2\pi, x) = u(t, x)$ for a.e. $(t, x) \in G\}$, equivalently W_B is the closure of the set $\{\Phi \in C^\infty(\mathbb{R} \times (0, \pi) : \Phi(t + 2\pi, x) = \Phi(t, x)$ and $\Phi(t, \cdot) \in C^\infty_0(0, \pi)\}$. As $a(u, u) \geq 1/2\|u\|^2_{H^1}$, we have

$$\|u\|^2_{W_B} = \frac{2}{\pi}\int_0^{2\pi}\int_0^\pi (u_x^2 + u_t^2)dx\, dt$$

and

$$\|u\|_{W_B^\infty} = \underset{0<t<2\pi}{\text{ess sup}}\|u_x(t, \cdot)\|_H + \underset{0<t<2\pi}{\text{ess sup}}\|u_t(t, \cdot)\|_H.$$

The bilinear from $B(\cdot, \cdot)$ takes the form

$$B(u, w) = \frac{2}{\pi}\int_0^{2\pi}\int_0^\pi (-u_t w_t + u_x w_x)dx\, dt.$$

The eigen vectors and eigen values of the operator A are

$$w_j = \sin jx,\ \lambda_j = j^2, \quad j = 1, 2, 3, \ldots$$

while the set $\Sigma = \{k^2 : k = 1, 1, 2, \ldots\}$. Hence $\lambda_j \in \Sigma$ for all $j = 1, 2, 3, \ldots$ which satisfies Theorem 2.25. We have

$$\begin{aligned}
X_0 &= \{U \in W_{\text{per}}(0, 2\pi) : B(u, w) = 0, \quad \forall w \in W_{\text{per}}(0, 2\pi)\},\\
X_1 &= \{u \in W_{\text{per}}(0, 2\pi) : (u, u_0)_{W_B} = 0. \quad \forall u_0 \in X_0\},\\
Y_1 &= \{f \in L^2(G) : (f, u_0)_{L^2(G)} = 0, \quad \forall u_0 \in X_0\},\\
\text{and } Y_0 &= \{f \in L^2(G) : (f, h)_{L^2(G)} = 0, \quad \forall h \in Y_1\}.
\end{aligned}$$

We also note that

$$\begin{aligned}
X_0^\infty &= \{u \in X_0 : \|u\|_{W^\infty} < \infty\},\\
X_1^\infty &= \{u \in X_1 : \|u\|_{W^\infty} < \infty\}\\
W^\infty_{\text{per}} &= X_0^\infty + X_1^\infty.
\end{aligned}$$

We recall $X_0 = X_0^\infty$ by Lemma 2.12.

We take $X_n = \text{span}\,\{\cos kt \sin kx,\ \sin kt \cos kx\}_{k=1}^{n}$ for $n = 1, 2, \ldots$

We define P, Q, P_n, Q_n as before.

Let $a < 0 < b$ where both a and b may be finite or infinite. Then for $g \in C^3(a, b)$, there are monotone nondecreasing functions k_1, k_2, k_3 such that $m = 1, 2, 3$, $|g^m(u)| \leq k_m(r)$, $u \in (a, b)$, $|u| \leq r < \max(|a|, b)$.

For $u \in W_{\text{per}}^\infty$, obviously

$$|u(t, x)| \leq \frac{\pi}{\sqrt{2}} \|u\|_{W^\infty} \text{ a.e. } (t, x) \in G. \tag{2.115}$$

For $r > 0$, let

$$\mathfrak{B} = \left\{ u \in W_{\text{per}}^\infty : \|u\|_{W^\infty} < \frac{\sqrt{2}}{\pi} r \right\}$$

$$u \in \mathfrak{B} \Rightarrow |u(t, x)| < r \text{ a.e. } (t, x) \in G.$$

We assume $g \in C^3(-\infty, \infty)$, $g(0) = 0$ and $g(u)$ satisfies the boundary condition whenever u does. In the following $g[\cdot]$ is the Nemytsky operator generated by g. For the proof of the following simple results (A), (B) and (C), see Appendix I.

(A) $g : \mathfrak{B} \to W_{\text{per}}^\infty(W_{\text{per}})$ is continuous. In particular $g : \mathfrak{B} \to W_{\text{per}}^\infty$ is Lipschitz, i.e.

$$\|g(u) - g(v)\|_{W^\infty} \leq c(r)\|u - v\|_{W^\infty} \quad \text{for all } u, \in \mathfrak{B}$$

where $c(r)$ is a constant depending only on r appearing in the definition of $\mathfrak{B}$.

(B) $g : \mathfrak{B} \to W_{\text{per}}^\infty$ is continuously (Fréchet) differentiable and in particular,

$$\sup_{\|h\|_{W^\infty}} \|(g'(u) - g'(v)) \cdot h\|_{W^\infty} \leq c_1(r)\|u - v\|_{W^\infty} \quad \text{for all } u, v \in \mathfrak{B}$$

where $c_1(r)$ is a constant depending only on r appearing in the definition of $\mathfrak{B}$.

(C) For all

$$\begin{aligned} u \in X_0,\ u \in C^{0,1/2}(G) \text{ and } \|u\|_{1/2} = \|u\|_{C^{0,1/2}(G)} & \\ &= \sup_{(\alpha,\beta)\neq(0,0)} \frac{|u(t+\alpha, x+\beta) - u(t, x)|}{(\alpha^2 + \beta^2)^{1/4}} \\ &\leq c\|u\|_W \end{aligned}$$

where c is a constant.

The following result is also well known (e.g. see Rabinowitz (1967) and Brézis, Coron, and Nirenberg (1980)).

(D) With

$$L = \left(\frac{\partial^2}{\partial t^2} - \frac{\partial^2}{\partial x^2}\right),$$

the corresponding linear map K as defined in subsection 3 has the property that $K : Y_1 \cap L^2(G) \to Y_1 \cap L^2(G)$ is a compact mapping (i.e. compact from $L^2(G)$-topology to $L^2(G)$-topology). We are now in a position to prove our Theorem 2.27.

Proof of Theorem 2.27. We first note for $u \in X_0$, $v \in W_{\text{per}}$, $(u, v)_W = (u, v)_W - B(u, v) = 2(u_t, v_t)_{L^2(G)}$. Also $P_nP : W_{\text{per}} \to X_n$ is an orthogonal projection of W_{per} onto X_n. We first prove that for each $n = 1, 2, \dots$ there is a point $u_n \in X_n$ such that $P_nP(g(u_n) + f) = 0$. To this end we define $T_n : X_n \to X_n$ by $T_n(u) = P_nP(g(u) + f)$. Then for $u, v \in X_n$,

$$\begin{aligned}(T_n(u) - T_n(v), u - v)_W &= (P_nP(g(u) - g(v)), u - v)_W = (g(u) - g(v), u - v)_W \\ &= \frac{4}{\pi}\int_0^{2\pi}\int_0^{\pi}(g'(u)u_t - g'(v)v_t)(u - v)_t dx\, dt\end{aligned}$$

as by (A), $g(u) - g(v) \in W_{\text{per}}$. Hence with $v = 0$

$$(T_n(u) - T_n(0), u)_W = \frac{4}{\pi}\int_0^{2\pi}\int_0^{\pi} g'(u)u_t^2 dx\, dt \geq \beta\|u\|_W^2.$$

Hence by the result of variational inequalities (e.g. [Kinderlehrer and Stampacchia (1980), p. 14]) there exists $u_n \in X_n$ such that $T_n(u_n) = 0$, i.e. $P_nP(g(u_n) + f) = 0$.

By using monotonicity of g we can easily prove that u_n is unique, a fact which we will not require.

Next we prove that $\{u_n \in X_n : n = 1, 2, \dots\}$ obtained above is W^∞-norm bounded:

$$0 = (T_n(u_n), u_n) = (g(u_n) + f, u_n)_W = \frac{4}{\pi}\int_0^{2\pi}\int_0^{\pi}(g'(u_n)(u_n)_t + f_t)(u_n)_t dx\, dt.$$

From this we obtain

$$\begin{aligned}\beta\|u_n\|_W^2 &\leq \frac{4}{\pi}\int_0^{2\pi}\int_0^{\pi}(g'(u_n)(u_n)_t^2 dx\, dt \leq \left|\frac{4}{\pi}\int_0^{2\pi}\int_0^{\pi} f_t(u_n)_t dx\, dt\right| \\ &\leq \frac{2}{\sqrt{\pi}}\|f_t\|_{L^2(G)}\|u_n\|_W.\end{aligned}$$

Thus $\{\|u_n\|_W\}$ is bounded and hence by Lemma 2.12 $\{\|u_n\|_{W^\infty}\}$ is bounded. In fact

$$\|u_n\|_{W^\infty} \leq \beta^{-1}\frac{2}{\sqrt{\pi}}C_1\|f_t\|_{L^2(G)}$$

for all n where C_1 is the constant of Lemma 2.12.

Now taking $r > \sqrt{2\pi}\beta^{-1}C_1\|f_t\|_{L^2(G)}$ and defining $\mathfrak{B}$ by

$$\mathfrak{B} = \left\{u \in W^{\infty}_{\text{per}} : \|u\|_{W^\infty} < \frac{\sqrt{2}}{\pi}r\right\},$$

we show that $g : \mathfrak{B} \to W^{\infty}_{\text{per}}$ satisfies all the conditions of Theorem 2.26.

We have proved above that for $n = 1, 2, \ldots,$ there exist points $u_n \in X_n$ satisfying $P_nPg[u_n] = 0$. Thus (b)′ of Theorem 2.26 is fulfilled. The condition (e)′ of Theorem 2.26 follows from (A) and (B) mentioned above. Indeed by (B) $\|g'(u_n) \cdot u - g'(0) \cdot u\|_{W^\infty} \le c_1(r)\|u_n\|_{W^\infty}$ for all $u \in W^{\infty}_{\text{per}}$. Hence $\|g'(u_n) - g'(0)\| \le c_1(r)\|u_n\|_{W^\infty}$ and therefore $\|g'(u_n)\| \le \|g'(0)\| + c_1(r)\|u_n\|_{W^\infty}$. Thus $\{\|g'(u_n)\|\}$ is bounded as $\{\|u_n\|_{W^\infty}\}$ is so. Similarly by (A) $\|g(u_n)\|_{W^\infty} = \|g(u_n) - g(0)\|_{W^\infty} \le C(r)\|u_n\|_{W^\infty}$. Thus condition (f)′ of Theorem 2.26 is satisfied. So we are left with the conditions (c)′ and (d)′ of Theorem 2.26 to verify.

Let T_n be the restriction of $P_nPg'(u_n)$ to X_n. Clearly T_n maps X_n into X_n. Let $u \in X_n$ and $T_nu = 0$. Then $P_nPg'(u_n) \cdot u = 0$. By Lemma 2.13 $Q_nQg'(u_n) \cdot u = 0$ and hence for all $v \in X_n$

$$\int_0^{2\pi}\int_0^{\pi} g'(u_n)uv\,dx\,dt = 0.$$

Taking $v = u$ we have

$$\beta\|u\|^2_{L^2(G)} \le \int_0^{2\pi}\int_0^{\pi} g'(u_n)u^2 dx\,dt = 0$$

which implies that $u = 0$.

Thus T_n is an injective linear mapping of X_n into X_n. But since X_n is finite dimensional, T_n is a linear homeomorphism of X_n onto X_n. Thus we have verified the condition (c)′ of Theorem 2.26.

Finally let $T_n^{-1}h = u \in X_n$ i.e. $Q_nQg'(u_n)u = P_nPg'(u_n)u = h$. Then

$$\begin{aligned}\beta\|u\|^2_{L^2(G)} &\le (g'(u_n)u,\, u)_{L^2(G)} = (Q_nQg'(u_n)u,\, u)_{L^2(G)} \le (h,\, u)_{L^2(G)}\\ &\le \|h\|_{L^2(G)}\|u\|_{L^2(G)}\end{aligned}$$

as $((I - Q_nQ)g'(u_n), u)_{L^2(G)} = 0$. Thus $\|T_n^{-1}h\|_{L^2(G)} = \|u\|_{L^2(G)} \le \beta^{-1}\|h\|_{L^2(G)}$. Therefore, $\|T_n^{-1}\| \le \beta^{-1}$ and the condition (d)′ is verified (noting that in finite dimensional spaces norm topologies are equivalent).

Hence by Theorem 2.26 there exist $\varepsilon_0 > 0$ and $\delta_0 > 0$ such that for each $n = 1, 2, \ldots,$ there exists a unique continuous mapping $\Phi_n : (-\varepsilon_0, \varepsilon_0) \to S_{\delta_0}(u_n, W^{\infty}_{\text{per}})$ (unit sphere with centre at u_n and radius δ_0) with the property that $\Phi_n(0) = u_n$ and $F_n(\varepsilon, \Phi_n(\varepsilon)) = 0$ for all $\varepsilon \in (-\varepsilon_0, \varepsilon_0)$ i.e. for each $\varepsilon \in (-\varepsilon_0, \varepsilon_0)$, $\Phi_n(\varepsilon)$ is a solution of the approximate equation $L_nu = \varepsilon(g[u] + f)$ where $L_nu = Lu + (P - P_nP)u$ and $Lu = \partial^2u/\partial t^2 - \partial^2u/\partial x^2$. Also since $\{\|u_n\|_{W^\infty}\}$ is bounded, we have from Corollary 2.26.1 that for each $\varepsilon \in (-\varepsilon_0, \varepsilon_0)$ the sequence $\{\|\Phi_n(\varepsilon)\|_{W^\infty}\}$ is bounded and therefore by Lemma 2.12 $\{\|\Phi_n(\varepsilon)\|_W\}$ is also bounded. Clearly by (2.115) $\{\|\Phi_n(\varepsilon)\|_{L^\infty(G)}\}$ is bounded and repeating the same argument we see with the aid of (A) that $\{\|g(\Phi_n(\varepsilon)\|_{L^\infty(G)}\}$ is bounded.

Let $\varepsilon \in (-\varepsilon_0, \varepsilon_0)$ be arbitrary. Then by what we have proved above a subsequence of $\{\Phi_n(\varepsilon)\}$ converges weakly to $\Phi(\varepsilon) \in W_{\text{per}} \cap L^2(G)$. And by Rellich compactness theorem (a further subsequence) converges to $\Phi(\varepsilon)$ strongly in $L^2(G)$ We claim that $\Phi(\varepsilon)$ is a solution of our problem (2.112)–(2.114).

We first note that since $\Phi_n(\varepsilon)$ is a solution of the approximate equation $L_n u = \varepsilon(g[u] + f)$, we have by (2.103) that for each $n = 1, 2, \ldots$

$$(I - P)\Phi_n(\varepsilon) = \varepsilon K(I - Q)[g(\Phi_n(\varepsilon)) + f]. \tag{2.116}$$

As $\{\|g(\Phi_n(\varepsilon)\|_{L^\infty(G)}\}$ is bounded, we see that $\{\|(I - Q)[g(\Phi_n(\varepsilon)) + f]\|_{L^2(G)}\}$ is bounded. Hence by (D) and (2.116) $\{(I - P)\Phi_n(\varepsilon)\}$ has a subsequence which converges in $L^2(G)$-norm.

[It is also clear from what we have proved before that before that $\{\|P(\Phi_n(\varepsilon))\|_{H^1(G)}\}$ is bounded and it follows from (C) that $\{\|P(\Phi_n(\varepsilon))\|_{1/2}\}$ is bounded. Hence by a well known compactness theorem $\{P\Phi_n(\varepsilon)\}$ has a subsequence which converges in $L^2(G)$-norm. Thus we see that $\{\Phi_n(\varepsilon)\} = \{P(\Phi_n(\varepsilon)) + (I - P)\Phi_n(\varepsilon)\}$ has a subsequence which converges to $\Phi(\varepsilon)$ in $L^2(G)$-norm.]

We have $\|\Phi_n(\varepsilon)\|_{W^\infty} \leq C$ for some constant C and for all n. We have already mentioned that $\|u\|_{W^\infty} \leq \sqrt{2}/\pi \Rightarrow |u(t, x)| \leq r$ a.e. $(t, x) \in G$. Hence taking $K_1 = \max\{|g'(u)| : |u| \leq (\pi/\sqrt{2}C\}$ where K_1 has the meaning as defined earlier in this section, we have for all $v \in L^2(G)$ (using mean value theorem),

$$\left|\int_0^{2\pi}\int_0^{\pi}(g(\Phi_n(\varepsilon)) - g(\Phi(\varepsilon)))\, v\, dx\, dt\right| \leq K_1 \int_0^{2\pi}\int_0^{\pi} |\Phi_n(\varepsilon) - \Phi(\varepsilon)|\, |v|\, dx\, dt$$

$$\leq K_1\|\Phi_n(\varepsilon) - \Phi(\varepsilon)\|_{L^2(G)}\|v\|_{L^2(G)} \to 0 \quad \text{as} \quad n \to \infty.$$

Thus $g(\Phi_n(\varepsilon)) \to g(\Phi(\varepsilon))$ weakly in $L^2(G)$. Again since $\Phi_n(\varepsilon)$ is a solution of the approximate equation for each n, we have from (2.105)

$$Q_n Q[g(\Phi(\varepsilon)) + f] = 0 \quad \text{for all} \quad n.$$

Now in the same way as in Corollary 2.26.2 we prove that $Q[g(\Phi(\varepsilon)) + f] = 0$. Let n_0 be a fixed but arbitrary positive integer. Then

$$Q_{n_0} Q[g(\Phi_m(\varepsilon)) + f] = 0 \quad \text{for all} \quad m \geq n_0,$$

i.e.

$$(g(\Phi_m(\varepsilon)) + f, v)_{L^2(G)} = 0 \quad \text{for all} \quad v \in X_{n_0} \quad \text{and all} \quad m \geq n_0.$$

Hence $(g(\Phi(\varepsilon)) + f,\, v)_{L^2(G)} = (g(\Phi(\varepsilon)) - g(\Phi_m(\varepsilon)),\, v)_{L^2(G)}$ for all $v \in X_{n_0}$ and all $m \geq n_0$. Letting $m \to \infty$, we obtain

$$(g(\Phi(\varepsilon)) + f,\, v)_{L^2(G)} = 0 \quad \text{for all} \quad v \in X_{n_0}$$

and hence $(g(\Phi(\varepsilon)) + f,\, v)_{L^2(G)} = 0$ for $v \in X_n$, $n = 1, 2, \ldots$.

Thus $P[g(\Phi(\varepsilon)) + f] = Q[g(\Phi(\varepsilon)) + f] = 0$ (by Lemma 2.13). Thus the condition (2.102) is fulfilled.

Now since $\Phi_n(\varepsilon)$ converges weakly to $\Phi(\varepsilon)$ in $L^2(G) \cap W_{\text{per}}$, $P(\Phi_n(\varepsilon)) \to P(\Phi(\varepsilon))$ weakly in $W_{\text{per}} \cap L^2(G)$. Also by Remark 2.16 (or by the Theorem in Rabinowitz (1967), $K(I - Q)[g(\Phi_n(\varepsilon)) + f]$ converges weakly to $K(I - Q)[g(\Phi(\varepsilon)) + f]$ in $W_{\text{per}} \cap L^2(G)$. Now letting $n \to \infty$ in (2.116) we obtain

$$(I - P)\Phi(\varepsilon) = \varepsilon K(I - Q)[g(\Phi(\varepsilon)) + f].$$

Thus the condition (2.101) is satisfied. $\Phi(\varepsilon)$ is therefore a solution of the problem (2.112)–(2.114). Finally since $\varepsilon \in (-\varepsilon_0, \varepsilon)$ is arbitrary, we have proved that the problem (2.112)–(2.114) has a solution for each $\varepsilon \in (-\varepsilon_0, \varepsilon_0)$. □

Appendix I

Proof of (A).

$g[u] = g(u(t, x))$ is certainly a well-defined measureable function. Also $g[u]$ satisfies the boundary conditions whenever $u \in \mathfrak{B}$.

Now

$$\begin{aligned} \|g(u(t, \cdot)\|_V &= \left\{2/\pi \textstyle\int_0^\pi |g'(u(t, x))|^2 |u_x(t, x)|^2\right\}^{\frac{1}{2}} \\ &\leq k_1(r)\, \|u(t, \cdot)\|_V \quad \text{(a.e.) } t \in (0, 2\pi). \end{aligned}$$

Similarly $\|\partial/\partial t[g(u(t, \cdot)]\|_H \leq k_1(r)\, \|u_t(t, \cdot)\|_H$. Thus $\|g(u)\|_{W^\infty} \leq k_1(r)\, \|u\|_{W^\infty} < \infty$ and therefore $g : \mathfrak{B} \to W^\infty_{\text{per}}$. We now prove the continuity. To this end we first note that (a.e.) $(t, x) \in G$,

$$\left|\frac{\partial}{\partial x}[g(u) - g(v)]\right| \leq |g'(u) - g'(v)|\, |u_x| + |g'(v)|\, |(u - v)_x|\,.$$

Now for $\mu, \nu \in (-r, r)$, there exists z between μ and ν such that $|g'(\mu) - g'(\nu)| \leq |g''(z)|\, |\mu - \nu|$. Thus for the constant $c = \pi/\sqrt{2}$, by what has been noted just before the result A we have, (a.e.) $(t, x) \in G$,

$$\begin{aligned} \left|\frac{\partial}{\partial x}[g(u) - g(v)]\right| &\leq k_2(r)\, |u - v|\, |u_x| + k_1(r)\, |(u - v_x)| \\ &\leq c k_2(r)\, \|u - v\|_{W^\infty}\, |u_x| + k_1(r)\, |(u - v)_x|\,. \end{aligned}$$

Thus we obtain, for almost all $t \in (0, 2\pi)$,

$$\begin{aligned}\|g(u(t,\cdot)) - g(v(t,\cdot))\|_V &\le 2[ck_2(r)\,\|u-v\|_{W^\infty}\,\|u(t,\cdot)\|_V]^2 \\ &\quad + 2[k_1(r)\,\|u(t,\cdot) - v(t,\cdot)\|_V]^2 \\ &\le 2\left([ck_2(r)]^2 + [k_1(r)]^2\right)\|u-v\|^2_{W^\infty}.\end{aligned}$$

A similar estimate is obtained for the term $\left\|\frac{\partial}{\partial t}[g(u(t,\cdot)) - g(v(t,\cdot))]\right\|_H$. This proves that $g : \mathfrak{B} \longrightarrow W^\infty_{\text{per}}$ is continuous. □

Proof of (B).

Let $u \in \mathfrak{B}$. For $h \in \mathfrak{B}$ with $u + h \in \mathfrak{B}$ we consider

$$q(t, x; u, h) = g(u+h) - g(u) - g'(u)h.$$

Then for almost all $(t, x) \in G$ we have

$$\begin{aligned}\frac{\partial q}{\partial x} &= g'(u+h)(u+h)_x - g'(u)u_x - g''(u)u_x h - g'(u)h_x \\ &= [g'(u+h) - g'(u)](u+h)_x - g''(u)u_x h \\ &= g''(u+\sigma h)h(u+h)_x - g''(u)u_x h,\end{aligned}$$

where $\sigma = \sigma(t, x)$ satisfies $0 < \sigma < 1$.

Since for any $\mu, / \nu \in (-r, r)$ with $\mu + \nu \in (-r, r)$ we have

$$|g''(\mu+\nu) - g''(\mu)| \le k_3(r)|\nu|,$$

it follows from the preceding equality that (a.e.) $(t, x) \in G$

$$\begin{aligned}\left|\frac{\partial q}{\partial x}\right| &\le k_3(r)|h|^2|u_x| + k_2(r)\,|h|\,|h_x| \\ &\le c^2 k_3(r)\,\|h\|^2_{W^\infty}\,|u_x| + ck_2(r)\,\|h\|_{W^\infty}\,|h_x|.\end{aligned}$$

Hence for almost all $t \in (0, 2\pi)$,

$$\begin{aligned}\|q(t,\cdot,u,h)\|^2_V &\le 2\left[c^2k_3(r)\,\|h\|^2_{W^\infty}\,\|u(t,\cdot)\|_V\right]^2 \\ &\quad + 2\left[ck_2(r)\,\|h\|_{W^\infty}\,\|h(t,\cdot)\|_V\right]^2 \\ &\le 2c^2\left([crk_3(r)]^2 + [k_2(r)]^2\right]\|h\|^4_{W^\infty}.\end{aligned}$$

A similar estimate holds for $\left\|\frac{\partial}{\partial t}[q(t,\cdot,u,h)]\right\|_H$.

Combining the two estimates we obtain

$$\|q(u, h)\|_{W^\infty} \le c(r)\, \|h\|^2_{W^\infty}$$

where $c(r)$ is a constant depending only on r and the functions k_2 and k_3. This shows that g is Fréchet differentiable at each $u \in \mathfrak{B}$.

It remains to prove that the map $u \longrightarrow g'(u)$ is continuous from $\mathfrak{B}$ to $B(W^\infty_{\text{per}}, W^\infty_{\text{per}})$, the space of bounded linear maps of W^∞_{per} into itself. To this end let $h \in W^\infty_{\text{per}}$. Then (a.e.) $(t, x) \in G$,

$$\begin{aligned}\left|\frac{\partial}{\partial x}[g'(u).h - g'(v).h]\right| &= |g''(u)u_x h - g''(v)v_x h + (g'(u) - g'(v))h_x| \\ &\le k_3(r)|u-v||u_x||h| + |g''(v)||(u-v)_x||h| + k_2(r)|u-v|h_x| \\ &\le c^2 k_3(r)\|u-v\|_{W^\infty}\|h\|_{W^\infty}|u_x| + ck_2(r)\|h\|_{W^\infty}|(u-v)_x| \\ &\quad + ck_2(r)\,\|u-v\|_{W^\infty}\,|h_x|\,.\end{aligned}$$

Thus for almost all $t \in (0, 2\pi)$

$$\begin{aligned}&\|g'(u(t,\cdot))h(t,\cdot) - g'(v(t,\cdot))h(t,\cdot)\|^2_V \\ &\quad \le 3c^3\left([crk_3(r)]^2 + 2[k_2(r)]^2\right)\|u-v\|_{w^\infty}\|h\|_{W^\infty}.\end{aligned}$$

A similar estimate holds for $\left\|\frac{\partial}{\partial t}[g'(u).h - g'(v).h]\right\|_H$.

From these estimates we deduce that

$$\sup_{\|h\|_{W^\infty}-1} \|g'(u)h - g'(v)h\| \le c(r)\,\|u-v\|_\infty$$

where $c(r)$ is a constant depending on r only. This completes the proof. □

Proof of (C).

Let $u = \sum_{k=1}^\infty (a_k \cos kt + b_k \sin kt)\sin kx \in X_0$.

Then we have $\|u\|^2_W = 2\pi \sum_{k=1}^\infty k^2(a_k^2 + b_k^2) < \infty$.

For $\alpha, \beta > 0$ we define

$$\begin{aligned}\Delta u(t, x) &= u(t+\alpha, x+\beta) - u(t, x) \\ &= u(t+\alpha, x+\beta) - u(t, x+\beta) + u(t, x+\beta) - u(t, x).\end{aligned}$$

By using Hölder's inequality,

$$\begin{aligned}|u(t+\alpha,\, x+\beta)-u(t,x+\beta)| &\le \sum_{k=1}^{\infty}(|a_k|\,(|\cos k(t+\alpha)-\cos kt|)) \\ &\quad + |b_k|\,(|\sin k(t+\alpha)-\sin kt|) \\ &\le \left(\sum_{k=1}^{\infty}k^2a_k^2\right)^{\frac{1}{2}}\left(\sum_{k=1}^{\infty}\frac{|\cos k(t+\alpha)-\cos kt|^2}{k^2}\right)^{\frac{1}{2}} \\ &\quad + \left(\sum_{k=1}^{\infty}k^2b_k^2\right)^{\frac{1}{2}}\left(\sum_{k=1}^{\infty}\frac{|\sin k(t+\alpha)-\sin kt|^2}{k^2}\right)^{\frac{1}{2}}.\end{aligned}$$

We note that

$$\begin{aligned}|\cos k(t+\alpha)-\cos kt|^2 &\le \tfrac{1}{2}\left(\left|e^{ik\alpha}-1\right|^2+\left|e^{-ik\alpha}-1\right|^2\right) \\ &= \tfrac{1}{2}\left[|\cos k\alpha-1+i\sin k\alpha|^2+|\cos k\alpha-i-1\sin k\alpha|\right]=2(1-\cos k\alpha)\end{aligned}$$

Similary we can show that $|\sin k(t+\alpha)-\sin kt|^2 \le 2(1-\cos k\alpha)$.

We know from Fourier series analysis

$$\frac{\pi}{2}|s|-\frac{s^2}{4}=\sum_{k=1}^{\infty}\frac{1-\cos ks}{k^2}, \quad -2\pi \le s \le 2\pi.$$

Hence for $|\alpha| \le 2\pi$

$$\begin{aligned}|u(t+\alpha,\, x+\beta)-u(t,\, x+\beta)| &\le \tfrac{1}{\pi}\,\|u\|_W\left(\textstyle\sum_{k=1}^{\infty}\frac{2(1-\cos\alpha)}{k^2}\right)^{\frac{1}{2}} \\ &\le \tfrac{1}{\pi}\,\|u\|_W\left(\pi|\alpha|-\tfrac{\alpha^2}{2}\right)^{\frac{1}{2}}.\end{aligned}$$

In the same fashion we can get estimate for $|\beta| \le 2\pi$

$$|u(t,x+\alpha)-u(t,x)| \le \frac{1}{2}\,\|u\|_W\left(\pi|\beta|-\frac{\beta^2}{2}\right)^{\frac{1}{2}}.$$

Thus for $|\alpha|$, $|\beta| \le 2\pi$

$$|\Delta u| \le \frac{1}{\pi}\,\|u\|_W\left[\left(\pi|\alpha|-\frac{\alpha^2}{2}\right)^{\frac{1}{2}}+\left(\pi|\beta|-\frac{\beta^2}{2}\right)^{\frac{1}{2}}\right].$$

Using this estimate we obtain for $|\alpha|$, $|\beta| \le 2\pi$

$$\frac{|\Delta u(t,x)|^2}{(\alpha^2+\beta^2)^{\frac{1}{2}}} \le \frac{1}{\pi^2}\,\|u\|_W\left[\frac{2(|\alpha|+|\beta|)-(\alpha^2+\beta^2)}{(\alpha^2+\beta^2)^{\frac{1}{2}}}\right] \le c\,\|u\|_W$$

where c is a constant independent of α and β. □

2.7 Set-Valued Contractions

Let X be a non-empty set. We will denote the family of all subsets of X by 2^X. We will also use the following notation:

$CB(X) = \{A \subset X : A \text{is non-empty closed and bounded}\}$ for a metric space (X, ρ) and $K(X) = \{A \subset X : A \text{is compact}\}$ for a topological space X.

Let (X, ρ) be a metric space and let for any $A \in CB(X)$ and $r > 0$, $N_r(A) = \{x \in X : \delta(x, A) < r\}$, where $\delta(x, A) = \inf\{\rho(x, a) : a \in A\}$ = the distance of x from A. Then for any pair $A, B \in CB(X)$, if we define $h(A, B) = \inf\{r : A \subset N_r(B) \text{ and } B \subset N_r(A)\}$, then we can easily verify that h is a metric on $CB(X)$. h is called the Hausdorff metric in $CB(X)$.

Let $A, B \in K(X)$ and define $d(A, B) = \sup\{\delta(x, B) : x \in A\}$. We can equivalently show that $h(A, B) = d(A, B) \vee d(B, A)$ is the Hausdorff metric on $K(X)$. In this case $(K(X), h)$ is called the space of fractals (see M. Barnsley (Barnsley (1988, p. 43)).

A mapping $T : X \to 2^X$ is called a set-valued mapping if for $x \in X$, $T(x) \in 2^X$.

Definition 2.3 Let (X, ρ_1) and (Y, ρ_2) be metric spaces. A set-valued mapping $T : X \to CB(Y)$ is said to be a Lipschitz mapping with Lipschitz constant α if $h(T(x), T(y)) \leq \alpha\rho_1(x, y)$, where $\alpha > 0$ and h is the Hausdorff metric in $CB(Y)$. If the Lipschitz constant α is < 1, then T is said to be a set-valued contraction with contraction constant α.

The following fixed point theorem is due to Nadler (1969).

Theorem 2.28 *Let (X, ρ) be a complete metric space and $T : X \to CB(X)$ is a set-valued contraction with contraction constant α. Then T has a fixed point $u \in X$ i.e., $u \in T(u)$.*

Proof. Let $x_0 \in X$ be an arbitrary point. We choose a point $x_1 \in T(x_0)$. Since $T(x_0), T(x_1) \in CB(X)$ and $\epsilon = \alpha > 0$, we have by definition of Hausdorff metric h on $CB(X)$, $T(x_0) \subset N_r(T(x_1))$ and $T(x_1) \subset N_r(T(x_0))$, where $r = h(T(x_0), T(x_1)) + \alpha$. Now since $x_1 \in T(x_0) \subset N_r(T(x_1))$, we have $\delta(x_1, T(x_1)) = \inf\{\rho(x_1, x) : x \in T(x_1)\} < r$. Hence there exists $x_2 \in T(x_1)$ such that $\rho(x_1, x_2) \leq r = h(T(x_0), T(x_1)) + \alpha$.

Now since $T(x_1), T(x_2) \in CB(X)$, $x_2 \in T(x_1)$ and $\epsilon = \alpha^2 > 0$, we have by the same argument as above a point $x_3 \in T(x_2)$ such that

$$\rho(x_2, x_3) \leq h(T(x_1), T(x_2)) + \alpha^2.$$

We continue this process to obtain a sequence $\{x_n\}_{n=1}^{\infty}$ of points of X such that $x_{n+1} \in T(x_n)$ and

$$\rho(x_n, x_{n+1}) \leq h(T(x_{n-1}), T(x_n)) + \alpha^n \tag{2.117}$$

for all $n \geq 1$.

Now from (2.117) and the definition of set-valued contraction we obtain

$$\begin{aligned}\rho(x_n, x_{n+1}) &\le h(T(x_{n-1}), T(x_n)) + \alpha^n \le \alpha\rho(x_{n-1}, x_n) + \alpha^n \\ &\le \alpha[h(T(x_{n-2}), T(x_{n-1})) + \alpha^{n-1}] + \alpha^n \\ &\le \alpha^2\rho(x_{n-2}, x_{n-1}) + 2\alpha^n \le \cdots \\ &\le \alpha^n\rho(x_0, x_1) + n\alpha^n \quad \text{for all} \quad n \ge 1. \end{aligned} \tag{2.118}$$

Hence using (2.118) we have

$$\begin{aligned}\rho(x_n, x_{n+j}) &\le \rho(x_n, x_{n+1}) + \rho(x_{n+1}, x_{n+2}) + \cdots + \rho(x_{n+j-1}, x_{n+j}) \\ &\le \alpha^n\rho(x_0, x_1) + n\alpha^n + \alpha^{n+1}\rho(x_0, x_1) + n + 1\alpha^{n+1} + \cdots \\ &\quad + \alpha^{n+j-1}\rho(x_0, x_1) + (n + j - 1)\alpha^{n+j-1} \\ &= (\textstyle\sum_{i=n}^{n+j-1} \alpha^i)\rho(x_0, x_1) + \sum_{i=n}^{n+j-1} i\alpha^i\end{aligned}$$

for all $n, j \ge 1$.

From (2.118) it follows that $\{x_n\}_{n=1}^{\infty}$ is a Cauchy sequence in (x, ρ). Since (x, ρ) is complete, $\{x_n\}_{n=1}^{\infty}$ converges to some point $x_0 \in X$. Now, from the relation $h(T(x_n), T(x_0)) \le \alpha\rho(x_n, x_0)$, it follows that $\{T(x_n)\}$ converges to $T(x_0)$ in $CB(X)$. Since $x_n \in T(x_{n-1})$ for all $n \ge 1$, it follows that $x_0 \in T(x_0)$. □

Definition 2.4 Given $\epsilon > 0$, a metric space (X, ρ) is said to be ϵ-chainable if and only if given $u, v \in X$ there is an ϵ-chain from u and v, i.e., there exists a finite set of points $x_0, x_1, \ldots, x_n$ with $u = x_0$ and $v = x_n$ such that $\rho(x_{i-1}, x_i) < \epsilon$ for all $i = 1, 2, \ldots, n$.

A set-valued mapping $T : (X, \rho) \to CB(X)$ is said to be (ϵ, α)-uniformly locally contractive with $\epsilon > 0$ and $0 < \alpha < 1$ provided that $h(T(x), T(y)) \le \alpha\rho(x, y)$ whenever $x, y \in X$ with $\rho(x, y) < \epsilon$.

Theorem 2.29 *Let (X, ρ) be a complete ϵ-chainable metric space and $T : X \to K(X)$ an (ϵ, α)-uniformly locally contractive set-valued mapping. Then T has a fixed point.*

Proof. For any $x, y \in X \times X$, we define

$$\rho_\epsilon(x, y) = \inf\{\sum_{i=1}^{n} \rho(x_{i-1}, x_i) : x_0 = x, x_1, \cdots, x_n = y \text{ is an } \epsilon\text{-chain from } x \text{ to } y\}.$$

We can easily verify that ρ_ϵ is a metric on X. ρ_ϵ also satisfies

(2.7.3) $$\rho(x, y) \le \rho_\epsilon(x, y) \quad \text{for all} \quad x, y \in X;$$

and

(2.7.4) $$\rho(x, y) = \rho_\epsilon(x, y) \quad \text{for all} \quad x, y \in X; \quad \text{with} \quad \rho(x, y) < \epsilon.$$

(2.7.3) follows from the triangle inequality:

$$\rho(x,y) \leq \rho(x,x_1) + \rho(x_1,x_2) + \cdots + \rho(x_{n-1},x_n).$$

It also follows from (2.7.3), (2.7.4) and the completeness of (X,ρ) that (X,ρ_ϵ) is complete. Let h_ϵ be the Hausdorff metric on $K(X)$ derived from ρ_ϵ. We can easily see that if $A,B \in K(X)$ such that $h(A,B) < \epsilon$ then $h_\epsilon(A,B) = h(A,B)$. Now we will show that $T : X \to K(X)$ is a set-valued contraction mapping of (X,ρ_ϵ) into $(K(X),h_\epsilon)$ with contraction constant α. Let $x,y \in X$ and $x_0 = x, x_1, \ldots, x_n = y$ be an ϵ-chain from x to y. Since $\rho(x_{i-1},x_i) < \epsilon$ for all $i = 1,2\ldots,n$, we have

$$h(T(x_{i-1}),T(x_i)) \leq \alpha\rho(x_{i-1},x_i) < \epsilon \quad \text{for all} \quad i = 1,2\ldots,n.$$

Hence

$$h_\epsilon(T(x),T(y)) \leq \sum_{i=1}^{n} h_\epsilon(T(x_{i-1}),T(x_i)) = \sum_{i=1}^{n} h(T(x_{i-1}),T(x_i)) \leq \alpha.$$

Now since $x_0 = x, x_1, \ldots, x_n = y$ is an arbitrary ϵ-chain, it follows that $h_\epsilon(T(x),T(y)) \leq \alpha\rho_\epsilon(x,y)$ for all $x,y \in X$. Thus T is a set-valued contraction mapping of (X,ρ_ϵ) into $(K(X),h_\epsilon)$ with contraction constant α. Hence by Theorem 2.28, T has a fixed point. □

Remark 2.20 Similar results for a single-valued mapping was first obtained by Edelstein (1961).

Definition 2.5 Let X and Y be topological spaces and $T : X \to 2^Y$ a set-valued mapping with $T(x) \neq \emptyset$ for each $x \in X$, i.e.,$T : X \to 2^Y \setminus \{\emptyset\}$. T is said to be upper semi-continuous at $x_0 \in X$ if given an open set G containing $f(x_0)$, there exists an open neighborhood $U(x_0)$ of x_0 such that $T(U(x_0) \subset G$, where for any subset A of X, $T(A) = \cup_{x\in A}T(x)$. T is said to be upper semi-continuous if T is upper semi-continuous at each point $x \in X$.

A set-valued mapping $T : X \to 2^Y \setminus \{\emptyset\}$ is said to be lower semi-continuous at $x_0 \in X$ if given an open set G in Y with $T(x_0) \cap G \neq \emptyset$, there exists an open neighborhood $U(x_0)$ of x_0 such that $T(x) \cap G \neq \emptyset$ for each $x \in U(x_0)$. T is said to be lower semi-continuous if T is lower semi-continuous at each point $x \in X$.

Lemma 2.18 *Let X and Y be non-empty sets and $T : X \to 2^Y \setminus \{\emptyset\}$ a set-valued mapping. Then for any non-empty set A of Y,*

$$X \setminus \{\cup_{y\in A}T^{-1}(y)\} = \{x \in X : T(x) \subset Y \setminus A\}.$$

Proof. Let u belong to the left-hand side (of the above expression). Then $u \notin T^{-1}(y)$ for any $y \in A$. This implies that $y \notin T(u)$ for any $y \in A$. Thus $T(u) \subset Y\setminus A$ which implies that u belongs to the right-hand side.

Next let u belong to the right-hand side. Then $T(u) \subset Y \setminus A$. It follows that $u \notin T^{-1}(y)$ for any $y \in A$. This implies that $u \notin \cup_{y\in A}T^{-1}(y)$; i.e., u belongs to the left-hand side. □

For any subset A of Y, let $T_+(A) = \{x \in X : T(x) \cap A \neq \emptyset\}$.

Lemma 2.19 *For any subset A of Y, $T_+(A) = \cup_{y\in A} T^{-1}(y)$.*

Proof. Let $x \in A$. Then $T(x) \cap A \neq \emptyset$. Let $y \in T(x) \cap A$. This implies that $x \in T^{-1}(y)$ with $y \in A$, i.e., $x \in T^{-1}(y) \subset \cup_{y\in A} T^{-1}(y)$. Next, let $u \in \cup_{y\in A} T^{-1}(y)$. Then $u \in T^{-1}(y)$ for some $y \in A$, i.e., $y \in T(u)$ with $y \in A$, i.e., $u \in T_+(A)$. □

Theorem 2.30 *Let X and Y be topological spaces and $T : X \to 2^Y \setminus \{\emptyset\}$ a set-valued mapping. Then the following statements are equivalent:*

- *(a) T is upper semi-continuous;*
- *(b) For each open set G in Y, $T^+(G) = \{x \in X : T(x) \subset G\}$ is open in X;*
- *(c) For each closed set F in Y, $T^{-1}(F) = \cup_{y\in F} T^{-1}(y)$ is closed in X, where $T^{-1}(y) = \{x \in X : y \in T(x)\}$;*
- *(d) For each $x \in X$ and every net $\{x_\delta : \delta \in D\}$ in X converging to x, and each open set G in Y with $T(x) \subset G$, $T(x_\delta) \subset G$ eventually, i.e., $T(x_\delta) \subset G$ for all $\delta_0 \geq \delta$ for some $\delta_0 \in D$.*

Proof. First let (a) hold. Let G be an open set in Y and $x_0 \in T^+(G)$. By upper semicontinuity of T at x_0, there exists an open neighborhood $U(x_0)$ of x_0 such that $T(U(x_0)) \subset G$. Hence $U(x_0) \subset T^+(G)$ and hence $T^+(G)$ is an open set. Thus (a) implies (b).

Now let $T^+(G)$ be open for every open set G in Y. Let $x_0 \in X$ and G be an open set containing $T(x_0)$. $T^+(G)$ is an open neighborhood of x_0 and $T(T^+(G)) \subset G$. Hence T is upper semicontinuous at x_0. Since x_0 is arbitrary, (b) implies (a). That $(b) \iff (c)$ is evident from Lemma 2.18.

Now we prove that (b) implies (d). Let $\{x_\delta : \delta \in D\}$ be a net converging to $x \in X$ and G an open subset of Y with $T(x) \subset G$. Then by (b), $T^+(G)$ is open and $x \in T^+(G)$. Since $x_\delta \to x$, $x_\delta \in T^+(G)$ eventually. Hence $T(x_\delta) \subset G$ eventually.

Finally, we prove that (d) implies (b). Let H be an open set in Y. If possible, let $T^+(H)$ be not open. Then there is a point $x_0 \in X$ such that $x_0 \in T^+(H)$ is not an interior point of $T^+(H)$. Let $D_0 = \mathcal{N}(x_0)$ be the system of all open neighborhoods of x_0. Then D_0 ordered partially by inclusion is a directed set. We choose $x_\delta \in D_0$ such that $x_\delta \notin T^+(H)$. This is possible as x_0 is not an interior point of $T^+(H)$. Evidently $\{x_\delta : \delta \in D_0\}$ is a net converging to x_0 and $T(x_0) \subset H$. Hence by (d), $T(x_\delta) \subset H$ eventually, which contradicts the fact that $x_\delta \notin T^+(H)$ for all $\delta \in D_0$. □

Theorem 2.31 *Let X and Y be topological spaces and $T : X \to 2^Y \setminus \{\emptyset\}$ a set-valued mapping. Then the following statements are equivalent:*

- *(a) T is lower semi-continuous;*
- *(b) For each open set G in Y, $T_+(G)$ is open in X;*
- *(c) For each closed set F in Y, $T^+(F)$ is closed in X;*

(d) *For each $x \in X$ and each net $\{x_\delta : \delta \in D\}$ in X converging to x, and each open set G in Y with $T(x) \cap G \neq \emptyset$, $T(x_\delta) \cap G \neq \emptyset$, i.e., $T(x_\delta) \cap G \neq \emptyset$ for all $\delta \geq \delta_0$ for some $\delta_0 \in D$.*

Proof. Let (a) hold. Let G be an open set in Y and $x_0 \in T_+(G)$. By the lower semi-continuity of T at x_0, there exists an open neighborhood $U(x_0)$ of x_0 such that $T(x) \cap G \neq \emptyset$ for each $x \in U(x_0)$, i.e., $U(x_0) \subset T_+(G)$. Hence $T_+(G)$ is open in X. Now by virtue of Lemma 2.19, $\cup_{y \in G} T^{-1}(y)$ is open. Thus (a) implies (b). We now suppose that (b) holds. Let $x_0 \in X$ and G be an open set in Y such that $T(x_0) \cap G \neq \emptyset$. Then $x_0 \in T_+(G)$. Hence by virtue of Lemma 2.19 and (b), $T_+(G) = \cup_{y \in G} T^{-1}(y)$ is an open neighborhood of x_0 in X. It follows that T is is lower semi-continuous at x_0. Thus (b) implies (a).

Now that (b) $\iff$ (c) follows from the Lemma 2.18. Finally, by giving similar argument as given in Theorem 2.30 we can prove that (b) implies (d). □

Theorem 2.32 (a) *Let X and Y be topological spaces with Y a T_3 space and $T : X \to 2^Y \setminus \{\emptyset\}$ a set-valued upper semi-continuous mapping with closed values. Then the graph $T = G(T) = \{(x, y) \in X \times Y : y \in T(x)\}$ is closed.*

(b) *Let X and Y be topological spaces with Y compact and $T : X \to 2^Y \setminus \{\emptyset\}$ a set-valued mapping with closed graph (i.e., $G(T)$ is closed). Then T is upper semi-continuous.*

Proof. (a) Let $\{(x_\delta, y_\delta) : \delta \in D\}$ be a net in $G(T)$ converging to (x, u). If possible, let $(x, u) \notin G(T)$, i.e., $u \notin T(x)$. Since $T(x)$ is closed and Y is T_3, there exist open sets G_1 containing u and G_2 containing $T(x)$ with $G_1 \cap G_2 = \emptyset$. Now since T is upper semi-continuous, by Theorem 2.30 (d), $T(x_\delta) \subset G_2$ eventually. But since $y_\delta \in T(x_\delta)$ for each $\delta \in D$, $y_\delta \in G_2$ ventually. This contradicts the fact that $y_\delta \to u$ as $u \in G_1$ and $G_1 \cap G_2 = \emptyset$.

(b) If possible, let T be not upper semi-continuous at a point $x \in X$. Let $\{x_\delta : \delta \in D\}$ be a net converging to x. Then there must exist, by Theorem 2.30 (d), at least one open set G in Y with $f(x) \subset G$ such that $T(x_\delta) \not\subset G$ eventually. We can choose a subnet $\{x_{\delta'} : \delta' \in D'\}$ of the net $\{x_\delta : \delta \in D\}$ such that $T(x_{\delta'}) \not\subset G$ for each $\delta' \in D'$. For this we can select $u_{\delta'}$ from each $T(x_{\delta'})$ such that $u_{\delta'} \notin G'$. Now since G' is compact, $\{u_{\delta'} : \delta' \in D'\}$ has a subnet $\{u_{\delta''} : \delta'' \in D''\}$ converging to a point $u \in G'$. Clearly, $\{(x_{\delta''}, u_{\delta''} : \delta'' \in D''\}$ is a net in $G(T)$ which converges to $(x, u) \notin G(T)$ as $u \notin T(x) \subset G$. □

Theorem 2.33 *Let X and Y be topological spaces, $T : X \to K(Y)$ a set-valued upper semi-continuous mapping and K a compact subset of X. Then $T(K) = \cup_{x \in K} T(x)$ is a compact subset of Y.*

Proof. Let $\{G_\alpha : \alpha \in I\}$ be an open covering of $T(K)$. Then for $\alpha \in I$, there exists an open set H_α in Y such that $G_\alpha = T(K) \cap H_\alpha$. For each $x \in K$, $T(x)$ being compact is covered by a finite number of H_α, say $H_{\alpha_1}, H_{\alpha_2}, \ldots, H_{\alpha_n}$ with $\bigcup_{i=1}^n H_{\alpha_i} \supset T(x)$.

We set $H_x = \bigcup_{i=1}^n H_{\alpha_i}$. Then $\{T^+(H_x) : x \in K\}$ is an open covering of K. Since K is compact, there is a finite subcovering $T^+(H_{x_1}), T^+(H_{x_2}), \ldots, T^+(H_{x_n})$ of K. It follows that $\{H_{x_i} : i = 1, 2, \ldots, n\}$ cover $T(K)$ and $\bigcup_{i=1}^n G_{x_i} = \bigcup_{i=1}^n (H_{x_i} \cap (K)) = T(K)$. Hence $\{G_{x_i} : i = 1, 2, \ldots, n\}$ is a subcover $T(K)$. □

Theorem 2.34 *Let X and Y be topological spaces, $T : X \to 2^Y \setminus \{\emptyset\}$ a set-valued upper semi-continuous (or lower semi-continuous) mapping, K is a connected subset of Y and $T(x)$ is a connected subset of Y for each $x \in K$. Then $T(K)$ is a connected subset of Y.*

Proof. If possible, we suppose that $T(K)$ is not connected. Then there are two disjoint non-empty open subsets G_1 and G_2 of $T(K)$ such that $T(K) = G_1 \cap G_2$. Hence there exist two non- empty open subsets H_1 and H_2 of Y such that $G_1 = H_1 \cap T(K)$ and $G_2 = H_2 \cap T(K)$. Thus $T(K) \subset H_1 \cup H_2$. By upper semi-continuity of T, $T^+(H_1)$ and $T^+(H_2)$ are open sets of X. Let $x \in K$, then $T(x) \subset H_1 \cup H_2$. But since $T(x)$ is connected, it follows that $T(x)$ is contained in either H_1 or in H_2. Thus $K \subset T^+(H_1) \cup T^+(H_2)$. Obviously, $T^+(H_1) \cap T^+(H_2) = \emptyset$ and $K \cap T^+(H_1) \neq \emptyset$ and $K \cap T^+(H_2) \neq \emptyset$. Thus K is not connected, which is a contradiction. Hence $T(K)$ must be connected. □

For the proof in the case of lower semi-continuity, we refer to Klein and Thompson (1984, p. 90).

Theorem 2.35 *Let X, Y and Z be topological spaces, and $T_1 : X \to 2^Y \setminus \{\emptyset\}$ and $T_2 : Y \to 2^Z \setminus \{\emptyset\}$ are set-valued upper semi-continuous mappings. Then the set-valued mapping $T : X \to 2^Z \setminus \{\emptyset\}$ defined by $T = T_2 \circ T_1 = T_2(T_1(x))$ is upper semi-continuous.*

Proof. Let G be an open subset of G. Then $T^+(G) = (T_2 \circ T_1)^+(G) = \{x \in X : (T_2 \circ T_1)(x) \subset G\} = \{x \in X : T_1(x) \subset T_2^+(G)\} = T_1^+[T_2^+(G)]$ is an open subset of X. Hence T is upper semi-continuous. □

2.7.1 *End Points*

Definition 2.6 For set-valued mapping $T : X \to 2^X \setminus \{\emptyset\}$, a point $x_0 \in X$ us said to be en end point of T if $T(x_0) = \{x_0\}$. Let X be a topological space. Then an upper semi-continuous set-valued mapping $T : X \to 2^X \setminus \{\emptyset\}$ with closed values is said to be a topological contraction if, for each non-empty closed subset A of X with $T(A) = A$, A is a singleton set, i.e., A is an end point of T.

Theorem 2.36 *Let X be a compact Hausdorff topological space and $T : X \to 2^X \setminus \{\emptyset\}$ a set-valued topological contraction. Then T has a unique end point $x_0 \in X$ such that $\{x_0\} = \bigcap_{n=0}^\infty T^n(X)$, where $T^0(X) = X$ and $T^n(X) = T(T^{n-1}(X))$ for $n = 1, 2, \ldots$.*

Proof. For each $n = 0, 1, 2, \ldots$, let $F_n = T^n(X)$. Since T is upper semi-continuous with closed (and hence compact) values, by Theorem 2.33 F_n is compact for each

n and moreover, $\{F_n\}$ is decreasing by virtue of the fact that $T(X) \subset X$. Hence it follows that $F = \bigcap_{n=0}^{\infty} F^n \neq \emptyset$ and is a closed subset of X. Also it is clear that $T(F) \subset F$. We will now prove that $F \subset T(F)$. To this end we take any point $x \in F$ and consider the set $A_n = T^{-1}(x) \cap F_n$ for each $n = 0, 1, 2, \ldots$, where $T^{-1}(x) = \{y \in X : x \in T(y)\}$. Then $A_n \neq \emptyset$ for each n. Indeed, since $x \in \bigcap_{n=0}^{\infty} F^n \subset T^{n+1}(X) = F_{n+1}$ there exists $u \in F_n$ such that $x \in T(u)$. Thus $u \in T^{-1}(x) \cap F_n$. Now since T is upper semi-continuous with closed values, it follows that $G(T) =$ the graph of T is closed. Hence $T^{-1}(x)$ is a closed subset of X and A_n is compact due to the compactness of X. Hence the decreasing sequence $\{A_n\}_{n=0}^{\infty}$ has non-empty intersection, i.e., $\bigcap_{n=0}^{\infty} A_n \neq \emptyset$.

Now we take a point $z \in \bigcap_{n=0}^{\infty} A_n$, i.e., $z \in T^{-1}(x) \cap (\bigcap_{n=0}^{\infty} F^n)$. Then $x \in T(z) \subset T(\bigcap_{n=0}^{\infty} A_n) \subset T(\bigcap_{n=0}^{\infty} F_n) = T(F)$. Thus we have proved $F \subset T(F)$. Hence $F = T(F)$. Since F is also closed and T is a set-valued topological contraction, we have $F = \{x_0\}$ for some $x_0 \in X$. Thus x_0 is an end point. The uniqueness of the end point follows from the fact that any other end must belong to F. □

Corollary 2.36.1 *Let X be a Hausdorff topological space, $T^{n_0} : X \to 2^X \setminus \{\emptyset\}$ a set-valued topological contraction and $T^{n_0}(X)$ is compact for some integer n_0. Then there is a unique end point x_0 of T such that $\{x_0\} = \bigcap_{n=0}^{\infty} T^n(X)$.*

Proof. Let $\hat{X} = T^{n_0}(X)$ and $\hat{T} = T^{n_0}$. Then $\hat{T} : \hat{X} \to 2^{\hat{X}} \setminus \{\emptyset\}$. By Theorem 2.36, there is a unique end point $x_0 \in X$ such that

$$\{x_0\} = \bigcap_{n=n_0}^{\infty} T^n(X) = \bigcap_{n=0}^{\infty} T^n(X).$$ □

Remark 2.21 We note that if we assume T to be a set-valued contraction instead of T^{n_0}, keeping the other assumption intact, the Corollary still holds as the set F remain unaltered.

Corollary 2.36.2 *Let X be a Hausdorff compact topological space, $T_i : X \to 2^X \setminus \{\emptyset\}$ a set-valued topological contraction and for $i = 1, 2, \ldots, n$, and $T : X \to 2^X$ the set-valued mapping defined by $T(x) = \cup_{i=1}^{n} T_i(x)$. Then for each subset S of X with $T(S) = S$, S contains all the end points of T_i, $i = 1, 2, \ldots, n$.*

Proof. For any open set G in X, $T^+(G) = \{x \in X : T(x) \subset G\} = \bigcup_{i=1}^{n} \{x \in X : T_i(x) \subset G\} \bigcup_{i=1}^{n} T^+(G)$ is open as each $T_i^+(G)$ is open due to the upper semi-continuity of T_i by Theorem 2.30. Hence by the same Theorem 2.30, T is upper semi-continuous. It is easy to see that T has closed values. Let S be a non-empty closed subset with $T(S) = S$ (such an S exists by the argument given in Theorem 2.36, see Theorem 2.41). Then clearly, $T_i(S) \subset S$ for $i = 1, 2, \ldots, n$. Now for $i = 1, 2, \ldots, n$ let $S_i = \bigcap_{n=0}^{\infty} T_i^n(S)$. Then, as is shown in Theorem 2.36, $T_i(S_i) = S_i$, $i = 1, 2, \ldots, n$. Now since for $i = 1, 2, \ldots, n$, T_i is a set-valued contraction, S_i is a singleton set, i.e., $S_i = \{x\}$ for some $x_i \in X$. It also follows that $S_i \subset S$ for each $i = 1, 2, \ldots, n$. Thus the corollary is proved. □

Remark 2.22 We note from above corollary that the finite union of set-valued topological contractions is not necessarily a set-valued topological contraction.

Definition 2.7 Let X be a topological space. Then an upper semi-continuous mapping $T : X \to 2^X \setminus \{\emptyset\}$ with closed and connected values is said to be a topological contraction in weak sense if, for each non-empty closed connected subset A of X with $T(A) = A$, A is a singleton, i.e., A is an end point of T.

Theorem 2.37 *Let X be a compact connected Hausdorff topological space and $T : X \to 2^X \setminus \{\emptyset\}$ be a set-valued contraction in the weak sense. Then T has a unique end point $x_0 \in X$ such that $\{x_0\} = \bigcap_{n=0}^{\infty} T^n(X)$, where $T^0(X) = X$ and $T^n(X) = T(T^{n-1}(X))$.*

Proof. The proof is similar to that of Theorem 2.36. We repeat it. For each $n = 0, 1, 2, \ldots$, let $F_n = T^n(X)$. Then for each $n = 1, 2, \ldots$, F_n is compact and connected by Theorems 2.33 and 2.34. Since $\{F_n\}$ is a decreasing sequence, it follows that $F = \bigcap_{n=0}^{\infty} F^n$ is a non-empty compact connectedsubset of X. Clearly, $T(F) \subset F$. Now we prove $F \subset T(F)$. Let $x \in F$ and consider the set $A_n = T^{-1}(x) \cap F_n$ for each $n = 0, 1, 2, \ldots$, where $T^{-1}(x) = \{y \in X : x \in T(y)\}$. Repeating the argument given in the proof of theorem 2.36, we obtain that $A_n \neq \emptyset$ for each n and $\bigcap_{n=0}^{\infty} A^n \neq \emptyset$. Let $z \in \bigcap_{n=0}^{\infty} A^n$, i.e., $z \in T^{-1}(x) \cap F_n$. Then $x \in T(z) \subset T(\bigcap_{n=0}^{\infty} A^n) \subset T(\bigcap_{n=0}^{\infty} F^n) = T(F)$. Thus $T(F) = F$. Since F is compact and connected, and T is a set-valued contraction in the weak sense, F is a singleton, say, x_0. Hence $T(x_0) = \{x_0\}$ is an end point. If u is any other end point, then $u \in \bigcap_{n=0}^{\infty} T^n(X) = x_0$. Thus x_0 is the unique end point. □

We can consider a single-valued mapping $T : X \to Y$ as a set-valued mapping by Tx as $\{Tx\}$ for each $x \in X$. As an application of Theorem 2.36 we have the following:

Corollary 2.37.1 *Let X be a Hausdorff compact space and $T : X \to X$ a set-valued topological contraction. Then T has a unique fixed point $x_0 \in X$ and, furthermore, $\{x_0\} = \bigcap_{n=0}^{\infty} T^n(X)$ and for each $x \in X$, the sequence $\{T^n(x)\}$ converges to the unique fixed point x_0.*

Proof. By Theorem 2.36 there is a point $x_0 \in X$ such that $\bigcap_{n=0}^{\infty} T^n(X) = \{x_0\}$ and x_0 is a fixed point of T. As $\{T^n(x)\}$ is a decreasing sequence of non-empty compact sets, it follows that for each $x \in X$, $T^n(x) \to x_0$. □

The following result which can be viewed as somewhat converse of our above corollary is due to Williamson and Janos (1987) which we state without proof.

Theorem 2.38 *Let $\Phi : S \to S$ be a continuous mapping of the compact metrizable space S into itself, with the property that $\bigcap_{n=1}^{\infty} \{\Phi^n(S)\}$ is a singleton set. Then given α with $0 < \alpha < 1$, there exists a metric ρ^*, with topology identical to the original one, such that $\rho^*(\Phi(x), \Phi(y)) \leq \rho^*(x, y)$ for all $x, y \in S$.*

While Bessaga has proved the following converse of Banach contraction principle which is stated below without proof.

Theorem 2.39 *Let $T : X \to X$ be a mapping of a non-empty set X into itself such that each iteration T^n, $n = 1, 2, \ldots$, has a unique fixed point. Then for each α with $0 < \alpha < 1$, there exists a complete metric ρ_α on X such that $\rho_\alpha(T(x), T(y)) \leq \alpha\rho_\alpha(x, y)$.*

2.8 Iterated Function Systems (IFS) and Attractor

In (Barnsley (1988)) a finite set $\{T_i : i = 1, 2, \ldots, n\}$ of single-valued contraction mappings $T_i : X \to X$ of a complete metric space (X, ρ) into itself with contraction constant λ_i with $0 < \lambda_i < 1$ has been said to be a hyperbolic iterated function system (IFS) $(X, T_i, i = 1, 2, \ldots, n)$. If $\lambda = \max\{\lambda_i : i = 1, 2, \ldots, n\}$, then the single-valued mapping $T : K(X) \to K(X)$ of the metric space $(K(X), h)$ into itself defined by

$$T(A) = T_1(A) \cup T_2(A) \cdots \cup T_n(A), \quad A \in K(X),$$

can easily be seen to be a contraction mapping with contraction constant λ (see Barnsley (1988, Lemma 5, p. 81)), where h is the Hausdorff metric corresponding to the metric ρ. λ is called the contraction constant of IFS.

It is also well-known that if (X, ρ) is a complete metric space, then the $K(X)$, the space of fractals, is a complete metric space with respect to the Hausdorff metric h (see e.g., Theorem 1 in Barnsley (1988), p. 37 or Theorem 4.3.9 in Klein and Thompson (1984), p. 45).

Hence the following theorem (Theorem 1 in Barnsley (1988), p. 82) is apparent from contraction mapping principle:

Theorem 2.40 *Let $\{X, T_i, i = 1, 2, \ldots, n\}$ be a hyperbolic iterated function system with contraction constant α as described above. Then the mapping $T : K(X) \to K(X)$ defined by*

$$T(B) = T_1(B) \cup T_2(B) \cup \cdots \cup T_n(B), \quad B \in K(X)$$

has a unique fixed point $A \in K(X)$, i.e.,

$$A = T(A) = T_1(A) \cup T_2(A) \cup \cdots \cup T_n(A)$$

and is given by

$$A = \lim_{n \to \infty} T^n(B) \quad \text{for any} \quad B \in K(X).$$

The fixed point $A \in K(X)$ is called the attractor (fractal) of the IFS.

In Remark 2.21 we have pointed out that the finite union of set-valued topological contractions is not a set-valued topological contraction. We cannot have a theorem as beautiful as Theorem 2.40. Nevertheless we can state the following theorem in a restricted situation.

Theorem 2.41 *Let X be a Hausdorff compact topological space, $T_i : X \to 2^X \setminus \{\emptyset\}$ a set-valued contraction mapping for $i = 1, 2, \dots, n$ and $T : X \to 2^X$ the set-valued mapping defined by*

$$T(x) = \cup_{i=1}^{n} T_i(x), \quad x \in X$$

$$(\textit{hence } \ T(A) = \cup_{i=1}^{n} T(A)).$$

Then there exists a non-empty closed (hence compact) subset F of X such that $T(F) = F = \bigcap_{n=0}^{\infty} T^n(X)$, where $T^0(X) = X$ and $T^n(X) = T(T^{n-1}(X))$.

Proof. The mapping T is upper semi-continuous (see the proof of Theorem 2.36) and has clear closed values. For each $n = 0, 1, 2, \dots$, let $F_n = T^n(X)$. Since T is upper semi-continuous, F_n is compact for each n and $\{F_n\}$ is decreasing as $T(X) \subset X$. Hence $F = \bigcap_{n=0}^{\infty} F_n \neq \emptyset$. Clearly, $T(F) \subset F$. Now repeating exactly the same argument of Theorem 2.36, we can show that $F \subset T(F)$. Thus $F = T(F)$ and $F = \bigcap_{n=0}^{\infty} F_n = \bigcap_{n=0}^{\infty} T^n(X)$.

We might call F to be the attractor of the system of set-valued topological contractions $\{T_i : i = 1, 2, \dots, n\}$. To justify our doing so, we can go back to IFS $\{X, T_i, i = 1, 2, \dots, n\}$ with contraction constant $\lambda = \max\{\lambda_i, i = 1, 2, \dots, n\}$. Let us consider the mapping $T : K(X) \to K(X)$ defined by

$$T(B) = \bigcup_{i=1}^{n} T_i(B), \quad B \in K(X).$$

Let

$$H_0 = T^0(K(X)), \ H_1 = T(K(X)), \dots, H_n = T^n(K(X)) = T(T^{n-1}(K(X))).$$

If A is the unique fixed point of the mapping $T : K(X) \to K(X)$ of Theorem 2.40 then clearly, $A \subset \bigcap_{n=0}^{\infty} H_n$. Moreover, if we assume that for some N, H_N is a bounded subset of the complete metric space $(K(X), h)$, then $\delta(H_{N+1}) \le \lambda\delta(H_N)$, where h is the Hausdorff metric and $\delta(A)$ is the diameter of a set $A \subset K(X)$.

Now since $\{\overline{H_{N+k}}\}_{k=1}^{\infty}$ is a decreasing sequence of non-empty closed sets with $\delta(\overline{H_{N+k}}) \le \lambda^k \delta(H_N) \to 0$ as $k \to \infty$, by Cantor Intersection Theorem $\bigcap_{n=0}^{\infty} \overline{H_n} = \bigcap_{k=1}^{\infty} \overline{H_{N+k}} = \{B\}$ where B is a single point in $K(X)$. Hence it follows that $A = B$ and

$$A = \bigcap_{n=0}^{\infty} H_n = \bigcap_{n=0}^{\infty} T^n(K(X)).$$

Thus in some sense F of Theorem 2.41 has analogy with the attractor A of IFS. □

Our next theorem which was first proved in Tarafdar and Výborný (1976) (also appeared in Tarafdar and Yuan (1997b) and Yuan (1999)) provides the existence, uniqueness and algorithm of end point of set-valued mappings. The rest of this section deals with materials which appeared in Tarafdar and Yuan (1997b) and partly in Tarafdar (1996b). Reference of Tarafdar and Yuan (1997b) is regrettably missing in Yuan (1999).

Definition 2.8 Let (X, ρ) be a metric space. Then a set-valued mapping $T : X \to 2^X \setminus \{\emptyset\}$ is called a generalized contraction if there exists a real number λ with $0 < \lambda < 1$ such that $\delta(T(A)) \leq \lambda\delta(A)$ for each non-empty closed bounded subset A of X with $T(A) \subset A$, where as before $T(A) = \cup_{a \in A} T(a)$ and $\delta(A) = \sup\{\rho(x, y) : x, y \in A\}$ is the diameter of A.

Given $x \in X$, let us define $T(x) = T^1(x)$, $T^2(x) = T(T^1(x))$, $\ldots, T^n(x) = T(T^{n-1}(x)) = \bigcup_{u \in T^{n-1}(x)} T(u)$, for each $n = 1, 2, \ldots$.

A sequence $\{x_n\}_{n\in\mathbb{N}}$ with $x_n \in T(x^{n-1}) \subset T^{n-1}(x)$ is called a generalized of iterates with respect to x. Note that for a single-valued mapping $T : X \to X$, given $x \in X$, we have a unique sequence of iterates with respect to x. This is not the case for the set-valued mapping T.

Theorem 2.42 *Let F be a non-empty closed bounded subset of a complete metric space (X, ρ) and $T : F \to 2^F \setminus \{\emptyset\}$ a generalized contraction. Then T has a unique fixed point $x_0 \in F$, which is also an end point such that for each $x \in X$, every generalized sequence of iterates with respect to x converges to x_0.*

Proof. We set $F_0 = F$ and define $F_n = \overline{T(F_{n-1})}$ for $n = 1, 2, \ldots$. Let $D = \bigcap_{i=1}^{\infty} F_i$. Then we can easily see that the following holds:

(i) $F_0 \supset F_1 \supset F_2 \supset \cdots \supset F_n \supset \cdots$;

(ii) $T(F_n) \subset F_n, \ \ n = 0, 1, 2, \ldots$;

(iii) $T(D) \subset D$.

To see (iii) let $u \in T(D)$. Let k be a positive integer. Since $u \in T(D)$, there exists $x \in D$ such that $u \in T(x)$.

Now since $x \in F_k$, $T(x) \subset T(F_k) \subset M_k$ by (ii). Hence $u \in T(x) \subset M_k$. Hence $u \in M_k$ for all k, i.e., $u \in D$. As T is a generalized contraction, we have

$$\begin{aligned}
\delta(F_n) = \delta(\overline{T(F_{n-1})}) &= \delta(T(F_{n-1})) \\
&\leq \lambda\delta(F_{n-1}) = \lambda\delta(\overline{T(F_{n-2})}) = \lambda\delta(T(F_{n-2})) \\
&\leq \lambda^2\delta(F_{n-2}) \\
&\leq \cdots \leq \lambda^{n-1}\delta(F_1) \to 0 \quad \text{as} \quad n \to \infty
\end{aligned}$$

Hence in view of (i) $D = \bigcap_{i=0}^{\infty} F_i$ is, by Cantor Intersection Theorem, a single point, say, x_0. Thus it follows from (iii) that $T(x_0) = x_0$.

We now prove the uniqueness of the end point. If possible, let $y = T(y)$ and $x \neq y$. Let $A = \{x_0, y\}$. Then $\delta(A) > 0$ and $T(A) = A$; $0 \neq \delta(T(A)) < \lambda\delta(A) =$

$\lambda\delta(T(A))$, which is impossible as $0 < \lambda < 1$. Also we can straightway see that $y \in D$ as $y = T(A)$. Thus $x_0 = y$.

Next, we see that x_0 is the unique fixed point point. Let $y' \in F$ such that $y' \in T(y')$. Since $y' \in T(y')$, we obtain

$$\text{(iv)} \qquad y' \in T(y') \subset T^2(y') \subset T^3(y') \subset \cdots .$$

We have also

$$\text{(v)} \qquad T^{n+1}(F_0) \subset F_n, \quad n = 0, 1, 2, \ldots .$$

By hypothesis $T(F_0) \subset F_0$. Hence $T^2(F_0) \subset T(F_0) \subset \overline{T(F_0)} = F_1$. Let $T^{k+1}(F_0) \subset F_k$. Then $T^{k+2}(F_0) \subset T(F_k) \subset \overline{T(F_k)} = F_{k+1}$. Thus by induction we obtain (v). Now by (iv) and (v) we can conclude that $y' \in D$. Hence $y' = x_0$.

Finally, let $x \in F$ and $\{x^n\}_{n\in\mathbb{N}}$ a generalized sequence of iterates with respect to x. To show that $x_n \to x_0$ as $n \to \infty$, we note that

$$x_{n+1} \in T^n(x) \subset T^n(F_0) \subset F_{n-1} \quad \text{by (iv)}.$$

Now since $\delta(F_n) \to 0$ as $n \to \infty$, we can show that$\{x_n\}$ is a Cauchy sequence in the complete metric space (F, ρ). Thus $x_n \to x$ in F for some $x \in F$.

Let n_0 be an arbitrary positive integer. Then we show that x is a limit point of F_{n_0}. Since F_{n_0} is a closed set, $x \in F_{n_0}$. Thus $x \in D$. Hence $x = x_0$. □

It is interesting to note that no continuity is assumed in the above Theorem. The following example due to Tarafdar and Vyborny (Tarafdar and Výborný (1976)) elucidates this fact.

Example 2.3 We define the mapping $T : [0, 2] \to [0, 2]$ by

$$T(t) = \begin{cases} \frac{t}{2} + \frac{1}{2}, & \text{if } 0 \le t \le 1, \\ \frac{t}{2} - \frac{1}{2}, & \text{if } 1 < t \le 2. \end{cases}$$

We can easily verify that $T[0, 2] = [0, 1]$ and T is a generalized contraction which is discontinuous at $t = 1$ and, therefore, not a Banach contraction. $t = 1$ is the unique fixed point.

In what follows we now merely reproduce the materials of (Tarafdar and Yuan (1997b)).

2.8.1 *Applications*

A. End Points, Nucleous and Pareto Optimum

In the sequel of this section we will need the following notions and definitions (Tarafdar (1996b) and Tarafdar and Yuan (1997b)).

A non-empty subset P of a real Banach space V is called a cone if $\bar{P} = P$, $P + P \subset P$, $\mathbb{R}_+P \subset P$ and $P \cap (-P) = \{0\}$, where $\bar{P}$ is the closure of P and

$\mathbb{R}_+ = [0, \infty)$. Each cone P induces in V an ordering '$\preceq$' defined by $x \preceq y$ if and only if $y - x \in P$. This relation '$\preceq$' is evidently reflexive, antisymmetric and transitive. The pair (V, P) is called an *ordered Banach space* with the ordering '$\preceq$' induced by P (which is called the positive cone of V). The set $P^* = \{f \in V^* : f(x) \geq 0 \text{ for all } x \in P\}$ is called the *dual cone*, where V^* is the continuous dual of V, i.e., P^* is the set of order preserving continuous linear functional on V.

A cone P of V is said to be *normal* if and only if there exists a positive number e such that if $x, y \in P$ with $\|x\| \geq 1$ and $\|y\| \geq 1$, then $\|x + y\| \geq e$ (for other equivalent definitions, see Knaster (1928)). In what follows, we always assume that V denotes an ordered Banach space and we write $x \prec y$ if $y - x \in P \setminus \{0\}$.

Definition 2.9 Let X be a non-empty convex subset of a Hausdorff real topological vector space. A mapping $F : X \to V$ is said to be *order or cone convex* if $F(\lambda x + \mu y) \preceq \lambda F(x) + \mu F(y)$ for all $x, y \in X$ and $\lambda \geq 0$, $\mu \geq 0$ with $\lambda + \mu = 1$, i.e. for such x, y, λ, μ we have

$$\lambda F(x) + \mu F(y) - F(\lambda x + \mu y) \in P.$$

In what follows, a Banach space V always means an ordered Banach space that is ordered by a normal cone C in the sense that $x \preceq y$ if and only if $y - x \in C$.

Lemma 2.20 *Let X be a non-empty convex subset of a Hausdorff real topological vector space and V an ordered Banach space. A mapping $F X \to V$ is cone convex if and only if the function $f \cdot F$ is convex for all $f \in P^*$.*

Proof. If F is cone convex, then $f \cdot F$ is convex for $f \in P^*$ as f is order preserving. Next, let $f \cdot F$ be convex for each $f \in V^*$. If possible, let for some $x, y \in X$ and $\lambda \geq 0$, $\mu \geq 0$ with $\lambda + \mu = 1$, $[\lambda F(x) + \mu F(y) - F(\lambda x + \mu y)] \notin P$. Then by a consequence of the separation theorem (e.g., see Schaefer (1971)), there exists an $f \in P^*$ such that $f[\lambda F(x) + \mu F(y) - F(\lambda x + \mu y)] < 0$ which implies that $f \cdot F$ is not convex. □

Let X be a topological space. A mapping $F : X \to V$ is said to be *order or cone lower* (respectively, *upper) semi-continuous* if $f \cdot F$ is lower (respectively, upper) semi-continuous for each $f \in P^*$.

Lemma 2.21 *If P is a normal cone of a real Banach space V and $x \in V$ with $x \succ 0$ (respectively, $\prec 0$), then there exists $f \in P^*$ such that $f(x) > 0$ (respectively, $f(x) < 0$).*

Proof. Let $x \succ 0$. Then $f(x) \geq 0$ for all $f \in P^*$. If possible, let $f(x) = 0$ for all $f \in P^*$. By Hanh-Banach extension theorem there exists $p \in V^*$ such that $p(x) \neq 0$. Now since P is a normal cone, $p = g - h$ where $g, h \in P^*$ (e.g., see Theorem 23.5, of Kelley and Namioka Kelley and Namioka (1963, p. 227)). Hence $p(x) = g(x) - h(x) = 0$ which is a contradiction. □

Definition 2.10 Let X be a non-empty set, V an ordered Banach space and $f : X \to V$ a (single-valued) mapping. Then we say that

(1) a point $a \in X$ is said to be a *Pareto maximal* (respectively, *Pareto minimal) optimum* of f if $f(y) \succeq f(a)$ (respectively, $f(y) \preceq f(a)$) for all $y \in X$ implies that $f(y) = f(a)$;

(2) if $a \in X$ is a Pareto maximal (respectively, minimal) optimum of f, then the set $N(f,a) = \{x \in X : f(a) = f(x)\}$ is called the (generalized) *Nucleolus* or the *indifference set* of f w.r.t. a.

We denote by $P(f)$ the set of all Pareto maximal (respectively, minimal) optima and $E(f)$ the set of all end points of f. It is easy to see that our definition for Pareto maximal (respectively, minimal) optimum is equivalent to the corresponding definitions introduced in (Tarafdar (1996b), pp. 2511–2519 and Tarafdar (1996a), pp. 2431–2439).

Also, the Pareto optima always means *Pareto maximal optima* in the remaining part of this section for simplicity, unless otherwise specified.

Definition 2.11 Let X be a non-empty set, V an ordered Banach space and $f : X \to V$ a (single-valued) mapping and $\phi : X \to 2^X$ a set-valued mapping. Then f is said to be

(1) *ϕ-monotone* if $x \in X$, $y \in \phi(x) \Rightarrow f(x) \preceq f(y)$; and

(2) strictly *ϕ-monotone* if $x \in X$, $y \in \phi(x)$ and $x \neq y \Rightarrow f(x) \prec f(y)$; (i.e., by Lemma 2.21, there exists $p \in P^*$ such that $pf(x) < pf(y)$ and $qf(x) \leq qf(y)$ for all $q \in P^*$).

It is clear that a ϕ-monotone mapping generalizes the ϕ-monotone vector function, introduced by Justman in (Justman (1978)) which, in turn, serves as the Lyapunov functions.

We note that for a given non-empty set X and an ordered Banach space V and for each mapping $f : X \to V$, there exist two set-valued mapping $\phi_f, \hat{\phi}_f : X \to 2^X$ such that f is ϕ_f-monotone and f is $\hat{\phi}_f$-strictly monotone, respectively, where $\phi_f, \hat{\phi}_f : X \to 2^X$ are defined by

$$(a) \qquad \phi_f(x) = \{y \in X : f(x) \preceq f(y)\}$$

and

$$(b) \qquad \hat{\phi}_f(x) = \{y \in X : f(x) \prec f(y)\}$$

for each $x \in X$.

Lemma 2.22 *Let X be a non-empty set and V an ordered Banach space. Suppose $f : X \to V$ is a (single-valued) mapping and $\phi : X \to 2^X$ a set-valued mapping. If f is strictly ϕ-monotone, then $N(f,a) \subset E(\phi)$ for each Pareto optimum a of f. Moreover, if f is ϕ-monotone, then each $z \in E(\phi_f)$ is a Pareto optimum of f.*

Proof. To see this, let $z \in N(f,a)$. If $z \notin E(\phi)$, then there exists $y \in X$ such that $y \neq z$ and $y \in \phi(z)$. The strict ϕ-monotone of f implies that $f(z) \preceq f(y)$ and $pf(z) < pf(y)$ for some $p \in P^*$ which contradicts the Pareto maximality of a of

f. It is clear that if $f : X \to V$ is a mapping, then each $z \in E(\phi_f)$ is a Pareto optimum of f. Indeed, if $z \in E(\phi_f)$ and $z \notin P(f)$, then there exists $y \in X$ with $x \neq y$ and $f(z) \preceq f(y)$. Thus, $y \in \phi_f(z)$ which is impossible since $\phi_f(z) = \{z\}$. □

Definition 2.12 Let X be a non-empty set and $\phi : X \to 2^X$ a set-valued mapping. Then a point $x_0 \in X$ is said to be a *maximal element* of ϕ if $\phi(x_0) = \emptyset$.

Lemma 2.23 *Let X be a non-empty set and V an ordered Banach space. Suppose $f : X \to V$ is a mapping. Then a point $x_0 \in X$ is a Pareto optimum of f if and only if $x_0 \in X$ is a maximal element of $\hat{\phi}_f$.*

Proof. It is easy to verify from the definitions. □

B. Set-Valued Dynamic Systems

Let X be a metric space with metric ρ. Following Aubin and Siegel (1980), a *set-valued dynamic system* F on X is a set-valued mapping $F : X \to 2^X$ which takes non-empty values. Any finite sequence $x^+ = x^0, x^1, \dots, x^n, \dots$, such that $x^{n+1} \in F(x^n)$ for each $n \in \mathbb{N}$ is called a *motion* of the system F at x_0. The set $\mathcal{F}(x^+) = \{x^n : n \in \mathbb{N} \text{ and } x^{n+1} \in F(x^n)\}$ is called the *trajectory* of this *motion* or *F-sequence* starting at x^0.

Let $U : X \to \mathbb{R}_+ \cup \{+\infty\}$ be a real-valued function with *Dom* $U \neq \emptyset$. A dynamic system $F : X \to 2^X$ is said to be *dissipative* (with respect to U) if for each $x \in X$,

$$U(y) + \rho(x, y) \leq U(x) \quad \text{for all } \; y \in F(x).$$

We first have the following fixed point theorem:

Lemma 2.24 *Let (X, ρ) be a complete metric space and $F : X \to 2^X$ an upper semi-continuous mapping with non-empty compact values such that the mapping $U'_F : X \to \mathbb{R}_+ \cup \{+\infty\}$ defined by*

$$(c) \quad U'_F(x) = \inf\{\sum_{n=0}^{\infty} \rho(x^n, x^{n+1}) | (x^n)_{n=0}^{\infty} \text{ is any } F\text{-sequence with } x^0 = x\}$$

for each $x \in X$, is proper (i.e., Dom $U'_F \neq \emptyset$). Then there exists a motion $\mathcal{F}(x^+)$ starting at some $x \in$ Dom U'_F which converges to a limit $\hat{x}$, which is a fixed point of F, i.e., $\hat{x} \in F(\hat{x})$.

Proof. Since F is upper semi-continuous with non-empty compact values, then for each $x \in X$, there exists $y \in F(x)$ such that $U'_F(x) \geq U'_F(y) + \rho(x, y)$ by Proposition 4 of (Aubin and Ekeland (1984), p. 242).

Noting that the graph of F is closed and *Dom* $U'_F \neq \emptyset$, by Proposition 2 of (Aubin and Ekeland (1984), p. 240), there is a motion $\mathcal{F}(x^+)$ starting at some $x \in$ *Dom* U'_F which converges to a limit $\hat{x}$, which is a fixed point of F, i.e., $\hat{x} \in F(\hat{x})$. □

In order to consider the existence of endpoints for set-valued dissipative dynamic systems which are lower semi-continuous, we need the following lemma.

Lemma 2.25 *Let (X, ρ) be a complete metric space and $F : X \to 2^X$ a set-valued dynamic system. Suppose that F is dissipative with respect to the lower semi-continuous function $U : X \to \mathbb{R}_+ \cup \{+\infty\}$. Then F has an endpoint $\hat{x}$ in X, i.e., $F(\hat{x}) = \{\hat{x}\}$.*

Proof. This is Proposition 1.3 of (Aubin and Ekeland (1984), p. 248). □

Lemma 2.26 *Let (X, ρ) be a metric space and $F : X \to 2^X$ a set-valued dynamic system. Then the function $U_F : X \to \mathbb{R}_+ \cup \{+\infty\}$ defined by*

$$(d) \quad U_F(x) = \sup\{\sum_{n=0}^{\infty} \rho(x^n, x^{n+1}) | (x^n)_{n=0}^{\infty} \quad \textit{is any} \quad F-\textit{sequence with} \quad x^0 = x\}$$

for each $x \in X$, satisfies the following inequality:

$$(*) \qquad U_F(y) + \rho(x, y) \leq U_F(x) \quad \textit{for all} \quad y \in F(x)$$

for each $x \in X$. Moreover, if U is any other non-negative valued function which satisfies the inequality $(*)$, then $U_F(x) \leq U(x)$ for all $x \in X$.

Proof. It is clear that for each $x \in X$, $U_F(y) + \rho(x, y) \leq U_F(x)$ for all $y \in F(x)$ by the definition of U_F. Moreover, it is trivial to verify that U_F is the smallest non-negative-valued function which satisfies the property $(*)$. Indeed, let U be any non-negative-valued function satisfying $(*)$. Now for any F-sequence, $(x^n)_{n=0}^{\infty}$ starting at $x^0 = x$. Noting that $U(x^n) \geq \rho(x^n, x^{n+1}) + U(x^{n+1})$ and $U(x^n) \geq 0$ for all $n = 0, 1, \ldots, \infty$, we have that $U(x) \geq \sum_{n=0}^{\infty} \rho(x^n, x^{n+1})$. Thus, $U_F(x) \leq U(x)$ by the definition of U_F. □

Lemma 2.27 *Let (X, ρ) be a metric space and $F : X \to 2^X$ a lower semi-continuous set-valued mapping with non-empty values. Then the function $U_F : X \to \mathbb{R}_+ \cup \{+\infty\}$ defined by formula (d) in Lemma 2.26 is lower semi-continuous. (Here F is lower semi-continuous at x_0 in the sense that if $x_n \to x_0$ and $y_0 \in F(x_0)$, then there exists $y_n \in F(x_n)$, $n = 1, 2, \ldots$, such that $y_n \to y_0$.)*

For the completeness we include the proof due to Maschler and Peleg in (Maschler and Peleg (1976), Lemma 2.1, p. 987).

Proof. Let $x \in X$ and $U_F(x) \geq r$. We need to show that $\liminf_{k\to\infty} U_F(u_k) \geq r$, whenever $u_k \in X$ for $k = 1, 2, \ldots$ and $\lim_{k\to\infty} u_k = x$.

Let $\epsilon > 0$. Then there exist points $x_0 = x, x_1 \in F(x_0), \ldots, x_T \in F(x_{T-1})$ such that

$$\sum_{t=0}^{T-1} \rho(x_{t+1}, x_t) \geq r - \epsilon.$$

Now since F is lower semi-continuous, there exist $T+1$ sequences $(u_{i,k})$, $i = 0, 1, \ldots, T$, where $u_{0,k} = u_k$, $u_{1,k} \in F(u_{0,k}), \ldots, u_{T,k} \in F(u_{T-1,k})$, $k = 1, 2\ldots$, such that

$$\lim_{k\to\infty} u_{t,k}) = x_t, \quad t = 0, 1, \ldots, T.$$

From above it is clear that

$$\liminf_{k\to\infty} U_F(u_k) \geq \lim_{k\to\infty} \sum_{t=0}^{T-1} \rho(u_{t+1,k}, u_{t,k}) \geq r - \epsilon.$$

Since ϵ is arbitrary, the lemma follows. □

Now we have the following result which is due to Maschler and Peleg in (Maschler and Peleg (1976), Remark 2.8, p. 988):

Theorem 2.43 *Let (X, ρ) be a complete metric space and $F : X \to 2^X$ a lower semi-continuous mapping with non-empty values. Suppose that the function $U_F : X \to \mathbb{R}_+ \cup \{+\infty\}$ defined by formula (d) is proper, i.e., Dom $U_F \neq \emptyset$. Then F has an endpoint $\hat{x} \in X$, i.e., $F(\hat{x}) = \{\hat{x}\}$.*

Proof. Note that F is lower semi-continuous, so that the function U_F is lower semi-continuous by Lemma 2.27. Then the conclusion follows from Lemma 2.25. □

For the convenience of our discussion, we shall call the function U_F defined in the formula (*d*) of Lemma 2.26, the *function induced* by F in the rest part of this section.

As an immediate consequence of Theorem 2.43, we have the following result due to Aubin and Ekeland in (Aubin and Ekeland (1984), Corollary 10, p. 246).

Theorem 2.44 *Let (X, ρ) be a complete metric space and let the dissipative dynamic system $F : X \to 2^X$ be lower semi-continuous. Then F has an endpoint $\hat{x}$ in X, i.e., $F(\hat{x}) = \{\hat{x}\}$.*

Proof. Since F is dissipative, there exists a function $U : X \to \mathbb{R}_+ \cup \{+\infty\}$ such that for each $x \in X$

$$U(y) + \rho(x, y) \leq U(x) \quad \text{for all} \quad y \in F(x)$$

and *Dom* $U \neq \emptyset$. Now, let $U_F : X \to \mathbb{R}_+ \cup \{+\infty\}$ be the induced function of F, i.e.,

$$U_F(x) = \sup\{\sum_{n=0}^{\infty} \rho(x^n, x^{n+1}) | (x^n)_{n=0}^{\infty} \text{ is any } F-\text{sequence with } x^0 = x\}$$

for each $x \in X$. By Lemma 2.26, $U_F(x) \leq U(x)$ for all $x \in X$. Hence, *Dom* $U_F \neq \emptyset$ because *Dom* $U \neq \emptyset$. Since F is lower semi-continuous with non-empty values, U_F is lower semi-continuous by Lemma 2.27. Therefore, the conclusion follows from Theorem 2.43 and we complete the proof. □

Now as applications of the previous results, we have the following theorems on the existence of Pareto optima in metric spaces.

Theorem 2.45 *Let (X, ρ) be a complete metric space, V an ordered Banach space and $f : X \to V$ a single-valued mapping. Suppose f is cone upper semi-continuous and the function $U_{\hat{\phi}_f} : X \to [0, \infty)$ which is defined from the function $\hat{\phi}_f$ by formula (d) of Lemma 2.26, is proper (i.e., Dom $U_{\hat{\phi}_f} \neq \emptyset$), where $\hat{\phi}_f$ is a mapping defined by formula (b) in the last section. Then $P(f) \neq \emptyset$.*

Proof. In order to prove that $P(f) \neq \emptyset$, by Lemma 2.23 it is sufficient to show that the mapping $\hat{\phi}_f$ has a maximal element. If it were false, then $\hat{\phi}_f(x) \neq \emptyset$ for each $x \in X$. Since f is cone upper semi-continuous, for each $y \in X$ the set $\hat{\phi}_f^{-1}(y) = \{x \in X : f(x) \prec f(y)\}$ is open in X, and therefore the mapping $\hat{\phi}_f$ is lower semi-continuous with non-empty values. By Theorem 2.43, there exists $x_0 \in X$ such that $x_0 = \hat{\phi}_f(x_0)$ which is impossible since $x \notin \hat{\phi}_f(x)$ for each $x \in X$. Thus, there must exist some $x \in X$ such that $\hat{\phi}_f(x) = \emptyset$, so that $x \in P(f)$ and the proof is completed. □

Theorem 2.46 *Let (X, ρ) be a closed bounded subset of a complete metric space, V an ordered Banach space and $f : X \to V$ a mapping. Suppose the induced set-valued mapping $\phi_f : X \to 2^X$ is a generalized contraction, where ϕ_f is a mapping defined by formula (a) before in the last section. Then $P(f) \neq \emptyset$.*

Proof. By assumption, the mapping ϕ_f is a generalized contraction, so that there exists $x_0 \in X$ such that $x_0 = \phi_f(x_0)$ by Theorem 2.42. Then $x_0 \in P(f)$ follows from Lemma 2.22. □

Corresponding to Theorem 2.36, we also have the following theorem.

Theorem 2.47 *Let (X, ρ) be a compact topological space, V an ordered Banach space and $f : X \to V$ a mapping. Suppose the set-valued mapping $\phi_f : X \to 2^X$ is a set-valued topological contraction, where ϕ_f is a mapping defined by formula (a). Then $P(f) \neq \emptyset$.*

Proof. By assumption, the mapping ϕ_f is a topological contraction, so that there exists $x_0 \in X$ such that $x_0 = \phi_f(x_0)$ by Theorem 2.36. Then $x_0 \in P(f)$ follows from Lemma 2.22 and we complete the proof. □

Theorems 2.45, 2.46 and 2.47 give the existence of Pareto optima in metric spaces.

We will now prove the existence of Pareto optimum in topological vector spaces.

Definition 2.13 Let X be a convex subset of a vector space E and V a Banach space. A mapping $f : X \to V$ is said to be *quasi-cone concave* if for each $x \in X$, the set $\{y \in X : f(y) \succ f(x)\}$ is convex.

Before we prove the existence of the Pareto maximum for the mapping f in topological vector spaces, we also need the following fixed point theorem (Tarafdar (1998)) the proof of which will be given in Chapter 4.

The Fixed Point Theorem

Let X be a non-empty convex subset of a Hausdorff topological vector space E. Suppose $T : X \to 2^X$ is a set-valued mapping with non-empty convex values such that

(i) for each fixed $y \in X$, $T^{-1}(y) = \{x \in X : y \in T(x)\}$ contains a relatively open set O_y of X (where O_y may be empty for some $y \in X$);

(ii) $\bigcup_{y \in X} O_y = X$; and

(iii) there exists a non-empty subset X_0 which is contained in a non-empty compact and convex subset X_1 of X such that the set $D = \bigcap_{x \in X_0} O_x^c$ is either empty or compact (where O_x^c denotes the complement of O_x in X).

Then T has a fixed point.

Now we have the following theorem:

Theorem 2.48 *Let X be a non-empty convex subset of a topological vector space E and V a Banach space. Suppose $f : X \to V$ is cone upper semi-continuous and cone quasi-concave. Furthermore, there exists a non-empty compact and convex subset X_0 of X such that the set $\bigcap_{y \in X_0} \{x \in X : f(x) \not\prec f(y)\}$ is either empty or compact. Then $P(f) \neq \emptyset$.*

Proof. We claim that the mapping $\hat{\phi}_f$ has a maximal element. Suppose if it were false, then $\hat{\phi}_f(x) \neq \emptyset$ for each $x \in X$ and $(\hat{\phi}_f)^{-1}(y) = \{x \in X : f(x) \prec f(y)\}$ is open since f is cone upper semi-continuous. Note that $\hat{\phi}_f(x)$ is convex since f is cone quasi-concave. Thus, all hypotheses of the above 'Fixed Point Theorem' are satisfied. Hence, by the above 'Fixed Point Theorem', there exists $x_0 \in X$ such that $x_0 \in \hat{\phi}_f(x_0)$, which is impossible by the definition of $\hat{\phi}_f$. Hence, there must exist $x \in X$ such that $\hat{\phi}_f(x) = \emptyset$. Thus $x \in P(f)$. □

As an immediate consequence of Theorem 2.48, we have the following corollary:

Corollary 2.48.1 *Let X be a non-empty compact subset of a topological vector space E and V an ordered Banach space. Suppose $f : X \to V$ is cone upper semi-continuous and cone quasi-concave. Then $P(f) \neq \emptyset$.*

C. The Stability of Pareto Optima

In this subsection, as applications of the results in the previous subsection, we shall study the stability of Pareto optima for the mapping f which takes values in ordered Banach spaces.

Definition 2.14 (Maschler and Peleg (1976)) Let Q be non-empty subset of a topological space X and $\phi : X \to 2^X$ a set-valued mapping with non-empty values. The set Q is called *stable* with respect to (in short w.r.t.) ϕ if for each open neighborhood U of Q, there exists a neighborhood V of Q such that if $x \in V$ and $\{x_n\}_{n\in\mathbb{N}}$ is a ϕ-sequence with $x_0 = x$, then $x_n \in U$ for $n = 1, 2, \ldots$ (or, equivalently, $\phi^n(V) \subset U$ for $n = 1, 2, \ldots$).

We first have the following lemma:

Lemma 2.28 *The set Q is stable w.r.t. ϕ if and only if for each open neighborhood U of Q, there exists another non-empty open subset $V \subset Q$ such that $\phi(V) \subset V$. Moreover, if Q is closed and stable w.r.t. ϕ and X is regular, then $\phi(Q) \subset Q$.*

Proof. If Q is stable and U is an open neighborhood of Q, then there is a neighborhood W of U such that $\phi^n(W) \subset U$ for $n \in \mathbb{N}$. Let $V = \bigcup_{n\in\mathbb{N}} \phi^n(W)$. Then V satisfies the stated properties. The converse is clear. Now suppose Q is stable and closed. Let $\{U_i : i \in I\}$ be the family of all non-empty open neighborhoods of Q, then $\phi(Q) \subset \bigcap_{i\in I} U_i$. Since Q is a closed subset of X, we claim that $\bigcap_{i\in I} U_i = Q$. It suffices to prove that $\bigcap_{i\in I} U_i \subset Q$. If not, there exists $w_0 \in \bigcap_{i\in I} U_i$ and $w_0 \notin Q$. As X is regular, and Q is closed, there exist a non-empty open neighborhood $O(w_0)$ of w_0 and a non-empty open neighborhood $\hat{U}$ of Q such that $O(w_0) \cup U = \emptyset$. But $x_0 \in \bigcap_{i\in I} U_i$, so that $x_0 \in \hat{U}$ as $\{U_i\}_{i\in I}$ is the family of all non-empty open neighborhoods of Q. Thus, $\bigcap_{i\in I} U_i \subset Q$. Therefore, we have $\phi(Q) \subset Q$. □

Theorem 2.49 *Let $f : X \to V$ be a mapping from a compact metric space (X, ρ) into an Ordered Banach space V. Let $N(f, a)$ be a generalized nucleolus of f. Suppose the following conditions hold:*

(i) f is ϕ-monotone;

(ii) f is cone upper semi-continuous on X; and

(iii) f is continuous on $N(f, a)$.

Then $N(f, a)$ is stable with respect to ϕ.

Proof. First we note that by the cone upper semi-continuity of f, the set $N(f, a) = \{x \in X : f(a) \preceq f(x)\}$ is a closed subset of X. Let $U \supset N(f, a)$ be an open subset of X. Then $S = X \setminus U$ is compact. Now for each $y \in S$, there exists by Lemma 2.21, a $p \in P^*$ and a positive integer $n \in \mathbb{N}$ such that $pf(y) < pf(a) + \frac{1}{n}$. Let

$$U_{p,n}(y) = \{x \in X : pf(x) < pf(a) + \frac{1}{n}\}$$

for each $p \in P^*$ and $n \in \mathbb{N}$. Then by the upper semi-continuity of pf at $a \in X$, $U_{p,n}(y)$ is an open set containing y. Thus $\{U_{p,n}(y)\}_{p\in P^*, n\in\mathbb{N}}$ is an open covering of S. As S is compact, there is a finite subcover, say, $U_{p_1,n_1}(y_1)$, $U_{p_2,n_2}(y_2), \ldots$, $U_{p_m,n_m}(y_m)$, i.e., $\bigcup_{i=1}^m U_{p_i,n_i}(y_i) \supset S$. For each $i = 1, 2, \ldots, m$, set

$$V_{p_i,n_i} = \{x \in X : p_i f(x) > p_i f(a) + \frac{1}{n_i}\}$$

which is, by the continuity of f on $N(f,a)$, an open set containing $N(f,a)$. Let $V = \bigcap_{i=1}^{m} V_{p_i,n_i}$. Then V is a neighborhood of $N(f,a)$ and is contained in U. To see this, let $y \in V$. If $y \notin U$, then $y \in S$ which would be a contradiction. Now if $x \in V$ and $\{x_n\}$ is a ϕ-sequence with $x_0 = x$, then by (i), we have for each $i = 1, 2, \ldots, m$,

$$p_i f(x_n) \geq p_i f(x_{n-1}) \geq \cdots \geq p_i f(x_0) > p_i f(a) + \frac{1}{n_i}.$$

Hence, $x_n \in V \subset U$, where $n = 1, 2, \ldots$. Thus $N(f,a)$ is stable. □

As applications of Theorem 2.49, we have the following theorems:

Theorem 2.50 *Let (X,ρ) be a compact complete metric space, V an ordered Banach space and $f : X \to V$ a single-valued mapping which is continuous on $P(f)$. Suppose f is cone upper semi-continuous on X and the induced function $U_{\hat{\phi}_f}$ of $\hat{\phi}_f$ from X to $[0,\infty)$ is proper, i.e., Dom $U_{\hat{\phi}_f} \neq \emptyset$, where $\hat{\phi}_f$ is a mapping defined by formula (b) of Definition* 2.11. *Then $N(f,a)$ is stable with respect to $U_{\hat{\phi}_f}$ for each $a \in P(f)$.*

Proof. By Theorem 2.45, $P(f) \neq \emptyset$. Since X is compact, the conclusion follows from Theorem 2.49. □

By employing Theorems 2.46, 2.47 and Corollary 2.48.1 and the same arguments as used in the proof of Theorem 2.50, we have the following theorems:

Theorem 2.51 *Let (X,ρ) be a compact subset of a complete metric space, V an ordered Banach space and $f : X \to V$ (single-valued) upper semi-continuous on X and continuous on $P(f)$. Suppose the set-valued mapping $\phi_f : X \to 2^X$ is a generalized contraction mapping, where ϕ_f is a mapping defined by formula (a) of Definition* 2.11. *Then $N(f,a)$ is stable with respect to U_{ϕ_f} for each $a \in P(f)$.*

Theorem 2.52 *Let (X,ρ) be a compact subset of a metric space, V an ordered Banach space and let $f : X \to V$ be upper semi-continuous and continuous on $P(f)$. Suppose the set-valued mapping $\phi_f : X \to 2^X$ is a topological contraction mapping. Then $N(f,a)$ is stable with respect to U_{ϕ_f} for each $a \in P(f)$.*

Theorem 2.53 *Let (X,ρ) be a non-empty compact subset of a topological vector space E and V be an ordered Banach space. Suppose $f : X \to V$ is cone upper semi-continuous and cone concave and f is continuous on $P(f)$. Then $N(f,a)$ is stable with respect to $\hat{\phi}_f$ for each $a \in P(f)$, where $\hat{\phi}_f$ is a mapping defined by formula (b) of Definition* 2.11.

2.9 Large Contractions

The materials of this section is taken from Watson (1998). Recently there has been some interesting advances in solving the following implicit Darboux problem with values in a Banach space E:

$$\begin{aligned} &\frac{\partial^2 z}{\partial x \partial y} = g\left(x, y, z, \frac{\partial^2 z}{\partial x \partial y}\right) && \text{for } (x,y) \in \Omega, \\ &z(x,0) = 0 && \text{for } x \geq 0, \\ &z(0,y) = 0 && \text{for } y \geq 0. \end{aligned} \tag{DP}$$

where $\Omega = \mathbb{R}^+ \times \mathbb{R}^+$ and $g : \Omega \times E \times E \to E$.

In (Wójtowicz (1997)) the author extends the results of (Rzepecki (1986)) by weakening the condition that the map

$$g(x, y, u, \cdot) : E \to E \tag{2.119}$$

be a contraction with constant $k < 1/2$. The author shows that it need only be a contraction with constant $k < 1$. Also in Wójtowicz (1997), the uniform continuity required in Rzepecki (1986) is weakened to continuity.

It is of interest therefore, to investigate to what extent the contractive condition can be weakened further. Using a fixed point theorem for so-called *Large Contractions* in Burton (1996), we are able to transform the implicit problem above to an explicit problem with right-hand side $f(x, y, u)$ a selection of

$$F(x, y, u) = \{z \in E : z = g(x, y, u, z)\}.$$

The explicit problem may then be solved by the same means as in Wójtowicz (1997). This is possible at the expence of the weaker continuity condition. We assume $g(\cdot, \cdot, \cdot, z)$ is continuous uniformly with respect to $z \in E$.

2.9.1 *Large Contractions*

The definition of large contractions is due to Burton (1996).

Definition 2.15 A mapping F from a metric space (M, d) into itself is said to be a Large Contraction if F is contractive, (i.e. for all $x \neq y$ $d(F(x), F(y)) < d(x, y)$) and for all $\varepsilon > 0$, there exists $0 \leq \delta(\varepsilon) < 1$ such that $d(x, y) \geq \varepsilon$ implies $d(F(x), F(y)) \leq \delta d(x, y)$.

We rephrase the Fixed Point Theorem, Theorem 1 of Burton (1996), in terms suitable for our goal. For $F : M \to M$, we shall say a subset $K \subset M$ is *invariant* under F (or simply invariant) if $F(K) \subset K$.

Theorem 2.54 *Burton (1996) Let (M, d) be a complete metric space and $F : M \to M$ a large contraction. If M contains a bounded invariant subset, then F has a unique fixed point.*

For a different application of Theorem 2.54 to an implicit integral equation, we refer to Burton (1996).

2.9.2 *The Transformation*

Let $\Omega(a,b) = [0,a] \times [0,b]$ and E a Banach space. Assume the following:

(i) $g : \Omega \times E \times E \to E$ is such that $g(\cdot,\cdot,\cdot,z)$ is continuous uniformly in $z \in E$;
(ii) the map $g(x,y,u,\cdot) : E \to E$ is a large contraction;
(iii) for each $(a,b) \in \Omega$ there exists $m(a,b) \geq 0$ such that for each $(x,y,u) \in \Omega(a,b) \times E$, there exists $K \subset B_{m(a,b)}(0) \subset E$ such that K is invariant under $g(x,y,u,\cdot)$.

Assumptions (ii) generalizes the assumption 2^0 of Wójtowicz (1997) in that a contraction with constant $k < 1$ is a large contraction. Assumption (iii) is not exactly the same as the corresponding assumption 3^0 of Wójtowicz (1997), although the assumptions 2^0 and 3^0 together are encompassed by (iii). This is evident from the discussion in Section 2 of Wójtowicz (1997).

Theorem 2.55 *Suppose Assumptions* (i)–(iii) *are satisfied. Then for each* $(x,y,u) \in \Omega \times E$ *there exists a unique* $f(x,y,u) \in E$ *such that*
(1) $f(x,y,u) = g(x,y,u,f(x,y,u))$;
(2) $f : \Omega \times E \to E$ *is continuous;*
(3) *for each* $(a,b) \in \Omega$ *there exists* $m(a,b) \geq 0$ *such that* $||f(x,y,u)|| \leq m(a,b)$ *for all* $(x,y) \in \Omega(a,b)$ *and* $u \in E$.

Proof. Fix $(x,y,u) \in \Omega \times E$ and let $F(z) = g(x,y,u,z)$. Then F is a large contraction. By Assumption (iii) and Theorem 2.54, F has a unique fixed point $f(x,y,u)$ so 1 is satisfied. Moreover for any $(a,b) \in \Omega$ and any $(x,y,u) \in \Omega(a,b) \times E$, there exists $m(a,b) \geq 0$ such that $||f(x,y,u)|| \leq m(a,b)$.

Now we show f is continuous. Let (x_n, y_n, u_n) be a sequence converging to (x_0, y_0, u_0). Then

$$f(x_n, y_n, u_n) = v_n = g(x_n, y_n, u_n, v_n) \tag{2.120}$$

and we claim $v_n \to v_0 = f(x_0, y_0, u_0)$. Suppose not. Then there exists $\varepsilon_0 > 0$ and a sequence $N_k \to \infty$ as $k \to \infty$ such that for each k there exists $n, m > N_k$ with $||v_n - v_m|| \geq \varepsilon_0$. By definition of a large contraction, there exists $\delta(\varepsilon_0) < 1$ such that

$$\begin{aligned}
\varepsilon_0 &\leq ||v_n - v_m|| \\
&\leq ||g(x_n, y_n, u_n, v_n) - g(x_m, y_m, u_m, v_m)|| \\
&\leq ||g(x_n, y_n, u_n, v_n) - g(x_n, y_n, u_n, v_m)|| \\
&\quad + ||g(x_n, y_n, u_n, v_m) - g(x_m, y_m, u_m, v_m)|| \\
&\leq \delta(\varepsilon_0)||v_n - v_m|| \\
&\quad + ||g(x_n, y_n, u_n, v_m) - g(x_m, y_m, u_m, v_m)||.
\end{aligned}$$

Thus

$$(1 - \delta(\varepsilon_0))||v_n - v_m|| \leq ||g(x_n, y_n, u_n, v_m) - g(x_m, y_m, u_m, v_m)||$$

and as $||v_n - v_m|| \geq \varepsilon_0$, it follows that

$$(1 - \delta(\varepsilon_0))\varepsilon_0 \leq ||g(x_n, y_n, u_n, v_m) - g(x_m, y_m, u_m, v_m)||. \tag{2.121}$$

From the continuity of $g(\cdot, \cdot, \cdot, z)$ uniformly in z, choosing k sufficiently large the right-hand side of (2.121) may be made arbitrarily small. This is a contradiction as $(1 - \delta(\varepsilon_0))\varepsilon_0$ is a fixed number greater than zero. Hence f is continuous and the proof is complete. □

Now a solution of (DP) is a solution of the explicit problem:

$$\begin{aligned} \frac{\partial^2 z}{\partial x \partial y} &= f(x, y, z) && \text{for } (x, y) \in \Omega, \\ z(x, 0) &= 0 && \text{for } x \geq 0, \\ z(0, y) &= 0 && \text{for } y \geq 0, \end{aligned} \tag{EDP}$$

where f is given as in Theorem 2.55.

2.9.3 *An Existence Theorem*

One further assumption is required for g. By $\alpha(X)$ where X is a subset of the Banach space E, we mean the Kuratowski measure of noncompactness (see Banaś and Goebel (1980).

Assume

$$\text{(iv)} \qquad \alpha(g(A \times X \times Y)) \leq h(\alpha(X))$$

where $h : \mathbb{R}^+ \to \mathbb{R}^+$ is a function and A, X, Y are bounded subsets, $A \subset \Omega$, and $X, Y \subset E$.

Theorem 2.56 *Suppose g satisfies* (i)–(iv) *where $h : \mathbb{R}^+ \to \mathbb{R}^+$ is continuous, nondecreasing function such that the scalar inequality*

$$0 \leq u(x, y) \leq \int_0^x \int_0^y h(u(s, t))dsdt, \quad (x, y) \in \Omega \tag{2.122}$$

has only a trivial solution. Then problem (DP) has a solution.

Proof. From assumptions (iii), (iv) and the construction of f, it is clear that

$$\alpha(f(A \times X)) \leq h(\alpha(X)) \tag{2.123}$$

for bounded subsets $A \subset \Omega$ and $X \subset E$.

Let $C(\Omega; E)$ be the space of continuous functions from Ω to E with the topology of uniform convergence on compact subsets of Ω.

A solution of (EDP) is a fixed point of the operator $T : C(\Omega; E) \to C(\Omega; E)$ defined by

$$T(z)(x,y) = \int_0^x \int_0^y f(s,t,z(s,t))dsdt. \tag{2.124}$$

The function f satisfies all the conditions required in the proof of Theorem 3 of Wójtowicz (1997). Hence by this theorem, T has a fixed point and the proof is complete. □

By careful analysis of Theorem 4 of Wójtowicz (1997), it can be seen that the conclusion of Theorem 4 is true in the situation examined here. This means that under the assumptions of Theorem 2.56, the set of all solutions of the implicit Darboux problem (DP) is an R_δ set (i.e., is homeomorphic to the intersection of a decreasing sequence of compact absolute retracts).

2.10 Random Fixed Point and Set-Valued Random Contraction

The aim of this section is to present a random fixed point theorem due to Itoh (1977). All the results of this section are from Itoh (1977). Throughout this section (X, ρ) will denote a Polish space, i.e., a separable metric space, $(T, \mathcal{A})$ a measurable space and $\mathcal{B}$ the σ-algebra of Borel sets of X generated by the metric topology of (X, ρ). A mapping $F : T \to 2^X$ is called weakly $\mathcal{A}$-measurable if for any open set B of X, $F^{-1}(B) \in \mathcal{A}$. In this section by measurable we will mean weakly measurable.

A single valued mapping $U : T \to X$ is said to be a measurable selection of a measurable mapping $F : T \to 2^X$ if for each $t \in T$, $U(t) \in F(t)$.

Definition 2.16 **(Random Fixed Point)** Let $F : T \times X \to CB(X)$ be a set-valued mapping such that for each $x \in X$, $F(., x)$ is $\mathcal{A}$-measurable. Then a measurable mapping $U : T \to X$ is called a random fixed point of F if $U(t) \in F(t, U(t))$ for all $t \in T$.

We will need the following lemmas for details of which we refer to (Himmelberg (1975) and Kuratowski and Ryll-Nardzewski (1965)).

Lemma 2.29 *Let $\{F_n\}$ be a sequence of measurable mappings, $F_n : T \to CB(X)$ and $F : T \to CB(X)$ a mapping such that for each $t \in T$, $h(F_n(t), F(t)) \to 0$ as $n \to \infty$, where h is the Hausdorff metric with respect to the metric ρ introduced in Section 2.7. Then F is measurable.*

Proof. By a theorem of Himmelberg (1975, Theorem 3.5), it will suffice to show that for each $x \in X$, the real-valued function defined on T by $t \to d(x, F(t))$ is measurable, where as before $d(x, F(t)) = \inf\{\rho(x, y) : y \in F(t)\}$. It follows from the definition of Hausdorff metric h that for any $A, B \in CB(X)$,

$$|d(x, A) - d(x, B)| \le h(A, B).$$

Hence $|d(x, F_n(t)) - d(x, F(t))| \le h(F_n(t), F(t))$. Thus by the hypothesis, for each $t \in T$, $d(x, F_n(t)) \to d(x, F(t))$ as $n \to \infty$. Therefore, $d(x, F(\cdot))$ being the pointwise limit of measurable functions $\{d(x, F_n(\cdot))\}$, is measurable. □

Lemma 2.30 *Let $F : T \times X \to CB(X)$ be a set-valued mapping such that for each $t \in T$, $F(t, \cdot)$ is $k(t)$-Lipschitz and for each $x \in X$, $F(\cdot, x)$ is measurable. Then for any measurable mapping $U : T \to X$, the mapping $G : T \to CB(X)$ defined by $G(t) = F(t, U(t)), t \in T$, is measurable.*

Proof. Since X is separable, there is a countable dense subset $X_1 = [x_n : n \in \mathbb{N}]$. Hence $\bar{X_1} = X$.

For each n, let $B_{1n} = \{x \in X : \rho(x, x_1) \le \frac{1}{n}\}$, and

$$B_{in} = \{x \in X : \rho(x, x_i) \le \frac{1}{n}\} - \bigcup_{j=1}^{i-1} B_{jn}, i = 2, 3 \dots.$$

Then $\{B_{in}\}$ is a countable partition of X, i.e., $B_{in} \in \mathcal{B}$, $\bigcup_{i=1}^{\infty} B_{in} = X$ and if $i \neq j$, $B_{in} \cap B_{jn} = \emptyset$.

For each $n = 1, 2, \dots$, we define the set-valued mapping

$$F_n : T \times X \to CB(X)$$

by $F_n(t, x) = F(t, x_i)$ if $t \in T$ and $x \in B_{in}$. It can be easily seen that F_n is well-defined as x belongs to a unique B_{in}. Then for any open set B of X we have

$$\begin{aligned}\{(t, x) \in T \times X : F_n(t, x) \cap B \neq \emptyset\} &= \textstyle\bigcup_{i=1}^{\infty}\{t \in T : F(t, x_i) \cap B \neq \emptyset\} \\ &= \textstyle\bigcup_{i=1}^{\infty}\{t \in T : F(t, x_i) \cap B \neq \emptyset\} \\ &\quad \times B_{in} \in \mathcal{A} \times \mathcal{B},\end{aligned}$$

where $\mathcal{A} \times \mathcal{B}$ is the product of σ-algebra on $T \times X$. Hence for each n, F_n is $\mathcal{A} \times \mathcal{B}$-measurable. For each $t \in T$, $x \in X$, there exists an i such that $x \in B_{in}$ and

$$h(F_n(t, x), F(t, x)) = h(F(t, x_i), F(t, x)) \le k(t)\rho(x_i, x) \le \frac{k(t)}{n}.$$

Thus $h(F_n(t, x), F(t, x)) \to 0$ as $n \to \infty$. Hence by Lemma 2.29, F is measurable. Also the mapping $g : T \to T \times X$ defined by $g(t) = (t, U(t)), t \in T$ is measurable in the sense that $g^{-1}(\mathcal{A} \times \mathcal{B}) \subset \mathcal{A}$. Now it follows that for every open set B of X,

$$G^{-1}(B) = g^{-1}(\{(t, x) \in T \times X : F(t, x) \cap B \neq \emptyset\}) \in \mathcal{A}.$$

Thus G is measurable. □

Lemma 2.31 *Let Y be a metric space, $f : T \times X \to Y$ a mapping such that for any $t \in T$, $f(t, \cdot)$ is continuous and for any $x \in X$, $f(\cdot, x)$ is measurable. Let $F : T \to 2^X$ be a set-valued measurable mapping such that for each $t \in T$, $F(t)$ is non-empty, closed and U an open subset of Y. Then the set-valued mapping $G : T \to 2^X$ defined by $G(t) = \{x \in F(t) : f(t, x) \in U\}, t \in T$ is measurable.*

Proof. By a well-known theorem (see (Castains (1969)), or (Himmelberg (1975, Theorem 5.6)), or (Wagner (1977))) which states that if $F : T \to 2^X$ is non-empty closed valued (weakly) measurable set-valued mapping, then there exists a countable family $\{U_n\}$ of measurable selections of F such that for each $t \in T$, $cl[U_n(t) : n = 1, 2 \cdots] = F(t)$ and $T \cap \{t : U_i(t) \in F(t)\}$ is measurable for each $i = 1, 2 \cdots$, where $clA = \bar{A}$ denotes the closure of the subset A of X. Since F satisfies the conditions of that theorem, we have such a countable family $\{U_n(t)\}$. Let B be any open subset of X, then

$$\begin{aligned} G^{-1}(B) &= \{t \in T : f(t, x) \in U \text{ for some } \ x \in F(t) \cap B\} \\ &= \{t \in T : f(t, U_n(t)) \in U, u_n(t) \in B \text{ for some } \ n\} \\ &= \bigcup_{n=1}^{\infty} \{t \in T : f(t, U_n(t)) \in U\} \bigcap u_n^{-1}(B). \end{aligned}$$

Then it follows (e.g., as in Theorem 6.5 of (Himmelberg (1975))) that $\{t \in T : f(t, u_n(t)) \in U\} \in \mathcal{A}$ for each n. Hence $G^{-1}(B) \in \mathcal{A}$ and G is measurable. □

Lemma 2.32 *Let $F, G : T \to CB(X)$ be measurable set-valued mapping, $U : T \to X$ a measurable selection of F, $r : T \to (0, \infty)$ a measurable function. Then there is a measurable selection $v : T \to X$ of G such that for each $t \in T$,*

$$\rho(u(t), v(t)) \leq h(F(t), G(t)) + r(t).$$

Proof. By the same reasons as in the previous theorem there exist a countable family $\{u_n\}$ of measurable selections of F and a countable family $\{v_n\}$ of measurable selections of G such that for each $t \in T$, $cl[u_n(t) : n = 1, 2, \ldots,] = F(t)$ and $cl[v_n(t) : n = 1, 2, \ldots,] = G(t)$. It follows from the definition of Hausdorff metric that for each $t \in T$,

$$h(F(t), G(t)) = \max\{\sup_i \inf_j \rho(u_i(t), v_j(t), \sup_j \inf_i \rho(u_i(t), v_j(t))\},$$

hence the real-valued function $h(F(\cdot), G(\cdot))$ on T is measurable. We now define $f : T \times X \to \mathbb{R}$ and $G_1 : X \to 2^X$ by

$$f(t, x) = \rho(u(t), x) - h(F(t), G(t)) - r(t)$$

and

$$G_1(t) = \{x \in G(t); f(t, x) < 0\}.$$

By Lemma 2.31, G_1 is measurable and by definition of Hausdorff metric, $G_1(t)$ is non-empty for all $t \in T$. Thus the mapping $G_2 : T \to CB(X)$ defined by $G_2(t) = clG_1(t)$ is measurable and by Kuratowski and Ryll-Nardzewski theorem (Kuratowski and Ryll-Nardzewski (1965, p. 398), also see Wagner (1977, Theorem 4.1)) has a measurable selection $v : T \to X$ which satisfies the requirement of the lemma. □

The following theorem is the random version of Nadler's fixed point theorem (Nadler (1969)) of set-valued contraction mapping which we have considered in Section 2.7.

Theorem 2.57 *Let $F : T \times X \to CB(X)$ be a mapping with the properties that for each $x \in X$, $F(\cdot, x)$ is measurable and for each $t \in T$, $F(t, \cdot)$ is a $k(t)$-contraction (i.e. contraction mapping with contraction constant $k(t)$), where $k : T \to [0, 1)$ is a measurable function. Then F has a random fixed point.*

Proof. We set $A_1 = \{t \in T : 0 < k(t)\}$ and $A_2 = T - A_1$. Then $A_1, A_2 \in \mathcal{A}$. First we prove the theorem on A_1. Let $v_0 : A_1 \to X$ be a measurable mapping. Then by Lemma 2.30, the mapping $G(\cdot) = F(\cdot, v(\cdot)) : A_1 \to CB(X)$ is measurableand has thus a measurable selection $v_1 : A_1 \to X$ of $F(\cdot, v(\cdot))$ by (Kuratowski and Ryll-Nardzewski (1965)). Hence by Lemma 2.32 there is a measurable selection $v_2 : A_1 \to X$ such that for any $t \in T$,

$$\rho(v_1(t), v_2(t)) \leq h(F(t, v_0(t)), F(t, v_1(t)) + k(t).$$

Again by Lemma 2.32, we have a measurable selection $v_3 : A_1 \to X$ of $F(\cdot, v_2(\cdot))$ such that for any $t \in T$,

$$\rho(v_2(t), v_3(t)) \leq h(F(t, v_1(t)), F(t, v_2(t)) + [k(t)]^2.$$

Continuing this process, we have by induction a sequence of measurable mappings $v_n : A_1 \to X$ such that for each $t \in T$, $v_n(t) \in F(t, v_{n-1}(t))$ and

$$\rho(v_n(t), v_{n+1}(t)) \leq h(F(t, v_{n-1}(t)), F(t, v_n(t)) + [k(t)]^n.$$

Let $t \in A_1$ be arbitrary but fixed. Then by the argument as (2.117) we have

$$\rho(v_n(t), v_{n+1}(t)) \leq [k(t)]^n \rho(v_0(t), v_1(t)) + n[k(t)]^n,$$

for all $n \geq 1$. Hence as in (2.118), we have

$$\rho(v_n, v_{n+j}) \leq (\Sigma_{i=n}^{n+j-1}[k(t)]^i \rho(v_0, v_1) + \Sigma_{i=n}^{n+j-m} i[k(t)]^i$$

for all $n, j = 1, 2, \ldots$. This by virtue of $0 < k(t) < 1$ shows that $\{v_n(t)\}$ is a Cauchy sequence in X. Hence $\{v_n(t)\}$ converges to a point $v(t) \in X$. Now it follows that for each n,

$$\begin{aligned}\rho(v(t), F(t, v)) &\leq \rho(v(t), v_n(t)) + \rho(v_n(t), F(t, v(t))\\ &\leq \rho(v(t), v_n(t)) + h(F(t, v_{n-1}(t)), F(t, v(t)) \text{as } \ v_n(t) \in F(t, v_{n-1}(t))\\ &\leq \rho(v(t), v_n(t)) + k(t)\rho(v_{n-1}(t), v(t)) \to 0 \text{as } n \to \infty.\end{aligned}$$

Hence $\rho(v(t), F(t, v(t)) = 0$. As $F(t, v(t))$ is closed, $v(t) \in F(t, v(t))$. Also the mapping $v : A_1 \to X$ being the pointwise limit of measurable mappings $\{v_n\}$, is measurable.

Now, we consider the case for A_2. For each $t \in A_2$ and every $x, y \in X$, we have

$$h(F(t,x), F(t,y)) \leq k(t)\rho(x,y) = 0.$$

Thus we can set $F(t,x) = F_0(t)$ for all $t \in A_2$, $x \in X$, where $F_0 : A_2 \to CB(X)$ is measurable. Hence again by Kuratowski and Ryll-Nardzewski (1965), we obtain a measurable selection $w : A_2 \to X$ of F_0. Then for any $t \in A_2$, $w(t) \in F_0(t) = F(t, w(t))$. We now define a mapping $U : T \to X$ by

$$U(t) = \begin{cases} v(t), & \text{if } t \in A_1 \\ w(t), & \text{if } t \in A_2. \end{cases}$$

It is clear that U is measurable and $U(t) \in F(t, w(t))$. In otherwords, U is a random fixed point of F. □

Chapter 3

Some Fixed Point Theorems in Partially Ordered Sets

3.1 Fixed Point Theorems and Applications to Economics

Partially Ordered Set

A nonempty set P with a partial order relation $\leq$ (reflexive, antisymmetric and transitive) is called a partially ordered set and is denoted by $(P, \leq)$. A subset C of $(P, \leq)$ is called a chain if, given any two elements x, y of C, either $x \leq y$ or $y \leq x$. An element $x \in P$ is called a maximal element if $x \leq y \Rightarrow x = y$.

Let A be a nonempty subset of $(P, \leq)$. An element $x \in P$ is called a lower (resp. an upper) bound of A if $x \leq a$ (resp. $a \leq x$) for all $a \in A$. A lower (resp. an upper) bound of A is called infimum (resp. supremum) of A if it is greater than or equal to (resp. less than or equal to) each lower bound (resp. each upper bound of A). If the supremum or infimum of A exists, it is unique (see Simmons (1963, pp. 43–44)) and is, respectively denoted sup$A$ and inf$A$.

In Section 3.2 we have obtained slightly generalized versions of Knaster-Tarski and Tarski-Kantorovitch theorems and a set-valued analogue of Knaster-Tarski theorem. In Section 3.3 we have applied Knaster-Tarski theorem in proving the existence of Nash equilibrium point of games and equilibrium point of economy. Finally we have discussed about the Pareto optimum point of both games and economy.

3.2 Fixed Point Theorem on Partially Ordered Sets

Theorem 3.1 (Generalized Knaster-Tarski Theorem) *Let $\{\leq_\alpha, I \in I\}$ be a family of preorder relations (reflexive and transitive) defined on a nonempty set P, where I is an index set. Let $T : P \to P$ be isotone with respect to each $\leq_\alpha$, i.e. $T(x) \leq_\alpha T(y)$ whenever $x \leq_\alpha y$. Assume that there exists $x_0 \in P$ such that $x_0 \leq_\alpha T(x_0)$ for each $\alpha \in I$. Assume that the relation $\leq$ defined by $x \leq y$ if and only if $x \leq_\alpha y$ for each $\alpha \in I$ is antisymmetric. Further assume that for every chain in $\{x \in P : x_0 \leq x\}$ has a suppremum. Then the set of fixed points of T is nonempty, which includes a maximal element of P.*

Proof. Define a partial order relation on P by $x \leq y$ if and only if $x \leq_\alpha y$ for each $\alpha \in I$. By hypothesis $\leq$ is a partial order on P, and T is an isotone with respect to $\leq$ and every chain $\{x \in P : x_0 \leq x\}$ has an upper bound. Consider the partial order set $\hat{P} = \{x : x \leq T(x)\} \cap \{x : x_0 \leq x\}$. Then $P \neq \emptyset$ as $x_0 \in \hat{P}$. Now we can easily see that every chain C in $(\hat{P}, \leq)$ has a supremum. Clearly C is also a chain in $\{x \in P : x_0 \leq x\}$ and has thus a supremum u in P, i.e. $x \leq u$ for all $x \in C$. Now since T is isotone with respect to $\leq$, $x \leq T(x) \leq T(u)$ for all x in C, which imply that $T(u)$ is an upper bound of C and $u \in \hat{P}$. Hence by Zorn's lemma $\hat{P}$ has a maximal element μ. Now $\mu \leq T(\mu)$ as $\mu \in \hat{P}$, and hence $T(\mu) \leq T(T\mu)$ as T is isotone. Thus $T(\mu) \in \hat{P}$. Hence by the maximality of $\mu, T(\mu) = \mu$ and the proof is complete. □

Example 3.1 Let (X, τ) be a Hausdorff uniform topological space, where the topology τ is genrated by the family $\{\rho_\alpha\}$ of psuedometrics. Let X^+ be the set $X \times \prod_{\alpha \in I} T_\alpha$, where $T_\alpha = [0, \infty)$ for each $\alpha \in I$. For each $\alpha \in I$ we define a preorder $\leq_\alpha$ in X^+ by: for $(x, \{a_\alpha\})$ and $(y, \{b_\alpha\})$ in X^+, $(x, \{a_\alpha\}) \leq_\alpha (y, \{b_\alpha\})$ if and only if $a_\alpha - b_\alpha \geq \rho_\alpha(x, y)$. Now we define the relation $\leq$ on X^+ by: $(x, \{a_\alpha\}) \leq (y, \{b_\alpha\})$ if and only if $(x, \{a_\alpha\}) \leq_\alpha (y, \{b_\alpha\})$ for each $\alpha \in I$. The antisymmetry of $\leq$ follows from its definition and the fact that X is Hausdorff.

Corollary 3.1.1 (Knaster-Tarski Theorem) (Tarski (1955); see also Knaster (1928)). *Let $(P, \leq)$ be a partially ordered set and $T : P \to P$ an isotone mapping. Assume that there exists an $x_0 \in P$ such that $x_0 \leq T(x_0)$ and that every chain in $\{x \in P : x_0 \leq x\}$ has a supremum. Then T has at least one fixed point which is also a maximal element in P.*

Proof. Take $I = \{1\}$ and apply Theorem 3.1 □

Our next theorem is a set-valued analogue of Knaster-Tarski Theorem.

Theorem 3.2 *Let $(P, \leq)$ be a partially ordered set and $T : P \to 2^P$ a set-valued mapping such that*

(i) *$T(x) \neq \emptyset$ for each $x \in P$;*
(ii) *T is isotone; i.e. $u \leq v$, for all u, v with $u \in T(x)$ and $v \in T(y)$ whenever $x \leq y$ and $x \neq y$.*
(iii) *there exists $x_0 \in P$ such that $x_0 \leq u$ for all $u \in T(x_0)$.*

Further assume that every chain in $\{x \in P : x_0 \leq x\}$ has a supremum.
Then there is an end point μ of T, i.e. $T(\mu) = \{\mu\}$, which is also a maximal element of P.

Proof. Let $\hat{P} = \{x \in P : x \leq u$ for all $u \in T(x)\} \cap \{x \in P : x_0 \leq x\}$. Then $\hat{P} \neq \emptyset$ as $x_0 \in \hat{P}$ and every chain C in $\hat{P}$ has an upper bound. To see this we first note that C is also a chain in $\{x \in P : x_0 \leq x\}$ and has, therefore, a supremum $\overline{x}$ in $\{x \in P : x_0 \leq x\}$. Thus $x_0 \leq \overline{x}$ and $c \leq \overline{x}$ for each $c \in C$. Now $c \leq u$ for all $u \in T(c)$ as $c \in \hat{P}$. Hence with $c \neq \overline{x}$, $c \leq u \leq v$ for all $v \in T(\overline{x})$ as $u \in T(c)$ and $v \in T(\overline{x})$, $c \leq \overline{x}$ and T is isotone. Thus each $v \in T(\overline{x})$ is an upper bound of C and hence $\overline{x} \leq v$ for each $v \in T(\overline{x})$. Hence $\overline{x} \in \hat{P}$. Obviously $x_0 \leq \overline{x}$. Thus C has a supremum in $\hat{P}$. Hence by Zorn's lemma there is a maximal element μ in $\hat{P}$. Let $u \in T(\mu)$ be arbitrary. Since $\mu \in \hat{P}, \mu \leq u \in T(\mu)$. Since T is isotone, $u \leq v$ for all $v \in T(u)$ if $u \neq \mu$ and also we have $x_0 \leq \mu \leq u$. Thus $u \in \hat{P}$. Now by the maximality of μ, we have $\mu = u$. As u is arbitrary, we have $T(\mu) = \{u\}$. □

Theorem 3.3 *Let $(P, \leq)$ be a partially ordered set and $T : P \to 2^P$ be a set-valued mapping such that*

(*i*) *for each $x \in P, T(x)$ is nonempty and supremum of $T(x)$ exists and is contained in it;*
(*ii*) *there is a point $x_0 \in P$ such that $x_0 \leq \sup T(x_0)$.*
(*iii*) *T is isotone in the sense that $x \leq y \Rightarrow \sup T(x) \leq \sup T(y)$.*

Further assume that every chain in $\{x \in P : x_0 \leq x\}$ has a supremum.

Then there is a fixed point of T which is also a maximal element.

Proof. Let $\hat{P} = \{x \in P : x \leq \sup T(x)\} \cap \{x \in P : x_0 \leq x\}$. Then $\hat{P}$ is nonempty as $x_0 \in \hat{P}$. Let C be a chain in $\hat{P}$. Then clearly C is a chain in $\{x \in P : x_0 \leq x\}$ and has a supremum $\overline{x}$, say. Thus $c \leq \overline{x}$ for all $c \in C$. Then as $c \in \hat{P}$ and by virtue of isotone property of $T, c \leq \sup T(c) \leq \sup T(\overline{x})$ for all $c \in C$. Thus $\overline{x} \leq \sup T(\overline{x}) = \overline{u}$, say. Hence $\overline{x} \in \hat{P}$ as obviously $x_0 \leq \overline{x}$. Thus C has a supremum in $\hat{P}$. Hence by Zorn's lemma there is a maximal element μ in $\hat{P}$. As $\mu \in \hat{P}$, we have $\mu \leq \sup T(\mu) = \overline{v}$, say. Hence $\mu \leq \sup T(\mu) \leq \sup T(\overline{v})$. Thus $\overline{v} \leq \sup T(\overline{v})$. This implies $\overline{v} \in \hat{P}$. Now by the maximality, $\mu = \overline{v} = \sup T(\mu) \in T(\mu)$. □

Definition 3.1 Let $(P, \leq)$ be a partially ordered set. Then a mapping $T : P \to P$ is said to be continuous if, for each countable chain $\{x_i\}$ in P such that supremum of $\{x_i\} = \sup x_i$ exists, we have $T(\sup\{x_i\}) = \sup\{T(x_i)\}$. Note that a continuous mapping $T : P \to P$ is isotone. Indeed, if $x \leq y$, then $y = \sup\{x, y\}$ and thus by continuity of $T, T(y) = T(\sup\{x, y\}) = \sup\{T(x), T(y)\}$. Hence $T(x) \leq T(y)$.

Theorem 3.4 (Generalized Tarski-Kantorovitch Theorem (Kantorovitch (1939))) *Let $\{\underset{\alpha}{\leq}, \alpha \in I\}$ be a family of preorder relations defined in a nonempty set P, where I is an index set. Let $T : P \to P$ be continuous with respect to $\leq$, where $\leq$ is defined on P by $u \leq v$ if and only if $u \leq_\alpha v$ for each α and $\leq$ is antisymmetric. Assume that*

(*a*) *there exists $x_0 \in P$ such that $x_0 \leq_\alpha T_\alpha(x_0)$ for each $\alpha \in I$;*
(*b*) *each countable chain $\{x \in P : x_0 \leq x\}$ has a supremum in P.*

Then the set of fixed points of T is nonempty. Furthermore, $\mu = \sup_n T^n(x_0)$ with respect to $\leq$ is a fixed point of T and μ is the infimum of the set of fixed points of T in $\{x : x_0 \leq x\}$.

Proof. We define $\leq$ in P as in the statement of the theorem by $x \leq y$ if and only if $x \leq_\alpha y$ for each $\alpha \in I$. Clearly $\leq$ is partial order relation as $\leq$ is antisymmetric. Also T is isotone by continuity of T and $x_0 \leq T(x_0)$. By induction we have $T^n(x_0) \leq T^{n+1}(x_0)$ for each $n \geq 0$. Hence $\{T^n(x_0) : n = 1, 2, \dots\}$ is a countable chain in $\{x : x_0 \leq x\}$ and $\sup T^n(x_0) = \mu$ exists in P. Now as $T : (P, \leq) \to (P, \leq)$ is continuous, $T(\mu) = \sup T(T^n(x_0)) = \sup T^{(n+1)}(x_0) = \mu$. Thus μ is a fixed point.

Now we prove that μ is the infimum of fixed points of T in $\{x : x_0 \leq x\}$. Let $\overline{\mu}$ be another fixed point in $\{x : x_0 \leq x\}$. Since $x_0 \leq \overline{\mu}, T(x_0) \leq T(\overline{\mu}) = \overline{\mu}$. By induction $T^n(x_0) \leq \overline{\mu}$ for each n. Hence $\overline{\mu}$ is an upper bound of $\{T^n(x_0) : n = 1, 2, \dots\}$. Therefore, we conclude that $\mu \leq \overline{\mu}$. Thus $\mu = \inf(\overline{\mu} : \overline{\mu}$ is a fixed point of T in $\{x : x_0 \leq x\}$. □

Corollary 3.4.1 (Tarski-Kantorovitch Theorem) *Let $(P, \leq)$ be a partially ordered set and $T : P \to P$ a continuous mapping. Assume that there is an $x_0 \in P$ such that $x_0 \leq T(x_0)$ and that each countable chain in $\{x \in P : x_0 \leq x\}$ has a supremum.*
Then the conclusion of the above theorem holds.

Proof. Take $I = \{1\}$ and apply the above theorem. □

3.3 Applications to Games and Economics

Let I be a finite or an infinite (countable or uncountable) set of players or economic agents. For each $i \in I$, let $X_i \neq \emptyset$ be the strategy or choice set, $A_i : X = \prod_{I \in I} X_i \to X_i$ the i-th constraint correspondence and $U_i : X \to \mathbb{R}$ the i-th utility or payoff function. Following Debreu (1959), Shafer and Sonnenschein (1975), Mas-Colell (1974), Yannelis (1987), Gale and Mas-Collel (1975), Florenzano (1983), Khan and Yannelis (1991), Tarafdar (1991) and Tarafdar (1992) and many others, we will describe an abstract economy by $\{(X_i, A_i, U_i) : i \in I\}$. For each $i \in I$, let $P_i : X \to 2^{X_i}$ be the i-th preference correspondence. The relation between U_i and P_i can be exhibited by the following definition:

$$P_i(x) = \{y_i \in X_i : U_i(y_i, x_{-i}) > U_i(x)\},$$

where x_{-i} is the projection of x onto $X_{-i} = \Pi_{j \neq i} X_j$ and (y_i, x_{-i}) is the point of X whose i-th co-ordinate is y_i (see Tarafdar (1991) and Tarafdar (1992).

Given an economy $\mathcal{E} = \{(X, A_i, U_i) : i = I\}$, we can pass onto the economy $\{(X, A_i, P_i) : i \in I\}$. A point $\overline{x} \in X$ is called an equilibrium point or generalized Nash equilibrium point of the economy $\mathcal{E} = \{(X, A_i, U_i) : i \in I\}$ if for each $i \in I$,

$$U_i(\overline{x}) = U_i[\overline{x}_i, \overline{x}_{-i}] = \sup_{z_i \in A_i(\overline{x})} U_i[z_i, \overline{x}_{-i}].$$

When for each $i \in I, A_i(x) = X_i$ for each $x \in X$, the economy reduces to a game and the equilibrium point to a Nash equilibrium point of the game introduced by Nash in (Nash (1950)). Thus the point $\overline{x} \in X$ is a Nash equilibrium point of the game $\{(X, U_i) : i \in I\}$ if for each $i \in I$,

$$U_i(\overline{x}) = U_i[\overline{x}_i, \overline{x}_{-i}] = \sup_{z_i \in X_i} U_i[z_i, \overline{x}_{-i}].$$

Thus the equilibrium point of an economy is the natural extension of the Nash equilibrium point.

3.3.1 *Game*

Let us first concentrate on a game $\{(X, U_i) : i \in I\}$. We want to apply Knaster-Tarski theorem to prove the existence of a Nash equilibrium point of the game $\mathcal{G} = \{(X, U_i) : i \in I\}$. To this end for each $i \in I$, if we define an order $\leq_i$ in X by:

$$x, y \in X, x \leq_i y \Leftrightarrow U_i(x) \leq U_i(y),$$

then $\leq_i$ fails to be antisymmetric because of the level sets $U_i^{-1}(r) = \{x \subset X : U_i(x) = r, r \text{ is a real number}\}$. The elements in $U_i^{-1}(r)$ are called indifferent (Debreu (1959, p. 54)).

We overcome this difficulty in the following ways: For each $i \in I$, we consider the set $X(i)$ of disjoint classes $[x_t]_i = \{x \in X : U_i(x) = t\}$. Clearly $X = X(i) = \cup_{t \in \mathbb{R}}[x_t]_i$.

Now for each $i \in I$, we define an order $\leq_i$ on $X(i)$ by $[x_t]_i \leq_i [x_{t'}]_i \Leftrightarrow t \leq t'$. In otherwords, $U_i^{-1}(t) \leq_i U_i^{-1}(t') \Leftrightarrow t \leq t'$. It is trivial to see that $(X(i), \leq_i)$ is a partially ordered set.

We now define partial order $\leq$ on X. Let $x \in X$ be arbitrary. Then for each $i \in I, x$ will belong to one class $[x_t]_i = \{x \in X : U_i^{-1}[U_i(x)]\}$ with $t = U_i(x)$. We now set $[x] = \{[x_t]_i : i \in I\}$ and $t = U_i(x)$. Given $x, y \in X$, we define

$$[x] \leq [y] \Leftrightarrow [x_t]_i = \{x \in X : U_i^{-1}[U_i(x)]\}$$

$$\leq_i [x_{t'}]_i = \{x \in X : U_i^{-1}[U_i(y)]\} \Leftrightarrow t = U_i(x) \leq t' = U_i(y),$$

for all $i \in I$.

Now we make the following assumptions:

(A1) For each $x \in X$ and each $i \in I, m_i(x) = \sup_{z_i \in X_i} U_i[z_i, x_{-i}]$ exists and is attained. Let $[\hat{x}]_i$ denote the corresponding class $U_i^{-1}(m_i(x))$.

(A2) For each $x \in X$, there exists $\tilde{x} \in X$ such that $U_i(\tilde{x}) = m_i(x)$ for all $i \in I$.

(A3) We assume that given x and y,

$$[x] \leq [y] \Rightarrow [\tilde{x}] \leq [\tilde{y}]$$

i.e. $U_i(x) \leq U_i(y)$ for each $i \in I \Rightarrow m_i(x) \leq m_i(y)$ for each $i \in I$.

Let us consider the partially ordered set $P = \{[x] : x \in X\}$, and make the final assumption.

(A4) Every chain on $\{[x] \in P : [x_0] \leq [x]\}$ has a supremum for some $x_0 \in X$.

Theorem 3.5 *Under the assumptions (A1) to (A4), the game $\mathcal{G}$ has a Nash equilibrium.*

Proof. Consider the mapping $T : P \to P$ defined by $T([x]) = [\tilde{x}], x \in X$. Inview of assmption (A2), T is well defined. We can easily see from (A2) that $[x] \leq [\tilde{x}]$ for each $x \in X$. By assumption (A3) T is isotone. Note that to make (A2) meaningful (A1) is needed. Now by Knaster-Tarski Theorem (Corollary 3.1.1) T has a fixed point $[\overline{x}] \in P$, i.e. $[\overline{x}] = T[\overline{x}] = [\tilde{\overline{x}}]$ which is also a maximal element. Hence for each $i \in I, [\overline{x}]_i = [\tilde{\overline{x}}]_i \Leftrightarrow U_i[\overline{x}_i, \overline{x}_{-i}] = m_i(\overline{x}) = \sup_{z_i \in X_i} U_i[z_i, \overline{x}_{-i}]$. □

3.3.2 *Economy*

We will now consider the abstract economy $\mathcal{E} = \{(X_i, A_i, U_i) : i \in I\}$. For each $i \in I$, we consider the same partial order $\leq_i$ on $X(i)$ and the order $\leq$ on X as defined in Subsection 3.3.1. Also for each $x \in X$, we define $[x]$ as in Subsection 3.3.1. Then we make the following assumptions:

$(A1)'$ For each $x \in X$ and each $i \in I, m_{i,A_i}(x) = \sup_{z_i \in A_i(x)} U_i[z_i, x_{-i}]$ exists and is attained.

$(A2)'$ For each $x \in X$, there exists $\tilde{x} \in X$ such that for each $i \in I, U_i(\tilde{x}) = m_{i,A_i}(x)$.

$(A3)'$ Given x and $y, [x] \leq [y] \Rightarrow [\tilde{x}] \leq_i [\tilde{y}]_i$ i.e. $U_i(x) \leq U_i(y)$ for each $i \in I \Rightarrow m_{i,A_i}(x) \leq m_{i,A_i}(y)$ for each $i \in I$.

Now as before we consider the partially ordered set $P = \{[x] : x \in X\}$.

$(A4)'$ Every chain in $\{[x] \in P : [x_0] \leq [x]\}$ has a supremum for some $x_0 \in X$.

Theorem 3.6 *Under the assumptions $(A1)'$ to $(A4)'$ the economy $\mathcal{E}$ has an equilibrium point.*

Proof. Consider the mapping $T : P \to P$ defined by $T[x] = [\tilde{x}], x \in X$ and apply the same argument as that in the proof of Theorem 3.5. □

Remark 3.1 Note that the same symbols $\leq$ have been used to denote the partial order in X and in $\mathbb{R}$ but it will be clear from the context whenever it occurs.
The conditions $(A1)$ and $(A4)$ (resp. $(A1)'$ and $(A4)'$) will be guaranteed by assuming compactness on X. The conditions $(A3)$ and $(A3)'$ might lead to something new. Results with topological conditons are wellknown (e.g. see Fan (1966), and Browder (1968)).

3.3.3 *Pareto Optimum*

The terms, Pareto optimum, Pareto efficient, Pareto equilibrium are often found in articles on economics, games theory, psychology, investment theory, theory of finances, traffic problems, problems of human migration and other related areas. We will define Pareto point of the game $\mathcal{G}$ and economy $\mathcal{E}$ considered in the beginning of this section. First we see Pareto's own description of optimal ophelimity (commonly known as Pareto optimum). Pareto in his book, Manuel d' économie potitique Pareto (1909,1971, 1971, p. 261), described the economic equilibriuim and remarks that

"..., the members of a collectively enjoy maximum ophelimity in a certain position when it is impossible to find a way of moving from that position very slightly in such a manner that the ophelimity enjoyed by each of the individuals of that collectivity increases or decreases. That is to say, any small displacement in departing from that position necessarily has the effect of increasing the ophelimity which certain individuals enjoy, and decreasing that which others enjoy, of being agreeable to some and disagreeable to others."

Pareto equilibrium point or Pareto optimum described above will simply come to mean the maximal element with respect to an order which is reflexive and transitive, not necessarily antisymmetric. We first consider the game $\mathcal{G} : \{(X, U_i) : i \in I\}$ and $X = \Pi_{i \in I} X_i$. For each $i \in I$, we define an order $\leq_i$ by: given $x, y \in X, x \leq_i y \Leftrightarrow U_i(x) \leq U_i(y)$ and an order $\leq$ on X by: given $x, y \in X, x \leq y \Leftrightarrow x \leq_i y$ for each $i \in I$.

A maximal element $\overline{x} \in X$ with respect to $\leq$ in X is said to be a Pareto Optimum Point. It is clear that at any other point some players will loose or some others will gain in their profit.

Next, we consider the economy $\mathcal{E} = \{(X_i, A_i, U_i) : i \in I\}$. We assume at the outset that for each $i \in I$ and each $x \in X, m_{i,A_i}(x) = \sup_{z_i \in A_i(x)} U_i(z_i, x_{-i})$ exists and is attained. Now define the fuction h by $h(x) = m_{1,A_1}(x) + m_{2,A_2}(x) + \cdots + m_{n,A_n}(x)$ for all $x \in X$.

If I is finite, say, $I = \{1, 2, \ldots, n\}$, then we have the following propositions.

Proposition 3.1 *If for the game $\mathcal{G}$, the function $f(x) = u_1(x) + u_2(x) + \cdots + u_n(x), x \in X$ has and attains its maximum over X at $\overline{x} \in X$, then $\overline{x}$ is a Pareto optimum of $\mathcal{G}$.*

Proof. If $\overline{x}$ is not a maximal, then $\overline{x} \leq y$ for some $y \in X$ and $x \neq y$, i.e. $\overline{x} \leq_i y$ for each i and $\overline{x} <_i y$, i.e. $U_i(\overline{x}) < U_i(y)$ for some i. This means $f(\overline{x}) < f(y)$ which is a contradiction. □

Let $I = \{1, 2, \ldots, n\}$ and for each $i \in I$, $\hat{U}_i(x) = \sup_{z_i \in X_i} U_i(z_i, x_i)$i exists and is attained. Let for each $x \in X$, $g(x) = \hat{U}_1(x) + \hat{U}_2(x) + \cdots + \hat{U}_n(x)$.Then we have the have the following proposition.

Proposition 3.2 *If for a game $\mathcal{G}$, f attains maximum at some point $\overline{x}$ and $f(\overline{x}) = g(\overline{x})$, then $\overline{x}$ is a Pareto optimum and equilibrium point of $\mathcal{G}$.*

Proof. For all $i = 1, 2, \ldots, n$, $U_i(x) \leq \hat{U}_i(x)$. The propostion follows from this observation. □

Proposition 3.3 *Given an economy $\mathcal{E}$, if f attains maximum at some point $\overline{x}$ and $f(\overline{x}) = h(\overline{x})$, then $\overline{x}$ is both a Pareto optimum and an equilibrium point of $\mathcal{E}$.*

Proof. That $\overline{x}$ is a Pareto optimum follows from the same argument as in Proposition 3.3. For each $i = 1, 2, \ldots, n$, $U_i(x) \leq m_{i,A_i}(x)$ by definition of $m_{i,A_i}(x)$. Hence the proposition follows by virtue of this observation. □

Remark 3.2 At the end we would like to record the following idea to be pursued later: Let f attains its maximum at two points $\overline{x}$ and $\overline{y}$, then the Pareto optimum $\overline{x}$ is said to be more socially acceptable than the Pareto optmum $\overline{y}$ if $\min\{U_i(\overline{y}) : i = 1, 2, \ldots, n\} < \min\{U_i(\overline{x}) : i = 1, 2, \ldots, n\}$ and $\max\{U_i(\overline{y}) : i = 1, 2, \ldots, n\} > \max\{U_i(\overline{x}) : i = 1, 2, \ldots, n\}$.

For the Pareto optimum and economic equilibrium for private exchange economy see Debreu (1959) and Tarafdar (1996b), Tarafdar (1996a), Tarafdar (1995b) and Tarafdar (1995a).

3.3.4 *The Contraction Mapping Principle in Uniform Space via Kleene's Fixed Point Theorem*

We will now consider Kleene's fixed point theorem wich is very much similar to Theorem 3.4 in the sense that every countable chain is replaced by every increasing sequence. We can easily verify that in a partially orderd set the statement that every countable chain has a supremum is equivalent to the statement that every increasing sequence has a supremum. First we point out some direction of applicability of Kleene's fixed point theorem. In every aspect of nonlinear analysis, including the denotational semantics of programming languages fixed point theorems are found to be an indispensable tool. Partially ordered sets and Kleene's fixed point theorem on such sets are used in a great variety of semantic model (Bakker (1980) and Stoy (1977)). In the recent past, attempts have been made to replace partially ordered sets by metric spaces and Kleen's fixed point theorem by contraction mapping principle (Bakker and Zucker (1983) and Hanes and Arbib (1986)). On the other hand Baranga (Baranga (1991)) has obtained contraction mapping principle in metric space from Kleene's fixed point theorem. In what follows we have deduced the contraction mapping principle in uniform space via Kleene's fixed point theorem.

Definition 3.2 Let $(X, \leq)$ be a partially ordered set, where $\leq$ is a partial order (reflexive, antisymmetric and transitive) relation on the nonempty set X. For an increasing sequence $\{x_n\}_{n\in\mathbb{N}}$ of $(X, \leq)$ we will denote by $\vee\{x_n : n \in \mathbb{N}\}$ the supremum of $\{x_n\}$, if it exists. $(X, \leq)$ is ω-complete if each increasing sequence $\{x_n\}_{n\in\mathbb{N}}$ in X has a supremum in X.

A mapping $f : (X, \leq) \to (Y, \leq)$ of a partially ordered set $(X, \leq)$ into a partially ordered set $(Y, \leq)$ is said to be ω-continuous if for each increasing sequence $\{x_n\}_{n\in\mathbb{N}}$ in X such that $\vee\{x_n : n \in \mathbb{N}\}$ exists, also $\vee\{f(x_n) : n \in \mathbb{N}\}$ exists in Y and $f(\vee\{x_n : n \in \mathbb{N}\}) = \vee\{f(x_n) : n \in \mathbb{N}\}$. We can easily see that on ω-continuous mapping is increasing. We now state Kleene's fixed theorem.

Theorem 3.7 *Let $(X, \leq)$ be an ω-complete partially ordered set and $f : X \to X$ an ω-continuous mapping. If $x_0 \in X$ is such that $x_0 \leq f(x_0)$, then $x_0^+ = \vee\{f^n(x_0) : n \in \mathbb{N}\}$ has the following properties:*

(1) $f(x_0^+) = x_0^+$;
(2) $x_0^+ \geq x_0$ *and for each* $y \in X$, *if* $y \geq x_0$ *and* $f(y) \leq y$, *then* $x_0^+ \leq y$ *(i.e.* x_0^+ *is the least fixed point in the set* $\{y \in X : x_0 \leq y\}$.

Proof. For convenience let $x_0 = x$. Define inductively

$$x_1 = f(x), x_2 = f(x_1), \ldots, x_n = f(x_{n-1}), \ldots.$$

Then under given conditions it follows from $x \leq f(x)$ that the sequence $\{x_n\}$ is inceasing. Thus we have from the ω-continuity that

$$x_0^+ = \vee\{x_n\} = f(\vee\{x_{n-1}\}) = f(x_0^+).$$

To prove the second part we need only to note that by the given condition it follows that for each $n \in \mathbb{N}$, $x_n \leq y$, where x_n is as defined above. Hence $x_0^+ \leq y$. □

Corollary 3.7.1 *Let $(X, \leq)$ be an ω-complete partially ordered set and $f : X \to X$ an ω-continuous mapping. If $x_0 \in X$ is such that $y_0 = f^{N-1}(x_0) \leq f^N(x_0)$ for some positive integer N, then $y_0^+ = \vee\{f^n(y_0) : n \in \mathbb{N}\}$ has the following properties:*

(1) $f(y_0^+) = y_0^+$;
(2) $y_0^+ \geq y_0$ *and for each* $y \in X$, *if* $y \geq y_0$ *and* $f(y) \leq y$,

then $y_0^+ \leq y$ *(that is,* y_0^+ *is the least fixed point in the set* $\{y \in X : y_0 \leq y\}$.

Proof. Since $y_0 = T^{N-1}(x_0) \leq T^N(x_0) = T(y_0)$, the corollary follows from Theorem 3.7. □

This corollary bears an analogy with Remark 2.2. Indeed the Contraction mapping principle are closely related in that the contraction mapping principle in a complete Hausdorff uniform space and hence in a complete metric space can be derived from Kleene's fixed point theorem. This has been accomplished in what follows next.

Let (X, τ) be a uniform Hausdorff topological space where the uniform topology τ is generated by the family $\{\rho_\alpha : \alpha \in I\}$ of pseudometrics, I being an index set. Let X^+ denote the Cartesian product $X \times \prod_{\alpha\in I} T_\alpha$, where $T_\alpha = [0, \infty)$ for each $\alpha \in I$. Let us consider the relation $\leq$ in X^+ defined by: for $\{a_\alpha\}$ and $\{b_\alpha\} \in \prod_{\alpha\in I} T_\alpha$, $(x, \{a_\alpha\}) \leq (y, \{b_\alpha\})$ if and only if $a_\alpha - b_\alpha \geq \rho_\alpha(x, y)$ for each $\alpha \in I$.

Clearly $\leq$ is reflexive and transitive. Furthermore $\leq$ is antisymmetric. Indeed if $(x, \{a_\alpha\}) \leq (y, \{b_\alpha\})$ and $(y, \{b_\alpha\}) \leq (x, \{a_\alpha\})$, then $a_\alpha - b_\alpha \geq \rho_\alpha(x, y)$ and $b_\alpha - a_\alpha \geq \rho_\alpha(y, x)$, i.e. $\rho_\alpha(x, y) = 0$ for each $\alpha \in I$. Since X is Hausdorff, it follows that $x = y$. Then it follows that $a_\alpha = b_\alpha$ for each $\alpha \in I$. Thus $(X^+, \leq)$ is a partially ordered set.

Theorem 3.8 *Let $\{(x_n, \{k_{n,\alpha}\})\}_{n\in\mathbb{N}}$ be an increasing sequence in $(X^+, \leq)$. Then for each $\alpha \in I$,*

(*i*) *the sequence $\{k_{n,\alpha}\}_{n\in\mathbb{N}}$ is decreasing sequence;*

(*ii*) $\sum_{n=0}^{\infty} \rho_\alpha(x_n, x_{n+1})$ *converges;*

(*iii*) $\{x_n\}_{n\in\mathbb{N}}$ *is a ρ_α-Cauchy sequence in X.*

Proof. By definition of $\leq$ in $(X^+, \leq)$, for each $\alpha \in I$, $\{k_{n,\alpha}\}_{n\in\mathbb{N}}$ is a decreasing sequence in $[0, \infty)$ and hence converges. For each $\alpha \in I$, let $\lim_{n\to\infty} k_{\alpha,n} = k_\alpha$. For each $\alpha \in I$, $\sum_{i=0}^{n} \rho_\alpha(x_i, x_{i+1}) \leq k_{0,\alpha} - k_{(n+1),\alpha} \to k_{0,\alpha} - k_\alpha$ as $n \to \infty$. Now (ii) and (iii) are clear. □

Theorem 3.9 *Let $\{(x_n, \{k_{n,\alpha}\})\}_{n\in\mathbb{N}}$ be an increasing sequence in $(X^+, \leq)$. The least upper bound of this sequence exists if $\{x_n\}_{n\in\mathbb{N}}$ is τ-convergent. Furthermore, the least upper bound is $(x, \{k_\alpha\})$, where $x = \tau - \lim_{n\to\infty} x_n$ and for each $\alpha \in I$, $k_\alpha = \lim_{n\to\infty} k_{n,\alpha}$. If in addition, (X, τ) is complete and $(x', \{k'_\alpha\})$ is the least upper bound of $\{(x_n, \{k_{n,\alpha}\})\}_{n\in\mathbb{N}}$ then $\lim_{n\to\infty} x_n = x'$ and $\lim_{n\to\infty} k_{n,\alpha} = k'_\alpha$ for each $\alpha \in I$.*

Proof. First, let $x = \tau - \lim_{n\to\infty} x_n$ and $k_\alpha = \lim_{n\to\infty} k_{n,\alpha}$. Then for each $\alpha \in I$, $\rho_\alpha(x_n, x) \to 0$ as $n \to \infty$; $x_n \to x$ in ρ_α-topology. Now if m and n are two positive integers with $m \geq n$. Then for each $\alpha \in I$, $\rho_\alpha(x_n, x_m) \leq k_{n,\alpha} - k_{m,\alpha}$ as $\{(x_n, \{k_{n,\alpha}\})\}$ is increasing in $(X^+, \leq)$. Letting $m \to \infty$, we see that for each $\alpha \in I$, $\rho_\alpha(x_n, x) \leq k_{n,\alpha} - k_\alpha$. This implies that $(x, \{k_\alpha\})$ is an upper bound of $\{(x_n, \{k_{n,\alpha}\})\}_{n\in\mathbb{N}}$. Now let $(x', \{k'_\alpha\})$ be another upper bound of this sequence. Then it follows that $\rho_\alpha(x_n, x') \leq k_{n,\alpha} - k'_\alpha$ for each $\alpha \in I$. Letting $n \to \infty$, $\rho_\alpha(x, x') \leq k_\alpha - k'_\alpha$ for each $\alpha \in I$. Hence $(x, \{k_\alpha\}) \leq (x', \{k'_\alpha\})$. Thus $(x, \{k_\alpha\})$ is the least upper bound.

Next, let for each $\alpha \in I$, $k_\alpha = \lim_{n\to\infty} k_{n,\alpha}$ and $(x', \{k'_\alpha\})$ be the least upper bound of $\{(x_n, \{k_{n,\alpha}\})\}_{n\in\mathbb{N}}$ in $(X^+, \leq)$. Since (X, τ) is complete, by Theorem 3.8 (iii), $\{x_n\}$ converges to a point $x \in X$, i.e. $\tau - \lim_{n\to\infty} x_n = x$. Hence by what we have proved in the first part, $(x, \{k_\alpha\})$ is the least upper bound of $\{(x, \{k_{n,\alpha}\})\}_{n\in\mathbb{N}}$. Now by the antisymmetry of $\leq$, it follows that $(x, \{k_\alpha\}) = (x', \{k'_\alpha\})$. □

Corollary 3.9.1 *If (X, τ) is a complete Hausdorff uniform topological space, then $(X^+, \leq)$ is ω-complete.*

We recall that a mapping $f : (X, \tau) \to (X, \tau)$ of a uniform topological space (X, τ) into itself is called contraction if, for each $\alpha \in I$, there exists a constant λ_α with $0 < \lambda_\alpha < 1$ such that $\rho_\alpha(f(x), f(y)) \leq \lambda_\alpha \rho_\alpha(x, y)$ for all $x, y \in X$.

Let $f : (X, \tau) \to (X, \tau)$ be a contraction mapping of a uniform topological space (X, τ) into itself as defined above. We define $f^+ : X^+ \to X^+$ as

$$f^+(x, \{k_\alpha\}) = (f(x), \{\lambda_\alpha k_\alpha\}).$$

Theorem 3.10 *If (X, τ) is a complete Hausdorff uniform topological space, then f^+ is ω-continuous.*

Proof. Let $\{(x_n, \{k_{n,\alpha}\})\}_{n\in\mathbb{N}}$ be an increasing sequence in X^+ with $(x, \{k_\alpha\} = \vee\{(x_n, \{k_{n,\alpha}\}) : n \in \mathbb{N}\}$. For each $\alpha \in I$, $\rho_\alpha(f(x_n), f(x_{n+1})) \leq \lambda_\alpha \rho_\alpha(x_n, x_{n+1}) \leq \lambda_\alpha(k_{n,\alpha} - k_{n+1,\alpha})$ which implies that $(f(x_n), \{\lambda_\alpha k_{n,\alpha}\}) \leq (f(x_{n+1}), \{\lambda_\alpha k_{n+1,\alpha}\})$ in X^+, $\{(f(x_n), \{\lambda_\alpha k_{n,\alpha}\})\}_{n\in\mathbb{N}}$ is increasing in X^+. Thus $f^+(x_n, \{k_{n,\alpha}\}) \leq f^+(x_{n+1}, \{k_{n+1,\alpha}\})$, i.e. f^+ is increasing. Now since X is complete, f is continuous and f^+ is increasing, we obtain by applying Theorem 3.8,

$$\begin{aligned} f^+(x, \{k_\alpha\}) &= (f(x), \{\lambda_\alpha k_\alpha\}) = \vee\{(f(x_n), \{\lambda_\alpha k_{n,\alpha}\}) : n \in \mathbb{N}\} \qquad (3.1)\\ &= \vee\{f^+(x_n, \{k_{n,\alpha}\}) : n \in \mathbb{N}\} \end{aligned}$$

where $\tau - \lim_{n\to\infty} x_n = x$ and $\lim_{n\to\infty} k_{n,\alpha} = k_\alpha$ for each $\alpha \in I$. □

The following Banach contraction mapping principle in a complete Hausdorff uniform topological space has been obtained in (Tarafdar (1974), Theorem 2.1) and will now be deduced from Kleene's fixed point theorem.

Theorem 3.11 (Banach contraction mapping principle) *Let (X, τ) be a complete Hausdorff uniform topological space and $f : X \to X$ a contraction mapping. Then f has a unique fixed point x^* and $x^* = \tau - \lim_{n\to\infty} f^n(x_0)$ for each $x_0 \in X$.*

Proof. Let $(X^+, \leq)$ and $f^+ : X^+ \to X^+$ be as defined as above. Let $x_0 \in X$ be arbitrary but fixed. Then for each $\alpha \in I$, we can find a positive real number a_α such that $(1 - \lambda_\alpha)a_\alpha > \rho_\alpha(x_0, f(x_0))$ (note that if $\rho_\alpha(x_0, f(x_0)) = 0$, then any $a_\alpha > 0$ will do). Hence $(x_0, \{a_\alpha\}) \leq (f(x_0), \{\lambda_\alpha a_\alpha\})$, i.e. $(x_0, \{a_\alpha\}) \leq f^+(x_0, \{a_\alpha\})$.

Thus $f^+ : X^+ \to X^+$ satisfies all the conditions of Kleene's theorem. Hence by Kleenes theorem $(\bar{x}, 0) == \vee\{(f^+)^n(x_0, \{a_\alpha\}) : n \in \mathbb{N}\}$ is a fixed point of f^+. Hence $\bar{x} = \lim_{n\to\infty} f^n(x_0)$ is a fixed point of f. Furthermore, $f^+(y, \{b_\alpha\}) \leq (y, \{b_\alpha\})$, i.e. $(f(y)\{\lambda_\alpha b_\alpha\}) \leq (y, \{b_\alpha\})$ if and only if $b_\alpha = 0$ and $\rho_\alpha(f(y), y) = 0$ for each $\alpha \in I$, if and only if $b_\alpha = 0$ for each α and $f(y) = y$ as X is Hausdorff. If $y \in X$ is a fixed point of f, we can find for each $\alpha \in I$ $\hat{a}_\alpha > 0$ such that $(x_0, \{\hat{a}_\alpha\}) \leq (y, \{0\})$ and $(x_0, \{\hat{a}_\alpha\}) \leq f^+(x_0, \{\hat{a}_\alpha\})$. Hence by the least fixed point property it follows that $(\bar{x}, \{0\}) \leq (y, \{0\})$ which implies that $\bar{x} = y$. Thus $\bar{x}$ is the unique fixed point of f. □

3.3.5 *Applications on Double Ranked Sequence*

For the following result we refer to Bakker (1980):

Let $(P, \leq)$ be a ω-complete partially ordered set. If $(x_{n,m})_{n,m\in\mathbb{N}}$ is a double ranked sequence in P satisfying

$$x_{n,m} \leq x_{n,m+1} \qquad \text{and} \tag{3.2}$$

$$x_{n,m} \leq x_{n+1,m} \qquad \text{for all } m, n \in \mathbb{N}, \tag{3.3}$$

then $(\{x_{n,m} : n \in \mathbb{N}\})_{m\in\mathbb{N}}$ and $(\vee\{x_{n,m} : n \in \mathbb{N}\} : m \in \mathbb{N}\})_{n\in\mathbb{N}}$ are increasing sequences in P and

$$\begin{aligned}\vee\{x_{n,m} : m \in \mathbb{N}\} : n \in \mathbb{N}\} &= \vee\{\vee\{x_{n,m} : n \in \mathbb{N}\} : m \in \mathbb{N}\} \\ &= \vee\{x_{k,k} : k \in \mathbb{N}\}.\end{aligned} \tag{3.4}$$

With help of this result we now obtain the following applications on double ranked sequence on uniform topological space similar to those obtained in Baranga (1991) on metric space.

Theorem 3.12 *Theorem on iterated limits Let (X, τ) be a complete uniform Hausdorff topological space as above and $\{x_{n,m}\}_{n,m\in\mathbb{N}}$ be a double ranked sequence of points of X such that for each $\alpha \in I$, there is a double ranked sequence $\{a_{n,m,\alpha}\}$ satisfying*

(*i*) $a_{n,m,\alpha} - a_{n,m+1,\alpha} \geq \rho_\alpha(x_{n,m}, x_{n,m+1})$ *and*
(*ii*) $a_{n,m,\alpha} - a_{n+1,m,\alpha} \geq \rho_\alpha(x_{n,m}, x_{n+1,m})$ *for all* $n, m \in \mathbb{N}$.

Then both iterated limits $\lim_{n\to\infty}\lim_{m\to\infty} x_{n,m} = \lim_{m\to\infty}\lim_{n\to\infty} x_{n,m}$ *exist and are equal to* $\lim_{k\to\infty} x_{kk}$.

Proof. By Corollary 3.9.1, $(X^+, \leq)$ is ω-complete, where $X^+ = X \times \prod_{\alpha\in I} T_\alpha$, $T_\alpha\,[0, \infty)$ for each $\alpha \in I$ and $\leq$ is as defined before. Then by (i) and (ii) we have respectively

$$x_{n,m} \leq x_{n,m+1} \tag{3.5}$$

and $$x_{n,m} \leq x_{n+1,m} \qquad \text{for all } n, m. \tag{3.6}$$

Hence the theorem follows from the above result of Bakker (1980). □

Corollary 3.12.1 *Let (X, τ) be a complete Hausdorff uniform topological space and $\{x_{n,m}\}_{n,m\in\mathbb{N}}$ be a double ranked sequence in X such that for each $\alpha \in I$ both series $\sum_{n=0}^{\infty}\sum_{m=0}^{\infty} \rho_\alpha(x_{n,m+1}, x_{n,m})$ and $\sum_{m=0}^{\infty}\sum_{n=0}^{\infty} \rho_\alpha(x_{n+1,m}, x_{n,m})$ converge. Then both iterated limits $\lim_{n\to\infty}\lim_{m\to\infty} x_{n,m}$ and $\lim_{m\to\infty}\lim_{n\to\infty} x_{n,m}$ exist and are equal to $\lim_{k\to\infty} x_{kk}$.*

Proof. For each $\alpha \in I$, define

$$a_{n,m,\alpha} = \sum_{i=n}^{\infty}\sum_{j=m}^{\infty} \rho_\alpha(x_{i,j+1}, x_{i,j}) + \sum_{j=m}^{\infty}\sum_{i=n}^{\infty} \rho_\alpha(x_{i+1,j}, x_{i,j}).$$

It is easy to verify that (i) and (ii) of Theorem 3.11 are satisfied. Hence the Corollary follows from Theorem 3.11. □

We should point out that taking $I = \{1\}$, we obtain all results of Baranga (1991) in metric spaces as corollaries.

3.4 Lattice Theoretical Fixed Point Theorems of Tarski

In this section we reproduce the proofs of lattce theoretical fixed point theorems of Tarski. A partially ordered set $(X, \leq)$ is said to be a lattice if for any two points $x, y \in A$ there is a supremum $x \vee y$ and an infimum $x \wedge y$ in X. A lattice $(X, \leq)$ is called complete if every subset A of X has a supremum $\vee A$ and an infimum $\wedge A$ in X. For a complete lattice we use the following notation: $1 = \sup X = \vee X$ and $0 = \inf X = \wedge X$.

Given two elements a and b with $a \leq b$, we denote the interval $[a, b] = \{x \in X : a \leq x \leq b\}$ by:

$$[a, b] = [x \in X : a \leq x \leq b].$$

$([a, b], \leq]$ is clearly a lattice which is also complete if X is complete.)

Theorem 3.13 (Tarski (1955)) *Let $(X, \leq)$ be a complete lattice and $f : X \to X$ an increasing mapping (isotone) (i.e. $x \leq y \Rightarrow f(x) \leq f(y)$). Then the set P of all fixed points of f is nonempty. Furthermore $(P, \leq)$ is a complete lattice, in particular*

$$\vee P = \vee\{x \in X : f(x) \geq x\} \in P \text{ and}$$
$$\wedge P = \wedge\{x \in X : f(x) \leq x\} \in P.$$

$\vee P$ and $\wedge P$ will respectively be called the maximal and the minimal fixed point.

Proof. Let

$$u = \vee\{x \in X : f(x) \geq x\}. \tag{3.7}$$

Then $x \leq u$ for all $x \in X$ with $f(x) \geq x$. Now since f is increasing, $x \leq f(x) \leq f(u)$ whenever $f(x) \geq x$. Hence by virtue of (3.7), we obtain

$$u \leq f(u). \tag{3.8}$$

Thus $f(u) \leq f(f(u))$. This implies that $f(u) \in \{x \in X : f(x) \geq x\}$ and hence by (3.7), we have

$$f(u) \leq u. \tag{3.9}$$

Now from (3.8) and (3.9)) u is a fixed point of f and also from (3.7) we conclude $u = \sup P = \vee P$, i.e.

$$\vee P = \vee\{x \in X : f(x) \geq x\} = u \in P.$$

Now let

$$v = \wedge\{x \in X : f(x) \leq x\}. \tag{3.10}$$

Then $v \leq x$ for all $x \in X$ with $f(x) \leq x$. Since f is increasing, $f(v) \leq f(x) \leq x$ whenever $f(x) \leq x$. Thus we obtain

$$f(v) \leq v. \tag{3.11}$$

Hence $f(f(v)) \leq f(v) \Rightarrow f(v) \in \{x \in X : f(x) \leq x\}$ and, therefore,

$$v \leq f(v). \tag{3.12}$$

Hence $v \in P$ and repeating the same argument as in the first part we obtain

$$\wedge P = \wedge\{x \in X : f(x) \leq x\} = v \in P. \tag{3.13}$$

It remains to prove that $(P, \leq)$ is a complete lattice. To this end it suffices to show that for every nonempty subset A of P, $\vee A \in P$ and $\wedge A \in P$. We first consider the interval $[\vee A, 1]$ which is a complete lattice. Now for any $x \in A$, $x \leq \vee A$ and hence $x = f(x) \leq f(\vee A)$ for each $x \in A$. Thus $\vee A \leq f(\vee A)$. Now for any y with $\vee A \leq y \leq 1$, we have $\vee A \leq f(\vee A) \leq f(y) \leq 1$. Thus the restriction f' of f to $[\vee A, 1]$ is an increasing mapping of $[\vee A, 1]$ to $[\vee A, 1]$. Hence applying (3.13) to the complete lattice $[\vee A, 1]$, $w =$ the infimum of all fixed points of f' is itself a fixed point of f'. Obviously w is a fixed point of f, and in fact is the least fixed point of f which is an upper bound of all elements of A. Thus it follows that $w = \vee A \in P$. Similarly we can prove that $\wedge A \in P$. □

Theorem 3.14 (Generalized Lattice-Theoretical Fixed Point Theorem) *Let $(X, \leq)$ be a complete lattice and $\mathcal{F}$ a commutative family of increasing mappings of X into itself. Then the set P of common fixed points, i.e. $\{x \in X : f(x) = x$ for all $f \in \mathcal{F}\}$ is nonempty. Furthermore $(P, \leq)$ is a complete lattice, in particular,*

$$\vee P = \vee\{x \in X : f(x) \geq x \textit{ for all } f \in \mathcal{F}\} \in P$$

and

$$\wedge P = \wedge\{x \in X : f(x) \leq x \textit{ for all } f \in \mathcal{F}\} \in P.$$

Proof. Let

$$u = \vee\{x \in X : f(x) \geq x \text{ for all } f \in \mathcal{F}\}. \tag{3.14}$$

Then as in the preceding theorem we can show that

$$u \leq f(u) \text{ for every } f \in \mathcal{F}. \tag{3.15}$$

Hence for all $g \in \mathcal{F}$, we have from (3.15)

$$g(u) \leq gf(u) \text{ as } g \text{ is increasing}$$

and thus by the commutativity of $\mathcal{F}$,

$$g(u) \leq f(g(u)) \text{ for all } f \in \mathcal{F}.$$

Therefore, $g(u) \in \{x \in X : f(x) \geq x \text{ for all } f \in \mathcal{F}\}$ and consequently,

$$g(u) \leq u.$$

Hence we have proved that

$$f(u) \leq u \text{ for all } f \in \mathcal{F} \tag{3.16}$$

(3.14)–(3.16) show that $u \in P$ and in fact u is the least upper bound of all such fixed points, i.e.

$$u = \vee P = \vee\{x \in X : f(x) \geq x \text{ for all } f \in \mathcal{F}\} \in P.$$

The proof of the remaining part is analogous to that of the preceding theorem and is left out. □

A lattice $(X, \leq)$ is said to be continuously or completely ordered if it is complete and densely ordered if, for all $x, y \in X$ with $x < y$, there exists $z \in X$ with $x < z < y$. Given a mapping $f : (X, \leq) \to (X, \leq)$ and $A \subset X$, let $f^*(A) = \{f(a) : f(a) \geq a$ and $a \in A\}$. A mapping $f : A \subset X \to B \subset X$ is called quasi-increasing if satisfies the formulas:

$$f(\vee Z) \geq \wedge f^*(Z) \text{ and } f(\wedge Z) \leq \vee(f^*(Z)$$

for every nonempty subet Z of A. It is called quasi-decreasing if it satisfies the formulas

$$f(\vee Z) \le \vee f^*(Z) \text{ and } f(\wedge Z) \ge \wedge f^*(Z).$$

A mapping is called continuous if it is both quasi-increasing and quasi-decreasing.

Theorem 3.15 *Let $(X, \le)$ be a continuously and densely ordered lattice and $f, g : X \to X$ be two mappings such that f is quasi-increasing, g is quasi-decreasing and $f(0) \ge g(0)$ and $f(1) \le g(1)$. Then the set $P = \{x \in X : f(x) = g(x)\}$ is nonempty and moreover, $(P, \le)$ is a continuous, ordered lattice; in particular,*

$$\vee P = \vee\{x \in X : f(x) \ge g(x)\} \in P$$

and

$$\wedge P = \wedge\{x \in X : f(x) \le g(x)\} \in P.$$

Proof. Let A be any subset of X such that

$$f(x) \ge g(x) \quad \text{for } x \in A. \tag{3.17}$$

We will prove that $f(\vee A) \ge g(\vee A)$.

If possible, let us assume that

$$f(\vee A) < g(\vee A). \tag{3.18}$$

Then the hypothesis $f(0) \ge g(0)$ implies that

$$\vee A \ne 0. \tag{3.19}$$

Also since X is densely ordered, (3.18) implies that there exists an element $z \in X$ such that

$$f(\vee A) < z < g(\vee A). \tag{3.20}$$

Let us consider the set

$$D = \{x \in X : x \le \vee A \text{ and } g(x) \le z\}. \tag{3.21}$$

Then we have

$$\vee D \le \vee A \tag{3.22}$$

and

$$\vee g^*(D) \le z. \tag{3.23}$$

Now we prove that equality does not hold in (3.22). For, if possible let $\vee D = \vee A$. Then from (3.19) $\vee D \neq 0$ and hence $D \neq \emptyset$. Since g is assumed to be quasi-decreasing, it follows that $g(\vee A) = g(\vee D) \leq \vee g^*(D)$ and hence by (3.23), $g(\vee A) \leq z$ which contradicts (3.20). Hence we have

$$\vee D < \vee A. \tag{3.24}$$

Next let

$$E = \{x \in A : \vee D < x\}. \tag{3.25}$$

Then E is nonempty. Indeed if $E = \emptyset$, then for every $x \in A$, $x \leq \vee D$ for all $x \in A$, which would imply $\vee A \leq \vee D$ contradicting (3.24). By (3.25) and density of X we can prove that $\vee E = \vee A$. Since, by assumption, f is quasi-increasing, we obtain

$$f(\vee A) = f(\vee E) \geq \wedge f^*(E)$$

and hence by virtue of (3.20)

$$z > \wedge f^*(E).$$

Therefore, we must have some $a \in E$ such that

$$z > f(a), \text{ for otherwise } z \leq \wedge f^*(E).$$

Thus by (3.25) and (3.17)

$$\vee D < a, \ a \in A = \text{ and } g(a) \leq z.$$

Hence by (3.21), $a \in D$.

The formulas $\vee D < a$ and $a \in D$ are impossible.

Thus we have proved that (3.18) does not hold for any subset A of X, i.e.

$$f(\vee A) \geq g(\vee A) \text{ for every nonempty } A = \{x \in X : f(x) \leq g(x)\}. \tag{3.26}$$

In a similar way we can prove that for every nonempty subset (empty or nonempty) of $B = \{x \in X : f(x) \leq g(x)\}$,

$$f(\wedge B) \leq g(\wedge B). \tag{3.27}$$

Now let Y be a subset (empty or nonempty) of

$$P = \{x \in X : f(x) = g(x)\}.$$

Let

$$u = \vee\{x \in X : f(x) \geq g(x) \text{ and } x \leq \wedge Y\}. \tag{3.28}$$

We note if $Y = \emptyset$, then (3.28) reduces to

$$u = \vee\{x \in X : f(x) \geq g(x)\}. \tag{3.29}$$

By (3.26) and (3.27) we obtain

$$f(u) \geq g(u) \text{ and } f(\wedge Y) \leq g(\wedge Y), \tag{3.30}$$

latter inequality follows from the fact that

$$Y = \{x \in X \cap Y : f(x) = g(x)\}.$$

If $u = \wedge Y$, then from (3.30) we obtain

$$f(u) = g(u), \quad \text{i.e. } u \in P. \tag{3.31}$$

If $u \neq \wedge Y$, then from (3.28) we have $u < \wedge Y$.

Since X is densely ordered, we can write

$$u = \wedge\{x \in X : u < x \leq \wedge Y\}. \tag{3.32}$$

Also from (3.28) we see that for every element x of the set

$$\{x \in X : u < x \leq \wedge Y\}, \quad f(x) < g(x).$$

Hence from (3.28) and (3.32) we have

$$f(u) \leq g(u).$$

Thus from (3.30) we conclude

$$f(u) = g(u), \quad \text{i.e. } u \in P.$$

Therefore we have proved that

for each subset Y of P, if $u = \vee\{x \in X : f(x) \geq g(x) \text{ and } x \leq \wedge Y\}$, then $u \in P$. (3.33)

Similarly we can prove that

for each subset Y of P, if $v = \wedge\{x \in X : f(x) \leq g(x) \text{ and } x \geq \vee Y\}$, then $v \in P$. (3.34)

From (3.33) it follows that u is the largest element, of P which is a lower bound of all elements of Y. Thus u is the greatest lower bound or inf of Y in $(P, \leq)$. Similarly from (3.34) it follows that v is the least upper bound or sup of Y in $(P, \leq)$. Hence

$$(P, \leq) \text{ is a continuously ordered system.} \tag{3.35}$$

Finally taking $Y = \emptyset$, (3.33) and (3.34) yield respectively

$$\vee P = \vee\{x \in X : f(x) \geq g(x)\} \in P \tag{3.36}$$

and

$$\wedge P = \wedge\{x \in X : f(x) \leq g(x)\} \in P. \tag{3.37}$$

The proof is complete. □

Every increasing (resp. decreasing) mapping is evidently quasi-increasing (resp. quasi-decreasing). The identity mapping, $g(x) = x$ for each $x \in X$ and the constant mapping, $g(x) = c \in X$, are continuous, i.e. both quasi-increasing and quasi-decreasing. Hence taking an arbitrary increasing mapping f and the identity mapping g in the above theorem, we obtain Theorem 3.13 in a continuously and densely ordered systems. On the otherhand by taking g to be the constant function $g(x) = c$ on X, we obtain:

Theorem 3.16 (Generalized Weierstrass Theorem) *Let $(X, \leq)$ be a continuously and densely defined system $f : X \to X$ a quasi-increasing mapping of X into itself, c an element of X such that*

$$f(0) \geq c \geq f(1).$$

Then the set $P = \{x \in X : f(x) = c\}$ is nonempty and $(P, \leq)$ is a continuously ordered system.

In particular,

$$\vee P = \vee\{x \in X : f(x) \geq c\} \in P$$

and

$$\wedge P = \wedge\{x \in X : f(x) \leq c\} \in P.$$

An analogous theorem for quasi-decreasing mapping can be obtained from Theorem 3.15 by taking an arbitrary constant mapping for f.

3.5 Applications of Lattice Fixed Point Theorem of Tarski to Integral Equations

In this section we will apply the lattice fixed point theorem of Tarski to integral equations which was originally considered in shendge and Joshi (1982).

Let $(X, \|.\|)$ be a real Banach space. A mapping $T : X \to X$ is said to be a nonlinear contraction if there exists a continuous non-decreasing real function $\varphi : \mathbb{R}^+ \to R^+$ with $\varphi(r) < r$ for $r > 0$ such that

$$\|T(x) - T(y)\| \leq \varphi(\|x - y\|), \text{ for } x, y \in x.$$

Let P be a cone in X. Then $(X, \leq)$ is a partially ordered set where $\leq$ is the partial order relation induced by P. Let P be a closed and bounded subset of X such that $(P, \leq)$ is a complete lattice with respect to the same relation $\leq$ defined in X.

Lemma 3.1 *If $T : X \to X$ is a nonlinear contraction, then $(I - T)^{-1}$ exists where I is the identity on X.*

Proof. Let $y \in X$ be arbitrary but fixed. Define $A : X \to X$ by $A(x) = y + T(x)$, $x \in X$. Then $\|A(u)) - A(v)\| = \|T(u) - T(v)\| \leq \varphi(\|u - v\|)$for all $u, v \in X$.

Hence by Corollary 2.10.2 with $I = \{1\}$, A has a unique fixed point x of A, i.e. $x = A(x) = y + T(x)$, i.e. $(I - T)(x) = y$. □

Theorem 3.17 *Let $A, B : P \to P$ be two mappings such that*

(*i*) *A is a nonlinear contraction,*
(*ii*) *$(I - A)^{-1}B$ is increasing (isotone) on P, where I is identity on P, and*
(*iii*) *$A(x) + B(y) \in P$, whenever $x, y, \in P$ Then the set S of the solutions of the equation*

$$Ax + Bx = x, x \in P \tag{3.38}$$

is nonempty and, furthermore $(S, \leq)$ is a complete lattice.
Note that $(I - A)^{-1}$ exists by Lemma 3.1.

Proof. Set $T = (I - A)^{-1}B$. Then it can be easily seen that (6.1) is equivalent to $T(x) = x$. Further, by (ii) T is increasing. We will now prove that T maps P into itself. Let $y \in P$ be arbitrary but fixed. We define a mapping $T_y : P \to P$ by $Ty(x) = A(x) + B(y)$, $x \in P$. By (iii) Ty maps P into itself.

Now for any $u, v \in P$,

$$\|T_y(u) - T_y(v)\| = \|A(u) - A(v)\| \leq \varphi(\|u - v\|).$$

As in Lemma 3.1 there is a unique fixed point $x \in P$ of Ty, i.e. $x = T_y(x) = A(x) + B(y)$, i.e. $(I - A)(x) = B(y)$, i.e. $x = (I - A)^{-1}B(y) = T(y)$. Now the theorem follows from Theorem 3.13. □

We now apply Theorem 3.17 to the following mixed type of nonlinear integral equations.

$$x(t) = h(t) + \int_0^t k_1(t, s)f(s, x(s))ds + \int_0^T k_2(t, s)g(s, x(s))ds \tag{3.39}$$

where $0 \leq s \leq t \leq T$.

Assuming appropriate conditions on the functions involved in (3.39) we will prove the existence of the minimal and maximal solutions of (3.39) by applying Theorem 3.17.

Let C denote the space of real-valued bounded functions defined on the interval $I = [0, T]$, which are also Lebesgue integrable. C is a Banach space with the norm defined by

$$\|x\| = \sup_{t \in I} |x(t)|, \qquad x \in C.$$

In C we define an order relation $\leq$ as follows:

for $x, y \in C, x \leq y$ if and only if $x(t) \leq y(t)$ for all $t \in I$. Then $(C, \leq)$ is a complete lattice Birchoff (1967).

We now make the following assumptions.

(1) $f(s,x)$ and $g(s,x)$ are real functions on $I \times \mathbb{R}$, and each is monotone non-decreasing in x for each $s \in I$, where $\mathbb{R}$ is the set of all real numbers. f satisfies Lipschitz condition with Lipschitz constant L.

(2) The functions $f(s, \varphi(s))$ and $g(s, \varphi(s))$ are Lebesque integrable for every Lebesque function φ on I.

(3) The functions h,f, g, k_1 and k_2 occurring in (3.39) are bounded for s, $t \in I$ and $|x| < \infty$, with bounds H_0, N_1, N_2, K_1 and K_2 respectively.

(4) The functions k_1 and k_2 are non-negative on $I \times I$.

Theorem 3.18 *Under the assumptions* (1) *to* (4) *the integral equation* (3.39) *has minimal and maximal solutions on I, if $LK_1T < 1$.*

Proof. Define the subset P of C by

$$P = \{x \in C : \|x\| \leq M\}$$

where $M = H_0 + N_1K_1T + N_2K_2T$. Obviously P is a closed bounded subset of C. Since $(C, \leq)$ is a Dedikind complete lattice (see (Birchoff (1967), p. 361), $(P, \leq)$ is also a complete lattice. Next, we define the mappings A and B by

$$A(x(t)) = h(t) + \int_0^t k_1(t,s)f(s,x(s))ds, \; t \in I$$

$$B(y(t)) = \int_0^T k_2(t,s)g(s,y(s))ds, \; t \in I.$$

By virtue of (3) it follows that $A(x) + B(y) \in P$, whenever x, $y \in P$. Indeed, for each $t \in I$,

$$\begin{aligned} |(A(x) + B(y))(t)| &\leq |h(t)| + \int_0^t |k_1(t, s)|\,|f(s, x(s))|\, ds \\ &\quad + \int_0^T |k_2(t, s)|\,|g(s, y(s))|\, ds \\ &\leq H_0 + N_1K_1T + N_2K_2T. \end{aligned}$$

Hence $\|A(x) + B(y)\| \leq H_0 + N_1K_1T + N_2K_2T$. Thus $A(x) + B(x) \in P$, whenever x, $y \in P$.

From the monotone nondecreasing character of f and g assumed in (1) and the nonnegative property of k_1 and k_2 assumed in (4), it readily follows that both A

and B are increasing (isotone) on P. Now we prove that A is a strict contraction. By using (1) we have

$$|A(x)t)) - A(y(t))| \leq L \int_0^t k_1(t,x)\,|x(s) - y(s)|\,ds$$

Hence by (3) $\|A(x) - A(y)\| \leq \lambda \|x - y\|$, where $\lambda = LK_1T < 1$. Finally, since A and B are both (isotone) increasing and A is a strict contraction, it follows that $(I - A)^{-1}B$ is (isotone) increasing. Thus all the conditions of (3.17) are satisfied. Hence the set S of solutions of the equation (3.39) is nonempty and is a complete lattice. Consequently $\wedge S=$ the minimal solution and $\vee S=$ maximal solution exist.

3.6 The Tarski-Kantorovitch Principle and Theory of Iterated Function Systems[2]

In Theorem 3.11 we have obtained the Banach cotraction mapping principle in a complete uniform topological space via Tarski-Katorovitch-Kleene fixed point theorem. On the other hand an Iterated Function System, IFS of Chapter 2 originally due to Hutchinson (1981) and Barnsley (1988) involves a finite set of contraction mappings, one naurally inclines to ask whether the results of Hutchinson-Barnley can be obtained in a complete partially ordered set $(X, \leq)$ with a finite set of appropriate self mappings. To this end we will present in this section some results of Jachymski, Gajek, and Pokarowski (2000). Let X be a set and $f_1, \ldots, f_n$ be selfmaps of X. The theory of iterated function systems (abbr., IFS) deals with the following *Hutchinson–Barnsley operator*:

$$F(A) := \bigcup_{i=1}^{n} f_i(A) \qquad \text{for } A \subset X. \tag{3.40}$$

We have already seen in a section of Chapter 2 that, if (X, ρ) is a complete metric space and all the maps f_i are Banach contractions, then F is the Banach contraction on the family $K(X)$ of all nonempty compact subsets of X, endowed with the Hausdorff metric. Consequently, F has then a unique fixed point A_0 in $K(X)$, which is called a fractal or an attractor in the sense of Barnsley. Moreover, for any set A in $K(X)$, the sequence $(F^n(A))_{n=1}^{\infty}$ of iterations of F converges to A_0 with respect to the Hausdorff metric. For an arbitrary IFS a set A_0 such that $A_0 = F(A_0)$ is called *invariant with respect to the IFS* $\{f_i : i = 1, \ldots, n\}$ (cf. Lasota–Myjak Lasota and Myjak (1996a). If $n = 1$, then such an A_0 is said to be a *modulus set* for the map f_1 (cf. Kuczma Kuczma (1968, p. 13)).

In this section we consider some applications of the Tarski–Kantorovitch fixed–point theorem (in abbreviation T-K theorem) in the theory of IFS (see

[2]The authors are grateful to Jacek Jachymski, Leslaw Gajek and Piotr Pokarowski for sending a computer file of a paper which was of cosiderable help in the prepartation of this section.

Corollary 3.4.1). So we will use the partial ordering technique to obtain results on fixed points of the Hutchinson–Barnsley operator. The idea of treating fractals as Tarski's fixed points appeared earlier in papers of Soto-Andrade and Varela (1984) and Hayashi (1985), however, they considered a version of Tarski's theorem other than that studying in this section. We also study the T–K principle for the following operator F, introduced by Lasota and Myjak (1996b),

$$F(A) := \mathrm{cl}\left(\bigcup_{i=1}^{n} f_i(A)\right) \qquad \text{for } A \subset X, \tag{3.41}$$

where cl denotes the closure operator. Using an idea of Williams Williams (1971), we show that, in such a case, the T–K theorem yields the Hutchinson–Barnsley theorem for a class of the *Heine–Borel metric spaces*, that is, spaces in which every closed and bounded set is compact (cf. Williamson–Janos Williamson and Janos (1987)). Given sets X and Y, and a map $f : X \mapsto Y$, the sets $f^{-1}(\{y\})$ $(y \in Y)$ are called *fibres* of f (cf. Engelking (1977, p. 14)).

In the rest of this section, we assume that a compact or countably compact space is Hausdorff by the definition.

Remark 3.3 As we have already noted in section 3.1 that the assumption "every countable chain has a supremum" is equivalent to "every increasing sequence (p_n) (that is, $p_n \le p_{n+1}$ for $n \in \mathbb{N}$) has a supremum". Similarly, in the definition of $\le$-continuity, we may substitute increasing sequences for countable chains. Such a reformulated Theorem is identical with the Kleene fixed–point theorem (cf., e.g., Baranga (1991)).

Lemma 3.2 *Let $(P, \le)$ be a partially ordered set, in which every countable chain has a supremum and such that for any $p, q \in P$ there exists an infimum* $\inf\{p, q\}$. *Assume that for any increasing sequences $(p_n)_{n=1}^{\infty}$ and $(q_n)_{n=1}^{\infty}$,*

$$\inf\left\{\sup_{n\in\mathbb{N}} p_n,\ \sup_{n\in\mathbb{N}} q_n\right\} = \sup_{n\in\mathbb{N}} \inf\{p_n, q_n\}. \tag{3.42}$$

Let $F_1, \dots, F_n$ be $\le$-continuous selfmaps of P and define a map F by

$$F(p) := \inf\{F_1(p), \dots, F_n(p)\} \qquad \textit{for } p \in P.$$

Then F is $\le$-continuous.

Proof. For the sake of simplicity, assume that $n = 2$; then an easy induction shows that our argument can be extended to the case of an arbitrary $n \in \mathbb{N}$. By Remark 3.3, it suffices to prove that given an increasing sequence (p_n), $F(p) = \sup_{n\in\mathbb{N}} F(p_n)$, where $p := \sup_{n\in\mathbb{N}} p_n$. Since F_1 and F_2 are increasing, so is F. Thus the sequence $(F(p_n))$ is increasing and by hypothesis, it has a supremum. Then, by

(2.3) and $\leq$-continuity of F_1 and F_2,

$$\sup_{n\in\mathbb{N}} F(p_n) = \sup_{n\in\mathbb{N}} \inf\{F_1(p_n), F_2(p_n)\} = \inf\{\sup_{n\in\mathbb{N}} F_1(p_n), \sup_{n\in\mathbb{N}} F_2(p_n)\}$$
$$= \inf\{F_1(\sup_{n\in\mathbb{N}} p_n), F_2(\sup_{n\in\mathbb{N}} p_n)\} = F(\sup_{n\in\mathbb{N}} p_n),$$

which proves the $\leq$-continuity of F. □

The following example which is also from Jachymski et al. (2000) shows that there exists a partially ordered set $(P, \leq)$, in which every countable chain has a supremum and for any $p, q \in P$ there exists $\inf\{p, q\}$, but condition (3.42) does not hold. In fact, the set $(P, \leq)$ defined below is a *complete lattice*, that is, every subset of P has a supremum and an infimum.

Example 3.2 Let $C(\mathbb{R})$ be the family of all nonempty closed subsets of the real line and $P := C(\mathbb{R}) \cup \{\emptyset\}$. Endow P with the inclusion $\subseteq$. If $\{A_t : t \in T\} \subseteq P$, then $\inf_{t\in T} A_t = \bigcap_{t\in T} A_t$ and $\sup_{t\in T} A_t = \mathrm{cl}\left(\bigcup_{t\in T} A_t\right)$. Define

$$A_n := [0, 1 - \frac{1}{n}], \quad B_n := [1 + \frac{1}{n}, 2] \qquad \text{for } n \in \mathbb{N}.$$

Then (A_n) and (B_n) are increasing and

$$\inf\{\sup_{n\in\mathbb{N}} A_n, \sup_{n\in\mathbb{N}} B_n\} = \mathrm{cl}\left(\bigcup_{n\in\mathbb{N}} A_n\right) \cap \mathrm{cl}\left(\bigcup_{n\in\mathbb{N}} B_n\right) = \{1\},$$

whereas $\sup_{n\in\mathbb{N}} \inf\{A_n, B_n\} = \mathrm{cl}\left(\bigcup_{n\in\mathbb{N}}(A_n \cap B_n)\right) = \emptyset$, so (3.42) does not hold.

3.7 The Iterated Function Systems on $(2^X, \supset)$

Throughout this section X is an abstract set, 2^X denotes the family of all subsets of X, and $f, f_1, \ldots, f_n$ are selfmaps of X. We consider the partially ordered set $(2^X, \supset)$. So for $A, B \subset X$, $A \leq B$ means that B is a subset of A. A sequence $(A_n)_{n=1}^{\infty}$ is $\supset$-increasing if it is decreasing in the usual sense; moreover, $\sup_{n\in\mathbb{N}} A_n$ in $(2^X, \supset)$ coincides with the intersection $\bigcap_{n\in\mathbb{N}} A_n$.

Proposition 3.4 *Let $F(A) := f(A)$ for $A \subset X$ so that $F : 2^X \mapsto 2^X$. The following conditions are equivalent:*

(*i*) *F is $\supset$-continuous;*

(*ii*) *given a decreasing sequence $(A_n)_{n=1}^{\infty}$ of subsets of X,*

$$f\left(\bigcap_{n\in\mathbb{N}} A_n\right) = \bigcap_{n\in\mathbb{N}} f(A_n);$$

(*iii*) *all fibres of f are finite.*

In particular, (iii) holds if f is injective.

Proof. The equivalence (i)$\Longleftrightarrow$(ii) follows from Remark 3.3. To prove (ii)$\Longrightarrow$(iii) suppose, on the contrary, that (iii) does not hold. Then there exist a $y \in X$ and a sequence $(x_n)_{n=1}^{\infty}$ such that $y = f(x_n)$ and $x_n \neq x_m$ if $n \neq m$. Set $A_n := \{x_k : k \geq n\}$ for $n \in \mathbb{N}$. Clearly, $(A_n)_{n=1}^{\infty}$ is decreasing and $\bigcap_{n\in\mathbb{N}} A_n = \emptyset$. Simultaneously, $f(A_n) = \{y\}$ so that

$$\bigcap_{n\in\mathbb{N}} f(A_n) = \{y\} \neq \emptyset = f\left(\bigcap_{n\in\mathbb{N}} A_n\right),$$

which violates (ii).

To prove (iii)$\Longrightarrow$(ii) assume that a sequence $(A_n)_{n=1}^{\infty}$ is decreasing. It suffices to show that $\bigcap_{n\in\mathbb{N}} f(A_n) \subset f(\bigcap_{n\in\mathbb{N}} A_n)$. Let $y \in \bigcap_{n\in\mathbb{N}} f(A_n)$. Then there is a sequence $(x_n)_{n=1}^{\infty}$ such that $x_n \in A_n$ and $y = f(x_n)$, that is, the set $\{x_n : n \in \mathbb{N}\}$ is a subset of the fibre $f^{-1}(\{y\})$. Condition (iii) implies that there is an $x \in X$ and a subsequence $(x_{k_n})_{n=1}^{\infty}$ of $(x_n)_{n=1}^{\infty}$ such that $x_{k_n} = x$. Hence $x \in \bigcap_{n\in\mathbb{N}} A_{k_n}$. Since $(A_n)_{n=1}^{\infty}$ is decreasing, $\bigcap_{n\in\mathbb{N}} A_{k_n} = \bigcap_{n\in\mathbb{N}} A_n$ so $x \in \bigcap_{n\in\mathbb{N}} A_n$. Moreover, $y = f(x)$ and thus $y \in f(\bigcap_{n\in\mathbb{N}} A_n)$. □

As an application of Proposition 3.4, Corollary 3.4.1 and Lemma 3.2 we obtain the following result on invariant sets of IFS in the set–theoretical case.

Theorem 3.19 *Let F be defined by* (2.1)*. If for $i = 1, \ldots, n$ all fibres of the maps f_i are finite, then for each set $A \subset X$ such that $F(A) \subset A$, the set $\bigcap_{n\in\mathbb{N}} F^n(A)$ is invariant with respect to the* IFS *$\{f_1, \ldots, f_n\}$. In particular, the set $\bigcap_{n\in\mathbb{N}} F^n(X)$ is the greatest invariant set with respect to this* IFS*. Hence, the system $\{f_1, \ldots, f_n\}$ has a nonempty invariant set if and only if the set $\bigcap_{n\in\mathbb{N}} F^n(X)$ is nonempty.*

Proof. We will apply Corollary 3.4.1 for the partially ordered set $(2^X, \supset)$ and the operator F. Clearly, $(2^X, \supset)$ is a complete lattice. We verify condition (3.42). Let $(A_n)_{n=1}^{\infty}$ and $(B_n)_{n=1}^{\infty}$ be decreasing sequences of subsets of X. Then (3.42) is equivalent to the equality

$$\bigcap_{n\in\mathbb{N}} A_n \cup \bigcap_{n\in\mathbb{N}} B_n = \bigcap_{n\in\mathbb{N}} (A_n \cup B_n),$$

which really holds. Let $F_i(A) := f_i(A)$ for $A \subset X$ and $i = 1, \ldots, n$. By Proposition 3.4, all the maps F_i are $\supset$-continuous. Thus all the assumptions of Corollary 3.4.1 are satisfied.

To show that $\bigcap_{n\in\mathbb{N}} F^n(X)$ is the greatest invariant set, observe that if $A_0 = F(A_0)$, then $A_0 = F^n(A_0)$ so that $A_0 = \bigcap_{n\in\mathbb{N}} F^n(A_0)$. Since F is increasing, so are all its iterates F^n and hence, $F^n(A_0) \subset F^n(X)$, which implies that $A_0 \subset \bigcap_{n\in\mathbb{N}} F^n(X)$. The conclusion of Theorem 3.19 is obvious. □

Let us notice that if X is a finite set, then condition (iii) of Proposition 3.4 is automatically satisfied so, by Theorem 3.19, for each map $f : X \mapsto X$ the set

$\bigcap_{n\in\mathbb{N}} f^n(X)$ is a modulus set for f. It turns out that this property characterizes finite sets only, according to the following

Proposition 3.5 *The following conditions are equivalent:*

(*i*) *X is a finite set;*
(*ii*) *for each map $f: X \mapsto X$, the set $\bigcap_{n\in\mathbb{N}} f^n(X)$ is a modulus set for f.*

Proof. The implication (i)$\Longrightarrow$(ii) follows from Theorem 3.19. To prove (ii)$\Longrightarrow$(i) suppose, on the contrary, that X is infinite. Let X_0 be a countable subset of X. Without loss of generality we may assume that

$$X_0 = \{a, b\} \cup \bigcup_{n=1}^{\infty} \bigcup_{k=1}^{n} \{a_{nk}\}$$

where elements a, b and a_{nk} are distinct. Set

$$\begin{aligned} f(x) &:= b \quad \text{for } x \in (X \setminus X_0) \cup \{a, b\}; \\ f(a_{n1} &:= a \quad \text{for } n \in \mathbb{N}; \\ f(a_{nk}) &:= a_{n,k-1} \quad \text{for } n \geq 2 \text{ and } 2 \leq k \leq n. \end{aligned}$$

Then $b = f^n(b)$ and $a = f^n(a_{nn})$ so $\{a, b\} \subseteq \bigcap_{n\in\mathbb{N}} f^n(X)$. On the other hand, it is easily seen that $\bigcap_{n\in\mathbb{N}} f^n(X) \subseteq \{a, b\}$. Therefore, we get

$$f\left(\bigcap_{n\in\mathbb{N}} f^n(X)\right) = f(\{a, b\}) = \{b\} \neq \{a, b\} = \bigcap_{n\in\mathbb{N}} f^n(X),$$

which violates (ii). □

We note that condition (iii) of Proposition 3.4 is not necessary for the set $\bigcap_{n\in\mathbb{N}} f^n(X)$ to be a modulus set for f. This fact can be deduced from Proposition 3.6 and Example 3.3 given below.

Proposition 3.6 *Let (X, ρ) be a bounded metric space and $f: X \mapsto X$ be a Banach contraction with a contractive constant $h \in (0, 1)$. Then for each set $A \subset X$ (with not necessarily $f(A) \subset A$), $\bigcap_{n\in\mathbb{N}} f^n(A)$ is a modulus set for f.*

Proof. Let $A \subset X$. Clearly, if the set $\bigcap_{n\in\mathbb{N}} f^n(A)$ is empty, then it is a modulus set for f. If this set is nonempty, then the diameter, $\delta\left(\bigcap_{n\in\mathbb{N}} f^n(A)\right)$, can be estimated as follows:

$$\delta\left(\bigcap_{n\in\mathbb{N}} f^n(A)\right) \leq \delta\left(f^n(A)\right) \leq \delta\left(f^n(X)\right) \leq h^n\delta(X) \to 0 \quad \text{as } n \to \infty,$$

which implies that $\bigcap_{n\in\mathbb{N}} f^n(A) = \{a\}$ for some $a \in X$. Hence, to prove that $\bigcap_{n\in\mathbb{N}} f^n(A)$ is a modulus set for f, it suffices to show that a is a fixed point of f. Since $a \in f^n(A)$ for $n \in \mathbb{N}$, there is a sequence $(a_n)_{n=1}^{\infty}$ such that $a = f^n(a_n)$. Then

$$\rho\big(a, f(a)\big) = \rho\big(f^n(a_n), f^{n+1}(a_n)\big) \leq h^n\rho\big(a_n, f(a_n)\big) \leq h^n\delta(X) \to 0,$$

which implies that $a = f(a)$. □

Example 3.3 Let $X := [-1, 1]$, $\alpha \in (0, 1/3)$, $f(0) := 0$ and $f(x) := \alpha x^2 \sin(1/x)$ for $x \in X \setminus \{0\}$. Endow X with the euclidean metric. Since $|f'(x)| \leq 3\alpha < 1$, f is a Banach contraction, so the assumptions of Proposition 3.6 are satisfied. On the other hand, Theorem 3.19 is not applicable here, since the fibre $f^{-1}(\{0\})$ is infinite.

3.8 The Iterated Function Systems on $(C(X), \supset)$

Throughout this subsection X is a Hausdorff topological space and $C(X)$ denotes the family of all nonempty closed subsets of X, endowed with the inclusion $\supset$. We start with examining the countable chain condition in this case.

Proposition 3.7 *The following conditions are equivalent:*

(i) *every countable chain in $(C(X), \supset)$ has a supremum;*
(ii) *for every decreasing sequence $(A_n)_{n=1}^{\infty}$ of nonempty closed subsets of X, the intersection $\bigcap_{n \in \mathbb{N}} A_n$ is nonempty;*
(iii) *X is countably compact.*

Proof. (i)$\Longleftrightarrow$(ii) follows from Remark 3.3. For (ii)$\Longleftrightarrow$(iii), see Engelking (1977, Theorem 3.10.2). □

Recall that a space X is *sequential* if every sequentially closed subset A of X (that is, A contains limits of all convergent sequences of its elements) is closed. In particular, every first–countable space is sequential (cf. Engelking (1977, Theorem 1.6.14)). Our next result deals with $\supset$-continuity of the Hutchinson–Barnsley operator in such spaces. It is interesting to note that $\supset$-continuity is related with appropriate properties of fibres of f (similarly, as in the set–theoretical space; (cf. Proposition 3.4 and Theorem 3.23 of this section), which, however, leads directly to continuity with respect to topology, according to the following

Proposition 3.8 *Let X be a countably compact and sequential space, $f : X \mapsto X$ and $F(A) := f(A)$ for $A \subset X$. The following conditions are equivalent:*

(i) *$F(C(X)) \subset C(X)$ and F is continuous on $C(X)$ with respect to the inclusion $\supset$;*
(ii) *f is continuous on X with respect to the topology.*

Proof. This equivalence follows from Remark 3.3, the fact that for a decreasing sequence $(A_n)_{n=1}^{\infty}$ of sets in $C(X)$, $\sup_{n \in \mathbb{N}} A_n$ in $(C(X), \supset)$ coincides with $\bigcap_{n \in \mathbb{N}} A_n$, and Theorem 3.25 of this section. □

The following example shows that in Proposition 3.8 the assumption that X is a sequential space is needed. Also observe that there exist countably compact and sequential spaces, which are not compact as, for example, the space W_0 defined below.

Example 3.4 Let ω_1 denote the smallest uncountable ordinal number, W_0 be the set of all countable ordinal numbers and $W := W_0 \cup \{\omega_1\}$. It is known that W is a compact space (cf. Engelking (1977, Example 3.1.27)) and W_0 is countably compact, but not compact (cf. Engelking (1977, Example 3.10.16)). Moreover, W_0 is a first–countable space, hence sequential. Let $X := W_0 \times W$. Then X is countably compact as the Cartesian product of a countably compact space and a compact space (cf. Engelking (1977, Corollary 3.10.14)). Define a map f by

$$f(x_1, x_2) := (0, x_2) \qquad \text{for } (x_1, x_2) \in X.$$

Clearly, f is a continuous selfmap of X so (ii) of Proposition 3.8 holds. Let $A := \{(x_1, x_1) : x_1 \in W_0\}$. Since the space W is Hausdorff, A is a closed subset of X. On the other hand $f(A) = \{0\} \times W_0$ so $\mathrm{cl}(f(A)) = \{0\} \times W$. Hence condition (i) of Proposition 3.8 does not hold. So the operator F is not a selfmap of $C(X)$.

As an immediate consequence of Propositions 3.7 and 3.7, we obtain the following

Corollary 3.19.1 *Let X be a sequential space, f and F be as in Proposition* 3.8 *The following conditions are equivalent:*

(*i*) *$(C(X), \supset)$ and F satisfy the assumptions of the T-K principle;*
(*ii*) *X is countably compact and f is continuous on X.*

In view of Corollary 3.19.1 the following theorem on invariant sets with respect to IFS on a sequential Hausdorff space can be deduced from the T–K theorem for the family $(C(X), \supset)$.

Theorem 3.20 *Let X be a countably compact and sequential space, and $f_1, \ldots, f_n$ be continuous selfmaps of X. Let F be defined by* (2.1) *and $A_0 := \bigcap_{n\in\mathbb{N}} F^n(X)$. Then the set A_0 is nonempty and closed, $A_0 = F(A_0)$, and A_0 is the greatest invariant set with respect to the* IFS *$\{f_1, \ldots, f_n\}$. Moreover, if X is metrizable, then the sequence $(F^n(X))_{n=1}^{\infty}$ converges to A_0 with respect to the Hausdorff metric.*

Proof. Denote $F_i(A) := f_i(A)$ for $A \in C(X)$ and $i = 1, \ldots, n$. By Corollary 3.19.1, $(C(X), \supset)$ and F_i satisfy the assumptions of Corollary 3.4.1. Clearly, for $A \in C(X)$ the set $F(A)$ is closed as a finite union of closed sets. Moreover, condition (3.42) is satisfied here (cf. the proof of Theorem 3.19) so, by Lemma 3.2, F is $\supset$-continuous. Thus, by Corollary 3.4.1, the set A_0 is invariant with respect to $\{f_1, \ldots, f_n\}$. Since $F(X) \subset X$ and F is increasing, the sequence $(F^n(X))_{n=1}^{\infty}$ is decreasing. Therefore, if X is metrizable, then $(F^n(X))_{n=1}^{\infty}$ converges to A_0 with respect to the Hausdorff metric as a decreasing sequence of compact sets (cf. Edgar (1990, Proposition 2.4.7))]. □

The above theorem generalizes an earlier result of Leader (1982) for $n = 1$. We close this subsection with a result on $\supset$-continuity of the operator F defined by (3.41).

Proposition 3.9 *Let X be a countably compact and sequential space, $f : X \mapsto X$ and $F(A) := \mathrm{cl}(f(A))$ for $A \in C(X)$. The following conditions are equivalent:*

(i) *F is continuous on $C(X)$ with respect to the inclusion $\supset$;*
(ii) *f is continuous on X with respect to the topology.*

Hence, if F is $\supset$-continuous, then $F(A) = f(A)$ for $A \in C(X)$.

Proof. By Remark 3.3, the $\supset$-continuity of F on $C(X)$ means that given a decreasing sequence $(A_n)_{n=1}^{\infty}$ of nonempty closed subsets of X,

$$\mathrm{cl}\left(f\left(\bigcap_{n\in\mathbb{N}} A_n\right)\right) = \bigcap_{n\in\mathbb{N}} \mathrm{cl}\left(f(A_n)\right).$$

By Theorem 3.26 of this section ((i)$\Longleftrightarrow$(ii)), this condition is equivalent to the topological continuity of f. Then, by Theorem 3.25 of this section ((i)$\Longleftrightarrow$(ii)), for $A \in C(X)$ the image $f(A)$ is closed so $F(A) = f(A)$. □

3.9 The Iterated Function System on $(K(X), \supset)$

Throughout this subsection X is a Hausdorff topological space and $K(X)$ denotes the family of all nonempty compact subsets of X, endowed with the inclusion $\supset$. Then every countable chain in $(K(X), \supset)$ has a supremum. Let F be defined by (3.40) for $A \subset K(X)$. If we are to apply Corollary 3.4.1 then, without loss of generality, we may assume that the space X is compact (in particular, countably compact), because the assumption of Corollary 3.4.1 "there is an $X_0 \in K(X)$ such that $X_0 \supset F(X_0)$" implies that all the maps $f_i|_{X_0}$ (the restriction of f_i to X_0) are selfmaps of the same compact set. Thus we arrive at the case considered in the previous subsection, however, this time we need not assume that a space X is sequential, since each continuous map f on X is closed the operator F, becomes a selfmap of $K(X)$.

Theorem 3.21 *Let X be a compact space and $f_1, \ldots, f_n$ be continuous selfmaps of X. Let F be defined by (2.1) and $A_0 := \bigcap_{n\in\mathbb{N}} F^n(X)$. Then the set A_0 is nonempty and compact, $A_0 = F(A_0)$, and A_0 is the greatest invariant set with respect to the* IFS $\{f_1, \ldots, f_n\}$.

Proof. Let $F_i(A) := f_i(A)$ for $A \in K(X)$ and $i = 1, \ldots, n$. The $\supset$-continuity of F_i follows from Proposition 3.10 of of this section. By Lemma 3.2, F is $\supset$-continuous, so 3.4.1 is applicable. □

Theorem 3.22 *Let X be a topological space (not necessarily Hausdorff), $f_1, \ldots, f_n$ be continuous selfmaps of X and F be defined by* (3.40). *The following conditions are equivalent:*

(i) *there exists a nonempty compact set A_0 such that $F(A_0) = A_0$;*

(ii) there exists a nonempty compact set A such that $F(A) \subset A$.

Proof. Obviously, it suffices to show that (ii) implies (i). This follows immediately from Theorem 3.21 applied to the compact set A and the restrictions $f_i|_A$ of the maps f_i to the set A. □

We will show the usefulness of Theorem 3.22 in the theory of IFS. We will prove in the next corollary the original theorem of Hutchinson-Barnsley considered in Section 2.7 of Chapter 2 in the special case when X is a Heine-Borel metric space without using Hausdorff metric. We recall that a metric space (X, ρ) is Heine-Borel if every closed bounded subset of X is compact. The closed ball with centre at a point $x \in X$ and with radius r is denoted by $B(x, r)$.

Corollary 3.22.1 *Let (X, ρ) be a Heine–Borel metric space, $f_1, \ldots, f_n$ be Banach's contractions on X with contractive constants $h_1, \ldots, h_n$ in $(0, 1)$, and F be defined by (3.40). Then there exists a nonempty compact set A_0 such that $F(A_0) = A_0$.*

Proof. We use an idea of Williams (1971) (also cf. Hayashi (1985)). Since a Heine–Borel metric space is complete, each map f_i has a unique fixed point x_i by the Banach contraction principle. Let $A := B(x_1, r)$, a radius r will be specified later. Denote $h := \max\{h_i : i = 1, \ldots n\}$ and $M := \max\{\rho(x_i, x_1) : i = 1, \ldots, n\}$. If $x \in A$, then by the triangle inequality and the contractive condition

$$\begin{aligned} \rho(f_i x, x_1) &\le \rho(f_i x, f_i x_i) + \rho(x_i, x_1) \le h\rho(x, x_i) + M \\ &\le h(\rho(x, x_1) + \rho(x_1, x_i)) + M \le hr + (1 + h)M. \end{aligned} \tag{3.43}$$

Now if we set $r := [(1 + h)/(1 - h)]M$, then $hr + (1 + h)M = r$ so, by (3.43), $f_i(x) \in A$. Since A does not depend on an integer i, we may infer that $F(A) \subset A$. Clearly, by the Heine–Borel property, A is compact and the existence of the set A_0 follows from Theorem 3.22. □

3.10 Continuity of Maps on Countably Compact and Sequential Spaces

In the proof of Theorem 3.21 we used the following

Proposition 3.10 *Let X be a countably compact space, Y be a set and $f : X \mapsto Y$. If all fibres of f are closed, then given a decreasing sequence $(A_n)_{n=1}^{\infty}$ of closed subsets of X,*

$$f\left(\bigcap_{n \in \mathbb{N}} A_n\right) = \bigcap_{n \in \mathbb{N}} f(A_n).$$

Proof. Let $(A_n)_{n=1}^{\infty}$ be a decreasing sequence of closed subsets of X. It suffices to show that $\bigcap_{n \in \mathbb{N}} f(A_n) \subset f\left(\bigcap_{n \in \mathbb{N}} A_n\right)$. Let $y \in \bigcap_{n \in \mathbb{N}} f(A_n)$. Then there is a sequence $(a_n)_{n=1}^{\infty}$ such that $y = f(a_n)$ and $a_n \in A_n$. Thus the sets B_n defined by

$$B_n := A_n \cap f^{-1}(\{y\})$$

are nonempty, closed and $B_{n+1} \subset B_n$. By the countable compactness of X, there exists an $x \in \bigcap_{n\in\mathbb{N}} B_n$. Then $y = f(x)$ and $x \in \bigcap_{n\in\mathbb{N}} A_n$, which means that $y \in f\left(\bigcap_{n\in\mathbb{N}} A_n\right)$. □

The next result is a partial converse to Proposition 3.10.

Proposition 3.11 *Let X be a Hausdorff topological space, Y be a set and $f : X \mapsto Y$. If for every decreasing sequence $(A_n)_{n=1}^{\infty}$ of nonempty compact subsets of X, $f\left(\bigcap_{n\in\mathbb{N}} A_n\right) = \bigcap_{n\in\mathbb{N}} f(A_n)$, then all fibres of f are sequentially closed.*

Proof. Suppose, on the contrary, that there is a $y \in X$ such that the fibre $f^{-1}(\{y\})$ is not sequentially closed. Then there exist an $x \in X$ and a sequence $(x_n)_{n=1}^{\infty}$ such that $f(x_n) = y$ and $f(x) \neq y$. Set $A_n := \{x\} \cup \{x_k : \ k \geq n\}$. Then the sets A_n are compact, since X is Hausdorff, and $A_{n+1} \subset A_n$. Clearly, $x \in \bigcap_{n\in\mathbb{N}} A_n$. Suppose that $x' \in \bigcap_{n\in\mathbb{N}} A_n$ and $x' \neq x$. Then there is a subsequence $(x_{k_n})_{n=1}^{\infty}$ of $(x_n)_{n=1}^{\infty}$ such that $x_{k_n} = x'$. Simultaneously, $(x_{k_n})_{n=1}^{\infty}$ converges to x so $x = x'$ (since, in particular, X is a T_1-space), a contradiction. Therefore $\bigcap_{n\in\mathbb{N}} A_n = \{x\}$ so that

$$f\left(\bigcap_{n\in\mathbb{N}} A_n\right) = \{f(x)\} \neq \{f(x), y\} = \bigcap_{n\in\mathbb{N}} f(A_n),$$

which contradicts the hypothesis. □

As an immediate consequence of Propositions 3.10 and 3.11, we get the following

Theorem 3.23 *Let X be a countably compact and sequential space, Y be a set and $f : X \mapsto Y$. The following conditions are equivalent:*

(*i*) *all fibres of f are closed;*
(*ii*) *given a decreasing sequence $(A_n)_{n=1}^{\infty}$ of nonempty closed subsets of X, $f\left(\bigcap_{n\in\mathbb{N}} A_n\right) = \bigcap_{n\in\mathbb{N}} f(A_n)$;*
(*iii*) *given a decreasing sequence $(A_n)_{n=1}^{\infty}$ of nonempty compact subsets of X, $f\left(\bigcap_{n\in\mathbb{N}} A_n\right) = \bigcap_{n\in\mathbb{N}} f(A_n)$.*

Proof. The implication (i)$\Longrightarrow$(ii) follows from Proposition 3.10, (ii)$\Longrightarrow$(iii) is obvious and (iii)$\Longrightarrow$(i) follows from Proposition 3.11. □

Remark 3.4 Observe that under the assumptions of Theorem 3.23, the classes $C(X)$ and $K(X)$ need not coincide, so the equivalence (ii)$\Longleftrightarrow$(iii) is not trivial. For example, define X as the set of all countable ordinal numbers; then $X \in C(X) \setminus K(X)$ (cf. Example 3.4).

In the sequel we will need the following lemma (cf. Engelking (1977, Proposition 1.6.15)).

Lemma 3.3 *Let X be a sequential space, Y be a topological space and $f : X \mapsto Y$. Then f is continuous if and only if f is sequentially continuous, that is, given a sequence $(x_n)_{n=1}^{\infty}$ in X,*

$$f(\lim x_n) \subset \lim f(x_n).$$

Proposition 3.12 *Let X be a topological space, Y be a countably compact and sequential space and $f : X \mapsto Y$. Then f is sequentially continuous if and only if the graph of f is sequentially closed in the Cartesian product $X \times Y$.*

Proof. First let a sequence $(x_n, f(x_n))_{n=1}^{\infty}$ converge to (x, y) in $X \times Y$. Then $x \in \lim x_n$ and $\{y\} = \lim f(x_n)$ since Y is Hausdorff. By hypothesis,

$$f(x) \in f(\lim x_n) \subset \lim f(x_n) = \{y\},$$

which means that $f(x) = y$. Thus the graph of f is sequentially closed.

Next, suppose, on the contrary, that f is not sequentially continuous. Then there exist a sequence $(x_n)_{n=1}^{\infty}$ and an $x \in X$ such that $x \in \lim x_n$ and $f(x) \notin \lim f(x_n)$. Without loss of generality, we may assume, by passing to a subsequence if necessary, that there is a neighborhood V of $f(x)$ such that $f(x_n) \notin V$ for all $n \in \mathbb{N}$. Since Y is also sequentially compact (cf. Engelking (1977, Theorem 3.10.31)), there is a convergent subsequence $(f(x_{k_n}))_{n=1}^{\infty}$ of $(f(x_n))_{n=1}^{\infty}$. Set $y := \lim f(x_{k_n})$ (this limit is unique since Y is Hausdorff). Since $x \in \lim x_{k_n}$ and the graph of f is sequentially closed, we infer that $y = f(x)$, that is, $(f(x_{k_n}))_{n=1}^{\infty}$ converges to $f(x)$. This yields a contradiction, since $f(x_{k_n}) \notin V$ and $f(x) \in V$. □

The next result is a closed graph theorem for maps on sequential spaces.

Theorem 3.24 *Let X and Y be sequential spaces and Y be countably compact. For a map $f : X \mapsto Y$ the following conditions are equivalent:*

(i) f is continuous;
(ii) the graph of f is closed in $X \times Y$;
(iii) the graph of f is sequentially closed in $X \times Y$;
(iv) f is sequentially continuous.

Proof. That (i) implies (ii) follows from Engelking (1977, Corollary 2.3.22). (ii)⇒(iii) is obvious. (iii)⇒(iv) follows from Proposition 3.12 and finally, (iv)⇒(i) holds by Lemma 3.3. □

The main result in this subsection is the following theorem, which gives a characterization of continuity of maps on countably compact and sequential spaces. This result was obtained as a by-product of our study of continuity of the Hutchinson–Barnsley operator with respect to the inclusion $\supset$ (cf. Proposition 3.8).

Theorem 3.25 *Let X and Y be countably compact and sequential spaces. For a map $f : X \mapsto Y$ the following conditions are equivalent:*

(i) *f is continuous;*
(ii) *for every closed subset A of X, the image $f(A)$ is closed, and all fibres of f are closed;*
(iii) *for every closed subset A of X, the image $f(A)$ is closed, and given a decreasing sequence $(A_n)_{n=1}^{\infty}$ of nonempty closed subsets of X, $f\left(\bigcap_{n\in\mathbb{N}} A_n\right) = \bigcap_{n\in\mathbb{N}} f(A_n)$;*
(iv) *for every compact subset A of X, the image $f(A)$ is compact, and given a decreasing sequence $(A_n)_{n=1}^{\infty}$ of nonempty compact subsets of X, $f\left(\bigcap_{n\in\mathbb{N}} A_n\right) = \bigcap_{n\in\mathbb{N}} f(A_n)$.*

Proof. (i)$\Longrightarrow$(ii). Let A be a closed subset of X. Since X is sequentially compact (cf. Engelking (1977, Theorem 3.10.31)), so is A (cf. Engelking (1977, Theorem 3.10.33)). Hence and by continuity of f, the image $f(A)$ is sequentially compact (cf. Engelking (1977, Theorem 3.10.32)). In particular, $f(A)$ is sequentially closed, hence closed since Y is sequential. Since, in particular, Y is a T_1-space it is clear that the fibres of f are closed.

(ii)$\Longrightarrow$(iii) follows immediately from Theorem 3.23.

We give a common proof of the implications (iii)$\Longrightarrow$(i) and (iv)$\Longrightarrow$(i). By Theorem 3.24, it suffices to show that the graph of f is sequentially closed. Let a sequence $(x_n, f(x_n))_{n=1}^{\infty}$ converge to (x, y) in $X \times Y$. Since both X and Y are Hausdorff, we may infer that $x = \lim x_n$ and $y = \lim f(x_n)$. Set $A_n := \{x\} \cup \{x_k : k \geq n\}$ for $n \in \mathbb{N}$. The the sets A_n are compact (hence closed), $A_{n+1} \subseteq A_n$ and $\bigcap_{n\in\mathbb{N}} A_n = \{x\}$. By hypothesis, $\bigcap_{n\in\mathbb{N}} f(A_n) = f\left(\bigcap_{n\in\mathbb{N}} A_n\right) = \{f(x)\}$. Since both (iii) and (iv) imply that the set $f(A_n)$ is closed and $f(x_k) \in f(A_n)$ for $k \geq n$, we may infer that $y = \lim_{k\to\infty} f(x_k) \in f(A_n)$ so that $y \in \bigcap_{n\in\mathbb{N}} f(A_n) = \{f(x)\}$, that is, $y = f(x)$. This proves that the graph of f is sequentially closed.

We have shown that conditions (i), (ii) and (iii) are equivalent, and that (iv) implies (i). To finish the proof it suffices to show that (iii) implies (iv). Since (iii) implies the continuity of f, the first part of (iv) holds. The second part of (iv) follows immediately from (iii). □

Our last theorem gives another characterization of continuity. This result was obtained as a by product of our study of $\supset$-continuity of operator F defined by Lasota and Myjak (1996b) (cf. Proposition 3.9).

Theorem 3.26 *Let X and Y be countably compact and sequential spaces. For a map $f : X \mapsto Y$ the following conditions are equivalent:*

(i) *f is continuous;*
(ii) *given a decreasing sequence $(A_n)_{n=1}^{\infty}$ of nonempty closed subsets of X, $\mathrm{cl}\left(f(\bigcap_{n\in\mathbb{N}} A_n)\right) = \bigcap_{n\in\mathbb{N}} \mathrm{cl}(f(A_n))$;*

(*iii*) *given a decreasing sequence* $(A_n)_{n=1}^{\infty}$ *of nonempty compact subsets of* X, $\mathrm{cl}\left(f(\bigcap_{n\in\mathbb{N}} A_n)\right) = \bigcap_{n\in\mathbb{N}} \mathrm{cl}(f(A_n))$.

Proof. (i)$\Longrightarrow$(ii): Let $(A_n)_{n=1}^{\infty}$ be a decreasing sequence of nonempty closed subsets of X. Since the intersection $\bigcap_{n\in\mathbb{N}} A_n$ is closed, we may conclude by Theorem 3.25 ((i)$\Longrightarrow$(ii)) that all the sets $f(\bigcap_{n\in\mathbb{N}} A_n)$ and $f(A_n)$ ($n \in \mathbb{N}$) are closed. Therefore, (ii) follows immediately from condition (iii) of Theorem 3.25.

(ii)$\Longrightarrow$(iii) is obvious.

(iii)$\Longrightarrow$(i). By Theorem 3.24, it suffices to show that the graph of f is sequentially closed. We use the same argument as in the proof of (iv)$\Longrightarrow$(i) in Theorem 3.25. So let $x = \lim x_n$ and $y = \lim f(x_n)$. Set $A_n := \{x\} \cup \{x_k : k \geq n\}$. By (iii),

$$\bigcap_{n\in\mathbb{N}} \mathrm{cl}(f(A_n)) = \mathrm{cl}\left(f\left(\bigcap_{n\in\mathbb{N}} A_n\right)\right) = \mathrm{cl}\left(\{f(x)\}\right) = \{f(x)\}.$$

Since $y \in \mathrm{cl}(f(A_n))$ for all $n \in \mathbb{N}$, we may infer that $y = f(x)$, which proves that the graph of f is sequentially closed. □

3.11 Solutions of Impulsive Differential Equations

Some interesting results concerning the existence of periodic solutions to the following impulsive differential equation have appeared, (e.g., see Kaul (1995), and Bajo and Liz (1996)).

$$\frac{d}{dt}x(t) = f(t, x(t)) \quad \text{for } t \neq \tau(x(t)) \tag{3.44}$$

$$x(t^+) = x(t) + I(x(t)) \quad \text{for } t = \tau(x(t)). \tag{3.45}$$

These results are proved by using monotone iterative techniques and considering $f : [0,T] \times \mathbb{R} \to \mathbb{R}$ only. Our aim is to establish a comparison result for (3.44)–(3.45) where the equation is interpreted as a system in $\mathbb{R}$, and then to apply this in proving the existence of periodic solutions to (3.44)–(3.45).

By a solution of (3.44)–(3.45) we mean a left continuous function $x : [0,T] \to \mathbb{R}$ such that if t satisfies $\tau(x(t)) \neq t$, $x(t)$ is differentiable and (3.44) is satisfied; and if $x(t)$ satisfies $\tau(x(t)) = t$, then $\lim_{s\to t^+} x(s) =: x(t^+) = x(t) + I(x(t))$ so (3.45) is true. See Lakshmikantham, Bainov, and Simeonov (1989) for more details.

Now we introduce some notations and definitions. A cone K in $\mathbb{R}$ is a closed convex set such that $\lambda K \subseteq K$ for all $\lambda \geq 0$ and $-K \cap K = \{0\}$. Such a set defines a partial ordering $\leq$ on $\mathbb{R}$: $x \leq y$ if and only if $y - x \in K$. The relation $\geq$ is defined similarly. Elements of K are said to be positive. We shall assume throughout that K has nonempty interior and $\mathbb{R} = K - K$. By $x < y$ we mean $y - x \in K\backslash\{0\}$. For $a, b \in \mathbb{R}$, the order interval $[a,b]$ is defined as $\{x \in \mathbb{R} : a \leq x \leq y\}$, which is compact and convex.

For K a cone in $\mathbb{R}$, K^+ is defined as those continuous linear functionals with domain $\mathbb{R}$ which are nonnegative on K. A mapping f between two partially ordered sets A and B is said to be increasing if $x \le y$ implies $f(x) \le f(y)$, and strictly increasing if $x \le y$ and $x \ne y$ implies $f(x) \le f(y)$ and $f(x) \ne f(y)$.

Let $f : [0,T] \times \mathbb{R} \to \mathbb{R}$. By the expression $\Omega \subset \mathbb{R}$ *is forward invariant (or flow invariant) with respect to* f, we mean any solution $x(t)$ of the (possibly impulsive) differential equation $x' = f(t,x)$ (3.44)–(3.45) with $x(0) \in \Omega$ satisfies $x(t) \in \Omega$ for all $t \in [0,T]$.

3.11.1 *A Comparison Result*

We collect some basic facts that are needed in the main theorem as a series of lemmas. First the continuity assumptions on the functions are

C1. $f : [0,T] \times \mathbb{R} \to \mathbb{R}$ is continuous and Lipschitz with respect to $x \in \mathbb{R}$;

C2. $\tau : \mathbb{R} \to (0,\infty)$ is continuously differentiable, strictly increasing with respect to K and $\tau(x) < T$ for some x;

C3. $I : \mathbb{R} \to -K$ is continuously differentiable.

Unless otherwise stated, C1-C3 will be assumed throughout the rest of this section. The condition in the next lemma ensures that $t = \tau(x)$ does not define a solution to (3.44)–(3.45).

Lemma 3.4 *Under the additional assumption*

$$\text{A1.} \qquad \nabla\tau \cdot f(t,x) \le 1 \quad \forall (t,x) \in [0,T] \times \mathbb{R}$$

the impulsive problem has a unique solution $x(t)$ defined on $[0,T]$, and if $x(0) = x_0$ is such that the solution meets the surface S at least once, the solution meets the surface exactly once.

The proof may be found in Lakshmikantham et al. (1989), Theorem 1.3.2. The following lemma is a comparison result for ordinary differential equations without impulse. First though we need the

Definition 3.3 Let K be a cone and $\le$ the partial ordering induced by K. A mapping $f : [0,T] \times \mathbb{R} \to \mathbb{R}$ is said to be *quasi-monotone* if for any $t \in [0,T]$ and any $x, y \in \mathbb{R}$ with $x \ge y$, and any $x^* \in K^+$ such that $x^*(x-y) = 0$ we have $x^*(f(t,x) - f(t,y)) \ge 0$.

Lemma 3.5 *If $u(t)$ and $v(t)$ are solutions of the ordinary differential equation $x' = g(t,x)$ on the interval $[0,T]$ with $u(0) < v(0)$, g satisfies* C1 *and g is quasi-monotone, then $u(t) < v(t)$ for all $t \in [0,T]$.*

See (Deimling (1977), Theorem 5.4) for a proof.

Lemma 3.6 *Under the assumptions* C1 – C3 *and* A1, *if $u(t)$ and $v(t)$ are solutions of the impulsive system* (3.44)–(3.45) *with f quasimonotone, $u(0) < v(0)$, and $v(t)$ meets the surface at time t_2, then $u(t)$ meets the surface at $t_1 \in (0, t_2)$.*

Proof. Suppose the solution $u(t)$ does not meet the surface S in the interval $(0, t_2)$. Then from Lemma 3.5 $u(t) < v(t)$ for all $t \in [0, t_2]$. As τ is strictly increasing, $\tau(u(t)) < \tau(v(t))$ for all $t \in [0, t_2]$. Then $p(t) = t - \tau(u(t))$ satisfies $p(0) < 0$ and $p(t_2) = t_2 - \tau(u(t_2)) > t_2 - \tau(v(t_2)) = 0$. As p is continuous in $[0, t_2]$ (by assumption that u does not meet S in this interval) there exists a point $t_1 \in (0, t_2)$ such that $p(t_1) = 0$. That is, $u(t)$ meets S, a contradiction which proves the lemma. □

The following comparison result complements the corresponding results in Bajo and Liz (1996), Kaul, Lakshmikantham, and Leela (1994), and Lakshmikantham, Leela, and Kaul (1994). By $\mathrm{D}I^*$ we mean the derivative of $I^* : \mathbb{R} \to \mathbb{R}$.

Theorem 3.27 *Suppose* C1–C3, A1 *and the following is satisfied.*

A2. $I^*(x) = x + I(x)$ *is increasing ;*

A3. $f(t, \cdot)$ *is quasimonotone for all* $t \in [0, T]$;

A4. $(\mathrm{D}I^*)(x) f(t, x) \geq f(t, I^*(x))$.

Then if $u(t)$ *and* $v(t)$ *are solution of the impulsive differential equation* (3.44)–(3.45) *with* $u(0) \leq v(0)$, *then* $u(t) \leq v(t)$ *for all* $t \in [0, T]$.

Proof. If $u(0) = v(0)$ then by uniqueness, $u(t) = v(t)$ for all $t \in [0, T]$ so the conclusion holds.

Suppose $u(0) < v(0)$. From Lemma 3.5, $u(t) < v(t)$ for as long as it takes for either solution to meet the surface S. If neither solution meets S in $[0, T]$, there is nothing to prove. If $u(t)$ meets S at $t_1 \in (0, T]$ and $v(t)$ does not meet S, then as $I(x) \in -K$, we have

$$u(t_1^+) = u(t_1) + I(u(t_1)) \leq u(t_1) < v(t_1). \tag{3.46}$$

Lemma 3.5 ensures that $u(t) < v(t)$ for all $t \in [0, t_1]$ and since $u(t_1^+) < v(t_1) = v(t_1^+)$, it follows by Lemma 3.5 again that $u(t) < v(t)$ for all $t \in (t_1, T]$.

It remains to consider the case when both $u(t)$ and $v(t)$ meet S. By Lemma 3.6, $u(t)$ meets S at $t_1 \in (0, T)$ and $v(t)$ meets S at $t_2 \in (t_1, T]$. Then $u(t) < v(t)$ for all $t \in [0, t_1]$. By the argument above, $u(t_1^+) < v(t_1)$ and so $u(t) < v(t)$ for all $t \in [t_1, t_2]$.

We now show $u(t_2) \leq v(t_2^+)$.

From A2 we have

$$u(t_1^+) = I^*(u(t_1)) \leq I^*(v(t_1)).$$

Let $k(t) = I^*(v(t)) - u(t)$ for $t \in (t_1, t_2]$ and notice that $k(t)$ is the unique solution of the initial value problem $x'(t) = g(t, x)$ with initial condition $x(t_1^+) = I^*(v(t_1)) - u(t_1^+) \in K$ where

$$g(t, x) = (\mathrm{D}I^*)(v(t)) f(t, v(t)) - f(t, I^*(v(t)) - x).$$

By A4, $g(t,x) \geq f(t, I^*(v(t))) - f(t, I^*(v(t)) - x)$ and as f is quasimonotone, g satisfies the condition that for any $x^* \in K^+$ with $x^*(x) = 0$, then $x^*(g(t,x)) \geq 0$. This is a sufficient condition to ensure the solution $k(t)$ remains in K for all subsequent time (that is; K is forward invariant under g, (see Guo and Lakshmikantham (1988), Theorem 4.2.2). In particular, $k(t_2) = v(t_2) + I(v(t_2)) - u(t_2) \geq 0$ and so $u(t_2) \leq v(t_2^+)$.

If $u(t_2) = v(t_2^+)$ then $u(t) = v(t)$ for all subsequent time from uniqueness. Otherwise $u(t_2) < v(t_2^+)$ and Lemma 3.5 shows $u(t) < v(t)$ for all subsequent time. Thus the theorem is proved. □

Assumption A4 is analogous to, though slightly weaker than, condition H_3 of Theorem 2.1 of Bajo and Liz (1996) where the authors deal with upper and lower solutions in one dimension.

3.11.2 *Periodic Solutions*

In this section, we examine the existence of solutions to the impulsive boundary value problem.

$$\begin{aligned} \frac{d}{dt}x(t) &= f(t, x(t)) \qquad \text{for } t \neq \tau(x(t) \\ x(t^+) &= x(t) + I(x(t)) \qquad \text{for } t = \tau(x(t) \\ x(0) &= x(T). \end{aligned}$$

A common technique in proving the existence of periodic solutions for ordinary differential equations is to show the displacement operator $P(z)$ defined by $P(z) = x(T;0,z)$ (that is; $P(z)$ is the value of the solution x at time T which has initial data $x(0) = z$) has a fixed point, via an application of Brouwer's fixed point theorem. Of course, the domain Ω of P is required to be forward invariant so that P maps Ω into Ω. The continuity of the displacement operator P follows from the continuous dependence of solutions of the differential equation on the initial data.

The case is somewhat different when impulses are involved. Then the displacement operator need not be continuous as the following simple example illustrates. (See Lakshmikantham et al. (1989) for results on the continuous dependence of solutions on the initial data for impulsive differential equations.)

Example 3.5 Let $f(t,x) = 1/2$ for all $(t,x) \in [0,2] \times \mathbb{R}$, $I(x) = -2$ and $\tau(x) = x + 1$. Then for $P(z) = x(2;0,z)$, it follows that $\lim_{z\to 0^+} P(z) = 1$ and $\lim_{z\to 0^-} P(z) = -1$.

Our approach in Tarafdar and Watson (1999) is analogous to that discussed above, in that it will be shown that $P(z) = x(T;0,z)$ has a fixed point, by using fixed point theorem of Tarski (3.13).

It is clear that the order interval $[a,b] = \{x \in \mathbb{R} : a \leq x \leq b\}$, where $\leq$ is the partial ordering induced by a generating cone K (that is; $\mathbb{R} = K - K$), is a complete lattice (Guo and Lakshmikantham (1988)).

We impose the following necessary and sufficient condition on f so that $\Omega = [a, b]$ is forward invariant under f (see Hartman (1972), Theorem 2). Specifically

$$A5. \qquad f(t,x) \in T_{[0,1]\times\Omega}(t,x) \text{ for all } (t,x) \in [0,T] \times \Omega,$$

where $T_{[0,1]\times\Omega}(t,x)$ is defined as follows:

$$T_{[0,1]\times\Omega}(t,x) = \left\{ y \in \mathbb{R} : \liminf_{\lambda \to 0+} \frac{\operatorname{dist}((t+\lambda, x+\lambda y), [0,1]\times\Omega)}{\lambda} = 0 \right\}. \tag{3.47}$$

If the impulse satisfies

$$A6. \qquad x + I(x) \in \Omega \text{ for all } x \in \Omega$$

and Ω is forward invariant under f, it is clear that the solution $x(t)$ of (5.1)–(5.2) with $x(0) \in \Omega$ remains in Ω for all $t \in [0, T]$.

Hence, assuming A5 and A6, we have $P(z) = x(T; 0, z)$ is a well-defined function from Ω to Ω, and if the conditions of Theorem 3.27 are satisfied, P is increasing, so Tarski's theorem ensures the existence of a fixed point of P in Ω. Thus we have proved the following result:

Theorem 3.28 *Let f, τ and I satisfy the continuity conditions* C1–C3. *Further, suppose* A1–A4 *are satisfied and that* A5 *and* A6 *hold for some nonempty order interval $\Omega = [a, b]$. Then there exists a maximal and minimal periodic solution to the impulsive differential equation* (3.44)–(3.45).

The conclusion that there exists a maximal and minimal periodic solution follows from the fact that the set of fixed points of P is a complete lattice.

For this subsection a further reading of Crandall (1972), Deimling (1992) and Hristova and Bainov (1987) is suggested.

Chapter 4

Topological Fixed Point Theorems

4.1 Brouwer Fixed Point Theorem

All fixed theorems to be considered in this chapter will be built on a single fixed point theorem, the Brouwer fixed point theorem which will be proved a little later. First we will need to consider some topological concepts.

Definition 4.1 Let X and Y be topological spaces. Two continuous mappings $f, g : X \to Y$ are said to be *homotopic* if there is a continuous mapping $F : X \times [0, 1] \to Y$ such that

$$\begin{aligned} F(x, 0) &= f(x), \qquad x \in X, \\ F(x, 1) &= g(x), \qquad x \in X \end{aligned}$$

F is called the *homotopy* of f and g. Also f and g are called homotopic through the homotopy F.

A topological space X is said to be *contractible* if the identity map on X is homotopic to a constant mapping, that is, there is a continuous mapping $F : X \times [0, 1] \to X$ such that $F(x, 1) = x$ for all $x \in X$ and $F(x, 0) = x_0 \in X$ for all $x \in X$. X is said to be contractible to x_0.

A nonempty subset A of a topological space X is said to be a retract of X if there is a continuous mapping $f : X \to X$ of X into itself such that $f(X) \subset A$ and $f(a) = a$ for all $a \in A$. In the sequel $B^{n+1} = \{x = (x_1, x_2, \ldots, x_{n+1}) \in \mathbb{R}^{n+1} : \sqrt{x_1^2 + x_2^2 + \cdots + x_{n+1}^2} \leq 1\}$ will denote the closed unit ball of $\mathbb{R}^{n+1}$ and $S^n = \{x = (x_1, x_2, \ldots, x_{n+1}) \in \mathbb{R}^{n+1} : \sqrt{x_1^2 + x_2^2 + \cdots + x_{n+1}^2} = 1\}$, the boundary of B^{n+1}. S^n is also called the n-sphere.

A continuous function of $f : U \longrightarrow \mathbb{R}$ defined on an open subset U of $\mathbb{R}^n$ is said to be smooth, or C^∞ if it has continuous partial derivatives of all orders. A function $f : A \longrightarrow \mathbb{R}$ defined on an arbitrary nonempty subset A of $\mathbb{R}^n$ is smooth, or C^∞, if there are an open subset U of $\mathbb{R}$ with $A \subset U$ and a smooth function $g : U \to \mathbb{R}$ such that $g/A = f$, where g/A denotes the restriction of g to A. Lastly, a continuous function $f : U$ (or A) $\longrightarrow \mathbb{R}^m$ is said to be smooth if each co-ordinate function is smooth.

Definition 4.2 *A nonempty subset A of $\mathbb{R}^n$ is said to be smoothly contractible if there exists a smooth function $F : A \times [0, 1] \longrightarrow A$ such that $F(a, 1) = a$ for all $a \in A$ and $F(a, 0) = a_0 \in A$ for all $a \in A$. A nonempty subset A of a set $B \subset \mathbb{R}^n$ is said to be smooth retract of B if there exists a smooth function $f : B \longrightarrow B$ such that $f(B) \subset A$ and $f(a) = a$ for all $a \in A$.*

The following theorems are equivalent:

Theorem 4.1 *A smooth mapping $f : B^{n+1} \to B^{n+1}$ has a fixed point.*

Theorem 4.2 *S^n is not a smooth retract of B^{n+1}.*

Theorem 4.3 *S^n is not smoothly contractible.*

Proof. As above, let Theorem 4.2 hold. If possible let f have no fixed point. Since $x \neq f(x)$, the points x and $f(x)$ determines a unique line $y = x+m(x-f(x))$ which intersects S^n in exactly two points. Let $m(x)$ be the nonnegative value of m where the line meets S^n, that is the ray emanating from $f(x)$ passing through x hits S^n. Then $m(x)$ is the larger solution of the quadratic equation:

$$1 = \|x + m(x - f(x))\|^2 = \|x\|^2 + 2mx.(x - f(x)) + m^2 \|x - f(x)\|^2, \qquad (4.1)$$

that is

$$m(x) = \frac{x.(f(x) - x) + \left([x.(x - f(x))]^2 + (1 - \|x\|^2) \|x - f(x)\|^2\right)^{\frac{1}{2}}}{\|x - f(x)\|^2}. \qquad (4.2)$$

The discriminant of the quadratic equation of (4.1) never vanishes as $x \neq f(x)$.

Hence $m(x)$ being a composition of smooth functions is smooth. We can easily see from (4.2) that $m(x) = 0$ whenever $x \in S^n$. Now we define a mapping $g : B^{n+1} \longrightarrow B^{n+1}$ by $g(x) = x + m(x)(x - f(x))$, $x \in B^{n+1}$. Clearly g is smooth and $g(B^{n+1}) \subset S^n$, and $g(x) = x$ for each $x \in S^n$, that is, g is a smooth retraction of B^{n+1} into S^n, which is a retraction. Geometrically, we have the figure below

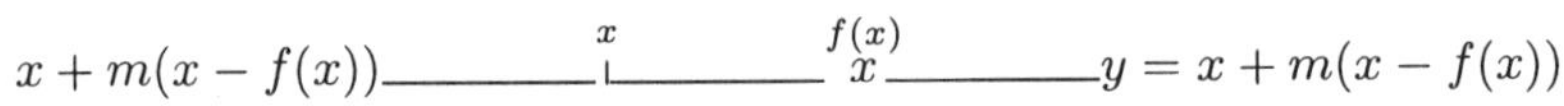

Next, let Theorem 4.1 hold. If possible, let Theorem 4.2 do not hold, that is, assume that there is a smooth retraction $f : B^{n+1} \longrightarrow B^{n+1}$. Now we define a smooth mapping $g : B^{n+1} \longrightarrow B^{n+1}$ by $g(x) = -f(x)$, $x \in B^{n+1}$. Clearly $g(x) \neq x$ for any $x \in B^{n+1}$, which violates 4.1. Now we prove that Theorem 4.2 implies Theorem 4.3. Let Theorem 4.2 hold. If possible, let S^n be smoothly contractible to a point $x_0 \in S^n$, i.e. there is a smooth mapping $F : S^n \times [0, 1] \longrightarrow S^n$ such that $F(x, 1) = x$ for all $x \in S^n$ and $F(x, 0) = x_0$ for all $x \in S^n$. We define a mapping

$f : B^{n+1} \longrightarrow S^n$ by

$$f(tx) = F(x,\, t),\ 0 \leq t \leq 1,\ x \in S^n. \tag{4.3}$$

Clearly f is a smooth retraction of B^{n+1} onto S^n, which contradicts Theorem 4.2. Finally we prove that Theorem 4.3 implies Theorem 4.2. Suppose that the Theorem 4.3 holds and, if possible, suppose that the Theorem 4.2 does not hold, that is, there exists a smooth retraction $f : B^{n+1} \longrightarrow B^{n+1}$ of B^{n+1} onto S^n. Then the mapping $F : S^n \times [0,\, 1] \longrightarrow S^n$ defined by (4.3) is clearly a smooth homotopy between the identity map on S^n to the constant map x_0, which contradicts Theorem 4.3. □

We assume the functions involved in the next two lemmas to be smooth.

Lemma 4.1 *Let $A = (a_{ij})$ be $n \times n$ matrix where each a_{ij} is a C^∞ function in $\mathbb{R}^n$, then for each $i = 1,\, 2,\, \ldots,\, n$,*

$$D_i \det A = D_i \begin{vmatrix} a_{11} & a_2 & \cdots & a_{1n} \\ a_{21} & a_{22} & \cdots & a_{2n} \\ \cdots & \cdots & \cdots & \cdots \\ a_{m1} & a_{m2} & \cdots & a_{mn} \\ \cdots & \cdots & \cdots & \cdots \\ a_{n1} & a_{n2} & \cdots & a_{nn} \end{vmatrix}$$

$$= \begin{vmatrix} D_i a_{11} & a_{12} & \cdots & a_{1n} \\ D_i a_{21} & a_{22} & \cdots & a_{2n} \\ \cdots & \cdots & \cdots & \cdots \\ D_i a_{n1} & a_{n2} & \cdots & a_{nn} \end{vmatrix} + \begin{vmatrix} a_{11} & D_i a_{12} & \cdots & a_{1n} \\ a_{21} & D_i a_{22} & \cdots & a_{2n} \\ \cdots & \cdots & \cdots & \cdots \\ a_{n1} & D_i a_{n2} & \cdots & a_{nn} \end{vmatrix}$$

$$+ \cdots + \begin{vmatrix} a_{11} & a_{12} & \cdots & D_i a_{1n} \\ a_{21} & a_{22} & \cdots & D_i a_{2n} \\ \cdots & \cdots & \cdots & \cdots \\ a_{n1} & a_{n2} & \cdots & D_i a_{nn} \end{vmatrix}, \ \text{where } D_i = \frac{\partial}{\partial x_i}$$

We will use both symbols D_i and $\frac{\partial}{\partial x_i}$ whenever it is convenient.

Proof. We first prove for 2×2 case. For each $i = 1,\, 2$

$$D_i \begin{vmatrix} a_{11} & a_{12} \\ a_{21} & a_{22} \end{vmatrix} = D_i[a_{11}a_{22} - a_{12}a_{21}]$$

$$= D_i a_{11} a_{22} + a_{11} D_i a_{22} - D_i a_{12} a_{21} - a_{12} D_i a_{21}$$

$$= \begin{vmatrix} D_i a_{11} & a_{12} \\ D_i a_{21} & a_{22} \end{vmatrix} + \begin{vmatrix} a_{11} & D_i a_{12} \\ a_{21} & D_i a_{22} \end{vmatrix}$$

We now assume that the lemma is true for the case $k \times k$.

Then for each $i = 1, 2, \ldots, k+1$,

$$D_i \begin{vmatrix} a_{11} & a_{12} & \cdots & a_{1\,k+1} \\ a_{21} & a_{22} & \cdots & a_{2\,k+1} \\ \cdots & \cdots & \cdots & \cdots \\ a_{k+1\,1} & a_{k+1\,2} & \cdots & a_{k+1\,k+1} \end{vmatrix}$$

$$= D_i a_{11} \cdot \begin{vmatrix} a_{22} & a_{23} & \cdots & a_{2\,k+1} \\ \cdots & \cdots & \cdots & \cdots \\ a_{k+12} & a_{k+13} & \cdots & a_{k+1\,k+1} \end{vmatrix} + a_{11} \left[\begin{vmatrix} D_i a_{22} & a_{23} & \cdots & a_{2\,k+1} \\ \cdots & \cdots & \cdots & \cdots \\ D_i a_{k+12} & a_{k+13} & \cdots & a_{k+1\,k+1} \end{vmatrix} \right.$$

$$\left. + \begin{vmatrix} a_{22} & D_i a_{23} & \cdots & a_{2\,k+1} \\ \cdots & \cdots & \cdots & \cdots \\ a_{k+1\,2} & D_i a_{k+1\,3} & \cdots & a_{k+1\,k+1} \end{vmatrix} + \cdots + \begin{vmatrix} a_{22} & a_{23} & \cdots & D_i a_{2\,k+1} \\ \cdots & \cdots & \cdots & \cdots \\ a_{k+1\,2} & a_{k+13} & \cdots & D_i a_{k+1\,k+1} \end{vmatrix} \right]$$

$$-D_i a_{12} \begin{vmatrix} a_{21} & a_{23} & \cdots & a_{2\,k+1} \\ \cdots & \cdots & \cdots & \cdots \\ a_{k+1\,1} & a_{k+1\,3} & \cdots & a_{k+1\,k+1} \end{vmatrix} - a_{12} \left[\begin{vmatrix} D_i a_{21} & a_{23} & \cdots & a_{2\,k+1} \\ \cdots & \cdots & \cdots & \cdots \\ D_i a_{k+1\,1} & a_{k+13} & \cdots & a_{k+1\,k+1} \end{vmatrix} \right.$$

$$\left. + \begin{vmatrix} a_{21} & D_i a_{23} & \cdots & a_{2\,k+1} \\ \cdots & \cdots & \cdots & \cdots \\ a_{k+1\,1} & D_i a_{k+1\,3} & \cdots & a_{k+1\,k+1} \end{vmatrix} + \cdots + \begin{vmatrix} a_{21} & a_{23} & \cdots & D_i a_{2\,k+1} \\ \cdots & \cdots & \cdots & \cdots \\ a_{k+1\,1} & a_{k+1\,3} & \cdots & D_i a_{k+1\,k+1} \end{vmatrix} \right]$$

$$+ \cdots + (-1)^{k+1} D_i a_{1\,k+1} \begin{vmatrix} a_{21} & a_{22} & \cdots & a_{2k} \\ \cdots & \cdots & \cdots & \cdots \\ a_{k+1\,1} & a_{k+1\,2} & \cdots & a_{k+1\,k} \end{vmatrix}$$

$$+ (-1)^{k+1} a_{1\,k+1} \left[\begin{vmatrix} D_i a_{21} & a_{22} & \cdots & a_{2k} \\ \cdots & \cdots & \cdots & \cdots \\ D_i a_{k+1\,1} & a_{k+1\,2} & \cdots & a_{k+1\,k} \end{vmatrix} \right.$$

$$\left. + \begin{vmatrix} a_{21} & D_i a_{22} & \cdots & a_{2k} \\ \cdots & \cdots & \cdots & \cdots \\ a_{k+1\,1} & D_i a_{k+1\,2} & \cdots & a_{k+1\,k} \end{vmatrix} + \cdots + \begin{vmatrix} a_{21} & a_{22} & \cdots & D_i a_{2k} \\ \cdots & \cdots & \cdots & \cdots \\ a_{k+1\,1} & a_{k+12} & \cdots & D_i a_{k+1\,k} \end{vmatrix} \right]$$

which can be easily seen to be

$$= \begin{vmatrix} D_i a_{11} & a_{12} & \cdots & a_{1\,k+1} \\ D_i a_{21} & a_{22} & \cdots & a_{2\,k+1} \\ \cdots & \cdots & \cdots & \cdots \\ D_i a_{k+1\,1} & a_{k+12} & \cdots & a_{k+1\,k+1} \end{vmatrix}$$

$$\begin{vmatrix} +a_{11} & D_i a_{12} & \cdots & a_{1\,k+1} \\ a_{21} & D_i a_{22} & \cdots & a_{2\,k+1} \\ \cdots & \cdots & \cdots & \cdots \\ a_{k+1\,1} & D_i a_{k+1\,2} & \cdots & a_{k+1\,k+1} \end{vmatrix}$$

$$+\cdots+\begin{vmatrix} a_{11} & a_{12} & \cdots & D_i a_{1\,k+1} \\ a_{21} & a_{22} & \cdots & D_i a_{2\,k+1} \\ \cdots & \cdots & \cdots & \cdots \\ a_{k+1\,1} & a_{k+1\,2} & \cdots & D_i a_{k+1\,k+1} \end{vmatrix}.$$

Thus the lemma is true for the case $(k+1)\times(k+1)$. □

Lemma 4.2 *Let $F(t, x) : B^{n+1}\times[0, 1] \longrightarrow \mathbb{R}^{n+1}$ be a smooth function. Then*

$$\sum_{l=1}^{n+1}\frac{\partial}{\partial x_l}\det Q_l = \frac{\partial}{\partial t}\det\left(\frac{\partial F_j}{\partial x_k}\right),$$

where $\left(\frac{\partial F_j}{\partial x_k}\right)$ is the Jacobian matrix of F with t being held as constant and Q_l is the matrix obtained from $\left(\frac{\partial F_j}{\partial x_k}\right)$ by replacing the l-th column by the partiul derivatives $\frac{\partial F_j}{\partial t}$, $1\le j\le n+1$.

Proof. First, we convince ourselves on the validity of this lemma for the case $F(t, x) : B^2\times[0, 1] \longrightarrow \mathbb{R}^2$.

$$\frac{\partial}{\partial t}\begin{vmatrix} \frac{\partial F_1}{\partial x_1} & \frac{\partial F_1}{\partial x_2} \\ \frac{\partial F_2}{\partial x_1} & \frac{\partial F_2}{\partial x_2} \end{vmatrix} = \begin{vmatrix} \frac{\partial^2 F_1}{\partial t\partial x_1} & \frac{\partial F_1}{\partial x_2} \\ \frac{\partial^2 F_2}{\partial t\partial x_1} & \frac{\partial F_2}{\partial x_2} \end{vmatrix} + \begin{vmatrix} \frac{\partial F_1}{\partial x_1} & \frac{\partial^2 F_1}{\partial t\partial x_2} \\ \frac{\partial F_2}{\partial x_1} & \frac{\partial^2 F_2}{\partial t\partial x_2} \end{vmatrix}$$

$$\sum_{l=1}^{2}\frac{\partial}{\partial x_l}\det Q_l = \frac{\partial}{\partial x_1}\det Q_1 + \frac{\partial}{\partial x_2}\det Q_2$$

$$= \frac{\partial}{\partial x_1}\begin{vmatrix} \frac{\partial F_1}{\partial t} & \frac{\partial F_1}{\partial x_2} \\ \frac{\partial F_2}{\partial t} & \frac{\partial F_2}{\partial x_2} \end{vmatrix} + \frac{\partial}{\partial x_2}\begin{vmatrix} \frac{\partial F_1}{\partial x_1} & \frac{\partial F_1}{\partial t} \\ \frac{\partial F_2}{\partial x_1} & \frac{\partial F_2}{\partial t} \end{vmatrix}$$

(By using Lemma 4.1)

$$= \begin{vmatrix} \dfrac{\partial^2 F_1}{\partial x_1 \partial t} & \dfrac{\partial F_1}{\partial x_2} \\ \dfrac{\partial^2 F_2}{\partial x_1 \partial t} & \dfrac{\partial F_2}{\partial x_2} \end{vmatrix} + \begin{vmatrix} \dfrac{\partial F_1}{\partial t} & \dfrac{\partial^2 F_1}{\partial x_1 \partial x_2} \\ \dfrac{\partial F_2}{\partial t} & \dfrac{\partial^2 F_2}{\partial x_1 \partial x_2} \end{vmatrix}$$

$$+ \begin{vmatrix} \dfrac{\partial^2 F_1}{\partial x_2 \partial x_1} & \dfrac{\partial F_1}{\partial t} \\ \dfrac{\partial^2 F_2}{\partial x_2 \partial x_1} & \dfrac{\partial F_2}{\partial t} \end{vmatrix} + \begin{vmatrix} \dfrac{\partial F_1}{\partial x_1} & \dfrac{\partial^2 F_1}{\partial x_2 \partial t} \\ \dfrac{\partial F_2}{\partial x_1} & \dfrac{\partial^2 F_2}{\partial x_2 \partial t} \end{vmatrix}$$

$$= \frac{\partial}{\partial t} \begin{vmatrix} \dfrac{\partial F_1}{\partial x_1} & \dfrac{\partial F_1}{\partial x_2} \\ \dfrac{\partial F_2}{\partial x_1} & \dfrac{\partial F_2}{\partial x_2} \end{vmatrix}$$

Thus the lemma is true in this case.

$$\sum_{l=1}^{n+1} \frac{\partial}{\partial x_l} \det Q_l = \frac{\partial}{\partial x_1} \begin{vmatrix} \dfrac{\partial F_1}{\partial t} & \dfrac{\partial F_1}{\partial x_2} & \dfrac{\partial F_1}{\partial x_3} & \cdots & \dfrac{\partial F_1}{\partial x_{n+1}} \\ \dfrac{\partial F_2}{\partial t} & \dfrac{\partial F_2}{\partial x_2} & \dfrac{\partial F_2}{\partial x_3} & \cdots & \dfrac{\partial F_2}{\partial x_{n+1}} \\ \cdots & \cdots & \cdots & \cdots & \cdots \\ \dfrac{\partial F_{n+1}}{\partial t} & \dfrac{\partial F_{n+1}}{\partial x_2} & \dfrac{\partial F_{n+1}}{\partial x_3} & \cdots & \dfrac{\partial F_{n+1}}{\partial x_{n+1}} \end{vmatrix}$$

$$+ \frac{\partial}{\partial x_2} \begin{vmatrix} \dfrac{\partial F_1}{\partial x_1} & \dfrac{\partial F_1}{\partial t} & \dfrac{\partial F_1}{\partial x_3} & \cdots & \dfrac{\partial F_1}{\partial x_{n+1}} \\ \dfrac{\partial F_2}{\partial x_1} & \dfrac{\partial F_2}{\partial t} & \dfrac{\partial F_2}{\partial x_3} & \cdots & \dfrac{\partial F_2}{\partial x_{n+1}} \\ \cdots & \cdots & \cdots & \cdots & \cdots \\ \dfrac{\partial F_{n+1}}{\partial x_1} & \dfrac{\partial F_{n+1}}{\partial t} & \dfrac{\partial F_{n+1}}{\partial x_3} & \cdots & \dfrac{\partial F_{n+1}}{\partial x_{n+1}} \end{vmatrix} + \cdots$$

$$+ \frac{\partial}{\partial x_{n+1}} \begin{vmatrix} \dfrac{\partial F_1}{\partial x_1} & \dfrac{\partial F_1}{\partial x_2} & \dfrac{\partial F_1}{\partial x_3} & \cdots & \dfrac{\partial F_1}{\partial t} \\ \dfrac{\partial F_2}{\partial x_1} & \dfrac{\partial F_2}{\partial x_2} & \dfrac{\partial F_2}{\partial x_3} & \cdots & \dfrac{\partial F_2}{\partial t} \\ \cdots & \cdots & \cdots & \cdots & \cdots \\ \dfrac{\partial F_{n+1}}{\partial x_1} & \dfrac{\partial F_{n+1}}{\partial x_2} & \dfrac{\partial F_{n+1}}{\partial x_3} & \cdots & \dfrac{\partial F_{n+1}}{\partial t} \end{vmatrix}$$

(by using the Lemma 4.1)

$$
\begin{aligned}
= & \begin{vmatrix}
\frac{\partial^2 F_1}{\partial x_1 \partial t} & \frac{\partial F_1}{\partial x_2} & \cdots & \frac{\partial F_1}{\partial x_{n+1}} \\
\frac{\partial^2 F_2}{\partial x_1 \partial t} & \frac{\partial F_2}{\partial x_2} & \cdots & \frac{\partial F_2}{\partial x_{n+1}} \\
\cdots & \cdots & \cdots & \cdots \\
\frac{\partial^2 F_{n+1}}{\partial x_1 \partial t} & \frac{\partial F_{n+1}}{\partial x_2} & \cdots & \frac{\partial F_{n+1}}{\partial x_{n+1}}
\end{vmatrix}
+
\begin{vmatrix}
\frac{\partial F_1}{\partial t} & \frac{\partial^2 F_1}{\partial x_1 \partial x_2} & \cdots & \frac{\partial F_1}{\partial x_{n+1}} \\
\frac{\partial F_2}{\partial t} & \frac{\partial^2 F_2}{\partial x_1 \partial x_2} & \cdots & \frac{\partial F_2}{\partial x_{n+1}} \\
\cdots & \cdots & \cdots & \cdots \\
\frac{\partial F_{n+1}}{\partial t} & \frac{\partial^2 F_{n+1}}{\partial x_1 \partial x_2} & \cdots & \frac{\partial F_{n+1}}{\partial x_{n+1}}
\end{vmatrix} \\
& + \cdots +
\begin{vmatrix}
\frac{\partial F_1}{\partial t} & \frac{\partial F_1}{\partial x_2} & \cdots & \frac{\partial F_1}{\partial x_1 \partial x_{n+1}} \\
\frac{\partial F_2}{\partial t} & \frac{\partial F_2}{\partial x_2} & \cdots & \frac{\partial F_2}{\partial x_1 \partial x_{n+1}} \\
\cdots & \cdots & \cdots & \cdots \\
\frac{\partial F_{n+1}}{\partial t} & \frac{\partial F_{n+1}}{\partial x_2} & \cdots & \frac{\partial F_{n+1}}{\partial x_1 \partial x_{n+1}}
\end{vmatrix}
+
\begin{vmatrix}
\frac{\partial^2 F_1}{\partial x_2 \partial x_1} & \frac{\partial F_1}{\partial t} & \cdots & \frac{\partial F_1}{\partial x_{n+1}} \\
\frac{\partial^2 F_2}{\partial x_2 \partial x_1} & \frac{\partial F_2}{\partial t} & \cdots & \frac{\partial F_2}{\partial x_{n+1}} \\
\cdots & \cdots & \cdots & \cdots \\
\frac{\partial^2 F_{n+1}}{\partial x_2 \partial x_1} & \frac{\partial F_{n+1}}{\partial t} & \cdots & \frac{\partial F_{n+1}}{\partial x_{n+1}}
\end{vmatrix} \\
& +
\begin{vmatrix}
\frac{\partial F_1}{\partial x_1} & \frac{\partial^2 F_1}{\partial x_2 \partial t} & \cdots & \frac{\partial F_1}{\partial x_{n+1}} \\
\frac{\partial F_2}{\partial x_1} & \frac{\partial^2 F_2}{\partial x_2 \partial t} & \cdots & \frac{\partial F_2}{\partial x_{n+1}} \\
\cdots & \cdots & \cdots & \cdots \\
\frac{\partial F_{n+1}}{\partial x_1} & \frac{\partial^2 F_{n+1}}{\partial x_2 \partial t} & \cdots & \frac{\partial F_{n+1}}{\partial x_{n+1}}
\end{vmatrix}
+ \cdots +
\begin{vmatrix}
\frac{\partial F_1}{\partial x_1} & \frac{\partial F_1}{\partial t} & \cdots & \frac{\partial^2 F_1}{\partial x_2 \partial x_{n+1}} \\
\frac{\partial F_2}{\partial x_1} & \frac{\partial F_2}{\partial t} & \cdots & \frac{\partial^2 F_2}{\partial x_2 \partial x_{n+1}} \\
\cdots & \cdots & \cdots & \cdots \\
\frac{\partial F_{n+1}}{\partial x_1} & \frac{\partial F_{n+1}}{\partial t} & \cdots & \frac{\partial^2 F_{n+1}}{\partial x_2 \partial x_{n+1}}
\end{vmatrix}
+ \cdots \\
& +
\begin{vmatrix}
\frac{\partial^2 F_1}{\partial x_{n+1} \partial x_1} & \frac{\partial F_1}{\partial x_2} & \cdots & \frac{\partial F_1}{\partial t} \\
\frac{\partial^2 F_2}{\partial x_{n+1} \partial x_1} & \frac{\partial F_2}{\partial x_2} & \cdots & \frac{\partial F_2}{\partial t} \\
\cdots & \cdots & \cdots & \cdots \\
\frac{\partial^2 F_{n+1}}{\partial x_{n+1} \partial x_1} & \frac{\partial F_{n+1}}{\partial x_2} & \cdots & \frac{\partial F_{n+1}}{\partial t}
\end{vmatrix}
+
\begin{vmatrix}
\frac{\partial F_1}{\partial x_1} & \frac{\partial^2 F_1}{\partial x_{n+1} \partial x_2} & \cdots & \frac{\partial F_1}{\partial t} \\
\frac{\partial F_2}{\partial x_1} & \frac{\partial^2 F_2}{\partial x_{n+1} \partial x_2} & \cdots & \frac{\partial F_2}{\partial t} \\
\cdots & \cdots & \cdots & \cdots \\
\frac{\partial F_{n+1}}{\partial x_1} & \frac{\partial^2 F_{n+1}}{\partial x_{n+1} \partial x_2} & \cdots & \frac{\partial F_{n+1}}{\partial t}
\end{vmatrix} \\
& + \cdots +
\begin{vmatrix}
\frac{\partial F_1}{\partial x_1} & \frac{\partial F_1}{\partial x_2} & \cdots & \frac{\partial^2 F_1}{\partial x_{n+1} \partial t} \\
\frac{\partial F_2}{\partial x_1} & \frac{\partial F_2}{\partial x_2} & \cdots & \frac{\partial^2 F_2}{\partial x_{n+1} \partial t} \\
\cdots & \cdots & \cdots & \cdots \\
\frac{\partial F_{n+1}}{\partial x_1} & \frac{\partial F_{n+1}}{\partial x_2} & \cdots & \frac{\partial^2 F_{n+1}}{\partial x_{n+1} \partial t}
\end{vmatrix}
\end{aligned}
$$

The first term of the first series + the second term of the second series + $\cdots$ + $(k+1)$th term of the last series make up

$$\frac{\partial}{\partial t}\begin{vmatrix} \dfrac{\partial F_1}{\partial x_1} & \cdots\cdots & \dfrac{\partial F_1}{\partial x_{n+1}} \\ \dfrac{\partial F_2}{\partial x_1} & \cdots\cdots & \dfrac{\partial F_2}{\partial x_{n+1}} \\ \cdots & \cdots & \cdots \\ \dfrac{\partial F_{n+1}}{\partial x_1} & \cdots\cdots & \dfrac{\partial F_{n+1}}{\partial x_{n+1}} \end{vmatrix}$$

and the rest of the terms cancel by pairs.

Thus we have proved the relation:

$$\frac{\partial}{\partial t}\det\left(\frac{\partial F_j}{\partial x_k}\right) = \sum_{l=1}^{n}\frac{\partial}{\partial x_l}\det Q_l. \qquad \square$$

Proof of Theorem 4.2. Let $f = (f_1, f_2, \dots, f_{n+1}) : B^{n+1} \longrightarrow B^{n+1}$ be an arbitrary smooth mapping. Let $V_f = \int\cdots\int_{B^{n+1}} \det\left(\frac{\partial f_j}{\partial x_k}\right) dx_1\cdots dx_{n+1}$ be the integral over B^{n+1} of Jacobian determinant of f. Let $g : B^{n+1} \longrightarrow B^{n+1}$ be another smooth map such that $f = g$ on S^n and $f(B^{n+1}) \subset S^n$.

We now define $F(t, x) : B^{n+1}\times[0, 1] \longrightarrow B^{n+1}$ by $F(t, x) = tf(x)+(1-t)g(x)$, $x \in B^{n+1}$, $0 \le t \le 1$, and $V(t) = tV_f + (1-t)V_g$, $0 \le t \le 1$, where V_g is integral over B_n of the Jacobian determinant of g. Then

$$\frac{dV}{dt} = \int\cdots\int_{B^{n+1}}\frac{\partial}{\partial t}\det\left(\frac{\partial F_j}{\partial x_k}\right)dx_1\cdots dx_{n+1}. \tag{4.4}$$

Since $f = g$ on S^n, we can easily see that if $x \in S^n$

$$\frac{\partial F_j}{\partial t}(t, x) = f_j(t, x) - g_j(t, x) = 0, \quad \text{for} \quad 1 \le j \le n+1.$$

Also it follows that $\det Q_l(t, x) = 0$, if $x \in S^n$ where Q_l has the meaning as defined in Lemma 4.2.

Also by Lemma 4.2 we have

$$\frac{\partial}{\partial t}\det\left(\frac{\partial F_j}{\partial x_k}\right) = \sum_{l=1}^{n+1}\frac{\partial}{\partial x_l}\det Q_l. \tag{4.5}$$

Now we set $B_l = B^{n+1} \cap \{x \in \mathbb{R}^{n+1} : x_l = 0\}$ and for $x \in B_l$, we define

$$\psi_l^+(x) = x + (0, \dots, (1-\sum_{j\neq l}x_j^2)^{\frac{1}{2}}, \dots, 0)$$

$$\psi_l^-(x) = x + (0, \dots, -(1-\sum_{j\neq l}x_j^2)^{\frac{1}{2}}, \dots, 0)$$

Now for each $l = 1, 2, \ldots, n$

$$\int \cdots \int_{B^{n+1}} \frac{\partial}{\partial x_l} \det Q_l(t, x) dx_1, \ldots, dx_{n+1}$$

$$= \int \cdots \int_{B_l} \int_{-(1-\sum_{j\neq l} x_j^2)^{\frac{1}{2}}}^{(1-\sum_{j\neq l} x_j^2)^{\frac{1}{2}}} \frac{\partial}{\partial x_l} Q_l dx_1, \ldots, dx_{n+1}$$

$$= \int \cdots \int_{B_l} \left[Q_l\left(t, \psi_l^+(x)\right) - Q_l\left(t, \psi_l^-(x)\right)\right] dx_1, \ldots, dx_{l-1}, dx_{l+1}, \ldots, dx_{n+1}$$

$$= 0 \text{ as } \psi_l^+(x), \psi_l^-(x) \in S^n.$$

It follows from (4.4) and (4.5) that $\frac{dV}{dt} = 0$ and hence $V_f = V_g$. Now if $g(x) = x$, $x \in B^{n+1}$ is the identity mapping, then

$$V_g > 0. \tag{4.6}$$

On the other hand, since $f(B^n) \subset S^n$, $\sum_{j=1}^{n+1} f_j^2 = 1$. Differentiating we obtain $\sum_{j=1}^{n+1} f_j \frac{\partial f_j}{\partial x_k} = 0$, $1 \leq k \leq n+1$. Hence at each point the vector $(f_1, f_2, \ldots, f_{n+1})$ is orthogonal to each of the vectors $\left(\frac{\partial f_1}{\partial x_k}, \ldots, \frac{\partial f_{n+1}}{\partial x_k}\right)$, $1 \leq k \leq n+1$. Hence the latter set of $(n+1)$ vectors are linearly dependent. Then it follows that the Jacobian determint vanishes and thus $V_f = 0$. Hecne in view of (4.6), f cannot coincide with the identity mapping g on S^n. □

We do not claim the orginality of the above proof. It is just a variant of the proof given in Gamlelin and Greene (1983, p. 170, Exercise 5), Gilbarg and Trudinger (1977, p. 236), and Dunford and Schwartz (1958). The proof in Gamlelin and Greene (1983, p. 168), by using differential forms and Stokes Theorem is elegant indeed. For the benefit of the readers unfamiliar with differential forms, the above proof is preferred.

Theorem 4.4 (Brouwer Fixed Point Theorem) *If $T : B^{n+1} \longrightarrow B^{n+1}$ is a continuous mapping of B^{n+1} into itself, then T has a fixed point.*

Proof. Let $T = (f_1, f_2, \ldots, f_{n+1})$. Then each f_i by the Stone-Weierstrass Theorem (see eg Royden (1970)), is a uniform limit of a sequence $\{f_i^k\}_{k=1}^{\infty}$ of smooth functions $f_i^k : B^{n+1} \longrightarrow [-1, 1]$ with $\left|f_i^k(x)\right| \leq |f_i(x)|$ for all $i = 1, 2, \ldots, n+1$, $x \in B^{n+1}$ and $k = 1, 2, \ldots$.

Let $T_k = (f_1^k, f_2^k, \ldots, f_{n+1}^k) : B^{n+1} \longrightarrow B^{n+1}$, for $k = 1, 2, \ldots$. Then we have $T = \lim_{k\to\infty} T_k$ uniformly. Now since Theorem 4.2 is equivalent to Theorem 4.1, it follows that for each k, T_k has a fixed point $x_k \in B^{n+1}$. Since B^{n+1} is compact, $\{x_k\}$ has a subsequence $\{x_{k_j}\}$ converging to a point $x \in B^{n+1}$. Hence

$$\begin{aligned}\|T(x) - x\| \leq \left\|T(x) - T(x_{k_j})\right\| + \left\|T(x_{k_j}) - T_{k_j}(x_{k_j}\right\| \\ + \left\|T_{k_j}(x_{k_j}) - x_{k_j}\right\| + \left\|x_{k_j} - x\right\| \to 0 \text{ as } j \to \infty.\end{aligned}$$

Therefore $T(x) = x$. □

4.1.1 *Schauder Projection*

Definition 4.3 (Schauder Projection) If $A = \{a_1, a_2, \ldots, a_n\}$ be a finite subset of a normed linear space $(E, \|\cdot\|)$. Then, for any real $\varepsilon > 0$ we set $A_\varepsilon = \cup \{S_\varepsilon(a_i) : i = 1, 2, \ldots, n\}$, where as before $S_\varepsilon(a_i) = \{x \in E : \|x - a_i\| < \varepsilon\}$. For each $i = 1, 2, \ldots, n$, we define the function $\lambda_i : A_\varepsilon \longrightarrow \mathbb{R}$ by $\lambda_i(x) = \max[0, \|x - a_i\|]$. The Schauder projection $p_\varepsilon : A_\varepsilon \longrightarrow \text{co } A$ is given by

$$p_\varepsilon(x) = \frac{\sum_{i=1}^{n} \lambda_i(x) a_i}{\sum_{i=1}^{n} \lambda_i(x)},$$

where co A denotes the convex hull of A.

We note that as for each $x \in A_\varepsilon$ belongs to at least one $S_\varepsilon(a_i)$, $\sum_{i=1}^{n} \lambda_i(x) \neq 0$ and, therefore p_ε is well defined.

Proposition 4.1 *If $A = \{a_1, a_2, \ldots, a_n\}$ is a subset of a convex set K of $(E, \|\cdot\|)$ and p_ε is the Schauder projection as defined above, then we have*

(a) *p_ε is compact mapping of A_ε into $\text{co}\, A \subset K$, i.e. p_ε is continuous and $p_\varepsilon(A_\varepsilon)$ is contained in a compact subset of E;*
(b) *$\|x - p_\varepsilon(x)\| < \varepsilon$ for all $x \in A_\varepsilon$.*

Proof. (a) is obvious. Now for each $x \in A_\varepsilon$,

$$\|x - p_\varepsilon(x)\| = \frac{1}{\sum_{i=1}^{n} \lambda_i(x)} \left\| \sum_{i=1}^{n} \lambda_i(x)[x - a_i] \right\| < \varepsilon.$$

Hence (b) follows. □

Theorem 4.5 *If X is a topological space, $K \subset (E, \|\cdot\|)$ is a convex set and $T : X \to K$ is a compact mapping, then for each $\varepsilon > 0$, there exist a finite subset $A = (a_1, a_2, \ldots, a_n) \subset T(X)$ and a finite dimensional mapping $T_\varepsilon : X \to K$ such that*

(i) *$\|T_\varepsilon(x) - T(x)\| < \varepsilon$ for each $x \in X$;*

and

(ii) *$T_\varepsilon(X) \subset \text{co}\, A \subset K$.*

Proof. By virtue of the fact that $\overline{T(X)}$ is compact, there is a finite subset $A = (a_1, a_2, \ldots, a_n) \subset T(X)$ with $T(X) \subset A_\varepsilon$ where A_ε is as defined before. Now we define $T_\varepsilon : X \to K$ as $T_\varepsilon = p_{\varepsilon} \circ T : X \longrightarrow K$ where $p_\varepsilon : A_\varepsilon \longrightarrow \operatorname{co} A$ is the Schauder projection. (i) follows from (b) of Proposition 4.1 while (ii) is obvious. □

Theorem 4.6 (Schauder fixed point theorem) *Let K be a convex subset of a normed linear space $(E, \|\cdot\|)$ and $T : K \to K$ a compact mapping. Then T has a fixed point.*

Proof. For each $\varepsilon > 0$, there is by Theorem 4.5 a mapping $T_\varepsilon : K \longrightarrow K$ such that

$$\|T_\varepsilon(x) - T(x)\| < \varepsilon \text{ for each } x \in K$$

and $T_\varepsilon(K) \subset \operatorname{co} A \subset K$ for some finite subset A of K. Now since $T_\varepsilon(\operatorname{co} A) \subset \operatorname{co} A$ and $\operatorname{co} A$ is homeomorphic to a finite-dimensional ball, T_ε has by Brouwer fixed point theorem a fixed point $x_\varepsilon \in \operatorname{co} A \subset K$. Thus for each $\varepsilon > 0$ we have $\|x_\varepsilon - T(x_\varepsilon)\| = \|T_\varepsilon(x_\varepsilon) - T(x_\varepsilon\| < \varepsilon$. Since $\overline{T(K)}$ is compact, we can find a subsequence $\{x_k\}$ converging to $x_0 \in K$ and satisfying

$$\|x_k - T(x_k)\| = \left\|T_{\frac{1}{k}}(x_k) - T(x_k)\right\| < \frac{1}{k}, \quad k = 1, 2, \ldots,$$

(taking $\varepsilon = \frac{1}{k}$).

Now as T is continuous, letting $K \to \infty$ we obtain $Tx_0 = x_0$. □

Corollary 4.6.1 *If K is a compact convex subset of a normed linear space $(E, \|\cdot\|)$ and $T : K \to K$ continuous mapping, then T has a fixed point.*

Proof. Clearly T is a compact mapping. The corollary follows from Theorem 4.6. □

Theorem 4.7 (Tychonoff fixed theorem) *If $T : K \longrightarrow K$ is a continuous mapping of a compact convex subset K of locally convex Hausdorff topological vector space E into K, then T has a fixed point.*

We delayed the proof until the next section. The proof is given after the proof of Corollary 4.14.1.

Theorem 4.8 (Markoff-Kakutani) *Let K be a cmpact convex subset of a locally convex Hausdorff topological vector space E and $\{T_\alpha : \alpha \in I\}$ a commuting family of continuous affine mappings of K into K. Then $\{T_\alpha : \alpha \in I\}$ has common fixed point.*

Proof. For each $\alpha \in I$, F_α, the set of fixed points of T_α is nonempty by Theorem 4.7. Also since T_α is continuous and affine, F_α is respectively a compact and convex subset of K. We need to show that $\cap\{F_\alpha : \alpha \in I\} \neq \emptyset$.

Since F_α is closed for each $\alpha \in I$, it would suffice to show that each finite intersection $\bigcap\limits_{i=1}^{n} F_{\alpha_i} \neq \emptyset$. We employ the method of induction on the number n of

F_{α_i}. The result is true for $n = 1$. We now suppose it to be true for $i = m < n$, i.e. $\bigcap_{i=1}^{m} F_{\alpha_i} \neq \emptyset$.

Since $\{T_\alpha : \alpha \in I\}$ is commuting, it follows that

$$F_{\alpha_{m+1}}\left[\bigcap_{i=1}^{m} F_{\alpha_i}\right] \subset \bigcap_{i=1}^{m} F_{\alpha_i}.$$

To see this let $x \in \bigcap_{i=1}^{m} F_{\alpha_i}$, then for each $i = 1, 2, \ldots, m$, $T_{\alpha_i}\left(T_{\alpha_{m+1}}(x)\right) = T_{\alpha_{m+1}}\left(T_{\alpha_i}(x)\right) = T_{\alpha_{m+1}}(x)$. Hence $T_{\alpha_{m+1}}(x) \in \bigcap_{i=1}^{m} F_{\alpha_i}$. Now as $\bigcap_{i=1}^{m} F_{\alpha_i}$ is a nonempty compact convex subset of E, it follows that $\bigcap_{i=1}^{m+1} F_{\alpha_i} \neq \emptyset$. Hence the result follows by induction. □

For further results and applications the readers are referred to Dugundji and Granas (1982).

4.1.2 *Fixed Point Theorems of Set Valued Mappings with Applications in Abstract Economy*

We will now consider the extension in Fan (1952) of the fixed point theorem of Kakutani in $\mathbb{R}^n$ and of Bohnenblust and Karlin (1950) in Banach space to locally convex Hausdorff topological vector space. Here we present the materials as we did in Tarafdar (1990b).

We begin with the following theorem.

Theorem 4.9 *Let K be a nonempty convex subset of normed linear space and $T : K \to 2^K$ an upper semicontinuous set valued mapping with nonempty compact convex values. If $X = \overline{T(K)}$ is compact and $X \subset K$, then there is a point $x_0 \in K$ such that $x_0 \in T(x_0)$, where $T(K) = \bigcup_{x \in K} T(x)$.*

For proof of this theorem see Dugundji and Granas (1978, p. 96, Theorem 11-3).

Corollary 4.9.1 *Let K be a nonempty compact convex subset of a normed linear space E and $T : K \to 2^K$ a upper semicontinuous set valued mapping with nonempty closed convex values. Then there exists a point $x_0 \in K$ such that $x_0 \in T(x_0)$.*

Proof. By Theorem 2.33 $T(K)$ is a compact set. Hence the corollary follows from the above theorem. □

Corollary 4.9.2 (Kakutani (1941)) *If K is a nonempty compact convex subset of $\mathbb{R}^n$ and $T : K \longrightarrow 2^K$ an upper semicontinuous set valued mapping with nonempty closed convex values, then there is a point $x_0 \in K$ such that $x_0 \in T(x_0)$.*

Lemma 4.3 (Fan (1952)) *Let X be a separated uniform topological space. Let A be a closed subset and B a compact subset of X. Then for each open entourage U of the uniformity of X, there is an open entourage V such that*

$$V(A) \cap V(B) \subset U(A \cap B),$$

where $W(D) = \{x \in X : (x, a) \in W$ for some $a \in D\}$ for a subset D of X and an entourage W of the uniformity.

Proof. For the sake of completeness we include the proof of Fan (1952). We can choose an open entourage U_1 such that $U_1 \circ U_1 \subset V$. Then $U_1(A \cap B)$ is an open subset of X. Let $K = B \cap [U_1(A \cap B)]^c$ which is a compact subset disjoint from the closed set A. Hence we can find an open entourage U_2 such that $K \cap U_2(A) = \phi$. This implies $B \cap U_2(A) \subset U_1(A \cap B)$. Indeed if $x \in B \cap U_2(A)$, $x \notin K$, ie. $x \in U_1(A \cap B)$. Now if V is an entourage such that $V = V^{-1} \subset U_1$ and $V \circ V \subset U_2$, then we can prove that $V(A) \cap V(B) \subset U(A \cap B)$. Indeed, let $x \in V(A) \cap V(B)$. Then $(a, x) \in V$ and $(b, x) \in V$ for some $a \in A$ and some $b \in B$. Since $V = V^{-1}$, $(a, b) \in V \circ V \subset U_2$. Hence $b \in B \cap U_2(A) \subset U_1(A \cap B)$. Thus there exists $c \in A \cap B$ such that $(c, b) \in U_1$. As we have already $(b, x) \in V$, we get $(c, x) \in U_1 \circ V \subset U_1 \circ U_1 \subset U$, ie. $x \in U(c) \subset U(A \cap B)$. □

We recall the following definition.

Let X and Y be two topological spaces. Then a set valued mapping $f : X \to 2^Y$ is said to be *upper semicontinuous* (*almost upper semicontinuous*) if for each $x_0 \in X$ and each open set U in Y with $f(x_0) \subset U$, there is an open set W with $x_0 \in W$ such that $f(x) \subset U$ ($f(x) \subset \overline{U}$) for all $x \in W$. Trivially an upper semicontinuous mapping is almost upper semicontinuous.

Theorem 4.10 *Let K be nonempty compact convex subset of a locally convex Hausdorff topological vector space E. Let $f, g : K \to 2^K$ be two set valued mappings with closed values such that*

(*i*) *f is almost upper semicontinuous and g is upper semicontinuous; and*
(*ii*) *for each $x \in X$, $g(x) \neq \emptyset$ and $\operatorname{co} g(x) \subset f(x)$.*

Then there exists a point $x_0 \in K$ such that $x_0 \in f(x_0)$.

Proof. For the sake of completeness we will repeat the argument of Fan (1952) wherever we find it necessary.

Let $\mathcal{B}$ be an open base of neighborhoods of 0 of E such that each $V \in \mathcal{B}$ is convex and symmetric, i.e. $V = -V$. For each $V \in \mathcal{B}$, we define the sets

$$F_V = \{x \in K : x \in f(x) + \overline{V}\}$$

and

$$G_V = \{x \in K : x \in \operatorname{co} g(x) + \overline{V}\},$$

where $\overline{V}$ denotes the closure of V.

Then by the condition of the theorem $G_V \subset F_V$ for each $V \in \mathcal{B}$. We first prove that $G_V \neq \emptyset$ for each $V \in \mathcal{B}$. To this end we consider a $V \in \mathcal{B}$ arbitrary but fixed. Since K is compact, there exists a finite number of points $x_1, x_2, \ldots, x_n$ in K such that $K \subset \bigcup_{i=1}^{n}(x_i + V)$. Let C be the closed convex hull of $\{x_1, x_2, \ldots, x_n\}$. For each $x \in C$, we define

$$g_V(x) = (g(x) + \overline{V}) \cap C.$$

Then clearly for each $x \in C$, $g_V(x)$ is a nonempty closed and hence compact subset of C. Now we prove that g_V is upper semicontinuous. Let $x_0 \in C$ and U be an open set in E such that $g_V(x_0) \subset U$. Since $g_V(x_0)$ is compact, we can find $V_1 \in \mathcal{B}$ such that $V_1 + g_V(x_0) \subset U$ (e.g. see Kelley and Namioka (1963, p. 351, 5.2 (vi))). Now by Lemma 4.3 we can find $V_2 \in \mathcal{B}$ such that

$$(V_2 + g(x_0) + \overline{V}) \cap (V_2 + C) \subset V_1 + [(g(x_0) + \overline{V}) \cap C].$$

Then it follows that

$$(V_2 + g(x_0) + \overline{V}) \cap C \subset V_1 + g_V(x_0) \subset U.$$

Now by the upper semicontinuity of g, there exists a neighborhood W of x_0 such that $g(x) \subset V_2 + g(x_0)$ for all $x \in W \cap K$.

Thus for all $x \in W \cap C$,

$$g_V(x) = (g(x) + \overline{V}) \cap C \subset [V_2 + g(x_0) + \overline{V}] \cap C \subset U.$$

Thus $g_V : C \to 2^C$ is an upper semicontinuous compact valued mapping. Now since C is a compact convex subset of finite dimensional subspace, the set valued mapping $h : C \to 2^C$ defined by $h(x) = \operatorname{co} g_V(x)$, $x \in C$ is upper semicontinuous and compact valued (e.g. see Nikaido (1968, Theorem 4.8 and Corollary to Theorem 2.9)). Hence by Kakutani's fixed point theorem there is a point $x_0 \in C$ such that

$$x_0 \in h(x_0) = \operatorname{co} g_V(x_0) = \operatorname{co}[(g(x_0) + \overline{V}) \cap C] \subset (\operatorname{co} g(x_0) + \overline{V}) \cap C,$$

i.e. $x_0 \in G_V$.

Since V is arbitrary, $G_V \neq \emptyset$ for each $V \in \mathcal{B}$. Hence F_V is nonempty for each $v \in \mathcal{B}$ as $G_V \subset F_V$.

In our next move, we prove that F_V is closed for each $V \in \mathcal{B}$. Again we consider a fixed but arbitrary $V \in \mathcal{B}$. We prove that $F_V^c = K \setminus F_V$ is open. Let $y \in F_V^c$. Then y is not contained in the closed set $f(y) + \overline{V}$. It is possible to find a $V' \in \mathcal{B}$ such that $(y + \overline{V}') \cap (f(y) + \overline{V} + \overline{V}') = \emptyset$. Now by almost upper semicontinuity of f, there is a $W \in \mathcal{B}$ such that $f(z) \subset \overline{f(y) + V'} \subset f(y) + \overline{V}'$ (as $f(y)$ is compact and $\overline{V}'$ is closed, $f(y) + \overline{V}'$ is closed, (e.g. see Kelley and Namioka (1963, p. 35, 5.2 (vii))) for all $z \in (y+W) \cap K$. We may assume $W \subset V'$. It then follows that for any $z \in (y+W) \cap K$, $z \notin f(z) + \overline{V}$, i.e. $z \notin F_V$ (for otherwise $z \in (y+W) \cap K \subset y + V'$ and $z \in f(z) + \overline{V} \subset f(y) + \overline{V} + \overline{V}'$ which leads to a contradiction). Thus we have proved that F_V^c is open. Since the finite intersection of members of $\mathcal{B}$ is again in $\mathcal{B}$, it follows that the family $\{F_V : V \in \mathcal{B}\}$ of closed sets has finite intersection property. Hence $\bigcap_{V \in \mathcal{B}} F_V \neq \emptyset$. Now it is easy to prove that for any point $x_0 \in \bigcap_{V \in \mathcal{B}} F_V$, $x_0 \in f(x_0)$. □

Lemma 4.4 *Let K be a nonempty compact convex subset of a locally convex Hausdorff topological vector space E and $g : K \to 2^K$ an almost upper semicontinuous set valued mapping. Then the set valued mapping $f : D \to 2^K$ defined by $f(x) = \overline{\text{co}} g(x)$, $x \in K$ is almost upper semicontinuous with nonempty closed values.*

Proof. Let U be an open set containing $f(x)$. Since $f(x)$ is a compact subset of E, we can find a convex open neighborhood N of 0 such that $f(x) + N \subset U$ (e.g. see Kelley and Namioka (1963, p. 35, 5.2. (vi)) and note that E is locally convex). Clearly $V = f(x) + N$ is convex open set containing $f(x)$ and $V \subset U$. Now since g is almost upper semicontinuous, there is an open set W containing x such that $g(y) \subset \overline{V}$ for every $y \in W \cap K$. Then as V is convex, $f(y) = \overline{\text{co}} g(y) \subset \overline{V} \subset \overline{U}$ for each $y \in W \cap K$. □

Corollary 4.10.1 *Let K be a nonempty compact convex subset of a locally convex Hausdorff topological vector space E and $g : K \to 2^K$ be an upper semicontinuous set valued mapping such that for each $x \in K$, $g(x)$ is a nonempty closed subset of K. Then there exists a point $x_0 \in D$ such that $x_0 \in \overline{\text{co}}\, g(x_0)$.*

Proof. We define the set valued mapping $f : K \to 2^K$ by $f(x) = \overline{\text{co}}\, g(x)$, $x \in K$. Then by Lemma 4.4, $f(x)$ is almost upper semicontinuous. Clearly the pair (f, g) satisfies all the conditions of Theorem 4.10. Hence there exists a point $x_0 \in K$ such that $x_0 \in f(x_0)$. □

Corollary 4.10.2 (Fan's fixed point theorem) *Let K be a nonempty compact convex subset of a locally convex Hausdorff topological vector space. If $f : K \to 2^K$ is an upper semicontinuous set valued mapping with closed values such that for each $x \in K$, $f(x)$ is a nonempty convex subset of K, then there exists a point $x_0 \in K$ such that $x_0 \in f(x_0)$.*

Proof. If we take $f = g$ in Theorem 4.10, the corollary is obtained. □

Theorem 4.11 *Let X be a topological and $\{Y_\alpha : \alpha \in I\}$ a family of compact Hausdorff spaces. If for each $\alpha \in I$, $f_\alpha : X \to 2^{Y_\alpha}$ is an upper semicontinuous (almost upper semicontinuous) set valued mapping with closed values, and $Y = \prod_{\alpha \in I} Y_\alpha$, then the set valued mapping $f : X \to 2^Y$ defined by*

$$f(x) = \prod_{\alpha \in I} f_\alpha(x), \quad x \in X$$

is upper semicontinuous (almost upper semicontinuous) with closed values.

Proof. Let $x_0 \in X$ be arbitrary and U be any open set in Y containing $f(x_0)$. If $f_\mu(x_0) = \emptyset$ for some $\mu \in I$, then by the upper semicontinuity of f_μ at x_0, we have an open set W containing x_0 such that $f_\mu(x) = \emptyset$ for all $x \in W$. Thus it follows that $f(x) = \prod_{\alpha \in I} f_\alpha(x) = \emptyset \subset$U $(\overline{U})$ for all $x \in W$. So we may assume that $f_\alpha(x_0) \neq \emptyset$ for all $\alpha \in I$. We first prove the theorem for the case when $I = \{1, 2, \dots, n\}$ is finite. Y_i can be regarded as a uniform space for each i (e.g. see Wilansky (1970, p. 222, Th.11.4.6)). Since $f_i(x_0)$ is a compact set in the uniform space Y_i and U is an open set containing $f(x_0) = \prod_{i=1}^n f_i(x_0)$, there exists for each i an open set U_i in Y_i such that $f_i(x_0) \subset U_i$ $(i = 1, 2, \dots, n)$ and $\prod_{i=1}^n U_i \subset U$. Now by the upper semicontinuity (almost upper semicontinuity) of each f_i, $i = 1, 2, \dots, n$, there exists an open neighborhood W_i of x_0 such that $f_i(x) \subset U_i$ $(\overline{U}_i)$ for $x \in W_i$. Hence $W = \bigcap_{i=1}^n W_i$ is the required open neighborhood of x_0 with $f(x) = \prod_{i=1}^n f_i(x) \subset \prod_{i=1}^n U_i \left(\prod_{i=1}^n \overline{U}_i\right) \subset U$ $(\overline{U})$. Thus we have proved theorem when I is finite. Now we assume I is arbitrary and $f_\alpha(x_0) \neq \emptyset$ for each $\alpha \in I$. Since U is the union of basic open sets in Y and $f(x_0 = \prod_{\alpha \in I} f_\alpha(x_0)$ is compact, there is a finite number of basic open sets $U^{(j)}$ in Y $(j = 1, 2, \dots, m)$ such that

$$f(x_0) = \prod_{\alpha \in I} f_\alpha(x_0) \subset \bigcup_{j=1}^m U^{(j)} \subset U \tag{4.7}$$

where the m sets $U^{(j)}$ are of the form $U^{(j)} = \prod_{i=1}^n U_{\nu_i}^{(j)} \times \prod_{\substack{\nu \in I \\ \nu \neq \nu_i}}$, by the definition of the product topology, where $U_{\nu_i}^{(j)}$ is open in Y_{ν_i}. It follows from (4.7) that

$$\prod_{i=1}^n f_{v_i}(x_0) \subset \bigcup_{j=1}^m \prod_{i=1}^n U_{\nu_i}^{(j)},$$

where the right hand side is an open set in $\prod_{i=1}^n Y_{\nu_i}$. Then by what we have proved for the case when I is finite, the mapping $x \to \prod_{i=1}^n f_{\nu_i}(x)$ is upper semicontinuous (almost upper semicontinuous). Thus there exists an open neighborhood W of x_0 such that

$$\prod_{i=1}^n f_{\nu_i}(x) \subset \bigcup_{j=1}^m \prod_{i=1}^n U_{\nu_i}^{(j)} \left(\overline{\bigcup_{j=1}^m \prod_{i=1}^n U_{\nu_i}^{(j)}}\right) \quad \text{for } x \in W.$$

It follows that

$$f(x) = \prod_{\alpha \in I} f_\alpha(x) \subset \bigcup_{j=1}^{m} U^{(j)} \subset U\left(\overline{\bigcup_{i=1}^{m} U^{(j)}} \subset \overline{U}\right). \qquad \square$$

The above proof for upper semicontinuous case is due to Fan (1952).

4.1.3 *Applications*

All topological spaces considered for the rest of this subsection are assumed to be Hausdorff.

Following Debreu (1952), Arrow and Debreu (1954) and Shafer and Sonnenschein (1975), we describe an abstract economy(or generalized game) with utility functions (or pay off) functions by $\mathcal{E} = \{\mathcal{X}_\alpha.\mathcal{A}_\alpha, \mathcal{U}_\alpha : \alpha \in \mathcal{I}\}$ where I is a finite or an infinite set of agents (or players) and for each $\alpha \in I$, X_α is the choice set (or strategy set), $A_\alpha : X = \prod_{\alpha \in I} X_\alpha \to 2^{X_\alpha}$ is the (budget) constraint correspondence, i.e. set valued mapping and $U_\alpha : X \to \mathbb{R}$ is the utility (or pay off) function and an economy with preference correspondences $\mathcal{E} = \{\mathcal{X}_\alpha, \mathcal{A}_\alpha, \mathcal{P}_\alpha : \alpha \in \mathcal{I}\}$, where I, X_α and A_α are as above and $P_\alpha : X \to 2^{X_\alpha}$ is the preference correspondnece for the agent $\alpha \in I$. Before going any further we first make clear the symbols and notations to be used throughout this section. X and $X_{-\alpha}$ will respectively denote the cartesian product $\prod_{\alpha \in I} X_\alpha$ and $\prod_{\substack{\beta \in I \\ \beta \neq \alpha}} X_\beta$ and $x_{-\alpha}$ will denote a generic element of $X_{-\alpha}$. We will also represent an element $x \in X$ by $\{x_\alpha\}$ where $x_\alpha\}$ is the projection of x onto X_α, i.e. x_α is the α-th co-ordinate of x. Thus we can write $x = \{x_\alpha\} = [x_\alpha, x_{-\alpha}]$ for each $\alpha \in I$.

A point $\overline{x} = \{\overline{x}_\alpha\}$ is called an *equilibrium point* of an abstract economy $\mathcal{E} = \{\mathcal{X}_\alpha, \mathcal{A}_\alpha, \mathcal{U}_\alpha : \alpha \in \mathcal{I}\}$ if for each $\alpha \in I$,

$$U_\alpha(\overline{x}) = U_\alpha[\overline{x}_\alpha, \overline{x}_{-\alpha}] = \sup_{z_\alpha \in A_\alpha(\overline{x})} U_\alpha[z_\alpha, \overline{x}_{-\alpha}].$$

It is worth noting that if $A_\alpha(x) = X_\alpha$ for each $x \in X$, the concept of an equilibrium point of the economy coincides with the well-known concept of Nash equilibrium point (1950), for in the latter case, $U_\alpha(\overline{x}) = U_\alpha[\overline{x}_\alpha, \overline{x}_{-\alpha}] = \sup_{z_\alpha \in X_\alpha} U_\alpha[z_\alpha, \overline{x}_{-\alpha}]$ for each $\alpha \in I$. An economy $\mathcal{E} = \{\mathcal{X}_\alpha, \mathcal{A}_\alpha, \mathcal{U}_\alpha : \alpha \in \mathcal{I}\}$ can be expressed as an economy of the form $\{X_\alpha, A_\alpha, P_\alpha : \alpha \in I\}$ if for each $\alpha \in I$, we define the correspondence $P_\alpha : X \to 2^{X_\alpha}$ by $P_\alpha(x) = \{y_\alpha \in X_\alpha : U_\alpha([y_\alpha, x_{-\alpha}]) > U_\alpha(x)\}$ for each $x = \{x_\alpha\} \in X$.

Now suppose that for each $\alpha \in I$, $P_\alpha : X \to 2^{X_\alpha}$ is defined as above. Then it is clear that $\overline{x}$ is an equilibrium point of the economy $\mathcal{E} = \{\mathcal{X}_\alpha, \mathcal{A}_\alpha, \mathcal{U}_\alpha : \alpha \in \mathcal{I}\}$ if and only if $P_\alpha(\overline{x}) \cap A_\alpha(\overline{x}) = \emptyset$ and $\overline{x}_\alpha \in A_\alpha(\overline{x})$ for each $\alpha \in I$.

Thus if the economy instead of being given by utility functions is given by preference relations, we can define an equilibrium point of an abstract economy $\mathcal{E} = \{\mathcal{X}_\alpha, \mathcal{A}_\alpha, \mathcal{P}_\alpha : \alpha \in \mathcal{I}\}$ as a point $\overline{x} = \{\overline{x}_\alpha\} \in X$ such that for each $\alpha \in I$, $\overline{x}_\alpha \in A_\alpha(\overline{x})$ and $P_\alpha(\overline{x}) \cap A_\alpha(\overline{x}) = \emptyset$.

The object in the rest of this section is to extend the following two theorems to the case of locally convex topological vector spaces.

Theorem A (Debrew). *Let $\mathcal{E} = \{X_i, A_i, U_i\}_{i=1}^N$ be an abstract economy (a game) such that for each $i = 1, 2, \ldots, N$*

(*i*) *X_i is a nonempty compact convex subset of $\mathbb{R}^\ell$;*

(*ii*) *$A_i : \prod_{i=1}^N X_i \to 2^{X_i}$ is a continuous correspondence such that for each $x \in X$, $A_i(x)$ is nonempty and convex;*

(*iii*) *$U_i : X \to \mathbb{R}$ is continuous on X and quasiconcave in x_i (i.e. $U_i(\cdot, x_{-i})$ is quasiconcave for each x_{-i}).*

Then $\mathcal{E}$ has an equilibrium point.

Theorem B. (Shafer and Sonnenschein (1975)). *Let $\{X_i, A_i, P_i\}_{i=1}^N$ be an abstract economy such that for each $i = 1, 2, \ldots, N$*

(*i*) *X_i is a nonempty compact convex subset of $\mathbb{R}^\ell$;*

(*ii*) *$A_i : X = \prod_{i=1}^N \to 2^{X_i}$ is a continuous correspondence such that for each $x \in X$, $A_i(x)$ is nonempty and convex;*

(*iii*) *$P_i : X \to 2^{X_i}$ has an open graph in $X \times X_i$ and for each $x = \{x_i\}_{i=1}^N$, $x_i \notin \mathrm{co}(P_i(x))$, where $\mathrm{co}\, A$ denotes the convex hull of A.*

Then $\mathcal{E}$ has an equilibrium point.

We should point out that our extension of Theorem B requires (1) I to be countable, (2) a strong irreflexivity of the preference correspondence, i.e. $x_\alpha \notin \overline{\mathrm{co}} P_\alpha(x)$ for each $x = \{x_\alpha\}$ (this seems to be unavoidable due to the pathological defect in the convex hull of a compact subset in an infinite dimensional space, i.e. the convex hull of a compact subset in an infinite dimensional space need not be compact, nor even closed, e.g. see Schaefer [(1966), p. 72] and lastly — but most importantly (3) an extension of Fan's fixed point theorem. The extension of Fan's fixed point theorem, i.e. our Theorem 4.1 overcomes the difficulty arising out of the following fact: If $g : K \to 2^K$ is an upper semicontinuous set valued mapping where K is a nonempty compact convex subset of a locally convex topological vector space, then the mapping $f : K \to 2^K$ defined by $f(x) = \overline{\mathrm{co}} g(x)$ is not necessarily upper semicontinuous. We have not made any distinction between a correspondence and a set valued mapping. Rather we have used both to conform to the existing literature.

4.1.4 *Equilibrium Point of Abstract Economy*

In this subsection we consider both kinds of economy described above, an abstract economy by utility functions and an abstract economy given by preference relations and prove the existence of an equilibrium point for either case.

Theorem 4.12 *Let $\mathcal{E} = \{\mathcal{X}_\alpha, \mathcal{A}_\alpha, \mathcal{U}_\alpha : \alpha \in \mathcal{I}\}$ be an abstract economy such that for each $\alpha \in I$*

(i) *X_α is a nonempty compact and convex subset of a locally convex topological vector space E_α;*

(ii) *$A_\alpha : X = \prod_{\alpha\in I} X_\alpha \to 2^{X_\alpha}$ is continuous correspondence so that for each $x \in X$, $A_\alpha(x)$ is a nonempty compact and convex subset of X_α;*

(iii) *$U_\alpha : X \to \mathbb{R}$ is continuous and is quasiconcave in x_α.*

Then there is an equilibrium point $\overline{x} = \{\overline{x}_\alpha\} \in X$ of the economy, i.e. for each $\alpha \in I$, $U_\alpha(\overline{x}) = U_\alpha[\overline{x}_\alpha, \overline{x}_{-\alpha}] = \sup_{z_\alpha\in A(\overline{x}_\alpha)} U_\alpha[z_\alpha, \overline{x}_{-\alpha}]$.

Proof. Let for each $\alpha \in I$, $F_\alpha(x) = \{y_\alpha \in X_\alpha : U_\alpha[y_\alpha, x_{-\alpha}] = \sup_{z_\alpha\in A_\alpha(x)} U_\alpha[z_\alpha, x_{-\alpha}]\}$. Since U_α is continuous and $A_\alpha(x)$ is compact, $F_\alpha(x) \neq \emptyset$ and clearly $F_\alpha(x) \subset A_\alpha(x)$. Thus for each $\alpha \in I$, $F_\alpha : X \to 2^{X_\alpha}$ is a set valued mapping. We will now prove that F_α has a closed graph. To this end we fix $\alpha \in I$ and let $\{(x^\delta, y^\delta_\alpha) : \delta \in D\}$ be a net in $X \times X_\alpha$ such that $x^\delta \to x$ and $y^\delta_\alpha \to y_\alpha$ and $y^\delta_\alpha \in F_\alpha(x^\delta)$, i.e. $U_\alpha[y^\delta_\alpha, x^\delta_{-\alpha}] = \sup_{z_\alpha\in A_\alpha(x^\delta)} U_\alpha[z_\alpha, x^\delta_{-\alpha}]$, i.e.

$$U_\alpha[y^\delta_\alpha, x^\delta_{-\alpha}] \geq U_\alpha[z_\alpha, x^\delta_{-\alpha}] \quad \text{for all} \quad z_\alpha \in A_\alpha(x^\delta). \tag{A}$$

Now since $y^\delta_\alpha \in F_\alpha(x^\delta) \subset A_\alpha(x^\delta)$, by upper semicontinuity of A_α, $y_\alpha \in A_\alpha(x)$. Now let $z_\alpha \in A_\alpha(x)$ be arbitrary. Then by Lower semicontinuity of A_α, there is $z^\delta_\alpha \in A_\alpha(x^\delta)$ such that $z^\delta_\alpha \to z_\alpha$. But from (A), we have $U_\alpha[y^\delta_\alpha, x^\delta_{-\alpha}] \geq U_\alpha[z^\delta_\alpha, x^\delta_{-\alpha}]$. Taking limit we obtain $U_\alpha[y_\alpha, x_{-\alpha}] \geq U_\alpha[z_\alpha, x_{-\alpha}]$. Thus we have proved that $U_\alpha[y_\alpha, x_{-\alpha}] = \sup_{z_\alpha\in A_\alpha(x)} U_\alpha[z_\alpha, x_{-\alpha}]$, i.e. $y_\alpha \in F_\alpha(x)$. Hence F_α has a closed graph and therefore $F_\alpha(x)$ is a closed subset of X_α for each $x \in X$.

Next we prove that $F_\alpha(x)$ is convex for each $x \in X$ and each $\alpha \in I$. Let $x \in X$ and $\alpha \in I$ be arbitrarily fixed. Let $y^1_\alpha, y^2_\alpha \in F_\alpha(x)$ and $\overline{y}_\alpha = \lambda y^1_\alpha + \mu y^2_\alpha$, $\lambda, \mu \geq 0$ and $\lambda + \mu = 1$. Then $U_\alpha[y^1_\alpha, x_{-\alpha}] = \sup_{z_\alpha\in A_\alpha(x)} U_\alpha[z_\alpha, x_{-\alpha}] = U_\alpha[y^2_\alpha, x_{-\alpha}]$. Suppose that $\overline{y}_\alpha \notin F_\alpha(x)$. Then there will exist $u_\alpha \in A_\alpha(x)$ such that $U_\alpha[u_\alpha, x_{-\alpha}] > U_\alpha[\overline{y}_\alpha, x_{-\alpha}]$. Let $u_0 = [u_\alpha, x_{-\alpha}]$. Since U_α is quasiconcave in α-th co-ordinate, the set $B = \{z_\alpha \in X_\alpha : U_\alpha[z_\alpha, x_{-\alpha}] > U_\alpha(u_0)\}$ is convex. Also by the continuity of U_α, $\overline{B} = \{z_\alpha \in X_\alpha : U_\alpha[z_\alpha, x_{-\alpha}] \geq U_\alpha(u_0)\}$ and is therefore convex. Now since $u_\alpha \in A_\alpha(x)$, it follows that $y^1_\alpha \in \overline{B}$ and $y^2_\alpha \in \overline{B}$. Hence $\overline{y}_\alpha \in \overline{B}$, i.e. $U_\alpha[\overline{y}_\alpha x_{-\alpha}] \geq U_\alpha(u_0) = U_\alpha[u_\alpha, x_{-\alpha}]$ which is a contradiction. □

Thus we have proved that for each $\alpha \in I$, the set valued mapping $F_\alpha : X \to 2^{X_\alpha}$ has closed graph and has closed convex value for each $x \in X$. Hence for each $\alpha \in I$, F_α is upper semicontinuous and hence by Theorem 4.11 of Fan (1952) the set valued mapping $F : X \to 2^X$ defined by $F(x) = \prod_{\alpha \in I} F_\alpha(x)$, $x \in X$ is upper semicontinuous and is evidently nonempty closed convex valued. Hence by fixed point theorem of Fan (here Corollary 4.10.2) there is a point $\overline{x} \in X$ such that $\overline{x} \in F(\overline{x})$. Now it is easy to see that this point $\overline{x}$ is an equilibrium point of the economy $\mathcal{E}$.

Corollary 4.12.1 *Let $\{X_\alpha : \alpha \in I\}$ be a family of nonempty compact convex sets, each in locally convex topological vector space E_α. Let for each $\alpha \in I$, $U_\alpha : X = \prod_{\alpha \in I} X_\alpha \to \mathbb{R}$ be a continuous function such that U_α is quasiconcave in x_α. Then there is a Nash equilibrium point.*

Proof. For each $\alpha \in I$, we define the set valued mapping $A_\alpha : X \to 2^{X_\alpha}$ by $A_\alpha(x) = A_\alpha[x_\alpha\, x_{-\alpha}] = X_\alpha$. Clearly A_α is continuous. Hence the Corollary follows from Theorem 4.12. □

Remark 4.1 More general result than Corollary 4.12.1 is known, and will be given in the appropriate place, for instance this result is known in Hausdorff topological vector space (see Theorems or Ma (1969) and Tan (1984)).

Theorem 4.13 *Let $\mathcal{E} = \{\mathcal{X}_\rangle, \mathcal{A}_\rangle, \mathcal{P}_\rangle : \rangle \in \mathcal{I}\}$ be an abstract economy, where I a countable set. Assume that for each $i \in I$,*

(a) X_i is a nonempty compact and convex subset of a locally convex metrizable space E_i;

(b) $A_i : X = \prod_{i \in I} X_i \to 2^{X_i}$ is a continuous correspondence such that for each $x \in X$, $A_i(x)$ is nonempty and convex;

(c) $P_i : X \to 2^{X_i}$ has an open graph in $X \times X_i$; and

(d) for each $x = \{x_i\} \in X$, $x_i \notin \overline{\text{co}} P_i(x)$.

Then there is an equilibrium point $\overline{x} \in X$ of $\mathcal{E}$.

Proof. As in Shafer and Sonnenschein (1975) we define for each $i \in I$ a continuous mapping $U_i : X \times X_i \to \mathbb{R}$ by $U_i(x, y_i) = \inf_{(u, z_i) \in G_i^c} \rho_i(P_i(x, y_i), (u, z_i))$ where G_i^c is the complement of the graph G_i of P_i and ρ_i is the metric in $X \times X_i$. Now for each $i \in I$, we define the set valued mapping $F_i : X \to 2^{X_i}$ by $F_i(x) = \{y_i, X_i : U_i(x, y_i) = \sup_{z_i \in X_i} U_i(x, z_i)\}$.

Since U_i is continuous and A_i being nonempty valued and upper semicontinuous is nonempty compact valued correspondence, it follows that $F_i(x)$ is nonempty for each $x \in X$ and by similar argument as given in Theorem 4.12 we can show that F_i has closed graph and is therefore upper semicontinuous. Hence by Theorem

4.11 the set valued mapping $F : X \to 2^X$ defined by $F(x) = \prod_{i \in I} F_i(x)$, $x \in X$ is upper semicontinuous. Thus by Corollary 4.10.1 there is a point $\overline{x} \in X$ such that $\overline{x} \in \overline{\text{co}}\, F(x) \subset \prod_{i \in I} \overline{\text{co}}\, F_i(\overline{x})$. Now we repeat the same argument of Shafer and Sonnenschein to show that $\overline{x}$ is an equilibrium point of the economy $\mathcal{E}$. Since $F_i(\overline{x}) \subset A_i(\overline{x})$ and $A_i(\overline{x})$ is closed and convex, $\overline{x} \in \prod_{i \in I} \overline{\text{co}}\, F_i(\overline{x}) \subset \prod_{i \in I} A_i(\overline{x})$. Thus $\overline{x}_i \in A_i(\overline{x})$ for each $i \in I$ where $\overline{x} = \{\overline{x}_i\}$. It remains to show that for each $i \in I$, $P_i(\overline{x}) \cap A_i(\overline{x}) = \emptyset$. If $z_i \in P_i(\overline{x}) \cap A_i(\overline{x})$, then $U_i(\overline{x}, z_i) > 0$. This implies that $U_i(\overline{x}, y_i) > 0$ for all $y_i \in F_i(\overline{x})$. Hence $z_i \in P_i(\overline{x}) \cap A_i(\overline{x})$ implies that $F_i(\overline{x}) \subset P_i(\overline{x})$. Thus $\overline{x}_i \in \overline{\text{co}}\, F_i(\overline{x}) \subset \overline{\text{co}}\, P_i(\overline{x})$ which contradicts (d). □

Remark 4.2 Since in a finite dimensional space, $\text{co}\, A$ of a compact subset A is compact, the condition (d) reduces to (iii) of Theorem B of Shafer and Sonnenschein. Thus the theorem is indeed a generalization of their theorem.

4.2 Fixed Point Theorems and KKM Theorems

In this section we will prove a series of fixed point and KKM theorems. Unless otherwise stated, all topological vector spaces throughout this chapter are assumed to be real and Hausdorff.

Theorem 4.14 (Tarafdar (1977)) *Let K be a nonempty compact convex subset of a topological vector space E. Let $T : K \to 2^K$ be a set valued mapping of K into 2^K such that*

(i) for each $x \in K$, $T(x)$ is a nonempty convex subset of K.

(ii) for each $y \in K$, $T^{-1}(y) = \{x \in K : y \in T(x)\}$ contains an open subset O_y of K (O_y may be empty);

and

(iii) $\bigcup \{O_y : y \in K\} = K$.

Then there exists a point $x_0 \in K$ such that $x_0 \in T(x_0)$.

Proof. Since K is compact, there exists by (iii) a finite family $\{y_1, y_2, \ldots, y_n\}$ such that $K = \bigcup_{i=1}^{n} O_{y_i}$. Let $\{f_1, f_2, \ldots, f_n\}$ be a partition of unity corresponding to this finite subcovering, i.e. each f_i, $i = 1, 2, \ldots, n$ is a real valued continuous function defined on K such that f_i vanishes outside O_{y_i}, $O \leq f_i(x) \leq 1$ for all $x \in K$ and $\sum_{i=1}^{n} f_i(x) = 1$ for each $x \in K$. □

We now define a mapping $p : K \to K$ by

$$p(x) = \sum_{i=1}^{n} f_i(x) y_i\,, \quad x \in K.$$

Obviously p maps K into K and is continuous. Also for each k with $f_k(x) \neq 0$, $x \in O_{y_k} \subset T^{-1}(y_k)$, i.e. $y_k \in T(x)$. As $T(x)$ is convex, it readily follows that $p(x) \in T(x)$ for each $x \in K$.

Now let S be the finite dimensional simplex spanned by $y_1, y_2, \ldots, y_n$. Then clearly p maps S into S. Also since E is Hausdorff vector space, the topology on S induced by the topology in E is Euclidean. Hence by the Brouwer fixed point theorem, there is point $x_0 = p(x_0) \in T(x_0)$. □

Corollary 4.14.1 (Brouwer, 1968)) *Let K be a nonempty convex compact subset of a topological vector space E and $T : K \to 2^K$ be set valued mapping of K into 2^K such that*

(a) for each $x \in K$, $T(x)$ is a nonempty convex subset of K;

and

(b) for each $y \in K$, $T^{-1}(y) = \{x \in K : y \in T(x)\}$ is open in K.

Then there exists a point $x_0 \in T(x_0)$.

Proof. Take $O_y = T^{-1}(y)$ of Theorem 4.14. Clearly $K = \bigcup_{y \in K} T^{-1}(y)$. To see this let $x \in K$. As $T(x) \neq \emptyset$, we can choose $y \in T(x)$. Hence $x \in T^{-1}(y)$. Thus the corollary follows from Theorem 4.14. □

We now prove the Tychonoff fixed point theorem.

Proof of Theorem 4.8. Let $\{p_\alpha : \alpha \in I\}$ be the family of seminorms which generates the topology of E. For each $\alpha \in I$, let

$$F_\alpha = \{x \in K : p_\alpha(x - T(x)) = 0\}.$$

Then since E is Hausdorff, x_0 is a fixed point of T if and only if $x_0 \in \bigcap_{\alpha \in I} F_\alpha$. Also since F_α is closed for each $\alpha \in I$ and K is compact, $\bigcap_{\alpha \in I} F_\alpha \neq \emptyset$ if and only if each finite intersection $F_{\alpha_1} \cap F_{\alpha_2} \cap \ldots \cap F_{\alpha_n} \neq \emptyset$. We prove that the later is nonempty. To this end it will suffice to prove that there is a point $u_0 \in k$ such that

$$\sum_{i=1}^{n} p_{\alpha_i}(u_0 - T(u_0)) \leq \sum_{i=1}^{n} p_{\alpha_i}(x - T(u_0)) \quad \text{for all } x \in K. \tag{4.8}$$

For then it would imply that $p_{\alpha_i}(u_0 - T(u_0)) = 0$ for $i = 1, 2, \ldots, n$ (taking $x = T(u_0)$), i.e. $u_0 \in \bigcap_{i=1}^{n} F_{\alpha_i}$. If possible, we suppose that (4.8) is false. Then we define the set valued mapping $G : K \to 2^K$ by

$$G(y) = \left\{ x \in K : \sum_{i=1}^{n} p_{\alpha_i}(y - T(y)) > \sum_{i=1}^{n} p_{\alpha_i}(x - T(y)) \right\},$$

$y \in K$ such that $G(y) \neq \emptyset$ for each $y \in K$. Also it is clear that $G(y)$ is convex for each $y \in K$. Moreover, it follows from the continuity of T and p_{α_i}, $i = 1, 2, \ldots, n$ that for each $x \in K$, $G^{-1}(x) = \{y \in K : x \in G(y)\} = \{y \in K : \sum_{i=1}^n p_{\alpha_i}(y - T(y)) > \sum_{i=1}^n p_{\alpha_i}(x - T(y))\}$ is an open set of K. Hence by Corollary 4.14.1 there is a point $u_0 \in G(u_0)$ which is absurd. Thus (4.8) is true. Therefore, $\bigcap_{\alpha \in I} F_\alpha \neq \emptyset$. □

Note that this proof also contains the proof of Corollary 4.7 as a special case and that unlike Dugundji and Granas (1982) the fixed point theorem of Browder has been used instead of KKM theorem.

We now prove the following lemma which was first proved by Fan (1961) as an extension of the well known finite dimensional result of Knaster-Kuratowski-Mazurkiewicz (1929) to topological vector space. We give here a different proof. In fact, it can be shown that the following lemma is equivalent to the Corollary 4.16.1 (see the proof of equivalence in Section 4.4).

Lemma 4.5 *Let X be a nonempty subset of a topological vector space E. For each $x \in X$, let a nonempty closed subset $F(x)$ be given such that*

(i) $F(x_0)$ is compact for some $x_0 \in X$;

and

(ii) for each finite subset $\{x_1, x_2, \ldots, x_n\}$ of X the convex hull of $\{x_1, x_2, \ldots, x_n\}$ is contained in the corresponding union $\bigcup_{i=1}^n F(x_i)$. Then $\bigcap_{x \in X} F(x) \neq \emptyset$.

Proof. In view of (i) it will suffice to show that for each finite set $\{x_1, x_2, \ldots, x_n\}$, $\bigcap_{i=1}^n F(x_i) \neq \emptyset$. If possible, we suppose that for some subset $\{x_1, x_2, \ldots, x_m\}$, $\bigcap_{i=1}^m F(x_i) = \emptyset$. Then for each $x \in S_m = \langle x_1, x_2, \ldots, x_m \rangle$, the convex hull of $\{x_1, x_2, \ldots, x_m\}$, the set $A(x) = \{y \in S_m \cap X : x \notin F(y)\}$ is not empty.

Indeed, at least one of the points x_i, $i = 1, 2, \ldots, m$ must be in $A(x)$ for otherwise $\bigcap_{i=1}^m F(x_i)$ would be nonempty. Now for each $y \in S_m$, $A^{-1}(y) = \{x \in S_m : y \in A(x)\} = \{x \in S_m : x \notin F(y)\} = [F(y)]^c \cap S_m$ is an open set in S_m. We set $O_y = A^{-1}(y)$. It can be easily seen that $\bigcup_{y \in S_m} O_y = S_m$. Indeed, let $x \in S_m$. Then since $A(x) \neq \emptyset$, we choose $y \in A(x)$. This implies that $x \in A^{-1}(y) = O_y$. Thus the assertion is made. Now we set up a set valued mapping $T : S_m \to S_m$ by $T(x) = \operatorname{co} A(x)$, $x \in S_m$. Then clearly T satisfies all the conditions of Theorem 4.14. Hence T has a fixed point x_0, i.e. $x_0 \in \operatorname{co} A(x_0)$. Hence there are points $y_1, y_2, \ldots, y_k$ in S_m such that $x_0 = \sum_{i=1}^k \lambda_i y_i$, $\sum_{i=1}^k \lambda_i = 1$, $\lambda_i \geq 0$, $y_i \in A(x_0)$,

$i = 1, 2, \ldots, k$. This implies that $x_0 \notin \bigcup_{i=1}^{k} F(y_i)$ which contradicts (ii). Hence have proved the theorem. □

4.2.1 *Duality in Fixed Point Theory of Set Valued Mappings*

Since each fixed point of a set valued mapping T is also a fixed point of the inverse mapping T^{-1} and vice-versa. It is possible to make a dual statement of a fixed point theorem to yield a fixed point of the inverse mapping. This simple phenomenon has been called the duality in fixed point theory in Tarafdar and Husain (1978) containing the next two results.

Corollary 4.14.1 has the following dual statement D1:

Theorem D1. *Let K be a nonempty subset of a linear topological vector space E. Let $T : K \to 2^E$ be a set valued mapping of K into 2^E such that*

(*i*) *$X = T(K)$ is a compact convex and $K \subset X$;*
(*ii*) *for each $y \in X$, $T^{-1}(y)$ is a nonempty convex subset of K;*

and

(*iii*) *for each $x \in K$, $T(x)$ is an open subset of X.*

Then there is a point $x_0 \in K$ such that $x_0 \in T(x_0)$.

Proof. We define a mapping $F : X \to 2^K$ by $F(x) = T^{-1}(x)$, $x \in X$ as $T^{-1}(x) \subset K \subset X$. Noting that $F^{-1}(y) = T(y)$ for each $y \in K$, it is easy to see that F satisfies the conditions of Corollary 4.14.1. Hence there is a fixed point $x_0 \in X$ such that $x_0 \in F(x_0) = T^{-1}(x_0)$. Thus $x_0 \in T(x_0)$ and $x_0 \in K$ as $F(x_0) \subset K$. □

Our next theorem is the dual of Theorem 4.8:

Theorem D2. *Let K be a nonempty compact subset of a locally convex Hausdorff topological vector space E. Let $T : K \to 2^E$ be a set valued mapping of K into 2^E such that*

(*i*) *$X = T(K)$ is compact convex and $K \subset X$;*
(*ii*) *T is upper semicontinuous;*
(*iii*) *$T(x)$ is nonempty and closed for each $x \in K$;*

and

(*iv*) *$T^{-1}(x)$ is a closed convex subset of K for each $x \in X$.*

Then there is a point $x_0 \in K$ such that $x_0 \in K$.

Proof. It is clear that the mapping $F : X \to 2^K$ defined by $F(x) = T^{-1}(x)$, $x \in X$ is nonempty valued. Since K and $T(K) = X$ are compact and T is upper semicontinuous with closed values, it follows by Theorem 2.32(a) (also Browder (1968), lemma, p. 285) that the graph of $T = G(T) = \{(x, y) : x \in K \text{ and } y \in T(x)\}$ is closed in $K \times X$. Now we prove that the graph of T^{-1} is closed. To this end, let $y_\delta \in T(K) = X$, $x_\delta \in T^{-1}(y_\delta)$ and $(y_\delta, x_\delta) \to (y, x)$. Hence $y_\delta \to y$,

$x_\delta \to x$ and $y_\delta \in T(x_\delta)$. Thus $y \in T(x)$ as $G(T)$ is closed, i.e. $x \in T^{-1}(y)$. Consequently $G(F) = G(T^{-1})$ is closed. T^{-1} having closed values is by Theorem 2.32(b) upper semicontinuous. Hence by Theorem 4.8 there is a point $x_0 \in X$ such that $x_0 \in F(x_0) = T^{-1}(x_0)$. Thus $x_0 \in T(x_0)$ and $x_0 \in K$ because $F(x_0) \subset K$. □

In the rest of this section we deal with the results obtained in Tarafdar (1982) and Fan (1966) and Fan (1984).

Lemma 4.6 *Let X be a nonempty convex subset of a topological vector E. To each point $x \in X$, let a nonempty subset $F(x)$ in E be given such that*

(a) *$x \in F(x)$ for each $x \in X$;*
(b) *$F(x_0)$ is compact for some $x_0 \in X$;*
(c) *for each $x \in X$, the set $A(x) = \{y \in X : x \notin F(y)\}$ is convex;*
(d) *for each $x \in X$, the intersection of $F(x)$ with any finite dimensional subspace of E is closed;*
(e) *for each $x \in X$, $F(x_0) \cap F(x)$ is closed.*

Then $\bigcap_{x \in X} F(x) \neq \emptyset$.

Remark 4.3 The conditions (a) and (c) together imply the condition (i) of Lemma 4.5. Thus if (d) and (e) are replaced by (ii) of Lemma 4.5 the above lemma would be a special case of Lemma 4.5.

Proof of Lemma 4.6. In view of (b) and (e) it would suffice to prove that $\bigcap_{i=1}^{n} F(x_i) \neq \emptyset$ for each finite subset $\{x_1, x_2, \ldots, x_n\}$ of X. If possible, let us assume that $\bigcap_{i=1}^{n} F(x_i) = \emptyset$. Then for each $x \in S$, the convex hull of $\{x_1, x_2, \ldots, x_n\}$ the set $B(x) = \{y \in S : x \notin F(y)\}$ is nonempty. Indeed, at least one of the x_i, $i = 1, 2, \ldots, n$ belongs to $B(x)$. Since S is convex, it follows from (c) that $B(x)$ is convex. Let us define a mapping $T : S \to 2^S$ by

$$T(x) = B(x) \text{ for each } x \in S.$$

Now $T^{-1}(x) = \{y \in S : x \in T(y)\} = \{y \in S : x \in B(y)\} = \{y \in S : y \notin F(x)\}$ is open in S by (d). Hence by Theorem 4.14.1 there is a point $x_0 \in S$ such that $x_0 \in T(x_0) = B(x_0)$. But this means that $x_0 \notin F(x_0)$ which contradicts (a). This proves the lemma. □

We now prove our fixed point theorems.

Theorem 4.15 *Let K be a nonempty convex subset of E. Let $T : K \to 2^K$ be a set valued mapping such that*

(a)′ *for each $x \in K$, $T(x)$ is a nonempty convex subset of K;*
(b)′ *for some $x_0 \in K$, the complement of $T^{-1}(x_0)$, in K, denoted by $[T^{-1}(x_0)]^c$ is compact;*

(c)′ *for each* $x \in K$, *the intersection of* $[T^{-1}(x)]^c$ *with any finite dimensional subsepace of* E *is closed;*

(d)′ *for each* $x \in K$, $[T^{-1}(x)]^c \cap [T^{-1}(x_0)]^c$ *is closed.*

Then there is a point $x_0 \in K$ *such that* $x_0 \in T(x_0)$.

Proof. If possible, let us assume that T has no fixed point, i.e. there is no point $x \in K$ such that $x \in T(x)$. This implies that there is no $x \in K$ such that $x \in T^{-1}(x)$. Thus we have (a) $x \in [T^{-1}(x)]^c$ for each $x \in K$; and (b) $[T^{-1}(x)]^c$ is nonempty for each $x \in K$.

We set $F(x) = [T^{-1}(x)]^c$ for each $x \in K$.

Now $A(x) = \{y \in K : x \notin F(y)\} = \{y \in K : x \notin [T^{-1}(y)]^c\} = \{y \in K : x \in T^{-1}(y)\} = T(x)$ which is convex by (a)′. Thus we have condition (c) of Lemma 4.6. Conditions (b), (d) and (e) of Lemma 4.6 follow from assumptions (b)′, (c)′ and (d)′ respectively. Hence there is a point $u \in K$ such that $u \in \bigcap_{x \in K} F(x)$, i.e. $u \in [T^{-1}(x)]^c$ for each $x \in K$ i.e. $u \notin T^{-1}(x)$ for any $x \in K$. However, $u \in K = \bigcup_{x \in K} T^{-1}(x)$ which is a contradiction. This proves the theorem. □

The following theorem is dual to the above theorem in the sense of Theorems D_1 and D_2.

Theorem 4.16 *Let* K *be a nonempty convex subset of* E. *Let* $T : K \to 2^K$ *be a multi-valued mapping such that*

(1) *for each* $x \in K$, $T(x)$ *be a nonempty subset of* K;

(2) *for some* $x_0 \in K$. $[T(x_0)]^c$ *is compact in* K;

(3) *for each* $x \in K$, $T^{-1}(x)$ *is convex (may be empty);*

(4) *for each* $x \in K$, *the intersection of* $[T(x)]^c$ *with any finite dimensional sub space of* E *is closed;*

(5) *for each* $x \in K$, $[T(x)]^c \cap [T(x_0)]^c$ *is closed;*

(6) $\bigcup_{x \in K} T(x) = K$.

Then there is a point $x_0 \in K$ *such that* $x_0 \in T(x_0)$.

Proof. As before, let us assume that there is no point $x \in K$ such that $x \in T(x)$. This implies (a) $x \in [T(x)]^c$ for each $x \in K$ and (b) $[T(x)]^c$ is nonempty for each $x \in K$.

We set $F(x) = [T(x)]^c$ for each $x \in K$.

Then $A(x) = \{y \in K : x \notin F(y)\} = \{y \in K : x \in T(y)\} = T^{-1}(x)$ which is convex by (3). Thus we have the condition (c) of Lemma 4.6 Conditions (2), (4) and (5) imply respectively conditions (b), (d) and (c) of Lemma 4.6. Hence there is a point $u \in K$ such that $u \in \bigcap_{x \in K} F(x) = \bigcap_{x \in K} [T(x)]^c$. This implies that $u \notin \bigcup_{x \in K} T(x)$ which is impossible by (6). Thus the theorem is proved. □

Corollary 4.16.1 *Let K be a nonempty convex subset of E and $T : K \to 2^K$ be a set valued mapping such that*

(i) *for each $x \in K$, $T(x)$ is a nonempty convex subset of K;*
(ii) *for each $x \in K$, $T^{-1}(x)$ is open in K;*
(iii) *for some $x_0 \in K$, $[T^{-1}(x_0)]^c$ is compact in K.*

Then there is a point $x_0 \in K$ such that $x_0 \in T(x_0)$.

Proof. Corollary follows from Theorem 4.15. □

Remark 4.4 This corollary generalizes the Corollary 4.14.1 of Browder (1968).

Corollary 4.16.2 *Let K be a nonempty convex subset of E and $T : K \to 2^K$ be a multi-valued mapping such that*

(i)′ *for each $x \in K$, $T(x)$ is a nonempty open subset of K;*
(ii)′ *for each $x \in K$, $T^{-1}(x)$ is convex (may be empty);*
(iii)′ *for some $x_0 \in K$, $[T(x_0)]^c$ is compact;*
(iv)′ $\bigcup_{x \in K} T(x) = K$.

Then there is a point $x_0 \in T$ such that $x_0 \in T(x_0)$.

Proof. This follows from Theorem 4.16. □

4.3 Applications on Minimax Principles

Theorem 4.17 (Fan's minimax principle) *Let K be a nonempty convex subset of E. Let $f(x, y)$ be a real valued function defined on $K \times K$ such that*

(i) *$f(x, x) \leq 0$ and $x \in K$;*
(ii) *for every $x \in K$, the set $A(x) = \{y \in K : f(x, y) > 0\}$ is convex;*
(iii) *there is a compact subset L of E and $x_0 \in L \cap K$ such that $f(x, x_0) > 0$ for all $x \in K$, $x \notin L$;*
(iv) *for every $y \in K$, we have*
 (1) *$f(x, y)$ is a lower semicontinuous function of x on the intersection of K with any finite dimensional subspace of E;*
 (2) *$f(x, y)$ is also a lower semicontinuous function of x on L.*

Then there exists a point $y_0 \in L$ such that $f(y_0, y) \leq 0$ for all $y \in K$.

Proof. For each $y \in K$, we set $F(y) = \{x : f(x, y) \leq 0\}$. It is easy to see that conditions (a), (c) and (d) of Lemma 4.6 follow from assumption (i), (ii) and (iv) respectively. $F(x_0)$ being a subset of L is compact by (iv) (2). Thus condition (b) of Lemma 4.6 holds. (e) follows also from (iv) (2). Hence by Lemma 4.6 there exist a point $y_0 \in \bigcap_{x \in K} F(x)$, i.e.

$$f(y_0, x) \leq 0 \quad \text{for all } x \in K. \qquad \square$$

Remark 4.5 This includes Fan's minimax principle (1966) and is different from the one of Brézis, Nirenberg and Stampacchia given later.

Although the following form of von Neumann and Sion minimax principle is a minor generalization of the one given by Brézis, Nirenberg and Stampacchia (1972), Proposition 1, we would like to include it as direct application of the minimax principle 4.17.

Theorem 4.18 (von Neumann–Sion minimax principle) *Let F be a Housdorff topological vector space and G be a vector space; let $A \subset F$ and $B \subset G$ be convex sets. Let $H(u, v)$ be a real valued function defined on $A \times B$ satisfying*

- (a)′ *for some $\tilde{v} \in B$ and some $\lambda > \sup_{v \in B} \inf_{u \in A} H(u, v)$, the set $P = \{u \in A : H(u, \tilde{v}) \leq \lambda\}$ is compact;*
- (b)′ *for each $v \in A$, $H(u, v)$ is a quasi-convex function of v on B and for each $v \in B$, $-H(u, v)$ is a quasi-convex function of u on A;*
- (c)′ *for each $v \in B$, $H(u, v)$ is a lower semicontinuous function of u on P and also a lower semicontinuous function of u on the intersection of A with any finite dimensional subspace of F;*
- (d)′ *for each $u \in A$, $-H(u, v)$ is a lower semicontinuous function of v on the intersection of B with any finite dimensional subspace of G.*

Then $\alpha = \sup_{v \in B} \inf_{u \in A} H(u, v) = \inf_{u \in A} \sup_{v \in B} H(u, v) = \beta$.

Proof. The same proof given in (H. Brézis and Stampacchia (1972)) with slight modification will do. We will maintain the notations given there. Obviously $\inf_{u \in A} H(u, v_0) \leq H(u_0, v_0) \leq \sup_{v \in B} H(u_0, v)$ for all $u_0 \in A$ and $v_0 \in B$. Thus $\alpha \leq \beta$. If possible, let us assume $\alpha < \beta$. We can choose a real number γ satisfying $\alpha < \gamma < \beta$, $\gamma \leq \lambda$.

Let $A(v) = \{u \in A : H(u, v) \leq \gamma\}$ and $B(u) = \{v \in B : H(u, v) \geq \gamma\}$. By choice of γ we have (1) $\bigcap_{v \in B} A(v) = \emptyset$ and (2) $\bigcap_{u \in A} B(u) = \emptyset$. We set $\overline{A}(v) = A(v) \cap P$ for each $v \in B$. Then by (c)′ $\overline{A}(v)$ is a closed subset of the compact subset P for each $v \in B$ and by (1) $\bigcap_{v \in B} \overline{A}(v) = \emptyset$. Hence we can find $v_1, v_2, \ldots, v_n \in B$ such that (3) $\bigcap_{i=1}^{n} \overline{A}(v_i) = \emptyset$. We note that $A(\tilde{v}) = \overline{A}(\tilde{v})$ as $\gamma \leq \lambda$. Consequently as $\gamma > \alpha$, we can assume $\lambda = \gamma$ and $\tilde{v}$ as one of v_i, $i = 1, 2, \ldots, n$. Let B' be the convex hull of $\{v_i, v_2, \ldots, v_n\}$. We now set $E = F \times R^n$ and $K = A \times B'$ where R^n is the usual n-dimensional Euclidean space. We define f on $K \times K$ by

$$f(x, y) = \min\{H(u, v') - \gamma, -H(u', v) + \gamma\}, \quad x = (u, v), \quad y = (u', v').$$

Obviously f satisfies (i) of 4.17 f satisfies (ii) of 4.17 by virtue of the quasiconvexity of H and $-H$ assumed in (b)$'$. We take $L = A(\tilde{v}) \times B'$. The lower semicontinuity of H on P assumed in (c)$'$ and the lower semicontinuity of $-H$ assumed in (d)$'$ imply the lower semicontinuity of f on L with respect to x for each fixed $y \in K$, i.e. (iv) (2) of 4.6 holds. For each fixed $y \in K$ the lower semicontinuity of f with respect to y on the intersection of K with any finite dimensional subspace of E follows from the corresponding lower semicontinuities of H and $-H$ assumed inn (c)$'$ and (d)$'$ (we recollect that minimum of two lower semicontinuous functions is lower semicontinuous), i.e. (iv) (1) of 4.17 holds. Finally we take $x_0 = (u_0, \tilde{v}) \in L \cap K$ for any $u_0 \in A$.

We can easily see that (iii) of 4.17 holds with this x_0.

Hence by Theorem 4.17 there is a point $y_0 = (u^0, v^0) \in K \cap L$ such that $f(y_0, y) \leq 0$ for all $y \in K$, i.e. for all $u \in A$, $v \in B'$, either $H(u^0, v) \leq \gamma$ or $\gamma \leq H(u, v^0)$. Let v be one of v_i. We can choose $v = v_i$ such that $u_0 \notin \overline{A}(v_i)$. This is possible as $\bigcap_{i=1}^{n} \overline{A}(v_i) = \emptyset$. Thus $H(u^0, v_i) > \gamma$ as $u^0 \in P$. Thus it follows $H(u, v^0) \geq \gamma$ for all $u \in A$, i.e. $v^0 \in \bigcap_{u \in A} B(u)$ which contradicts (2). □

4.3.1 *Applications on Sets with Convex Sections*

Theorem 4.19 *Let $K_1, K_2, \ldots, K_n$ be $n \geq 2$ nonempty convex sets, each in a topological vector space E_n, and let $K = \prod_{j=1}^{n} K_j$. Let $S_1, S_2, \ldots, S_n$ be n subsets of K having the following properties:*

(*a*) *Let $K_{-j} = \prod_{i \neq j} K_i$ and let us denote the points of K_{-j} by x_{-j}. For $j = 1, 2, \ldots, n$ and for each $x_{-j} \in K_{-j}$, the set $S_j(x_{-j}) = \{x_j \in K_j : [x_j, x_{-j}] \in S_j\}$ is a nonempty convex subset of K_j.*

(*b*) *For each $j = 1, 2, \ldots, n$ and for each point $x_j \in K_j$, the set $S_j(x_j) = \{x_{-j} \in K_{-j} : [x_j, x_{-j}] \in S_j\}$ is an open subset of K_{-j}.*

and

(*c*) *For some point $x^0 = (x_1^0, x_2^0, \ldots, x_n^0) \in K$, the complement of the set $\bigcap_{j=1}^{n} \{S_j(x_j^0) \times K_j)$ is compact in K. Then $\bigcap_{j=1}^{n} S_j \neq \emptyset$.*

Proof. For each $x \in K$, let $A(x) = \prod_{j=1}^{n} S_j(x_{-j})$ where x_{-j} is the natural projection of x on K_{-j}. By (a) $A(x)$ is a nonempty convex subset of K for each $x \in K$. We define a set valued mapping $T : K \to 2^K$ by $T(x) = A(x)$, $x \in K$.

Now $x \in T^{-1}(y)$ if $y \in T(x) = A(x) = \prod_{j=1}^{n} S_j(x_{-j})$, i.e. $y_j \in S_j(x_{-j})$ for each $j = 1, 2, \ldots, n$, i.e. $x_{-j} \in S_j(y_j)$ for $j = 1, 2, \ldots, n$, where $y = (y_1, y_2, \ldots, y_n)$. Hence $T^{-1}(y) = \bigcap_{j=1}^{n} \{S_j(y_j) \times K_j\}$ which is an open set by (b).

Final by (c) $[T^{-1}(x^0)]^c$ is compact in K. Hence by Corollary 4.16.1 there is a point $z \in K$ such that $z \in T(z)$, i.e. $z \in \prod_{j=1}^{n} S_j(z_{-j})$, $z_j \in S_j(z_{-j})$ for $j = 1, 2, \ldots, n$. Hence $z = [z_j, z_{-j}] \in S_j$ for $j = 1, 2, \ldots, n$. Thus $x \in \bigcap_{j=1}^{n} S_j$. □

Our next corollary is due to Fan (1966, Theorem 1) and afterwards by Brower (1968, Theorem 11) by different method.

Corollary 4.19.1 *Let $K_1, K_2, \ldots, K_n$ be $n \geq 2$ nonempty compact convex sets, each in a topological vector space E_n and let $K = \prod_{j=1}^{n} K_j$. Let $S_1, S_2, \ldots, S_n$ be n subsets of K having the properties (a) and (b) of Theorem 4.19. Then $\bigcap_{j=1}^{n} S_j \neq \emptyset$.*

Proof. The condition (c) of Theorem 4.19 follows automatically by the compactness of K_j, $j = 1, 2, n$. Hence the corollary follows from Theorem 4.19. □

Theorem 4.20 *Let $K_1, K_2, \ldots, K_n, K, K_{-j}, S_1, S_2, \ldots, S_n$ be as in Theorem 4.19 satisfying the following properties:*

(*i*) *For each $j = 1, 2, \ldots, n$ and each point $x_j \in K_j$, the set $S_j(x_j) = \{x_{-j} \in K_{-j} : [x_j, x_{-j}] \in S_j\}$ is a convex subset of K_{-j}.*
(*ii*) *For each $j = 1, 2, \ldots, n$ and each $x_{-j} \in K_{-j}$, the set $S_j(x_{-j}) = \{x_j \in K_j : [x_j, x_{-j}] \in S_j\}$ is a nonempty open subset of K_j.*
(*iii*) *For some point $x^0 = (x_1^0, x_2^0, \ldots, x_n^0) \in K$, the complement of the set $\prod_{j=1}^{n} S_j(x_{-j})$ is compact.*
(*iv*) *$\bigcup_{x \in K} A(x) = K$, where $A(x) = \prod_{j=1}^{n} S_j(x_{-j})$, x_{-j} being the natural projection of x on K_{-j}.*

Then $\bigcap_{j=1}^{n} S_j \neq \emptyset$.

Proof. By (ii) $A(x)$ is a nonempty open subset of K for each $x \in K$. We define the set valued mapping $T : K \to 2^K$ by $T(x) = A(x)$, $x \in K$. By the same argument as in Theorem 4.19, for each $y \in K$, $T^{-1}(y) = \bigcap_{j=1}^{n} \{S_j(y_i) \times K_j\}$ which is convex by (i). By (iii) the complement of $T(x_0)$ is compact and by (iv) $\bigcup_{x \in K} T(x) = K$. Hence by Corollary 4.16.2, there is a point $z \in K$ such that $z = T(z) = A(z)$, i.e. $z = [x_j, z_{ij}] \in S_j$ for all $j = 1, 2, \ldots, n$. Hence $z \in \bigcap_{j=1}^{n} S_j$. □

Corollary 4.20.1 *Let $K_1, K_2, \ldots, K_n$ be $n \geq 2$ nonempty compact convex sets, each in a topological vector space, $K = \prod_{j=1}^{n} K_j$ and $K_{-j} = \prod_{j \neq i}$, $j = 1, 2, \ldots, n$. Let $S_1, S_2, \ldots, S_n$ be n subsets of K satisfying (i), (ii) and (iv) of Theorem 4.20. Then $\bigcap_{j=1}^{n} S_j \neq \emptyset$.*

Proof. The condition (iii) of Theorem 4.20 is automatically satisfied due to the compactness of K_j, $j = 1, 2, \ldots, n$. Hence the corollary from Theorem 4.20. □

Theorem 4.21 *Let $K_1, K_2, \ldots, K_n$ and K be as in Theorem 4.19. Let $F_1, f_2, \ldots, f_n$ be n real valued functions defined on K satisfying properties:*

(a) *For each $j = 1, 2, \ldots, n$, and for each point $x_j \in K_j$. $f_j(x_j, x_{-j})$ is a lower semi-continuous function of x_{-j} on K_{-j}.*
(b) *For each $j = 1, 2, \ldots, n$ and for each $x_{-j} \in K_{-j}$, $f_j(x_j, x_{-j})$ is a quasi-concave function of x_i on K_j (i.e. for each real number t, the set $\{x_j \in K_j, f_j(x_j, x_{-j}) > t\}$ is a convex subset of K_j).*
(c) *Let $t_1, t_2, \ldots, t_n$ be n real numbers such that for each $j = 1, 2, \ldots, n$ and each point x_{-j} of K_{-j}, there exists a point $y_j \in K_j$ such that $f_j(y_j, x_{-j}) > t_j$.*
(d) *Further assume that for some point $x^0 = (x_1^0, x_2^0, \ldots, x_n^0) \in K$, the complement of the set $\bigcap_{j=1}^{n} \left[\{x_{-j} \in K_{-j} : f_j(x_j^0, x_{-j}) > t_j\} \times K_j\right]$ in K is compact.*

Then there is a point $u \in K$ such that $f_j(u) > t_j$ for all $j = 1, 2, \ldots, n$.

Proof. For $j = 1, 2, \ldots, n$, we define the subset S_j of K by $S_j = \{x \in K : f_j(x) > t_j\}$ which is nonempty by condition (c). The condition (d) says that the complement of $\bigcap_{j=1} \{S_j(x_j^0) \times K_j\}$ in K is compact. The rest of the proof follows from Theorem 4.19. □

Corollary 4.21.1 (Fan (1966) and also Brower (1968)) *Let $K_1, K_2, \ldots, K_n$, K and K_{-j} for $j = 1, 2, \ldots, n$ be as in Corollary 4.19.1. Let $f_1, f_2, \ldots, f_n$ be n real valued functions defined on K satisfying (a), (b) and (c) of Theorem 4.21.*

Then there is a point $u \in K$ such that $f_j(u) > t)j$ for all $j = 1, 2, \ldots n$.

Proof. For $j = 1, 2, \ldots n$, we define $S_j = \{x \in K : f_j(x) > t\}$ which is a nonempty subset by (b). the condition (c) is automatic due to the compactness of K_j, $j = 1, 2, \ldots, n$. the corollary follows either from Theorem 4.21 or from Corollary 4.19.1. □

Our next theorem is dual to Theorem 4.21.

Theorem 4.22 *Let $K_1, K_2, \ldots, K_n$, K and K_{-j}, $j = 1, 2, \ldots, n$ be as in Theorem 4.21. Let $f_1, f_2, \ldots, f_n$ be n real valued functions defined on K satisfying the properties:*

(*i*) *For each* $j = 1, 2, \ldots, n$ *and each* $x_j \in K_j$, $f(x_j, x_{-j})$ *is a quasi-concave function of* x_{ij} *on* K_{-j}.

(*ii*) *For each* $j = 1, 2, \ldots, n$ *and each* $x_{-j} \in K_{-j}$, $f_j(x_j, x_{-j})$ *is a lower semi-continuous function of* x_j *on* K_j.

(*iii*) *Let* $t_1, t_2, \ldots, t_n$ *be* n *real numbers such that for each* $j = 1, 2, \ldots, n$ *and each* $x_{-j} \in D_{-j}$, *there exists* $y_j \in K_j$ *such that* $f_j(y_j, x_{-j}) > t_j$.

(*iv*) *For some point* $x^0 = (x_1^0, x_2^0, \ldots, x_n^0) \in K$, *the complement of the set* $\prod_{j=1}^{n} \{x_j \in K_j : f_j(x_j, x_{-j}^0) > t_j\}$ *in* K *is compact.*

(*v*) $\bigcup_{x \in K} A(x) = K$, *where* $A(x) = \prod_{j=1}^{n} \{x_j \in K_j : f(x_j, x_{-j}) > t_j\}$, x_{-j} *being the projection of* x *on* K_{-j}.

Then there is a point $u \in K$ *such that* $f_j(u) > t_j$ *for all* $j = 1, 2, \ldots n$.

Proof. As before, for each $j = 1, 2, \ldots, n$ we define the subsets S_j of K by $S_j = \{x \in K : f_j(x) > t_j\}$ which is nonempty by (iii). The condition (iv) reduces to the condition that the complement of $\prod_{j=1}^{n} S_j(x_{-j}^0) = A(x_0)$ in K is compact. Now it is easy to see that the corollary follows from Theorem 4.20. □

Corollary 4.22.1 *Let* $K_1, K_2, \ldots, K_n$, K *and* K_{-j}, $j = 1, 2, \ldots, n$ *be as in Corollary* 4.19.1. *Let* f_1, f_2, f_n *be* n *real valued functions defined on* K *satisfying the properties* (*i*), (*ii*), (*iii*) *and* (*v*) *of Theorem* 4.22.

Then there is a point $u \in K$ *such that* $f_j(u) > t_j$ *for all* $j = 1, 2, \ldots, n$.

Proof. Noting that the condition (iv) of Theorem 4.22 is automatic, the corollary follows from Theorem 4.22. □

Remark 4.6 The materials from Lemma 4.6 to Theorem 4.18 and Theorems 4.19, 4.20 and 4.21 are due Tarafdar (1982) and Corollaries 4.20.1 and 4.22.1 are due to Tarafdar and Husain (1978).

4.4 More on Sets with Convex Sections

In our remaining consideration the index set I will be finite or infinite. The results in this section are more general then those of the previous section.

Theorem 4.23 (Fan (1984)) *Let* $\{E_\alpha : \alpha \in I\}$ *be a family of topological vector spaces, where* I *is a finite or an infinite index set. For each* $\alpha \in I$, *let* X_α *be a nonempty compact convex subset in* E_α. *Let* $X = \prod_{\alpha \in I} X_\alpha$ *and* $X_{-\alpha} = \prod_{\substack{\beta \in I \\ \beta \neq \alpha}} X_\beta$ *for each* $\alpha \in I$. *Let* $\{A_\alpha : \alpha \in I\}$ *and* $\{B_\alpha : \alpha \in I\}$ *be two families of subsets of* X *having the following properties:*

(*a*) *For each* $\alpha \in I$ *and each* $x_\alpha \in X_\alpha$, *the set*

$B_\alpha(x_\alpha) = \{x_{-\alpha} \in X_{-\alpha} : [x_\alpha, x_{-\alpha}] \in B_\alpha\}$ is open in X_α; and

(b) For each $\alpha \in I$ and each $x_{-\alpha} \in X_{-\alpha}$, the set $B_\alpha(x_{-\alpha}) = \{x_\alpha \in X_\alpha : [x_\alpha, x_{-\alpha}] \in B_\alpha\}$ is nonempty and the set $A_\alpha(x_{-\alpha}) = \{x_\alpha \in X_\alpha : [x_\alpha, x_{-\alpha}] \in A_\alpha\}$ contains the convex hull of $B_\alpha(x_{-\alpha})$. Then $\bigcap_{\alpha\in I} A_\alpha \neq \emptyset$.

Proof. For each $\alpha \in I$ and each $x_{-\alpha}$, there exists by (b) $x_\alpha \in B_\alpha(x_{-\alpha})$, i.e. $x_{-\alpha} \in B_\alpha(x_\alpha)$. Hence $\{B_\alpha(x_\alpha(: x_\alpha \in X_\alpha\}$ is an open covering of the compact set $X_{-\alpha}$. Consequently for each $\alpha \in I$, there is a finite set $\{x_{\alpha 1}, x_{\alpha 2}, \ldots, x_{an(\alpha)}\} \subset X_\alpha$ such that

$$X_{-\alpha} = \bigcup_{k=1}^{n(\alpha)} B_\alpha(x_{\alpha k}).$$

Let $\{f_{\alpha 1}, f_{\alpha 2}, f_{\alpha n(\alpha)}\}$ be a continuous partition of unity subordinate to this finite open subcovering of $X_{-\alpha}$, i.e.

$$f_{\alpha k}(x_{-\alpha}) = 0 \quad \text{for} \quad x_{-\alpha} \in X_{-\alpha} \setminus B_\alpha(x_{\alpha k}),$$

$$\sum_{k=1}^{n(\alpha)} f_{\alpha k}(x_{-\alpha}) = 1 \quad \text{for} \quad x_{-\alpha} \in X_{-\alpha},$$

and $0 \leq f_{\alpha k}(x_{-\alpha}) \leq 1$ for all $x_{-\alpha} \in X_{-\alpha}$. We now define a continuous mapping $p_\alpha : X_{-\alpha} \to X_\alpha$ by

$$p_\alpha(x_{-\alpha}) = \sum_{k=1}^{n(\alpha)} f_{\alpha k}(x_{-\alpha}) x_{\alpha k}, \quad x_{-\alpha} \in X_{-\alpha}.$$

Since $f_{\alpha k}(x_{-\alpha}) \neq 0$ implies $x_{-\alpha} \in B_\alpha(x_{\alpha k})$, i.e. $x_{ak} \in B_\alpha(x_\alpha)$. Hence it follows that $p_\alpha(x_{-\alpha}) \in$ convex hull of $B_\alpha(x_{-\alpha}) \subset A_\alpha(x_{-\alpha})$ by (b) for $\alpha \in I$ and $x_{-\alpha} \in X_{-\alpha}$.

For each $\alpha \in I$, let K_α be the convex hull of $\{x_{\alpha 1}, x_{\alpha 2}, \ldots, x_{n(\alpha)}\}$. Then $K_\alpha \supset X_\alpha$. Let F_α denote the vector subspace of E_α generated by K_α. F_α being finite dimensional is locally convex. Then $K = \prod_{\alpha\in I} K_\alpha$ is a compact convex subset of the locally convex Hausdorff topological vector space $\prod_{\alpha\in I} F_\alpha$. For each $\alpha \in I$, let $K_{-\alpha} = \prod_{\beta\in I\, \beta\neq\alpha} K_\beta$. Then clearly $K_{-\alpha} \subset X_{-\alpha}$ for each $\alpha \in I$.

Finally we define a continuous mappinga $q : K \to K$ in the following manner for each: $x = [x_\alpha, x_{-\alpha}] \in K_\alpha \times K_{-\alpha}$, $q(x) = \{p_\alpha(x_{-\alpha})\}_{\alpha\in I} \in \prod_{\alpha\in I} K_\alpha = K$. Clearly q is continuous. Hence by Tychonoff's fixed point Theorem 4.8 there is a fixed point $u \in K$ of q, i.e. $u = \{u_\alpha\}_{\alpha\in I} = q(u) = \{q_\alpha(u_{-\alpha})\}_{\alpha\in I}$. Thus for each $\alpha \in I$, we have $u_\alpha = p_\alpha(u_{-\alpha}) \in A_\alpha(u_{-\alpha})$. i.e. $u = [u_\alpha, u_{-\alpha}] \in A_\alpha$. Hence $\bigcap_{\alpha\in I} A_\alpha$. □

Corollary 4.23.1 (Ma (1969)) *Let $\{E_\alpha : \alpha \in I\}$, $\{X_\alpha : \alpha \in I\}$, X and $X_{-\alpha}$ be as in Theorem 4.19.1. Let $\{A_\alpha : \alpha \in I\}$ be a family of subsets of X satisfying the following conditions:*

(i) *for each $\alpha \in I$ and each $x_\alpha \in X_\alpha$, the set $A_\alpha(x_\alpha) = \{x_{-\alpha} \in X_{-\alpha} : [x_\alpha, x_{-\alpha}] \in A_\alpha\}$ is open in $X_{-\alpha}$;*

and

(ii) *for each $\alpha \in I$ and each $x_{-\alpha} \in X_{-\alpha}$, the set $A_\alpha(x_{-\alpha}) = \{x_\alpha \in X_\alpha : [x_\alpha, x_{-\alpha}] \in A_\alpha\}$ is nonempty and convex.*

Then $\bigcap_{\alpha\in I} A_\alpha \neq \emptyset$.

Proof. For each $\alpha \in I$, we take $A_\alpha = B_\alpha$ and apply Theorem 4.23 to obtain the corollary. □

We now prove the following extended version of Theorem 4.23.

Theorem 4.24 *Let $\{E_\alpha : \alpha \in I\}$ be as in Theorem 4.19.1. Let for each $\alpha \in I$, X_α be nonempty convex subset of E_α. Let $\{A_\alpha : \alpha \in I\}$ and $\{B_\alpha : \alpha \in I\}$ be two families of subsets of $X = \prod_{\alpha\in I} X_\alpha$ satisfying the properties (a) and (b) of Theorem 4.23. Further assume that (c) there is a nonempty compact convex subset K of X such that for every $x \in X \setminus K$, $x = [x_\alpha, x_{-\alpha}] \in X_\alpha \times X_{-\alpha}$, there exists $y \in K$, $[y_\alpha, y_{-\alpha}] \in X_\alpha \times X_{-\alpha}$ satisfying $[y_\alpha, x_{-\alpha}] \in B_\alpha$ for each$\alpha \in I$. Then $\bigcap_{\alpha\in I} A_\alpha \neq \emptyset$.*

Proof. For each $\alpha \in I$, let $P_\alpha : X \to X_\alpha$ be the projection of X on X_α. For each $\alpha \in I$. let $P_\alpha(K) = X'_\alpha$ which is nonempty compact and convex. Let $X' = \prod_{\alpha\in I} X'_\alpha$ and for each $\alpha \in I$, let $A'_\alpha = A_\alpha \cap X'$ and $B'_\alpha n = B)\alpha \cap X'$. We also set $X'_{-\alpha} = \prod_{\substack{\beta\in I\\ \beta\neq\alpha}} X'_\beta$ for each $\alpha \in I$.

Then for each $\alpha \in I$ and each $x_\alpha \in X'_\alpha$, it is easily seen that the set

$$B'_\alpha(x) = \{x_{-\alpha} \in X'_{-\alpha} : [x_\alpha, x_{-\alpha}] \in B'_\alpha\} = \{x_{-\alpha} \in X_{-\alpha} : [x_\alpha, x_{-\alpha}] \in B_\alpha\} \cap X_{-\alpha}]$$

which is relatively open in $x_{-\alpha}]$ by cndition (a). Also for each $\alpha \in I$ and each $x_{-\alpha} \in X'_{-\alpha}$, the set

$$B'_\alpha(x_{-\alpha}) = \{x_\alpha \in X'_\alpha : [x_\alpha, x_{-\alpha}] \in B'_\alpha\}$$

which is nonempty for the following reasons.

By condition (b) $B_\alpha(x_{-\alpha}) \neq \emptyset$. Let $x_\alpha \in B_\alpha(x_{-\alpha})$. If $x_\alpha \notin B'_\alpha(x_{-\alpha})$, it follows that $x \notin X'$ and hence $x \notin K$ as $K \subset X'$. By condition (c) there exists $y \in K$, $y = [y_\alpha, y_{-\alpha}] \in C_\alpha \times X_{-\alpha}$ such that $[y_\alpha, x_{-\alpha}] \in B_\alpha$ for each $\alpha \in I$. Thus as $y_\alpha \in P_\alpha(K) = X'_\alpha$ and $x_{-\alpha} \in X'_{-\alpha}$, $[y_\alpha, x_{-\alpha}] \in B_\alpha \cap X' = B'_\alpha$. Hence $y_\alpha \in B'_\alpha(x_{-\alpha})$. Therefore $B'(x_{-\alpha})$ is always nonempty.

Finally it can easily be seen that for each $\alpha \in I$ and $x_{-\alpha} \in X'_{-\alpha}$, the set

$$A'_\alpha(x_{-\alpha}) = \{x_\alpha \in X'_\alpha : [x_\alpha, x_{-\alpha}] \in A'_\alpha\} = \{x_\alpha \in X_\alpha : [x_\alpha, x_{-\alpha}] \in A_\alpha\} \cap X'_\alpha$$

contains the convex hull of the set

$$B'_\alpha(x_{-\alpha}) = \{x_\alpha \in X'_\alpha : [x_\alpha, x_{-\alpha}] \in B'_\alpha\} = \{x_\alpha \in X_\alpha : [x_\alpha, x_{-\alpha}] \in B_\alpha\} \cap X'_\alpha.$$

Hence the system $\{X'_\alpha, A'_\alpha, B'_\alpha\}$ satisfies all the conditions of Theorem 4.23. Therefore $\bigcap_{\alpha \in I} A'_\alpha \neq \emptyset$. Hence $\bigcap_{\alpha \in I} A_\alpha \neq \emptyset$. □

The following theorem, which includes a theorem of von Neumann (1937), which in turn implies the fundamental theorem of two-person zero sum games was proved by Fan (1952) and also by Browder (1968), while Theorems 4.24 and 4.27 have been taken from Tarafdar (1988) and Theorem 4.26 from Tarafdar and Husain (1978).

Theorem 4.25 *Let $\{K_\alpha : \alpha \in I\}$ be an indexed family of nonempty compact convex sets, each in a locally convex topological vector space E_α. Let $K = \prod_{\alpha \in I} K_\alpha$ and $K_{-\alpha} = \prod_{\substack{\beta \in I \\ \beta \neq \alpha}} K_\beta$.*

Let $\{S_\alpha : \alpha \in I\}$ be a corresponding indexed family of closed subsets of K having the property that for each $x = \{x_\alpha : \alpha \in I\} \in K$ and each $\alpha \in I$, the set $S_\alpha(x_{-\alpha}) = \{y_\alpha \in K_\alpha : [y_\alpha, x_{-\alpha}] \in S_\alpha\}$ is nonempty and convex.

Then $\bigcap_{\alpha \in I} S_\alpha \neq \emptyset$.

Proof. Here we give the proof as given by Browder (1968). We define a set valued mapping $T : K \to 2^K$ by

$$T(x) = T\{x_\alpha : \alpha \in I\} = \{y_\alpha : \alpha \in I\},\ x = \{x_\alpha : \alpha \in I\}$$

if and only if for each $\alpha \in I$, $y_\alpha \in S_\alpha(x_{-\alpha})$. Since for each $\alpha \in I$, $S_\alpha(x_{-\alpha})$ is nonempty convex and closed as S_α is closed, it follows that $T(x)$ is a nonempty compact convex subset of K for each $x \in K$. Now we show that T is upper semicontinuous. It will suffice to show that the graph $G(T)$ of T is closed in $K \times K$. If possible, we assume that $[x, y] \notin G(T)$. Then there exists an index $\beta \in I$ such that $y_\beta \notin S_\beta$, i.e. $[y_\beta, x_{-\beta}] \notin S_\beta$. Since S_β is compact, there is a neighborhood U of y_β in K_β and a neighborhood V of $x_{-\beta}$ in $K_{-\beta}$ such that $U \times V$ does not intersect S_β. We consider the neighborhoods U' of y and V' of x in K defined by $U' = U \times K_{-\beta}$ and $V' = K_\beta \times V$. It now follows that for any $\overline{x} \in V'$ and any $\overline{y} \in U'$, $\overline{y}_\beta \notin S_\beta(x_{-\beta})$. Hence $U' \times V'$ does not intersect $G(T)$. Thus $[x, y]$ is not a limit point of $G(T)$. It follows then that $G(T)$ is closed in $K \times K$ and T is upper semicontinuous. Hence by Tychonoff Theorem 4.8, there is a point $u \in K$ such that $u \in T(u)$, i.e. $\{u_\alpha, \alpha \in I\} = T(u_\alpha : \alpha \in I\}$, i.e. $u_\alpha \in S_\alpha(u_{-\alpha})$, i.e. $u = [u_\alpha, u_{-\alpha}] \in S_\alpha$ for each $\alpha \in I$. Hence $u \in \bigcap_{\alpha \in I} S_\alpha$. □

Our next theorem is dual to Theorem 4.25.

Theorem 4.26 *Let $\{K_\alpha : \alpha \in I\}$ be a family of nonempty compact convex sets, each in locally convex topological vector space E_α. Let $K = \prod_{\alpha\in I} K_\alpha$ and for each $\alpha \in I$, $K_{-\alpha} = \prod_{\substack{\beta\in I\\ \beta\neq\alpha}} K_\beta$. Let $\{S_\alpha : \alpha \in I\}$ be corresponding indexed family of closed subsets of K having the following properties:*

(a) *For each $x = x_\alpha : \alpha \in I\}$ and each $\alpha \in I$, the set $S_\alpha(x_{-\alpha}) = \{y_\alpha \in K_\alpha : [y_\alpha, x_{-\alpha}] \in S_\alpha\}$ is nonempty;*
(b) *For each $x = \{x_\alpha : \alpha \in I\}$ and each $\alpha \in I$, the set $S_\alpha(x_\alpha) = \{y_{-\alpha} \in K_{-\alpha} : [x_\alpha, y_{-\alpha}] \in S_\alpha\}$ is a convex subset of $K_{-\alpha}$*

and

(c) *Let for each $x = \{x_\alpha : \alpha \in I\} \in K$, $A(x) = \prod_{\alpha\in I} S_\alpha(x_{-\alpha})$. Assume that $\bigcup_{x\in K} A(x) = K$. Then $\bigcap_{\alpha\in I} S_\alpha \neq \emptyset$.*

Proof. Since K_α and S_α are compact for each $\alpha \in I$, it is easy to see that for each $x \in K$ and $\alpha \in I$, $S_\alpha(x_{-\alpha})$ is closed and, therefore, compact subset of K_α. Hence for each $x \in K$, $A(x)$ is a compact subset of K. Let us define a set valued mapping $T : K \to 2^K$ by

$$T(x) = A(x), \quad x \in K.$$

Thus $T(x)$ is compact and hence closed for each $x \in K$. That $T(x)$ is nonempty follows from the condition (a). Now we consider the set $T^{-1}(y)$, $y \in K$. $x \in T^{-1}(y)$ if and only if $y \in T(x) = A(x) = \prod_{\alpha\in I} S_\alpha(x_{-\alpha})$, i.e. if and only if $y_\alpha \in S)\alpha(x_{-\alpha})$ for each $\alpha \in I$, i.e. if and only if $x_{-\alpha} \in S_\alpha(y_\alpha)$ for each $\alpha \in I$. Hence $T^{-1}(y) = \bigcap_{\alpha\in I} \{S_\alpha(y_\alpha) \times K_\alpha\}$ which is a convex set by condition (b). Further since $K_{-\alpha}$ and S_α are conpact, it follows that for each $y_\alpha \in K_\alpha$, $S_\alpha(y_\alpha)$ is a closed set. Therefore $T^{-1}(y)$ is also closed for each $y \in K$. Now by repeating the same argument as in Theorem 4.25 we can prove that $G(T)$ is closed. Hence T is upper semicontinuous. Hence by Theorem 4.2.1 of section 4.2, there is a point $u \in K$ such that $u \in T(u)$, i.e. $\{u_\alpha : \alpha \in T\} = \{u_\alpha : \alpha \in I\}$, i.e. $u_\alpha \in S_\alpha(u_{-\alpha})$, i.e. $u = [u_\alpha, u_{-\alpha}] \in S_\alpha$ for each $\alpha \in I$. Thus $u \in \bigcap_{\alpha\in I} S_\alpha$. □

Using Theorem 4.25 we now prove the following extension.

Theorem 4.27 *For each $\alpha \in I$, let X_α be a nonempty convex subset of locally convex topological vector space E_α. Let $\{S_\alpha : \alpha \in I\}$ be a family of closed subsets of $X = \prod_{\alpha\in I} X_\alpha$ having the property that for each $\alpha \in I$ and each $x_{-\alpha} \in X_{-\alpha} = \prod_{\substack{\beta\in I\\ \beta\neq\alpha}} X_\beta$*

the set

$$S_\alpha(x_{-\alpha}) = \{x_\alpha \in X_\alpha : [x_\alpha, x_{-\alpha}] \in S_\alpha\}$$

is a nonempty convex set.

Furthermore, assume that there is a nonempty compact convex subset K of X such that for every $x \in X/K$, $px_\alpha, x_{-\alpha}] \in X_\alpha \times X_{-\alpha}$, there exists $y \in K$, $y = [y_\alpha, y_{-\alpha}] \in X_\alpha \times X_{-\alpha}$ satisfying $[y_\alpha, x_{-\alpha}] \in S_\alpha$ for each $\alpha \in I$.

Then $\bigcap_{\alpha \in I} S_\alpha \neq \emptyset$.

Proof. For each $\alpha \in I$, let $P_\alpha : X \to X_\alpha$ be the projection of X onto X_α. For each $\alpha \in I$, let $P_\alpha(K) = X'_\alpha$ which is a nonempty compact convex set in E_α. Let $X' = \prod_{\alpha \in I} X'_\alpha$ and for each $\alpha \in I$, let $S'_\alpha = S_\alpha \cap X'$. Then S'_α is a closed subset of X' for each $\alpha \in I$ as X' is compact and S_α is closed.

Now for each $\alpha \in I$ and each $x_{-\alpha} \in X_\alpha$, we can prove by giving similar arguments as those in Theorem 4.20.1 that the set $S'_\alpha(x_{-\alpha}) = \{x_\alpha \in X'_\alpha : [x_\alpha, x_{-\alpha}] \in S'_\alpha \neq \emptyset$. That the set $S'_\alpha(x_{-\alpha})$ is convex follows from the equality:

$$S'_\alpha(x_{-\alpha}) = \{x_\alpha \in X'_\alpha : [x_\alpha, x_{-\alpha}] \in S'_\alpha\} = \{x_\alpha \in X_\alpha : [x_\alpha, x_{-\alpha}] \in S_\alpha\} \cap X'_\alpha.$$

Now applying Theorem 4.25 to the system $\{X'_\alpha, S'_\alpha\}$ we have $\bigcap_{\alpha \in I} S'_\alpha \neq \emptyset$ and hence $\bigcap_{\alpha \in I} S_\alpha \neq \emptyset$. □

The following two Theorems 4.28 and 4.29 were stated by Ma (1969) in topological vector spaces while Browder (1968) proved these in locally convex topological vector spaces. Theorem 4.28 provides a generalization of Corollary 4.21.1 concerning a known result of Nash (1951) on the existence of equilibrium points from finite to infinite person games.

Theorem 4.28 *Let $\{X_\alpha : \alpha \in I\}$ be a family of nonempty compact convex sets, each in a topological vector space E_α, where I is a before a finite or a infinite index set and for each $\alpha \in I$, $X_{-\alpha} = \prod_{\substack{\beta \in I \\ \beta \neq \alpha}} X_\beta$. Let $\{f_\alpha : \alpha \in I\}$ be a family of real valued functions defined on $X = \prod_{\alpha \in I} X_\alpha$ satisfying the following properties:*

- (*a*) *For each $\alpha \in I$ and for each $x_\alpha \in X_\alpha$, $f_\alpha(x_\alpha, x_{-\alpha})$ is a lower semicontinuous function of $x_{-\alpha} \in X_{-\alpha}$.*
- (*b*) *For each $x_{-\alpha} X_{-\alpha}$, $f_\alpha(x_\alpha, x_{-\alpha})$ is quasi-concave function of x_α on X_α.*
- (*c*) *Let $\{t_\alpha : \alpha \in I\}$ be a family of real numbers such that for each $\alpha \in I$ and each $x_{-\alpha} \in X_{-\alpha}$, there exists a point $y_\alpha \in X_\alpha$ such that $f_\alpha(y_\alpha, x_{-\alpha}) > t_\alpha$. Then there is a point $u \in X$ such that $f_\alpha(u) > t_\alpha$.*

Proof. For each $\alpha \in I$, we define the subsets A_α of X by $A_\alpha = \{x \in X : F_\alpha(x) > t_\alpha\}$.

Then by virtue of (a), (b) and (c), we have the conditions (i) and (ii) of Corollary 4.23.1. Hence the theorem follows from Corollary 4.23.1. □

In order to prove the concluding theorems of this section we need the concept of uniform continuity of a function defined on topological vector and the following lemma.

A real valued function f defined on a nonempty subset X of a topological vector space E is called uniformly continuous if, given $\varepsilon > 0$, there exists an open neighborhood W of $O \in E$ such that $|f(x) - f(y)| < \varepsilon$ whenever $x - y \in W \cap X$.

Lemma 4.7 *A real valued continuous function f defined on a compact subset X of a topological vector space E is uniformly continuous. Although it is known, we include a proof for the benefit of the readers.*

Proof. Let $\varepsilon > 0$ be a real number. Then by the continuity of f for each $x \in X$, there exists a symmetric open neighborhood W_x of $O \in E$ such that $|f(x) - f(y)| < \varepsilon$ whenever $x - y \in W_x \cap X$. For each $x \in X$ we select a symmetric open neighborhood $\widehat{W}_x$ of O such that $\widehat{W}_x + \widehat{W}_x \subset W_x$. Clearly $\{(\widehat{W}_x + x) \cap X \ : \ x \in X\}$ is an open cover of X. Let $\{(\widehat{W}_{x_i} + x_i) \cap X \ : \ i = 1, 2, n\}$ be a finite open cover. Let $W = \bigcap_i^n W_{x_i}$. Now let $x, y \in X$ with $x - y = W$. Then there is $m = 1, 2, \ldots n$ such that $x \in \widehat{W}_{x_m} + x_m$. Now

$$y - x_m = y - x + x - x_m \in W + \widehat{W}_{x_m} \subset \widehat{W}_{x_m} + \widehat{W}_{x_m} \subset W_{x_m}.$$

Hence $|f(x) - f(y)| \leq |f(x) - f(x_m)| + |f(x_m - f(y)| < \frac{\varepsilon}{2} + \frac{\varepsilon}{2} = \varepsilon$. □

We now state and prove a theorem on the existence of equilibrium point.

Theorem 4.29 *Let $\{X_\alpha \ : \ \alpha \in I\}$ be a fmaily of nonempty compact convex sets, each in a topological vector space E_α, where I is as in Theorem 4.28. Let $\{f_\alpha \ : \ \alpha \in I\}$ be a family of continuous functions defined on $X = \prod_{\alpha \in I} X_\alpha$. If for each $\alpha \in I$ and for each $x_{-\alpha} \in S_{-\alpha}$, $f_\alpha(x_\alpha, x_{-\alpha})$ is a quasi-convex function of $x_\alpha \in X_\alpha$, where $X_{-\alpha}$ is as defined in Theorem 4.28.*

Then there is a point $u \in X$ such that for each $\alpha \in I$,

$$f_\alpha(u) = f_\alpha[u_\alpha, u_{-\alpha}] = \sup_{x_\alpha \in X_\alpha} f_\alpha[x_\alpha, u_{-\alpha}].$$

u is called an equilibrium point.

Proof. For each $\alpha \in I$ and each $x_{-\alpha} \in X_{-\alpha}$, let

$$g_\alpha(x_{-\alpha} = \sup_{y_\alpha \in X_\alpha} f_\alpha(y_\alpha, x_{-\alpha}).$$

By Lemma 4.7 f_α is uniformly continuous. Thus it follows that g_α is a real valued continuous function on $X_{-\alpha}$. For each $\varepsilon > 0$, let

$$H_\varepsilon = \{x \in X \ : \ f_\alpha(x) \geq g_\alpha(x_{-\alpha}) - \varepsilon, \ \text{for all } \alpha \in I\},$$

where $x_{-\alpha}$ is the projection of x on $X_{-\alpha}$. By the continuity of f_α, g_α and the projection mapping on $X_{-\alpha}$, it follows that H_ε is a compact subset of X. Also we note that H_ε decreases as ε decreases to zero. If we set $H_0 = \bigcap_{\varepsilon>0} H_\varepsilon$, then clearly H_0 is precisely the set, of all points which will satisfy the conclusion of the theorem. H_0 is nonempty if each H_ε is nonempty. It thus suffice to prove that H_ε is nonempty for each $\varepsilon > 0$. Now for fixed $\varepsilon > 0$, let

$$A_{\alpha,\varepsilon} = \{x \in X \,:\, f_\alpha(x_\alpha, x_{-\alpha}) > g_\alpha(x_{-\alpha}) - \varepsilon\},\ \alpha \in I.$$

For each $\alpha \in I$ and each $x_\alpha \in X_\alpha$, the set

$$\begin{aligned} A_{\alpha,\varepsilon} &= \{x_{-\alpha} \in X_{-\alpha} \,:\, [x_\alpha, x_{-\alpha}] \in A_{\alpha,\varepsilon}\} \\ &= \{x_{-\alpha} \in X_{-\alpha} \,:\, f_\alpha(x_\alpha, x_{-\alpha}) > g_\alpha(x_{-\alpha}) - \varepsilon\} \end{aligned}$$

is open, by virtue of the continity of f_α and g_α. Also for each $\alpha \in I$ and each $x_{-\alpha} \in X_{-\alpha}$, the set

$$\begin{aligned} A_{\alpha\in\varepsilon} &= \{y_\alpha \in X_\alpha \,:\, [y_\alpha, x_{-\alpha}] \in A_{\alpha,\varepsilon}\} \\ &= \{y_\alpha \in X_\alpha \,:\, f_\alpha(y_\alpha, x_{-\alpha}) > g_\alpha(x_{-\alpha}) - \varepsilon\} \end{aligned}$$

is convex and nonempty, by virtue of the quasi-convexity of the function $f_\alpha(y_\alpha, x_{-\alpha})$ in y_α and the definition of g_α. Hence by Corollary 4.23.1, there is a point $u \in \bigcap_{\alpha\in I} A_{\alpha,\varepsilon}$. It readily follows that $u \in H_\varepsilon$. The proof is complete. □

The following theorem which is similar to the above theorem is proved by Browder in locally convex topological vector space.

Theorem 4.30 (Browder (1968)) *Let $\{X_\alpha : \alpha \in I\}$ be a family of nonempty compact convex sets, each in a locally convex topological vector space E_α. Let X and $X_{-\alpha}$ for each $\alpha \in I$ be as defined in the previous theorem. Let $\{f_\alpha : \alpha \in I\}$ be a correspondingly indexed family of continuous real valued functions on X. Assume that for each $\alpha \in I$, $f \in \mathbb{R}$, and each $x_{-\alpha} \in X_{-\alpha}$, the set $\{y_\alpha \in X_\alpha : f_\alpha(y_\alpha, x_{-\alpha}) \geq t\}$ is a convex subset of X_α. Then there is a point $u \in X$ such that for each $\alpha \in I$,*

$$f_\alpha(u) = \sup_{y_\alpha \in X_\alpha} f_\alpha(y_\alpha, u_{-\alpha}).$$

Proof. For each $\alpha \in I$, let

$$A_\alpha = \{x \in X \,:\, f_\alpha \geq \sup_{y_\alpha \in X_\alpha} f(y_\alpha, x_{-\alpha})\},$$

where $x_{-\alpha}$ is the projection of x on $X_{-\alpha}$. Since X is compact and f_α is by Lemma is uniformly continuous, it follows that the function $g_\alpha(x_{-\alpha}) = \sup_{y_\alpha \in X_\alpha} f_\alpha(y_\alpha, x_{-\alpha})$ is continuous on $X_{-\alpha}$. It follows that A_α is a closed subset of X for each $\alpha \in I$. Now for each $\alpha \in I$ and each $u_{-\alpha} \in X_{-\alpha}$, the set $A_\alpha = (u_{-\alpha}) = \{v_\alpha \in X_\alpha : f_\alpha(v_\alpha, u_{-\alpha}) \geq g_\alpha(u_{-\alpha})\}$ is a nonempty closed subset of X_α which is convex by assumption. Hence by Theorem $\bigcap_{\alpha\in I} A_\alpha \neq \emptyset$, and if $u \in \bigcap_{\alpha\in I} A_\alpha$, then u satisfies the requirement of the theorem. □

Remark 4.7 The interested readers are encouraged to obtain the dual forms (see e.g. Husian and Tarafdar (1976)) and the noncompact version of the last two theorems.

4.5 More on the Extension of KKM Theorem and Ky Fan's Minimax Principle

In this section we will consider two extensions of Ky Fan's generalization (Theorem) of the well known classical finite dimensional result of Knaster-Kuratowski-Mazurkiewiéz (KKM) (1929), the first one by Brezis, Nirenberg and Stampacchia (1972) and the second one by Tarafdar and Thompson (1978). We will make a detailed comparison between the two and make applications.

Brézis, Nirenberg and Stampacchia's Extension of Ky Fan's Generalization of KKM Theorem

Lemma 4.8 *Let X be a nonempty subset of a topological vector space E. To each $x \in X$, let a nonempty subset $F(x)$ of E be given such that*

(i) $F(x_0) = L$ is compact for some $x_0 \in X$;
(ii) the convex hull of every finite subset $\{x_1, x_2, \ldots, x_n\}$ of X is contained in the corresponding union $\bigcup_{i=1}^{n} F(x_i)$;
(iii) for each $x \in X$, the intersection of $F(x)$ with any finite dimensional subspace is closed;
(iv) for every convex subset D of E the following equality holds

$$\left\{\overline{\bigcap_{x\in X\cap D} F(x)}\right) \cap D = \left(\bigcap_{x\in X\cap D} F(x)\right) \cap D.$$

Then $\bigcap_{x\in X} F(x) \neq \emptyset$.

The proof of this lemma is not given. The idea of the proof will be the same as that of Lemma 4.10.

Lemma 4.9 *Let X be a nonempty subset of E. To each $x \in X$, let a nonempty set $F(x)$ of E be given such that*

(a) $x \in F(x)$ for each $x \in F(x)$;
(b) $F(x_0)$ is compact for some $x_0 \in X$;

(c) *for each finite subset* $\{x_1, x_2, \ldots, x_n\}$ *of* X *and each* $x \in S_n = \langle x_1, x_2, \ldots, x_n \rangle =$ *the convex hull of* $\{x_1, x_2, \ldots, x_n\}$, *the set* $A(x) = \{y \in S_n \cap X : x \notin F(y)\}$ *has the property that whenever* $A(x)$ *is nonempty, it contains a nonempty convex subset* $H(x)$ *such that the set* $P(x) = \{y \in S_n : x \notin H(y)\}$ *is closed;*

(d) $F(x_0) \cap F(x)$ *is closed for each* $x \in C$.

Then $\bigcap_{x \in X} F(x) \notin \emptyset$.

Proof. In view of (b) and (d) it suffices to prove that $\bigcap_{i=1}^{n} F(x_i) \neq \emptyset$ for each finite subset $\{x_1, x_2, \ldots, x_n\}$ of X. On the contrary we suppose that for some finite subset $\{x_1, x_2, x_k\}$ of X we have $\bigcap_{i=1}^{k} F(x_i + \emptyset$. Then for each $x \in S_k = \langle x_1, x_2, \ldots, x_k \rangle$ the set $A(x) = \{y \in S_k \cap X : x \notin F(y)\}$ is nonempty. Indeed, at least one of the points x_i, $i = 1, 2, \ldots, k$ must be in $A(x)$, for otherwise $\bigcap_{i=1}^{k} F(x_i)$ would be nonempty. We now define a multi-valued mapping $T : S_k \to 2^{S_k}$ by $T(x) = H(x)$, $x \in S|k$; T is well defined by virtue of (c). Now for each $x \in S_k$, $T^{-1}(x) = \{y \in S_k : x \in T(y)\} = \{y \in S_k : x \in H(y)\} =$ complement of $P(x)$ in S_K which is an open set in S_k by condition (c) $P(x)$ being closed in S_k). Hence by the fixed point theorem of Browder there is a point $x_0 \in S_k$ such that $x_0 \in T(x_0)$. But then by definition of $T(x_0)$ we have $x_0 \notin F(x_0)$ which contradicts (a). Thus $\bigcap_{i=1}^{k} F(x_i) \neq \emptyset$. □

We are now in a position to prove our main lemma.

Tarafdar and Thompson's Extension of Ky Fan's Generalization of KKM Theorem

Lemma 4.10 *Let* X *be a nonempty subset of* E. *To each* $x \in X$, *let a nonempty subset* $F(x)$ *of* E *be given such that*

(α) $x \in F(x)$ *for each* $x \in X$;

(β) $\overline{F(x_0)} = L$ *is compact for some* $x_0 \in X$;

(γ) *for each finite subset* $\{x_1, x_2, \ldots, x_n\}$ *of* X *and each* $x \in S_n = \langle x_1, x_2, \ldots, x_n \rangle$ *the set* $A(x) = \{y \in S_n \cap X : x \notin F(y)\}$ *has the same property as laid down in (c) of Lemma 2.2.*

(δ) *for each* $x \in X$, *the intersection of* $F(x)$ *with any finite dimensional subspace is closed;*

(ω) *the Brezis-Nirenberg-Stampacchia condition holds, that is, for every convex subset* D *of* E *we have* $\overline{(\bigcap_{x \in X \cap D} F(x))} \cap D = (\bigcap_{x \in X \cap D} F(x)) \cap D$.

Then $\bigcap_{x \in X} F(x) \neq \emptyset$.

Proof. We may assume $x_0 = 0$. Let $(E_i)_{i\in I}$ be the class of all finite dimensional subspaces of E ordered by inclusion i.e. $i \geq j$ means $E_j \subset E_i$. Restricting to E_i the conditions of Lemma 4.9 apply to $X_i = X \cap E_i$ and $F_i(x) = F(x) \cap E_i$. Clearly (a) and (c) are satisfied and (b) and (d) follow from (β) and (δ). By Lemma 4.9 there is $u_i \in L \cap E_i$ satisfying

$$u_i \in F_i(x) \subset F(x) \quad \text{for every} \quad x \in X_i.$$

We now repeat the argument of Brezis Nirenberg, and Stampacchia (1972). Let $\phi_i = \bigcup_{j\geq i} \{u_j\}$ and so $u \in F(z)$ for $u \in \phi + i$ and $z \in x_i$ and hence $\phi_i \subset \bigcap_{x\in x_i} F(z)$.

Suppose $\tilde{x} \in \bigcap_{i\in I} \overline{\phi}_i$—which is non-empty since $\overline{\phi}_i \subset L$ is compact and let i_i be such that $\tilde{x} \in E_{i_0}$. For any $x \in X$ we can find $i \geq i_0$ such that $x \in E_i$. We have

$$\tilde{x} \in \overline{\phi}_i \cap E_i \subset \left(\overline{\bigcap_{z\in X_i} F(z)}\right) \cap E_i = \bigcap_{z\in X_i} F_i)(z)$$

by (ω). Therefore $\tilde{x} \in F_i(x) \subset F(x)$ and consequently $\tilde{x} \in \bigcap_{x\in X} F(x)$. □

Comparison between Lemma 4.8 and Lemma 4.10

(A) *If condition* (γ) *of Lemma* 4.10 *is strengthened to the condition:*

(γ)′: *for each* $x \in X_n = \langle x_1, x_2, \ldots, x_n \rangle$ *the set* $A(x) = \{y \in S_n \cap X : x \notin F(y)\}$ *is convex, then Lemma* 4.10 *follows from Lemma* 4.8.

To show this, it is enough to show that $(\gamma)'$ implies condition (*ii*) of Lemma 4.8. Let (α) hold and $\{x_1, x_2, \ldots, x_n\}$ be any finite subset of X. Suppose (*ii*) fails and $S_n = \langle x_1, x_2, \ldots, x_n = rangle \not\subseteq \bigcup_{i=1}^{n} F(x_i)$. Then there is $x \in S_n$ with $x \notin \bigcup_{i=1}^{n} F(x_i)$, $x = \sum_{i=1}^{n} \lambda_i x_i$, $\lambda_i \geq 0$, and $\sum_{i=1}^{n} \lambda_i = 1$. Sicne $x \notin F(x_i)$, $x_i \in A(x)$ for all $i = 1, 2, \ldots, n$, and hence $x = \sum_{i=1}^{n} \lambda_i x_i \in A(x)$ by $(\gamma)'$. This means that $x \notin F(x)$ contradicting (α). Thus (*ii*) of Lemma 4.8 and Lemma 4.10 follow from Lemma 2.1.

Remark 4.8 It is interesting to note that in this case we can take $H(x) = A(x)$ for each $x \in X$ since $P(x) = \{y \in S_n : x \notin H(y) = A(y)\} = \{y \in X_n : x \notin F(y)\}$ is automatically closed by (δ).

(B) *Lemma* 4.8 *applies to the following example although Lemma* 4.10 *does not apply.*

Let E be the plane R^2, $S = \{(u, v) \in R^2 : -1 \leq u, v \leq 1\}$, and $X = \{(u, v) \in S : |u| = |v| = 1\}$. For $x = (i, j) \in X$ set $F(x) = \{(u, v) \in R^2 : 0 \leq iu, jv \leq 1\}$.

Clearly Lemma 4.8 applies and by inspection $\bigcap_{x \in X} F(x) = \{0, 0\}$. That Lemma 4.10 does not apply can be seen as follows. For x in S let $A(x) = \{y \in X \cap X : x \notin F(y)\}$ so that for $x \neq (0, 0)$, $A(x)$ is a non-empty subset of X. Let $H(x)$ be a non-empty convex subset of $A(x)$ for $x \neq (0, 0)$. Suppose $H^{-1}(x) = \{y \in S : x \in H(y)\}$ is open in S for all x in S. Now $H^{-1}(x)$ is empty for x not in X and since $H(x)$ is a single element for $x \neq (0, 0)$ non-empty $H^{-1}(x)$ are disjoint. Now $\bigcup_{x \in X} H^{-1}(x) = S \setminus \{(0, 0)\}$ is connected which is a contradiction.

(C) *Lemma* 4.10 *applies to the following example although Lemma* 4.8 *does not apply.*

Let E be the reals, $F(-3) = \{x \in R : -3 \leq x \leq -2 \text{ or } |x| \leq 1\}$ and $F(3) = \{x \in R : 2 \leq x \leq 3 \text{ or } |x \leq 1\}$. Clearly Lemma 4.8 does not apply since $[-3, 3]$ is not a subset of $F(3) \cup F(-3)$. Now Lemma 4.10 applies since for x in $[-3, 3]$, $A(x) = \{y \in [-3, 3] \cap \{-3, 3\} : x \notin F(y)\}$ and we may choose

$$H(x) = \begin{cases} -3, & \text{for } x > 1 \\ 3, & \text{for } x < -1. \end{cases}$$

Then $H(x)$ is a convex subset of $A(x)$ and $H^{-1}(x)$ is open in $[-3, 3]$. The other conditions of Lemma 4.10 are clearly satisfied.

Applications

Theorem 4.31 (Minimax principle) *Let K be a non-enpty convex subset of E and $f(x, y)$ be a real valued function defined on $K \times K$ such that*

(*i*) *$f(x, x)(\leq 0$ for each $x \in K$;*

(*ii*) *for each finite subset $\{x_1, x_2, \ldots, x_n\}$ of K and $x \in S_n = \langle x_1, x_2, \ldots, x_n\rangle$ the set $A(x) = \{y \in S_n : f(x, y) > 0\}$ if non-empty contains a non-empty convex subset $H(x)$ such that the set*

$$P(x) = \{y \in S_n : x \notin H(y)\}$$

is closed;

(*iii*) *for each $y \in K$, $f(x, y)$ is a lower semicontinuous function of x on the intersection of K with any finite dimensional subspace of E;*

(*iv*) *there is a compact subset L of E and $y_0 \in L \cap K$ such that $f(x, y_0) > 0$ for $x \in K$, $x \notin L$;*

(*v*) *whenever $x, y \in K$ and x_δ is a net on K converging to x, then $f(x_\alpha(1 - t)x + ty) \leq 0$ for every $t \in [0, 1]$ implies $f(x, y) \leq 0$.*

Then there is a point $x_0 \in L \cap K$ such that

$$f(x_0, y) \leq 0 \quad \textit{for all} \quad y \in K.$$

In particular, $\inf_{x \in K} \sup_{y \in K} f(x, y) \leq 0$.

Proof. For each $z \in K$ we set

$$F(z) = \{x \in K : f(x, z) \leq 0\}.$$

For each finite subset $\{x_1, x_2, \ldots, x_n\}$ of K and $x \in S_n = \langle x_1, x_2, \ldots, x_n \rangle$ the set $A(x) = \{y \in S_n : x \notin F(y)\} = \{y \in S_n : f(x, y) >\}$ has the property (γ) of Lemma 4.10 by (ii). While (α), (δ) and (ω) of Lemma 4.10 follow from (i), (iii) and (v) respectively. Finally by (iv), $F(y_0)$ is compact and hence we have (β) of Lemma 4.10. Thus by Lemma 4.10 there is a point $x_0 \in L \cap K$ such that

$$x_0 \in \bigcap_{x \in K} F(x), \text{ that is, } f(x_0, y) \leq 0 \text{ for all } y \in K.$$

We note that $x_0 \in L$ by virtue of (iv). □

Corollary 4.31.1 (Brezis Nirenberg and Stampacchia (1972)) *Let K be a non-empty convex subset of E and $f(x, y)$ be a real valued function defined on $K \times K$ such that*

(i)′ $f(x, x) \leq 0$ for each $x \in K$;
(ii)′ for every $x \in K$, the set $\{y \in K : f(x, y) > 0\}$ is convex;
(iii)′ the condition (iii) of Theorem 4.31 holds;
(iv)′ the condition (iv) of Theorem 4.31 holds;
(v)′ the condition (v) of Theorem 4.31 holds.

Then there exists a point $x_0 \in L \cap K$ such that

$$f(x_0, y) \leq 0 \quad \text{for all} \quad y \in K.$$

Proof. As before we set

$$F(z) = \{x \in K : f(x, z) \leq 0\} \quad \text{for each} \quad z \in K.$$

The set $A'(x) = \{y \in K : f(x, y) > 0\}$ is convex by (ii)′. Hence for any finite subset $\{x_1, x_2, \ldots, x_n\}$ of K and $x \in S_n = \langle x_1, x_2, \ldots, x_n \rangle$ the set $A(x) = \{y \in S_n : f(x, y) > 0\}$ is convex. Now we choose $H(x) = A(x)$ for each $x \in K$. The set $P(x) = \{y \in S_n : x \in H(y)\}$ is closed by (iii)′ because of the reason given in the remark following **(A)**. Thus the conclusion of the corollary follows from Theorem 4.31. □

Corollary 4.31.2 (Ky Fan (1972)) *Let K be a non-empty compact convex subset of E and $f(x, y)$ be a real valued function defined on $K \times K$ such that*

(0) $f(x, x) \leq 0$ for each $x \in K$;
(00) for each $x \in K$, the set $\{y : f(x, y) > 0\}$ is convex;
(000) for each $y \in K$, $f(x, y)$ is a lower semicontinuous function of x on K.

Then there is a point $x_0 \in K$ such that $f(x_0, y) \leq 0$ for all $y \in K$.

Proof. This follows from Corollary 4.31.1. □

Let E be Hausdorff topological vectors space over the reals and K be a subset of E. Then a mapping A of K into E^* is called *pseudomonotone* if, whenever x_δ is a net in K converging to x with $\limsup(Ax_\delta,\, x_\delta - x) \leq 0$ then $\liminf(Ax_\delta,\, x_\delta - y) \geq (Ax,\, x - y)$. Here $(\cdot\,,\cdot)$ denotes the paring between E^* and E.

Corollary 4.31.3 (Brezis (1968), Corollary 29) *Let K be convex subset of E (over reals) and let $f(x,\, y) = (Ax, x-y)+\phi(x)-\phi(y)$ where A is a pseudo-monotone mapping from K into E^* and ϕ is a lower semicontinuous convex function. In addition we assume that A is continuous from any finite dimensional subspace of E to the weak topology of E^* and condition (iv)′ of Corollary* 4.31.1 *holds. Then there exists $x_0 \in L \cap K$ such that $(Ax_0,\, x_0 - y) + \phi(x_0) - \phi(y) \leq 0$ for all $y \in K$.*

Proof. The conditions (i)′, (ii)′ and (iii)′ of Corollary 4.31.1 follow immediately. To verify that (v)′ holds, let $x_\delta \to x$ and $f(x_\delta,\, (1-t)x + ty) \leq 0$ for each $t \in [0,\, 1]$. In particular $f(x_\delta,\, x) \leq 0$, so that

$$\langle Ax_\delta,\, x_\delta - x\rangle + \phi(x_\delta) - \phi(x) \leq 0.$$

Hence $\limsup\langle Ax_\delta,\, x_\delta - x\rangle \leq \limsup(\phi(x) - \phi(x_\delta))$ and thus $\limsup\langle Ax_\delta,\, x_\delta - y\rangle \geq \langle Ax,\, x - y\rangle$ for each $y \in K$. But we have also

$$f(x_\delta,\, y) = \langle Ax_\delta,\, x_\delta - y\rangle + \phi(x_\delta) - \phi(x) \leq 0.$$

Hence $\langle Ax,\, x - y\rangle + \phi(x) - \phi(y) \leq 0$. □

4.6 A Fixed Point Theorem Equivalent to the Fan–Knaster–Kuratowski–Mazurkiewicz Theorem

Fan (1984) has obtained a further generalization of the classical KKM Theorem (Knaster, Kuratowski, and Mazurkiewicz (1929)). Tarafdar (1987) has proved a fixed point theorem for setvalued mapping with the aid of an earlier theorem (Theorem 4.14) of him and has shown that this theorem is equivalent to Fan's Theorem. In this section we will present these two results.

Theorem 4.32 (Fan–Knaster–Kuratowski–Mazurkiwicz Theorem) *Let Y be a nonempty convex subset of a topological vector space and $\phi \neq X \subset Y$. For each $x \in X$, let $F(x)$ be a relatively closed subset of Y such that the convex hull of each finite subset $\{x_1, x_2, \ldots, x_n\}$ of X is contained in the corresponding union $\bigcup_{i=1}^{n} F(x_i)$. Then for each nonempty subset X_0 of X such that X_0 is contained in a compact convex subset of Y, $\bigcap_{x \in X_0} F(x) \neq \emptyset$. Furthermore, if for some such X_0 (i.e. X_0 is contained in a compact convex subset of Y) the nonempty set $\bigcap_{x \in X_0} F(x)$ is compact, then $\bigcap_{x \in X} F(x) \neq \emptyset$.*

We first prove the following fixed point theorem independently of Theorem 4.32 and show the equivalence of Theorem 4.32 with our fixed point Theorem 4.33 below:

Theorem 4.33 *Let X be a nonempty convex subset of a topological vector space. Let $F : X \to 2^X$ be a set valued mapping such that*

(i) *for each $x \in X$, $f(x)$ is a nonempty convex subset of X;*

(ii) *for each $y \in X$, $f^{-1}(y) = \{x \in X : y \in F(x)\}$ contains a relatively open subset O_y of X (O_y may be empty for some y):*

(iii) *$\bigcup_{x \in X} O_x = X$; and*

(iv) *there exists a nonempty $X_0 \subset X$ such that X_0 is contained in a compact convex subset X_1 of X and the set $D = \bigcap_{x \in X_0} O_x^c$ is compact, (D could be empty and as before O_x^c denotes the complement of O_x in X). Then there exists a point $x_0 \in X$ such that $x_0 \in f(x_0)$.*

Proof of Theorem 4.33. We first assume that $D = \emptyset$. In this case for each $x \in X_1$, $f(x) \cap X_1 \neq \emptyset$. Indeed if $f(x_0) \cap X_1 = \emptyset$ for some $x_0 \in X_1$, then for all $x \in X_1$, $x \notin f(x_0)$, i.e., $x_0 \notin f^{-1}(x) \supseteq O_x$. Thus $x_0 \in \bigcap_{x \in X_1} O_x^c \subseteq \bigcap_{x \in X_0} O_x^c = D$, which contradicts that $D = \emptyset$. Therefore we can define a set valued mapping $g : X_1 \to 2^{X_1}$ by $g(x) = f(x) \cap X_1$, so that $g(x)$ is a non-empty convex subset of X_1 for each $x \in X_1$. Now for each $y \in X_1$, $g^{-1}(y) = \{x \in X_1 : y \in g(x)\} = \{x \in X_1 : y \in f(x) \cap X_1\} = f^{-1}(y) \cap X_1$ contains the relatively open set $O_y^1 = O_y \cap X_1$ in X_1. Also since $\bigcap_{x \in X_0} O_x^c = \emptyset$, we have $\bigcup_{x \in X_0} O_x = X$ and hence $\bigcup_{x \in X_1} O_x = X$. Thus $\bigcup_{x \in X_1} O_x^1 = \bigcup_{x \in X_1} (O_x \cap X_1) = X_1$. Hence by our Theorem 4.14 there is a point $x_0 \in X_1$ such that $x_0 \in g(x_0) \subseteq f(x_0)$.

We now consider the case when D is a nonempty compact set. In this case we prove the theorem by contradiction. Let us assume that f has no fixed point. Then O_x^c is nonempty for each $x \in X$, for $O_x^c = \emptyset$ implies that $x \notin O_x^c$, i.e., $x \in O_x \subset f^{-1}(x)$, i.e. $x \in f(x)$. More generally, the convex hull of each finite subset $\{x_1, x_2, \dots, x_n\}$ of X is contained in the union $\bigcup_{i=1}^{n} O_{x_i}^c$. To see this let $x = \sum_{i=1}^{n} \lambda_i x_i \notin \bigcup_{i=1}^{n} O_{x_i}^c$ for some finite subset $\{x_1, x_2, \dots, x_n\}$ of X and $\lambda_i \geq 0$, $i = 1, 2, \dots, n$, with $\sum_{i=1}^{n} \lambda_i = 1$. This implies that $x \in O_{x_i} \subseteq f^{-1}(x_i)$ for each $i = 1, 2, \dots, n$. Hence $x_i \in f(x)$ for each $i = 1, 2, \dots, n$. However, since $f(x)$ is convex, $x \in f(x)$, contradicting our assumption. For convenience, we set $F(x) = O_x^c$ for each $x \in X$. In our next move we prove by applying Theorem 4.14 that for each finite subset $\{x_1, x_2, \dots, x_n\}$ of X, $\bigcap_{x \in K} F(x) \neq \emptyset$, where K is the convex hull of $X_1 \cup \{x_1, x_2, \dots, x_n\}$. Clearly K is a compact convex subset of X. If possible, we suppose that $\bigcap_{x \in K} F(x) = \emptyset$. Then we can define a set valued mapping $h : K \to 2^K$

by $h(y) = \{x \in K : y \notin F(x)\}$ such that $h(y)$ is nonempty for each $y \in K$. Now for $x \in K$, $h^{-1}(x) = \{y \in K : x \in h(y)\} = \{y \in K : y \notin F(x)\} = (F(x))^c \cap K = O_x \cap K = \hat{O}_x$, which is a relatively open set in K. We now define a set valued mapping $j : K \to 2^K$ by $j(x) =$ convex hull of $h(x)$ for each $x \in K$. Since $j(x) \supset h(x)$ for each $x \in K$, it follows that $j^{-1}(x) \supseteq h^{-1}(x) \supseteq \hat{O}_x$ for each $x \in K$. Also $\bigcap_{x\in K} F(x) = \emptyset$ implies that $\bigcup_{x\in K} O_x = X$. Hence $\bigcup_{x\in K} \hat{O}_x = \bigcup_{x\in K} (O_x \cap K) = K$. Therefore by Theorem 4.14 there exists $x_0 \in K$ such that $x_0 \in j(x_0) =$ convex hull of $h(x_0)$. This implies that there exist points $y_1, y_2, \ldots, y_m$ in K such that $y_i \in h(x_0)$ for $i = 1, 2, \ldots, m$, where $x_0 = \sum_{i=1}^{n} \lambda_i y_i$, $\lambda_i \geq 0$ for $i = 1, 2, \ldots, m$, and $\sum_{i=1}^{m} \lambda_i = 1$. This means that $x_0 \notin F(y_i)$ for $i = 1, 2, \ldots, m$, i.e., $x_0 = \sum_{i=1}^{m} \lambda_i y_i \notin \bigcup_{i=1}^{m} F(y)$, which contradicts our established fact that the convex hull of each finite subset $\{y_1, y_2, \ldots, y_n\}$ of X is contained in the corresponding union $= \bigcup_{i=1}^{m} F(y_i) = \bigcup_{i=1}^{m} O^c_{y_i}$.

Thus we have proved that $\bigcap_{x\in K} F(x) \neq \emptyset$. Hence $D \bigcap (\bigcap_{i=1}^{n} F(x_i)) \supseteq \bigcap_{x\in K} F(x) \neq \emptyset$ as $X_0 \cup \{x_1, x_2, \ldots, x_n\} \subseteq K$. What we have then proved above is that for each finite subset $\{x_1, x_2, \ldots, x_n\}$ of X, $\bigcap_{i=1}^{n} (D \cap F(x_i)) \neq \emptyset$. Now since D is compact and $F(x)$ is closed, $F(x) \cap D$ is compact for each $x \in X$. Hence $\bigcap_{x\in X} (F(x) \cap D) \neq \emptyset$ and, therefore, $\bigcap_{x\in X} F(x) = \bigcap_{x\in X} O^c_x \neq \emptyset$, which contradicts our condition (iii). Thus f must have a fixed point. □

Corollary 4.33.1 *Let X be a nonempty convex subset of a topological vector space. Let $f : X \to 2^X$ be a set valued mapping such that*

(i) *for each $x \in X$, $f(x)$ is a nonempty convex subset of X;*
(ii) *for each $y \in X$, $f^{-1}(y)$ contains a relatively open subset O_y of X;*
(iii) *$\bigcup_{x\in X} O_x = X$; and*
(iv) *there exists a point $x_0 \in X$ such that $O^c_{x_0}$ is compact.*

Then there exists a point $x \in X$ such that $x \in f(x)$.

Equivalence of Theorem 4.32 and Theorem 4.33

We first prove that Theorem 4.33 implies Theorem 4.32.

Let us assume that the conditions of Theorem 4.32 hold. If possible, suppose that $\bigcap_{x\in X} F(x) = \emptyset$. Then we can define a set valued mapping $f : Y \to 2^Y$ by $f(y) = \{x \in X : y \notin F(x)\}$. Clearly $f(y)$ is a nonempty subset of Y for each $y \in Y$. It also follows that for each $x \in Y$, $f^{-1}(x) = (F(x))^c = O_x$ is a relatively open set in Y. Let $g : Y \to 2^Y$ be the set valued mapping defined by $g(y) =$ convex hull of $f(y)$ for each $y \in Y$. Thus for each $y \in Y$, $g(y)$ is a nonempty convex subset of Y

and for each $x \in Y$, $g^{-1}(x) \supset f^{-1}(x) = O_x$. Also $\bigcap_{x\in X} F(x) = \emptyset$ implies $\bigcup_{x\in X} O_x = Y$ and hence $\bigcup_{x\in Y} O_x = Y$. Finally, $\bigcap_{x\in X_0} O_x^c = \bigcap_{x\in X_0} F(x) = D$ is compact. Hence by Theorem 4.33 there exists a point $x_0 \in X$ such that $x_0 \in g(x_0) =$ convex hull of $f(x_0)$. Now by giving an argument similar to that in the proof of Theorem 4.33 we can show that this contradicts the hypothesis of Theorem 4.32 that the convex hull of each finite subset $\{x_1, x_2, \dots, x_n\}$ of X is contained in the corresponding union $\bigcup_{i=1}^{n} F(x_i)$. Hence $\bigcap_{x\in X} F(x) \neq \emptyset$.

Now to show that Theorem 4.32 implies Theorem 4.33, we assume that the conditions of Theorem 4.33 hold.

For each $x \in X$, $F(x) = O_x^c$, which is a relatively closed set in X. Let us first consider the case when $D = \emptyset$. Then by taking $Y = X$ in Theorem 4.32 we must have a finite subset $\{x_1, x_2, \dots, x_n\}$ of X such that the convex hull of $\{x_1, x_2, \dots, x_n\}$ is not contained in the corresponding union $\bigcup_{i=1}^{n} F(x_i)$, for otherwise D will be nonempty by the first part of Theorem 4.32. This means that $x_0 = \sum_{i=1}^{n} \lambda_j x_i \notin F(x_i) = O_{x_i}^c$ for each $i = 1, 2, \dots, n$ and for some $\lambda_i \geq 0$, $i = 1, 2, \dots, n$ with $\sum_{i=1}^{n} \lambda_i = 1$. Thus $x_0 \in O_{x_i} \subset f^{-1}(x_i)$, i.e., $x_i \in f(x_0)$ for each $i = 1, 2, \dots, n$. Hence $x_0 \in f(x_0)$ as $f(x_0)$ is convex and Theorem 4.33 is proved in this case.

Finally, let $D \neq \emptyset$. If the convex hull of each finite subset $\{x_1, x_2, \dots, x_n\}$ of X is contained in the corresponding union $\bigcup_{i=1}^{n} F(x_i)$, then by Theorem 4.32 $\bigcap_{x\in X} O_x^c = \bigcap_{x\in X} F(x) \neq \emptyset$, which contradicts the condition (iii) of Theorem 4.33. Hence there must exist a finite subset $\{x_1, x_2, \dots, x_r\}$ of X such that the convex hull of $\{x_1, x_2, \dots, x_r\}$ is not contained in the corresponding union $\bigcup_{i=1}^{r} F(x_i)$. Now repeating the same argument as in the first case, we obtain a point $x_0 \in X$ such that $x_0 \in f(x_0)$. This completes the proof. □

We will have many opportunities of seeing the applications of Theorem 4.33 in the sequel but in the following theorem (Tarafdar (1989)) we note a simple application.

Theorem 4.34 *Let $X_1, X_2, \dots, X_n$ $n(\geq 2)$ be nonempty sets each in a Hausdorff topological vector space and let $X = \prod_{1\leq j\leq n} X_i$ Let $X_{-j} = \prod_{1\leq i\leq n, i\neq j} X_i$ and let us denote the points of X_{-j} by x_{-j}. Let $A_1, A_2, \dots, A_n$ and $B_1, B_2, \dots, B_n$ be subsets of X such that*

(a) *for each $j = 1, 2, \dots, n$ and for each $x_j \in X_j$ the set $B_j(x_j) = \{x_{-j} \in X_{-j} : [x_j, x_{-j}] \in B_j\}$ is open in X_{-j};*
(b) *for each $j = 1, 2, \dots, n$ and each $x_{-j} \in X_{-j}$, the set $B_j(x_{-j}) = \{x_j \in X_j :$*

$[x_j, x_{-j} \in B_j\}$ is nonempty;

(c) *for each $j = 1, 2, \ldots, n$ and $x_{-j} \in X_{-j}$, the set $A_j(x_{-j}) = \{x_j \in X_j : [x_j, x_{-j}] \in A_j\}$ contains the convex hull of $B_j(x_{-j})$;*

and

(d) *there is a nonempty compact convex subset K of X such that for every $x \in X \setminus K$, $x = [x_j, x_{-j}] \in X_j \times X_{-j}$, there exists a $y \in K$, $y = [y_j, y_{-j}] \in X_j \times X_{-j}$ satisfying $[y_j, x_{-j}] \in B_j$ for $j = 1, 2, \ldots, n$. Then $\bigcap_{i=1}^{n} A_j \neq \emptyset$.*

Proof. We define two set valued mappings $f : X \to 2^X$ and $g : X \to 2^X$ by $f(x) = \prod_{1\le j\le n} \operatorname{co} B_j(x_{-j})$ and $g(x) = \prod_{1\le j\le n} B_j(x_{-j})$ for each $x = [x_j, x_{-j}] \in X$, where $\operatorname{co} B_j(x_{-j})$ denotes the convex hull of $B_j(x_{-j})$. By condition (b), $f(x)$ and $g(x)$ is nonempty for each $x \in X$. Clearly $f(x)$ is convex and $g(x) \subset f(x)$ for each $x \in X$ by condition (c). For each point $y = (y_1, y_2, \ldots, y_n) \in X$ we consider the set $g^{-1}(y) = \{x \in X : y \in g(x)\}$. Now

$$x \in g^{-1}(y) \Leftrightarrow y \in g(x) = \prod_{1\le j\le n} B_j(x_{-j}) \Leftrightarrow y_j \in B_j(x_{-j})$$

for each $j = 1, 2, \ldots, n \Leftrightarrow x_{-j} \in B_j(y_j)$ for each $j = 1, 2, \ldots, n$. Thus for each $y \in X$,

$$g^{-1}(y) = \bigcap_{j=1}^{n} \{B_j(y_j) \times X_j\}$$

is an open set in X by virtue of the condition (a) $= O_y$, say. As $g(y) \subset f(y)$ for each $y \in X$, $O_y = g^{-1}(y) \subset f^{-1}(y)$ for each $y \in X$. Also $\bigcup_{x\in X} O_x = X$. [For let $y \in X$. Then $g(y) \neq \emptyset$. Let $x \in g(y)$. Obviously $y \in g^{-1}(x) = O_x$].

Now

$$x \in \cap_{y\in K} O_y^c = \cap_{y\in K} [g^{-1}(y)]^c \Leftrightarrow x \in [g^{-1}(y)]^c$$

for each $y \in K \Leftrightarrow x \notin g^{-1}(y)$ for each

$$y \in K \Leftrightarrow x \notin \bigcap_{j=1}^{n} B_j(y_j) \times X_j$$

for each $y = [y_j, y_{-j}] \in K \Leftrightarrow x \notin B_j(y_j) \times X_j$ for each $y = [y_j, y_{-j}] \in K$ and for some $j = 1, 2, \ldots, n, \ldots (*)$. On the the other hand by condition (d) for each $x \in X \setminus K$, $x = (x_j, x_{-j}) = X_j \times X_{-j}$, there exists $y \in K$, $y = (y_j, y_{-j})$ satisfying $(y_j, x_{-j}) \in B_j$ for $j = 1, 2, \ldots, n$. In other words

$$x \in \bigcap_{j=1}^{n} B_j(y_j)) \times x X_j.$$

This fact and (*) together imply that

$$\bigcap_{x\in K} O_x^c = \bigcap_{x\in K} [g^{-1}(x)]^c \subset K$$

Hence $\bigcap_{x\in K} O_x^c$ being closed subset of a compact set is compact. Hence all the conditions of Theorem 4.33 are fulfilled for the set valued mapping f. Thus by Theorem 4.33 there is a fixed point $x \in X$ of f, i.e.

$$x \in f(x) = \prod_{1\le j\le n} \operatorname{co} B_j(x_{-j}).$$

Hence by (c).

$$x \in \prod_{1\le j\le n} \operatorname{co} B_j(x_{-j}) \subseteq \prod_{1\le j\le n} A_j(x_{-j}),$$

i.e., $x_j \in A_j(x_{-j})$ for $j = 1, 2, \ldots, n$, i.e. $[x_j, x_{-j}] \in A_j$ for $j = 1, 2, \ldots, n$. Thus $x \in \bigcap_{j=1}^{n} A_j$. □

The original idea of introducing two families $A_1, A_2, \ldots, A_n$ and $B_1, B_2, \ldots, B_n$ is due to Dugundji and Granas (1978) and Ben-El-Mechaiekh, Deguire, and Granas (1982).

Further applications of this theorem may be found in Tarafdar (1986).

4.7 More on Fixed Point Theorems

We prove the following fixed point theorems of Ding and Tarafdar (1994) which differ from Theorem 4.33 in condition (iv).

As before all topological vector spaces are assumed to be Hausdorff in this section.

Theorem 4.35 *Let X be a nonempty convex subset of a topological vector space E and $F, G : X \to 2^X$ be such that*

(1) *for each $x \in X$, $F(x) \subset G(x)$;*
(2) *for each $y \in X$, $F^{-1}(y)$ is compactly open in X;*
(3) *there exist a nonempty compact convex subset X_0 of X and a nonempty compact subset K of X such that for $y \in X \setminus K$, there is an $x \in \operatorname{co}(X_0 \cup \{y\})$ with $y \notin \operatorname{cl}_C((X \setminus (\operatorname{co} G)^{-1}(x)) \cap C)$ for any nonempty compact subset C of X; and*
(4) *for each $x \in K$, $F(x) \neq \emptyset$, where $\operatorname{cl}_X(A)$ denotes the closure of A in X.*

Then there exists $\hat{y} \in X$ such that $\hat{y} \in \operatorname{co}((G(\hat{y}))$.

Proof. Suppose that the conclusion does not hold. Then for each $x \in X$, $x \notin \mathrm{co}((G(x))$. For each $y \in X$ and for each nonempty compact subset C of X, $(\mathrm{co}\, F)^{-1}(y) \cap C$ is open in C. Indeed, if $x \in (\mathrm{co}\, F)^{-1}(y) \cap C$, then $x \in C$ and $y \in \mathrm{co}(F(x))$. Let $y_1, \ldots, y_n \in F(x)$, $\lambda_1, \ldots, \lambda_n \geq 0$ with $\sum_{i=1}^{n} \lambda_i = 1$ such that $y = \sum_{i=1}^{n} \lambda_i y_i$. For each $i = 1, \ldots, n$, $F^{-1}(y_i) \cap C$ is open in C by (2) and $x \in F^{-1}(y_i) \cap C$. Let $U = \bigcap_{i=1}^{n} (F^{-1}(y_i) \cap C)$, then U is an open neighborhood of x in C. If $z \in U$, then $z \in C$ and $y_i \in F(z)$ for each $i = 1, \ldots, n$, so that $y = \sum_{i=1}^{n} \lambda_i y_i \in \mathrm{co}(F(z))$ and hence $z \in (\mathrm{co}\, F)^{-1}(y) \cap C$ for all $z \in U$. Therefore $(\mathrm{co}\, F)^{-1}(y) \cap C$ is open in C and $(\mathrm{co}\, F)^{-1}(y)$ is compactly open in X for each $y \in X$.

For each $x \in X$, let

$$T(x) = \mathrm{cl}_X(X \setminus (\mathrm{co}\, G)^{-1}(x)) \cap K,$$
$$S(x) = (X \setminus (\mathrm{co}\, F)^{-1}(x)) \cap K.$$

We shall prove that the family $\{T(x) : x \in X\}$ has the finite intersection property. Let $\{x_1, \ldots, x_n\}$ be any finite subset of X and let

$$D = \mathrm{co}(X_0 \cup \{x_1, \ldots, x_n\}),$$

then D is a compact convex subset of X. Define two mappings $T_0, S_0 : D \to 2^D$ by

$$T_0(x) = \mathrm{cl}_D((X \setminus (\mathrm{co}\, G)^{-1}(x)) \cap D),$$
$$S_0(x) = (X \setminus (\mathrm{co}\, F)^{-1}(x)) \cap D), \qquad \text{for each } x \in D.$$

We observe that

(a) for each $x \in D$, $S_0(x)$ is closed in D since $(\mathrm{co}\, F)^{-1}(x)$ is compactly open and D is compact;
(b) for each $x \in D$, $T_0(x) \subset S_0(x)$ by (1) and (a);
(c) for each $x \in D$, $T_0(x)$ is compact in D;
(d) T_0 is a KKM mapping, i.e. for any finite subset A of D, $\mathrm{co}(A) \subset \bigcup_{x \in A} T_0(x)$.

Indeed, it is enough to prove the mapping $T^* : D \to 2^D$ defined by

$$T^*(x) = (X \setminus (\mathrm{co}\, G)^{-1}(x)) \cap D \quad \text{for each } x \in D,$$

is a KKM mapping. If it were false, then there exist $\{u_1, \ldots, u_n\} \subset D$ and $\lambda_1, \ldots, \lambda_n \geq 0$ with $\sum_{i=1}^{n} \lambda_i = 1$ such that

$$\sum_{i=1}^{n} \lambda_i u_i \notin \bigcup_{i=1}^{n} T^*(u_i) = \bigcup_{i=1}^{n} (X \setminus (\mathrm{co}\, G)^{-1}(u_i)) \cap D.$$

It follows that

$$\sum_{i=1}^{n}\lambda_i u_i \in D\cap\bigcap_{i=1}^{n}(\operatorname{co}G)^{-1}(u_i)$$

and

$$u_j\in(\operatorname{co}G)(\sum_{i=1}^{n}\lambda_i u_i)\quad\text{for all}\quad j=1,\ldots,n.$$

Hence, we have

$$\sum_{i=1}^{n}\lambda_i u_i\in\operatorname{co}(G(\sum_{i=1}^{n}\lambda_i u_i))$$

which contradicts the fact that for each $x\in X$, $x\notin\operatorname{co}(G(x))$. therefore T^* is a KKM mapping and so T_0 is also a KKM mapping.

By Lemma 4.5, we have

$$\emptyset\neq\bigcap_{x\in D}T_0(x)=\bigcap_{x\in D}\operatorname{cl}_D((X\setminus(\operatorname{co}G)^{-1}(x))\cap D)\subset\bigcap_{x\in D}\operatorname{cl}_X(X\setminus(\operatorname{co}G)^{-1}(x))\cap D.$$

Take any $\overline{y}\in\bigcap_{x\in D}\operatorname{cl}_D((X\setminus(\operatorname{co}G)^{-1}(x))\cap D)$. Since $\operatorname{co}(X_0\cup\{\overline{y}\})\subset D$, by the assumption (3), we must have $\overline{y}\in K$ and hence

$$\overline{y}\in\bigcap_{i=1}^{n}\operatorname{cl}_X(X\setminus(\operatorname{co}G)^{-1}(x_i))\cap K=\bigcap_{i=1}^{n}T(x_i);$$

that is, the family $\{T(x):x\in X\}$ has the finite intersection property. By the compactness of K, $\bigcap_{x\in X}T(x)\neq\emptyset$. By (1) and (2), we have $T(x)\subset S(x)$ for each $x\in X$ and hence

$$\emptyset\neq\bigcap_{x\in X}S(x)=\bigcap_{x\in X}(X\setminus(\operatorname{co}F)^{-1}(x))\cap K.$$

Take any $\hat{y}\in\bigcap_{x\in X}(X\setminus(\operatorname{co}F)^{-1}(x))\cap K=K\cap(X\setminus\bigcup_{x\in X}(\operatorname{co}F)^{-1}(x))=K\setminus\bigcup_{x\in X}(\operatorname{co}F)^{-1}(x))$. But, by the assumption in (4) that for each $x\in K$, $F(x)\neq\emptyset$, we have

$$K\subset\bigcup_{x\in X}(\operatorname{co}F)^{-1}(x))$$

which is a contradiction. Therefore the conclusion must hold. □

Remark 4.9 Theorem 4.35 with $F=G$ improves Lemma 1 of Ding and Tan (1993) and Theorem 3″ of Ding and Tan (1992a) in the following two aspects: (1) the coercive condition in the assumption (3) of Theorem 4.35 is weaker than that in Lemma 1 of and Theorem 3″ of Ding and Tan (1993); (2) clearly, the condition for each $x \in K$, $F(x) \neq \emptyset$ is weaker than the condition for each $x \in X$, $F(x) \neq \emptyset$ since $K \subseteq X$.

The following result is an immediate consequence of Theorem 4.35.

Theorem 4.36 *Let X be a nonempty convex subset of a topological vector space E and $F, G : X \to 2^X$ be such that*

(1) *for each $x \in X$, $F(x) \subset G(x)$;*
(2) *for $y \in X$, $F^{-1}(y)$ is compactly open in X;*
(3) *there exist a nonempty closed and compact subset K of X and a point $x_0 \in X$ such that $X \setminus K \subset (\operatorname{co} G)^{-1}(x_0)$; and*
(4) *for each $x \in K$, $F(x) \neq \emptyset$.*

Then there exists $\hat{y} \in X$ such that $\hat{y} \in \operatorname{co}(G(\hat{y}))$.

Proof. Let $X_0 = \{x_0\}$, then X_0 is a nonempty compact convex subset of X and $x_0 \in \operatorname{co}(X_0 \cup \{y\})$ for any $y \in X \setminus K$. It follows from the condition (3) that $X \setminus (\operatorname{co} G)^{-1}(x_0) \subset K$ so that $\operatorname{cl}_C((X \setminus (\operatorname{co} G)^{-1}(x_0)) \cap C) \subset K$ for any nonempty compact subset C of X. Thus for each $y \in X \setminus K$, $y \notin \operatorname{cl}_C((X \setminus (\operatorname{co} G)^{-1}(x_0)) \cap C)$. The conclusion follows from Theorem 4.35. □

Remark 4.10 Theorem 4.36 is Theorem 2.4″ of (Tan and Yuan (1993)) and hence Theorem 4.35 improves and generalizes Theorems 2.4′, 2.4″ and 2.4‴ of (Tan and Yuan (1993)) and Theorem 1 of (Ding and Tan (1992b)).

We now prove that the following theorem is equivalent to Theorem 4.35.

Theorem 4.37 *Let X be a nonempty convex subset of a topological vector space E and $G : X \to 2^X$ be such that*

(1) *for each $y \in X$, $G^{-1}(y)$ contains a compactly open subset O_y of X (which may be empty);*
(2) *there exist a nonempty compact convex subset X_0 of X and a nonempty compact subset K of X such that for each $y \in X \setminus K$, there is an $x \in \operatorname{co}(X_0 \cup \{y\})$ with $y \notin \operatorname{cl}_C((X \setminus (\operatorname{co} G)^{-1}(x)) \cap C)$ for any nonempty compact subset C of X, and $K \subset \bigcup_{y \in X} O_y$.*

Then there exists $\hat{y} \in X$ such that $\hat{y} \in \operatorname{co}(G(\hat{y}))$.

Proof. We first prove that Theorem 4.35 ⇒ Theorem 4.37. We define $F : X \to 2^X$ by

$$F(x) = \{y \in X : x \in O_y\}, \quad x \in X,$$

By the last part of the condition (2), $F(x) \neq \emptyset$ for each $x \in K$. Thus the condition (4) of Theorem 4.35 holds. By the condition $F(x) \subset G(x)$ for each $x \in X$ which is precisely the condition (1). Again since $F^{-1}(y) = 0_y$ which is compactly open by (1) of this theorem. Lastly the condition (3) of Theorem 4.35 the same as the first part of condition 2 of Theorem 4.37. Hence the Theorem 4.37 follows from Theorem 4.35. Next we suppose that Theorem 4.37 holds. For each $y \in X$ we take $O_y = F^{-1}(y)$ where F is given in Theorem 4.35. Then we can easily see that Theorem 4.37 $\Rightarrow$ Theorem 4.35. □

Theorem 4.38 *Let X be a nonempty convex subset of a topological vector space E and $G: X \to 2^X$ be such that*

(1) *for each $y \in X$, $G^{-1}(y)$ contains a compactly open subset O_y of X (which may be empty),*
(2) *there exist a nonempty closed and compact subset K of X and a point $x_0 \in X$ such that $x_0 \in (\operatorname{co} G)(y)$ for all $y \in X \setminus K$,*
(3) $K \subset \bigcup_{y \in X} O_y$.

Then there exists a point $\hat{y} \in X$ such that $\hat{y} \in \operatorname{co}(G(\hat{y}))$.

Proof. Since the condition (2) of Theorem 4.38 implies that the condition (2) of Theorem 4.37 is satisfied. The conclusion holds from Theorem 4.37. □

Remark 4.11 Theorem 4.38 is Theorem 2.4'''' of Tan and Yuan (1993) and generalizes Theorems 2.4'''' and 2.4''''' of Tan and Yuan (1993), Theorem 2 of Ding and Tan (1992a) and the corresponding results of Tarafdar (1977), Mehta and Tarafdar (1987), Border (1985), Browder (1968) and Yannelis (1985).

The following fixed point theorem seems to be very useful in mathematical economics and related areas.

Theorem 4.39 (Tarafdar 1991). *Let $\{X_\alpha : \alpha \in I\}$ be a family of nonempty compact convex sets, each in a topological vector space E_α, where I is an indexing set. Let $X = \prod_{\alpha \in I} X_\alpha$. For each $\alpha \in I$, let $T_\alpha : X \to 2^{X_\alpha}$ be a set valued mapping such that*

(i) *for each $x \in X$, $T_\alpha(x)$ is a nonempty convex subset of X_α,*
(ii) *for each $x_\alpha \in X_\alpha$, $T_\alpha^{-1}(x_\alpha) = \{y \in X : x_\alpha \in T_\alpha(y)\}$ contains a relatively open subset O_{x_α} of X such that*

$$\bigcup_{x_\alpha \in X_\alpha} O_{x_\alpha} = X \qquad (O \text{ may be empty for some } x_\alpha),$$

Then there is a point $x \in X$ such that $x \in T(x) = \prod_{\alpha \in I} T_\alpha(x)$, i.e., $x_\alpha \in T_\alpha(x)$ for each $\alpha \in I$, where x_α is the projection of x onto X_α for each $\alpha \in I$.

Proof. We fix $\alpha \in I$. Since X is compact, by (ii) there is a finite set $\{x_\alpha^1, x_\alpha^2, \ldots, x_\alpha^n\}$ of points in X_α such that $\bigcup_{i=1} O_{x_\alpha^i} = X$. Let $\{f_\alpha^1, f_\alpha^2, \ldots, f_\alpha^n\}$ be a partition of unity corresponding to this finite subcovering $O_{x_\alpha^2}, O_{x_\alpha^2}, \ldots, O_{x_\alpha^n}$, i.e., each f_α^i vanishes outside $O_{x_\alpha}^i$, $0 \leqq f_\alpha^i(x) \leqq 1$ and $\sum_{i=1}^n f_\alpha^i(x) = 1$ for each $x \in X$.

Now we define a mapping $q_\alpha : X \to X_\alpha$ by

$$q_\alpha(x) = \sum_{i=1}^{n} f_\alpha^i(x)x_\alpha^i, \qquad x \in X.$$

Clearly q_α maps X into X_α and is continuous. Also for each $i = 1, 2, \ldots, n$ with $f_\alpha^i(x) \neq O$, $x \in O_{x_\alpha^i} \subset T_\alpha^{-1}(x_\alpha^i)$, i.e., $x_\alpha^i \in T_\alpha(x)$. But since $T_\alpha(x)$ is convex, it follows that $q_\alpha(x) \in T_\alpha(x)$ for each $x \in X$.

Let F_α be the linear hull of the set $\{x_\alpha^1, x_\alpha^2, \ldots, x_\alpha^n\}$. Then F_α is a locally convex (Hausdorff) topological vector space as it is finite dimensional. The convex hull S_α of the set $\{x_\alpha^1, x_\alpha^2, \ldots, x_\alpha^n\}$ is a compact subset of F_α. It is also clear that q_α maps X into S_α. If we do this for each $\alpha \in I$, we obtain a compact convex subset $S = \prod_{\alpha\in I} S_\alpha$ of the locally convex (Hausdorff) topological vector space $F = \prod_{\alpha\in I} F_\alpha$ and a mapping $q : S \to S$ defined by $q(x) = \prod_{\alpha\in I} q_\alpha(x)$ such that for each $x \in S$ and each $\alpha \in I$ $q_\alpha(x) \in T_\alpha(x)$.

Since for each $\alpha \in I$, q_α is a continuous ampping of X into S_α, it follows that q is a continuous mapping of S into itself. Hence by Tychonoff fixed point theorem, q has a fixed point, i.e., there is a point $x \in q(x) = \prod_{\alpha\in I} q_\alpha(x) \in \prod_\alpha T_\alpha(x)$, i.e., $x_\alpha \in T_\alpha(x)$ for each $\alpha \in I$, where x_α is the projection of x onto X_α. □

Remark 4.12 (1) If the set I consists of only one element, i.e., the cardinality of I is one, then Theorem 4.39 reduces to a theorem of Tarafdar (1977, Theorem 1), (in here Theorem 4.14) which contains the theorem of Browder (1968, Theorem 1), (in here Corollary 4.14.1) (in here Corollary 4.14.1).

(2) The condition (ii) in Theorem 4.39 above can be replaced by the stronger condition: for each $x_\alpha \in X_\alpha$, $T_\alpha^{-1}(x_\alpha)$ is relatively open in X_α.

To see this we take $O_{x_\alpha} = T_\alpha^{-1}(x_\alpha)$. Then

$$\bigcup_{x_\alpha\in X_\alpha} O_{x_\alpha} = \bigcup_{x_\alpha\in X_\alpha} T_\alpha^{-1}(x_\alpha) = X.$$

Let $x \in X$. Then by (i) we can find $y_\alpha \in T_\alpha(x)$, i.e., $x \in T_\alpha^{-1}(y_\alpha) = O_{y_\alpha}$.

We adapt the following notation. Let $\{X_\alpha : \alpha \in I\}$ be a family of non-empty sets, where I is an index set, finite or infinite (countable or uncountable). X will denote their cartesian product, i.e., $X = \prod_{\alpha\in I} X_\alpha$ and for each $\alpha \in I$, $X_\alpha = \prod_{\beta\in I,\beta\neq\alpha} X_\beta$. We also denote a generic element of X_α by $x_{-\alpha}$. Further, we represent each $x \in X$ by $x = [x_\alpha, x_\alpha]$ for each $\alpha \in I$ where x_α is the projection of x onto X_α. Finally, we also write $x = \{x_\alpha\}$ where x_α is the projection of x onto X_α for each $\alpha \in I$.

We apply our Theorem 4.39 to obtain a related theorem directly in the following remark.

Remark 4.13 Let $\{X_\alpha : \alpha \in I\}$ be a family of nonempty, compact convex sets, each in a topological vector space E_α. Let $\{A_\alpha\}$ and $\{B_\alpha\}$ be two families of subsets of X having the following properties:

(a) for each $\alpha \in I$ and each $x_\alpha \in X_\alpha$, the set $B_\alpha(x_\alpha) = \{x_{-\alpha} \in X_{-\alpha} : [x_\alpha, x_{-\alpha}] \in B_\alpha\}$ is open in $X_{-\alpha}$;

(b) for each $\alpha \in I$ and each $x_{-\alpha} \in X_{-\alpha}$, the set $B_\alpha(x_{-\alpha}) = \{x_\alpha \in X_\alpha : [x_\alpha, x_{-\alpha}] \in B_\alpha\}$ is nonempty and the set $A_\alpha(x_{-\alpha}) = \{x_\alpha \in X_\alpha : [x_\alpha, x_{-\alpha}] \in A_\alpha\}$ contains the convex hull of $B_\alpha(x_{-\alpha})$.

Then $\bigcap_{\alpha \in I} A_\alpha \neq \emptyset$.

Proof. For each $\alpha \in I$ and each $x = \{x_\alpha\}$ we set $T_\alpha(x) = \operatorname{co} B_\alpha(x_{-\alpha})$. Then for each $\alpha \in I$, $T_\alpha : X \to 2^{X_\alpha}$ is a set valued mapping satisfying the condition (i) of Theorem 4.39 [by the first part of condition (b)].

Now for each $\alpha \in I$ and each $y_\alpha \in X_\alpha$,

$$\begin{aligned} T_\alpha^{-1}(y_\alpha) &= \{x = \{x_\alpha\} \in X : y_\alpha \in T_\alpha(x)\} \\ &= \{x = \{x_\alpha\} \in X : y_\alpha \in \operatorname{co} B_\alpha(x_{-\alpha})\} \supseteq \{x = \{x_\alpha\} \in X : y_\alpha \in B_\alpha(x_{-\alpha})\} \\ &= \{x = \{x_\alpha\} \in X : x_{-\alpha} \in B_\alpha(y_\alpha)\} = X_\alpha \times B_\alpha(y_\alpha) = O_{y_\alpha}, \end{aligned}$$

say, which is an open set in X. Let $x = \{x_\alpha\} \in X$. Then by virtue of the first part of the condition (b) there is $y_\alpha \in B_\alpha(x_{-\alpha})$. Hence $x_{-\alpha} \in B_\alpha(y_\alpha)$. This implies that $x \in O_{y_\alpha}$. Thus for each $\alpha \in I$, $\bigcup_{y_\alpha \in X_\alpha} O_{y_\alpha} = X$ and hence the condition (ii) of Theorem 4.39 is fulfilled. Hence by the Theorem 4.39 there exists a point $x = \{x_\alpha\} \in X$ such that for each $\alpha \in I$, $x_\alpha \in T_\alpha(x) = \operatorname{co} B_{-\alpha}(x) \subset A_\alpha(x_{-\alpha})$, i.e., $x = [x_\alpha, x_{-\alpha}] \in A_\alpha$ for each $\alpha \in I$.

4.8 Applications of Fixed Point Theorems to Equilibrium Analysis in Mathematical Economics and Game Theory

Shafer and Sonnenschein (1975) extended the Debreu theorem on the existence of equilibrium in a generalized N-person game (Debreu (1952)) or an abstract economy (Arrow and Debreu (1954)). In essence, Shafer and Sonnenschein maintained the spirit of the pioneering works of Debreu, Arrow, Mas-Colell (1974) and Gale and Mas-Colell (1975), Gale and Mas-Colell (1979). Bewley (1972) proved the existence of equilibrium point with infinite dimensional commodity space. In recent years, many authors (e.g., Yannelis and Prabhakar (1983) and Toussaint (1984)) have proved the existence of equilibrium point of an abstract economy with infinite dimensional commodity space and infinite agents.

Following Debreu and Schafer and Sonnenschein we will describe an abstract economy or generalized qualitative game by $\mathcal{E} = \{\mathcal{X}_\alpha, \mathcal{A}_\alpha, \mathcal{U}_\alpha : \alpha \in \mathcal{I}\}$ where

I is finite or infinite (countable or uncountable) set of agents or players, and for each $\alpha \in I$, X_α is the choice set or strategy set; $A_\alpha : X = \prod_{\alpha \in I} X_\alpha \to 2^{X_\alpha}$ is the constraint correpsondence (set valued mapping) and $U_\alpha : X \to \mathbb{R}$ is the utility or pay off function. X_α will be a subset of a topological vector space for each $\alpha \in I$. We denote the product $\prod_{\beta \in I, \beta \neq \alpha}$ by X_α $X_{-\alpha}$ and a generic element of $X_{-\alpha}$ by $x_{-\alpha}$.

An abstract economy instead of being given by $\{X_\alpha, A_\alpha, U_\alpha : \alpha \in I\}$ may be given by $\mathcal{E} = \{\mathcal{X}_\alpha, \mathcal{P}_\alpha, \mathcal{A}_\alpha : \alpha \in \mathcal{I}\} : \alpha \in \mathcal{I}\}$ where for each $\alpha \in I$, $P_\alpha : X \to 2^{X_\alpha}$ is the preference correspondence. The relationship between the utility function U_α and the preference correspondence P_α can be exhibited by the definition

$$P_\alpha(x) = \{y_\alpha \in X_\alpha : U_\alpha([y_\alpha, x_{-\alpha}]) > U_\alpha(x)\},$$

where for each $\alpha \in I$, $x_{-\alpha}$ is the projection of x onto $X_{-\alpha}$ and $[y_\alpha, x_{-\alpha}]$ is the point of X whose αth coordinate is y_α.

In the case of the economy being given by $\mathcal{E} = \{\mathcal{X}_\alpha, \mathcal{A}_\alpha, \mathcal{U}_\alpha : \alpha \in \mathcal{I}\}$, a point $\overline{x} \in X$ is called an equilibrium point or a generalized Nash equilibrium point of the economy $\mathcal{E}$ if

$$U_\alpha(\overline{x}) = U_\alpha[\overline{x}_\alpha, \overline{x}_{-\alpha}] = \sup_{z_\alpha \in A_\alpha(\overline{x})} U_\alpha[z_\alpha, \overline{x}_{-\alpha}]$$

for each $\alpha \in I$ where $\overline{x}_\alpha$ and $\overline{x}_{-\alpha}$ are respectively projections of $\overline{x}$ onto X_α and $X_{-\alpha}$. In this case the equilibrium point is the natural extension of the equilibrium point introduced by Nash (1950). Now let $\mathcal{E} = \{\mathcal{X}_\alpha, \mathcal{A}_\alpha, \mathcal{U}_\alpha : \alpha \in \mathcal{I}\}$ be an abstract economy and let for each $\alpha \in I$, P_α be obtained as above. Then it can be easily checked that a point $\overline{x} \in X$ is an equilibrium point of $\mathcal{E}$ is and only iff for each $\alpha \in I$, $P_\alpha(\overline{x}) \cap A_\alpha(\overline{x}) = \emptyset$ and $\overline{x}_\alpha \in A_\alpha(\overline{x})$. Thus given an abstract economy $\mathcal{E} = \{\mathcal{X}_\alpha, \mathcal{P}_\alpha, \mathcal{A}_\alpha : \alpha \in \mathcal{I}\}$ we can define an equilibrium point of $\mathcal{E}$ as follows: A point $\overline{x} \in X$ is said to be an equilibrium point of the abstract economy $\mathcal{E} = \{\mathcal{X}_\alpha, \mathcal{P}_\alpha, \mathcal{A}_\alpha : \alpha \in \mathcal{I}\}$ if for each $\alpha \in I$, $P_\alpha(\overline{x}) \cap A_\alpha(\overline{x}) = \emptyset$ and $\overline{x}_\alpha \in A_\alpha(\overline{x})$ where $\overline{x}_\alpha$ is the projection of $\overline{x}$ onto X_α.

4.8.1 *Fixed Point and Equilibrium Point*

Given an abstract economy $\mathcal{E} = \{\mathcal{X}_\alpha, \mathcal{P}_\alpha, \mathcal{A}_\alpha : \alpha \in \mathcal{I}\}$, for each $x \in X$ we define $I(x) = \{\alpha \in I : P_\alpha(x) \cap A_\alpha(x) \neq \emptyset\}$. Assume that for each $x \in X$ and for each $\alpha \in I$, $x_\alpha \notin$ convex full of $P_\alpha(x)$.

For each $\alpha \in I$, we define the set valued mapping $T_\alpha : X \to 2^X$ by

$$T_\alpha(x) = \begin{cases} \operatorname{co} P_\alpha(x) \cap A_\alpha(x) & \text{if } \alpha \in I(x), \\ A_\alpha(x) & \text{if } \alpha \notin KI(x), \end{cases}$$

where $\operatorname{co} P_\alpha(x)$ denotes the convex hull of $P_\alpha(x)$.

Then it is easy to see that $\overline{x} \in X$ is an equilibrium point of the economy $\mathcal{E}$ if and only if $\overline{x}$ is the fixed point of the set valued mapping defined by $T(x) = \prod_{\alpha\in I} T_\alpha(x)$.

We recall the following notations:

For each $\alpha \in I$, X_α will denote a nonempty set in a topological vector space E_α, $P_\alpha\, X = \prod_{\alpha\in I} X_\alpha \to 2^{X_\alpha}$ a set valued mapping (preference correspondence) and $A_\alpha\, X \to 2^{X_\alpha}$ a set valued ampping (constraint correspondence).

$\{X_\alpha, P_\alpha : \alpha \in I\}$ will be called a qualitative game and $\{X_\alpha, P_\alpha, A_\alpha : \alpha \in I\}$ an abstract economy.

A point $x \in X$ is called a maximal element of the game $\{X_\alpha, P_\alpha : \alpha \in I\}$ if $P_\alpha(x) = \emptyset$ for each $\alpha \in I$ and a point $x \in \{x_\alpha\} \in X$ is called an equilibrium point of the abstract economy $\{X_\alpha, P_\alpha, A_\alpha : \alpha \in I\}$ if, for each $\alpha \in I$, $x_\alpha \in A_\alpha(x)$ and $P_\alpha(x) \cap A_\alpha(x) = \emptyset$.

Theorem 4.40 *Let $\mathcal{E} = \{\mathcal{X}_\alpha, \mathcal{P}_\alpha, \mathcal{A}_\alpha : \alpha \in \mathcal{I}\}$ be an abstract economy such that for each $\alpha \in I$, the following conditions hold:*

(i) X_α is compact and convex;

(ii) for each $x \in X$, $A_\alpha(x)$ is nonempty and convex valued;

(iii) for each $x_\alpha \in X$, $\{P_\alpha^{-1}(x_\alpha) \cup F_\alpha\} \cap A_\alpha^{-1}(x_\alpha)$ contains a relatively open subset O_{x_α} of X such that $\bigcup_{x_\alpha\in X_\alpha} O_{x_\alpha} = X$, where $F_\alpha = \{x \in X : P_\alpha(x) \cap A_\alpha(x) = \emptyset\}$;

(iv) for each $x = \{x_\alpha\} \in X$, $x_\alpha \notin \operatorname{co} P_\alpha(x)$.

Then $\mathcal{E}$ has the equilibrium point.

Proof. For each $\alpha \in I$, let $G_\alpha = \{x \in X : P_\alpha(x) \cap A_\alpha(x) \neq \emptyset\}$ and for each $x \in X$, let $I(x) = \{\alpha \in I : P_\alpha(x) \cap A_\alpha(x) \neq \emptyset\}$.

Now for each $\alpha \in I$, we define the set valued mapping $T_\alpha : X \to 2^{X_\alpha}$ by

$$T_\alpha(x) = \begin{cases} \operatorname{co} P_\alpha(x) \cap A_\alpha(x) & \text{if } \alpha \in I(x), \text{ i.e., if } x \in G_\alpha, \\ A_\alpha(x) & \text{if } \alpha \notin I(x). \end{cases}$$

Then for each $x \in X$, $T_\alpha(x)$ is nonempty and convex valued. Also for each $y_\alpha \in X_\alpha$, it can be easily checked that

$$\begin{aligned} T_\alpha^{-1}(y_\alpha) &= \left[\{(\operatorname{co} P_\alpha)^{-1}(y_\alpha) \cap A_\alpha^{-1}(y_\alpha)\} \cap G_\alpha\right] \cup \left[A_\alpha^{-1}(y_\alpha) \cap F_\alpha\right] \\ &\supset \left[\{P_\alpha^{-1}(y_\alpha) \cap A_\alpha^{-1}(y_\alpha)\} \cap G_\alpha\right] \cup \left[A_\alpha^{-1}(y_\alpha) \cap F_\alpha\right] \\ &= \left[P_\alpha^{-1}(y_\alpha) \cap A_\alpha^{-1}(y_\alpha)\right] \cup \left[F_\alpha \cap F_\alpha^{-1}(y_\alpha)\right] \\ &= \left[P_\alpha^{-1}(y_\alpha) \cup F_\alpha\right] \cap A_\alpha^{-1}(y_\alpha). \end{aligned}$$

The first inclusion follows from the fact that as $P_\alpha(x) \subset \operatorname{co} P_\alpha(x)$ for each $x \in X$, $P_x^{-1}(y_\alpha) \subset (\operatorname{co} P_\alpha)^{-1}(y_\alpha)$ for each $y_\alpha \in X_\alpha$. Hence by virtue of condition (iii) for each $y_\alpha \in X_\alpha$, $T_\alpha^{-1}(y_\alpha)$ contains an open subset O_{y_α} of X such that $\bigcup_{y_\alpha \in X_\alpha} O_{y_\alpha} = X$. Thus by Theorem 4.39 there exists a point $x = \{x_\alpha\} \in X$ such that $x_\alpha \in T_\alpha(x)$ for each $\alpha \in I$. By (iv) and definition of T_α, it follows that x is an equilibrium point of $\mathcal{E}$. □

Remark 4.14 It is easy to see that the condition (iii) of Theorem 4.40 can be replaced by the following stronger condition:

(iii)' For each $x_\alpha \in X_\alpha$, $\{P_\alpha^{-1}(x_\alpha) \cup F_\alpha\} \cap A_\alpha^{-1}(x_\alpha)$ is relatively open in X. For in this case for each $\alpha \in I$, $\bigcup_{x_\alpha \in X_\alpha} \{P_\alpha^{-1}(x_\alpha) \cup F_\alpha\} \cap A_\alpha^{-1}(x_\alpha) = X$.

To see this, let $x \in X$. If $x \notin F_\alpha$, we can find $y_\alpha \in P_\alpha(x) \cap A_\alpha(x)$, i.e., $x \in P_\alpha^{-1}(y_\alpha) \cap A_\alpha^{-1}(y_\alpha)$ and hence $x \in [P_\alpha^{-1}(y_\alpha) \cup F_\alpha] \cap A_\alpha^{-1}(y_\alpha)$. Let $x \in F_\alpha$. Since $A_\alpha(x) \neq \emptyset$ for each $x \in X$, we can find $y_\alpha \in A_\alpha(x)$, i.e., $x \in A_\alpha^{-1}(y_\alpha)$ and hence $x \in [P_\alpha^{-1}(y_\alpha) \cup F_\alpha] \cap A_\alpha^{-1}(y_\alpha)$.

Corollary 4.40.1 *Let $\mathcal{E} = \{\mathcal{X}_\alpha, \mathcal{P}_\alpha, \mathcal{A}_\alpha : \alpha \in \mathcal{I}\}$ be an abstract economy such that for each $\alpha \in I$ the following conditions hold:*

(i) *X_α is compact and convex;*
(ii) *for each $x \in X$, $A_\alpha(x)$ is nonempty and convex valued;*
(iii) *the set $G_\alpha = \{x \in X : P_\alpha(x) \cap A_\alpha(x) \neq \emptyset\}$ is a closed subset of X;*
(iv) *for each $y_\alpha \in X_\alpha$, $P_\alpha^{-1}(y_\alpha)$ is a relatively open subset in G_α and $A_\alpha^{-1}(y_\alpha)$ is a relatively open subset in X;*
(v) *for each $x = \{x_\alpha\} \in X$, $x_\alpha \notin \operatorname{co} P_\alpha(x)$ for each $\alpha \in I$.*

Then there is a equilibrium point of the economy $\mathcal{E}$.

Proof. Since $P_\alpha^{-1}(y_\alpha)$ is relatively open in G_α, $P_\alpha^{-1}(y_\alpha) = G_\alpha \cap U_\alpha$ for some open subset U_α of X,. Hence $P_\alpha^{-1}(y_\alpha) \cup F_\alpha = (G_\alpha \cap U_\alpha) \cup F_\alpha = X \cap (U_\alpha \cup F_\alpha)$. Thus $\{P_\alpha^{-1}(y_\alpha) \cup F_\alpha\} \cap A_\alpha^{-1}(y_\alpha) = (U_\alpha \cup F_\alpha) \cap A_\alpha{}^1(y_\alpha)$ is a relatively open subset of X as U_α, F_α and $A_\alpha^{-1}(y_\alpha)$ are open subsets of X. Now the corollary follows from Theorem 4.40 and Remark 4.14.

Theorem 4.41 *Let $\mathcal{G} = \{X_\alpha, P_\alpha : \alpha \in I\}$ be a qualitative game such that for each $\alpha \in I$, the following conditions hold:*

(i) *X_α is compact and convex;*
(ii) *for each $x_\alpha \in X_\alpha$, $\{P_\alpha^{-1}(x_\alpha) \cup F_\alpha\}$ contains a relatively open subset O_{x_α} of X such that $\bigcup_{x_\alpha \in X_\alpha} O_{x_\alpha} = X$, where*

$$F_\alpha = \{x \in X : P_\alpha(x) = \emptyset\};$$

(iii) *for each $x = \{x_\alpha\} \in X$, $x_\alpha \notin \operatorname{co} P_\alpha(x)$.*

Then there is a maximal element of the game $\mathcal{G}$.

Proof. For each $\alpha \in I$, if we define the set valued mapping $A_\alpha : X \to 2^{X_\alpha}$ by $A_\alpha(x) = X_\alpha$, $x \in X$, then the theorem will follow from Theorem 4.40. □

Remark 4.15 The condition (ii) in Theorem 4.41 can be replaced by the following stronger condition:

$(ii)'$ For each $x_\alpha \in X)\alpha$, the set $\{P_\alpha^{-1}(x_\alpha) \cup F_\alpha\}$ is a relatively open subset of X. For this case for each $\alpha \in I$, $\bigcup_{x_\alpha \in X_\alpha} \{P_\alpha^{-1}(x_\alpha) \cup F_\alpha\} = X$.

To see this let $x \in X$. If $x \notin F_\alpha$, then $P_\alpha(x) \neq \emptyset$. Hence we can find $y_\alpha \in P_\alpha(x)$, i.e., $x \in P_\alpha^{-1}(y_\alpha)$. Thus we can conclude that $x \in P_\alpha^{-1}(y_\alpha) \cup F_\alpha$.

Corollary 4.41.1 *Let $\mathcal{G} = \{X_\alpha, P_\alpha : \alpha \in I\}$ be a qualitative game such that for each $\alpha \in I$ the following conditions hold:*

(i) *X_α is compact and convex;*
(ii) *the set $G_\alpha = \{x \in X : P_\alpha(x) \neq \emptyset\}$ is a closed subset of X;*
(iii) *for each $y_\alpha \in X_\alpha$, $P_\alpha^{-1}(y_\alpha) = \{x \in G_\alpha : y_\alpha \in P_\alpha(x)\}$ is relatively open in G_α;*
(iv) *for each $x = \{x_\alpha\} \in X$, $x_\alpha \notin \operatorname{co} P_\alpha(x)$.*

Then there is a maximal element of $\mathcal{G}$.

Proof. Since $P_\alpha^{-1}(y_\alpha)$ is relatively open in G_α,

$$P_\alpha^{-1}(y_\alpha) = G_\alpha \cap U_\alpha \text{ for some open subset } U_\alpha \text{ of } X.$$

Thue $P_\alpha^{-1}(y_\alpha) \cup F_\alpha = (G_\alpha \cap U_\alpha) \cup F_\alpha = (G_\alpha \cup F_\alpha) \cap (U_\alpha \cup F_\alpha) = X \cap (U_\alpha \cup F_\alpha) = O_{y_\alpha}$, say, which is open subset of X as $F_\alpha = X \setminus G_\alpha$ is an open subset of X.

The corollary now follows from Theorem 4.41 and Remark 4.15. □

Definition 4.4 Let A be a subset of a topological space X. We shall denote by $\operatorname{int}_X(A)$ the interior of A in X and by $\operatorname{cl}_X(A)$ the closure of A in X. A is said to be compactly open (resp. closed) in X if for each nonempty compact subset C of X, $A \cap C$ is open (resp. closed) in C. If A is a subset of a vector space, we shall denote by $\operatorname{co}(A)$ the convex hull of A. If A is a nonempty subset of a topological vector space E and $S, T : A \to 2^E$ are correspondences, then $\operatorname{co} T, T \cap S : A \to 2^E$ are correspondences defined by $(\operatorname{co} T)(x) = \operatorname{co}(T(x))$ and $(T \cap S)(x) = T(x) \cap S(x)$ for each $x \in A$, respectively. If X and Y are topological spaces and $T : X \to 2^Y$ is a correspondence, the graph of T is the set $\operatorname{Gr}(T) = \{(x, y) \in X \times T : y \in T(x)\}$ and the correspondence $\overline{T} : X \to 2^Y$ is defined by $\overline{T}(x) = \{y \in Y : (x, y) \in \operatorname{cl}_{X \times Y}(\operatorname{Gr}(T))\}$ (the set $\operatorname{cl}_{X \times Y}(\operatorname{Gr}(T))$ is called the adherence of the graph of T) and $\operatorname{cl} T : 2^Y$ is defined by $(\operatorname{cl} T)(x) = \operatorname{cl}_Y(T(x))$ for each $x \in X$. It is easy to see that $(\operatorname{cl} T)(x) \subset (\overline{T})(x)$ for each $x \in X$.

The following notions are slightly general than the corresponding notions due to Ding and Tan (1992c), Ding and Tan (1992a), Ding and Tan (1993) (also see

Tan and Yuan (1993)). Let X be a topological space, Y be a nonempty subset of a vector space $E, \theta : X \to E$ be a map and $\varphi : X \to 2^Y$ be a correspondence. Then (1) φ is said to be of class $L_{\theta,F}$ if (a) if for each $x \in X$, $\operatorname{co}\varphi(x) \subset Y$ and $\theta(x) \notin \operatorname{co}\varphi(x)$ and (b) there exists a correspondence $\psi : X \to 2^Y$ such that for each $x \in X$, $\psi(x) \subset \varphi(x)$ and for each $y \in Y$, $\psi^{-1}(y) = \{x \in X : y \in \psi(x)\}$ is compactly open in X and $\{x \in X : \varphi(x) \neq \emptyset\} = \{x \in X : \psi(x) \neq \emptyset\}$; (2) (φ_x, ψ_x, N_x) is a $L_{\theta,F}$-majorant of φ at x if $\varphi_x, \psi_x : X \to 2^Y$ and N_x is an open neighborhood of x in X such that (a) for each $z \in N_x$, $\varphi(z) \subseteq \varphi_x(z)$ and $\theta(z) \notin \operatorname{co}\varphi_x(z)$, (b) for each $z \in X$, $\psi_x(z) \subseteq \varphi_x(z)$ and $\operatorname{co}\varphi_x(z) \subseteq Y$ and (c) for each $y \in Y$, $\psi_x^{-1}(y)$ is compactly open in X; (3) φ is said to be $L_{\theta,F}$-majorized if for each $x \in X$ with $\varphi(x) \neq \emptyset$, there exists an $L_{\theta,F}$-majorant (φ_x, ψ_x, N_x) of φ at x such that for any nonempty finite subset A of the set $\{x \in X : \varphi(x) \neq \emptyset\}$, we have

$$\left\{z \in \bigcap_{x\in A} N_x : \bigcap_{x\in A} \operatorname{co}\varphi_x(z) \neq \emptyset\right\} = \left\{z \in \bigcap_{x\in A} N_x : \bigcap_{x\in A} \operatorname{co}\psi_x(z) \neq \emptyset\right\}.$$

It is clear that every correspondence of class $L_{\theta,F}$ is $L_{\theta,F}$-majorized. If for each $y \in Y$, $\psi_x^{-1}(y)$ are open in X, then these notions reduce to the corresponding notions of Ding and Tan (1992c), if for each $x \in \{x \in X : \varphi(x) \neq \emptyset\}$, $\varphi_x(z) = \psi_x(z)$ for all $z \in X$, then these notions reduce to the corresponding notions of Ding and Tan (1993). Hence these notions in turn generalize the corresponding notions introduced by Ding et al. (1992) and Tulcea (1986). In what follows, we shall deal mainly with either the case (I) $X = Y$ and is a nonempty convex subset of a topological vector space E and $\theta = I_X$, the identity map on X, or the case (II) $X = \prod_{i\in I} X_i$ and $\theta = \pi_j : X \to X_j$ is the projection of X onto X_j and X_j is a nonempty convex subset of a topological vector space. In both cases (I) and (II), we shall write L_F in place of $L_{\theta,F}$.

4.8.2 *Existence of Maximal Elements*

We recall the more general definition of maximal element.

Let X be a topological space and $T : X \to 2^X$ be a correspondence. Then a point $x_0 \in X$ is said to be a maximal element of T if $T(x_0) = \emptyset$.

The following result is Lemma 3.1 of Tan and Yuan (1993) and the proof can be found in Yuan (1999, p. 247).

Lemma 4.11 *Let X be a regular topological vector space and Y be a nonempty subset of a vector space E. Let $\theta : X \to E$ and $P : X \to 2^Y$ be $L_{\theta,F}$-majorized. If each open subset of X containing the set $B = \{x \in X : P(x) \neq \emptyset\}$ is paracompact, then there exists a correspondence $\varphi : X \to 2^Y$ of class $L_{\theta,F}$ such that $P(x) \subseteq \varphi(x)$ for each $x \in X$.*

Theorem 4.42 *Let X be a nonempty convex subset of a topological vector space*

E and $G : X \to 2^X$ be of class L_F. Suppose that there exist a nonempty compact convex subset X_0 of X and a nonempty compact subset K of X such that for each $y \in X \setminus K$, there is an $x \in \text{co}(X_0 \cup \{y\})$ with $y \notin \text{cl}_C((X \setminus (\text{co}\, G)^{-1}(x)) \cap C)$ for any nonempty compact subset C of X. Then there exists an $\hat{x} \in K$ such that $G(\hat{x}) = \emptyset$, i.e. $\hat{x}$ is a maximal element of G.

Proof. Since G is of class L_F, we have

(a) for each $x \in X$, $x \notin \text{co}(G(x))$;
(b) there exists a correspondence $F : X \to 2^X$ such that
  (i) for each $x \in X$, $F(x) \subset G(x)$,
  (ii) for each $y \in X$, $F^{-1}(y)$ is compactly open in X,
  (iii) $\{x \in X : F(x) \neq \emptyset\} = \{x \in X : G(x) \neq \emptyset\}$.

Suppose $G(x) \neq \emptyset$ for each $x \in K$. By (iii), we have $F(x) \neq \emptyset$ for each $x \in K$. It follows from Theorem 4.35 that there exists $\overline{x} \in X$ such that $\overline{x} \in \text{co}(G(\overline{x}))$, which contradicts (a). Hence there must exist an $\hat{x} \in K$ such that $G(\hat{x}) = \emptyset$. □

Remark 4.16 Theorem 4.42 improves Theorem 3 and 4 of Ding and Tan (1993) and Theorem 3.2 of Tan and Yuan (1993).

Theorem 4.43 *Let X be a nonempty paracompact convex subset of a topological vector space E and $P : X \to 2^X$ be L_F-majorized. Suppose that there exist a nonempty compact convex subset X_0 of X and a nonempty compact subset K of X such that for each $y \in X \setminus K$, there is an $x \in \text{co}(X_0 \cup \{y\})$ with $y \notin \text{cl}_C((X \setminus (\text{co}\, P)^{-1}(x)) \cap C)$ for any nonempty compact subset C of X. Then there exists an $\hat{x} \in K$ such that $P(\hat{x}) = \emptyset$.*

Proof. Suppose that the conclusion does not hold. Then by the coercive condition, $P(x) \neq \emptyset$ for all $x \in X$ and hence the set $\{x \in X : P(x) \neq \emptyset\} = X$ is paracompact. By Lemma 4.11, there exists a correspondence $\varphi : X \to 2^X$ of class L_F such that $P(x) \subset \varphi(x)$ for each $x \in X$. Note that for each $y \in X \setminus K$, $y \notin \text{cl}_C((X \setminus (\text{co}\, P)^{-1}(x)) \cap C)$ implies $y \notin \text{cl}_C((X \setminus (\text{co}\, \varphi)^{-1}(x)) \cap C)$. By Theorem 4.42, there exists a point $\hat{x} \in K$ such that $\varphi(\hat{x}) = \emptyset$ so that $P(\hat{x}) = \emptyset$ which is a contradiction. Therefore, there exists a point $\hat{x} \in K$ such that $P(\hat{x}) = \emptyset$. □

Remark 4.17 Theorem 4.43 improves and generalizes Theorem 1 of Ding and Tan (1993). Note that the coercive condition of Theorem 3.3 of Tan and Yuan (1993) implies the coercive condition of Theorem 4.40 and hence Theorem 4.43 also generalizes Theorem 3.3 of Tan and Yuan (1993), Theorem 5 of Ding and Tan (1992c) and in turn generalizes Corollary 1 of Borglin and Keiding (1976), Theorem 2.2 of Toussaint (1984), Theorem 2 of Tulcea (1986), Theorem 5.1 and Corollary 5.1 of Yannelis-Prabhakar (1984) and Theorem 2 of Yannelis (1987).

4.8.3 *Equilibrium Existence Theorems*

Let I be a (possibly infinite) set of agents. For each $i \in I$, let its choice or strategy set X_i be a nonempty subset of a topological vector space. Let $X = \prod_{i\in I} X_i$. For each $i \in I$, let $P_i : X \to 2^{X_i}$ be a correspondence. Following the notion of Gale and Mas-Colell (1978), the collection $\Gamma = (X_i, P_i)_{i\in I}$ will be called a qualitative game. A point $x \in X$ is said to be an equilibrium of the game Γ if $P_i(x) = \emptyset$ for all $i \in I$. For each $i \in I$, let A_i be a subset of X_i. Then for each fixed $k \in I$, we define

$$\prod_{j\in I, j\neq k} A_j \otimes A_k = \{x = (x_i)_{i\in I} : x_i \in A_i \text{ for all } i \in I\}.$$

Some authors use two constraint correspondences. In their context an abstract economy (= generalized game) is a family of quadruples $\Gamma = (X_i; A_i, B_i, P_i)_{i\in I}$ where I is a (finite or infinite) set of agents (players) such that for each $i \in I$, X_i is a nonempty subset of a topological vector space and $A_i, B_i : X = \prod_{i\in I} X_i \to 2^{X_i}$ are constraint correspondences and $P_i : X \to 2^{X_i}$ is a preference correspondence. When $I = \{I, \ldots, N\}$ where N is a positive integer, $\Gamma = (X_i; A_i, B_i, P_i)_{i\in I}$ is also called an N-person game. An equilibrium of Γ is a point $x \in X$ such that for each $i \in I$, $x_i \in \overline{B}_i(x)$ and $A_i(x) \cap P_i(x) = \emptyset$. We remark that when $\overline{B}_i(x) = \mathrm{cl}_{X_i}(B_i(x))$ (which is the case when B_i has a closed graph; in particular, when $\mathrm{cl}\, B_i$ is upper semicontinuous with closed values), the definition of an equilibrium point coincides with that of Ding et al. (1992) and Ding and Tan (1992a), Ding and Tan (1993); and if in addition, $A_i = B_i$ for each $i \in I$, the definition of an equilibrium point coincides with the standard definition; e.g. in Borglin and Keiding (1976), Yannelis and Prabhakar (1983), Yannelis-Prabhakar (1984), Tulcea (1986), and Tulcea (1988).

As an application of Theorem 4.42, we obtain the following equilibrium existence theorem for an one-person game.

Theorem 4.44 *Let X be a nonempty convex subset of a topological vector space. Let $A, B, P : X \to 2^X$ be such that*

(1) *for each $x \in X$, $\mathrm{co}(A(x)) \subset \overline{B}(x)$;*

(2) *for each $y \in X$, $A^{-1}(y)$ is compactly open in X;*

(3) *$A \cap P$ is of class L_F;*

(4) *there exist a nonempty compact convex subset X_0 of X and a nonempty compact subset K of X such that for each $y \in X \setminus K$, there is an $x \in \mathrm{co}(X_0 \cup \{y\})$ with $y \notin \mathrm{cl}_C((X \setminus (\mathrm{co}(A \cap P))^{-1}(x)) \cap C)$ for any non-empty compact subset C of X and for each $x \in K$, $A(x) \neq \emptyset$.*

Then there exists a point $\hat{x} \in K$ such that $\hat{x} \in \overline{B}(\hat{x})$ and $A(\hat{x}) \cap P(\hat{x}) = \emptyset$.

Proof. Let $M = \{x \in X : x \notin \overline{B}(x)\}$, then M is open in X. Define $\varphi : X \to 2^X$ by

$$\varphi(x) = \begin{cases} A(x) \cap P(x), & \text{if } x \notin M, \\ A(x), & \text{if } x \in M. \end{cases}$$

Since $A \cap P$ is of class L_F, for each $x \in X$, $x \notin \mathrm{co}(A(x) \cap P(x))$ and there exists a correspondence $\beta : X \to 2^X$ such that (a) for each $x \in X$, $\beta(x) \subset A(x) \cap P(x)$; (b) for each $y \in X$, $\beta^{-1}(y)$ is compactly open in X and (c) $\{x \in X : \beta(x) \neq \emptyset\} = \{x \in X : A(x) \cap P(x) \neq \emptyset\}$. Now define $\psi : X \to 2^X$ by

$$\psi(x) = \begin{cases} \beta(x), & \text{if } x \notin M, \\ A(x), & \text{if } x \in M. \end{cases}$$

Then clearly for each $x \in X$, $\psi(x) \subset \varphi(x)$ and $\{x \in X : \psi(x) \neq \emptyset\} = \{x \in X : \varphi(x) \neq \emptyset\}$ by (c). For each $y \in X$, it is easy to see that $\psi^{-1}(y) = (M \cup \beta^{-1}(y)) \cap A^{-1}(y)$ and is compactly open in X by (2) and (b). For each $x \in X$, if $x \in M$, then $x \notin \overline{B}(x)$, it follows from (1) that $x \notin \mathrm{co}(\varphi(x))$; if $x \notin M$, $x \notin \mathrm{co}(A(x) \cap P(x)) = \mathrm{co}(\varphi(x))$ since $x \notin \mathrm{co}(A(x) \cap P(x))$ for all $x \in X$. This shows that φ is of class L_F. By (4) and the definition of φ, for each $y \in X \setminus K$, there is an $x \in \mathrm{co}(X_0 \cup \{y\})$ such that $y \notin \mathrm{cl}_C((X \setminus (\mathrm{co}\,\varphi)^{-1}(x)) \cap C)$ for any nonempty compact subset C of X. By Theorem 4.42, there exists a point $\hat{x} \in K$ such that $\varphi(\hat{x}) = \emptyset$. Since for each $x \in K$, $A(x) \neq \emptyset$ and by (4), for each $y \in X \setminus K$, $A(y) \neq \emptyset$. Hence $A(x) \neq \emptyset$ for all $x \in X$ so that we must have $\hat{x} \in \overline{B}(\hat{x})$ and $A(\hat{x}) \cap P(\hat{x}) = \emptyset$. □

We note that Theorem 4.44 improves Theorem 4 of Ding and Tan (1992c).

As an application of Theorem 4.43, we shall prove the following equilibrium existence theorem for a noncompact qualitative game.

Theorem 4.45 *Let $\Gamma = (X_i, P_i)_{i \in I}$ be a qualitative game such that $X = \prod_{i \in I} X_i$ is paracompact. Suppose the following conditions are satisfied:*

(1) *for each $i \in I$, X_i is a nonempty convex subset of a topological vector space;*
(2) *for each $i \in I$, $P_i : X \to 2^{X_i}$ is L_F-majorized;*
(3) $\bigcup_{i \in I} \{x \in X : P_i(x) \neq \emptyset\} = \bigcup_{i \in I} \mathit{int}_X \{x \in X : P_i(x) \neq \emptyset\}$;
(4) *there exist a nonempty compact convex subset X_0 of X and a nonempty compact subset K of X such that for each $y \in X \setminus K$, there is an $x \in \mathrm{co}(X_0 \cup \{y\})$ with $y \notin \mathrm{cl}_C((X \setminus (\mathrm{co}\, P_i)^{-1}(x_i)) \cap C)$ for each $i \in I$ and for any nonempty compact subset C of X.*

Then Γ has an equilibrium point in K.

Proof. For each $x \in X$. Let $I(x) = \{i \in I : P_i(x) \neq \emptyset\}$. For each $i \in I$, define

$P'_i : X \to 2^X$ by $P'_i(x) = \prod_{j\in I, j\neq i} X_j \otimes P_i(x)$. Furthermore, define $P : X \to 2^X$ by

$$P(x) = \begin{cases} \bigcap_{i\in I(x)} \mathrm{co}(P'_i(x)), & \text{if } I(x) \neq \emptyset, \\ \emptyset, & \text{if } I(x) = \emptyset. \end{cases}$$

Then for each $x \in X$, $P(x) \neq \emptyset$ if and only if $I(x) \neq \emptyset$. By (2), (3) and the similar argument as in the proof of Theorem 7 of Ding-Tan (1992(b)) (also see the proof of Theorem 4.2 of Tan-Yuan (1993)), we can prove that P is L_F-majorized (i.e. see Yuan (1999), pp. 251–252 for a detailed proof which, with slight modification will serve our purpose). By (4), for each $y \in X \setminus K$, there exists an $x \in \mathrm{co}(X_0 \cup \{y\})$ such that for any nonempty compact subset C on X,

$$\begin{aligned} y \notin \bigcup_{i\in I} \mathrm{cl}_C((X \setminus (\mathrm{co}\, P_i)^{-1}(x_i)) \cap C) \\ = \mathrm{cl}_C\left(\bigcup_{i\in I} (X \setminus (\mathrm{co}\, P_i)^{-1}(x_i)) \cap C\right) \\ = \mathrm{cl}_C\left(\left(X \setminus \bigcap_{i\in I} (\mathrm{co}\, P_i)^{-1}(x)_i)\right) \cap C\right), \end{aligned}$$

and $P_i(y) \neq \emptyset$ for each $y \in X \setminus K$ and $i \in I$ and hence $I(y) = I$ for each $y \in X \setminus K$. By the definition of P'_i and P, we have

$$\begin{aligned} \mathrm{co}(P(y)) &= \bigcap_{i\in I} \mathrm{co}(P'_i(y)) \\ &= \prod_{i\in I} \mathrm{co}(P_i(y)). \end{aligned}$$

It follows that

$$\begin{aligned} (\mathrm{co}\, P)^{-1}(x) &= \{y \in X : x \in \mathrm{co}(P(y))\} \\ &= \{y \in X : x \in \prod_{i\in I} \mathrm{co}(P_i(y))\} \\ &= \{y \in X : y \in (\mathrm{co}\, P_i)^{-1}(x_i) \text{ for each } i \in I\} \\ &= \bigcap_{i\in I} (\mathrm{co}\, P_i)^{-1}(x_i). \end{aligned}$$

Thus, we have that for each $y \in X \setminus K$, there exists an $x \in \mathrm{co}(X_0 \cup \{y\})$ such that $y \notin \mathrm{cl}_C((X \setminus (\mathrm{co}\, P)^{-1}(x)) \cap C)$ for any nonempty compact subset C of X. By Theorem 4.43, there exists a point $\hat{x} \in K$ such that $P(\hat{x}) = \emptyset$. This implies $I(\hat{x}) = \emptyset$ and therefore $P_i(\hat{x}) = \emptyset$ for each $i \in I$. □

Remark 4.18 Theorem 4.45 improves Theorem 3 of Ding-Tan (1993). It is easy to see that the coercive condition (d) of Theorem 4.2 of Tan-Yuan (1993) implies the coercive (4) of Theorem 4.45. Hence Theorem 4.404 also generalizes Theorem 4.2 of Tan-Yuan (1993), Theorem 7 of Ding-Tan (1992b). In Theorem 4.45, if X_i is compact for each $i \in I$, then $X = \prod_{i\in I} X_i$ is also compact. By Letting $X_0 = K = X$, the condition (4) of Theorem 4.45 is satisfied trivially. Hence Theorem 4.45 also generalizes Theorem 2.4 of Toussaint (1998) and Proposition 3 of Tulcea (1986) in several aspects which in turn generalize the fixed point theorem of Gale and Mas-Colell (1978).

As an application of Theorem 4.45, we shall show the following equilibrium existence theorem for a noncompact abstract economy with an infinite number of agents.

Theorem 4.46 *Let $\Gamma = (X_i, A_i, B_i, P_i)_{i\in I}$ be an abstract economy such that $X = \prod_{i\in I} X_i$ is paracompact. Suppose that the following conditions are satisfied:*

(1) *for each $i \in I$, X_i is a nonempty convex subset of a topological vector space;*
(2) *for each $i \in I$ and for each $x \in X$, $A_i(x)$ is nonempty and $\mathrm{co}(A_i(x)) \subset B_i(x)$;*
(3) *for each $i \in I$ and for each $y \in X$, $A_i^{-1}(y)$ is compactly open in X;*
(4) *for each $i \in I$, $A_i \cap P_i$ is L_F-majorized;*
(5) *for each $i \in I$, $E_i = \{x \in X : (A_i \cap P_i)(x) \neq \emptyset\}$ is open in X;*
(6) *there exist a nonempty compact convex subset X_0 of X and a nonempty compact subset K of X such that for each $y \in X \setminus K$, there is an $x \in \mathrm{co}(X_0 \cup \{y\})$ with $y \notin \mathrm{cl}_C((X \setminus (\mathrm{co}(A_i \cap P_i))^{-1}(x_i)) \cap C)$ for each $i \in I$ and for any nonempty compact subset C of X.*

Then Γ has an equilibrium point in K.

Proof. For each $i \in I$, let $F_i = \{x \in X : x_i \notin \overline{B}_i(x)\}$, then F_i is open in X. For each $i \in I$, define $Q_i : X \to 2^{X_i}$ by

$$Q_i(x) = \begin{cases} (A_i \cap P_i)(x), & \text{if } x \notin F_i, \\ A_i(x), & \text{if } x \in F_i. \end{cases}$$

We shall prove that the qualitative game $\Gamma = (X_i, Q_i)_{i\in I}$ satisfies all hypotheses of Theorem 4.45. For each $i \in I$, we have that the set

$$\begin{aligned} \{x \in X : Q_i(x) \neq \emptyset\} &= \{x \in F_i : Q_i(x) \neq \emptyset\} \cup \{x \in X \setminus F_i : Q_i(x) \neq \emptyset\} \\ &= F_i \cup \{x \in X \setminus F_i : (A_i \cap P_i)(x) \neq \emptyset\} \\ &= F_i \cup [(X \setminus F_i) \cap E_i] \\ &= F_i \cup E_i \end{aligned}$$

is open in X and hence the condition (3) of Theorem 4.45 is satisfied. By (4), for each $x \in E_i$, there exist an open neighborhood N_x of x in X and correspondences $\psi_x, \varphi_x : (X \to 2^{X_i}$ such that

(a) for each $z \in N_x$, $(A_i \cap P_i)(z) \subset \varphi_x(z)$ and $z_i \notin \text{co}(\varphi_x(z))$;
(b) for each $z \in X$, $\psi_x(z) \subset \varphi_x(z)$;
(c) for each $y \in X$, $\psi_x^{-1}(y)$ is compactly open in X;
(d) for each nonempty finite set $A \subset E_i$,

$$\left\{z \in \bigcap_{x \in A} N_x : \bigcap_{x \in A} \text{co}(\psi_x(z)) \neq \emptyset\right\}$$
$$= \left\{z \in \bigcap_{x \in A} N_x : \bigcap_{x \in A} \text{co}(\varphi_x(z)) \neq \emptyset\right\}.$$

Now for each $x \in X$ with $Q_i(x) \neq \emptyset$, let

$$M(x) = \begin{cases} F_i, & \text{if } x \in F_i, \\ N_x, & \text{if } x \notin F_i, \end{cases}$$

and define correspondences $\Phi_x, \Psi_x : X \to 2^{X_i}$ by

$$\Phi_x(z) = \begin{cases} \varphi_x(z), & \text{if } z \notin F_i, \\ A_i(z), & \text{if } z \in F_i, \end{cases}$$

$$\Psi_x(z) = \begin{cases} \psi_x(z), & \text{if } z \notin F_i, \\ A_i(z), & \text{if } z \in F_i. \end{cases}$$

Then for each $x \in X$ with $Q_i(x) \neq \emptyset$, $M(x)$ is an open neighborhood of x such that

(i) for each $z \in M(x)$, $Q_i(z) \subset \Phi_x(z)$ and $z_i \notin \text{co}(\Phi_x(z))$ by (2) and (a);
(ii) for each $z \in X$, $\Psi_x(z) \subset \Phi_x(z)$ by (b);
(iii) for each $y \in X_i$,
$$\begin{aligned}\Psi_x^{-1}(y) &= \{z \in X \setminus F_i : y \in \Psi_x(z)\} \cup \{z \in F_i : y \in \Psi_x(z)\} \\ &= \{z \in X \setminus F_i : y \in \psi_x(z)\} \cup \{z \in F_i : y \in A_i(z)\} \\ &= [(X \setminus F_i) \cap \psi_x^{-1}(y)] \cup (F_i \cap A_i^{-1}(y)) \\ &= [F_i \cup \psi_x^{-1}(y)] \cap A_i^{-1}(y)\end{aligned}$$
is compactly open in X by (3), (c) and F_i being open in X.

Now let A be a finite subset of $\{x \in X : Q_i(x) \neq \emptyset\}$, then $A = A_1 \cup A_2$ where $A_1 = \{x \in A : x \in F_i\}$ and $A_2 = \{x \in A : x \notin F_i\}$.

Case 1. If $A_1 = \emptyset$, then by (d),

$$\{z \in \bigcap_{x\in A} M(x) : \bigcap_{x\in A} \text{co}(\Psi_X(z)) \neq \emptyset\}$$

$$= \{z \in \bigcap_{x\in A_2} M(x) \cap F_i : \bigcap_{x\in A_2} \text{co}(\Psi_x(z)) \neq \emptyset\}$$

$$\cup \{z \in \bigcap_{x\in A_2} M(x) \setminus F_i : \bigcap_{x\in A_2} \text{co}(\Psi_x(z)) \neq \emptyset\}$$

$$= \{z \in \bigcap_{x\in A_2} M(x) \cap F_i : A_i(z) \neq \emptyset\}$$

$$\cup\, z \in \bigcap_{x\in A_2} M(x) \setminus F_i : \bigcap_{x\in A_2} \text{co}(\psi_x(z)) \neq \emptyset\}$$

$$= \{z \in \bigcap_{x\in A_2} M(x) \cap F_i : A_i(z) \neq \emptyset\}$$

$$\cup \{z \in \bigcap_{x\in A_2} M(x) \setminus F_i : \bigcap_{x\in A_2} \text{co}(\varphi_x(z)) \neq \emptyset\}$$

$$= \{z \in \bigcap_{x\in A_2} M(x) \cap F_i : \bigcap_{x\in A_2} \text{co}(\Phi_x(z)) \neq \emptyset\}$$

$$\cup \{z \in \bigcap_{x\in A_2} M(x) \setminus F_i : \bigcap_{x\in A_2} \text{co}(\Phi_x(z)) \neq \emptyset\}$$

$$= \{z \in \bigcap_{x\in A} M(x) : \bigcap_{x\in A} \text{co}(\Phi_x(z)) \neq \emptyset\}.$$

Case 2. If $A_1 \neq \emptyset$, then

$$\{z \in \bigcap_{x\in A} M(x) : \bigcap_{x\in A} \text{co}(\Psi_x(z) \neq \emptyset\}$$

$$= \{z \in \bigcap_{x\in A_1} M(x) \cap \bigcap_{x\in A_2} M(x) : \bigcap_{x\in A} \text{co}(\Psi_x(z) \neq \emptyset\}$$

$$= \{z \in F_i \cap \bigcap_{x\in A_2} M(x) : \bigcap_{x\in A} \text{co}(\Psi_x(z)) \neq \emptyset\}$$

$$= \{z \in F_i \cap \bigcap_{x\in A_2} M(x) : \bigcap_{x\in A} \text{co}(\Phi_x(z)) \neq \emptyset\}$$

$$= \{z \in \bigcap_{x\in A} M(x) : \bigcap_{x\in A} \text{co}(\Phi_x(z)) \neq \emptyset\},$$

since $\Psi_x(z) = \Phi_x(z) = A_i(z)$ for each $z \in F_i$. This shows that for each $i \in I$, Q_i is L_F-majorized.

Finally, by (6) there exist a nonempty compact convex subset X_0 of X and a nonempty compact subset K of X such that for each $y \in X \setminus K$, there is an $x \in \text{co}(X_0 \cup \{y\})$ with $y \notin \text{cl}_C((X \setminus (\text{co}\, Q_i)^{-1}(x_i)) \cap C)$ for each $i \in I$ and for any nonempty compact subset C of X. By Theorem 4.43, there exists a point $\hat{x} \in K$ such that $Q_i(\hat{x}) = \emptyset$ for all $i \in I$. By (2) and the definition of Q_i, this implies that for each $i \in I$, $\hat{x}_i \in \overline{B}_i(\hat{x})$ and $A_i(\hat{x}) \cap P_i(\hat{x}) = \emptyset$. □

Remark 4.19 Note that the notion of L_C-majorized correspondences in Ding-Tan (1993) is a special case of the notion of L_F-majorized correspondences. Thus Theorem 4.46 improves Theorem 4 of Ding and Tan (1993) and in turn generalizes Theorem 3 of Tulcea (1986) (which is also Theorem 4 of Tulcea (1988)).

Theorem 4.47 *Under the hypotheses of Theorem* 4.46, *if the coercive condition* (6) *is replaced by the following coercive condition*

(6)′ *there exist a nonempty closed compact subset K of X and a point $x^0 = (x_i^0)_{i \in I} \in X$ such that $x_i^0 \in \text{co}(A_i(y) \cap P_i(y))$ for all $i \in I$ and for all $y \in X \setminus K$,*

then Γ has an equilibrium point in K.

Proof. We prove that the coercive condition (6)′ implies the condition (6) of Theorem 4.46. By (6)′, we have that for each $i \in I$,

$$X \setminus K \subset (\text{co}(A_i \cap P_i))^{-1}(x_i^0)$$

and hence

$$X \setminus (\text{co}(A_i \cap P_i))^{-1}(x_i^0) \subseteq K.$$

Since K is closed and compact, we have

$$\text{cl}_C((X \setminus (\text{co}(A_i \cap P_i))^{-1}(x_i^0)) \cap C) \subseteq K \cap C \subseteq K$$

for each nonempty compact subset C of X. Now, let $X_0 = \{x^0\}$, then X_0 is a nonempty compact convex subset of X and $x^0 \in \text{co}(X_0 \cup \{y\})$ for all $y \in X$. Hence, there exist a nonempty compact convex subset X_0 of X and a nonempty compact subset K of X such that for each $y \in X \setminus K$, there is an $x^0 \in \text{co}(X_0 \cup \{y\})$ with $y \notin \text{cl}_C((X \setminus (\text{co}(A_i \cap P_i))^{-1}(x_i^0)) \cap C)$ for all $i \in I$ and for any nonempty compact subset C of X, i.e. the condition (6) of Theorem 4.46 is satisfied. The conclusion holds from Theorem 4.46. □

Remark 4.20 Theorem 4.47 generalizes Theorem 4.3 of Tna-Yuan in (1993) to $A_i \cap P_i$ being L_F-majorized for each $i \in I$. Hence Theorem 4,8,18 positively answers the open question presented by Tan-Yuan in (1993). Theorem 4.47 also improves and generalizes Theorem 8 of Ding-Tan (1992b) in several aspects which in turn generalizes Theorem 3 of Tulcea in (1986) and Theorem 4 of Tulcea in (1988).

Corollary 4.47.1 *Let $\Gamma = (X_i, A_i, B_i, P_i)_{i\in I}$ be an abstract economy such that $X = \prod_{i\in I} X_i$ is paracompact. Suppose the following conditions are satisfied:*

(1) *for each $i \in I$, X_i is a nonempty convex subset of a topological vector space:*
(2) *for each $i \in I$ and for each $x \in X$, $A_i(x)$ is nonempty, $\text{co}(A_i(x)) \subset \overline{B}_i(x)$ and $x_i \notin \text{co}(P_i(x))$;*
(3) *for each $i \in I$ and for each $y \in X_i$, $A_i^{-1}(y)$ and $P_i^{-1}(y)$ are open in X;*
(4) *there exist a nonempty compact convex subset X_0 of X and a nonempty compact subset K of X such that for each $y \in X \setminus K$, there is an $x \in \text{co}(X_0 \cup \{y\})$ with $y \notin \text{cl}_C((X \setminus (\text{co}(A_i \cap P_i))^{-1}(x_i)) \cap C)$ for each $i \in I$ and for any nonempty compact subset C of X.*

Then Γ has an equilibrium point in K.

Proof. Since $\{x \in X : (A_i \cap P_i)(x) \neq \emptyset\} = \bigcup_{y\in X_i} (A_i^{-1}(y) \cap P_i^{-1}(y))$, by (3), the conditions (3) and (5) of Theorem 4.8.16 are satisfied. Since for each $i \in I$ and for each $y \in X_i$, $(A_i \cap P_i)^{-1}(y) = A_i^{-1}(y) \cap P_i^{-1}(y)$ is open in X, for given any $x \in \{x \in X : (A_i \cap P_i)(x) \neq \emptyset\}$, let $N_x = X$, $\Psi_x = \phi_x = A_i \cap P_i$, then it is easy to see that the condition (4) of Theorem 4.46 is also satisfied. The conclusion holds from Theorem 4.46. □

Remark 4.21 Note that for any $x_i \in X_i$, $(\text{co}(A_i \cap P_i))^{-1}(x_i)$ is open in X by the condition (3) of Corollary 4.8.20 and hence $\text{cl}_C((X \setminus (\text{co}(A_i \cap P_i))^{-1}(x_i)) \cap C) = (X \setminus \text{co}(A_i \cap P_i))^{-1}(x_i) \cap C$. It follows that $y \notin \text{cl}_C((X \setminus (\text{co}(A_i \cap P_i))^{-1}(x_i)) \cap C)$ implied $x_i \in \text{co}(A_i(y) \cap P_i(y))$. Corollary 4.47.1 improves Corollary 1 of Ding-Tan in (1993). Since the coercive condition (v) of Corollary 4.4 of Tan-Yuan (1993) implies the coercive condition (4), therefore Corollary 4.47.1 also generalizes Corollary 4.4 of Tan-Yuan (1993), Corollary 1 of Ding-Tan (1992b), Corollary 2 of Tulcea (1986) (also Corollary in (1988), Theorem 2.5 of Toussaint (1984) and Theorem 6.1 of Yannelis-Prabhakar (1983).

In the following, we shall employ the approximation technique used by Tulcea in (Tulcea (1986)). As an application of Theorem 4.45, we have the following existence theorem of approximate equilibrium point for an abstract economy.

Theorem 4.48 *Let $\Gamma = (X_i, A_i, B_i, P_i)_{i\in I}$ be an abstract economy such that $X = \prod_{i\in I} X_i$ is paracompact. Suppose that the following conditions are satisfied:*

(1) *for each $i \in I$, X_i is a nonempty convex subset of a topological vector space E_i;*
(2) *for each $i \in I$, A_i is lower semicontinuous such that for each $x \in X$, $A_i(x)$ is nonempty and $\text{co}(A_i(x)) \subset \overline{B}(x)$;*
(3) *for each $i \in I$, $L_i \cap P_i$ is L_F-majorized;*
(4) *for each $i \in I$, $E^i = \{x \in X : (A_i \cap P_i)(x) \neq \emptyset\}$ is open in X;*

(5) *there exist a nonempty compact convex subset* X_0 *of* X *and a nonempty compact subset* K *of* X *such that for each* $y \in X \setminus K$, *there is an* $x \in \mathrm{co}(X_0 \cup \{y\})$ *with* $y \notin \mathrm{cl}_C((X \setminus (\mathrm{co}(A_i \cap P_i))^{-1}(x_i)) \cap C)$ *for each* $i \in I$ *and for any nonempty compact subset* C *of* X.

Then given $V = \prod_{i \in I} V_i$ *where fore each* $i \in I$, V_i *is a convex open neighborhood of zero in* E_i, *there exists a point* $\hat{x}_V = (\hat{x}_{V_i})_{iii} \in K$ *such that for each* $i \in I$, $\hat{x}_{V_i} \in \overline{B}_{V_i}(\hat{x}_V)$ *and* $A_i(\hat{x}_V) \cap P_i(\hat{x}_V) = \emptyset$, *where* $B_{V_i}(x) = (B_i(x) + V_i) \cap X$ *for each* $i \in I$ *and for each* $x \in X$.

Proof. Let $V = \prod_{i \in I} V_i$ be given. For each $i \in I$, define the correspondences $A_{V_i}, B_{V_i} : X \to 2^{X_i}$ by

$$A_{V_i}(x) = (\mathrm{co}(A_i(x)) + V_i) \cap X_i,$$
$$B_{V_i}(x) = (B_i(x) + V_i) \cap X_i \quad \text{for each} \quad x \in X.$$

Then, by (2) and Lemma 4.1 of Chang (1990) (or see (1986, p. 7), A_{V_i} has an open graph in $X \times X_i$ which implies that A_{V_i} has open lower sections (see (1976), pp. 265–266), i.e. $(A_{V_i})^{-1}(y)$ is open in X for each $y \in X_i$. For each $i \in I$, define the correspondence $Q_{V_i} : X \to 2^{X_i}$ by

$$Q_{V_i}(x) = \begin{cases} (A_i \cap P_i)(x), & \text{if } x \notin F_{V_i}, \\ A_{V_i}(x), & \text{if } x \in F_{V_i}, \end{cases}$$

where $F_{V_i} = \{x \in X : x_i \notin \overline{B}_{V_i}(x)\}$. Clearly, F_{V_i} is open in X. By using same argument as in the proof of Theorem 4.46, we can prove that the qualitative game $\Gamma = (X_i, Q_{V_i})_{i \in I}$ satisfies that for each $i \in I$, Q_{V_i} is L_F-majorized. Note that by the defination of Q_{V_i}, we have that for each $i \in I$ and for each $y \in X$, $A_i(y) \cap P_i(y) \subset A_{V_i}(y) \subset Q_{V_i}(y)$ and so for each $i \in I$, for each $x_i \in X_i$ and for any nonempty compact subset C of X, $\mathrm{cl}_C((X \setminus (\mathrm{co}\, Q_{V_i})^{-1}(x_i)) \cap C) \subset \mathrm{cl}_C((X \setminus (\mathrm{co}(A_i \cap P_i))^{-1}(x_i)) \cap C)$. Hence the condition (5) implies that there exist a nonempty compact convex subset X_0 and a nonempty compact subset K such that for each $y \in X \setminus K$, there is an $x \in \mathrm{co}(X_0 \cup \{y\})$ with $y \notin \mathrm{cl}_C((X \setminus (\mathrm{co}\, Q_{V_i})^{-1}(x_i)) \cap C)$. Hence all hypotheses of Theorem 4.45 are satisfied. By Theorem 4.45 there exists a point $\hat{x}_V = (\hat{x}_{V_i})_{i \in I} \in K$ such that $Q_{V_i}(\hat{x}_V) = \emptyset$ for all $i \in I$. Since for each $i \in I$ and for each $x \in X$, $A_{V_i}(x) \neq \emptyset$ by (2), we must have $\hat{x}_{V_i} \in \overline{B}_{V_i}$ and $A_i(\hat{x}_V) \cap P_i(\hat{x}_V) = \emptyset$ for each $i \in I$. □

Remark 4.22 Theorem 4.48 generalizes Theorem 5.2 of Tan-Yuan in (1993) in the following two aspects: (a) for each $i \in I$, $A_i \cap P_i$ is L_F-majorized; (b) the coercive condition (5) is weaker than the coercive condition (e) of Theorem 5.2 of Tan-Yuan (1993). Hence Theorem 4.48 answers in the affirmative the open question presented by Tan-Yuan in (1993).

The following result is Lemma 5.3 of Tan-Yuan in (1993).

Lemma 4.12 *Let X be a topological space, Y be a nonempty subset of a topological vector space E, $\mathcal{B}$ be a neighborhood base at zero in E and $B : X \to 2^Y$. For each $V \in \mathcal{B}$, let $B_V : X \to 2^Y$ be defined by $B_V(x) = (B(x) + V) \cap Y$ for each $x \in X$. If $\hat{x} \in X$ and $\hat{y} \in Y$ are such that $\hat{y} \in \bigcap_{V\in\mathcal{B}} B_V(x)$, then $\hat{y} \in \overline{B}(\hat{x})$.*

We shall now show the following equilibrium existence theorem for an abstract economy in locally convex topological vector space.

Theorem 4.49 *Let $\Gamma = (X_i, A_i, B_i, P_i)_{i\in I}$ be an abstract economy such that $X = \prod_{i\in I} X_i$ is paracompact. Suppose the following conditions are satisfied:*

(1) *for each $i \in I$, X_i is a nonempty convex subset of a locally convex topological vector space E_i;*
(2) *for each $i \in I$, A_i is lower semicontinuous such that for each $x \in X$, $A_i(x)$ is nonempty and $\mathrm{co}(A_i(x)) \subset \overline{B}_i(x)$;*
(3) *for each $i \in I$, $A_i \cap P_i$ is L_F-majorized;*
(4) *for each $i \in I$, the set $E^i = \{x \in X : (A_i \cap P_i)(x) \neq \emptyset\}$ is open in X;*
(5) *there exist a nonempty compact convex subset X_0 of X and a nonempty compact subset K such that for each $y \in X \setminus K$, there is an $x \in \mathrm{co}(X_0 \cup \{y\})$ with $y \notin \mathrm{cl}_C((X \setminus (\mathrm{co}(A_i \cap P_i))^{-1}(x_i)) \cap C)$ for each $i \in I$ and for any nonempty compact subset C of X.*

Then Γ has an equilibrium point in K.

Proof. For each $i \in I$, let $\mathcal{B}_i$ be the collection of all open convex neighborhoods of zero in E_i and let $\mathcal{B} = \prod_{i\in I} \mathcal{B}_i$. Given $V = \prod_{i\in I} V_i \in \mathcal{B}$ where $V_i \in \mathcal{B}_I$ for each $i \in I$. By Theorem 4.48, there exists an $\hat{x}_V \in K$ such that $\hat{x}_{V_i} \in \overline{B}_{V_i}(\hat{x}_V)$ and $A_i(\hat{x}_V) \cap P_i(\hat{x}_V) = \emptyset$ for each $i \in I$, where $B_{V_i}(x) = (B_i(x) + V_i) \cap X_i$ for each $x \in X$. It follows that the set

$$D_V = \{x \in K : x_i \in \overline{B}_{V_i}(x) \text{ and } A_i(x) \cap P_i(x) = \emptyset \text{ for all } i \in I\}$$

is a nonempty closed subset of K by (4) and hence D_V is compact.

Now we prove that the family $(D_V)_{V\in\mathcal{B}}$ has the finite intersection property. Let $\{V^1, \dots, V^n\}$ be any finite subset of $\mathcal{B}$. For each $k = 1, \dots, n$, let $V^k = \prod_{i\in I} V_i^k$ where $V_i^k \in \mathcal{B}_i$ for each $i \in I$; let $V = \prod_{i\in I}(\bigcap_{k=1}^n V_i^k)$ then $D_V \neq \emptyset$. Obviously, $D_V \subset \bigcap_{k=1}^n D_V k$ so that $\bigcap_{k=1}^n D_V k \neq \emptyset$. Therefore the family $\{D_V : V \in \mathcal{B}\}$ has the finite intersection property. Since K is compact, $\cap\{D_V : V \in \mathcal{B}\} \neq \emptyset$. Now take any $\hat{x} \in \cap\{D_V : V \in \mathcal{B}\}$, then for each $i \in I$, $\hat{x}_i \in \overline{B}_{V_i}(\hat{x})$ for each $V_i \in \mathcal{B}_i$ and $A_i(\hat{x}) \cap P_i(\hat{x}) = \emptyset$. By Lemma 4.12, for each $i \in I$, $\hat{x}_i \in \overline{B}_i(\hat{x})$. □

Remark 4.23 Theorem 4.49 improves and generalizes Theorem 5.4 of Tan-Yuan in (1993) to see that for each $i \in I$, $A_i \cap P_i$ is L_F-majorized and hence Theorem 4.49 positively answers the open problem presented by Tan-Yuan in (1993).

Corollary 4.49.1 *Let $\Gamma = (X_i, A_i, B_i, P_i)_{i\in I}$ be an abstract economy such that $X = \prod_{i\in I} X_i$ is paracompact. Suppose the following conditions are satisfied:*

(1) *for each $i \in I$, A_i is a nonempty comvex subset of a locally convex topological vector space;*
(2) *for each $i \in I$, A_i has an open graph (resp., is lower semicontinuous) such that for each $x \in X$ $A_i(x)$ is nonempty and $\mathrm{co}(A_i(x)) \subset \overline{B}_i(x)$;*
(3) *for each $i \in I$, P_i is lower semicontinuous (resp., has an open graph);*
(4) *for eahc $i \in I$, $A_i \cap P_i$ is L_F-majorized;*
(5) *there exist a nonempty compact convex subset X_0 of X and a nonempty compact subset K of X such that for each $y \in X \setminus K$, there is an $x \in \mathrm{co}(X_0 \cup \{y\})$ with $y \notin \mathrm{cl}_C((X \setminus (\mathrm{co}(A_i \cap P_i))^{-1}(x_i)) \cap C)$ for each $i \in I$ and for any nonempty compact subset C of X.*

Then Γ has an equilibrium point in K.

Proof. Since A_i has an open graph (resp., is lower semicontinuous) and P_i is lower semicontinuous (resp., has an open graph), the correspondence $A_i \cap P_i : X \to 2^{X_i}$ is also lower semicontinuous by Lemma 4.2 of Yannelix (1987), so that the set $E_i = \{x \in X : (A_i \cap P_i)(x) \neq \emptyset\}$ is open in X. By Theorem 4.49, Γ has an equilibrium point in K.

Corollary 4.49.2 *Let $\Gamma = (X_i, A_i, B_i, P_i)_{iii}$ be an abstract economy such that*

(1) *for each $i \in I$, X_i is a nonempty compact comvex subset of a locally convex topological vector space;*
(2) *for each $i \in I$ A_i has an open graph (resp., is lower semicontinuous) such that for each xiX, $A_i(x)$ is nonempty and $\mathrm{co}(A_i(x)) \subset \overline{B}_i(x)$;*
(3) *for each $i \in I$, P_i is lower semicontinuous (resp., has an open graph);*
(4) *for each $i \in I$, $A_i \cap P_i$ is L_F-majorized.*

Then Γ has an equilibrium point in X.

Proof. By Corollary 4.49.1 with $X_0 = K = \prod_{iii} X_i$, the conclusion holds. □

Remark 4.24 Corollaries 4.49.1 and 4.49.2 generalize Corollaries 5.5 and 5.6 of Tan-Yuan (1993) to L_F-majorized correspondences and generalize Corollary 3 of Borglin-Keiding (1976), Theorem 4.1 of Chang (1990) and Theorem of Shafer-Sonnenschein (1975) in several aspects. Corollary 5.2 positively also answers the open problem presented by Tan-Yuan in (1993).

Finally we note that the results of Sections 4.7 and 4.8 have been taken from Ding and Tarafdar (1994).

4.9 Fixed Point of ψ-Condensing Mapping, Maximal Elements and Equilibria

In this section we shall introduce the ψ-condensing and ψ_α-condensing mapping and prove to existence of fixed point of such mappings. We will also prove the existence of maximal element of games and equilibria of abstract economies.

Let C denote a lattice with a least element 0. We first recall the following definitions due to Fitzpatrick and Petryshyn (1974).

Definition 4.5 Let X be a locally convex topological vector space. The mapping $\psi : 2^X \to C$ is said to be a measure of noncompactness provided the followings hold for all $A, B \in 2^X$:

(i) $\psi(A) = 0$ if and only if A is relatively compact, i.e. $\overline{A}$ = closure of A is compact;
(ii) $\psi(\overline{\text{co}}\, A) = \psi(A)$;
(iii) $\psi(A \cup B) = \max(\psi(A), \psi(B))$.

It follows from (3) that if $A \subset B$, then $\Psi(A) \leq \Psi(B)$. The above notion is a generalization of the set-measure of noncompactness (Kuratowski (1920)) and the ball-measure of noncompactness (Sadovskii (1972)) defined either in terms of a family of seminorms when X is a locally convex topological vector space or of a single norm when X is a Banach space. For more details we refer the readers to reference Fitzpatrick and Detryshyh (1974).

Let K be anonempty subset of X. A set valued mapping $T : K \to 2^K$ values is called ψ-condensing provided that if $\Omega \subset K$ and $\psi(T(\Omega)) \geq \psi(\Omega)$, then Ω is relatively compact, where $\psi : 2^X \to C$ is a measure of noncompactness.

Note that if $T : D \to 2^X$ is a compact mapping (i.e., $T(D)$ is precompact), then T is Ψ-condensing for any measure of noncompactness Ψ. Various Ψ-condensing mappings which are not compact have been considered in Borisovich et al. (1980), Gohberg et al. (1947), Nussabauss (1971), Detryshyh et al. (1974), Reich (1972), Sadovskii (1972), etc. Moreover, when the measure of noncompactness Ψ is either the set-measure of noncompactness or ball-measure of noncompactness, Ψ-condensing mappings are called condensing mappings.

Definition 4.6 Let $\{X_\alpha : \alpha \in I\}$ be an indexed family of nonempty sets, each in locally convex topological vector space E_α, where I is a finite or an infinite index set. For each $\alpha \in I$, let $T_\alpha : X = \prod_{\alpha \in I} X_\alpha \to 2^{X_\alpha}$ be a set valued mapping and $\psi_\alpha : 2^{X_\alpha} \to C$ a measure of noncompactness in X_α. Then for $\alpha \in I$, T_α is said to be ψ_α-condensing provided that if $\Omega \subset X$ and $\psi_\alpha(T_\alpha(\Omega)) \geq \psi_\alpha(P_\alpha(\Omega))$, then $P_\alpha(\Omega)$ is relatively compact in X_α.

Note that if $I = \{1\}$, then $P_{\{1\}} = I$ = the identity on X and this definition reduces to the Definition 4.5 above.

Definition 4.7 Let X be a nonempty. A set valued mapping $T : X \to 2^X$ is said to acyclic (e.g., see Bergstrom (1975, p. 403) if for each $n \in \mathbb{N}$ with $x_{i+1} \in T(x_i)$ implies that $x_i \notin T(x_n)$.

Theorem 4.50 *Let $\{X_\alpha : \alpha \in I\}$ be a family of nonempty closed convex subsets, each in a locally convex topological vector space E_α, where I is a finite or an infinite index set. For each $\alpha \in I$, let $T_\alpha : X = \prod_{\alpha\in I} X_\alpha \to 2^{X_\alpha}$ be a ψ_α-condensing mapping with nonempty values where ψ_α is ameasure of noncompactness in X_α.*

Then there exist nonempty compact convex subset K of X and compact convex subset K_α of X_α for each $\alpha \in I$ such that $K = \prod_{\alpha\in I} K_\alpha$ and $T_\alpha : K \to 2^{K_\alpha}$ for each $\alpha \in I$.

Proof. Let $x_0 \in X$ and $\mathcal{F}$ be the family of closed convex subsets C of X contains x_0 and for each $\alpha \in I$, $\mathcal{F}_\alpha$ be the family c_α of closed convex subsets of X_α such that
(*) $\quad C = \prod_{\alpha\in I} C_\alpha$ and $T_\alpha : C \to 2^{C_\alpha}$ for each $\alpha \in I$.
Clearly $\mathcal{F} \neq \emptyset$ and $\mathcal{F} \neq \emptyset$ for each $\alpha \in I$ as $X \in \mathcal{F}$ and $X_\alpha \in \mathcal{F}$ with $X = \prod_{aii} X_\alpha$. Now let $K = \cap\{C : C \in \mathcal{F}\}$ and $K_\alpha = \cap\{C_\alpha : C_\alpha \in \mathcal{F}_\alpha\}$ closed covnex subsets of X and X_α respectively as $x_0 \in K$ and $P_\alpha x_0 \in K_\alpha$. Now we first note that $K = \prod_{\alpha\in I} K_\alpha$. Let $x \in K$. Then $x \in C = \prod_{\alpha\in I} C_\alpha$ for all $C \in \mathcal{F}$ and all $C \in \mathcal{F}$ and $C_\alpha \in \mathcal{F}_\alpha$ satisfying (*), i.e. $P_\alpha x \in C_\alpha$ for each $\alpha \in I$. It implies that $P_\alpha x \in K_\alpha$ for each $\alpha \in I$. Thus $x \in \prod_{\alpha\in I} K_\alpha$. Next, let $x \in \prod_{\alpha\in I} K_\alpha$. Then for each $\alpha \in I$, $P_\alpha x \in C_\alpha$, i.e. $x \in \prod_{\alpha\in I} C_\alpha = C$. This implies that $x \in K$. Thus $K = \prod_{\alpha\in I} K_\alpha$.

Next, we prove that $T_\alpha : K \to 2^{K_\alpha}$ for each $\alpha \in I$. Let $x \in K$. Then $x \in C = \prod_{\alpha\in I} C_\alpha$ where C and C_α satisfy (*). Hence by (*), $T_\alpha(x) \subset C_\alpha$ for $C_\alpha \in \mathcal{F}_\alpha$, $T_\alpha(x) \subset K_\alpha$. Thus $T_\alpha : K \to 2^{K_\alpha}$. It remains to prove that K and K_α are compact for each $\alpha \in I$.

Let us consider the set valued mapping $T : X \to 2^X$ defined by $T(x) = \prod_{\alpha\in I} T_\alpha(x)$. Then clearly we have $T : K \to 2^K$. Now let $\widehat{K} = \overline{\text{co}}\,(\{x_0\} \cup T(K))$, then clearly $\widehat{K} \subset K$ as $x_0 \in K$ $T(K) \subset K$ and K is closed and convex, where $T(K) \cup \{T(x) : x \in K\}$. For each $\alpha \in I$, let $\widehat{K}_\alpha = P_\alpha \widehat{K}$. Then $\widehat{K}_\alpha \subset P_\alpha K = K_\alpha$. Now let $\hat{\widehat{K}} = \prod_{\alpha\in I} P_\alpha \widehat{K} = \prod_{\alpha\in I} \widehat{K}_\alpha$. Then clearly $\hat{\widehat{K}}$ is closed convex, $x_0 \in \hat{\widehat{K}}$ and $\hat{\widehat{K}} \subset K$. Now if $x \in \hat{\widehat{K}}$, then $T(x) \subset T(K)$ as $\hat{\widehat{K}} \subset K$. Thus $T(x) \subset \widehat{K}$. Hence $T_\alpha(x) = P_\alpha T(x) \subset P_\alpha \widehat{K} = \widehat{K}_\alpha$. Hence $T_\alpha : \hat{\widehat{K}} \to 2^{\widehat{K}_\alpha}$. This implies that $\hat{\widehat{K}} = K$ and $P_\alpha \widehat{K} = \widehat{K}_\alpha = K_\alpha$, for each $\alpha \in I$. Finally let α be arbitrary but fixed. Then by using the continuity of P_α, we can easily check that

$$\begin{aligned} P_\alpha \widehat{K} &= P_\alpha \overline{\text{co}}\,(\{x_0\} \cup T(K)) \\ &= \overline{\text{co}}\,(\{P_\alpha(x_0)\} \cup P_\alpha T(K)) = \overline{\text{co}}\,(\{P_\alpha(x_0)\} \cup T_\alpha(K)) . \end{aligned}$$

Hence

$$\begin{aligned}\psi_\alpha(K_\alpha) &= \psi_\alpha(P_\alpha\widehat{K}) = \psi_\alpha\left(\overline{\text{co}}\{P_\alpha(x_0)\} \cup T_\alpha(K)\right)\\ &= \psi_\alpha\left(\{P_\alpha(x_0)\} \cup T_\alpha(K)\right) = \max\{\psi_\alpha\{(P_\alpha(x_0)\},\, \psi_\alpha(T_\alpha(K))\}\\ &= \psi_\alpha\left(T_\alpha(K)\right).\end{aligned}$$

This shows that $K_\alpha = P_\alpha K$ is compact. Now since $\alpha \in I$ is arbitrary, K_α is compact for each $\alpha \in I$ and hence by Tychonov Theorem K is compact. □

Taking $I = \{1\}$ we obtain the following corollary (Lemma 2.1, of Mehta et al (1996)).

Corollary 4.50.1 *Let X be a nonempty closed convex subset of a locally convex topological vector space E and $T : X \to 2^X$ be a ψ-condensing mapping with nonempty convex values where $\psi : 2^E \to C$ is a measure of noncompactness. Then there is a nonempty compact convex subset K of X such that $T : K \to 2^K$.*

Theorem 4.51 *Let $\{X_\alpha : \alpha \in I\}$ and X be as in Theorem 4.50 and for each $\alpha \in I$, $T_\alpha L\, X \to 2^{X_\alpha}$ be a ψ-condensing upper semicontinuous mapping with nonempty closed values. Then the mapping $T : X \to 2^X$ defined by $T(x) = \prod_{\alpha\in I} T_\alpha(x)$ has a fixed point.*

Proof. By Theorem 4.50 there exist a nonempty compact convex set $K \subset X$ and a compact convex set K_α for each $\alpha \in I$ such that $K = \prod_{\alpha\in I} K_\alpha$ and $T_\alpha : K \to 2^{K_\alpha}$ for each $\alpha \in I$. It is now trivial to see that $T_\alpha : K \to 2^{K_\alpha}$ is upper semicontinuous and $T_\alpha(x)$ is nonempty compact convex subset of K_α for each $x \in K$. Now by Theorem 4, the mapping $T : K \to 2^K$ defined by $T(x) = \prod_{\alpha\in I} T_\alpha(x)$, $x \in K$ is upper semicontinuous with compact comvex values. Hence by Theorem 4 of Fan, there exists a point $x_0 \in K \subset X$ such that $x_0 \in T(x_0) = \prod_{\alpha\in I} I_\alpha(x_0)$. □

Specializing $I = \{1\}$, we have the following corollary.

Corollary 4.51.1 *Let X be a nonempty closed convex subset of locally convex topological vector space E. Let the mapping $T : X \to 2^X$ be upper semicontinuous and ψ-condensing with nonempty closed convex values. Then T has a fixed point x_0, i.e. $x_0 \in T(x_0)$.*

Theorem 4.52 *Let $\{X_\alpha : \alpha \in I\}$ and X be as in Theorem 4.50. For each $\alpha \in I$, let $T_\alpha : X \to 2^{X_\alpha}$ be a ψ_α-condensing mapping such that*

(*i*) *for each $x \in X$, $T_\alpha(x)$ is a nonempty convex of X_α;*
(*ii*) *for each $x_\alpha \in X_\alpha$, $T_\alpha^{-1}(x_\alpha)$ contains a compactly open subset O_{x_α} of X such that $\bigcup_{x_\alpha\in X_\alpha} O_{x_\alpha} = X$ (O_{x_α} may be empty for some x_α).*

Then there is a point $x \in X$ such that $x \in T(x) = \prod_{\alpha\in I} T_\alpha(x)$.

Proof. By Theorem 4.50 there exist a nonempty compact convex subset K of X and a compact convex subset K_α of X_α such that $K = \prod_{\alpha\in I} K_\alpha$ and $T : K \to 2^{K_\alpha}$. By (i), $T_\alpha(x)$ is a nonempty convex subset of K_α for each $x \in K$.

By (ii), for each $x_\alpha \in K_\alpha \subset X$, there exists a compactly open set O_{x_α} of X and hence relatively open set $\hat{O}_{x_\alpha} = O_{x_\alpha} \cap K$ of K. Also if $x \in K$, by (ii) $x \in O_{x_\alpha}$ for some $x_\alpha \in X_\alpha$. Hence $x \in O_{x_\alpha} \cap K = O_{x_\alpha}$. This shows that $K \subset \bigcup_{x_\alpha\in K_\alpha} \hat{O}_{x_\alpha}$. Hence $K = \bigcup_{x_\alpha\in K_\alpha} \hat{O}_{x_\alpha}$. Thus all the conditions of Theorem 4.39 are fulfilled. Therefore by Theorem 4.39 there exists a point $x \in T(x) = \prod_{\alpha\in I} T_\alpha(x)$. □

Now if we take $I = \{1\}$, we obtain the following corollary.

Corollary 4.52.1 *Let X be a nonempty closed convex subset of a locally convex topological vector space E. Let $X \to 2^X$ be a ψ-condensing mapping with nonempty convex values such that for each $y \in X$, $T^{-1}(y)$ contains compactly open subset O_y of X with $\bigcup_{y\in X} O_y = X$. Then T has a fixed point.*

Corollary 4.52.1 contains Theorem 4.8.5 of Yuan (1999) as a special case. Corollaries 4.51.1 and 4.52.1 generalize the corresponding results on fixed point theorems in locally convex topological vector spaces of Dugundji and Granas (1982), Reich (1972), Smart (1974), Istratescu (1981), Zeidler (1985) and others.

Let X and Y be topological spaces. A correspondence $T . X \to 2^Y$ is said to be (1) quasi-regular if (a) it has open lower sections, (b) $T(x)$ is non-empty and convex for each $x \in X$, and (c) $\overline{T}(x) = \text{cl}_X T(x)$ for each $x \in X$; (2) the correspondence $T : X \to 2^Y$ is said to be regular if it is quasi-regular and has an open graph.

If X is a set, Y is a subset of a vector space, and $F : X \to 2^Y$ is such that for each $x \in X$, $\text{co}\, F(x) \subset Y$, then the mapping $\text{co}\, F : X \to 2^Y$ is defined by $(\text{co}\, F)(x) = \text{co}\, F(x)$ for each $x \in X$. If $\{X + i : i \in I\}$ and $Y_i : i \in I\}$ are collections of sets and $F_i \prod_{j\in I} X_j \to 2^{Y_i}$ is a set-valued mapping for each $i \in I$, then the mapping $\prod_{i\in I} F_i : \prod_{i\in I} X_i \to 2^{\prod_{i\in I} Y_i}$ is defined by $(\prod_{i\in I} F_i)(x) = \prod_{i\in I} F_i(x)$ for each $x \in \prod_{i\in I} X_i$. We note that if X is a topological space, Y is a topological vector space, and $FL\, X \to 2^Y$ is lower semicontinuous, it is clear that $\text{co}\, F$ is lower semicontinuous.

We need notation used in Tulcea (1988). Let X be a non-empty set, Y a non-empty subset of topological vector space E, and $F : X \to 2^Y$ a set-valued mapping. A family $(f_j)_{j\in J}$ of correspondences between X and Y, indexed by a non-empty filtering set J (we denote by '$\leq$' the order relation in J) is an upper approximating family for F if:

(A_I) $F(x) \subset f_j(x)$ for all $x \in X$ and all $j \in J$;

(A_{II}) for each $j \in J$ there is $j^* \in J$ such that for each $h \geq j^*$ and $h \in J$, $f_h(x) \subset f_j(x)$ for each $x \in X$;

(A_{III}) for each $x \in X$ and $V \in \mathcal{B}$, where $\mathcal{B}$ is the fundamental system of zero of the topological vector space E, there is $j_{x,\nu} \in J$ such that $f_h(x) \subset F(x) + V$ if $h \in J$ and $j_{x,V} \leq h$.

From (A_I)–(A_{III}), it is easy to deduce that

(A_{IV}) for each $x \in X$ and $l \in J$,

$$F(x) \subset \bigcap_{j \in J} f_j(x) = \bigcap_{k \leq j,\, k \in J} f_j(x) \subset \overline{F(x)} \subset \overline{F}(x).$$

Let X be a topological space, Y a non-empty subset of a vector space E, $\theta : X \to E$ be a single-valued mapping, and $\phi : X \to 2^Y$ be a correspondence. Then (1) ϕ is said to be of class $L_{\theta,C}$ if (a) for each $x \in X$, $\operatorname{co}\phi(x) \subset Y$ and $\theta(x) \notin \operatorname{co}\phi(x)$, and (b) for each $y \in Y$, $\phi^{-1}(y)$ is compactly open in X; (2) (ϕ_x, N_x) is an $L_{\theta,C}$-majorant of ϕ at x if $\phi_x : X \to 2^Y$ and N_x is an open neighborhood of x in X such that (a) for each $z \in N_x$, $\phi(z) \subset \phi_x(z)$ and $\theta(z) \notin \operatorname{co}\phi_x(z)$, (b) for each $z \in X$, $\operatorname{co}\phi_x(z) \subset T$, and (c) for each $y \in Y$, $\phi_x^{-1}(y)$ is compactly open in X; (3) ϕ is said to be $L_{\theta,C}$-majorized if for each $x \in X$ with $\phi(x) \neq \emptyset$, there exists an $L_{\theta,C}$-majorant (ϕ_x, N_x) of ϕ at x.

It is clear that every correspondence of class $L_{\theta,C}$ is $L_{\theta,C}$-majorized. We note that our notions of the correspondence ϕ being of class $L_{\theta,C}$ and $L_{\theta,C}$-majorized correspondence generalize the notions of correspondence of class $\mathcal{L}_\theta^*$ and $\mathcal{L}_\theta^*$-majorized correspondence, respectively introduced by Ding *et al.* which in turn generalized the notions of $\phi \in C(X, Y, \theta)$ and C-majorized correspondence, respectively, introduced by Tulcea. In this paper, we shall deal mainly with either the case (I) $X = Y$ and is a non-empty convex subset of the topological vector space E and $\theta = I_X$, the identity mapping on X; or the case (II) $X = \prod_{i \in I} X_i$ and $\theta = \pi_j$; $X \to X_j$ is the jth projection of X onto X_j and $Y = X_j$ is a non-empty convex subset of a topological vector space. In both cases (I) and (II), we shall write L_C in place of $L_{\theta,C}$.

For the proof of the following two lemmas we refer to Yuan (1999, Theorem 4.7.2 and Lemma 4.7.3, pp. 300–301), (also see Klein and Thompson (1984, Theorem 7.3.10, p. 86).

Lemma 4.13 *Let X be a topological space and Y a normal space. If $F_1, F_2 : X \to 2^Y$ are upper semicontinuous at $x \in X$ and have closed values, then $F_1 \cap F_2$ is upper semicontinuous at x.*

Lemma 4.14 *Let X and Y be two topological space and A be a nonempty closed (resp., open) subset of X. If $F_1 : X \to 2^Y$ and $F_2 : A \to 2^Y$ are lower semicontinuous (resp. upper semicontinuous) mappings such that $F_2(x) \subset F_1(x)$ for all*

$x \in A$, then the mapping $F : X \to 2^Y$ defined by

$$F(x) = \begin{cases} F_1(x), & \text{if } x \notin A \\ F_2(x), & \text{if } x \in A \end{cases}$$

is lower semicontinuous (resp. upper semicontinuous).

We should point out that Lemma 6.1 of Yannelis and Prabhakav (1983) is a particular case of Lemma 4.14 above.

We now extend over the Theorems which are an extension of a theorem of Debreu and Shafer and Sonnenschein (1975).

Theorem 4.53 *Let $\mathcal{E} = \{X_\alpha, A_\alpha, U_\alpha : \alpha \in I\}$ be an abstract economy such that for each $\alpha \in I$.*

(i) X_α is a nonempty closed convex subset of a locally convex vector space E_α.

(ii) $A_\alpha : X = \prod_{\alpha \in I} X_\alpha \to 2^{X_\alpha}$ is a continuous ϕ_α-condensing correspondence such that for each $x \in X$, $A_\alpha(x)$ is nonempty closed convex subset of X_α;

(iii) $U_\alpha : X \to \mathbb{R}$ is continuous and quasiconcave in x_α.

Then there is an equilibrium point $\overline{x} = \{\overline{x}_\alpha\} \in X$ of the economy, i.e. $U_\alpha(\overline{x}) = U_\alpha[x_\alpha, \overline{x}_\alpha] = \sup_{Z_\alpha \in A_\alpha(\overline{x})} U[z_\alpha, \overline{x}_{-\alpha}]$.

Proof. By Theorem 4.50 there exist a nonempty compact convex subset K of X and for each $\alpha \in I$, a compact convex K_α of X_α such that $K = \prod_{\alpha \in I} K_\alpha$ and $A_\alpha : K \to 2^{K_\alpha}$ for each $\alpha \in I$. Now for each $\alpha \in I$, let $\hat{U}_\alpha = U \setminus K =$ the restriction of U_α to K. Then clearly $\hat{U}_\alpha : K \to \mathbb{R}$ is continuous and quasiconcave in x_α. Thus the truncated economy $\hat{\xi} = \{K_\alpha, A_\alpha, \hat{U}_\alpha : \alpha \in I\}$ has an equilibrium point $\overline{x} \in K$ of $\hat{\xi}$ by Theorem.

Now $U_\alpha(\overline{x}) = \hat{U}_\alpha(\overline{x}) - \hat{U}_\alpha[\overline{x}_\alpha, \overline{x}_{-\alpha}] = \sup_{z_\alpha \in (\overline{x})} \hat{U}_\alpha[\overline{x}_\alpha, x_{-\alpha}] = \sup_{z_\alpha \in A_\alpha(x)} \times U_\alpha[z_\alpha . \overline{x}_{-\alpha}]$ for each $\alpha \in I$. □

Theorem 4.54 *Let $\xi = \{X_i, A_i, P_i : i \in I\}$ be an abstract economy, where I is a countable index set. Assume that for each $i \in I$,*

(a) X_i is s nonempty closed convex subset of a locally convex metrizable space E_i;

(b) $A_i : X = \prod_{i \in I} X_i \to 2^{X_i}$ is a continuous ϕ_i-condensing correspondence such that for each $x \in X$, $A_i(x)$ is nonempty and convex, where ψ_i is a measure of noncompactness in X_i;

(c) $P_i : X \to 2^{X_i}$ has a compactly open graph in $X \times X_i$;

(d) for each $x = \{x_i\} \in X$, $x_i \notin \overline{co}\, P_i(x)$ for all $i \in I$.

Then the exists an equilibrium point $\overline{x} = \{x_i\} \in X$ of ξ, i.e. $\overline{x}_i \in A_i(\overline{x})$ and $P_i(\overline{x}) \cap A_i(\overline{x}) = \phi$ for each $i \in I$.

Proof. By Theorem 4.50 there exist a nonempty compact convex subset K of X and a compact convex subset K_i of X_i for each $i \in I$ such that $K = \prod_{i \in J} K_i$ and $A_i : K \to 2^{K_i}$. For each $i \in I$, let us define $\hat{P}_i : K \to 2^{K_i}$ by $\hat{P}(x) = P_i(x) \cap K_i$, $x \in K$.

Now for each $i \in I$,

$$\begin{aligned} G(\hat{P}_i) &= \text{ the graph of } \hat{P}_i \\ &= \{(x, u_i) : u_i \in \hat{P}_i(x) \subset K_i, x \in K\} \\ &= \{(x, u_i) : u_i \in P_i(x) \subset X_i, x \in X\} \cap K \times K_i \\ &= G(P_i) \cap (K \times K_i) \text{ which is open by (c).} \end{aligned}$$

Thus the truncated economy $\xi = \{K_i, A_i, \hat{P}_i : i \in I\}$ satisfies all the condition of Theorem. Hence there is a point $\overline{x} = \{\overline{x}_i\} \in K$ such that for each $i \in I$, $\overline{x}_i \in A_i(\overline{x})$ and $\hat{P}_i(\overline{x}) \cap A_i(\overline{x}) = \emptyset$. Now $\hat{P}_i(\overline{x}) \cap A_i(\overline{x}) = \phi$ implies that $P_i(\overline{x})A_i(\overline{x}) = \emptyset$. For if $y_i \in P_i(\overline{x}) \cap A_i(\overline{x})$, then $y_i \in A_i(\overline{x}) \subset K_i$ as $\overline{x} \in K$ and thus $y_i \in P_i(\overline{x}) \cap K_i = \hat{P}(\overline{x})$ which would be contraction. Hence $\overline{x} \in K$ is an equilibrium point of ξ. □

Theorem 4.55 *Let $\Gamma = (X_i, P_i)_{i \in I}$ be a qualitative game such that $X = \prod_{i \in I} X_i$. Suppose the following conditions are satisfied:*

(a) *X_i is a non-empty compact and convex subset of a topological vector space for each $i \in I$;*
(b) *$P_i : X \to 2^{X_i}$ is L_C-majorized for each $i \in I$.*
(c) $\bigcup_{i \in I} \{x \in X : P_i(x) \neq \emptyset\} = \bigcup_{i \in I} \int_X \{x \in X : P_i(x) \neq \emptyset\}$.

Then Γ has an equilibrium point in X.

Proof. For each $i \in I$, X_i is a non-empty compact and convex subset of the topological vector space E_i. By Theorem 3 of Ding and Tan (1993), there exists a point $x \in X$ such that $P_i(x) = \emptyset$ for all $i \in I$. □

We shall use the approximation technique inspired by Chang (1990) and Tulcea (1988) to obtain equilibrium existence theorems for generalized games in which the constraint correpsondences are lower semicontinuous instead of having lower open sections. The outline of ideas is as follows: for given a generalized game $\Gamma = (X_i; A_i, B_i; P_i)_{i \in I}$, we first construct an associated approximate generalized game $\Gamma_V = (X_i; (A_V)_i, (B_V)_i; (P_V)_i)_{iii}$ for each non-empty open neighborhood V of zero in locally convex topological vector space. Then, for the associated approximate generalized game $\Gamma_V = (X_i; (A_V)_i, (B_V)_i; (P_V)_i)_{i \in I}$, there exists an associated simple qualitative game $\mathcal{I}_V = (X_i; (Q_V)_i)_{iii}$ which exhibits the same equilibrium points as the approximate generalized game $\Gamma_V = (X_i; (A_V), (B_V)_i; (P_V)_i)_{i \in I}$ by Theorem 4.55. Finally by employing our approximate Lemma 4.12 and finite intersection property, the result is deduced.

Theorem 4.56 *Let $\mathcal{G} = (X_i; A_i, B_i; P_i)_{i\in I}$ be a generalized game and $X = \prod_{i\in I} X_i$. Suppose the following conditions are satisfied:*

(a) *for each $i \in I$, X_i is a non-empty compact and convex subset of a locally convex Hausdorff topological vector space E_i;*
(b) *for each $i \in I$, $A_i : X \to 2^{X_i}$ is lower semicontinuous such that for each $x \in X$, $A_i(x)$ is non-empty and $\operatorname{co} A_i(x) \subset B_i(x)$;*
(c) *for each $i \in I$, $A_i \cap P_i$ is L_C-majorized;*
(d) *for each $i \in I$, the set $E^i = \{x \in X : (A_i \cap P_i)(x) \neq \emptyset\}$ is open in X.*

Then $\mathcal{G}$ has an equilibrium point in X, i.e. there exists a point $\hat{x} = (\hat{x}_i)_{i\in I} \in X$ such that for each $i \in I$, $\hat{x}_i \in \overline{B}_i(\hat{x})$ and $A_i(\hat{x}) \cap P_i(\hat{x}) = \emptyset$.

Proof. Let $V = \prod_{i\in I} V_i$ be given where for each $i \in I$, V_i is an open convex neighborhood of zero in E_i. Fix any $i \in I$ and define $A_{V_i}, B_{V_i} : X \to 2^{X_i}$ by $A_{V_i}(x) = (\operatorname{co} A_i(x) + V_i) \cap X_i$ and $B_{V_i}(x) = (B_i(x) + V_i) \cap X_i$ for each $x \in X$. By (a), A_i is lower semicontinuous so that $\operatorname{co} A_i$ is also lower semicontinuous by Proposition 2.6 of (1956, p. 366). It follows from Lemma 4.1 of (1990, p. 7) that A_{V_i} has an open graph in $X \times X$. Now let $F_{V_i} = \{x \in X : x \in\notin \overline{B_{V_i}}(x)\}$. Then F_{V_i} is open in X. Define the mapping $Q_{V_i} : X \to 2^{X_i}$ by

$$Q_{V_i}(x) = \begin{cases} (A_i \cap P_i)(x), & \text{if } x \notin F_{V_i}, \\ A_i(x), & \text{if } x \in F_{V_i}. \end{cases}$$

We shall prove that the qualitative game $\mathcal{I} = (X_i, Q_{V_i})_{i\in I}$ satisfies all conditions of Theorem 4.55. First we note that for each $i \in I$, the set

$$\begin{aligned} &\{x \in X : Q_{V_i}(x) \neq \emptyset\} \\ &\quad = F_{V_i} \cap \{x \in X : A_i(x) \neq \emptyset\} \cup \{x \in X \setminus F_{V_i} : A_i(x) \cap P_i(x) \neq \emptyset\} \\ &\quad = F_{V_i} \cup \{x \in X \setminus F_{V_i} : A_i(x) \cap P_i(x) \neq \emptyset\} \quad \text{(by (b))} \\ &\quad = F_{V_i} \cup \left((X \setminus F_{V_i}) \cap E^i\right) = F_{V_i} \cup E^i \end{aligned}$$

is open in X by (d). Let $x \in X$ be such that $Q_{V_i}(x) \neq \emptyset$. We consider the following two cases:

Case 1. $x \in F_{V_i}$. Let $\psi_x = A_{V_i}$ and $N_x = F_{V_i}$. Then N_x is an open neighborhood of x in X such that (i) $Q_{V_i}(z) \subset \psi_x(z)$ and by (b), $z_i \notin \operatorname{co}\psi_x(z)$ fir each $z \in N_x$; (ii) $\operatorname{co}\psi_x(z) \subset X_i$ for all $y \in X_i$ since A_{V_i} has an open graph. Therefore, ψ_x is an L_C-majorant of Q_{V_i} at x.

Case 2. $x \notin F_{V_i}$. Since $Q_{V_i}(x) = (A_i \cap P_i)(x) \neq \emptyset$ and $A_i \cap P_i$ is L_C-majorized, there exist an open neighborhood N_x of x in X and a correspondence $\phi_x : X \to 2^{X_i}$ such that (i) $(A_i \cap P_i)(z) \subset \phi_x(z)$ and $z_i \notin \operatorname{co}\phi_x(z)$ for each $z \in N_x$; (ii) $\operatorname{co}\phi_x(z) \subset X_i$ for each $z \in X$; and (iii) $\phi_x^{-1}(y)$ is compactly open in X for each $y \in X_i$. Define

$\psi_x : X \to 2^{X_i}$ by

$$\psi_x(z) = \begin{cases} A_{V_i}(z) \cap \phi_x(z), & \text{if } z \notin F_{V_i}, \\ A_{V_i}(z), & \text{if } z \in F_{V_i}. \end{cases}$$

Note that as $(A_i \cap P_i)(z) \subset \phi_x(z)$ and $A_i(z) \subset A_{V_i}(z)$ for each $z \in N_x$, we have $Q_{V_i}(z) \subset \psi_x(z)$ and $\operatorname{co}\psi_x(z) \subset X_i$. It is easy to see that $z_i \notin \operatorname{co}\psi_x(z)$ for all $z \in X$. Moreover, for any $y \in X_i$, the set

$$\begin{aligned} \psi_x^{-1}(y) &= \{z \in X : y \in \psi_x(z)\} \\ &= \{z \in X \setminus F_{V_i} : y \in \psi_x(z)\} \cup \{z \in F_{V_i} : y \in \psi_x(z)\} \\ &= \{z \in X \setminus F_{V_i} : y \in A_{V_i}(z) \cap \phi_x(z)\} \cup \{z \in F_{V_i} : y \in A_{V_i}(z)\} \\ &= \left[(X \setminus F_{V_i}) \cap A_{V_i}^{-1}(y) \cap \phi_x^{-1}(y)\right] \cup \left[F_{V_i} \cap A_{V_i}^{-1}(y)\right] \\ &= \left[\phi_x^{-1}(y) \cup F_{V_i}\right] \cap A_{V_i}^{-1}(y) \end{aligned}$$

is compactly open in X. Therefore, ψ_x is an L_C-majorant of Q_{V_i} at point x.

Hence Q_{V_i} is an L_C-majorized correspondence. Now by our assumption, the set $\{x \in X : Q_{V_i}(x) \neq \emptyset\} = F_{V_i} \cup \{x \in X \setminus F_{V_i} : (A_i \cap P_i)(x) \neq \emptyset\} = F_{V_i} \cup E^i$ is open in X by condition (c). Therefore all hypotheses of Theorem 4.55 are satisfied, so that by Theorem 4.55, there exists a point $x_V = (x_{V_i})_{i \in I} \in X$ such that $Q_{V_i}(x_V) = \emptyset$ for all $i \in I$. Since for each $i \in I$, $A_i(x)$ is non-empty, we must have $x_{V_i} \in \overline{B_{V_i}}(x_V)$ and $A_i(x_V) \cap P_i(x_V) = \emptyset$.

For each $i \in I$, let $\mathcal{B}_i$ be the collection of all open convex neighborhoods of zero in E_i and $\mathcal{B} = \prod_{i \in I} \mathcal{B}_i$. Given any $V \in \mathcal{B}$, let $V = \prod_{j \in I} V_j$, where $V_j \in \mathcal{B}_j$ for each $j \in I$. By the argument above, there exists a $\hat{x}_V \in X$ such that $\hat{x}_{V_i} \in \overline{B_{V_i}}(\hat{x}_V)$ and $A_i(\hat{x}_V) \cap P_i(\hat{x}_V) = \emptyset$ for each $i \in I$, where $B_{V_i}(x) = (B_i(x) + V_i) \cap X_i$ for each $x \in X$. It follows that the set $Q_V := \{x \in K : x_i \in \overline{B_{V_i}}(x) \text{ and } A_i(x) \cap P_i(x) = \emptyset\}$ is a non-empty and closed subset of X by (d).

Now we want to prove $\{W_V\}_{V \in \mathcal{B}}$ has the finite intersection property. Let $\{V_1, \dots, V_n\}$ be any finite subset of $\mathcal{B}$. For each $i = 1, \dots, n$, let $V_i = \prod_{j \in I} V_{ij}$ where $V_{ij} \in \mathcal{B}_i$ for each $j \in I$; let $V = \prod_{j \in I}(\bigcap_{i=1}^{n} V_{ij})$, then $Q_V \neq \emptyset$. Clearly $Q_V \subset \bigcap_{i=1}^{n} Q_{V_i}$ so that $\bigcap_{i=1}^{n} Q_{V_i} \neq \emptyset$. Therefore the family $\{Q_V : V \in \mathcal{B}\}$ has the finite intersection property. Since X is compact, $\bigcap_{V \in B} Q_V \neq \emptyset$. Now take any $\hat{x} \in \bigcap_{V \in \mathcal{B}} Q_V$, then for each $i \in I$, $\hat{x}_i \in \overline{B_{V_i}}(\hat{x})$ for each $V_i \in \mathcal{B}_i$ and $A_i(\hat{x}) \cap P_i(\hat{x}) = \emptyset$. By Lemma 4.12, we must have for each $i \in I$, $\overline{B_i}(\hat{x})$. □

As a consequence of Theorem 4.56, we have the following:

Corollary 4.56.1 *Let $\mathcal{G} = (X_i, A_i, B_i, P_i)_{i \in I}$ be a generalized game such that $X = \prod_{i \in I} X_i$ is paracompact. Suppose the following conditions are satisfied:*

(a) *for each* $i \in I$, X_i *is a non-empty compact and convex subset of a locally convex Hausdorff topological vector space* E_i;

(b) *for each* $i \in I$ *and for each* $x \in X$, $A_i(x)$ *is non-empty and* $\operatorname{co} A_i(x) \subset B_i(x)$;

(c) *for each* $i \in I$, A_i *has an open graph in* $X \times X_i$ *(resp., is lower semicontinuous) and* P_i *is lower semicontinuous (resp., has an open graph in* $X \times X_i$*)*;

(d) *for each* $i \in I$, $A_i \cap P_i$ *is* L_C*-majorized.*

Then there exists $\overline{x} \in X$ *such that for each* $i \in I$, $A_i(\overline{x}) \cap P_i(\overline{x}) = \emptyset$ *and* $\overline{x}_i \in A_i(\overline{x})$.

Proof. For each $i \in I$, since A_i has an open graph in $X \times X_i$ (resp., is lower semicontinuous) and P_i is lower semicontinuous (resp., has open graph in $X \times X_i$) the ampping $A_i \cap P_i : X \to 2^{X_i}$ is also lower semicontinuous(e.g. see Lemma 4.1 of (1983, p. 103), so that the set $E^i = \{x \in X : A_i(x) \cap P_i(x) \neq \emptyset\}$ is open in X. Therefore all conditions of Theorem 4.56 are satisfied and the conclusion follows. □

Now, by Theorem 4.56 and the approximation Theorem 3 of Tulcea (1988, p. 280), we are able to give the existence theorem of the generalized game $\mathcal{G} = (X_i; A_i; P_i)_{i \in I}$ in which the constraint correspondences are upper semicontinuous and preferences are lower semicontinuous instead of having open lower sections or open graphs.

Theorem 4.57 *Let* $\mathcal{G} = (X_i, A_i, P_i)_{i \in I}$ *be a generalized game and* $X = \prod_{i \in I} X_i$. *Suppose the following conditions are satisfied for each* $i \in I$:

(a) X_i *is a non-empty compact and convex subset of a locally convex topological vector space* E_i;

(b) $A_i : X \to 2^{X_i}$ *is upper semicontinuous with non-empty closed and convex values;*

(c) $P_i : X \to 2^{X_i}$ *is lower semicontinuous and* L_C*-majorized;*

(d) *for each* $i \in I$, $E^i = \{x \in X; (A_i \cap P_i)(x) \neq \emptyset\}$ *is open in* X.

Then there exists $\overline{x} \in X$ *such that for each* $i \in I$, $A_i(\overline{x}) \cap P_i(\overline{x}) = \emptyset$ *and* $\overline{x}_i \in A_i(\overline{x})$.

Proof. For each $i \in I$, X_i is non-empty compact and convex, so that $X = \prod_{i \in I} X_i$ is also compact and convex. Given any $i \in I$, since A_i is upper semicontinuous, Theorem 3 of Tulcea (1988, p. 280) implies that there exists a family $(B_{ij})_{j \in J}$ indeed by filtering set J, consisting of regular correspondences between X and X_i, such that both $(B_{ij})_{j \in J}$ and $(\overline{B_{ij}})_{j \in J}$ are upper approximating of families for B_i.

The game $\mathcal{G}_j = (X_i; (B_{ij})_i, (\overline{B_{ij}})_i; P_i)_{i \in I}$ satisfies the hypotheses of Theorem 4.56 (hence also Corollary 4.56.1) for each $j \in J$. Hence $\mathcal{G}_j$ has an equilibrium $x^{j^*} \in X$ for each fixed $j \in J$ such that $B_{ij}(x^{j^*}) \cap P_i(x^{j^*}) = \emptyset$, and $\pi_i(x^{j^*}) \in \overline{B_{ij}}(x^{j^*})$ for an $i \in I$.

Let $\mathcal{U}$ be an ultrafilter on J finer than the filter sections of J. Since $(x^{j^*})_{j \in J} \subset X$, $\overline{x} = \lim_{j, \mathcal{U}} x^{j^*}$. Then for all $i \in I$, $\pi_i(\overline{x}) = \lim_{j, \mathcal{U}} (x^{j^*})_i$. For each $i \in I$, we know that

$A_i(x) \subset B_{ij}(x)$ for all $x \in X$ and $j \in J$. Since for each $j \in J$, $A_i(x^{j^*}) \cap P_i(x^{j^*}) = \emptyset$ for all $i \in I$. By condition (d), it implies that $A_i(\overline{x}) \cap P_i(\overline{x}) = \emptyset$ for all $i \in I$.

For each $j \in J$, x^{j^*} is an equilibrium point of the generalized game $\mathcal{G}_j$ and B_{ij} is regular. Therefore $\operatorname{cl} B_{ij}(x^{j^*}) = \overline{B_{ij}}(x^{j^*})$ and $(x^{j^*})_i \in \operatorname{cl}(B_{ij}(x^{j^*})) = \overline{B_{ij}}(x^{j^*})$. Since $\overline{B_{ij}}$ has a closed graph, $(\overline{x}, (\overline{x})_i) \in$ Graph $\overline{B_{ij}}$ for every $i \in I$. But for each $i \in I$, $(B_{ij})_{i \in I}$ is the upper approximating family of the correspondence A_i, by the property (A_{IV}) of the upper approximating family, for each $x \in X$. $\bigcap_{j \in J} B_{ij}(x)\overline{A_i}(x)$. We also know that A_i has a closed graph by (b), so that for each $i \in I$, $A_i(\overline{x}) \cap P_i(\overline{x}) = \emptyset$ and $\pi_i(\overline{x}) \in A_i(\overline{x})$. □

We note that Theorem 4.57 generalizes Theorem 5 of Tulcea (1988, p. 284) to the extent that the preference correspondence P_i is L_C-majorized instead of $\mathcal{E}_i$-majorized. Theorem 4.57 also generalizes the Theorem 4.54 of Shafer and Sonnenschein (1975, p. 374) and a corresponding result due to Tian.

The following example shows that the "for each $i \in I$, the set $E^i = \{x \in X : (A_i \cap P_i)(x) \neq \emptyset\}$ is open in X" in condition (d) of Theorem 4.57 is essential.

Example 4.1 Let $I = \{1\}$ and $X = [0, 1]$. Define $A, P : X \to 2^X$ by

$$A(x) = \begin{cases} [1/2, 1], & \text{if } x \in [0, 1/2), \\ [0, 1] & \text{if } x = 1/2, \\ [0, 1/2], & \text{if } x \in (1/2, 1], \end{cases}$$

and

$$P(x) = \begin{cases} \emptyset, & \text{if } x = 0, \\ [0, x), & \text{if } x \in (0, 1]. \end{cases}$$

Then A is upper semicontinuous with non-empty closed convex values and the fixed point set of A is a singleton $\{1/2\}$. The correspondence P has convex values with open lower sections since for each $y \in [0, 1]$, $P^{-1}(y) = (y, 1]$ is open in X. Therefore A, P, and X satisfy all conditions except that "$E = \{x \in [0, 1] : A(x) \cap P(x) \neq \emptyset\} = [1/2, 1]$" is closed in $[0, 1]$. But $A(1/2) \cap P(1/2) \neq \emptyset$, i.e., the generalized game $\mathcal{G} = ([0, 1], A, P)$ has no equilibrium point since $A(1/2) \cap P(1/2) \neq \emptyset$.

By Theorem 4.57, we can also obtain the following well-known Fan-Glicksberg fixed point theorem:

Corollary 4.57.1 *Let X be a convex compact subset of a locally convex topological vector space E and $A : X \to 2^X$ be upper semicontinuous with non-empty closed and convex values for each $x \in X$. Then A has a fixed point.*

Proof. By taking $I = \{1\}$, and $P_i(x) = \emptyset$ for each $x \in X$ in Theorem 4.57 the conclusion follows. □

Remark 4.25 So far, we have probed the existence theorems of equilibria for generalized games with compact and infinite dimensional strategy spaces, in infinite numbers of agents, and nontotal-nontransitive constraint and preference correspondences which may not have open graphs or open lower (upper) sections. Since it is well known that if a correspondence has an open graph, then it has open upper and lower sections and thus the correspondences with open lower sections are lower sections semicontinuous. However, a continuous correspondence does not hold open lower (or upper) sections properties in general (e.g., see Yannelis and Prabhakar (1983, p. 237)). The results of this section generalize many results in the existing literature by relaxing the openness of graphs or lower (upper) sections of constraint correspondences.

By Remark 4.25, Theorem 4.57 (also Corollary 4.57.1) not only shows that Theorem 6.1 of Yannelis and Prabhakar (1983, p. 242) can be extended to non-metrizable subsets without introducing additional assumption and hence the question raised by Yannelis and Prabhakar is answered in the affirmative. In fact, some of the assumptions of their question can be further weakened.

Since we also know that in the infinite settings, the set of feasible allocations generally is not compact in any topology of the commodity spaces, it is necessary to consider the existence of equilibria for generalized games in which the strategy spaces which may not be compact. This is done by strengthening the assumptions on the preference or constraint correspondences which enables one to remove altogether the compactness (or paracompactness) assumptions on the strategy spaces in the following sections.

The work on the rest of this chapter is mainly due to E. U. Tarafdar.

Noncompact Economy

Theorem 4.58 *Let $\mathcal{G} = (X_i, A_i, B_i, P_i : i \in I)$ be a generalized game and $X = \prod_{i\in I} X_i$ suppose the following conditions are satisfied:*

- $(a)'$ *for each $i \in I$, X_i is a nonempty closed convex of a locally convex topological vector space E_i;*
- $(b)'$ *for each $i \in I$, $A_i : X \to 2^{X_i}$ is lower semicontinuous such that for each $x \in X$, $A_i(x)$ is nonempty and $\operatorname{co} A_i(x) \subset B_i(x)$ and for each $i \in I$, B_i is ψ_i-condensing, where ψ_i is a measure of noncompactness in X_i;*
- $(c)'$ *for each $i \in I$, $A_i \cap P_i$ is L_c-majorized;*
- $(d)'$ *for each $i \in I$, the set $E^i = \{x \in X : (A_i \cap P_i)(x) \neq \emptyset\}$ is compactly open in X.*

Then $\mathcal{G}$ has an equilibrium point $\overline{x} \in X$, i.e. there exists a point $\overline{x} = \{\overline{x}_i\} \in X$ such that for each $i \in I$, $\overline{x}_i \in \overline{B}_i(x)$ and $A_i(\overline{x}) \cap B_i(\overline{x}) = \emptyset$.

Proof. Since B_i is nonempty valued for each $i \in I$, we have by Theorem 4.50 there exist a nonempty compact convex subset K of X and for each $i \in I$, a compact convex subset K of X such that $K = \prod_{i\in I} K_i$ and $B_i : K \to 2^{K_i}$ for each $i \in I$.

Clearly $A_i : k \to 2^{k_i}$ for each $i \in I$. As $\operatorname{co} A_i(x) \subset B_i(x)$, $x \in X$. Let us define $\hat{P}_i : k \to 2^{k_i}$ for each $i \in I$, by $\hat{P}_i(x) = P_i(x) \cap K_i$, $x \in k$.

We now consider the generalized game $\hat{\mathcal{G}} = (k_i, A_i, B_i, \hat{P}_i : i \in I\}$. The conditions (a) and (b) of Theorem 4.56 are immediate for $\hat{G}$. It is only routine to check that for each $i \in I$, $A_i \cap \hat{P}_i$ is L_C-majorized. Thus the condition (c) of Theorem 4.56 holds for $\hat{\mathcal{G}}$. Also for each $i \in I$, $\hat{E} = \{x \in k : (A_i \cap \hat{P}_i(x) \neq \emptyset\} = \{x \in X : (A_i \cap P_i)(x) \neq \emptyset\} \cap k$ is open in k. Thus by Theorem 4.56 there exists a point $\overline{x}$ such that $\overline{x}_i \in \overline{B}_i(x)$ and $A_i(\overline{x}) \cap \hat{P}_i(x) = 0$ for each i. We complete the proof by noting that $A_i(\overline{x}) \cap \hat{P}_i(x) = \emptyset$ implies that $A_i(\overline{x}) \cap P_i(\overline{x}) = \emptyset$. As $A_i(\overline{x}) \subset k_i$. □

Exactly in the same way by using the Corollary 4.56.1 we can prove the following theorem:

Theorem 4.59 *Let $\mathcal{G} = (X_i, A_i, B_i, P_i, i \in I)$ be a generalized game such that $X = \prod_{i\in I} X_i$ is a paracompact. Assume that the following conditions are satisfied:*

(a)′ *for each $i \in I$, X_i is a nonempty closed convex subset of a locally convex topological vector space E_i;*
(b)′ *for each $i \in I$ and for each $x \in X$, $A_i(x)$ is nonempty and $\operatorname{co} A_i(x) \subset B_i(x)$ and for each $i \in I$, B_i is ψ_i-condensing;*
(c)′ *for 4each $i \in I$, A_i has a compactly open graph in $X \times X_i$ (resp. is lower semicontinuous) and P_i is lower semicontinuous (resp. has a compactly open graph in $X \times X_i$);*
(d)′ *for each $i \in I$, $A_i \cap P_i$ is L_c-majorized.*

Then $\mathcal{G}$ has an equilibrium point $\overline{x} \in X$.

Theorem 4.60 *Let $\mathcal{G} = (X_i, A_i, P_i : i \in I\}$ be a generalized game and $X = \prod_{i\in I} X_i$. Assume that for each $i \in I$ the following conditions hold:*

(a)′ *X_i is a nonempty closed convex subset of a locally convex topological vector space E_i;*
(b)′ *$A_i : X \to 2^{X_i}$ is upper semicontinuous with nonempty closed convex values and ψ_i-condensing;*
(c)′ *$P_i : X \to 2^{X_i}$ is lower semicontinuous and L_C-majorized;*
(d)′ *for each $i \in I$, $E^i\{x \in X : (A_i \cap P_i)(x) \neq \emptyset\}$ is compactly open in X.*

Then there exists a point $\overline{x} = \{\overline{x}_i\} \in X$ such that for each $i \in I$, $\overline{x}_i(\overline{x})$ and $A_i(\overline{x}) \cap B_i(\overline{x}) = \emptyset$.

Proof. Since for each $i \in I$, A_i is nonempty valued and ψ_i-condensing, there exist by Theorem 4.50 a nonempty compact convex subset k of X and for each $i \in I$, a compact convex subset k_i of X_i such that $k = \prod_{i\in I} j_i$ and $A_i : k \to 2^{k_i}$. We can easily see that $A_i : k \to 2^{k_i}$ is upper semicontinuous with nonempty closed convex values by (b)$'$. Thus (a) and (b) of Theorem 4.57 are satisfied by $(k,\, k_i,\, A_i : i \in I)$. For each $i \in I$, we define as before $\hat{P}_i : k \to 2^{k_i}$ by $\hat{P}_i(x) = P_i(x) \cap k_i$, $x \in k$ and verify that (c) and (d) of Theorem 4.57 are satisfied by $\hat{\mathcal{G}} = \{k_i,\, A_i,\, P_i : i \in I\}$. Hence by Theorem 4.57 there exists $\overline{x} = \{\overline{x}_i\} \in k \subset X$ such that for each $i \in I$, $\overline{x}_i \in A_i(\overline{x})$ and $A_i(\overline{x}) \cap \hat{P}_i(\overline{x}) \neq \emptyset$ and hence $A_i(\overline{x}) \cap P_i(\overline{x}) = \emptyset$ as $\overline{A}_i(\overline{x}) \subset k_i$. □

4.9.1 *Equilibrium on Paracompact Spaces*

In the last three decades, the classical Arrow-Debreu's existence theorem of Walrasian equilibria (1984) has been generalized in many directions. In finite dimensional spaces, Gale and Mas-Colell (1978) proved the existence of a competitive equilibrium without the assumptions of total or transitive preference correspondences. Shafer and Sonnenschein (1975) obtain results in the same direction and they proved the Arrow-Debreu Lemma for abstract economies for the case where preference correspondences may not be total or transitive. For the infinite dimensional strategy spaces and finite or infinite many players, the existence results of equilibria for generalized games were given by Aubin and Ekeland (1984), Bewley (1972), Border (1985), Chang (1990), Ding-Tan (1993), Ding *et al.* (1992), Flam (1979), Florenzano (1983), Khan and Vohra (1979), Khan and Papageorigou (1987), Kim-Richter (1986), Kim *et al.* (1989), Lassonde and Schenkel (1992), Tarafdar and Mehta (1987), Tian (1992), Toussiant (1984, 1988), Yannelis-Prabhakar (1983), etc. All existence theorems mentioned above, however, are obtained by assuming that the constraint and preference correspondences have open graphs or have open lower (or upper) open sections. Besides, in most of these models, the strategy sets are assumed to be compact in topological vector spaces. These are restricted assumptions since it is well known that if a correspondence has an open graph, then it has open upper and lower sections and thus the correspondences with open lower sections are lower semicontinuous. However, a continuous correspondence does not hold open lower (or upper) sections properties in general. Moveover, we also know that in the infinite settings, the set of feasible allocations is generally not compact in any topology of commodity spaces. The motivations for economists interested in setting forth conditions for the existence of equilibria come from the importance of generalized games (also called abstract economy) in the study of markets and other general games and from the restrictions of the existing theorems.

In this subsection, by the approximate theorem for the upper semicontinuous correspondence of Tulcea (1988), we give the existence theorems of equilibria for non-compact generalized games in which constraint correspondences are upper semicontinuous instead of having lower (upper) open sections or open graph in locally convex topological vector spaces. Moreover in our framework, strategy spaces may be infinite-dimensional and non-compact; the number of players may be uncountable and preference correspondences may be non-total or non-transitive. Thus our results generalize many of the existence theorems of equilibria in generalized games by relaxing the compactness of strategy spaces and the continuity of constraint correspondences. In particular, we answer the question raised by Yannelis and Prabhakar (1983, p. 243) affirmatively with weaker assumptions. As applications, the Fran-Glicksberg fixed point theorem and an existence theorem for constrained game are derived. These results generalize the corresponding results due to Aubin and Ekeland (1874), Chang (1990), Shafer-Sonnenschein (1975), Toussaint (1984), Tulcea (1986, 1988), Yannelis-Prabhakar (1983) and others.

If X and Y are topological spaces, we recall that: (1) $F : X \to 2^Y$ is said to be lower semicontinuous (respectively, upper semicontinuous) if for each closed (respectively, open) subset C of Y, the set $\{x \in X : F(x) \subset C\}$ is closed (respectively, open) in X; (2) F is said to have compactly open lower (respectively, upper) sections if $F^{-1}(y) := \{x \in X : y \in F(x)\}$ is compactly open for each $y \in Y$ (respectively, $F(x)$ is compactly open in Y for each $x \in X$) and (3) F is said to be compact if for each $x \in X$, there exists a neighborhood V_x at x in X such that $F(V_x) = \bigcup_{z \in V_x} F(z)$ is relatively compact in Y. If X is a subset of a topological vector space E, X is said to have the property (K) if for every compact subset B of X, the convex hull of B is relatively compact in X.

Let X and Y be topological spaces. A correspondence $T : X \to 2^Y$ is said to be quasi-regular if:

(1) it has open lower sections, i.e. for each $y \in Y$, $T^{-1}(y)$ is open in X,
(2) $T(x)$ is non-empty and convex for each $x \in X$,
(3) $\overline{T}(x) = \text{cl}_Y\, T(x)$ for all $x \in X$.

The correspondence T is said to be regular if it is quasi-regular and has an open graph. Let I be a (possibly infinite) set of players. For each $i \in I$, let its choice or strategy set X_i be a non-empty subset of a topological vector space and $X = \prod_{i \in I} X_i$. A generalized game (or an abstract economy) is a family of quadruples $\Gamma = (X_i; A_i, B_i; P_i)_{i \in I}$ where I is a (finite or infinite) set of players (agents) such that for each $i \in I$, X_i is a non-empty subset of a topological vector space and $A_i, B_i : X = \prod_{j \in I} X_j \to 2^{X_i}$ are constraint correspondences and $P_i : X \to 2^{X_i}$ is a preference correspondence. An equilibrium of Γ is a point $\hat{x} \in X$ such that for each $i \in I$, $\hat{x}_i = \pi_i(\hat{x}) \in \overline{B}_i(\hat{x})$ and $A_i(\hat{x}) \cap P_i(\hat{x}) = \emptyset$ where $\pi_i : X_i \to X_i$ is the i-th projection of X onto X_i. We remark that when $\overline{B}_i(\hat{x}) = \text{cl}_{X_i} B_i(\hat{x})$ (which is the case when B_i has a closed graph in $X \times X_i$; in particular, when $\text{cl}\, B_i$ is upper semicontinuous with closed values), our definition of an equilibrium point coincides with that of Ding *et al.* (1992); and if in addition, $A_i = B_i$ for each $i \in I$, our definition of an equilibrium point coincides with standard definition; e.g. in Borglin-Keiding (1976), Tulcea (1986) and Yannelis-Prabhakar (1983).

Note that our generalized game model is more general than the one given by Borglin-Keiding (1976) (see also Tulcea (1986) and Yannelis-Prabhakar (1983)) in the sense that the constraint mapping has been split into two parts A and B due to the fact that the "small" constraint mapping may have not enough fixed points but a "big" constraint mapping B does so.

The following example illustrates that how our generalized game model does work but the former one given by Borglin-Keiding *et al.* does not.

Example 4.2 Let $X = [0, 1]$ and define the constraint mapping $A : X \to 2^X$ by

$$A(x) = \begin{cases} [0, 1-x), & \text{if } x \in (0, 1]; \\ \{1\}, & \text{if } x = 0. \end{cases}$$

Define the preference mapping $P : X \to 2^X \cup \{\emptyset\}$ by

$$P(x) = \begin{cases} (0, 1], & \text{if } [0, 1); \\ \emptyset, & \text{if } x = 1. \end{cases}$$

Then the fixed point set of A is $(0, 1/2)$. It is also clear that $x \notin P(x)$ and P has open lower sections (due to the fact that $P^{-1}(y) = \emptyset$ if $y = 1$ and $P^{-1}(y) = (y, 1]$ if $y \in [0, 1)$). Let $B : X \to 2^X$ be defined by $B(x) = A(x)$ for each $x \in X$. Note that the mapping $\overline{B} : X \to 2^X$ is such that $\overline{B}(x) = [0, 1-x]$ for each $x \in X$ and $1/2$ is also a fixed point of B so that $1/2$ is an equilibrium point of the generalized game $\Gamma = ([0, 1]; A, G; P)$ (in the sense of Yannelis-Prabhakar (1983)) since $1/2 \notin A(1/2)$ even though $A(1/2) \cap P(1/2) = \emptyset$.

We first need the following existence theorem of equilibria for generalized games:

Theorem 4.61 *Let* $\Gamma = (x_i, A_i, B_i, P_i)_{i \in I}$ *be a generalized game such that* $X = \prod_{i \in I} X_i$ *is paracompact. Suppose that the following conditions are satisfied:*

(a) *for each* $i \in I$, X_i *is a non-empty convex subset of a topological vector space,*
(b) *for each* $i \in I$ *and for each* $x \in X$ $A_i(x)$ *is non-empty and* $conv A_i(x) \subset B_i(x)$,
(c) *for each* $i \in I$ *and for each* $y \in X_i$, $A_i^{-1}(y)$ *and* $P_i^{(1)}(y)$ *are open in* X,
(d) *for each* $i \in I$ *and for each* $x \in X$, $X_i \notin conv P_i(x)$;
(e) *there exist a non-empty closed and compact subset* K *of* X *and* $x^0 = (x_i^0)_{i \in I} \in X$ *such that* $x_i^0 \in conv(A_i(y) \cap P_i(y))$ *for each* $i \in I$ *and for all* $y \in X \setminus K$.

Then Γ *has an equilibrium in* K.

Proof. Since $\{x \in X : (A_i \cap P_i)(x) \neq \emptyset\} = \bigcup_{y \in X_i} (A_i^{-1}(y) \cap P_i^{-1}(y))$ is open by (c), the conditions (c) and (e), all hypotheses of Theorem 4.3 of Tan and Yuan (1993) are satisfied. By Theorem in Tan and Yuan (1993), the conclusion follows. □

Theorem 4.61 generalizes Theorem 2.5 of Toussaint in (1984), Corollary 2 of Tulcea in (Tulcea (1986)) (also Corollary 2 in (Tulcea (1988))) and Theorem 6.1 in Yannelis and Prabhakar (1983) to non-compact case.

Let X be a non-empty set, Y a non-empty subset of topological vector space E and $F : X \to 2^Y$. A family $(f_i)_{j\in J}$ of correspondences between X and Y, indexed by a nonempty filtering set J (we denote by $\leq$ the order relation in J) is an upper approximating family for F (Tulcea (1988), p. 269) if

(A_I) $F(x) \subset f_j(x)$ for all $x \in X$ and all $j \in J$,

(A_{II}) for all $j \in J$ there is $j^* \in J$ such that for each $h \geq j^*$ and $\in J$, $f_h(x) \subset f_j(x)$ for each $x \in X$,

(A_{III}) for each $x \in X$ and $V \in \mathcal{B}$ is a base for the zero neighborhood in E, there is $j_{x,V} \in J$ such that $f_h(x) \subset F(x) + V$ if $h \in J$ and $j_{x,V} \geq h$.

From (A_I)–(A_{III}), it is easy to deduce that:

(A_{IV}) for each $x \in X$ and $k \in J$, $F(x) \subset \bigcap_{j\in J} f_j(x) = \bigcap_{k\leq j,\, k\in J} f_j(x) \subset \operatorname{cl} F(x) \subset \overline{F}(x)$.

By Theorem 3 and its Remark of Tulcea (1988, pp. 280–282), we have the following fact:

Lemma 4.15 *Let $(X_i)_{i\in I}$ be a family of paracompact space and let $(Y_i)_{i\in I}$ be a family of set such that for each $i \in I$, Y_i is a non-empty convex subset of a locally convex Hausdorff topological vector space E_i and Y_i has the property (K). For each $i \in I$, let $F_i : X_i \to 2^{Y_i}$ be such that F_i is compact and upper semicontinuous with compact convex values. Then there is a common filtering set J (independent of $i \in I$) such that for each $i \in I$, there is a family $(f_{ij})_{j\in J}$ of correspondences between X_i and Y_i with the following properties:*

(a) for each $j \in J$ is regular,

(b) $(f_{ij})_{j\in J}$ and $(\overline{f_{ij}})_{j\in J}$ are upper approximating families for F_i,

(c) for each $j \in J$, $\overline{f_{ij}}$ is continuous if Y_i is compact.

Remark 4.26 In the statement of above Lemma, Tulcea (1988, Theorem 3, p. 270)) assumed that Y_i is also closed. In his proof, the hypothesis "Y is closed" actually is not needed.

4.9.2 *Equilibria of Generalized Games*

In this section, by the approximation Lemma 4.15 and Theorem 4.61, we shall give the existence theorem of equilibria for the non-compact generalized game $\Gamma = (X_i;\, A_i,\, B_i;\, P_i)_{i\in I}$ in which the constraint correspondences are upper semi-continuous instead of having open lower or upper sections.

Theorem 4.62 *Let $\Gamma = (X_i, A_i, B_i, P_i)_{i\in I}$ be a generalized game such that $X = \prod_{i\in I} X_i$ is paracompact. Suppose the following conditions are satisfied for each $i \in I$:*

(a) X_i *is a non-empty convex subset of locally convex Hausdorff topological vector space* E_i *and* X_i *has the property* (K),

(b) $A_i,\, B_i : X \to 2^{X_i}$ *is such that* B_i *is compact and upper semicontinuous with nonempty compact convex values and* $A_i(x) \subset B_i(x)$ *for each* $x \in X$,

(c) $P_i : X \to 2^{X_i}$ *has compactly open lower sections and for each* $x \in X$, $x_i \notin conv\, P_i(x)$;

(d) *the set* $E^i = \{x \in X : (A_i \cap P_i)(x) \neq \emptyset\}$ *is open in* X,

(e) *there exist a non-empty compact subset* K *of* X *and* $x^0 \in X$ *for each* $y \in X \setminus K$ *with* $x_i^0 \in conv(A_i(y) \cap P_i(y))$.

Then there exists $\overline{x} \in K$ *such that for each* $i \in I$, $A_i(\overline{x}) \cap P_i(\overline{x}) = \emptyset$ *and* $\overline{x} \in \overline{B}_i(\overline{x})$.

Proof. By Lemma 4.15 there is a common filtering set J such that for every $i \in I$, there exists a family $(B_{ij})_{j\in J}$ of regular correspondences between X and X_i, such that both $(B_{ij})_{j\in J}$ and $(\overline{B_{ij}})_{j\in J}$ are the upper approximating of families for B_i.

Let $j \in J$ be arbitrarily fixed. The game $\Gamma_j = (X_{ij};\, B_{ij},\, \overline{B_{ij}};\, P_i)_{j\in J}$ satisfies all hypotheses of Theorem 4.60. Hence Γ_j has an equilibrium $\overline{x}^j \in K$ such that $B_{ij}(\overline{x}^j) \cap P_i(\overline{x}^j) = \emptyset$, and $\pi_i(\overline{x}^j \in \overline{B_{ij}}(x\overline{x}^j)$ for all $i \in I$.

Since $(\overline{x}^j)_{j\in J}$ is a net in the compact set K, without loss of generality we may assume that $(\overline{x}^j)_{j\in J}$ converges to $x^* \in K$. Then for each $i \in I$, $\pi_i(x^*) = \lim_{j\in J} \pi_i(\overline{x}^j)$. Noting that for every $j \in J$ and $x \in X$, $A_i(x) \subset B_i(x) \subset B_{ij}(x)$, we have $A_i(\overline{x}^j) \cap P_i(\overline{x}^j) = \emptyset$ for all $i \in I$. By condition (d), for every $i \in I$, $A_i(x^*) \cap P_i(x^*) = \emptyset$. As $\overline{B_{ij}}$ has closed graph, $(x^*,\, x_i^*) \in \text{Graph}\,\overline{B_{ij}}$ for every $i \in I$. For each $i \in I$, since $(\overline{B_{ij}})_{j\in J}$ is also an upper approximation family for B_i, $\bigcap_{j\in J} \overline{B_{ij}}(x) \subset \overline{B}_i(x)$ for each $x \in X$ so that $(x^*,\, x_i^*) \in \text{Graph}\,\overline{B_i}$. Therefore, for each $i \in I$, $A_i(x^*) \cap P_i(x^*) = \emptyset$ and $\pi_i(x^*) \in \overline{B_i}(x^*)$. □

We remark that Theorem 4.62 generalizes Theorem 2.5 of Toussaint (1984, p. 103), Theorem 4.1 of Chang (1990, p. 246), Theorem 6.1 of Yannelis-Prabhakar (1983) to non-compact generalized games and the constraint correspondences need not have open lower sections. In particular, Theorem 4.62 answers the question raised by Yannelis and Prabhakar (1983, 243) in the affirmative with weaker conditions. In Theorem 4.62, let $A_i = B_i$ for each $i \in I$ and we have the following:

Theorem 4.63 *Let* $\Gamma = (X_i,\, A_i,\, P_i)_{i\in J}$ *be a generalized game such that* $X = \prod_{i\in I} X_i$ *is paracompact. Suppose the following conditions are satisfied for each* $i \in I$:

(a) X_i *is a non-empty convex subset of a locally convex Hausdorff topological vector space* E_i *and* X_i *has the property* (K),

(b) $A_i : X \to 2^{X_i}$ *is compact and upper semicontinuous with non-empty compact and convex values for each* $x \in X$,

(c) $P_i : X \to 2^{X_i}$ *has compactly open lower sections and for each* $x \in X$, $x_i \notin conv\,P_i(x)$;

(d) *the set* $E^i = \{x \in X : (A_i \cap P_i)(x) \neq \emptyset\}$ *is open in* X,

(e) *there exist a non-empty compact subset* K *of* X *and* $x^0 \in X$ *such that for each* $y \in X \setminus K$ *with* $x_i^0 \in conv(A_i(y) \cap P_i(y))$.

Then there exists $\overline{x} \in K$ *such that for each* $i \in I$, $A_i(\overline{x}) \cap P_i(\overline{x}) = \emptyset$ *and* $\overline{x}_i \in A_i(\overline{x})$.

If X_i is compact and closed convex in Theorem 4.63, we have:

Corollary 4.63.1 *Let* $\Gamma = (X_i; A_i; P_i)_{i \in I}$ *be a generalized game and* $X := \prod_{i \in I} X_i$. *Suppose the following conditions are satisfied for each* $i \in I$:

(a) X_i *is a non-empty closed compact convex subset of locally convex Hausdorff topological vector space* E_i,
(b) $A_i : X \to 2^{X_i}$ *is upper semicontinuous with non-empty compact and convex values for each* $x \in X$,
(c) $P_i : X \to 2^{X_i}$ *has open lower sections and* $x_i \notin conv\, P_i(x)$ *for each* $x \in X$,
(d) *the set* $E^i = \{x \in X : (A_i \cap P_i)(x) \neq \emptyset\}$ *is open in* X.

Then there exists an $\overline{x} \in K$ *such that for each* $i \in I$, $A_i(\overline{x}) \cap P_i(\overline{x}) = \emptyset$ *and* $\overline{x}_i \in A_i(\overline{x})$.

Corollary 4.63.1 generalizes the Theorem of Shafer-Sonnenschein (1975, p. 374) in the following ways: (1) I is uncountable or countable infinite instead of finite; (2) for each $i \in I$, the locally convex Hausdorff topological vector space is infinite dimensional instead of being finite dimensional; (3) for each $i \in I$, A_i is upper semicontinuous instead of continuous and (4) P_i has open lower sections instead of an open graph.

As an application of Corollary 4.63.1, we obtain the well-known Fan-Glicksberg's fixed point theorem (see (1952) or (1952) for upper semicontinuous correspondence in locally convex topological vector spaces.

Corollary 4.63.2 *Let* X *be a convex compact subset of a locally convex topological vector space and let* $A : X \to 2^X$ *be upper semicontinuous with non-empty closed and convex values for each* $x \in X$. *Then* A *has a fixed point.*

Proof. Let $I = \{1\}$ and $P_i = \emptyset$ for each $x \in X$ in Corollary 4.63.1. The conclusion follows from Corollary 4.63.1. □

The following example shows that the condition (d) "for each $i \in I$, the set $E^i = \{x \in X : (A_i \cap P_i)(x) \neq \emptyset\}$ is open in X" in Theorem 4.63 is essential.

Example 4.3 Let $I = \{1\}$ and $X = [0, 1]$. Define $A, P : X \to 2^X$ by

$$A(x) = \begin{cases} [1/2, 1], & \text{if } x \in [0, 1/2), \\ [0, 1], & \text{if } x = 1/2, \\ [0, 1/2], & \text{if } x \in (1/2, 1]. \end{cases}$$

and

$$P(x) = \begin{cases} \emptyset, & \text{if } x = 0, \\ [0,\, x) & \text{if } x \in (0,\, 1]\,. \end{cases}$$

Then A is upper semicontinuous with non-empty closed convex values and the fixed point set of A is the singleton set $\{1/2\}$. The correspondence P has convex values with open lower sections, since for each $y \in [0,\, 1]$, $P^{-1}(y) = (y,\, 1]$ which is open in X. Therefore A, P and X satisfy all conditions of Theorem 4.62 except that $E = \{x \in [0,\, 1] : A(x) \cap P(x) \neq \emptyset\} = [1/2,\, 1]$ is closed but not open in $[0,\, 1]$. However, $A(1/2) \cap P(1/2) \neq \emptyset$, i.e. the generalized game $\Gamma = (0,\, 1]\,;\, A;\, P)$ has no equilibrium point.

4.9.3 *Applications*

In this section, as applications of Theorem 4.63, we shall give the existence theorem of equilibria for constrained games in locally convex spaces. For simplicity, we only consider the compact constraint generalized games.

Let $I = \{1,\, 2, \ldots, N\}$. Each player i chooses a strategy x_i in a subset X_i of a locally convex topological vector space E_i. Denote by X the (Cartesian) product $\prod_{j \in I,\, j \neq i} X_j$. Denote by x and x_{-i} an element of X and X_{-i} respectively. Each player i has a payoff (utility) function $u_i : X \to \mathbb{R} \cup \{-\infty,\, +\infty\}$. Given x_{-i} (the strategies of others), the choice of the i-th player is retricted to a non-empty compact and convex set $A_i(x_{-i}) \subset X_i$, the *feasible strategy set*; the i-th player chooses $x_i \in A_i(x_{-i})$ so as to minimize $u_i(x_{-i},\, x_i)$ over $A_i(x_{-i})$, where $(x_{-i},\, x_i)$ is the point $y = (y_j)_{j \in J}$ such that $y_i = x_{-i}$ and $y_i = x_i$. The family $\mathcal{G}(X_i;\, A_i;\, u_i)_{i=1}^N$ is then called a constrained N-person game and an equilibrium for $\mathcal{G}$ is an $x^* \in X$ such that $x_i^* \in A_i(x_{-i}^*)$ and $u_i(x^*) \leq u_i(x_{-i}^*,\, x_i)$ for all $x_i \in A_i(x_{-i}^*)$ (e.g. $u_i(x^*) = \inf_{x_i \in A_i(x_{-i}^*)} u_i(x_{-i}^*,\, x_i)$) for each $i = 1,\, 2, \ldots, N$.

Note that if $A_i(x_{-i}) = X_i$ for each $i = 1,\, 2, \ldots, N$, the constrained N-person game reduced to the conventional game $\mathcal{G} = (X_i;\, u_i)_{i \in I}$ and its equilibrium is called Nash equilibrium.

Theorem 4.64 *Let $\mathcal{G} = (X_i;\, A_i;\, U_i)_{i=1}^N$ be a constrained game and $X = \prod_{i=1}^N X_i$ is a non-compact convex subsets of a locally convex topological vector space E_i for each $i = 1,\, 2, \ldots, N$. Suppose the following conditions are satisfied:*

(a) *the correspondence $A : X \to 2^X$ defined by $A(x) = \prod_{i=1}^N A_i(x_{-i})$ for each $x = (x_{-i},\, x_i) \in X$ is upper semicontinuous with closed convex values,*

(b) *the function* $\psi : X \times X \to \mathbb{R} \cup \{-\infty, +\infty\}$ *defined by*

$$\psi(x, y) = \sum_{i=1}^{N} [u_i(x_{-i}, x_i) - u_i(x_{-i}, y_i)]$$

for each $(x, y) \in X \times X$ *is such that for each* $y \in X$, $x \mapsto \psi(x, y)$ *is lower semicontinuous on* X, *where* $x = (x_{-i}, x_i)$ *and* $y = (y_{-i}, y_i)$,

(c) *for each* $x \in X$, $x \notin conv(\{y \in X : \psi(x, y) > 0\})$,

(d) *the set* $\{x \in X : \sup_{y \in A(x)} \psi(x, y) > 0\}$ *is open in* X,

Then there exists $x^* \in X$ *such that for each* $i = 1, 2, \ldots, N$,

$$x_i^* \in A_i(x_{-i}^*) \quad \text{and} \quad u_i(x^*) \leq \inf_{x_i \in A(x_{-i}^*)} u_i(x_{-i}^*, x_i).$$

Proof. Define $P : X \to 2^X$ by $P(x) = \{y \in X : \psi(x, y) > 0\}$ for each $x \in X$. Note that (b) implies that P has open lower sections in X. By (c), $x \notin \text{conv} P_i(x)$ for each $x \in X$. The condition (d) implies that the set $\{x \in X : A(x) \cap P(x) \neq \emptyset\}$ is open in X. Therefore $\mathcal{G} = (X; A; P)$ satisfies all the hypotheses of Theorem 4.62 with $A = B$. By Theorem 4.62, there exists an $x^* \in X$ such that $x^* \in A(x^*)$ and $A(x^*) \cap P(x^*) = \emptyset$. Since

$$\{x \in X : A(x) \cap P(x) \neq \emptyset\} = \{x \in X : \sup_{y \in A(x)} \psi(x, y) > 0\},$$

it follows $x^* \in A(x^*)$ and $\sup_{y \in A(x^*)} \psi(x^*, y_i) \leq 0$.

For each $i = 1, 2, \ldots, N$, and $y_i \in A_i(x_{-i}^*)$, let $y = (x_{-i}^*, y_i)$. Then $y \in A(x^*)$ so that $(u_i(x^*) - u_i(x_{-i}^*, y_i)) = \sum_{i=1}^{N} [u_i(x^*) - u_i(x_{-i}^*, y_i)] = \psi(x^*, y) \leq \sup_{y \in (x^*)} \psi(x^*, y) \leq 0$.

Therefore $(u_i(x^*) - u_i(x_{-i}^*, y_i)) \leq 0$ for all $y_i \in A_i(x_{-i}^*)$. Hence x^* is an equilibrium point of the constrained game $\mathcal{G} = (X_i; A_i; u_i)_{i=1}^N$. □

Theorem 4.64 generalizes the corresponding result of Aubin and Ekeland (1984, pp. 350–351) in the sense that the feasible correspondence A_i is upper semicontinuous instead of being continuous.

4.10 Coincidence Points and Related Results, an Analysis on *H*-Spaces

Let X and Y be two nonempty sets. Let $T : X \to 2^Y$ and $S : Y \to 2^X$ be two set valued mappings. Then a point $(x_0, y_0) \in X \times Y$ is said to be a coincidence point if $y_0 \in T(x_0)$ and $x_0 \in S(y_0)$.

Note that the problem on the existence of a coincidence point is equivalent to that of fixed point of the composition. To see this, let $U = S_0 T$ be defined by

$$U(x) = ST(x) = U\{S(z) : z \in T(x)\}, \quad x \in X.$$

Then $x_0 \in U(x_0)$ implies that there exists a $y_0 \in T(x_0)$ such that $x_0 \in S(y_0)$ i.e. (x_0, y_0) is a coincidence point while $(x_0, y_0) \in X \times Y$ is a coincidence point implies that x_0 is a fixed point of $S_0 T$ as defined above.

The importance of coincidence phenomenon was displayed by Von Neumann (1937) in proving his well-known minimax inequality. Since then, many authors have contributed and enriched the field, e.g. Eilenberg and Montgomery (1946), Kakutani (1941), Nash (1950), Debreu (1952), Fan (1952), Kneser (1952), Sion (1958), Gorniewic and Granas (1981), Gale (1955), Nikaido (1967), Browder (1984), Ko and Tan (1986), Ding and Tarafdar (1993) and Tarafdar and Watson (1998) to name a few of many.

A set valued mapping $T : X \to 2^X$ is said to be H-KKM if, for each finite subset A of X, $H - \text{co}\, A \subset \bigcup_{x \in A} T(x)$. We should point out that in Bardaro and Ceppitelli (1988) $T : X \to 2^X$ is called H-KKM if, for each finite subset A of X, $F_A \subset \bigcup_{x \in A} T(x)$. Thus if T is H-KKM in our sense, then T is H-KKM in the sense of Bardaro and Ceppitelli. In here by H-KKM we mean by our sense.

We will need the following result of Ho (1987) in the sequel.

Lemma 4.10 *Let X be a topological space such that for every subset J of $\{0, 1, \ldots, n\}$ there is a nonempty contractible subset F_J of X having the property that $F_J \subset F_{J'}$ whenever $J \subset J'$. Then there is a continuous mapping $g : \Delta_n \longrightarrow X$ such that $g(\Delta_J) \subset F_J$ for each subset J of $\{0, 1, \ldots, n\}$, where Δ_n is the standard n-dimensional simplex with vertices $e_0, e_1, \ldots, e_n$, $\{e_0, e_1, \ldots, e_n\}$ is the canonical basis of $\mathbb{R}^{n+1}$ and for any subset J of $\{0, 1, \ldots, n\}$ $\Delta_J (\subset \Delta_n)$ is the convex hull of the vertices $\{e_j : j \in J\}$.*

Proof. We include the same proof as given by Horvath (1987, Th.1) (see also Lemma 2.2.5, Yuan (1999)). For each $i \in \{0, 1, \ldots, n\}$ we choose a point $x_i \in F_{\{i\}}$. Let us assume that for each subset J of cardinality $\leq k$ we have constructed a function $\sigma_J : \Delta_J \longrightarrow X$ such that $\sigma_J(\Delta_J) \subset F_J$ and $\sigma_J = \sigma_{J'}$ on $\Delta_J \cap \Delta_{J'}$ if $J \cap J' \neq \emptyset$. Now let $J = \{i_0, i_1, \ldots, i_k\}$ be a subset of $\{0, 1, \ldots, n\}$ with cardinality $k + 1$ and let $J_a = J \setminus \{i_a\}$. We have $J = J_0 \cup \cdots \cup J_k$ is the boundary of the simplex Δ_J. For each a we have by assumption a continuous mapping $\sigma_{J_a} : \Delta_{J_a} \longrightarrow F_{J_a}$. Since these mappings coincide on the intersection of the faces, we obtain a continuous mapping

$$\tilde{\sigma}_J : \Delta_{J_0} \cup \cdots \cup \Delta_{J_k} \longrightarrow F_{J_0} \cup \cdots \cup F_{J_k} \subset F_J.$$

Since F_J is contractible, $\tilde{\sigma}_J$ can be continuously extended to the simplex Δ_J. Thus for each subset J of $\{0, 1, \ldots, n\}$ with cardinality $k + 1$ we have constructed a mapping $\sigma_J : \Delta_J \longrightarrow F_J$ such that

$$\sigma_J/\Delta_J \cap \Delta_{J'} = \sigma_{J'}/\Delta_J \cap \Delta_{J'}$$

whenever $\Delta_J \cap \Delta_{J'} \neq \emptyset$.

Hence by a finite number of step we obtain a continuous mapping $g : \Delta_n \longrightarrow X$ with $g(\Delta_J) \subset F_J$. □

Definition 4.8 A nonempty subset of a topological space X is called acyclic if all of its reduced Cech homology groups over the rationals vanish. In particular, any contractible space is acyclic, and hence any convex or star-shaped set in a topological vector space is acyclic.

Readers are warned not to be confused with Definition 4.7 of an acyclic set valued mapping.

Shioji (1988, p. 188) gave a somewhat new proof of the following result of Eilenberg and Montgomery (1946) by using the Theorem 6.3 of Gorniewicz (1975). Since we are restricted by the length of the tools, we state the result as a lemma and leave it without proof.

Lemma 4.17 *Let Δ_n be an n-dimensional simplex with the Euclidean topology and W a compact topological space. Let $\psi : W \to \Delta_n$ be a single valued continuous mapping and $T : \Delta_n \to 2^W$ be a set valued mapping with nonempty compact acylic values. Then there exists a point $x_0 \in \Delta_n$ such that $x_0 \in \psi_0 T(x_0) = \psi(T(x_0))$.*

Theorem 4.65 *Let X be a contractible space and Y a compact Hausdorff topological vector space. Let $A : X \to 2^Y$ be upper semicontinuous with closed contractible values. Suppose that $B : Y \to 2^X$ is such that:*

(a) *$B^{-1}(x)$ contains an open set O_x (which may be empty) such that $\bigcup_{x\in X} O_x = Y$; and*

(b) *for each open set Q in Y, the set $\bigcap_{y\in Q} B(y)$ is empty or contractible.*

Then there exists $w_0 \in X$ and $z_0 \in Y$ such that $w_0 \in B(z_0)$ and $z_0 \in A(w_0)$.

Proof. We first show that there exist a n-simplex Δ_N and two functions $f : \Delta_N \to X$ and $\psi : Y \to \Delta_N$ such that $f(\psi(y)) \in B(y)$ for all $y \in Y$.

As Y is compact and $\bigcup_{x\in X} O_x = y$, there exists a finite subset $\{x_0, \ldots, x_n\}$ of X such that $\bigcup_{i=0}^{n} O_{x_i} = Y$. For each nonempty subset J of $N := \{0, \ldots, n\}$, we define

$$F_J = \begin{cases} \cap\{B(y) : y \in \cap_{j\in J} O_{x_j}, & \text{if } \bigcap_{h\in J} O_{x_j} \neq \emptyset, \\ X, & \text{otherwise.} \end{cases}$$

Note that $y \in \bigcap_{j\in J} O_{x_j}$ implies $\{x_j : j \in J\} \subset B(y)$ and so using (b), if $\bigcap_{j\in J} O_{x_j} \neq \emptyset$, then $F_J = \cap\{B(y) : y \in \bigcap_{j\in J} O_{x_j}\}$ is nonempty and contractible. Moreover, it is clear that $F_J \subset F_{J'}$, whenever $\emptyset \neq J \subset J' \subset \{0, \dots, n\}$. Thus, F satisfies all the hypotheses of Lemma 4.16, and so there is a continuous function $f : \Delta_N \to X$ such that $f(\Delta_N) \subset F_J$ for all $J \subset N$. Let $\psi_i : i \in N\}$ be a continuous partition of unity subordinate to the open covering $\{O_{x_i} : i \in N\}$; that is, for each $i \in N$, $\psi_i : Y \to [0, 1]$ is continuous, $\{y \in Y : \psi_i(y) \neq 0\} \subset O_{x_i}$ such that $\sum_{i=0}^{n} \psi_i(y) = 1$ for all $y \in Y$. Define $\psi : Y \to \Delta_N$ by $\psi(y) = (\psi_0(y), \dots, \psi_n(y))$ for each $y \in Y$. Then $\psi(y) \in \Delta_{J(y)}$ for all $\in\in Y$, where $Y(y) = \{i \in \{0, 1, \dots, n\} : \psi(y) \neq \emptyset\}$. Therefore, $f(\psi(y)) \in f(\Delta_{J(y)}) \subset F_{J(y)} \subset B(y)$.

It is clear the composition $A \circ f : \Delta_N \to 2^Y$ is upper semicontinuous with closed, contractible values and $\psi : Y \to \Delta_N$ is continuous. Lemma 4.17 assures the existence of an $x_0 \in \Delta_N$ such that $x_0 \in \psi(A(f(x_0)))$. Defining $w_0 = f(x_0)$, we have $w_0 = f(x_0) \in f(\psi(A(w_0)))$ and so we can choose a $z_0 \in A(w_0)$ such that $w_0 = f(\psi(z_0)) \in B(z_0)$. □

Remark 4.27 Theorem 4.65 is a generalization of Theorem 1 of Tarafdar and Yuan (1994).

We now prove the following theorem.

Theorem 4.66 *Let X be a complete convex subset of a locally convex Hausdorff topological vector space E, and Y be convex set in Hausdorff topological vector space H. Let $g : Y \to 2^X$ be upper semicontinuous with nonempty compact convex values and $f : X \to 2^Y$ have nonempty convex values such that:*

(i) for each $y \in Y$, $f^{-1}(y)$ contains an open set $O_y \subset X$ (which may be empty for some $y \in Y$);

(ii) $\bigcup_{y\in Y} O_y = X$;

(iii) there exists a compact convex set $Y_1 \subset Y$ and $Y_0 \subset Y_1$ such that $D = \bigcap_{y\in Y_0} O_y^c$ is compact or empty (here O_y^c denotes the complement of O_y in X).

Then there exists an $x_0 \in X$ such that $g^{-1}(x_0) \cap f(x_0) \neq \emptyset$.

Proof. First, suppose $D = \emptyset$. in this case, $f(x) \cap Y_1 \neq \emptyset$ for all $x \in \overline{\text{co}}\, g(Y_1)$. To see this, suppose there is an $x_0 \in \overline{\text{co}}\, g(y_1)$ such that $f(x_0) \cap Y_1 = \emptyset$. Then for each $y \in Y_1$, $y \notin f(x_0)$, i.e., $x_0 \notin f^{-1}(y)$ and as $O_y \subset f^{-1}(y)$, $x_0 \in O_y^c$ for all $y \in Y_1$. So $x_0 \in \bigcap_{y\in Y_1} O_y^c \subset \bigcap_{y\in Y_0} O_y^c = D$, which contradicts our initial assumption that $D = \emptyset$. Thus, we may define a nonempty set-valued mapping $h : \overline{\text{co}}\, g(Y_1) \to 2^{Y_1}$ by $h(x) = f(x) \cap Y_1$. Then $h(x)$ is convex and $h^{-1}(y) = \{x \in \overline{\text{co}}\, g(Y_1) : y \in h(x)\} = \{x \in \overline{\text{co}}\, g(y_1) : y \in f(x) \cap Y_1\} = f^{-1}(y) \cap \overline{\text{co}}\, g(Y_1) \supset O_y \cap \overline{\text{co}}\, g(Y_1) = O_y^1$ which is relatively open in $\overline{\text{co}}\, g(Y_1)$. Moreover, $\bigcup_{y\in Y_1} O_y^1 = \overline{\text{co}}\, g(Y_1)$ since $\bigcap_{y\in Y_0} O_y^c = \emptyset$. Thus by Lemma 4.16, there exists an $x_0 \in \overline{\mathrm{co}}\, g(Y_1)$ such that $h(x_0) \cap g^{-1}(x_0) \neq \emptyset$, that is, $f(x_0) \cap g^{-1}(x_0) \neq \emptyset$.

Next suppose $D \neq \emptyset$, and is compact. In this case, we argue by contradiction, so suppose there is $x_0 \in X$ such that $f(x_0) \cap g^{-1}(x_0) \neq \emptyset$. If $O_y^c = \emptyset$, for some $y \in Y$ as $f^{-1}(y)^c \subset O_y^c$, then $f^{-1}(y) = X$. So $y \in f(x)$ for all $x \in X$ and choosing $z \in g(y)$, we have $y \in F(z)$, that is, $y \in f(z) \cap g^{-1}(z)$, which is a contradiction. Thus, $O_y^c \neq \emptyset$ for all $y \in Y$. Moreover, the sets O_y^c satisfy the following: for any finite subset $\{y_1, \ldots, y_n\} \subset Y$, $g(\mathrm{co}\{Y_1, \ldots, y_n\}) \subset \bigcup_{i=1}^{n} O_{y_i}^c$. To prove this, suppose $z \in g(\mathrm{co}\{y_1, \ldots, y_n\})$ but $z \notin \bigcup_{i=1}^{n} O_{y_i}^c$. Then $z \in \bigcap_{i=1}^{n} O_y$, and so $y_i \in f(z)$ for all $i = 1, \ldots, n$. The set $f(z)$ is convex so $\mathrm{co}\{y_1, \ldots, y_n\} \subset f(z)$ and as $z \in g(\mathrm{co}\{y_1, \ldots, y_n\})$, we may choose $w \in \mathrm{co}\{y_1, \ldots, y_n\}$ such that $z \in g(w)$. Then $w \in f(z)$, and this contradicts the assumption that there is no coincidence point. Thus, $g(\mathrm{co}\{y_1, \ldots, y_n\}) \subset \bigcup_{i=1}^{n} O_y^c$ for any finite subset $\{y_1, \ldots, y_n\}$ of Y.

For our next move, let $\{y_1, \ldots, y_n\} \subset Y$ be arbitrary and define $K = \overline{\mathrm{co}}(\{y_1, \ldots, y_n\} \cup Y_1)$, which is compact and convex. We claim $\bigcap_{y \in K} O_y^c \neq \emptyset$. Indeed if this is not the case, the set-valued mapping $h : \overline{\mathrm{co}}\, g(K) \to 2^K$ defined by $h(x) = \{y \in K : x \notin O_y^c\}$ is nonempty for all $x \in \overline{\mathrm{co}}\, g(K)$. Also, $h^{-1}(y) = \{x \in \overline{\mathrm{co}}\, g(K) : x \notin O_y^c\} = \overline{\mathrm{co}}\, g(K) \cap O_y = \hat{O}_y$ and $\hat{O}_y$ is relatively open in $\overline{\mathrm{co}}\, g(K)$. Define $j(x) = \overline{\mathrm{co}}\, h(x)$. Then $\hat{O}_y \subset h^{-1}(y) \subset j^{-1}(y)$. As $\bigcap_{y \in K} O_y^c = \emptyset$, $\bigcup_{y \in K} O_y = X$, and so $\bigcup_{y \in K} \hat{O}_y = \overline{\mathrm{co}}\, g(K)$. By Theorem 4.65 there exists an $x_0 \in \overline{\mathrm{co}}\, g(K)$ such that $j(x_0) \cap g^{-1}(x_0) \neq \emptyset$. That is, there exists an $x_0 \in \overline{\mathrm{co}}\, g(K)$ and $w_0 \in K$, such that $w_0 \in j(x_0)$ and $x_0 \in g(w_0)$. So $w_0 = \sum_{i=1}^{m} \lambda_i w_i$, where $w_i \in h(x_0)$ and $\sum_i \lambda_i = 1$, $\lambda_i \geq 0$. From the definition of h, we have $x_0 \notin O_{w_i}^c$ for all $i = 1, \ldots, m$, that is, $x_0 \notin \bigcup_{i=1}^{m} O_{w_i}^c$. But $x_0 \in g(w_0) \subset g(\overline{\mathrm{co}}\{w_1, \ldots, w_m\}) \subset \bigcup_{i=1}^{m} O_{w_i}^c$, a contradiction which proves that $\bigcap_{y \in K} O_y^c \neq \emptyset$. So $[\bigcap_{y \in Y_1} O_y^c] \cap [\bigcap_{i=1}^{n} o_{y_i}^c] \neq \emptyset$, which implies $\bigcap_{i=1}^{n} (D \cap O_{y_i}^c) \neq \emptyset$. As D is compact and $O_{y_i}^c \cap D$ is closed with the finite intersection property, $\bigcap_{y \in Y} (O_y^c \cap D) \neq \emptyset$ and so $\bigcap_{y \in Y} O_y^c \neq \emptyset$, contradiction to (ii). Thus, the proof is complete. □

Remark 4.28 Due to the pathological fact that the closed convex hull of a compact set in an arbitrary topological vector space need not be compact, it is necessary that we assume X is locally convex and complete. (Therefore, our Theorem 4.66 is not a strict generalization of Theorem 4.65, although it extends Theorem 4.65 in the sense that a coincidence point is generally not a fixed-point for either of the mappings involved.)

Corollary 4.66.1 *Let X and Y be as in Theorem 4.66. Let $g : Y \to 2^X$ be upper semicontinuous with nonempty compact convex values and $f : X \to 2^Y$ have nonempty convex values such that:*

(i) *for each $y \in Y$, $f^{-1}(y)$ contains an open set $O_y \subset X$ (which may be empty for some $y \in Y$);*

(ii) $\bigcup_{y\in Y} O_y = X$;

(iii) *there exists a point $y_0 \in Y$ such that $O^c_{y_0}$ is compact or empty.*

Then there exists an $x_0 \in X$ such that $g^{-1}(x_0) \cap f(x_0) \neq \emptyset$.

Proof. Take $Y_0 = Y_1 = \{y_0\}$ in Theorem 4.66. □

In the following, we employ Theorem 4.66 to prove a minimax inequality generalising Theorem 1 of Ha (1987).

Theorem 4.67 *Let X and Y be as in Theorem 4.66. Let $f : Y \to 2^X$ be upper semicontinuous with compact convex values and $g : Y \times X \to \mathbb{R}$ such that:*

(i) *for all $y \in Y$, $x \mapsto g(y, x)$ is lower semicontinuous;*

(ii) *for all $x \in X$, $y \mapsto g(y, x)$ is quasi-concave; and*

(iii) *there exists a nonempty compact convex set $Y_1 \subset Y$ and $Y_0 \subset Y_1$ such that for all $z \in X \setminus f(Y_1)$ there exists a $w \in Y_0$ such that $g(w, z) > \sup_{\substack{y\in Y\\ u\in f(y)}} g(y, u)$.*

Then

$$\inf_{x\in X} \sup_{y\in Y} g(y, x) \leq \sup_{y\in Y//u\in f(y)} g(y, u).$$

Proof. We begin by noting that Condition (iii) makes no sense when $\sup_{\substack{y\in Y\\ u\in f(y)}} g(y, u) = \infty$, though then the conclusion is trivially satisfied. Without loss of generality, we may assume

$$r := \sup_{\substack{y\in Y\\ u\in f(y)}} g(y, u) < \infty.$$

Suppose, for a contradiction, that $\inf_{x\in X} \sup_{y\in Y} g(y, x) > r$. Then the set-valued mapping $h : X \to 2^Y$ defined by

$$h(x) = \{y \in Y : g(y, x) > r\}$$

is nonempty with convex values. Furthermore, $h^{-1}(y) = \{x \in X : g(y, x) > r\} = O_y$ is open as $g(y, \cdot)$ is lower semicontinuous. It is easily seen that $\bigcup_{y\in Y} O_y = X$. Finally, we show $\bigcap_{y\in Y} O^c_y$ is compact or empty. For $z \in X \setminus f(Y_1)$, employing (iii), we have the existence of a $w \in Y_0$ such that $g(w, z) > r$; that is there is a $w \in Y_0$ such that $z \notin O_w^c$. This implies $z \notin O_w^c$ and so $\bigcap_{y \in Y_0} O_y^c \subset f(Y_1)$. Theorem 4.66 implies the existence of a $y_0 \in Y$ such that $f(y_0) \cap h^{-1}(y_0) \neq \emptyset$. Let w_0 be a member of this intersection. Then $g(y_0, w_0) > r$ and $w_0 \in f(y_0)$, a contradiction to the definition of r. The proof is complete. □

To begin with we explain the notion of a H-space introduced by Horvath (1983, 1984, 1987) in some way and further developed by Bardaro and Cappitelli (1988) and (1989) and related concepts on H-spaces.

Definition 4.9 Let (X, τ) be a topological space and $\mathcal{F}(X)$ the family of all nonempty finite subsets of X. Let $\{F_A\}$ be a family of nonempty contractible subsets of X indexed by $A \in \mathcal{F}(X)$ such that $F_A \subset F_{A'}$ whenever $A \subset A'$. The pair $(X, \{F_A\})$ is called an H-space.

Examples

1. Let E be a topological vector space. For each nonempty finite subset $A = \{x_1, x_2, x_n\}$ of E, we set $F_A = \text{co}\{x_1, x_2, \ldots, x_n\}$. Then $(E, \{F_A\})$ is an H-space.
2. Let X be a contractible topological space. Then $(X, \{F_A\})$ is an H-space, where for each $A \in \mathcal{F}(X)$, $F_A = X$.
3. Let X be a nonempty convex subset of a topological vector space E, Y a topological space and $f : X \to Y$ a continuous bijection. For each $A \in \mathcal{F}(Y)$, $\text{co}[f^{-1}(A)]$ is a compact subset of X. Thus $f : \text{co}[f^{-1}(A)] \to f(\text{co}[f^{-1}(A)])$ is an homeomorphism. Now for each $A \in \mathcal{F}(Y)$, let $F_A = f(\text{co}[f^{-1}(A)])$. Clearly F_A is contractible (see 4 below). Then $(Y\{F_A\})$ is a H-space. Y could be a torus, Möbius band or Klien bottle and hence an H-space is not necessarily contractible.

The above examples were given in Horvath (1991) and more examples can be found there.

4. Let $(X, \{F_A\})$ be a H-space and Y an homeomorphic image of X. Then Y is also an H-space. Let $h : X \to Y$ be a homeomorphism. For any $A \in \mathcal{F}(Y)$, let $G_A = h_0 F_{h^{-1}(A)}$. Clearly $A, B \in \mathcal{F}(Y)$ and $A \subset B$ imply $G_A \subset G_B$. Cince $F_{h^{-1}(A)}$ is contractible to a point $y_0 \in F_{h^{-1}(A)}$, there exists a continuous mapping $F : F_{h^{-1}(A)} \times [0, 1] \longrightarrow F_{h^{-1}(A)}$ such that $F(y, 1) = y$ for all $y \in F_{h^{-1}(A)}$ and $F(y, 0) = y_0$ for all $y \in F_{h^{-1}(A)}$. Clearly the mapping $G : G_A \times [0, 1] \longrightarrow G_A$ defined by $G(x, t) = hF(h^{-1}(x), t)$, $x \in G_A$ is continuous and has the properties $G(x, 1) = h_0 h^{-1}(x) = x$ for all $x \in G_A$ and $H(x, 0) = h_0 G(x, 0) = h(y_0) = l_0$, say. Thus G_A is contractible. Hence (Y, G_A) is an H-space.

Given an H-space $(X, \{F_A\})$, a nonempty subset D of X is called

(i) H-convex if $F_A \subset D$ for each finite subset A of D;

(ii) weakly H-convex if $F_A \cap D$ is nonempty and contractible for each finite subset A of D and

(iii) compactly open (closed) if $D \cap B$ is open (closed) in B for each compact subset B of X. Also a subset K of X is called H-compact if, for every finite subset A of X, there exists a compact, weakly H-convex subset D of X such that $K \cup A \subset D$.

Throughout the rest of this chapter by a finite subset we will always mean nonempty finite subset.

Let $(X, \{F_A\})$ be an H-convex space. Then given a nonempty subset K of X, we define the H-convex hull of K, denoted by $H - \operatorname{co} K$ as

$$H - \operatorname{co} K = \cap\{D \subset X : D \text{ is } H\text{-convex and } D \supset K\}.$$

$H - \operatorname{co} K$ is H-convex. Indeed if A is a finite subset of $H - \operatorname{co} K$, then for every H-convex subset D of X with $D \supset K$, we have $H - \operatorname{co} K \subset D$ and thus $A \subset D$. Hence as D is H-convex, $F_A \subset D$ and hence $F_A \subset H - \operatorname{co} K$. It also follows that $H - \operatorname{co} K$ is the smallest H-convex subset containing K.

The following lemmas are proved in Tarafdar (1990):

Lemma 4.18 *Let $(X, \{F_A\})$ be a H-space and K be a nonempty subset of X. Then $H - \operatorname{co} K = \cup\{H - \operatorname{co} A : A$ is a finite subset of $K\}$.*

Proof. Let A be a finite usbset of K. Then $H - \operatorname{co} A$ is the smallest H-convex subset containing A and $H - \operatorname{co} K$ is the smallest H-convex subset containing K. Thus it follows that $H - \operatorname{co} A \subset H - \operatorname{co} K$. Hence $\cup\{H - \operatorname{co} A : A$ is a finite subset of $K\} \subset H - \operatorname{co} K$.

Next, let $\cup\{H - \operatorname{co} A : A$ is a finite subset of $K\} = L$. Then L contains K as a subset and we prove that L is H-convex.

Let $B = \{x_1, x_2, \ldots, x_n\}$ bc a finite subset of L. Then there are finite subsets $A_1, A_2, \ldots, A_n$ of K such that $x_i \in H - \operatorname{co} A_i$, $i = 1, 2, \ldots, n$. Obviously $A' = \bigcup_{i=1}^{n} A_i$ is a finite subset of K, and $x_i \in H - \operatorname{co} A'$ for $i = 1, 2, \ldots, n$. Therefore, as $H - \operatorname{co} A'$ is H-convex, $F_B \subset F_{A'} \subset H - \operatorname{co} A' \subset L$. Thus L is an H-convex subset containing K. Hence $H - \operatorname{co} K \subset \cup\{H - \operatorname{co} A : A$ is a finite subset of $K\}$. □

Now let D_α be an H-convex subset of X_α for each $i \in I$; then $D = \prod_{i \in I} D_\alpha$ is an H-convex subset of X. To see this let A be a finite subset of D. Then for each $i \in I$, $A_\alpha = P_\alpha(A)$ is a finite subset of D_α and $F_{A_\alpha} \subset D_\alpha$ as D_α is H-convex. Hence $F_A = \prod_{i \in I} F_{A_\alpha} \subset \prod_{i \in I} A_\alpha = D$.

Then we have proved the following:

Lemma 4.19 *The product of any number of H-spaces is an H-space and the product of H-convex subsets is H-convex.*

Proof. Let $\{(X_\alpha, \{F_{A_\alpha}\}) : i \in I\}$ be a family of H-spaces where I is a finite or an infinite index set for each nonempty subset A of $X = \prod\{X_\alpha : i \in I\}$, we set $F_A = \prod\{F_{A_\alpha} : i \in I\}$, where for each $i \in I$, $A_\alpha = P_\alpha(A)$ and $P_\alpha : X \to X_\alpha$ is the projection of X onto X_α. Since for each $i \in I$, F_{A_α} is contractible, it is easy to see that F_A is contractible. Indeed, if for each $i \in I$, F_{A_α} is contractible to $x_\alpha^0 \in X_\alpha$ through the homotopy $h_\alpha : F_{A_\alpha} \times [0, 1] \to F_{A_\alpha}$, that is h_α is continuous, $h_\alpha(x_\alpha, 1) = x_\alpha$ for each $x_\alpha \in F_{A_\alpha}$ and $h_\alpha(x_\alpha, 0) = x_\alpha^0$ for each $x_\alpha \in F_{A_\alpha}$. Then the mapping $h : F_A \times [0, 1] \to F_A$ defined by $h(x, t) = \prod_{i\in I} h_\alpha(x_\alpha, t)$ is clearly a homotopy map and F_A is contractible to $\prod_{i\in I} x_\alpha^0 \in X$, where $P_\alpha(x) = x_\alpha$. Moreover if A and B are two nonempty subsets of X with $A \subset B$, then for each $i \in I$, $P_\alpha(A) \subset P_\alpha(B)$, that is, $A_\alpha \subset B_\alpha$ and consequently $F_{A_\alpha} \subset F_{B_\alpha}$. Hence we have $F_A = \prod_{i\in I} F_{A_\alpha} \subset \prod_{i\in I} F_\beta = F_B$. Thus $(X, \{F_A\})$ is an H-space.

Next, let D_α be an H-convex subset of X_α for each $i \in I$, then $D = \prod_{i\in I} D_\alpha$ is an H-convex subset of X.. To see this, let A be a nonempty finite subset of D. Then for each $i \in I$, $A_\alpha = P_\alpha(A)$ is a finite subset of D_α and $F_{A_\alpha} \subset D_\alpha$ as D_α is H-convex. Hence $F_A = \prod_{i\in I} F_{A_\alpha} \subset \prod_{i\in I} D_\alpha$. □

We will also need the following definitions.

Definitions. A nonempty subset D of an H-space is called

(i) weakly H-convex if $F_A \cap D$ is nonempty and contractible for each finite subset A of D and
(ii) compactly open (closed) if $D \cap B$ is open (closed) in B for each compact subset B of X.

Also a subset K of X is called H-compact if, for every finite subset A of X, there exists a compact, weakly H-convex subset D of X such that $K \cup A \subset D$.

A set valued mapping $T : X \to 2^X$ is said to be T-KKM (see Park (1992)) if for each finite subset A of X, $H - \mathrm{co}\, A \subset \bigcup_{x\in A} T(x)$.

We would point out that (Bardaro and Ceppitelli (1988)), T is called H-KKM if for each finite subset A of X, $F_A \subset \bigcup_{x\in A} T(x)$. Thus if T is T-KKM then T is H-KKM.

The following lemma is due to Horvath (1991).

Lemma 4.20 *Let $(X, \{F_A\})$ be an H-space and $\{H_i\}_{i=0}^n$ be a family of closed nonempty subsets of X. If there exists a subset $\{x_0, x_1, \ldots, x_n\}$ of X such that for each nonempty subset J of $\{0, 1, \ldots, n\}$, $F_{\{x_j\}j\in J} \subset \bigcup_{j\in J} H_j$.*

Proof. By Lemma 4.16 there is continuous mapping $g : \Delta_n \to X$ such that $g(\Delta_J) \subset F_J$ for each subset J of $\{0, 1, \ldots, n\}$, where F_J denotes $F_{\{x_j\}j\in J}$.

For each $i = 0, 1, \ldots, n$, let $S_i = g^{-1}\left(F_{\{0,1,\ldots,n\}} \cap H_i\right)$, so that S_i is a closed subset of the simplex Δ_n.

For each nonempty subset J of $\{0, 1, \ldots, n\}$, we have $\bigcup\limits_{j\in J} S_j = g^{-1}\Bigg(F_{\{0,1,\ldots,n\}} \cap (\bigcup\limits_{j\in J} H_j)\Bigg) \supset g^{-1}(F_{\{0,1,\ldots,n\}} \cap F_J) = g^{-1}(G_J) \supset \Delta_J$. Hence $\text{co}\{e_j : j \in J\} \subset \bigcup\limits_{j\in J} S_j$. Thus by the classical Knaster-Kuratowski-Mazurkiewicz Theorem $\bigcap\limits_{i=0}^{n} S_i \neq \emptyset$. Let $u \in \bigcap\limits_{i=0}^{n} S_i$. Then $g(u) \in \bigcap\limits_{i=0}^{n} \{F_{\{0,1,\ldots,n\}} \cap H_i\}$. Hence $\bigcap\limits_{i=0}^{n} H_i \neq \emptyset$. □

Lemma 4.21 (Horvath (1989)) *Let $(X, \{F_A\})$ be a compact H-space and $G : X \to 2^X$ a set valued mapping with closed values and $F_A \subset \bigcup\limits_{x\in A} G(x)$ for then $\bigcap\limits_{x\in X} G(x) \neq \emptyset$.*

Proof. Since $G(x)$ is closed and X is compact, by Lemma $\bigcap\limits_{x\in A} G(x) \neq \emptyset$, for each finite subset A of X. Hence $\bigcap\limits_{x\in X} G(x) \neq \emptyset$. □

The following theorem is proved by Bardaro and Ceppitelli (1988).

Theorem 4.68 *Let $(X, \{F_A\})$ be an H-space and $T : X \to 2^X$ an H-KKM set valued mapping such that*

(a) *for $x \in X$, $T(x)$ is compactly closed;*
(b) *there is a compact subset L of X and an H-compact subset K of X such that for every weakly H-convex subset D with $K \subset D \subset X$, we have*

$$\bigcap_{x\in D} (T(x) \cap D) \subset L.$$

Then

$$\bigcap_{x\in X} T(x) \neq \emptyset.$$

Proof. It suffices to prove that $\bigcap\limits_{x\in X} \{T(x) \cap L\} \neq \emptyset$. Since by (a) $T(x) \cap L$ is closed in the compact set L, it suffices to prove that $\bigcap\limits_{x\in A} (T(x) \cap L) \neq \emptyset$ for each finite subset A of X. Let $A \subset X$ be a finite set and $X_0 \subset X$ be a compact weakly H-convex set such that $K \cup A \subset X_0$. By (b), $\bigcap\limits_{x\in X_0} (T(x) \cap X_0) \subset L$ and thus $\bigcap\limits_{x\in A} (T(x) \cap L) \supset \bigcap\limits_{x\in X_0} (T(x) \cap X_0)$. Therefore, it is sufficient to show that $\bigcap\limits_{x\in X_0} (T(x) \cap X_0) \neq \emptyset$. To this end we now consider the set valued $G : X_0 \to 2^{X_0}$ defined by $G(x) = T(x) \cap X_0$, $x \in X_0$. Since T is H-KKM, it easily follows that G is H-KKM in the H-space $(X_0, \{F_{A\cap X_0} \cap X_0\})$ by the Lemma 4.21 $\bigcap\limits_{x\in X_0} G(x) = \bigcap\limits_{x\in X_0} (T(x) \cap X_0) \neq \emptyset$. □

In what follows we prove that this theorem is equivalent to the following fixed point theorem:

Theorem 4.69 *Let $(X.\{F_A\})$ be an H-space and $f : X \to 2^X$ be a set-valued mapping such that*

(i) *for each $x \in X$, $f(x)$ is non-empty and H-convex;*

(ii) *for each $y \in X$, $f^{-1}(y) = \{x \in X : y \in f(x)\}$ contains a compactly open subset O_y of X (O_y could be empty for some y);*

(iii) *$\bigcup_{x\in X} O_x = X$; and*

(iv) *there exists a compact subset L of X and an H-compact subset K of X such that for every weakly H-convex subset D with $K \subset D \subset X$, we have*

$$\cap_{x\in D}(O_x^c \cap D) \subset L,$$

where O_x^c denotes the complement of O_x in X.

Then there is a point $x_0 \in X$ such that $x_0 \in F(x_0)$.

Proof. We first prove that Theorem 4.68 implies Theorem 4.69. Let the conditions of Theorem 4.69 hold. For each finite subset A of X, we set $T(x) = O_x^c$. If for each finite subset A of X, $H - \operatorname{co} A \subset \bigcup_{x\in A} T(x)$, then for each finite subset A of X, $F_A \subset \bigcup_{x\in A} T(x)$ as $H - \operatorname{co} A$ is an H-convex subset. Thus the set-valued mapping $T : X \to 2^X$ would satisfy all the conditions of Theorem 4.68 and hence $\bigcap_{x\in X} T(x) \neq \emptyset$ which would contradict the condition (iii). Hence there must exist at least one finite subset A of X such that $H - \operatorname{co} A \not\subset \bigcup_{x\in A} T(x)$, that is, there exists a point $y \in H - \operatorname{co} A$ such that $y \notin \bigcup_{x\in A} T(x)$, that is, $y \in [T(x)]^c$ for each $x \in A$, that is, $y \in O_x \subset f^{-1}(x)$ for each $x \in A$. Hence $x \in f(y)$ foreach $x \in A$, that is $A \subset f(y)$. But as $f(y)$ is H-convex, $H - \operatorname{co} A \subset f(y)$ which implies that $y \in F(y)$.

Next we prove that Theorem 4.69 implies Theorem 4.68. Assume that the conditions of Theorem 4.68 hold. If possible, suppose that $\bigcap_{x\in X} T(x) = \emptyset$. Then we can define a set-valued mapping $g : mX \to 2^X$ by $g(y) = \{x \in X : y \notin T(x)\}$. Clearly $g(y)$ is a nonempty subset of X for each $y \in Y$. Also for each $x \in X$, $g^{-1}(x) = (T(x))^c \setminus O_x$, say which is open subset of X. Let $f : X \to 2^X$ be the set-valued mapping defined by $f(y) = H - \operatorname{co} g(y)$ for each $y \in X$. Thus for each $y \in X$, $f(y)$ is an H-convex subset of X with $g(y) \subset f(y)$, and for each $x \in X$, $f^{-1}(x) \supset g^{-1}(x) = O_x$. Moreover, $\bigcap_{x\in X} T(x) = \emptyset$ implies $\bigcup_{x\in X} O_x = X$. Finally, $\bigcap_{x\in D} (O_x^c \cap D) = \bigcap_{x\in D} (T(x) \cap D) \subset L$. Hence the mapping f satisfies the conditions of the Theorem 4.69. Thus there exists a point $x_0 \in X$ such that $x_0 \in f(x_0) = H - \operatorname{co} g(x_0)$, that is, there is by Lemma 4.18 a finite subset $A = \{x_1, x_2, \ldots, x_n\}$ of $g(x_0)$ such that $x_0 \in H - \operatorname{co} A \subset f(x_0)$. But $x_i \in g(x_0)$, $i = 1, 2, \ldots, \Rightarrow x_0 \notin$

$T(x_i)$, $i = 1, 2, \ldots, n$, that is, $x_0 \notin \bigcup_{i=1}^{n} T(x_i)$, that is, $H - \operatorname{co} A \not\subset \bigcup_{x \in A} T(x)$ which contradicts that T is H-K.K.M. This proves our assertion. □

Our next theorem generalizes a theorem of Fan (Theorem 16, (1984)), Browder (1968) and the author (1989).

Theorem 4.70 *Let $X_1, X_2, \ldots, X_n$ be $n \geq 2$ H-spaces and let $X = \prod_{j=1}^{n} X_j$. Let $\{A_j\}_{j=1}^{n}$ and $B_j\}_{j=1}^{n}$ be two families of subsets of X having the following properties:*

(a) *Let $\widehat{X}_j = \prod_{i \neq j} X_i$ and let $\hat{x}_j$ denote a generic element of $\widehat{X}_j$. For each $j = 1, 2, \ldots, n$ and for each point $\hat{x}_j \in \widehat{X}_j$, the set $B_j(\hat{x}_j) = \{x_j \in X_j : [x_j, \hat{x}_j] \in B_j\}$ is nonempty and the set $A_j(\hat{x}_j) = \{x_j \in X_j : [x_j, \hat{x}_j] \in A_j\}$ contains the H-convex hull of $B_j(\hat{x}_j)$.*

(b) *For each $j = 1, 2, \ldots, n$ and for each point $x_j \in X_j$, the set $B_j(x_j) = \{\hat{x}_j \in \widehat{X}_j : [x_j, \hat{x}_j] \in B_j\}$ is compactly open in $\widehat{X}_j$.*

(c) *There exists an H-compact subset X_0 of X such that $\bigcap_{x \in X_0} O_x^c$ is compact where $O_x = \bigcap_{j=1}^{n} \{B_j(x_j) \times X_j\}$ and x_j is the projection of x into X_j for each $j = 1, 2, \ldots, n$.*

Then $\bigcap_{j=1}^{n} A_j \neq \emptyset$.

Proof. We define two set-valued mappings $f : X \to 2^X$ and $g : X \to 2^X$ by $f(x) = \prod_{j=1}^{n} H - \operatorname{co} B_j(\hat{x}_j)$ and and $g(x) = \prod_{j=1}^{n} B(\hat{x}_j)$ for each $x = [x_j, \hat{x}_j] \in X$ where x_j and $\hat{x}_j$ are respectively the projections of x into X_j and $\widehat{X}_j$. Clearly for $x \in X$, by Lemma 2 $f(x)$ is H-convex, and by (a) $g(x) \neq \emptyset$ and $f(x) \supset g(x)$. For each $y \in X$, we consider the set $g^{-1}(y) = \{x \in X : y \in G(x)\}$. Now $x \in G^{-1}(y) \Leftrightarrow y = (y_1, y_2, \ldots, y_n) \in g(x) = \prod_{j=1}^{n} B_j(\hat{x}_j) \Leftrightarrow y_j \in B_j(hx_j)$ for each $j = 1, 2, \ldots, n \Leftrightarrow \hat{x}_j \in B_j(y_j)$ for each $j = 1, 2, \ldots, n$. Thus for each $y \in X$, $g^{-1}(y) = \bigcap_{j=1}^{n} \{B_j(y_j) \times X_j\} = O_y$, which is compactly open. To show this it would suffice that $B_j(y_j) \times X_j$ is compactly open. Let K be a compact subset of X. Let $\hat{P}_j(K) = \widehat{K}_j$ and $P_j(K) = K_j$ where $\hat{P}_j$ and P_j are respectively the projections of X onto $\widehat{X}_j$ and X_j. Then $\widehat{K}_j$ and K_j are compact subsets of $\widehat{X}_j$ and X_j respectively and $(B_j(y_j) \times X_j) \cap (\widehat{K}_j \times K_j) = \left(B_j(y_j) \cap \widehat{K}_j\right) \times K_j$. This shows that $(B_j(y_j) \times X_j)$ is open in $\widehat{K}_j \times K_j$ by virtue of (b). Now since $\widehat{K}_j \times K_j \subset K$, it follows that $B_j(y_j) \times X_j$ is open in K. Now since $g(x) \subset f(x)$ for each $x \in x$, it follows that for each $y \in X$, $f^{-1}(y)$ contains a compactly open subset $g^{-1}(y) = O_y$.

Furthermore $\bigcup_{x \in X} O_y = X$. [For let $x \in X$. Since $g(x) \neq \emptyset$, $g(x)$ contains a point $y \in X$. Thus $x \in g^{-1}(y) = O_y$.] Finally by (e) there exists an H-compact subset X_0 of X such that $\bigcap_{x \in X_0} O_x^c = L$ is compact. Clearly with this pair (X_0, L) the condition (iv) of Theorem 4.69 is satisfied. Thus by Theorem 4.69 there exists a point $x \in X$ such that

$$x \in f(x) = \prod_{j=1}^{n} H - \operatorname{co} B(\hat{x}_j) \subset \prod_{j=1}^{n} A_j(\hat{x}_j)$$

by (a), that is, $x_j \in A_j(\hat{x}_j)$ for $j = 1, 2, \ldots, n$, that is $[x_j, \hat{x}_j] \in A_j$ for $j = 1, 2, \ldots, n$. Thus $x \in \bigcap_{j=1}^{n} A_j$. □

Remark 4.29 The theorem which is dual, in the sense of Tarafdar and Husain (1978), to the above theorem can similarly be stated and proved.

Bardaro and Ceppitelli (1988) proved some generalisations of Fan's minimax inequalities in Riesz space. We prove a variant of one of these (Bardaro and Ceppitelli (1988), Theorem 3) by means of our Theorem 4.69.

Let (E, C) be a Riesz space, where C is the positive cone, provided with a linear, order compatible topology (for example, see Fremlin (1974)) and $\overset{\circ}{C}$, the interior of C is assumed to be nonempty.

Theorem 4.71 *Let $(X, \{F_A\})$ be an H-space and $f, g : X \times X \to (E, C)$ two functions such that with a given $\lambda \in E$ the following conditions hold:*

(a) *$g(x, y) \leq f(x, y)$ for all $x, y \in X$;*
(b) *$f(x, x) \notin \lambda + \overset{\circ}{C}$ for all $x \in X$;*
(c) *for every $y \in X$, the set $\{x \in X : f(x, y) \in \lambda + \overset{\circ}{C}\}$ is H-convex;*
(d) *for every $x \in X$, the set $\{y \in X : g(x, y) \in \lambda + \overset{\circ}{C}\}$ is compactly open;*
(e) *there exists an H-compact subset X_0 of X such that $\{y \in X : g(x, y) \notin \lambda + \overset{\circ}{C}$, for each $x \in X_0\}$ is a compact subset of X.*

Then the set $S = \{y : g(x, y) \notin \lambda + \overset{\circ}{C}$ for all $x \in X\}$ is a nonempty compactly closed subset of X.

Proof: For each $x \in X$, let $F(x) = \{y \in X : f(x, y) \notin \lambda + \overset{\circ}{C}\}$ and $G(x) = \{y \in X : g(x, y) \notin \lambda + \overset{\circ}{C}\}$. Then by (d), for each $x \in X$, $G(x)$ is compactly closed. it is clear that $S = \cap_{x \in X} G(x)$ and S is compactly closed. So we need to show that $S \neq \emptyset$. If possible, let $S = \emptyset$. Then for each $y \in X$, the set $h(y) = \{x \in X : y \notin G(x)\} = \{x \in X : g(x, y) \in \lambda + \overset{\circ}{C}\}$ is non-empty. Hence for each $y \in X$, the set

$$k(y) = \{x \in X : f(x, y) \in \lambda + \overset{\circ}{C}\} \supset h(y) = \{x \in X : g(x, y) \in \lambda + \overset{\circ}{C}\}.$$

The last inclusion follows from the inclusion $G(x)^c \subset F(x)^c$ which in turn follows from (b). [To set this let $y \notin G(x)$, that is, $g(x, y) \in \lambda + \overset{\circ}{C}$. Then there is a neighborhood $V$of O in E such that $g(x, y) + V \subset \lambda + \overset{\circ}{C}$. Now $g(x, y) \leq f(x, y) \Rightarrow \lambda < g(x, y) + v \leq f(x, y) + v$ for each $v \in V$. Thus $f(x, y) + V \subset \lambda + \overset{\circ}{C}$, that is $y \notin F(x)$]. Now for each $x \in X$,

$$h^{-1}(x) = \{y \in X : x \in h(y)\} = \{y \in X : g(x, y) \in \lambda + \overset{\circ}{C}\} = O_x,$$

say, is compactly open by (d). Thus for the set-valued mapping $k : X \to 2^X$, $k(y)$ is nonempty and H-convex (by (c)) and for each $x \in X$, $k^{-1}(x)$ contains a compactly open subset $O_x = h^{-1}(x)$. [That $h^{-1}(x) \subset k^{-1}(x)$ follows from the fact that $h(x) \subset k(x)$]. Also $\bigcup_{x \in X} h^{-1}(x) = \bigcup_{x \in X} O_x = X$. [To see this let $y \in X$. Since $h(y) \neq \emptyset$, we can assume $x \in h(y)$. Then $y \in h^{-1}(x) = O_x$]. Finally

$$(e) \Rightarrow \bigcap_{x \in X_0} O_x^c = \bigcap_{x \in X_0} \{y \in X : g(x, y) \notin \lambda + \overset{\circ}{C}\} = L,$$

say, in compact. Thus the pair (L, X_0) satisfies the condition (iv) of Theorem 4.69 for the mapping k. Hence this mapping $k : X \to 2^X$ fulfils all the conditions of Theorem 4.69 and, therefore, there is a point $x_0 \in X$ such that $x_0 \in k(x_0)$, that is, $f(x_0, x_0) \in \lambda + \overset{\circ}{C}$ which contradicts (b). Thus we have proved the theorem. □

Remarks. In the same way we can deduce Theorem 4 and Corollary 1 of (Bardaro and Ceppitelli (1988)) from our Theorem 4.69. Theorem 4.71 here includes a theorem of Allen (1977) and also of Tarafdar (1986). The results of this section are from Tarafdar and Watson (1998) and Tarafdar (1999).

We shall return to some new coincidence theorems which generalize some recent results in the literature.

Theorem 4.72 *Let K be a nonempty compact subset of a topological space X and $(Y, \{F_Z\})$ and H-space. Let $G : X \to 2^Y$ and $T : Y \to ka(K)$ be set-valued mapping such that*

(i) T is u.s.c. on Y,
(ii) for each $x \in X$, $G(x)$ is H-convex and for each $y \in Y$, $G^{-1}(y)$ contains a compactly open subset O_y of X (O_y may be empty for some y) such that $K \subset \bigcup_{y \in Y} O_y$, where $ka(K)$ denotes the family of all nonempty compact subsets of K.

Then there exist $x_0 \in K$ and $y_0 \in Y$ such that $x_0 \in T(y_0)$ and $y_0 \in G(x_0)$.

Proof. Since O_y is compactly open for each $y \in Y$ and K is compact, by (ii), there exists a finite subset $\{y_0, y_1, \ldots, y_n\}$, let $F_j = F_{\{y_j\}_{j \in J}}$. Then clearly $F_J \subset F_{J'}$ whenever $J \subset J'$. By Lemma 4.16, there is a continuous mapping $g : \Delta_n \to Y$

such that $g(\Delta_J) \subset F_J$ for each nonempty subset J of $\{0, 1, \ldots, n\}$. By (i), $T : Y \to ka(K)$ is u.s.c. and hence the composition mapping $T \circ g : \Delta_n \to ka(K)$ is also u.s.c. Now let $\{f_0, f_1, \ldots, f_n\}$ be a partition of unity subordinate to the open covering $\{O_{y_i} \cap K\}_{i=0}^n$. Define a mapping $f : K \to \Delta_n$ by

$$f(x) = \sum_{i=0}^{n} f_i(x) e_i \quad \text{for each} \quad x \in K.$$

Clearly, f is continuous. By Lemma 4.17 there exists a point $x^* \in \Delta_n$ such that $x^* \in f(T \circ g(x^*))$ so that there exists a point $x_0 \in T \circ g(x^*) \subset K$ such that $x^* = f(x_0) = \sum_{i=0}^{n} f_i(x_0) e_i$. Let $J(x_0) = \{i \in \{0, 1, \ldots, n\} : f_i(x_0) \neq 0\}$, then

$$x^* = \sum_{i \in J(x_0)} f_i(x_0) e_i \in \Delta_{J(x_0)}$$

and for each $i \in J(x_0)$, $x_0 \in O_{y_i} \cap K \subset O_{y_i} \subset G^{-1}(y_i)$. It follows that $y_i \in G(x_0)$ for each $i \in J(x_0)$. Since $G(x_0)$ is H-convex, we have

$$g(x^*) \in g(\Delta_{J(x_0)}) \subset F_{J(x_0)} \subset G(x_0).$$

Let $y_0 = g(x^*)$. Then we have $x_0 \in T(y_0)$ and $y_0 \in G(x_0)$. □

Corollary 4.72.1 *Let K be a nonempty compact subset of a topological space X and $(Y, \{F_A\})$ an H-space. Let $G : X \to 2^Y$ be a set valued mapping such that for each $x \in X$, $G(x)$ is H-convex and for each $y \in Y$, $G^{-1}(y)$ contains a compactly open subset O_y of X (O_y may be empty for some y) such that $K \subset \bigcup_{y \in Y} O_y$. Then for any continuous mapping $t : Y \to K$, there exists a point $y_0 \in Y$ such that $y_0 \in G(t(y_0))$.*

Proof. Define a mapping $T : Y \to 2^K$ by

$$T(y) = \{t(y)\} \quad \text{for each} \quad y \in Y.$$

Then $T : Y \to ka(K)$ is u.s.c. By Theorem 4.72, there exist $x_0 \in K$ and $y_0 \in Y$ such that $x_0 \in T(y_0) = \{t(y_0)\}$ and $y_0 \in G(x_0)$. Hence we must have $y_0 \in G(t(y_0))$. □

Remark 4.30 Corollary 4.72.1 is a Corollary 2.2 of Tarafdar (1992).

Theorem 4.73 *Let X be a Hausdorff locally convex topological vector space, $(Y, \{F_A\})$ be a compact H-space, $T : Y \to 2^X$ be u.s.c. with closed values and $g : Y \to X$ be continuous such that*

(i) for each $y \in Y$, $T(y) \cap g(Y)$ is a non-empty acyclic space,
(ii) for each $x \in g(Y)$, $\lambda > 0$ and any continuous semi-norm p on X, the set $\{y \in Y : p(g(y) - x) < \lambda\}$ is H-convex.

Then there exists $y_0 \in Y$ such that $g(y_0) \in T(y_0)$.

Proof. Assume that the conclusion is not true. Then $g(y) \notin T(y)$ for all $y \in Y$. Then for each $y \in Y$, the origin O of X to the closed set $g(y) - T(y)$ and hence there exist $\delta_y > 0$ and a continuous seminorm o_p on X such that $p_y(g(y) - u) > 2\delta_y$ for all $u \in T(y)$. Now by the upper semicontinuity of T and continuity of g, there exists $N(y)$ of y in Y such that $p_y(g(z) - v) > \delta_y$ for all $z \in N(y)$ and $v \in T(z)$. Since Y is compact and $\{N(y) : y \in Y\}$ is an open cover of Y, there exists a finite subcover $\{N(y_i) : i = 1, 2, \ldots, n\}$ of Y. let $p = \max\{p_{y_i} : i = 1, 2, \ldots, n\}$ and $\lambda = \min\{\delta_{y_i} : i = 1, 2, \ldots, n\}$. Then p is a continuous seminorm on X such that

$$p(g(y) - x) > \lambda \text{ for all } y \in Y \text{ and } x \in T(y). \tag{4.9}$$

(The idea of construction of p and λ is borrowed from Ha (1985).)

Clearly $g(Y)$ is a compact subset of X. Define a mapping $T^* : Y \to 2^{g(Y)}$ by

$$T^*(y) = T(y) \cap g(Y).$$

By Theorem 3.1.8 of Aubin and Ekeland (1984) and the condition (i), $T^* : Y \to ka(g(Y))$ is u.s.c.. Define $G : g(Y) \to 2^Y$ by

$$G(x) = \{y \in Y : p(g(y) - x) < \lambda\}, \quad \forall x \in g(Y).$$

By the condition (ii), for each $x \in g(Y)$, $G(x)$ is H-convex. It follows from the continuity of p that for each $y \in Y$, $G^{-1}(y) = \{x \in g(Y) : p(g(y) - x) < \lambda\}$ is an open subset of $g(Y)$. For each $x \in g(Y)$, there is some $y_1 \in Y$ such that $x = g(y_1)$ and hence $y_1 \in G(x)$ and $x \in G^{-1}(y_1)$. Hence $g(Y) = \bigcup_{y \in Y} G^{-1}(y)$. By Theorem 4.72 there exist $x_0 \in g(Y)$ and $y_0 \in Y$ such that $x_0 \in T^*(y_0) = T(y_0) \cap g(Y)$ and $y_0 \in G(x_0)$. Hence, we have $p(g(y_0) - x_0) < \lambda$ and $x_0 \in T(y_0)$ which contradicts (4.9). Therefore there exists a point $y_0 \in Y$ such that $g(y_0) \in T(y_0)$. □

Theorem 4.74 *Let X be a Hausdorff locally convex topological vector space, $(Y, \{\Gamma_A\})$ be a compact H-space. $T : Y \to 2^X$ be u.s.c. with closed values and $g : Y \to X$ be continuous such that*

(i) *for each $y \in Y$, $g^{-1}(T(y))$ is a non-empty acyclic set.*
(ii) *for each closed convex subset C of X, $g^{-1}(C)$ is an H-convex subset of Y.*

Then there exists a point $y_0 \in Y$ such that $g(y_0) \in T(y_0)$.

Proof. Assume that the conclusion does not hold, then, by an argument similar to that in the proof of Theorem 4.73 there exist $\lambda > 0$ and a continuous semi-norm p on X such that

$$p(g(y) - x) > \lambda, \text{ for all } y \in Y \text{ and } x \in T(y).$$

Define mappings $T^* : G = Y \to 2^Y$ by

$$T^*(y) = g^{-1}(T(y)) \text{ for each } y \in Y,$$
$$G(y) = \{z \in Y : p(g(z) - g(y)) < \lambda\} \text{ for each } y \in Y.$$

Since g is continuous and T is u.s.c., it is easy to see that T^* has a closed graph. Note that Y is compact, so by Corollary 3.1.9 of Aubin and Ekeland (1984), T is u.s.c. with compact values. By (i), $T^* : Y \to ka(X)$ is u.s.c. For each $y \in Y$ and $A = \{y_1, \dots, y_n\} \subset G(y)$, let $u_i = g(y_i)$, $i = 1, \dots, n$, and hence $y_i \in g^{-1}u_i$, $i = 1, \dots, n$ and $A \subset g^{-1}[\text{co}(u_1, \dots, u_n)]$. By (ii), $g^{-1}[\text{co}(u_1, \dots, u_n)]$ is H-convex and so $\Gamma_A \subset g^{-1}[\text{co}(u, \dots, u_n 0]$. For any $z \in F_A$, there exist $\lambda_i \geq 0$, $i = 1, \dots, n$ with $\sum_{i=1}^{n} \lambda_i = 1$ such that $g(z) = \sum_{i=1}^{n} \lambda_i u_i$. Note that $y_i \in G(y)$ for $i = 1, \dots, n$. It follows that

$$
\begin{aligned}
p(g(z) - g(y)) &= p\left(\sum_{i=1}^{n} \lambda_i u_i - g(y)\right) = p\left(\sum_{i=1}^{n} \lambda_i (g(y_i) - g(y))\right) \\
&\leq \sum_{i=1}^{n} \lambda_i p(g(y_i) - g(y)) < \lambda
\end{aligned}
$$

and hence $F_A \subset G(y)$ and $G(y)$ is H-convex. By the continuity of p and g, for each $z \in Y$,

$$G^1(z) = \{y \in Y : p(g(z) - g(y)) < \lambda\}$$

is open in Y. For each $y \in Y$, we have $y \in G(y)$ and hence $Y = \bigcup_{y \in Y} G(y)$. By Theorem 4.72 there exist z_0, $y_0 \in Y$ such that $y_0 \in T^*(z_0) = g^{-1}(T(z_0))$ and $z_0 \in G(y_0)$. Hence we have $g(y_0) \in T(z_0)$ and $p(g(z_0) - g(y_0)) < \lambda$ which contradicts 4.10. Hence there exists a point $y_0 \in Y$ such that $g(y_0) \in T(y_0)$. □

Remark 4.31 Theorem 4.74 improves Theorem 2 of Ha (1985), Theorem 1 of Fan (1952) and Theorem 2 of Fan (1961).

Applications

Let X and Y be two topological spaces and $T : X \to 2^Y$ be a preference correspondence. We recall that point $x \in X$ is said to be a maximal element of the preference correspondence T uf $T(x) = \emptyset$. The existence theorems of maximal elements have become an important tool in proving the equilibrium existence of abstract economics or generalized games, see for example, Borglin and Keiding (1976), Yannelis and Prabhakar (1983), Ding and Tan (1992, 1993) and Tarafdar (1991).

Theorem 4.75 *Let K be a nonempty compact subset of a topological space X and $(Y, \{F_A\})$ an H-space. Let $G : X \to 2^Y$ and $T : Y \to 2^K$ be two correspondences. Suppose that*

(*i*) *T is u.s.c. such that for each $y \in Y$, $T(y)$ is either an empty set or a closed acyclic set,* (*ii*) *for each* $y \in Y$, *some* $Q^{-1}(y)$ *contains a compactly open subset* O_y *of* X (O_y *may be empty for some* y) *such that* $K \subset \bigcup_{y \in Y} O_y$ *where* $Q(x) = H - \text{co}(G(x))$, *the* H*-convex hull of* $G(x)$, *(see Tarafdar (1990)), for each* $x \in X$,

(*iii*) *for each* $(x, y) \in K \times Y$, $x \in T(y)$ *implies* $y \notin Q(x)$.

Then either T *has a mzximal element in* Y *or* Q *has a maximal element in* K.

Proof. Assume that both T and Q do not have maximal elements. Then, by (i), $T : Y \to ka(K)$ is u.s.c. By (ii), for each $x \in X$, $Q(x)$ is H-convex and for each $y \in Y$, $Q^{-1}(y)$ contains a compactly open subset O_y of X such that $K \subset \bigcup_{y \in Y} O_y$. By Theorem 4.72, there exist $x_0 \in K$ and $y_0 \in Y$ such that $x_0 \in T(y_0)$ and $y_0 \in Q(x_0)$ which contradicts the condition (iii). The conclusion must hold. □

Corollary 4.75.1 *Let* $(X, \{F_A\})$ *be a compact* H *and* $G : X \to 2^X$ *a preference correspondence such that*

(*i*) *for each* $x \in X$, $x \notin H - \text{co}(G(x))$

(*ii*) *for each* $y \in X$, $Q^{-1}(y)$ *contains a compactly open subset* O_y *of* X (O_y *may be empty for some* y) *such that* $X = \bigcup_{y \in Y} O_y$, *where* $Q(x) = H - \text{co}(G(x))$ *for each* $x \in X$.

Then G *has a maximal element in* X.

Proof. By letting $X = Y = K$ and $T(x) = \{x\}$ for each $x \in X$ in Theorem 4.75, the conclusion follows from Theorem 4.75.

Remark 4.32 The results beginning from Theorem 4.72 to Corollary 4.75.1 are taken from Ding and Tarafdar (1994).

4.11 Applications to Mathematical Economics: An Analogue of Debreu's Social Equilibrium Existence Theorem

In our consideration a social system might have a finite or an infinite agent and each agent has a range of actions from which he/she chooses one. However his/her choice is not totally free in the sense that this choice is to be taken from a subset determined by actions of other agents. When the action of every agent is known, the outcome of the social activity is known. The preferences of each agent yield his/her complete ordering of outcomes and each one chooses his/her best possible action from his/her restricted subset. Debreu (1959) considers a social system consisting of m number of agents. Each agent $i = 1, 2, \ldots, m$ operates on a contractible polyhedron (a homeomorphic image of a geometric polyhedron) in $\mathbb{R}^n$. Under this setting Debreu (1959) has presented an existence theorem with general conditions under which there is an equilibrium for such a social system, i.e. an outcome of the social system where the action of each agent belongs to his/her restricted subset and no agent has incentive to consider another action. This basic result has now become a classical result and has generated a deep interest in the work of mathematical economist worldwide in the field of equilibrium analysis. As Debreu (1952) has used a particular case of the fixed point theorem of Eilenbery and Montgomery (1946), or even the more general result of Begle (1950), a contractible polyhedron for each agent was appropriate for his choice set. Since we are going to use the fixed point theorems in locally H-convex spaces, the choice set X_α for each agent $\alpha \in I$ will appropriately assumed to be a locally H-convex compact space.

In this section the authors consider a social system in which there could be theoretically a finite or an infinite agents characterized by $i \in I$, I being an index set. Each agents $i \in I$ operates on a locally H-convex space $(X, \{F_{A_i}\})$ which contains a locally convex topological vector space as a special case and will be defined shortly. It will be proved that homeomorphic image of an H-space is an H-space and a H-convex subset (a concept which will be defined shortly) is a topologal, i.e. remain invariant under at topological mapping (homeomorphism). Thus the non-linear concepts of economics are well-suited in H-space. In sharp contrast to this, one will need linear homeomorphism in topological vector spaces.

The α-th agent chooses an action from X_α, his choice set which is a locally H-convex compact Hausdorff space. For each $\alpha \in I$, the function $f_\alpha : \prod_{\alpha\in I} \to \overline{\mathbb{R}} = [-\infty, \infty]$ denotes the payoff function. Given $x_{-\alpha} \in X_{-\alpha} = \prod_{\beta\neq\alpha}^{\beta\in I} X_\beta$, the choice of the α-th agent is restricted to a nonempty compact set $A_\alpha(x_{-\alpha}) \subset X_\alpha$, $A_\alpha : X_{-\alpha} \to X_\alpha$ is the constraint mapping. The α-th agent chooses x_α in $A_\alpha(x_{-\alpha})$ so as to maximize $f_\alpha(x_\alpha, x_{-\alpha})$, assumed to be continuous in x_α on $A_\alpha(x_{-\alpha})$. As before, $x^* \in X$ is an equilibrium point if for each $\alpha \in I$, $x^*_\alpha \in A_\alpha(x^*_{-\alpha})$ and $f_\alpha(x^*) = \sup\limits_{x_\alpha \in A_\alpha(x^*_{-\alpha})} f_\alpha(x_\alpha, x^*_\alpha)$.

Theorem 4.76 *For each $\alpha \in I$, let X_α be as above, $A_\alpha : X_{-\alpha} \to 2X_\alpha$ set valued mapping with closed graph G_α and nonempty compact values, and f_α a continuous function from G_α to the completed real line such that $\varphi_\alpha(x_{-\alpha}) = \sup\limits_{x_\alpha\in A_\alpha(x_{-\alpha})} f_\alpha(x_\alpha, x_{-\alpha})$ is continuous. Further assume that for each $\alpha \in I$, and $x_{-\alpha} \in X_{-\alpha}$, the set $M_{x_{-\alpha}} = \{x_\alpha \in A_\alpha(x_{-\alpha}) : f_\alpha(x_\alpha, x_{-\alpha}) = \varphi_\alpha(x_{-\alpha})\}$ is either*

(i) contractible (more generally acyclic);
(ii) H-convex in X_α.

Then there is an equilibrium point.

Proof. By theorem $X = \prod_{\alpha\in I} X_\alpha$ is a Hausdorff locally H-convex space. Now we define the set valued mapping $\varphi : X \to 2^X$ by

$$\varphi(x) = \prod_{\alpha\in I} M_{x_{-\alpha}}, \qquad x = \{x_\alpha\}_{\alpha\in I} \in X.$$

Under (i) $\varphi(x)$ is nonempty contractible (or more generally acyclic) for each $x \in X$ as $M_{x_{-\alpha}}$ is so for each $\alpha \in I$.

Under (ii) $\varphi(x)$ is nonempty H-convex for each $x \in X$, the product of H-convex sets being evidently H-convex.

Now just as in Debreu (1952), for each $\alpha \in I$, we define in $X_\alpha \times X_{-\alpha}$, the subset of X,

$$\begin{aligned} M_\alpha &= \{(x_\alpha, x_{-\alpha}) : x_\alpha \in M_{x_{-\alpha}}\} \\ &= \{(x_\alpha, x_{-\alpha}) \in G_\alpha : f_\alpha(x_\alpha, x_{-\alpha}) = \varphi_\alpha(x_{-\alpha})\} \end{aligned}$$

which is a closed set by virtue of the fact that G_α is closed and f_α and φ_α are continuous.

The graph G of φ is the subset of $X \times X$, and

$$\begin{aligned} G &= \{(x, x') : x' \in \phi(x)\} \\ &= \{(\{x_\alpha\}, \{x'_\alpha\}) : x'_\alpha \in M_{x_{-\alpha}} \text{ for all } \alpha \in I\} \\ &= \{(x, x') : (x'_\alpha, x_{-\alpha}) \in M_\alpha \text{ for all } \alpha \in I\}. \end{aligned}$$

Now since M_α is closed for each α, the subset $N_\alpha = \{(x,x') : (x'_\alpha, x_{-\alpha}) \in M_\alpha\}$ is closed for each $\alpha \in I$.

It then follows that $G = \bigcap_{aii} N_\alpha$ is closed.

Hence by Theorem under (i) and Theorem under (ii), there is a fixed point $\overline{x} \in X$ of ϕ, i.e. $\overline{x} \in \phi(\overline{x}) = \prod_{\alpha \in I} M_{\overline{x}_{-\alpha}}$, where $x = \{\overline{x}_\alpha\}$. Hence $\overline{x}_\alpha \in M_{\overline{x}_\alpha} = \{x_\alpha \in A_\alpha(\overline{x}_{-\alpha}) : f_\alpha(x_\alpha, \overline{x}_{-\alpha}) = \varphi_\alpha(\overline{x}_{-\alpha}\}$ for each $\alpha \in I$, i.e. $\overline{x}_\alpha \in A_\alpha(\overline{x}_{-\alpha})$ and

$$f_\alpha(\overline{x}) = f_\alpha(\overline{x}_\alpha, \overline{x}_{-\alpha}) = \phi_\alpha(\overline{x}_{-\alpha}) = \sup_{x_\alpha \in A_\alpha(\overline{x}_{-\alpha})} f_\alpha(x_\alpha, \overline{x}_{-\alpha})$$

for each $\alpha \in I$. Thus $\overline{x} \in X$ is an equilibrium point. We should point out that theorem can be proved in locally H-convex spaces. □

Chapter 5

Variational and Quasivariational Inequalities in Topological Vector Spaces and Generalized Games

5.1 Simultaneous Variational Inequalities

Since the appearance of Minty(1962, 1963), Hartman and Stampacchia (1966) and Browder (1965b), the theory of monotone (nonlinear) operators in general and the variational inequality in particular have generated a tremendous interest amongst mathematicians. This is because of the wide applicability of the variational inequalities in nonlinear elliptic boundary value problems, obstacle problems, complementarity problems, mathematical programming, mathematical economics, optimizations and in many other areas. Papers which concern differential equations and variational inequalities are too many to cite. We will only cite Browder (1965b), Hartman and Stampacchia (1966), Kinderlehrer and Stampacchia (1980), Browder (1970) and Pascali and Sburlan (1978). Readers will find further references cited in these. In Karamardian (1972) it was shown that the problem of complementarity can be reduced to that of variational inequality, while the relationship between the variational inequality and mathematical programming was shown in Mancino and Stampacchia (1972), between the variational inequality and convex functions by Rockafeller (1970) and Moreau (1966), and between the variational inequality and and the equilibrium point of Walrasian economy in Riesz spaces in Aliprantis and Brown (1993).

5.1.1 *Variational Inequalities for Single Valued Functions*

Throughout this subsection X will denote a nonempty convex subset of a real Hausdorff topological vector space E (preferably a locally convex Hausdorff topological vector space whenever we are considering a nonlinear mapping $T : X \to E^*$, for otherwise we might be in the trivial case $E^* = \{0\}$, E^* being the continuous dual of E. For a subset A of X, A^c will denote the complement of A in X, i.e. $A^c = X \setminus A$ and $\overline{A}$ or $\operatorname{cl} A$ the closure of A in E. We will use both notations for closure as our need demands. Our fixed point theorem will play an important role in this subsection. Here we deal with the materials of Husain and Tarafdar (1994), Husain and Tarafdar (1996), Tarafdar (1990a) and Tarafdar (1977). At the outset we should mention that to prove the main theorem on variational inequalities for single valued function we will need our fixed point theorem of set-valued mapping.

Definition 5.1 A pair of real valued functions f and g defined on $X \times X$ is said to be monotone or a monotonic pair if

$$f(x, y) + g(y, x) \geq 0$$

for all $x, y \in X$, and strictly monotone if in addition the equality

$$f(x, y) + g(y, x) = 0$$

implies $x = y$.

A function $f : X \times X \to \mathbb{R}$ is said to be hemicontinuous if if the function

$$k(t) = f(x + t(y - x), y)$$

of the real variable t is lower semicontinuous on X as $t \downarrow 0$ for arbitrary given vectors x and y of X.

Remark 5.1 If we take $f = g$ in the above definition, we obtain a monotone function as defined in Tarafdar (1990a) and Mosco (1976).

Definition 5.2 A single valued (possibly nonlinear) mapping $A : X \to E^*$ is said to be monotone if

$$\langle Ax - Ay, x - y \rangle \geq 0$$

for all $x, y \in X$ and strictly monotone if

$$\langle Ax - Ay, x - y \rangle = 0$$

implies $x = y$. Here $\langle \cdot, \cdot \rangle$ denotes the pairing between E^* and E. The mapping $A : X \to E^*$ is called hemicontinuous if A is continuous from the line segment of X to the weak topology of E^*.

Remark 5.2 To see the connection between the Definitions 5.1 with $f = g$ and 5.2, let $A : \to E^*$ be a monotone mapping according to Definition 5.2. Then the function $f : X \times X \to \mathbb{R}$ is defined by

$$f(x, y) = \langle A(x), x - y \rangle \qquad \text{for } x, y \in X \tag{5.1}$$

is a monotone function according to Definition 5.1 with $f = g$. It is also easy to see that f is hemicontinuous according to Definition 5.1 if A is hemicontinuous according to Definition 5.2.

We give some examples:

Example 5.1 Let X be a nonempty subset of a real Hilbert space $(H, \langle \cdot, \rangle)$, then the functions $f, g : X \to \mathbb{R}$ defined by

$$f(x, y) = \alpha \|x - y\|^2 - \langle x, y \rangle$$

and

$$g(x, y) = \langle x, y \rangle,$$

$x, y \in X$, (α is a positive real number) are a monotonic pair.

Example 5.2 If X is a nonempty subset of a normed linear space $(E, \|\cdot\|)$ and $f : X \to \mathbb{R}$ is a function, then for each positive number α, the function

$$f_\alpha(x, y) = f(x) - f(y) + \alpha \|x - y\|,$$

$x, y \in X$ is a strictly monotone function.

Example 5.3 If X is as in Example 5.2 and $A : X \to E^*$ is monotone (strictly monotone) mapping, then the function $f : X \times X \to \mathbb{R}$ defined by

$$f(x, y) = \langle Ax, x - y \rangle,$$

$x, y \in X$ is a monotone (strictly) function.

Example 5.4 Let X be a nonempty convex subset of $\mathbb{R}^n$ and $f : X \to \mathbb{R}$ is a continuously differentiable convex (strictly convex) function, then the $F : X \times X \to \mathbb{R}$ defined

$$F(x, y) = \langle \Delta f(x), x - y \rangle,$$

$x, y \in X$ is monotone (strictly monotone) (see Pascali and Sburlan (1978, p. 16)).

Let $\varphi : E \to \mathbb{R}$ be a differentiable function defined on a Banach space E and let φ'_x denote the (Gateaux) derivative at $x \in E$, i.e.

$$\varphi'_x(y) = \lim_{t \to 0} \frac{\varphi(x + ty) - \varphi(x)}{t}.$$

Thus φ'_x is a bounded linear functional on E for each $x \in E$ and the mapping $x \to \varphi'_x$ is a mapping of E into E^*. We now state the following Proposition first explicitly given and stated by Russian mathematicians (e.g. see Browder (1970)).

Proposition 5.1 *Let φ'_x be as above, then the function $f : E \times E \to \mathbb{R}$ defined by*

$$f(x, y) = \langle \varphi'_x, x - y \rangle$$

is monotone if and only if φ is convex.

Proof. From what we said in Example 5.3 it will suffice to prove that the operator $A = \varphi'$ is monotone if and only if φ is convex. We should point out that our proof from this point onward is almost the same as that of Browder (1970, p.4). First let φ be convex. Then for each α with $0 \leq \alpha \leq 1$, we have

$$\varphi[y + \alpha(x - y)] - \varphi(y) \leq \alpha[\varphi(x) - \varphi(y)] \qquad \text{for } x, y \in E \tag{5.2}$$

Dividing by α and letting $\alpha \to 0^+$ we have

$$\langle {\varphi'}_y, x - y \rangle \leq \varphi(x) - \varphi(y). \tag{5.3}$$

Interchanging x and y we have similarly

$$\langle {\varphi'}_x, y - x \rangle \leq \varphi(y) - \varphi(x). \tag{5.4}$$

Adding (5.3) and (5.4) we obtain

$$\langle {\varphi'}_x - {\varphi'}_y, x - y \rangle \geq 0.$$

Next, we assume that $A = \varphi'$ is monotone. For given $x, y \in X$, we construct the function

$$p(\lambda) = \varphi[\lambda x + (1-\lambda)y] - \lambda\varphi(x) - (1-\lambda)\varphi(y) \qquad \text{for } 0 \leq \lambda \leq 1. \tag{5.5}$$

It suffices to show that $p(\lambda) \leq 0$ for all $0 \leq \lambda \leq 1$. If possible, suppose this is not true. Then as p is differentiable with $p(0) = p(1) = 0$, it follows that $p(\lambda)$ must have maximum at a point $\eta \in (0,1)$ with $p'(\eta) = 0$. Let λ be a point in [0,1] with $\eta \leq \lambda$. Then using the monotonicity of A,

$$\begin{aligned} p'(\lambda) - p'(\eta) &= \langle A(\lambda x + (1-\lambda)y) - A(\eta x + (1-\eta)y), x - y \rangle \\ &= (\lambda - \eta)^{-1} \langle A(\lambda x + (1-\lambda)y) - A(\eta x + (1-\eta)y), [\lambda x + (1-\lambda)y] \\ &\quad -[\eta x + (1-\eta)y] \rangle \\ &\geq 0. \end{aligned}$$

Thus p does not decease for $\lambda \geq \eta$. Hence it follows that $p(\eta) \leq p(1) = 0$, and $p(\lambda) \leq 0$ for all λ for $0 \leq \lambda \leq 1$. This is a contradiction. The proof is complete. □

5.1.2 *Solutions of Simultaneous Nonlinear Variational Inequalities*

Let X and E be as in the previous subsection.

We consider the following problem. We assume that

$$\varphi : X \to (-\infty, \infty] \text{ and } \varphi \not\equiv \infty, \tag{5.6}$$

φ is convex and lower semicontinuous on E,

$$f, g : X \times X \to \mathbb{R} \qquad \text{with } f(x,x) = g(x,x) = 0 \text{ for each } x \in X \tag{5.7}$$

such that

(*i*) for each $x \in X$, $f(x,.)$ and $g(x,.)$ are both concave and upper semicontinuous on E;

(*ii*) the pair (f, g) is monotone and both f, g are hemicontinuous.

We are interested to find the set of points $x_0 \in X$ which simultaneously satisfy the inequalities

$$\varphi(x_0) + f(x_0, y) \leq \varphi(y) \qquad \text{for all } y \in X \tag{5.8}$$

and

$$\varphi(x_0) + g(x_0, y) \leq \varphi(y) \qquad \text{for all } y \in X. \tag{5.9}$$

i.e. we want to find the set of common solutions of the variational inequalities (5.8) and (5.9). Such a solution x_0 is called a solution of simultaneous variational inequalities.

In what follows the following lemma will play an important role.

Lemma 5.1 *Assume that φ is defined as in* (5.6) *and $f, g : X \times X \to \mathbb{R}$ are a monotonic pair of functions. Then if x_0 is a solution of the inequality* (5.8), *x_0 is a solution of the inequality*

$$\varphi(x_0) \leq \varphi(y) + g(y, x_0) \qquad \text{for all } y \in X \tag{5.10}$$

and if x_0 is a solution of the inequality (5.9), *then x_0 is a solution of the inequality*

$$\varphi(x_0) \leq \varphi(y) + f(y, x_0) \qquad \text{for all } y \in X. \tag{5.11}$$

If f and g are as defined in (5.7) *satisfying* (5.7) (*i*) *and are hemicontinuous, and φ is defined as in* (5.6), *then if x_0 is a solution of the inequality* (5.10), *x_0 is a solution of the inequality* (5.9) *and if x_0 is a solution of the inequality* (5.11), *x_0 is a solution of the inequality* (5.8).

Proof. Let x_0 be a solution of (5.8), i.e. $\varphi(x_0) + f(x_0, y) \leq \varphi(y)$ for all $y \in X$. Adding $g(y, x_0)$ to both sides and using monotonicity of the pair (f, g), we get $\varphi(x_0) \leq \varphi(y) + g(y, x_0)$ for all $y \in X$,i,e. x_0 is a solution of (5.10). Similarly if x_0 is a solution of (5.9), then adding $f(y, x_0)$ to both sides and using monotonicity of the pair (f, g), we obtain that x_0 is a solution of (5.11).

Next let x_0 be a solution of (5.11), i.e. $\varphi(x_0) \leq \varphi(y) + f(y, x_0)$ for all $y \in X$. If possible, let x_0 be not a solution of (5.8). Then there must exist point $y \in X$ with $\varphi(y) < \infty$ such that

$$\varphi(x_0) + f(x_0, y) > \varphi(y) \tag{5.12}$$

Now since X is convex, $y_t = (1 - t)x_0 + ty \in X$ for all $t \in [0, 1]$. By our assumption made in (5.6) and by virtue of hemicontinuity of f, it follows that the function $\varphi(y_t) + f(y_t, y)$ of real variable t is lower semicontinuous as $t \downarrow 0^+$. Hence by (5.12), there must a real $\bar{t}$ such that

$$\varphi(y_t) + f(y_t, y) > \varphi(y) \qquad \text{for all } t \in (0, \bar{t}]. \tag{5.13}$$

Moreover as x_0 is solution of (5.11), we have

$$\varphi(y_t) + f(y_t, x_0) \geq \varphi(x_0) \qquad \text{for all } t \in [0,1]. \tag{5.14}$$

Now multiplying (5.13) by t and (5.14) by $(1-t)$ and adding, and using convexity of φ and concavity of $f(y_t, .)$, we obtain $\varphi(y_t) + f(y_t, y_t) > \varphi(y_t)$ which will imply that $\varphi(y_t) < \infty$ and $f(y_t, y_t) > 0$ which in turn contradict (5.7). Hence x_0 must be a solution of (5.8). Similarly we can prove that if x_0 is a solution of (5.10), then x_0 is a solution of (5.9). □

Theorem 5.1 *If φ is as defined in* (5.6) *and f, g are as defined in* (5.7) *satisfying* (5.7) *(i) and (ii), then x_0 is a common solution of the variational inequalities* (5.8) *and* (5.9) *if and only if x_0 is either a solution of* (5.8) *or a solution of* (5.9).

Proof. The theorem follows from Lemma 5.1. Indeed, if x_0 is a solution of (5.8), by the first part of the lemma x_0 is a solution of (5.10) and by the second part of the lemma is a solution of (5.9). Similarly if x_0 is a solution of (5.9), then it is a solution of (5.8). □

Theorem 5.2 *Let $\varphi : X \to (-\infty, \infty]$ and $f, g : X \times X \to (-\infty, \infty)$ be as in* (5.6) *and* (5.7) *respectively and (f, g) satisfy* (5.7) *(i) and (ii) and X be closed. Then for each $x \in X$, the sets $F(x) = \text{cl}\{y \in X : \varphi(y) + f(y,x) \leq \varphi(x)\}$ and $G(x) = \text{cl}\{y \in X : \varphi(y) + g(y,x) \leq \varphi(x)\}$ are subsets of X. Furthermore assume that there is a nonempty subset X_0 contained in a compact convex subset X_1 of X such that one of the following conditions holds:*

(a) *the set $L = \bigcap_{x \in X_0} F(x)$ is compact;*
(b) *the set $M = \bigcap_{x \in X_0} G(x)$ is compact;*
(c) *the set $P = \bigcap_{x \in X_0} B(x)$ is compact, where $B(x) = \{y \in X : \varphi(y) \leq \varphi(x) + f(x,y)\}$;*
(d) *the set $Q = \bigcap_{x \in X_0} C(x)$ is compact, where $C(x) = \{y \in X : \varphi(y) \leq \varphi(x) + g(x,y)\}$.*

Then the set S of simultaneous solutions of the variational inequalities (5.8) *and* (5.9) *is a nonempty compact convex subset of X.*

Proof. It is easily seen that for each $x \in X$, by lower semicontinuity of φ and $-f(x,.)$ in E, the set $B(x) = \{y \in X : \varphi(y) - f(x,y) \leq \varphi(x)\} = \{y \in X : \varphi(y) \leq \varphi(x) + f(x,y)\}$ is a closed subset of X. Again by the monotonicity of the pair (f,g), for each $x \in X$, the set $\{y \in X : \varphi(y) + g(y,x) \leq \varphi(x)\} \subseteq \{y \in X : \varphi(y) \leq \varphi(x) + f(x,y)\} = B(x)$. Hence for each $x \in X$, the set $G(x) = \text{cl}\{y \in X : \varphi(y) + g(y,x) \leq \varphi(x)\} \subseteq B(x)$ and is, therefore, a closed subset of X. Similarly, for each $x \in X$, by the lower semicontinuity of φ and $-g(x,.)$ in E, it follows that the set $C(x) = \{y \in X : \varphi(y) \leq \varphi(x) + g(x,y)\}$ is a closed subset of X and by the monotonicity of the pair (f,g), the set $\{y \in X : \varphi(y) + f(y,x) \leq \varphi(x)\} \subseteq C(X)$. Hence for each $x \in X$, $F(x) = \text{cl}\{y \in X : \varphi(y) + f(y,x) \leq \varphi(x)\} \subseteq C(x)$ and is, therefore, a closed subset of X. Next, we first prove the existence of the solution of the simultaneous variational inequalities (5.8) and (5.9) under the condition (b). To this end we consider the following condition:

(*) for each $x \in X$, there exists a point $y \in X$ such that $\varphi(y) + f(y,x) < \varphi(x)$.

The condition (*) may or may not hold. We will prove that in either case the solution of the simultaneous variational inequalities (5.8) and (5.9) exists. First, let us assume that the condition (*) does not hold. This precisely means that there exists $x_0 \in X$ such that $\varphi(y) + f(y,x_0) \geq \varphi(x_0)$ for all $y \in X$, i.e. x_0 is a solution of the inequality (5.11) and hence by Lemma 5.1 x_0 is is a solution of the inequality (5.8). Therefore by Theorem 5.1 x_0 is a solution of the simultaneous variational inequalities (5.8) and (5.9). Next, we assume that the condition (*) holds. By virtue of the Theorem 5.1, it would suffice to prove the existence of a solution of the inequality (5.9). If possible,we suppose that there is no solution of the inequality (5.9). This then implies that for each $y \in X$, the set $T(y) = \{x \in X : \varphi(y) + g(y,x) > \varphi(x)\}$ is nonempty. Thus T defines a set-valued mapping of X into 2^X. Also by convexity of φ and concavity of $g(y,.)$, it follows that $T(y)$ is a convex subset of X for each $y \in X$. Now for $x \in X$,

$$\begin{aligned}
T^{-1}(x) &= \{y \in X : x \in T(y)\} \\
&= \{y \in X : \varphi(y) + g(y,x) > \varphi(x)\} \\
&= \{y \in X : \varphi(y) + g(y,x) \leq \varphi(x)\}^c \supseteq [\text{cl}\{y \in X : \varphi(y) + g(y,x) \leq \varphi(x)\}]^c \\
&= [G(x)]^c = O_x,
\end{aligned}$$

say, which is a relatively open subset of X. Further by condition (b), $M = \bigcap_{x \in X_0} G(x) = \bigcap_{x \in X_0} {O_x}^c$ is compact. Lastly since (*) holds, it follows that $\bigcup_{x \in X} [B(x)]^c = \bigcup_{x \in X} \{y \in X : \varphi(y) > \varphi(x) + f(x,y)\} = X$. To see this let, for each $x \in X$, let $W(x) = \{y \in X : \varphi(y) + f(y,x) < \varphi(x)\}$ which is nonempty by (*). Hence $W : X \to 2^X$ is a set-valued mapping with $W(x) \neq \emptyset$ for each $x \in X$.Thus $X = \bigcup_{x \in X} W^{-1}(x) = \bigcup_{x \in X} \{y \in X : x \in W(y)\} = \bigcup_{x \in X} \{y \in X : \varphi(x) + f(x,y) < \varphi(y)\} = \bigcup_{x \in X} [B(x)]^c$. Hence the set-valued mapping $T : X \to 2^X$ satisfies all the conditions of a fixed point theorem. Therefore, there exists a point $x_0 \in X$ such that $x_0 \in T(x_0)$, $\varphi(x_0) + f(x_0,x_0) > \varphi(x_0)$ which implies that $\varphi(x_0) < \infty$ and $f(x_0,x_0) > 0$, contradicting an assumption made in (5.7). This contradiction proves the assertion. Similarly we can prove the existence of the simultaneous variational inequalities (5.8) and (5.9) under the condition (a) by replacing f by g in the condition (*) and g by f in the arguments that follow. It remains to show that S is a compact convex subset of X. For each $x \in X$, let $H(x) = \{y \in X : \varphi(y) + f(y,x) \leq \varphi(x)\}$ and $K(x) = \{y \in X : \varphi(y) + g(y,x) \leq \varphi(x)\}$, i.e. $\operatorname{cl} H(x) = F(x)$ and $\operatorname{cl} K(x) = G(x)$. It is then clear that $S = \bigcap_{x \in X} [H(x) \cap K(x)]$. Also by Lemma 5.1, it follows that $S = \bigcap_{x \in X} B(x) \cap C(x)$. But as we have already shown that for each $x \in X$, $G(x) \subseteq B(x)$ and $F(x) \subseteq C(x)$, it is obvious that for each $x \in X$, $F(x) \cap G(x) \subseteq B(x) \cap C(x)$. Thus $S = \bigcap_{x \in X} [H(x) \cap K(x)] \subseteq [B(x) \cap C(x)] = S$ being a closed subset of L or M according as we have (a) or (b), is a compact subset of X. We conclude the proof under (a) or (b) by observing that due to the convexity of φ and concavity of $f(x,.)$ and $g(x,.)$, the sets $B(x)$ and $C(x)$ are convex for each $x \in X$. Finally we finish the proof by noting that the condition (c) implies the condition (b) and the condition (d) implies the condition (a). This follows from the fact that in the beginning of the proof we have proved that for each $x \in X$, $F(x) \subseteq C(x)$ and $G(x) \subseteq B(x)$. □

Remark 5.3 From the proof it is clear that S is compact convex subset of L or M according as we have (a) and (d) or (b) and (c).

Corollary 5.2.1 *Let $\varphi : X \to (-\infty, \infty]$ and $f, g : X \times X \to (-\infty, \infty)$ be as defined in* (5.6) *and* (5.7) *respectively,* (f, g) *satisfy* (5.7) *(i) and (ii) and X be closed.*

Further assume that there is a nonempty subset X_0 contained in a compact convex subset X_1 of X such that one of the following holds:

- *(i) for each $y \in X \setminus X_1$, there exists $x \in X_0$ with $\varphi(x) < \infty$ such that $\varphi(y) + f(y,x) > \varphi(x)$;*
- *(ii) same as (i) with g in place of f.*
- *(i)′ for each $y \in X \setminus X_1$, there exists $x \in X_0$ with $\varphi(x) < \infty$ such that $\varphi(y) > \varphi(x) + g(x,y)$;*
- *(ii)′ same as (i)′ with f in place of g.*

Then the set of solutions of simultaneous variational inequalities (5.8) *and* (5.9) *is a nonempty compact convex subset of X.*

Proof. First suppose that (i) holds. Then for each $y \in X \setminus X_1$, there exists $x \in X_0$ such that $\varphi(y) + f(y,x) > \varphi(x)$. This implies that $\bigcap_{x \in X_0} H(x) \subseteq X_1$, where $H(x) = \{y \in X : \varphi(y) + f(y,x) \leq \varphi(x)\}$. By notation of Theorem 5.2, $\operatorname{cl}(H(x)) = F(x)$ for each $x \in X$. Since X_1 is closed, it follows that $\bigcap_{x \in X_0} \operatorname{cl}(H(x)) = \bigcap_{x \in X_0} F(x) \subseteq X_1$. Hence $\bigcap_{x \in X_0} F(x)$ being a closed subset of a compact set X_1 is compact, i.e. the condition (a) of Theorem 5.2 holds. Thus in this case the corollary now follows from Theorem 5.2. In the same way we can show that the condition (ii) implies the condition (b) of Theorem 5.2. It is also easy to see that $(i)'$ implies that

$\bigcap_{x \in X_1} C(x) \subseteq X_1$, where $C(x) = \{y \in X : \varphi(y) \leq \varphi(x) + g(x,y)\}$ for each $x \in X$. It follows that $\bigcap_{x \in X_0} C(x)$ is compact, i.e. the condition (d) of Theorem 5.2 holds. Finally, in the same way we can show that $(ii)'$ implies (c) of Theorem 5.2. □

Corollary 5.2.2 *Let $\varphi : X \to (-\infty, \infty]$ and $f, g : X \times X \to (-\infty, \infty)$ be as defined in* (5.6) *and* (5.7) *respectively, (f,g) satisfy* (5.7) *(i) and (ii) and X be closed. Further assume that one of the following conditions holds:*

(*i*) *there exists a point $x_0 \in X$ such that the set $F(x_0) = \mathrm{cl}\{y \in X : \varphi(y) + f(y,x_0) \leq \varphi(x_0)\}$ is compact;*

(*ii*) *there exists a point $x_0 \in X$ such that the set $G(x_0) = \mathrm{cl}\{y \in X : \varphi(y) + g(y,x_0) \leq \varphi(x_0)\}$ is compact;*

(*iii*) *there exists a point $x_0 \in X$ such that the set $B(x_0) = \{y \in X : \varphi(y) \leq \varphi(x_0) + f(x_0,y)\}$ is compact;*

(*iv*) *there exists a point $x_0 \in X$ such that the set $C(x_0) = \{y \in X : \varphi(y) \leq \varphi(x_0) + g(x_0,y)\}$ is compact;*

(*v*) *there exists a point x_0 contained in a compact convex subset X_1 of X such that $\varphi(x_0) < \infty$ and $\varphi(y) + f(y,x_0) > \varphi(x_0)$ for all $y \in X \setminus X_1$;*

(*vi*) *there exists a point x_0 contained in a compact convex subset X_1 of X such that $\varphi(x_0) < \infty$ and $\varphi(y) + g(y,x_0) > \varphi(x_0)$ for all $y \in X \setminus X_1$;*

(*vii*) *there exists a point x_0 contained in a compact convex subset X_1 of X such that $\varphi(x_0) < \infty$ and $\varphi(y) > \varphi(x_0) + f(x_0,y)$ for all $y \in X \setminus X_1$;*

(*viii*) *there exists a point x_0 contained in compact convex subset X_1 of X such that $\varphi(x_0) < \infty$ and $\varphi(y) > \varphi(x_0) + g(x_0,y)$ for all $y \in X \setminus X_1$.*

Then the set of solutions of the simultaneous variational inequalities (5.8) *and* (5.9) *is a nonempty compact convex subset of X.*

Proof. The corollary under (i) to (iv) follows directly from Theorem 5.2 and (v) to (viii) from Corollary 5.2.1. □

Next result was first proved in Tarafdar (1990a).

Corollary 5.2.3 *If $\varphi : X \to (-\infty, \infty]$ is as defined in* (5.6) *and f is a function as defined in* (5.7) *satisfying* (5.7) *(i) and (ii) (as a pair (f,f)) and X is closed, then for each $x \in X$, the set*

$$F(x) = \mathrm{cl}\{y \in X : \varphi(y) + f(y,x) \leq \varphi(x)\}$$

is a closed subset of X. Further assume that there is a nonempty subset X_1 of X such that one of the following conditions holds:

(*A*) *same as (a) of Theorem* 5.2*;*

(*B*) *same as (c) of Theorem* 5.2*;*

(*C*) *same as (i) of Corollary* 5.2.1*;*

(D) *same as* (ii)′ *of Corollary* 5.2.1.

Then the set of solutions of the variational inequality (5.8) *is a nonempty compact convex subset of* X.

Proof. The proof of the corollary under condition (A) and (B) follows from Theorem 5.2 by taking by taking $f = g$, whereas that conditions (C) and (D) imply (d) of Theorem 5.2 can easily be checked (see e.g. the proof of Theorem 1.2 in Tarafdar (1990a)). □

Theorem 5.3 *If* $f, g : X \times X \to \mathbb{R}$ *are a strictly monotonic pair of functions and* $x_1 \in X$ *and* $x_2 \in X$ *are respectively two solutions of the variational inequalities* (5.8) *and* (5.9), *then* $x_1 = x_2$, *i.e. the solution of the simultaneous variational inequalities, if exists, must be unique.*

Proof. Since $x_1 \in X$ is a solution of (5.8), $\varphi(x_1) + f(x_1, x_2) \leq \varphi(x_2)$. Again since $x_2 \in X$ is a solution of (5.9), we have $\varphi(x_2) + g(x_2, x_1) \leq \varphi(x_1)$. Adding these two we obtain $f(x_1, x_2) + g(x_2, x_1) \leq 0$. On the other hand by monotonicity we have $f(x_1, x_2) + g(x_2, x_1) \geq 0$. Thus $f(x_1, x_2) + g(x_2, x_1) = 0$, which by strict monotonicity implies $x_1 = x_2$. □

Corollary 5.3.1 *Let* X *be a closed convex subset of* E *and* $T : X \to E^*$ *be a monotone and hemicontinuous mapping according to Definition* 5.2. *Further assume that* X *contains a nonempty subset* X_0 *contained in a compact convex subset* X_1 *of* X *such that one of the followings holds:*

(α) *the set* $\bigcap_{x \in X_0} \mathrm{cl}\{y \in X : \langle T(y), y - x\rangle \leq 0\}$ *is compact;*

(β) *the set* $\bigcap_{x \in X_0} \{y \in X : \langle T(x), x - y\rangle \geq 0\}$ *is compact;*

(γ) *for each* $y \in X \setminus X_1$, *there exists* $x \in X_0$ *such that* $\langle T(y), y - x\rangle > 0$;

(δ) *for each* $y \in X \setminus X_1$, *there exists* $x \in X_0$ *such that* $\langle T(x), x - y\rangle < 0$.

Then the set of solutions of the variational inequality, i.e. the set of points $x_0 \in X$ *satisfying:*

$$\langle T(x_0), x_0 - y\rangle \leq 0 \qquad \text{for all } y \in X, \tag{5.15}$$

is a nonempty compact convex subset of X.

Proof. We define the real valued function $f(x, y)$ on $X \times X$ by$f(x, y) = \langle T(x), x - y\rangle$, $x, y \in X$ and $\varphi(x) \equiv 0$ on X. Then the corollary follows from Theorem 5.2 and Corollary 5.2.1. □

Corollary 5.3.2 *Let* X *and* T *be as in Corollary* 5.3.1. *Further assume that one of the following conditions holds:*

(α)′ *there exists a point* $x_0 \in X$ *such that the set* $\mathrm{cl}\{y \in X : \langle T(y), y - x_0\rangle \leq 0\}$ *is compact;*

$(\beta)'$ *there exists a point* $x_0 \in X$ *such that the set* $\{y \in X : \langle T(x_0), x_0 - y\rangle \geq 0\}$ *is compact;*

$(\gamma)'$ *there exists a point* x_0 *contained in a compact convex subset* X_1 *of* X *such that* $\langle T(y), y - x_0\rangle > 0$ *for all* $y \in X \setminus X_1$;

$(\delta)'$ *there exists a point* x_0 *contained in a compact convex subset* X_1 *of* X *such that* $\langle T(x_0), x_0 - y\rangle < 0$ *for all* $y \in X \setminus X_1$.

Then the conclusion of Corollary 5.3.1 *holds.*

Proof. We define f and φ as in the proof of Corollary 5.3.1 and apply Corollary 5.2.2 with (i), (iii), (v) and (vii) respectively for $(\alpha)'$, $(\beta)'$, $(\gamma)'$ and $(\delta)'$. □

Corollary 5.3.3 *Let* X *be a compact convex subset of* E. *Let* $T : X \to E^*$ *be a monotone and hemicontinuous mapping of* X *into* E^*. *Then the set of the variational inequality* (5.15) *is a nonempty compact convex subset of* X.

Proof. We take $X = X_1 = X_0$. Then the set $\bigcap_{x\in X} \{y \in X : \langle T(x), x - y\rangle \geq 0\}$ being the intersection of closed sets is a closed subset of the compact set X and is, therefore, compact. Hence the corollary follows from the Corollary 5.3.1 (β). □

Remark 5.4 Corollary 5.3.3 includes the results obtained by Tarafdar (1977), Browder (1965b) and Hartman and Stampacchia (1966) with operator $C(u) \equiv 0$.

Corollary 5.3.4 *Let* X *be a closed convex subset of a reflexive Banach space* V *and* $T : X \to V^*$ *be a monotone and hemicontinuous mapping. Let* $v^* \in V^*$ *be given. Assume that* X *contains a nonempty subset* X_0 *contained in a closed bounded convex subset* X_1 *of* X *such that one of the following holds:*

(*i*) *the set* $\bigcap_{x\in X_0} \text{cl}\{y \in X : \langle T(y), y - x\rangle \leq \langle v^*, y - x\rangle\}$ *is bounded;*

(*ii*) *the set* $\bigcap_{x\in X_0} \langle T(x), x - y\rangle \geq \langle v^*, x - y\rangle\}$ *is bounded;*

(*iii*) *for each* $y \in X\setminus X_1$, *there exists* $x \in X_0$ *such that* $\langle T(y), y-x\rangle > \langle v^*, y-x\rangle$;

(*iv*) *for each* $y \in X\setminus X_1$, *there exists* $x \in X_0$ *such that* $\langle T(x), x-y\rangle < \langle v^*, x-y\rangle$.

Then the set of solutions of the variational inequality

$$\langle T(x), y - x\rangle > \langle v^*, y - x\rangle \qquad \text{for all } y \in X \tag{5.16}$$

is a closed convex bounded subset of X.

Proof. We define $f : X \times X \to \mathbb{R}$ by $f(x, y) = \langle T(x) - v^*, x - y\rangle$, and $\varphi \equiv 0$ on X. Then f is monotone and hemicontinuous and satisfies all the conditions of Corollary 5.2.3 with V equipped with weak topology. Conclusion of the Corollary follows from Corollary 5.2.3. □

Corollary 5.3.5 *Let* X, V, T *and* v^* *be as in Corollary* 5.3.4. *Assume that any one of the following conditions holds:*

(*i*) *there exists a point* $x_0 \in X$ *such that the set* $\{y \in X : \langle T(y), y - x_0 \rangle \leq \langle v^*, y - x_0 \rangle\}$ *is bounded;*

(*ii*) *there exists a point* $x_0 \in X$ *such that the set* $\{y \in X : \langle T(x_0), x_0 - y \rangle \geq \langle v^*, x_0 - y \rangle\}$ *is bounded;*

(*iii*) *there exists a point* x_0 *contained in a closed bounded convex subset* X_1 *of* X *such that* $\langle T(y), y - x_0 \rangle > \langle v^8, y - x_0 \rangle$ *for all* $y \in X \setminus X_1$*;*

(*iv*) *there exists a point* x_0 *contained in a closed bounded convex subset* X_1 *of* X *such that* $\langle T(x_0), x_0 - y \rangle < \langle v^*, x_0 - y \rangle$ *for all* $y \in X \setminus X_1$.

Then the conclusion of the Corollary 5.3.4 *holds.*

Proof. The proof follows immediately from Corollary 5.3.4 with $X_0 = \{x_0\}$. □

Remark 5.5 A mapping $T : X \to E^*$ is said to be coercive at $x_0 \in X$ if there is a compact subset B of E such that $x_0 \in B \cap X$ and $\langle T(y), y - x_0 \rangle > 0$ for all $y \in X \setminus B$. Thus if $E = V$ and T is coercive at x_0, then the condition (iii) above holds.

5.1.3 *Application to Nonlinear Boundary Value Problem for Quasilinear Operator of Order 2m in Generalized Divergence Form*

As an application of our results of previous subsection we consider the existence of solutions of the boundary value problem for a quasilinear operator of oder $2m$ in generalized divergence form as formulated in Browder (1970). Before we proceed further we make the following well-known observation.

If $T : V \to V^*$ is a mapping of a Banach space V into V^* and x_0 is a solution of the variational inequality

$$\langle T(x_0), y - x_0 \rangle \geq 0 \qquad \text{for all } y \in V \tag{5.17}$$

then $\langle T(x_0), z \rangle \geq 0$ for all $z \in V$, for we can take $y = z + x_0$. Hence $\langle T(x_0), -z \rangle \geq 0$ for all $z \in V$. Thus $\langle T(x_0), z \rangle = 0$ for all $z \in V$. Hence inequality (5.17) becomes equality. In fact if x_0 is an interior point of X, we can show that the inequality (5.15) becomes equality.

Here we adapt the notations of Browder (1970) and Pascali and Sburlan (1978). Let Ω be a domain in $\mathbb{R}^N (N \geq 2)$ and $x = (x_1, x_2, \cdots, x_N)$ denote an element of Ω. Let $D^\alpha = D_1^{\alpha_1} D_2^{\alpha_2} \cdots D_N^{\alpha_N}$, where $D_j = \frac{\partial}{\partial x_j}$ and $\alpha = (\alpha_1, \alpha_2, \cdots, \alpha_N)$ is an N-tuples of nonnegative integers with $|\alpha| = \alpha_1 + \alpha_2 + \cdots + \alpha_N$. Let $\mathbb{R}^{N_m}$ be the vector space whose elements are of the form $\zeta = \{\zeta_\alpha : |\alpha| \leq m\}$. Then a nonlinear generalized divergence operator of order $2m$ has the form

$$A(u) = \sum_{|\alpha| \leq m} (-1)^{|\alpha|} D^\alpha A_\alpha(x, \zeta(u(x))) \tag{5.18}$$

where $\zeta(u) = \{D^\alpha u : |\alpha| \leq m\}$ and A_α is a real valued function on $\Omega \times \mathbb{R}^{N_m}$ for each α with $|\alpha| \leq m$. The direct method of the calculus of variations involves in obtaining solutions of the variational boundary value problem of the Euler-Lagrange equation (5.18). In here we are interested in Browder's approach to the problem as in Browder (1970) and Tarafdar (1990a). Let V be a closed subspace of $W^{m,p}(\Omega)$ such that

$$W_0^{m,p}(\Omega) \subseteq V \subseteq W^{m,p}(\Omega) \qquad \text{with } 1 < p < \infty. \tag{5.19}$$

Here $W^{m,p}(\Omega)$ is the Banach space of all L^P functions u on Ω (with respect to Lebesgue N-measure) whose distribution derivative up to order m are also L^p functions. $W^{m,p}(\Omega)$ becomes an uniformly convex (hence reflexive) Banach space with respect to the norm

$$\|u\|_{m,p}^p = \sum_{|\alpha| \leq m} \|D^\alpha u\|_{L^p}^p \qquad \text{for } 1 < p < \infty. \tag{5.20}$$

The subspace V in which we are interested must include the testing functions $C_c^\infty(\Omega)$ with compact supports in Ω and must be closed in $W^{m,p}(\Omega)$. Hence it must include the smallest such space $W_0^{m,p}(\Omega)$, the closure of the testing functions in $W^{m,p}(\Omega)$, i.e. $W_0^{m,p}(\Omega) \subseteq V \subseteq W^{m,p}(\Omega)$ (for details see Browder (1970), or Pascali and Sburlan (1978) and see Adams (1975) for general informations on Sobolev spaces).

Let V be a closed subspace of $W^{m,p}(\Omega)$ as in (5.19). If for each α, $A_\alpha(.,\zeta(u(.))) \in L^q$ for every $u \in W^{m,p}(\Omega)$ with $\frac{1}{p} + \frac{1}{q} = 1$, the operator A is assigned its generalized Dirichlet form

$$a(u,v) = \sum_{|\alpha| \leq m} \int_\Omega A_\alpha(x, \zeta(u(x))) D^\alpha v(x) dx \qquad \text{for all } u, v \in V. \tag{5.21}$$

The operator A defined in (5.18) is said to be of variational type if the boundary conditions are implicitly verified by the constraint

$$u \in V \qquad \text{and } \langle Au, v \rangle = a(u,v) \qquad \text{for all } v \in V. \tag{5.22}$$

Here $\langle .,. \rangle$ denotes the pairing between V^* and V. The choice of V and together with (5.22) determines the boundary conditions in the manner as explained in Browder (1970) (see also Pascali and Sburlan (1978, p. 273)):

$V = W_0^{m,p}$ and (5.22) yield generalized Dirichlet boundary conditions, while $V = W^{m,p}$ and (5.22) yield generalized Neumann boundary conditions. Any other choice of V with $W_0^{m,p}(\Omega) \subseteq V \subseteq W^{m,p}(\Omega)$ and (5.22) give what are called the generalized boundary conditions of mixed type.

Now as in Browder (1970) we make some assumptions to make precise the meaning of the solution of (5.18).

Assumption (A). Each $A_\alpha : \Omega \times \mathbb{R}^{N_m} \to \mathbb{R}$ satisfies Caratheodory conditions (i.e. for each fixed $\zeta \in \mathbb{R}^{N_m}$, $A_\alpha(x,\zeta)$ is measurable in x with respect to Lebesgue measure in Ω and for almost all $x \in \Omega$, $A_\alpha(x,\zeta)$ is continuous in ζ on $\mathbb{R}^{N_m}$ and there exists a real number p with $1 < p < \infty$ such that

$$|A_\alpha(x,\zeta)| \le C|\zeta|^{p-1} + g(x) \tag{5.23}$$

where $C > 0$ is a constant and $g \in L^q(\Omega)$ with $\frac{1}{p} + \frac{1}{p} = 1$. (For any unexplained term or gap in the proof of this problem please see Pascali and Sburlan (1978, pp. 272–276)).

By virtue of condition (A), for each $u \in W^{m,p}(\Omega)$, $A_\alpha(x,\zeta(u(x)))$ is measurable function of x and belongs to $L^q(\Omega)$. Also it follows from *Hölders'* inequality that $a(u,v)$ is well defined for all $u, v \in W^{m,p}(\Omega)$ and satisfies an inequality of the form:

$$|a(u,v)| \le \phi(\|u\|_{m,p})\ \|v\|_{m,p}\,, \tag{5.24}$$

where $\phi : \mathbb{R}^+ \to \mathbb{R}^+$ is a continuous function defined by $\phi(r) = cr^{p-1} + \|g\|_q$ and $c > 0$ is a constant (see Pascali and Sburlan (1978, p. 275) for details). Thus by virtue of (5.24) for each $u \in V$ with $W_0^{m,p}(\Omega) \subseteq V \subseteq W^{m,p}(\Omega)$, $a(u,.)$ is a continuous linear functional, say, $T(u)$ on V. Hence by (5.24) we can define an operator (possibly nonlinear) $T\,V \to V^*$ by

$$a(u,v) = \langle T(u), v\rangle \qquad \text{for all } v \in V. \tag{5.25}$$

The mapping $T : V \to V^*$ is determined by the generalized Dirichlet form $a(u,v)$ and the closed subspace V of $W^{m,p}(\Omega)$. Let $f \in V^*$ be given. Then $u_0 \in V$ is said to be a solution of the variational boundary value problem for $A(u) = f$ if

$$a(u_0,v) = \langle T(u_0), v\rangle = \langle f, v\rangle \qquad \text{for all } v \in V, \tag{5.26}$$

where $A(u)$ is the quasilinear operator in generalized divergence form given in (5.18). Hence by the observation made in the beginning of this subsection, given $f \in V^*$, $u_0 \in V$ is a solution of the variational boundary value problem $A(u) = f$ if and only if $u_0 \in V$ is a solution of the variational inequality

$$\langle T(u_0), v - u_0\rangle \ge \langle f, v - u_0\rangle \qquad \text{for all } v \in V. \tag{5.27}$$

The condition (A) also implies that the Nemitsky operator $F_\alpha = A_\alpha(.,\zeta(u))$ is continuous from $W^{m,p}(\Omega)$ into $L^q(\Omega)$ for each α (see Pascali and Sburlan (1978, p. 166 and p. 275)). Now it follows that $T : V \to V^*$ is continuous. Indeed, let $\{u_n\}$ be a sequence in V converging to u in norm of V. Then

$$\begin{aligned} \|(T(u_n) - T(u)\| &= \sup_{\|v\|_{m,p}\le 1} \langle T(u_n) - T(u), v\rangle \\ &\le \sup_{\|v|_{m,p}\le 1} \sum_{|\alpha|\le m} \int_\Omega |A_\alpha(x,\zeta(u_n)) - A_\alpha(x,\zeta(u))|\,|D^\alpha v|dx \to 0 \end{aligned}$$

as $n \to \infty$. Thus $T : V \to V^*$ is continuous and hemicontinuous.

Assumption (B). For each $x \in \Omega$ and each pair $\zeta, \zeta' \in \mathbb{R}^{N_m}$,

$$\sum_{|\alpha|\leq m} [A_\alpha(x,\zeta) - A_\alpha(x,\zeta')]\,[\zeta_\alpha - \zeta'_\alpha] \geq 0.$$

The condition (B) implies that the operator T is monotone.

$$\begin{aligned}\langle T(u) - T(v),\, u - v\rangle &= \langle T(u),\, u-v\rangle - \langle T(v),\, u-v\rangle\\ &= a(u,\, u-v) - a(v,\, u-v) = a(u-v,\, u-v)\\ &= \int_\Omega \sum_{|\alpha|\leq m} [A_\alpha(x,\zeta(u)) - A_\alpha(x,\zeta(v))]\,[\zeta_\alpha(u) - \zeta_\alpha(v)]\,dx\\ &\geq 0.\end{aligned}$$

Theorem 5.4 *Let $A(u)$ be a quasilinear operator of order $2m$ in generalized divergence form satisfying (A) and (B) on an open set Ω of $\mathbb{R}^N$. Let V be a closed subspace of $W^{m,p}(\Omega)$ satisfying* (5.19). *Let $f \in V^*$ be given. Further assume that either*

(C) V contains a nonempty subset U_0 contained in a closed bounded convex subset U_1 of V such that any one of the following conditions holds:

(i) the set $\bigcap_{u\in U_0} \mathrm{cl}\{v \in V : \sum_{|\alpha|\leq m} \int_\Omega [A_\alpha(x,\, \zeta(v)) - f][\zeta_\alpha(v) - \zeta_\alpha(u)]\,dx \leq 0\}$ is bounded;

(ii) the set $\bigcap_{u\in U_0} \{v \in V : \sum_{|\alpha|\leq m} \int_\Omega [A_\alpha(x,\, \zeta(u)) - f][\zeta_\alpha(u) - \zeta_\alpha(v)]\,dx \geq 0\}$ is bounded;

(iii) for each $v \in V \setminus U_1$, there exists $u \in U_0$ such that $\sum_{|\alpha|\leq m} \int_\Omega [A_\alpha(x,\, \zeta(v)) - f][\zeta_\alpha(v) - \zeta_\alpha(u)]\,dx > 0$; and

(iv) for each $v \in V \setminus U_1$, there exists $u \in U_0$ such that $\sum_{|\alpha| lem} \int_\Omega [A_\alpha(x,\, \zeta(u)) - f][\zeta_\alpha(u) - \zeta_\alpha(v)]\,dx < 0$;

or (D) any one of the following conditions holds:

(i)′ there exists a point $u_0 \in V$ such that the set $\{v \in V : \sum_{|\alpha|\leq m} \int_\Omega [A_\alpha(x,\, \zeta(v)) - f][\zeta_\alpha(v) - \zeta_\alpha(u_0)]\,dx \leq 0\}$ is bounded;

(ii)′ there exists a point $u_0 \in V$ such that the set $\{v \in V : \sum_{|\alpha|\leq m} \int_\Omega [A_\alpha(x,\, \zeta(u_0)) - f][\zeta_\alpha(u_0) - \zeta_\alpha(v)]\,dx \geq 0\}$ is bounded;

(iii)′ there exists a point u_0 contained in a closed bounded convex subset U_1 of V such that $\sum_{|\alpha|\leq m} \int_\Omega [A_\alpha(x,\, \zeta(v)) - f][\zeta_\alpha(v) - \zeta_\alpha(u_0)]\,dx > 0$ for all $v \in V\setminus U_1$;

(iv)′ there exists a point u_0 contained in a closed bounded convex subset U_1 of V such that
$\sum_{|\alpha|\leq m} \int_\Omega [A_\alpha(x,\, \zeta(u_0)) - f][\zeta_\alpha(u_0) - \zeta_\alpha(v)]\,dx < 0$
for all $v \in V \setminus U_1$.

Then the variational boundary value problem for $A(u) = f$ with respect to V in the sense as explained above has a solution and the set of such solutions is a closed bounded convex subset of V.

Proof. The mapping $T : V \to V^*$ has been proved to be monotone and hemicontinuous. We take $V = X$ in the Corollary 5.3.4 and Corollary 5.3.5 and note that (C) (i), (ii), (iii) and (iv) imply respectively (i), (ii), (iii) and (iv) of Corollary 5.3.4 and (D) $(i)'$, $(ii)'$, $(iii)'$ and $(iv)'$ imply respectively $(i)'$, $(ii)'$, $(iii)'$ and $(iv)'$ of Corollary 5.3.5. Thus the Theorem follows from Corollary 5.3.4 and Corollary 5.3.5. □

Remark 5.6 It can be easily seen that any coercivity condition such as in Browder (1970) or Pascali and Sburlan (1978) will imply all the conditions in (C) of our theorem.

5.1.4 *Minimization Problems and Related Results*

In this subsection we consider the minimization problem:

$$\inf_{x \in X} f(x) \tag{5.28}$$

where X is a nonempty convex subset of a real Hausdorff topological vector space E which will be assumed to be locally convex in the case when $E^* \neq 0$ is considered. We apply the results of the previous subsection to prove the existence of the solution of the problem (5.28). We recall that for a nonlinear mapping $A : X \to E^*$, a point $x_0 \in X$ is called a solution of the variational inequality if

$$\langle A(x_0), x_0 - y \rangle \leq 0 \qquad \text{for all } y \in X, \tag{5.29}$$

where as before $\langle .,. \rangle$ denotes the pairing between E^* and E. For a function $F : X \times X \to \mathbb{R}$, as before a point $x_0 \in X$ is called a solution of the variational inequality if

$$F(x_0, y) \leq 0 \qquad \text{for all } y \in X. \tag{5.30}$$

That (5.29) is a special case of (5.30) can be seen by defining $F : X \times X \to \mathbb{R}$ to be

$$F(x, y) = \langle A(x), x - y \rangle \qquad \text{for all } x, y \in X. \tag{5.31}$$

Now $x_0 \in X$ is a solution of (5.30) with F as defined in (5.31) if and only if x_0 is a solution of (5.29). If $f : X \subseteq \mathbb{R}^n \to \mathbb{R}$ is a continuously differentiable function and $A = \nabla f$, the gradient of f, then it is well known (see Pascali and Sburlan (1978, p. 15-16)) that $x_0 \in X$ is a solution of the minimization problem (5.28), then x_0 is a solution of the variational inequality (5.29). If f is convex, then any solution of (5.29) is solution of (5.28). It is trivially always the case that $x_0 \in X$ is a solution of the minimization problem (5.28) if and only if x_0 is a solution of the variational inequality (5.30) with $F : X \times X \to \mathbb{R}$ defined by $F(x, y) = f(x) - f(y), x, y \in X$.

Theorem 5.5 *Let X be a closed convex subset of a Banach $U = V^*$ (i.e. the continuous dual of a Banach space V) and $f : X \times X \to \mathbb{R}$ a monotone and hemicontinuous function satisfying the conditions:*

(i) for each $x \in X, f(x,x) = 0$;
(ii) for each $x \in X, f(x,.)$ is concave and upper semicontinuous on X.

Then for each $x \in X$, the set $F(x) = \omega^*$-closure of $\{y \in X : f(y,x) \leq 0\}$ is a subset of X.

Further assume that there is a nonempty set X_0 contained in a closed bounded convex subset X_1 of X such that one of the following conditions holds:

(A) the set L defined in (a) of Theorem 5.2 is bounded;
(B) the set $K = \bigcap_{x \in X_0} B(x)$ is bounded, where $B(x) = \{y \in X : f(x,y) \geq 0\}$;
(C) for each $y \in X \setminus X_1$, there exists $x \in X_0$ such that $f(y,x) > 0$;
(D) for each $y \in X \setminus X_1$, there exists $x \in X_0$ such that $f(x,y) < 0$.

Then the set of solutions of the variational inequality (5.8) with $\varphi \equiv 0$ is a nonempty ω^*-compact convex subset of X.

Proof. Equip U with the ω^*-topology (weak topology) of V. Then as X is convex, it is ω^*-closed. Also by the known result (e.g. see Pascali and Sburlan (1978, p. 23)), for each $x \in X, F(x,.)$ is ω^*-upper semicontinuous and the Banach-Alaoglu theorem, X_1 is ω^*-compact. Hence the result follows from Corollary 5.2.3 with $\varphi(x) \equiv 0$. □

Remark 5.7 It is easy to see that the above theorem holds if X is a reflexive Banach space.

Corollary 5.5.1 *Let X be a closed convex subset of a Banach space $U = V^*$ and $f : X \times X \to \mathbb{R}$ a monotone and hemicontinuous function satisfying (i) and (ii) of Theorem 5.5. Further assume that there exists a point $x \in X$ such that one of the following conditions holds:*

(i) the set $F(\overline{x}) = \omega^*$-closure of the set $\{y \in X : f(y,\overline{x}) \leq 0\}$ is bounded;
(ii) the set $B(\overline{x}) = \{y \in X : f(\overline{x},y) \geq 0\}$ is bounded;
(iii) $\overline{x}$ is contained in a closed bounded convex subset X_1 of X such that $f(y,\overline{x}) > 0$ for all $y \in X \setminus X_1$;
(iv) $\overline{x}$ is contained in closed bounded convex subset X_1 of X such that $f(\overline{x},y) < 0$ for all $y \in X \setminus X_1$.

Then the set of solutions of the variational inequality (5.8) is a nonempty ω^*-compact convex subset of X.

Proof. Taking $\{\overline{x}\} = X_0 = X_1$, the corollary follows from Theorem 5.5 under conditions (i) and (ii) and taking $\{\overline{x}\} = X_0$, the corollary follows from Theorem 5.5 under conditions (iii) and (iv). □

Corollary 5.5.2 *Let X be closed convex subset of a Banach space $U = V^*$ and $g : X \to \mathbb{R}$ a convex lower semicontinuous function. Further assume that either of the conditions (a) or (b) holds:*

(a) *there exists a nonempty subset X_0 contained in a closed bounded convex subset X_1 of X such that one of the following conditions holds:*
   (i) *the set $K = \bigcap_{x \in X_0} B(x)$ is bounded, where $B(x) = \{y \in X : g(x) \geq g(y)\}$;*
   (ii) *for each $y \in X \setminus X_1$, there exists $x \in X_0$ such that $g(y) > g(x)$.*

(b) *there exists a point $\overline{x} \in X$ such that the one of the following conditions holds:*
   (i) *the set $B(\overline{x}) = \{y \in X : g(\overline{x}) \geq g(y)\}$ is bounded;*
   (ii) *$\overline{x}$ is contained in a closed bounded convex subset X_1 of X such that $g(\overline{x}) < g(y)$ for all $y \in X \setminus X_1$.*

Then the set of solutions of the minimization problem (5.28) is a nonempty ω^*-compact convex subset of X.

Proof. Define $f(x,y) = g(x) - g(y), x, y \in X$. Then $f(x,y)$ is a monotone and hemicontinuous function satisfying (i) and (ii) of Theorem 5.5. Hence the corollary follows from Theorem 5.5 and Corollary 5.5.1. □

Corollary 5.5.3 *Let X be a closed convex subset of a Banach space $U = V^*$ and $g : X \to \mathbb{R}$ a convex lower semicontinuous function on X. Further assume that*

$$\frac{g(x)}{\|x\|} \to \infty \qquad \text{as } \|x\| \to \infty.$$

Then the set of solutions of minimization problem (5.28) is a nonempty ω^*-compact convex subset of $X \bigcap B_R$ for a sufficiently large $R > 0$, where $B_R = \{x \in X : \|x\| \leq R\}$.

Proof. We define $f(x,y) = g(x) - g(y), x, y \in X$. Then f is a monotone function satisfying (i) and (ii) of Theorem 5.5. Now $\lim_{\|x\| \to \infty} \frac{g(x)}{\|x\|} = \infty$ implies $\lim_{\|x\| \to \infty} \frac{g(x) - g(x_0)}{\|x\|} = \infty$, where $x_0 \in X$ is a particular point. Hence there exists $R > g(x_0)$ such that $f(x, x_0) = g(x) - g(x_0) > 0$ for every $\|x\| > R$. Hence the condition (ii) of Corollary 5.5.2 is satisfied. Thus the Corollary follows from Corollary 5.5.2. □

Remark 5.8 Note that in the proof of Corollaries 5.5.2 and 5.5.3 we have used the fact that the lower semicontinuity of g implies the the lower semicontinuity of the function $k(t) = f(x + t(y - x), y) = g(x + t(y - x)) - g(y)$ of real variable t as $t \downarrow 0$ i.e. hemicontinuity of f.

5.1.5 *Extension of a Karamardian Theorem*

In this subsection we extend a theorem of Karamardian (1971). Throughout this subsection X will be assumed to be a nonempty closed convex subset of a locally convex (real) Hausdorff topological vector space E.

The lemma which is somewhat known is proved here for the sake of completeness.

Lemma 5.2 *If $T : X \to E^*$ is a continuous mapping and if there is a nonempty subset X_0 contained in a compact convex subset X_1 of X such that*

(i) $\bigcap_{x \in X_0} \{x \in X : \langle T(x), y - x\rangle \geq 0\}$ *is compact,*

then there is a point $x_0 \in X$ such that

$$\langle T(x_0), y - x_0\rangle \geq 0 \qquad \text{for all } y \in X$$

and the set of all such x_0 is a compact subset of X.

Proof. If the assertion of the lemma is false, then for each $x \in X$, there must exist one $y \in X$ such that $\langle T(x), y - x\rangle < 0$. Thus we can define a set valued mapping $F : X \to 2^X$ by setting $F(x) = \{y \in X : \langle T(x), y - x\rangle < 0\}, x \in X$. Then for each $x \in X, F(x)$ is nonempty and evidently convex.

Also for each $y \in X, F^{-1}(y) = \{x \in X : y \in F(x)\} = \{x \in X : \langle T(x), y - x\rangle < 0\} = O_y$,say, which is a relatively open set as T is continuous (see, e.g. Browder (1967, Lemma 1). Obviously , $\bigcup_{y \in X} F^{-1}(y) = \bigcup_{y \in X} O_y = X$. (To see this let $x \in X$. Since $F(x) \neq \emptyset$, there is a point $y \in F(x)$ and so $x \in F^{-1}(y)$). Furthermore, $\bigcap_{y \in X_0} O^c{}_y = \bigcap_{y \in X_0} \{x \in X : \langle T(x), y - x\rangle \geq 0\}$ is compact. Hence by a fixed point theorem, there is a point $x_0 \in X$ such that $x_0 \in F(x_0)$, i.e. $\langle T(x_0), x_0 - x_0\rangle < 0$ which is absurd. Hence the assertion of the lemma must be true. The rest follows from the continuity of T. □

Remark 5.9 The following condition:

"There exists a nonempty set X_0 contained in a compact convex subset X_1 of X such that for each $y \in X \setminus X_1$, there exists $x \in X_0$ such that $\langle T(y), x - y\rangle < 0$"

implies (i) of the Lemma above. We can easily see this by verifying that under the above condition the set $D = \bigcap_{y \in X_0} \{x \in X : \langle T(x), y - x\rangle \geq 0\} \subseteq X_1$. Thus D being a closed subset of a compact set is compact.

Our next result which we write also as a lemma appeared in Tarafdar (1990a) (also partly in Tarafdar (1986)).

Lemma 5.3 *Let $T : X \to E^*$ be a monotone and hemicontinuous mapping. Further assume that there is a nonempty subset X_0 contained in a compact convex subset X_1 of X such that one of the following conditions holds:*

(a) *the set* $\bigcap_{x \in X_0} \text{cl}\{y \in X : \langle T(y), y - x\rangle \leq 0\}$ *compact;*

(*b*) *the set* $\bigcap_{x\in X_0} \{y \in X : \langle T(x), x-y\rangle \geq 0\}$ *is compact;*
(*c*) *for each* $y \in X \setminus X_1$, *there exists* $x \in X_0$ *such that* $\langle T(y), y-x\rangle > 0$;
(*d*) *for each* $y \in X \setminus X_1$, *there exists* $x \in X_0$ *such that* $\langle T(x), x-y\rangle < 0$.

Then the set of points $x_0 \in X$ *satisfying the variational inequality* $\langle T(x_0), y-x_0\rangle \geq 0$ *for all* $y \in X$ *is a nonempty convex subset of* X.

Proof. We define the function $f : X \times X \to \mathbb{R}$ by $f(x,y) = \langle T(x), x-y\rangle, x, y \in X$ and apply Corollary 5.2.3 with $\varphi(x) \equiv 0$ on X. □

Theorem 5.6 *Let* X *be a closed convex cone of* E *and* $T : X \to E^*$ *either a continuous mapping or a monotone hemicontinuous mapping. Further assume that there is a nonempty set* X_0 *contained in a compact convex subset* X_1 *of* X *such that for each* $y \in X \setminus X_1$, *there exists* $x \in X_0$ *such that* $\langle T(y), y-x\rangle > 0$.

Then there exists $\overline{x} \in X$ *such that* $T(\overline{x}) \in X^*$ *and* $\langle T(\overline{x}), \overline{x}\rangle = 0$, *where*

$$X^* = \{f \in E^* : f(x) = \langle f, x\rangle \geq 0\} \qquad \textit{for all } x \in X,$$

i.e. X^* *is the positive dual cone of* X.

Proof. By Lemma 5.2 or Lemma 5.3 according as T continuous or monotone hemicontinuous, the set of solutions of the variational inequality $\langle T(x), y-x\rangle \geq 0$ for all $y \in X$ is a nonempty subset of X. Now we prove that each solution $\overline{x} \in X$ of the variational inequality satisfies the requirement of the theorem. Since $\langle T(\overline{x}), y-\overline{x}\rangle \geq 0$ for all $y \in X$, $\langle T(\overline{x}), \overline{x}\rangle \leq 0$, as $0 \in X, X$ being a cone. Also taking $y = 2\overline{x}$, $\langle T(\overline{x}), \overline{x}\rangle \geq 0$. Thus $\langle T(\overline{x}), \overline{x}\rangle = 0$. Now as $\langle T(\overline{x}), y-\overline{x}\rangle \geq 0$ for all $y \in X$, we have $\langle T(\overline{x}), y\rangle \geq \langle T(\overline{x}, \overline{x}\rangle = 0$ for all $y \in X$, i.e. $T(\overline{x}) \in X^*$. □

Remark 5.10 If $E = \mathbb{R}^n$ and T is continuous, Theorem 5.6 reduces to a theorem of Karamardian (1971) and is,therefore, an extension of Karamardian theorem.

Remark 5.11 Some of the results which we have obtained in this section have generalized in Singh, Tarafdar, and Watson (1996) by replacing monotone pair of functions by pseudomontotone pair with the following definition:

The real valued functions f and g defined on $X \times X$ are said to be pseudomonotone or pseudomonotonic pair if $f(x,y) \geq 0$ whenever $g(y,x) \leq 0$ for all $x, y \in X$.

Since the proofs of those results involve no new technique, we did not consider these results. We have,however, considered some special results on pseudomonotone mappings of Singh, Watson, and Srivastava (1997) in the next section.

5.2 Variational Inequalities for Setvalued Mappings

It is well-known that variational inequality theory does not only have many important applications in partial differential equations such as free boundary problems and so on (e.g., see Baiocchi and Capelo (1984), but it also has been successfully used in the study of operations research, mathematical programming and optimization theory (e.g., see Aubin and Ekeland (1984). Due to the development of set-valued analysis, the study of variational inequalities has been under much attention recently, for example, see Ding and Tan (1990), Harker and Pang (1990), Husain and Tarafdar (1994), Granas (1990), Karamolegos and Kravvaritis (1992), Kravvaritis (1979) (1979), Mosco (1976), Shih and Tan (1985), Shih and Tan (1988b), Shih and Tan (1988a), Tarafdar and Yuan (1994) and many others whose names are not mentioned here. In this section we present the results obtained in Tan-Tarafdar-Yuan (1999) on the existence of solutions for variational inequalities and quasi-variational inequalities of set-valued mappings either in simultaneous form or in implicit form as applications of Ky Fan's KKM mapping principle in Fan (1961) (we can use our fixed point theorem as well) and Fan-Glicksberg fixed point theorem (see Fan (1952) and Glicksberg (1952)). Precisely, we shall establish the existence of solutions for simultaneous variational inequalities in subsection 2. Then implicit variational inequality and implicit quasi-variational inequality in which set-valued mappings are monotone (resp., upper semicontinuous) will be investigated in subsection 3 (resp., in subsection 4). These results either generalize or improve corresponding ones given in recent literatures.

We shall denote by $\mathbb{R}$ and $\mathbb{N}$ the set of real numbers and the set of natural numbers, respectively. Let X be a nonempty set. For convenience, throughout this section we shall denote by 2^X the family of all non-empty subsets of X. If X is a topological space (resp., a non-empty subset of a topological vector space) , we shall denote by $K(X)$ (resp., $KC(X)$) the family of all non-empty compact subsets of X (resp., the family of all non-empty compact and convex subsets of X). If X is a subset of a vector space E, then coX denotes the convex hull of X in E. Let $f : X \to 2^{\mathbb{R}}$ be a (set-valued) mapping. For each $x \in X$, let $\inf f(x) := \inf\{z : z \in f(x)\}$. Let E^* be the dual space of a complex Hausdorff topological vector space E and X be a non-empty subset of E. We shall denote by $\langle w, x\rangle$ the dual pair between E^* and E for $w \in E^*$ and $x \in E$ and by $Re\langle w, x\rangle$ the real part of the complex number $\langle w, x\rangle$.

Definition 5.3 A mapping $T : X \to 2^{E^*}$ is said to be monotone if for each $x, y \in X$, $Re\langle u - v, x - y\rangle \geq 0$ for all $u \in T(x)$ and $v \in T(y)$.

Throughout this section E denotes a given Hausdorff topological vector space unless otherwise specified.

Definition 5.4 Let X be a non-empty convex subset of E, f, $g : X \times X \longrightarrow 2^{\mathbb{R}}$, $f_1 : X \longrightarrow 2^{\mathbb{R}}$, $h : X \to \mathbb{R} \cup \{-\infty, +\infty\}$ and $H : X \to 2^{E^*}$. Then

(1) $\{f, g\}$ is said to be a monotone pair if for each x, $y \in X$, $u + w \geq 0$ for each $u \in f(x, y)$ and $w \in g(y, x)$; f is said to be monotone if the pair $\{f, f\}$ is monotone. In particular, when f is single-valued, we recover the notion of monotone pair reduces to that of a monotone mapping defined by Mosco (1976) (see Tarafdar (1990a) and also Husain and Tarafdar (1994)).

(2) f is said to be hemicontinuous if for each $x, y \in X$, the mapping $k : [0, 1] \longrightarrow 2^{\mathbb{R}}$ defined by $k(t) := f((1-t)x + ty, y)$ for all $t \in [0,1]$ is such that for each given $s \in \mathbb{R}$ with $f(x,y) \subset (s, +\infty)$, there exists $t_0 \in (0,1]$ such that $f((1-t)x + ty, y) \subset (s, +\infty)$ for all $t \in (0, t_0)$. We note that if f is single-valued, our definition of hemicontinuity reduces to the classical one given by Mosco (1976), i.e., the function $t \mapsto f(x + t(y-x), y)$ from $[0,1]$ to $\mathbb{R}$ is lower semicontinuous as $t \downarrow 0$.

(3) f_1 is said to be concave if for each $n \in \mathbb{N}$, $x_1, \cdots, x_n \in X$ and nonnegative $\lambda_1, \cdots, \lambda_n$ with $\Sigma_{i=1}^n \lambda_i = 1$ and for each $u \in f_1(\Sigma_{i=1}^n \lambda_i x_i)$, there exist $v_i \in f_1(x_i)$ for $i = 1, \cdots, n$ such that $u \geq \Sigma_{i=1}^n \lambda_i v_i$.

(4) h is said to be lower semicontinuous (resp., upper semicontinuous if for each $\lambda \in \mathbb{R}$, the set $\{x \in X : h(x) \leq \lambda\}$ (resp., $\{x \in X : h(x) \geq \lambda\}$ is closed in X.

(5) H is said to be w^*- demicontinuous if for each $x \in X$, $\lambda \in \mathbb{R}$ and $z \in E$ with $H(x) \subset \{p \in E^* : Re\langle p, z\rangle > \lambda\}$, there exists an open neighborhood N of x in X such that $H(y) \subset \{p \in E^* : Re\langle p, z\rangle > \lambda\}$ for all $y \in N$.

Example 5.5 Let X be a non-empty convex subset of a Banach space $(E, \|\cdot\|)$ and $\psi : X \to \mathbb{R} \cup \{+\infty\}$ be a convex function. We may assume its subdifferential $\partial\psi(x)$ exists for some $x \in X$ (e.g., if ψ is lower semicontinuous and convex by Theorem 5.4.3 of Aubin and Ekeland (1984, p. 262), i.e.,

$$\partial\psi(x) := \{p \in E^* : \psi(x) - \psi(z) \leq Re\langle p, x - z\rangle \text{ for all } z \in X\}.$$

Then the mapping $A : X \to 2^{E^*}$ defined by $A(x) := \partial\psi(x)$ for each $x \in X$ is a monotone mapping. Define $f : X \times X \to 2^{\mathbb{R}}$ by $f(x,y) := \{Re\langle u, x-y\rangle : u \in A(x)\}$ for each $x \in X$. It is clear that f is a monotone mapping. For each fixed positive real number β, define $g : X \times X \to 2^{\mathbb{R}}$ by

$$g_\beta(x,y) := \{Re\langle u, x-y\rangle : u \in A(x)\} + \beta\|x-y\|\}$$

for each $(x,y) \in X \times X$. Then it is obvious that $\{f, g_\beta\}$ is a monotone pair.

Definition 5.5 Let X and Y be two topological spaces, $F : X \to 2^Y$ and $G : X \to 2^{\mathbb{R}}$. Then (a) F is said to be upper semicontinuous (in short, USC) (resp., lower semicontinuous (in short, LSC)) if for each $x \in X$ and for each open set U in Y with $F(x) \subset U$ (resp., $F(x) \cap U \neq \emptyset$), there is an open neighborhood N of x in X such that $F(y) \subset U$ (resp., $F(y) \cap U \neq \emptyset$) for all $y \in N$; (b) the graph of F is the set $\{(x,y) \in X \times Y : y \in F(x)\}$ and (c) G is lower (resp., upper) demicontinuous if for each $x \in X$ and $s \in \mathbb{R}$ with $G(x) \subset (s, \infty)$ (resp., $G(x) \subset (-\infty, s)$), there is an open neighborhood N of x in X such that $G(y) \subset (s, \infty)$ (resp., $G(y) \subset (-\infty, s)$) for all $y \in N$. We note that (i) if G is USC, then G is both lower demicontinuous and upper demicontinuous, (ii) when $X \subset E$, $Y = E^*$ and E^* is equipped with the w^*-topology, if F is USC, then F is w^*- demicontinuous and (iii) when G is single-valued, the notions of lower demicontinuity (resp., upper demicontinuity) and lower semicontinuity (resp., upper semicontinuity) coincide.

Example 5.6 Define $F : [0, \infty) \to 2^{\mathbb{R}}$ by $F(x) = \{x\}$ if $x \geq 1$ or $x = 0$ and $F(x) = [x, 1/x]$ if $0 < x < 1$. Define $G : (-\infty, 0] \to 2^{\mathbb{R}}$ by $G(x) = \{x\}$ if $x \leq -1$ or $x = 0$ and $G(x) = [1/x, x]$ if $-1 < x < 0$. Then it is easy to see that (1) F is both lower demicontinuous and w^*- demicontinuous but not upper semicontinuous and not upper demicontinuous and (2) G is both upper demicontinuous and w^*-demicontinuous but not upper semicontinuous and not lower demicontinuous.

For each non-empty subset A of E and each $r > 0$, let $U(A; r) := \{w \in E^* : \sup_{x \in A} |\langle w, x\rangle| < r\}$. Let $\delta(E^*, E)$ be the topology on E^* generated by the family $\{U(A; r) : A$ is a non-empty bounded subset of E and $r > 0\}$ as a base for the neighborhood system at 0. Then E^*, when equipped with the topology $\delta(E^*, E)$ becomes a locally convex topological vector space. The topology $\delta(E^*, E)$ is called the strong topology on E^*.

5.2.1 *Simultaneous Variational Inequalities*

For the sake of typing convenience we use ψ instead of φ unlike previous section. Let X be a non-empty convex subset of E, $\psi : X \to \mathbb{R}$ and $f, g : X \times X \to \mathbb{R}$. One of the interesting problem is to find a point $x_0 \in X$ which simultaneously satisfies the following inequalities:

$$\psi(x_0) + f(x_0, y) \leq \psi(y) \qquad \text{for all } y \in X \tag{5.32}$$

and

$$\psi(x_0) + g(x_0, y) \leq \psi(y) \qquad \text{for all } y \in X \tag{5.33}$$

i.e., to find a common solution for both variational inequalities (5.32) and (5.33) above. This is the so-called existence problem for solutions of simultaneous variational inequalities and this problem has been studied by Husain and Tarafdar (1994) as given in in the beginning of this chapter. In this subsection, we shall study the existence of solutions for the simultaneous variational inequality problem in the set-valued setting. We first need the following:

Lemma 5.4 *Let f, $g : X \times X \longrightarrow 2^{\mathbb{R}}$.*

(1) *Suppose $\{f, g\}$ is a monotone pair and $x, y \in X$. If* $\inf f(x, y) \leq 0$, *then* $\inf g(y, x) \geq 0$.

(2) *Suppose f is hemicontinuous and for each $x \in X$,* $\inf f(x, x) \leq 0$ *and $y \mapsto f(x, y)$ is concave. If $x_0 \in X$ is such that* $\inf f(y, x_0) \geq 0$ *for all $y \in X$, then* $\inf f(x_0, y) \leq 0$ *for all $y \in X$.*

Proof. (1) If $\inf f(x,y) \leq 0$, then for any $\epsilon > 0$, there exists $u \in f(x,y)$ such that $u < \epsilon$. As $\{f,g\}$ is a monotone pair, for each $w \in g(y,x)$, we have $u + w \geq 0$, so that $w \geq -u > -\varepsilon$. Thus $\inf g(y,x) \geq -\varepsilon$, which implies that $\inf g(y,x) \geq 0$ as $\varepsilon > 0$ is arbitrary.

(2) Assume that $\inf f(y, x_0) \geq 0$ for all $y \in X$, but $\inf f(x_0, y_0) > 0$ for some $y_0 \in X$. Let $s \in \mathbb{R}$ be such that $\inf f(x_0, y_0) > s > 0$. Let $U := (s, \infty)$. Then $f(x_0, y_0) \subset U$. Since f is hemicontinuous, there exists $t_0 \in (0, 1)$ such that $f(z_t, y_0) \subset U$ for all $t \in (0, t_0)$, where $z_t := (1-t)x_0 + ty_0$ for each $t \in [0, 1]$. As $y \mapsto f(z_{t_0}, y)$ is concave, for each $u \in f(z_{t_0}, (1-t)x_0 + ty_0)$, there exist $v_1 \in f(z_{t_0}, x_0)$ and $v_2 \in f(z_{t_0}, y_0)$ such that $u \geq (1-t_0)v_1 + t_0 v_2 >> (1-t_0) \cdot s + t_0 \cdot 0) = (1-t_0)s$ as $\inf f(z_{t_0}, x_0) \geq 0$ by assumption. Hence $\inf f(z_{t_0}, z_{t_0}) = \inf f(z_{t_0}, (1-t)x_0 + ty_0) \geq (1-t_0)s > 0$, which contradicts the assumption that $\inf f(x, x) \leq 0$ for each $x \in X$. □

As an application of Lemma 5.4 2.1, we have the following:

Theorem 5.7 *Let* $f, g : X \times X \longrightarrow 2^{\mathbb{R}}$ *be such that*

(*i*) $\{f, g\}$ *is a monotone pair;*

(*ii*) *for each* $x \in X$, $\inf f(x, x) \leq 0$ *and* $\inf g(x, x) \leq 0$;

(*iii*) f, g *are hemicontinuous;*

(*iv*) *for each* $x \in X$, *the mappings* $y \mapsto f(x, y)$ *and* $y \mapsto g(x, y)$ *are concave.*

Then $x_0 \in X$ *is a solution of the following simultaneous variational inequalities*

$$\begin{cases} \inf f(x_0, y) \leq 0 & \text{for all } \ y \in X \\ \inf g(x_0, y) \leq 0 & \text{for all } \ y \in X \end{cases}$$

if and only if that x_0 *is either a solution of the variational inequality:*

$$\inf f(x_0, y) \leq 0 \quad \text{for all } \ y \in X \tag{5.34}$$

or, a solution of the following variational inequality

$$\inf g(x_0, y) \leq 0 \quad \text{for all } \ y \in X \tag{5.35}$$

Proof. We only need to prove the sufficiency. Suppose $\inf f(x_0, y) \leq 0$ for all $y \in X$. By Lemma 5.4 (1), $\inf g(y, x_0) \geq 0$ for all $y \in X$. By Lemma 5.4 (2), $\inf g(x_0, y) \leq 0$ for all $y \in X$. Similarly, if $\inf g(x_0, y) \leq 0$ for all $y \in X$, then by Lemma 5.4, $\inf f(x_0, y) \leq 0$ for all $y \in X$. □

As an immediate consequence of Theorem 5.7, we have the following result which is Theorem 5.7, i.e. Theorem of Husain and Tarafdar (1994):

Corollary 5.7.1 *Let* X *be a non-empty convex subset of* E *and* $\psi : X \to \mathbb{R}$ *a convex function. Suppose that* $f, g : X \times X \to \mathbb{R}$ *satisfy:*

(1) $\{f, g\}$ *is a monotone pair;*

(2) *for each* $x \in X$, $f(x,x) = g(x,x) = 0$; *and*
(3) *for each fixed* $x \in X$, *both* $f(x,\cdot)$ *and* $g(x,\cdot)$ *are concave; and* f *and* g *are hemicontinuous.*

Then there exists $x_0 \in X$ *is a common solution of both* (5.32) *and* (5.33) *if and only if* x_0 *is either a solution of* (5.32) *or a solution of* (5.33).

Proof. Define $f, g : X \times X \to \mathbb{R}$ by

$$\hat{f}(x,y) := \psi(x) + f(x,y) - \psi(y)$$

and

$$\hat{g}(x,y) := \psi(x) + g(x,y) - \psi(y)$$

for each $(x,y) \in X \times X$. Applying Theorem 5.7 to $\hat{f}$ and $\hat{g}$, the conclusion follows. □

In what follows, we shall prove some sufficient conditions which guarantee that either the inequality (5.34) or (5.35) has a solution. In order to do so, we need the following:

Lemma 5.5 *Let* $g : X \longrightarrow 2^{\mathbb{R}}$ *be lower demicontinuous. Then the mapping* $G : X \longrightarrow \mathbb{R} \cup \{-\infty\}$ *defined by* $G(x) := \inf g(x)$ *for each* $x \in X$ *is lower semicontinuous.*

Proof. Let $\lambda \in \mathbb{R}$ be given. Suppose $\{x_\alpha\}_{\alpha\in\Gamma}$ is a net in X and $x_0 \in X$ such that $\inf g(x_\alpha) \le \lambda$ for all $\alpha \in \Gamma$ and $x_\alpha \longrightarrow x_0$. Suppose $\inf g(x_0) > \lambda$. Choose any $s \in \mathbb{R}$ such that $\inf g(x_0) > s > \lambda$. Let $U := (s, \infty)$, then $g(x_0) \subset U$. Since g is lower demicontinuous, there exists an open neighborhood N of x_0 in X such that $g(x) \subset U$ for all $x \in N$. But then there exists $\alpha_0 \in \Gamma$ such that $x_\alpha \in N$ for all $\alpha \ge \alpha_0$. Hence $g(x_{\alpha_0}) \subset U$ so that $\inf g(x_{\alpha_0}) \ge s > \lambda$ which is a contradiction. Therefore we must have $\inf g(x_0) \le \lambda$. This shows that the set $\{x \in X : \inf g(x) \le \lambda\}$ is closed in X. Thus G is lower semicontinuous. □

Let X be a non-empty subset of a vector space V and $F : X \to 2^V$. W recall that F is said to be a KKM mapping (e.g., see Fan (1961)) if $co\{x_i : i = 1, \cdots, n\} \subset \cup_{i=1}^n F(x_i)$ for each $x_1, \cdots, x_n \in X$ and $n \in \mathbb{N}$.

We shall also need the following simple observation:

Lemma 5.6 *Let* V *be a vector space and* X *a non-empty convex subset of* V. *Suppose* $f : X \times X \longrightarrow 2^{\mathbb{R}}$ *is such that*

(*i*) *for each* $x \in X$, $\inf f(x, x) \le 0$;
(*ii*) *for each* $x \in X$, $y \mapsto f(x, y)$ *is concave.*

Define $F : X \longrightarrow 2^X$ *by* $F(w) = \{x \in X : \inf f(x, w) \leq 0\}$ *for each* $w \in X$. *Then* F *is a KKM-mapping.*

Proof. Suppose not, then there exist $n \in \mathbb{N}$, $w_1, \ldots, w_n \in X$ and $\lambda_1, \ldots, \lambda_n > 0$ with $\sum_{i=1}^n \lambda_i = 1$ such that $\sum_{i=1}^n \lambda_i w_i \notin \bigcup_{j=1}^n F(w_j)$. It follows that $\inf f(\sum_{i=1}^n \lambda_i w_i, w_j) > 0$ for all $j = 1, \ldots, n$. Let $s \in \mathbb{R}$ be such that $\min_{1\leq j\leq n} \inf f(\sum_{i=1}^n \lambda_i w_i, w_j) > s > 0$. Since $y \mapsto f(\sum_{i=1}^n \lambda_i w_i, y)$ is concave by (ii), for each $u \in f(\sum_{i=1}^n \lambda_i w_i, \sum_{j=1}^n \lambda_j w_j)$, there exist $v_j \in f(\sum_{i=1}^n \lambda_i w_i, w_j)$ for $j = 1, \ldots, n$ such that $u \geq \sum_{j=1}^n \lambda_j v_j > s$. Thus $\inf f(\sum_{i=1}^n \lambda_i w_i, \sum_{j=1}^n \lambda_j w_j) \geq s > 0$, which contradicts (i). Hence F must be a KKM-mapping. □

Theorem 5.8 *Let* X *be a non-empty closed convex subset of* E. *Suppose* $f : X \times X \longrightarrow 2^{\mathbb{R}}$ *is such that*

(*i*) *for each* $x \in X$, $\inf f(x, x) \leq 0$,

(*ii*) *for each* $x \in X$, $y \mapsto f(x, y)$ *is concave,*

(*iii*) *for each* $y \in X$, $x \mapsto f(x, y)$ *is lower demicontinuous,*

(*iv*) *there exist a non-empty compact subset* B *of* X *and* $w_0 \in B$ *such that*

$$\inf f(x, w_0) > 0 \text{ for all } x \in X \setminus B.$$

Then the set $S := \{x \in X : \inf f(x, w) \leq 0$ *for all* $w \in X\}$ *is a non-empty compact subset of* B.

Proof. Define $F : X \to 2^X$ by

$$F(w) := \{x \in X : \inf f(x, w) \leq 0\}$$

for each $w \in X$. By(i), $F(w) \neq \emptyset$ for all $w \in X$, so that F is well-defined. By (iii) and Lemma 5.5, for each $w \in X$, the set $F(w)$ is closed in X. By (iv), $F(w_0)$ is a closed subset of B so that $F(w_0)$ is compact. By (i), (ii) and Lemma 5.6, F is a KKM-mapping. By Ky Fan's KKM mapping principle, $\bigcap_{w\in X} F(w) \neq \emptyset$. Thus $S = \bigcap_{w\in X} F(w)$ is a non-empty compact subset of B. □

Lemma 5.7 *Let* X *be a non-empty closed convex subset of* E. *Suppose* $g : X \longrightarrow 2^{\mathbb{R}}$ *and let* $W := \{x \in X : \inf g(x) \geq 0\}$. *Then* (*a*) W *is closed in* X *if* g *is LSC and* (*b*) W *is convex if* g *is concave.*

Proof. (a) If W were not closed in X, then there would exist a net $\{x_\alpha\}_{\alpha\in\Gamma}$ in X and $x_0 \in X$ such that $x_\alpha \longrightarrow x_0$, and $\inf g(x_\alpha) \geq 0$ for all $\alpha \in \Gamma$ but $\inf g(x_0) < 0$.

Let $s \in \mathbb{R}$ be such that $\inf g(x_0) < s < 0$ and $U := (-\infty, s)$. Then $g(x_0) \cap U \neq \emptyset$. Since g is LSC, there exists an open neighborhood N of $x_0 \in X$ such that $g(x) \cap U \neq \emptyset$ for all $x \in N$. As $x_\alpha \longrightarrow x$, there exists $\alpha_0 \in \Gamma$ such that $x_\alpha \in N$ for all $\alpha \geq \alpha_0$. Thus $g(x_{\alpha_0}) \cap U \neq \emptyset$ so that $\inf g(x_{\alpha_0}) < s < 0$, which is a contradiction. Thus W is closed in X.

(b) Suppose $x, y \in W$ and $\lambda \in (0, 1)$, then $\inf g(x) \geq 0$ and $\inf g(y) \geq 0$. Since g is concave, for each $u \in g(\lambda x + (1 - \lambda)y)$, there exist $v_1 \in g(x)$ and $v_2 \in g(y)$ such that $u \geq \lambda v_1 + (1 - \lambda)v_2 \geq 0$. Thus $\inf g(\lambda x + (1 - \lambda)y) \geq 0$ and we have $\lambda x + (1 - \lambda)y \in W$. Therefore W is convex. □

Theorem 5.9 *Let X be a non-empty closed convex subset of E and $f : X \times X \longrightarrow 2^{\mathbb{R}}$ be such that*

(*i*) *for each $x \in X, \inf f(x, x) \leq 0$,*

(*ii*) *for each $x \in X, y \mapsto f(x, y)$ is concave and LSC,*

(*iii*) *f is hemicontinuous,*

(*iv*) *there exist a non-empty compact $B \subset X$ and $w_0 \in B$ such that*

$$\inf f(x, w_0) > 0 \text{ for all } x \in X \setminus B;$$

(*v*) *f is monotone.*

Then the set $S := \{x \in X : \inf f(x, w) \leq 0 \text{ for all } w \in X\}$ is a non-empty compact convex subset of B.

Proof. Define $F, G, H : X \longrightarrow 2^X$ by

$$\begin{aligned} F(w) &= \{x \in X : \inf f(x, w) \leq 0\}, \\ G(w) &= cl_X F(w), \\ H(w) &= \{x \in X : \inf f(w, x) \geq 0\}, \end{aligned}$$

for each $w \in X$. Then by (i), (ii) and Lemma 5.6, F is a KKM-mapping so that G is also a KKM-mapping. Note that by (iv), $F(w_0) \subset B$ so that $G(w_0) \subset B$ and $G(w_0)$ is compact.

Again by Ky Fan's KKM mapping principle, $\bigcap_{w \in X} G(w) \neq \emptyset$. By (ii) and Lemma 5.7(a), for each $w \in X$, $H(w)$ is closed and convex.

To complete the proof, it sufficient to show that

$$S = \bigcap_{w \in X} F(w) = \bigcap_{w \in X} G(w) = \bigcap_{w \in X} H(w).$$

Indeed, if $w \in X$ and $x \in F(w)$, then $\inf f(x, w) \leq 0$ so that by (v) and Lemma 5.4 (1), $\inf f(w, x) \geq 0$. It follows that $x \in H(w)$. Hence $F(w) \subset H(w)$ so that $G(w) \subset H(w)$. Therefore $\bigcap_{w \in X} F(w) \subset \bigcap_{w \in X} G(w) \subset \bigcap_{w \in X} H(w)$.

On the other hand, if $x \in \bigcap_{w\in X} H(w)$, then $\inf f(w,, x) \geq 0$ for all $w \in X$. Thus by (i), (ii), (iii), (v) and Lemma 5.4(2), we have $\inf f(x, w) \leq 0$ for all $w \in X$. Thus $x \in \bigcap_{w\in X} F(w)$. Therefore $\bigcap_{w\in X} H(w) \subset \bigcap_{w\in X} F(w)$. Hence we have $S = \bigcap_{w\in X} F(w) = \bigcap_{w\in X} G(w) = \bigcap_{w\in X} H(w)$. □

5.2.2 *Implicit Variational Inequalities — The Monotone Case*

Let C be a non-empty subset of E and C_1 a non-empty subset of C. Suppose $f : C_1 \times C \times C \to \mathbb{R}$ and $g : C_1 \times C \to \mathbb{R}$ are such that $f(u, v, v) \geq 0$ for all $u \in C_1$ and $v \in C$. Mosco in Mosco (1976) had investigated the following so called implicit variational inequality problems:
Find a vector $v \in C_1$ such that

$$g(v, v) \leq f(v, v, w) + g(v, w) \text{ for all } w \in C. \tag{5.36}$$

In this subsection, it is our goal to study the existence of solutions for implicit variational inequality and implicit quasivariational inequalities which are variant forms of the implicit variational inequality (5.36) above. Indeed, as applications of Theorem 5.9 and by combining Fan-Glicksberg fixed point theorem, we shall provide some sufficient conditions to guarantee the existence of variational and quasivariational inequalities in their implicit forms, and in which the set-valued mappings are monotone.

As an application of Theorem 5.9, we have the following variational inequality:

Theorem 5.10 *Let X be a non-empty closed convex subset of E and $T : X \longrightarrow 2^{E^*}$ be monotone such that*

(*i*) *for each $x \in X$, $T(x)$ is w^*-compact;*
(*ii*) *T is w^*-upper semicontinuous from line segments in X to the weak*-topology $\sigma(E^*, E)$ on E^**
(*iii*) *there exist a non-empty weakly compact subset B of X and $w_0 \in B$ such that*

$$\underset{u}{\in} T(x) \to \inf Re\langle u, x - w_0\rangle > 0 \quad \textit{for all } x \in X \setminus B.$$

Then the set $S := \{y \in X : \underset{w}{\in} Ty \to \inf Re\langle w, y - x\rangle \leq 0$ for all$x \in X\}$ is a non-empty weakly compact convex subset of B.

Proof. Define $f : X \times X \longrightarrow 2^{\mathbb{R}}$ by

$$f(x, y) = \{Re\langle u, x - y\rangle : u \in Tx\}$$

for each $x, y \in X$. Then we have

(1) f is monotone as T is monotone.
(2) For each $x, y \in X$, $f(x, y)$ is a non-empty compact subset of $\mathbb{R}$;

(3) For each $x \in X$, $f(x,x) = \{0\}$ so that $\inf f(x, x) \leq 0$.

(4) For each $x \in X$, the mapping $y \mapsto f(x, y)$ is concave. Indeed, for each $n \in \mathbb{N}$, $y_1, \cdots, y_n \in X$ and $\lambda_1, \cdots, \lambda_n \in [0,1]$ with $\Sigma_{i=1}^n \lambda_i = 1$ and for each $s \in f(x, \sum_{i=1}^n \lambda_i y_i)$, there exists $u \in Tx$ such that $s = Re\langle u, x - \sum_{i=1}^n \lambda_i y_i \rangle$. But then $Re\langle u, x - y_i \rangle \in f(x, y_i)$ for each $i = 1, 2, \ldots, n$ and

$$s = Re\langle u, x - \sum_{i=1}^{n} \lambda_i y_i \rangle = \sum_{i=1}^{n} Re\langle u, x - y_i \rangle.$$

Therefore $y \mapsto f(x, y)$ is concave.

(5) For each $x \in X$, the mapping $y \mapsto f(x, y)$ is weakly LSC; i.e., the mapping $y \mapsto f(x,y)$ is LSC when X is equipped with the relative weak topology. Indeed, let $y_0 \in X$ and $U \subset \mathbb{R}$ be open such that $f(x, y_0) \cap U \neq \emptyset$. Then there exists $u \in Tx$ such that $Re\langle u, x - y_0 \rangle \in U$. For each fixed $x \in X$ and $u \in T(x)$, as $y \mapsto Re\langle u, x - y \rangle$ is weakly continuous, there exists a weakly open neighborhood N of y_0 in X such that $Re\langle u, x - y \rangle \in U$ for all $y \in N$, so that $f(x, y) \cap U \neq \emptyset$ for all $y \in N$. Thus $y \mapsto f(x, y)$ is weakly LSC.

(6) f is hemicontinuous. Indeed, fix any $x, y \in X$ and define $k : [0, 1] \longrightarrow X$ by $k(t) = f((1-t)x + ty, y)$ for each $t \in [0, 1]$. Let $U = (s, \infty)$ where $s \in \mathbb{R}$ be such that $f(x, y) \subset U$. Note that $f(x, y)$ is compact as Tx is weak*-compact. Let $r_0 = \inf f(x,y)$. Then $r_0 > s$. Set $r := (r_0 + s)/2, t_1 := (r - s)/r$ and $V := (r, \infty)$. Then $t_1 \in (0,1)$, $f(x,y) \subset V$ and $(1-t)V \subset U$ for all $t \in (0, t_1)$. Let $W = \{w \in E^* : Re\langle w, x - y \rangle > r\}$, then W is w^*-open and $T(x) \subset W$. By (*ii*), there exists $t_0 \in (0, t_1)$ such that $T((1-t)x + ty) \subset W$ for all $t \in (0, t_0)$. Thus for each $u \in T((1-t)x + ty)$ and $t \in (0, t_0)$, we have

$$U \supset (1-t)V \supset (1-t)Re\langle u, x - y \rangle = Re\langle u, ((1-t)x + ty) - y \rangle.$$

Therefore $U \supset f((1-t)x + ty, y)$ for all $t \in (0, t_0)$. Hence f is hemicontinuous.

(7) By (iii), there exists a non-empty weakly compact subset B of X and $w_0 \in B$ such that

$$\inf f(x, w_0) = \inf_{u \in Tx} Re\langle u, x - w_0 \rangle > 0.$$

for all $x \in X \setminus B$.

Now equip E with weak topology, then all hypotheses of Theorem 5.9 are satisfied. Thus

$$\begin{aligned} S &= \{x \in X : \inf f(x, w) \leq 0 \text{ for all } w \in X\} \\ &= \{x \in X : \inf_{u \in Tx} Re\langle u, x - w \rangle \leq 0 \text{ for all } w \in X\} \end{aligned}$$

is a non-empty weakly compact convex subset of B. □

As an application of Theorem 5.10, we have the following result which is Theorem 1 of Shih and Tan (1988b):

Corollary 5.10.1 *Let $(E, \|\cdot\|)$ be a reflexive Banach space and X a non-empty closed convex subset of E. Suppose $T: X \longrightarrow 2^{E^*}$ is monotone such that each $T(x)$ is a weakly compact subset of E^* and T is upper semicontinuous from line segments in X to the weak topology of E^*. Assume that there exists $x_0 \in X$ such that*

$$\lim_{\substack{\|y\| \to \infty \\ y \in X}} \inf_{w \in Ty} Re\langle w, y - x_0 \rangle > 0. \tag{5.37}$$

Then there exists $\hat{y} \in X$ such that

$$\inf_{w \in T\hat{y}} Re\langle w, \hat{y} - x \rangle \le 0 \quad \text{for all } x \in X.$$

Proof. By (5.37), there exist $M > 0$ and $R > 0$ with $||x_0|| \le R$ such that $\inf_{w \in Ty} Re\langle w, y - x_0 \rangle > M$ for all $y \in X$ with $||y|| > R$. Let $B := \{x \in X : ||x|| \le R\}$. Then B is a non-empty weakly compact (and convex) subset of X such that $\inf_{w \in Ty} Re\langle w, y - x_0 \rangle > 0$ for all $x \in X \setminus B$. It is easy to see that all hypotheses of Theorem 5.10 are satisfied so that the conclusion follows. □

We note that under the assumptions in Corollary 5.10.1, the conditions "T is upper semicontinuous from line segments in X to the weak topology of E" and "T is w^*-demicontinuous from line segments in X to the w^*-topology of E" are equivalent (see e.g., [1, Theorem 10, p.128]).

As a second application of Theorem 5.9, we have the following implicit variational inequality:

Theorem 5.11 *Let E be locally convex, X be a non-empty compact convex subset of E and $g : X \times X \times X \longrightarrow K(\mathbb{R})$ be such that*

(i) For each $u, x \in X$, $\inf g(u, x, x) \le 0$.

(ii) For each $u, x \in X$, the mapping $y \mapsto g(u, x, y)$ is concave.

(iii) For each $u \in X$, the mapping $(x, y) \mapsto g(u, x, y)$ is monotone and hemicontinuous.

(iv) For each $x \in X$, the mapping $(u, y) \mapsto g(u, x, y)$ is LSC.

Then the set $W := \{u \in X : \inf f(u, u, w) \le 0$ for all $w \in X\}$ is a non-empty compact subset of X.

Proof. For each fixed $u \in X$, define $f_u : X \times X \longrightarrow 2^{\mathbb{R}}$ by

$$f_u(x, y) = g(u, x, y)$$

for each $x, y \in X$. Then f_u satisfies all hypotheses in Theorem 5.9 so that the set

$$\begin{aligned} S(u) &= \{x \in X : \inf f_u(x, w) \le 0 \text{ for all } w \in X\} \\ &= \{x \in X : \inf g(u, x, w) \le 0 \text{ for all } w \in X\} \end{aligned}$$

is a non-empty compact convex subset of X and S is thus a mapping from X to $K(X)$. We shall show that S has a closed graph. Indeed, let $(x_\alpha)_{\alpha\in\Gamma}$ be a net in X and $y_\alpha \in S(x_\alpha)$ for all $\alpha \in \Gamma$ such that $x_\alpha \longrightarrow x_0 \in X$ and $y_\alpha \longrightarrow y_0 \in X$. Note that for each $\alpha \in \Gamma$, $\inf g(x_\alpha, y_\alpha, w) \leq 0$ for all $w \in X$. Let $w \in X$ be given and fix an arbitrary $\alpha \in \Gamma$. Since $g(x_\alpha, y_\alpha, w)$ is compact, there exists $u_\alpha \in g(x_\alpha, y_\alpha, w)$ such that $u_\alpha = \inf g(x_\alpha, y_\alpha,, w) \leq 0$. Since $(y, z) \mapsto g(x_\alpha, y, z)$ is monotone, for each $v \in g(x_\alpha, w, y_\alpha)$, we have $u_\alpha + v \geq 0$ so that $v \geq -u_\alpha \geq 0$. Thus $\inf g(x_\alpha, w, y_\alpha) \geq 0$. As $w \in X$ is arbitrarily given, $\inf g(x_\alpha, w, y_\alpha) \geq 0$ for all $w \in X$. By (iv) and Lemma 5.7, for each $w \in X$, the set $\{(x, y) \in X \times X : \inf g(x, w, y) \geq 0\}$ is closed. It follows that $\inf g(x_0, w, y_0) \geq 0$ for all $w \in X$. By Lemma 5.4(2), $\inf g(x_0, y_0, w) \leq 0$ for all $w \in X$ which shows that $y_0 \in S(x_0)$. Hence S has a closed graph so that S is upper semicontinuous. Now by Fan-Glicksberg fixed point theorem (e.g., see Fan (1952) or Glicksberg (1952)), there exists $\hat{x} \in X$ such that $\hat{x} \in S(\hat{x})$, i.e., $\inf g(\hat{x}, \hat{x}, w) \leq 0$ for all $w \in X$ so that $W \neq \emptyset$. To complete the proof, it remains to show that W is a closed subset of X. Suppose $\{u_\alpha\}_{\alpha\in\Gamma}$ is a net in W such that $u_\alpha \longrightarrow u_0 \in X$. Then $\inf g(u_\alpha, u_\alpha, w) \leq 0$ for all $w \in X$. Now by the same argument as above (with $y_\alpha = x_\alpha = u_\alpha$ for all $\alpha \in \Gamma$ and $x_0 = y_0 = u_0$), $\inf g(u_0, u_0, w) \leq 0$ for all $w \in X$. Thus $u_0 \in S(u_0)$ so that $u_0 \in W$. Therefore W is closed in X. □

As an application of Theorem 5.11, we have the following implicit quasi-variational inequality:

Theorem 5.12 *Let E be locally convex, X be a non-empty compact convex subset of E, $S : X \longrightarrow KC(X)$ be continuous and $g : X \times X \times X \longrightarrow 2^{\mathbb{R}}$ be such that*

(i) *For each $u, x \in X$, $\inf g(u, x, x) \leq 0$.*

(ii) *For each $u, x \in X$, the mapping $y \mapsto g(u, x, y)$ is concave and for each $y \in X$, the mapping $u \mapsto g(u, y, u)$ is concave.*

(iii) *For each $u \in X$, the mapping $(x, y) \mapsto g(u, x, y)$ is monotone and hemicontinuous.*

(iv) *For each $x \in X$, the mapping $(u, y) \mapsto g(u, x, y)$ is LSC.*

(v) *The mapping $(u, x) \mapsto g(u, x, u)$ is LSC.*

Then (a) there exists $\hat{y} \in X$ such that

$$\begin{cases} \hat{y} \in S(\hat{y}) \\ \inf g(\hat{y}, \hat{y}, w) \leq 0 \quad \text{for all } w \in S(\hat{y}) \end{cases}$$

and (b) the set

$$\{y \in X : y \in S(y) \text{ and } \inf g(y, y, w) \leq 0 \text{ for all } w \in S(y)\}$$

is a non-empty compact subset of X.

Proof. (a) Define $F : X \longrightarrow KC(X)$ by

$$F(u) = \{y \in S(u) : \inf g(y, y, w) \leq 0 \text{ for all } w \in S(u)\}$$

for each $u \in X$. Let $u \in X$ be given. By Theorem 5.11, $F(u)$ is non-empty and compact. We shall now show that $F(u)$ is also convex. Let $x, y \in F(u)$ and $\lambda \in (0, 1)$ be given. As $x, y \in S(u)$ and $S(u)$ is convex, $\lambda x + (1-\lambda)y \in S(u)$. Since $\inf g(x, x, w) \leq 0$ and $\inf g(y, y, w) \leq 0$ for all $w \in S(u)$, $\inf g(x, w, x) \geq 0$ and $\inf g(y, w, y) \geq 0$ for all $w \in S(u)$ by (iii) and Lemma 5.4 (1). It follows that inf $g(\lambda x + (1-\lambda)y, w, \lambda x + (1-\lambda)y) \geq 0$ for all $w \in S(u)$ by (ii) and Lemma 5.7. By Lemma 5.4(2), $\inf g(\lambda x + (1-\lambda)y, \lambda x + (1-\lambda)y, w) \leq 0$ for all $w \in S(u)$. Thus $\lambda x + (1-\lambda)y \in F(u)$. Hence $F(u)$ is also convex. This shows that F is well-defined.

Now we shall show that F has a closed graph. Indeed, let $((x_\alpha, y_\alpha))_{\alpha\in\Gamma}$ be a net in $X \times X$ and $(x_0, y_0) \in X \times X$ be such that $(x_\alpha, y_\alpha) \longrightarrow (x_0, y_0)$ and $y_\alpha \in F(x_\alpha)$ for all $\alpha \in \Gamma$. Since $y_\alpha \in S(x_\alpha)$ for each $\alpha \in \Gamma$, $y_0 \in S(x_0)$ as S is USC. Now fix an arbitrary $w_0 \in S(x_0)$. Since S is LSC, there is a net $(w_\alpha)_{\alpha\in\Gamma}$ in X with $w_\alpha \in S(x_\alpha)$ for all $\alpha \in \Gamma$ such that $w_\alpha \longrightarrow w_0$. Since $\inf g(y_\alpha, y_\alpha, w_\alpha) \leq 0$ for all $\alpha \in \Gamma$, by (iii) and Lemma 5.4 (1), we have $\inf g(y_\alpha, w_\alpha, y_\alpha) \geq 0$ for all $\alpha \in \Gamma$. By (v) and Lemma 5.7, $\inf g(y_0, w_0, y_0) \geq 0$. Since $w_0 \in S(x_0)$ is arbitrary, we have $\inf g(y_0, w, y_0) \geq 0$ for all $w \in S(x_0)$. By (ii), (iii) and Lemma 5.4(2), it follows that $\inf g(y_0, y_0, w) \leq 0$ for all $w \in S(x_0)$ so that $y_0 \in F(x_0)$. Thus F has a closed graph and hence F is USC.

By Fan-Glicksberg fixed point theorem again, there exists $\hat{y} \in X$ such that $\hat{y} \in F(\hat{y})$; i.e.,

$$\begin{cases} \hat{y} \in S(\hat{y}) \\ \inf g(\hat{y}, \hat{y}, w) \leq 0 \quad \text{for all } w \in S(\hat{y}). \end{cases}$$

(b) By (a), the set $\{y \in X : y \in S(y) \text{ and } \inf g(y, y, w) \leq 0 \text{ for all } w \in S(y)\}$ is non-empty; it is also compact by following the same argument as in the proof of Theorem 5.11. □

We would like to remark that our results in this subsection unify and generalize corresponding results in the literatures given in Aubin and Ekeland (1984), Baiocchi and Capelo (1984), Harker and Pang (1990), Husain and Tarafdar (1994), Mosco (1976), Shih and Tan (1985) and Shih and Tan (1988a).

5.2.3 *Implicit Variational Inequalities — The USC Case*

Parallel to the ideas used in previous subsection and as application of Theorem 5.8 instead of Theorem 5.9, we can also study the existence of solutions for implicit variational and implicit quasi-variational inequalities in which real set-valued mappings are upper semicontinuous instead of being monotone. First we have the following implicit variational inequality:

Theorem 5.13 *Let E be locally convex, X be a non-empty compact convex subset of E and $g : X \times X \times X \longrightarrow K(\mathbb{R})$ be such that*

(i) *For each $u \in X$, $\inf g(u, x, x) \leq 0$.*

(ii) *For each $u, x \in X$, the mapping $y \mapsto g(u, x, y)$ is concave.*

(iii) *For each $y \in X$, the mapping $(u, x) \mapsto g(u, x, y)$ is lower demicontinuous.*

Then the set $W := \{u \in X : \inf f(u, u, w) \leq 0$ for all $w \in X\}$ is a non-empty compact subset of X.

Proof. For each fixed $u \in X$, define $f_u : X \times X \longrightarrow 2^{\mathbb{R}}$ by

$$f_u(x, y) = g(u, x, y)$$

for each $x, y \in X$. Then f_u satisfies all hypotheses in Theorem 5.8 so that the set

$$\begin{aligned} S(u) &= \{x \in X : \inf f_u(x, w) \leq 0 \text{ for all } w \in X\} \\ &= \{x \in X : \inf g(u, x, w) \leq 0 \text{ for all } w \in X\} \end{aligned}$$

is a non-empty compact convex subset of X and S is thus a mapping from X to $K(X)$. We shall now show that S has a closed graph. Indeed, let $(x_\alpha)_{\alpha \in \Gamma}$ be a net in X and $y_\alpha \in S(x_\alpha)$ for each $\alpha \in \Gamma$ such that $x_\alpha \longrightarrow x_0 \in X$ and $y_\alpha \longrightarrow y_0 \in X$. Note that for each $\alpha \in T$, $\inf g(x_\alpha, y_\alpha, w) \leq 0$ for all $w \in X$. By (iii) and Lemma 5.5, for each $w \in X$, the mapping $(x, y) \mapsto \inf g(x, y, w)$ is lower semicontinuous. It follows that $\inf g(x_0, y_0, w) \leq 0$ for all $w \in X$ so that $y_0 \in S(x_0)$. Thus S has a closed graph and hence is USC. Now by Fan-Glicksberg fixed point theorem, there exists $\hat{x} \in X$ such that $\hat{x} \in S(\hat{x})$, i.e., $\inf g(\hat{x}, \hat{x}, w) \leq 0$ for all $w \in X$. This shows that $\hat{x} \in W$ so that the set W is non-empty. Moreover, by (iii) and Lemma 5.5, the set W is closed in X and hence is compact. □

So far, we have established some existence theorems of solutions for implicit variational inequalities and quasi-variational inequalities as applications of Fan-Glicksberg fixed point theorem. However, we can also study variational inequalities as applications of existence theorems of equilibria for generalized games (resp., abstract economics). Some results in this direction have been given in Tarafdar and Yuan (1994). In what follows, we shall use that method to prove an implicit quasi-variational inequality (Theorem 5.14 below). We need the following result which is a special case of Theorem 5 of Tulcea (1988) (see also Yuan (1998):

Lemma 5.8 *Let E be locally convex, X be a non-empty compact convex subset of E, $A : X \longrightarrow KC(X)$ be USC and $P : X \longrightarrow 2^X \cup \{\emptyset\}$ be such that*

(i) *For each $y \in X$, the set $P^{-1}(y) := \{x \in X : y \in P(x)\}$ is open in X.*

(ii) *For each $x \in X$, $x \notin coP(x)$.*

(iii) *The set $\{x \in X : A(x) \cap P(x) \neq \emptyset\}$ is open in X.*

Then there exists $\hat{x} \in X$ such that $\hat{x} \in A(\hat{x})$ and $A(\hat{x}) \cap P(\hat{x}) = \emptyset$.

We shall now apply Lemma 5.8 to prove the following implicit quasi-variational inequality:

Theorem 5.14 *Let E be locally convex, X be a non-empty compact convex subset of E, $S : X \longrightarrow KC(X)$ be continuous (i.e., S is both LSC and USC on X) and $f : X \times X \longrightarrow 2^{\mathbb{R}}$ be lower demicontinuous such that*

(i) *For each $x \in X$, $\inf f(x, x) \leq 0$.*

(ii) *For each $x \in X$, $y \mapsto f(x, y)$ is concave.*

Then there exists $u \in X$ such that

$$\begin{cases} u \in S(u) & \textit{and} \\ \inf f(u, w) \leq 0, & \textit{for all } w \in S(u). \end{cases}$$

Proof. Define $P : X \longrightarrow 2^X \cup \{\emptyset\}$ by

$$P(x) = \{y \in X : \inf f(x, y) > 0\}$$

for each $x \in X$. We then have:

(1) For each $y \in X$, the set $P^{-1}(y)$ is open in X by Lemma 5.5 as $x \mapsto f(x, y)$ is lower demicontinuous.

(2) For each $x \in X$, $x \notin coP(x)$. Indeed, suppose there exists $x_0 \in X$ such that $x_0 \in coP(x_0)$. Let $y_1, \ldots, y_n \in P(x_0)$, $\lambda_1, \ldots, \lambda_n > 0$ with $\sum_{i=1}^{n} \lambda_i = 1$ be such that $x_0 = \sum_{i=1}^{n} \lambda_i y_i$. As $y \mapsto f(x_0, y)$ is concave, for each $u \in f(x_0, x_0) = f(x_0, \sum_{i=1}^{n} \lambda_i y_i)$, there exist $u_i \in f(x_0, y_i)$ for $i = 1, \ldots, n$ such that $u \geq \sum_{i=1}^{n} \lambda_i u_i \geq \sum_{i=1}^{n} \lambda_i \inf f(x_0, y_i)$. Then $\inf f(x_0, x_0) \geq \sum_{i=1}^{n} \lambda_i \inf f(x_0, y_i) > 0$, which contradicts (i). Hence $x \notin coP(x)$ for all $x \in X$.

(3) The set $\{x \in X : S(x) \cap P(x) \neq \emptyset\}$ is open in X. Indeed, suppose $S(x_0) \cap P(x_0) \neq \emptyset$. Let $y_0 \in S(x_0) \cap P(x_0)$. Then $y_0 \in S(x_0)$ and $\inf f(x_0, y_0) > 0$. Let $s \in \mathbb{R}$ be such that $\inf f(x_0, y_0) > s > 0$ and $U := (s, \infty)$. Since f is lower demicontinuous and $f(x_0, y_0) \subset U$, there exist open neighborhoods N_1 of x_0 in X and V of y_0 in X such that $f(x, y) \subset U$ for all $(x, y) \in N_1 \times V$. Since $V \cap S(x_0) \neq \emptyset$ and S is LSC, there exists an open neighborhood N_2 of x_0 in X such that $V \cap S(x) \neq \emptyset$ for all $x \in N_2$. Let $N := N_1 \cap N_2$. Then N is an open neighborhood of x_0 in X. Suppose $x \in N$ is given. As $V \cap S(x) \neq \emptyset$, we may take any $y \in V \cap S(x)$; then $f(x, y) \subset U$ so that $\inf f(x, y) \geq s > 0$ and hence $y \in P(x) \cap S(x)$. Thus $S(x) \cap P(x) \neq \emptyset$ for all $x \in N$. Therefore the set $\{x \in X : S(x) \cap P(x) \neq \emptyset\}$ is open in X.

Now by Lemma 5.8, there exists $\hat{y} \in X$ such that $\hat{y} \in S(\hat{y})$ and $S(\hat{y}) \cap P(\hat{y}) = \emptyset$, i.e.,

$$\begin{cases} \hat{y} \in S(\hat{y}) \\ \inf f(\hat{y}, w) \leq 0 \text{ for all } w \in S(\hat{y}). \end{cases}$$ □

Lemma 5.9 *Let X be a non-empty and bounded subset of E and $T : X \longrightarrow K(E^*)$ be USC, where E^* is equipped with the strong topology. Define $f : X \times X \longrightarrow 2^{\mathbb{R}}$ by*

$$f(x, y) = \{Re\langle u, x - y\rangle : u \in Tx\} \quad \textit{for all } x, y \in X.$$

Then f is USC.

Proof. Let $x_0, y_0 \in X$ and $U \subset \mathbb{R}$ be open such that

$$\{Re\langle u, x_0 - y_0\rangle : u \in Tx_0\} = f(x_0, y_0) \subset U.$$

Note that the mapping $(u, z) \mapsto \langle u, z\rangle$ is (jointly) continuous on $(X - X) \times E^*$. Thus for each $u \in Tx_0$, there exist a strongly open neighborhood V_u of u and an open neighborhood M_u of x_0 in X and an open neighborhood N_u of y_0 in X such that

$$\{Re\langle v, w - z\rangle : v \in V_u, w \in M_u, z \in N_u\} \subset U.$$

Since $Tx_0 \subset \bigcup_{w \in Tx_0} V_u$ and Tx_0 is strongly compact, there exist $u_1, \ldots, u_n \in Tx_0$ such that $Tx_0 \subset \bigcup_{i=1}^{n} V_{u_i}$. Since T is USC, there exists an open neighborhood M_1 of x_0 in X such that $Tx \subset \bigcup_{i=1}^{n} V_{u_i}$ for all $x \in M_1$. Let $M_{x_0} := M_1 \cap \bigcap_{i=1}^{n} M_{u_i}$ and $N_{y_0} := \bigcap_{i=1}^{n} N_{u_i}$. Then M_{x_0} and N_y are open neighborhoods of x_0 and y_0 in X, respectively. Now suppose $x \in M_{x_0}$, $y \in N_{y_0}$, and $u \in Tx$ are given. Let $i_0 \in \{1, \ldots, n\}$ be such that $u \in V_{u_{i_0}}$. As $x \in M_1 \cap M_{u_{i_0}}$ and $y \in N_{u_{i_0}}$, $Re\langle u, x - y\rangle \in U$. It follows that $f(x, y) \subset U$ for all $x \in M_{x_0}$ and $y \in N_{y_0}$. Therefore f is USC. □

By combining both Theorem 5.14 and Lemma 5.9, we have the following result which is Theorem 4 in Shih and Tan (1985):

Corollary 5.14.1 *Let E be locally convex, X be a non-empty compact convex subset of E, $S : X \longrightarrow KC(X)$ be continuous and $T : X \longrightarrow K(E^*)$ be USC, where E^* is equipped with the strong topology. Then exists $\hat{y} \in X$ such that*

$$\begin{cases} \hat{y} \in S(\hat{y}), \\ \inf_{w \in T\hat{y}} Re\langle w, \hat{y} - x\rangle \leq 0 \quad \textit{for all } x \in S(\hat{y}). \end{cases}$$

Proof. Define $f : X \times X \longrightarrow 2^{\mathbb{R}}$ by

$$f(x,\, y) = \{Re\langle u,\, x - y\rangle : u \in Tx\}$$

for each $x,\, y \in X$. By Lemma 5.9, f is USC. Now the conclusion follows from Theorem 5.14. □

Finally, we have the following implicit quasi-variational inequality:

Theorem 5.15 *Let E be locally convex, X be a non-empty compact convex subset of E, $S : X \longrightarrow KC(X)$ be continuous and $g : X \times X \times X \longrightarrow 2^{\mathbb{R}}$ be such that*

(i) For each $u,\, x \in X$, $\inf g(u,\, x,\, x) \leq 0$.

(ii) For each $u, y \in X$, the mapping $w \mapsto g(u, y, w)$ is concave.

(iii) g is lower demicontinuous on $X \times X \times X$.

(iv) For each $(u, w) \in X \times X$, the mapping $y \mapsto inf\, g(u, y, w)$ is convex.

Then (a) there exists $\hat{y} \in X$ such that

$$\begin{cases} \hat{y} \in S(\hat{y}) \\ \inf g(\hat{y},\, \hat{y},\, w) \leq 0 \ \ \textit{for all } w \in S(\hat{y}) \end{cases}$$

and (b) the set

$$\{y \in X : y \in S(y) \ \textit{and} \ \inf g(y,\, y,\, w) \leq 0 \ \textit{for all } w \in S(y)\}$$

is a (non-empty) compact subset of X.

Proof. Define $F : X \longrightarrow KC(X)$ by

$$F(u) = \{y \in S(u) : \inf g(u,\, y,\, w) \leq 0 \quad \text{for all } w \in S(u)\}$$

for each $u \in X$. By Theorem 5.8, F is non-empty valued. Now we shall show that F has a closed graph. Indeed, let $((x_\alpha,\, y_\alpha))_{\alpha\in\Gamma}$ be a net in $X \times X$, $(x_0,\, y_0) \in X \times X$ such that $(x_\alpha,\, y_\alpha) \longrightarrow (x_0,\, y_0)$ and $y_\alpha \in S(x_\alpha)$ for each $\alpha \in \Gamma$. Then $y_0 \in S(x_0)$ since S is USC. Now fix an arbitrary $w_0 \in S(x_0)$. Since S is LSC, there is a net $(w_\alpha)_{\alpha\in\Gamma}$ in X with $w_\alpha \in S(x_\alpha)$ for all $\alpha \in \Gamma$ such that $w_\alpha \longrightarrow w_0$. Note that $\inf g(x_\alpha,\, y_\alpha,\, w_\alpha) \leq 0$ for all $\alpha \in \Gamma$. By (iii) and Lemma 5.5, $\inf g$ is jointly lower semicontinuous. It follows that $\inf g(x_0,\, y_0,\, w_0) \leq 0$. As $w_0 \in S(x_0)$ is arbitrary, $y_0 \in F(x_0)$. Thus F has a closed graph. It follows that for each $u \in X$, $F(u)$ is closed in X and is therefore compact, and is also convex by (iv). Therefore F is well-defined. Moreover, as X is compact and F has a closed graph, F is USC. By Fan-Glicksberg fixed point theorem again, there exists $\hat{y} \in X$ such that $\hat{y} \in F(\hat{y})$; i.e.,

$$\begin{cases} \hat{y} \in S(\hat{y}) \\ \inf g(\hat{y},\, \hat{y},\, w) \leq 0 \ \ \text{for all } w \in S(\hat{y}). \end{cases}$$

Thus the proof is completed. □

Before we conclude this section, we would like to note that the results established in this paper can be applied to study many nonlinear problems such as nonlinear operators, nonlinear optimization, complementarity problems and so on by using those ideas which have been illustrated by Aubin and Ekeland (1984), Baiocchi and Capelo (1984), Granas (1990), Harker and Pang (1990), Husain and Tarafdar (1994), Karamolegos and Kravvaritis (1992), Kravvaritis (1979) (1979), Mosco (1976) and references therein.

5.3 Variational Inequalities and Applications

The aim of this section is to prove further results for pseudo-monotone operators and derive a few interesting results as corollaries as in Singh, Tarafdar, and Watson (1997) and Singh-Tarafdar-Watson (1998).

Let E stand for a real locally convex Hausdorff topological vector space and X a nonempty convex subset of E with $E^* \neq \{0\}$.

We need the following definitions.

Definition 5.6 Let $T : X \to E^*$ be a nonlinear map. T is said to be h-pseudomonotone if $\langle Tx, x-y\rangle + h(x) - h(y) \geq 0$ whenever $\langle Ty, x-y\rangle + h(x) - h(y) \geq 0$ for all $x, y \in X$, where $h : X \to R$ is a function. One can readily see that a monotone mapping h-pseudomonotone for each function $h : X \to R$. The mapping $T : X \to E^*$ is hemicontinuous if T is continuous from the line segment of X to the weak topology of E^*.

A point $x_0 \in X$ is said to be a solution of the variational inequality if

$$\langle Tx_0, y - x_0\rangle \geq 0 \text{ for all } y \in X.$$

We will be interested in the following:
Find $u \in X$ such that

$$\langle T(u), v - u\rangle + hv - hu \geq 0 \text{ for all } v \in X$$

where $T : X \to E^*$ is a nonlinear mapping and $h : X \to R$ is lower semicontinuous and convex function.

First we give the following lemma which is analogous to well-known Minty's lemma.

Lemma 5.10 *If X is a nonempty convex subset of a topological vector space E and $T : X \to E^*$ is a h-pseudomonotone and hemicontinuous, then $u \in X$ is a solution of*

$$\langle T(u), v - u\rangle + h(v) - h(u) \geq 0 \textit{ for all } v \in X \tag{5.38}$$

if and only if $u \in X$ is a solution of

$$\langle T(v), v-u\rangle + h(v) - h(u) \geq 0 \textit{ for all } v \in X, \tag{5.39}$$

where $h : X \to R$ is a convex function.

Proof. Let $u \in X$ be a solution of (5.38). Then by h-pseudomonotonicity of T, we have

$$\langle T(v), v-u\rangle + h(v) - h(u) \geq 0 \text{ for all } v \in X.$$

Now assume that u satisfies (5.39) and let $v \in X$ be arbitrary. Then $v_t = (1-t)u + tv \in X$ for all $t \in (0,1)$ since X is convex. Hence,

$$\langle T(v_t), v_t - u\rangle + h(v_t) - h(u) \geq 0.$$

So,

$$\langle T(v_t), t(v-u)\rangle + t(h(v) - h(u)) \geq 0.$$

If $0 < t < 1$, then

$$\langle T(v_t), v-u\rangle + h(v) - h(u) \geq 0.$$

As $t \to 0, v_t \to u$, we get $\langle T(u), v-u\rangle + h(v) - h(u) \geq 0$. □

Now, we state the following:

Theorem 5.16 *Let X be a nonempty closed convex subset of a real Hausdorff topological vector space E with $E^* \neq \{0\}$. Let $T : X \to E^*$ be a h-pseudomonotone and hemicontinuous map and $h : X \to R$ a lower semicontinuous and convex function. Further, assume that there exists a nonempty set X_0 contained in a compact, convex subset X_1 of X such that the set*

$$D = \cap_{u \in X_0}\{v \in X : \langle T(u), u - v\rangle + h(u) - h(v) \geq 0\}$$

is either empty or compact.

Then there exists a $u_0 \in X$ such that

$$\langle T(u_0), v - u_0\rangle + h(v) - h(u_0) \geq 0 \textit{ for all } v \in X.$$

Proof. Consider the following condition:

(*) for each $v \in X$ there exists a $u \in X$ such that $\langle T(u), u-v\rangle + h(u) - h(v) < 0$.

Then condition (*) may or may not hold. In either case, we will prove the existence of a $u_0 \in X$ satisfying

$$\langle T(u_0), v - u_0\rangle + h(v) - h(u_0) \geq 0 \quad \text{for all } v \in X.$$

First suppose that (*) does not hold. This means that there exists at least one $v_0 \in X$ such that

$$\langle T(u), u - v_0 \rangle + h(u) - h(v_0) \geq 0 \quad \text{for all } u \in X,$$

i.e., $v_0 \in X$ is a solution of (5.39). Then by Lemma 5.10, $v_0 \in X$ is a solution of (5.38).

Next, suppose that (*) holds. If possible, suppose that there is no solution of (5.38) under condition (*).

Then for each $u \in X$, the set

$$F(u) = \{v \in X : \langle T(u), v - u \rangle + h(v) - h(u) < 0\}$$

must be nonempty. It also follows from the convexity of h that the set $F(u)$ is convex for each $u \in X$. Thus, $F : X \to 2^X$ is a set-valued map with $F(u)$ nonempty and convex for each $u \in X$.

Now for each $u \in X$,

$$F^{-1}(u) = \{v \in X : u \in F(v)\} = \{v \in X : \langle T(v), u - v \rangle + h(u) - h(v) < 0\}.$$

Hence, for each $u \in X$, $(F^{-1}(u))^c = \{v \in X : \langle T(v), u - v \rangle + h(u) - h(v) \geq 0\} \subset \{v \in X : \langle T(u), u - v \rangle + h(u) - h(v) \geq 0\}$ (by h-pseudomonotonicity of T) $= G(u)$ which is a relatively closed subset of X since h is lower semicontinuous.

Hence, for each $u \in X, F^{-1}(u) \supset (G(u))^c = 0_u$ say, which is a relatively open subset of X.

Now by the condition (*), we can easily see that $\cup_{u \in X} 0_u = X$. (Indeed, if $v \in X$, by (*) there exists a $u \in X$ such that $v \in (G(u))^c = 0_u$. Thus, $v \in \cup_{u \in X} 0_u$. Hence, $\cup_{u \in X} 0_u = X$.)

Finally, $D = \cap_{u \in X_0} G(u) = \cap_{u \in X_0} 0_u^c$ is compact or empty by the given condition.

Hence, by fixed point Theorem 4.33, there exists a $u \in X$ such that $u \in Fu$, i.e.,

$$\langle T(u), u - u \rangle + h(u) - h(u) < 0$$

which is impossible. Hence there is a solution in this case as well. □

Now we give a few results that are special cases to Theorem 5.16.

Corollary 5.16.1 *Let $T : X \to E^*$ be monotone and hemicontinuous, $h : X \to R$ convex and lower semicontinuous. Further, assume that there exists a nonempty set X_0 contained in a compact convex subset X_1 of X such that $D = \cap_{u \in X_0} \{v \in X : \langle T(u), u - v \rangle + h(u) - h(v) \geq 0\}$ is either empty or compact. Then there is a $u \in X$ satisfying* (5.38).

Remark 5.12 Corollary 5.16.1 contains a well-known result of Tarafdar (1986).

Corollary 5.16.2 *Let X be a compact, convex subset of E and $T : X \to E^*$ h-pseudomonotone and hemicontinuous. Suppose $h : X \to R$ is lower semicontinuous and convex. Then there is a $u \in X$ satisfying* (5.38).

Remark 5.13 If we take $T = A - B$ where A is a monotone map and B is anti-monotone and both are hemicontinuous, then we derive a result due to Siddiqi, Ansari, and Kazmi Siddiqi and Kazmi (1994). Here we need only two conditions, lower semicontinuity and convexity of the function h.

Now we give the following, known as the complementarity problem.

Theorem 5.17 *Let X be a nonempty closed convex cone in E with $E^* \neq \{0\}$. Let $T : X \to E^*$ be pseudo-monotone and hemicontinuous. Further assume that there exists a nonempty set X_0 contained in a compact convex subset X_1 of X such that the set*

$$D = \cap_{u \in X_0} \{v \in X : \langle T(u), u - v\rangle \geq 0\}$$

is either empty or compact.

Then there is a $u_0 \in X$ such that

$$T(u_0) \in X^* \text{ and } \langle T(u_0), u_0\rangle = 0,$$

where X^* is positive dual cone of X.

Proof. By Theorem 5.16, there is a $u_0 = u \in X$ such that $\langle Tu, v - u\rangle \geq 0$ for all $v \in X$. Let $v = 2u$. Then $\langle T(u), u\rangle \geq 0$. Also, if $v = 0$, then $\langle Tu, u\rangle \leq 0$. Thus, $\langle T(u), u\rangle = 0$. Also, since $\langle T(u), v\rangle \geq \langle T(u), u\rangle \geq 0, T(u) \in X^*$. Hence, the result. □

5.3.1 *Application to Minimization Problems*

We will consider the following problem:

$$\inf_{x \in E} f(x)$$

where E is a normed linear space and f is the sum of two real valued functions $g, h : E \to (-\infty, \infty]$.

Before we consider this problem, we need the following notations and results.

We recall that a function $g : E \to (-\infty, \infty]$ is said to be Gateaux differentiable at an interior point $x_0 \in \text{Dom } f \equiv$ domain of f if

$$Dg(x_0)(x) = \lim_{t \to 0^+} \frac{g(x_0 + tx) - g(x_0)}{t}$$

exists for each $x \in E$ and the map

$$x \to Dg(x_0)(x) = \langle Dg(x_0), x\rangle = \langle g'(x_0), x\rangle$$

is continuous and linear. In this case, $Dg(x_0)$ is called the gradient of g at x_0.

Definition 5.7 A set-valued mapping $T : E \to E^*$ is monotone if for all $x, y \in E$

$$\langle p - q, x - y\rangle \geq 0 \text{ whenever } p \in T(x) \text{ and } q \in T(y).$$

In Section 5.1 of this chapter we have have proved the following result (see Proposition 5.1).

Theorem 5.18 *A Gateaux differentiable function $g : E \to (-\infty, \infty)$ is convex if and only if the mapping $T : E \to E^*$ defined by*

$$Tx = g'(x) = Dg(x)$$

is monotone.

Definition 5.8 Let a function $g : E \to (-\infty, \infty]$ be given. Then the subdifferentiable of g at $x_0 \in E$ denoted by $\partial g(x_0)$ is defined by

$$\partial g(x_0) = \{p \in E^* : g(x_0) - g(x) \leq \langle p, x_0 - x\rangle \text{ for all } x \in E\}.$$

$\partial g(x_0)$ may be empty. g is said to be subdifferentiable at $x_0 \in E$ if $\partial g(x_0) \neq \emptyset$.

Remark 5.14 If $f : E \to (-\infty, \infty)$ is a subdifferentiable function, then it is a well-known fact that the set-valued mapping $T : E \to E^*$ defined by $T(x) = \partial f(x)$ is monotone. Indeed, if $p \in T(x), q \in T(y)$ and $x, y \in E$, then

$$\begin{aligned}\langle p - q, x - y\rangle &= \langle p, x - y\rangle + \langle q, y - x\rangle \\ &\geq f(x) - f(y) + f(y) - f(x) = 0.\end{aligned}$$

The following proposition is obtained in an article of Tarafdar and Yuan in a more general case which will be considered in appropriate section.

Proposition 5.2 *Let $f = g + h$ be the sum of two functions $g, h : E \to (-\infty, \infty]$ such that g is subdifferentiable. Then $\overline{x} \in E$ minimizes f if there exists $p \in \partial g(\overline{x})$ such that*

$$\langle p, \overline{x}\rangle + h(\overline{x}) - h(x) < 0 \textit{ for all } x \in E.$$

Proof. Let $p \in \partial g(\overline{x})$ such that

$$\langle p, \overline{x}\rangle + h(\overline{x}) - h(x) \leq 0 \text{ for all } x \in E.$$

Then $p \in \partial g(\overline{x})$,

$$\begin{aligned}f(\overline{x}) - f(x) &= g(\overline{x}) - g(x) + h(\overline{x}) - h(x) \\ &\leq \langle p, \overline{x} - x\rangle + h(\overline{x}) - h(x) \text{ for all } x \\ &\leq 0 \quad \text{for all } x \in E.\end{aligned}$$

□

Theorem 5.19 *Let E be a normed linear space and let $f = g+h$ be the sum of two functions $g, h : E \to (-\infty, \infty]$ such that h is convex and lower semicontinuous and g is convex and both Gateaux differentiable and subdifferentiable. Further, assume* *that there exists a nonempty subset K_0 contained in a compact convex subset K_1 of E such that the set*

$$D = \cap_{x \in K_0}\{y \in K : \langle Dg(x), x - y\rangle + h(x) - h(y) \geq 0\}$$

is either empty or compact.

Then there exists a point $x_0 \in E$ which minimizes f, i.e., $f(x_0) = \inf_{x \in E} f(x)$.

Proof. Since g is convex and both Gateaux differentiable and subdifferentiable, it follows that $\partial f(x) = \{Df(x)\}$ (e.g., see Aubin (1979, p. 219). Now we define the set-valued mapping $B : E \to E^*$ by

$$B(x) = Df(x) = f'(x), x \in E.$$

By Remark 5.14 and also Theorem 5.18, B is monotone. Also B is continuous by the definition of Gateaux differentiability.

Hence, it follows that B is pseudo-monotone and hemicontinuous. Hence, by Theorem 5.16, there exists a point $x_0 \in E$ such that $\langle B(x_0), x_0 - x\rangle + h(x_0) - h(x) \leq 0$ for all $x \in E$, i.e., $\langle Df(x_0), x_0 - x\rangle + h(x_0) - h(x) \leq 0$ for all $x \in E$.

Hence, by Proposition 5.2, x_0 minimizes f on E. □

5.4 Duality in Variational Inequalities

In this section E will denote a real locally convex Hausdorff topological vector space and E^* its continuous dual with the duality $\langle .,. \rangle$.

The idea of duality in variational inequalities is due to Mosco (1972) where he expounded this principle for an injective mapping and made a comment in a remark that it is possible to do the same thing for a set-valued mapping. In this section we present this principle for a set-valued mapping.

Many variational problems with unilateral constraints for partial differential operators can be formulated as a problem of variational inequalities (e.g. see the initial works in Stampacchia (1964), Lions-Stampacchia (1965), Lions-Stampacchia (1967), Lions (1969), and Duvaut-Lions (1971) and the extensive bibliography that can found in these).

Let A : Dom $(A) \to 2^{E^*}$ be a set-valued mapping and let $f : E \to (-\infty, \infty]$ be a lower semicontinuous convex function with $f \not\equiv \infty$, where $D(A) =$ Dom (A) is the domain of A.

Then (p, u) with $u \in D(A), p \in A(u)$ is said to satisfy the variational inequality if

$$\langle p, v - u \rangle \geq f(u) - f(v) \qquad \text{for all } v \in E. \tag{5.40}$$

We will consider an equation dual to (5.40) involving A^{-1} and the Fenchel conjugate f^* of f, i.e. the lower semicontinuous convex function f^* defined on E^* and defined by

$$f^*(v^*) = \sup\{\langle v^*, v \rangle - f(v) : v \in E\} \qquad \text{for } v^* \in E^*; \tag{5.41}$$

(see Fenchel (1949), Brondsted (1964), and Moreau (1962)).

Let ∂f denotes the subdifferential of f. We identify with its canonical image into the bidual E^{**} and consider the mapping $\partial f^*{}_0 : E^* \to 2^E$ defined by

$$\partial f^*{}_0(v^*) = \partial f^*(v^*) \cap X \qquad \text{for all } v^* \in E^*, \tag{5.42}$$

where $\partial f^* : E^* \to 2^{E^{**}}$ is the subdifferential of f^*.

Lemma 5.11 *With f as above, $\partial f^*{}_0$ is the inverse of the set-valued mapping $\partial f : E \to 2^{E^*}$, i.e. $p \in \partial f(u)$ if and only if $u \in \partial f^*{}_0(p)$.*

Proof. The proof is well-known (e.g. see Moreau (1966, Section 10.6)). However we will indicate the proof. Let $p \in \partial f(u)$. Then for all $v \in E$, $f(u) - f(v) \le \langle p, u - v \rangle \le \langle p, u \rangle - \langle p, v \rangle$. Thus $\langle p, v \rangle - f(v) \le \langle p, u \rangle - f(u)$ for all $v \in E$. Hence

$$f^*(p) \le \langle p, u \rangle - f(u) \tag{5.43}$$

Also it follows from (5.41) that for each $v^* \in E^*$, we have

$$\langle v^*, v \rangle - f(v) \le f^*(v^*) \qquad \text{for all } v \in E.$$

This implies that for each $v^* \in E^*$,

$$\langle v^*, u \rangle - f(u) \le f^*(v^*).$$

Hence we have

$$-\langle v^*, u \rangle + f(u) \ge -f^*(v^*). \tag{5.44}$$

Now adding (5.43) and (5.44) we obtain
$f^*(p) - f^*(v^*) \le \langle p, u \rangle - \langle v^*, u \rangle = \langle u, p - v^* \rangle$ for all $v^* \in E^*$. Hence $u \in \partial f^*{}_0(p)$. Similarly we can prove the other way round. □

To gain symmetry we construct a new set-valued mapping $A' : \text{Dom}\,(A') \to 2^E$ defined by:

$$A'(v^*) = -A^{-1}(-v^*) = -\{u \in E : -v^* \in A(u)\}, \tag{5.45}$$

where $D(A') = -\text{range}(A) = -\bigcup_{u \in D(A)} A(u)$.

We now consider the dual variational inequality:
With $u \in A'(u^*)$ and $u^* \in D(A')$, the pair (u, u^*) is said to satisfy the following variational inequality if

$$\langle u, v^* - u^* \rangle \ge f^*(u^*) - f^*(v^*) \qquad \text{for all } v^* \in E^*. \tag{5.46}$$

(5.40) and (5.46) could be viewed as dual of one to another.

Now we prove the following theorem:

Theorem 5.20 *(p,u) satisfies (5.40) if and only if (u,u^*) with $u^* = -p \in -A(u)$ satisfies (5.46). Moreover, (5.40) and (5.46) hold if and only if $u^* \in -A(u)$ or $u \in -A'(u^*)$ and the following is satisfied*

$$f(u) + f^*(u^*) = \langle u^*, u \rangle. \tag{5.47}$$

Proof. If (p,u) is a solution of (5.40), then we have (A): for all $v \in E$,

$$f(u) - f(v) \le \langle p, v-u \rangle = \langle -p, u-v \rangle \Longleftrightarrow -p \in \partial f(u)$$

and $p \in A(u)$ (it was first noted in Lescarret (1965), see also Rockafeller (1970)). Then by Lemma 5.11 $u \in \partial f^*{}_0(-p)$ with $p \in A(u)$. Let us set $u^* = -p \in -A(u)$ which implies that $u \in A^{-1}(-u^*) = -(-A^{-1}(-u^*)) = -A'(u^*)$. Thus we have $u^* \in D(A')$ and we already have $u \in \partial f^*{}_0(-p)$, i.e. $u \in \partial f^*{}_0(u^*)$, i.e. by what we said before $u \in [-A'(u^*)] \cap \partial f^*{}_0(u^*)$. Hence applying (A) to f^* and A' we conclude that (u,u^*) is a solution of (5.46). Similarly we reverse the argument.

Finally we prove that (5.47) holds under the given conditions.

Let u, u^* and p be as in the theorem. Then by definition of f^*, we have

$$\langle u^*, v \rangle - f(v) \le f^*(u^*) \qquad \text{for all } v \in E.$$

Hence

$$\langle u^*, u \rangle - f(u) \le f^*(u^*). \tag{5.48}$$

Also since (p,u) is a solution of (5.40), we have

$$f(u) - f(v) \le \langle p, v-u \rangle \qquad \text{for all } v \in E.$$

Hence

$$\langle -p, v \rangle - f(v) \le -f(u) - \langle p, u \rangle \qquad \text{for all } v \in E.$$

Thus

$$f^*(-p) \le -f(u) - \langle p, u \rangle. \tag{5.49}$$

Noting $-p = u^*$ it now follows from (5.48) and (5.49) that $f^*(u^*) + f(u) = \langle u^*, u \rangle$. □

Corollary 5.20.1 *Let $A : D(A) \to E^*$ be an injective mapping from E into E^*. Then a vector $u \in E$ is a solution of the variational inequality*

$$u \in D(A), \langle A(u), v-u \rangle \ge f(u) - f(u) \qquad \textit{for all } v \in E, \tag{5.50}$$

if and only if the vector $u^* = -A(u)$ of E^* is a solution of the dual variational inequality

$$u^* \in D(A'), \langle A'(u^*), v^* - u^* \rangle \ge f^*(u^*) - f(v^*) \qquad \textit{for all } v^* \in E^*. \tag{5.51}$$

Moreover, (5.50) *and* (5.51) *hold if and only if* $u^* = -A(u)$, *or* $u = -A'(u^*)$ *and the following identity is satisfied*

$$f(u) + f^*(u^*) = \langle u^*, u \rangle. \tag{5.52}$$

Proof. The proof follows from Theorem 5.20 noting that A' is also injective and $p = A(u)$. □

5.4.1 *Some Auxiliary Results*

Let K be a nonempty closed convex subset of E and δ_K the indicator function of K, i.e.,

$$\delta_K(v) = 0 \text{ if } v \in K,\ \delta_K(v) = +\infty \text{ if } v \in E \setminus K.$$

δ_K is a lower semicontinuous function on E and its conjugate is the support function σ_K of K,

$$\sigma_K(v^*) = \sup\{\langle v^*, v \rangle : v \in K\},\ v^* \in E^*.$$

If $f = \delta_K$, then (5.40) and (5.46) become respectively

$$p \in A(u) \text{ with } u \in D(A) \cap K, \langle p, v - u \rangle \geq 0 \qquad \text{for all } v \in K, \tag{5.53}$$

and

$$u \in A'(u^*) \text{ with } u^* \in D(A'), \langle u, v^* - u^* \rangle \geq \sigma_K(u^*) - \sigma_K(v^*) \qquad \text{for all } v^* \in E^*. \tag{5.54}$$

The following corollary is now immediate from Theorem 5.20.

Corollary 5.20.2 *If* $f = \delta_K$, *then* (p, u) *is solution of* (5.53) *if and if* (u^*, u) *with* $u^* = -p \in -A(u)$ *is a solution of* (5.54). *Moreover,* (5.53) *and* (5.54) *hold if and only if* $u^* \in -A(u)$ *or* $u \in -A'(u^*)$ *and the identity*

$$u \in D(A) \cap K, \sigma_K(u^*) = \langle u^*, u \rangle \tag{5.55}$$

holds.

Now specializing to the injective mapping A : Dom $(A) \to E^*$ and taking $f = \delta_K$, (5.50) and (5.51) take respectively the forms

$$)u \in D(A) \cap K, \langle A(u), v - u \rangle \geq 0 \qquad \text{for all } v \in K, \tag{5.56}$$

and

$$u^* \in D(A'), \langle A'(u^*), v^* - u^* \rangle \geq \sigma_K(u^*) - \sigma_K(v^*) \qquad \text{for all } v^* \in E^*. \tag{5.57}$$

The following corollary are now immediate from 5.20.1:

Corollary 5.20.3 *A vector u is a solution of* (5.56) *if and only if $u^* = -A(u)$ is a solution of* (5.57). *Moreover,* (5.56) *and* (5.57) *hold if and only if $u^* = -A(u)$, or $u = -A'(u^*)$, and*

$$u \in D(A) \cap K, \sigma_K(u^*) = \langle u^*, u \rangle. \tag{5.58}$$

Now we consider special form of K defined by

$$K = v_0 + H = \{v \in E : v = v_0 + w, w \in H\}, \tag{5.59}$$

where H is a closed convex cone in E with vertex at the origin and $v_0 \in E$ is a given vector. The support function σ_K of this K is given by
$\sigma_K(v^*) = \langle v^*, v_0 \rangle$ if $v^* \in H^*$, $\sigma_K(v^*) = \infty$ if $v^* \notin H^*$,
where

$$H^* = \{v^* \in E* : \langle v^*, v \rangle \leq 0, \text{ for all } v \in H\}$$

is the polar cone of H, i.e. the nonpositve dual cone of H.

For K as defined in (5.59), inequalities (5.53) and (5.54) become respectively

$$p \in A(u) \text{ with } u \in D(A) \cap (v_0 + H), \langle p, v - u \rangle \geq 0 \qquad \text{for all } v \in v_0 + H, \tag{5.60}$$

and

$$u \in A'(u^*) \text{ with } u^* \in D(A'), \langle u + v_0, v^* - u^* \rangle \geq 0 \qquad \text{for all } v^* \in H^*. \tag{5.61}$$

Now just like our Corollary 5.20.2, we have the following corollary

Corollary 5.20.4 *If K is as defined in* (5.59) *and $f = \delta_K$, then (p, u) is a solution of* (5.60) *if and only if (u^*, u) with $u^* = -p \in -A(u)$ is a solution of* (5.61). *Moreover,* (5.60) *and* (5.61) *hold if and only $u^* \in_A (u)$ or $u \in -A'(u^*)$ and the identity*

$$u \in D(A) \cap (v_0 + H), u^* \in H^*, \langle u^*, u - v_0 \rangle = 0. \tag{5.62}$$

holds.

Further specializing to injective mapping A and $K = v_0 + H$ we consider the followings:

$$u \in D(A) \cap (v_0 + H), \langle A(u), v - u \rangle \geq 0 \qquad \text{for all } v \in (v_0 + H); \tag{5.63}$$

and

$$u^* \in D(A') \cap H^*, \langle A'(u^*) + v_0, v^* - u^* \rangle \geq 0 \qquad \text{for all } v^* \in H^*. \tag{5.64}$$

Corollary 5.20.5 *If A is an injective mapping, $f = \delta_K$ and K is as in* (5.59), *then $u \in K$ is a solution of* (5.63) *if and only if $u^* = -A(u)$ is a solution of* (5.64).

Moreover, (5.63) *and* (5.64) *hold if and only if* $u^* = -A(u)$ *or* $u = -A'(u^*)$, *and the following*

$$u \in D(A) \cap (v_0 + H), u* \in H*, \langle u^*, u - v_0 \rangle = 0. \tag{5.65}$$

holds.

Proof. Noting that $p = A(u)$ as A and henceA' is injective, the corollary follows from Theorem 5.20 or more specifically from Corollary 5.20.2. □

The following remarks are from Mosco (1972). For the benefit of the readers we include these here:

Remark 5.15 If E is reflexive, then it is well-known that $f^{**} = f$ and $A'' = A$ (e.g. see Moreau (1966, Section 6)). Hence in this (5.40) and (5.46) are dual of each other.

Remark 5.16 If A is the subdifferential of a lower semicontinuous convex function h on E, Theorem 5.20 is closely related to Fenchel's duality theorem (see Fenchel (1949), Moreau (1966), and Rockafeller (1970)). A typical example of such phenomenon can be seen in the classical potential theory, where the capacity of a set can be defined both in terms of the Dirichlet integral and the energy integral, which are conjugate of one to another over an appropriate Sobolev space. In the results of this section, this dual characterization is preserved for the capacities relative to non-symmetric second order elliptic partial operators as studied in Stampacchia (1964), and Stampacchia (1965) (see Matzeu (1972)).

Remark 5.17 The system involving (5.65) where $u = -A'(u^*)$, can be seen as an infinite analogue of the so-called complementarity systems that occur, for example, in convex programming and game theory (e.g. see Cottle and Dantzig (1968) and Cottle (1966)). In fact, if $E \simeq E^* \simeq \mathbb{R}^n$ and $-H$ is the nonnegative orthant $\{v = (v_i) : v_i \geq 0, i = 1, 2, \cdots, n\}$ of $\mathbb{R}^n$, then the system (5.65) reduces to

$$u^* \geq 0, v \geq 0, \langle u^*, v \rangle = 0, v = A'(u^*) + v_0$$

which is a complementarity system in the sense of references quoted above.

Remark 5.18 The explicit consideration of the dual variational inequality can be useful in dealing with problems of regularity, stability or numerical approximation of solutions. We refer to the works of Fusciardietal for some applications in this direction.

We now show how the results of previous section can be used to derive the dual result.

Theorem 5.21 *Let* $f : E \to \mathbb{R}$ *be a lower semicontinuous convex function and* $A : D(A) \to E^*$ *be a* f*-pseudomonotone and hemicontinuous mapping, where* $D(A)$

is a closed convex subset of E. *Further assume that there exists a nonempty subset* X_0 *contained in compact convex subset* X_1 *of* $D(A) = X$ *such that the set*

$$D = \bigcap_{u \in X_0} \{v \in X : \langle A(u), u - v \rangle + f(u) - f(v) \geq 0\}$$

is either empty or compact.

Then there exists $u_0 \in D(A)$ *such that*

$$\langle A(u_0), v - u_0 \rangle = h(u) - h(v) \text{ for all } v \in D(A) \tag{5.66}$$

and (u, u^*) *with* $u^* = -p = -A(u_0)$ *satisfies*

$$u \in A'(u^*), \text{ and } u^* \in D(A'), \langle u, v^* - u^* \rangle \geq f^*(u^*) - f^*(v^*) \qquad \text{for all } v^* \in E^*, \tag{5.67}$$

where $A' : D(A') \to 2^E$ *is defined by:*

$$A'(v^*) = -A^{-1}(-v^*) = -\{u \in E : -v^* = A(u)\}$$

and $D(A') = -(A)$. *Furthermore, the following identity holds:*

$$f(u_0) + f^*(u^*) = \langle u^*, u_0 \rangle.$$

Proof. Apply Theorem 5.21 and Theorem 5.20. □

In the remaining sections of Chapter 5, we shall present some important results with their proofs after thorough revisions and considerations of different research articles of Chowdhury and Tan, Ding, Kim and Tan, Kim and Tan, Shih and Tan, Tarafdar and Yuan, and of Tarafdar and Mehta in variational inequalities, generalized variational inequalities, generalized quasi-variational inequalities and in generalized games.

5.5 A Variational Inequality in Non-Compact Sets with Some Applications

We shall assume in all results of this section that E is a Hausdorff topological vector space. The following Fan-Browder fixed point theorem (Browder (1967)) is essential in convex analysis and also the basic tool in proving many variational inequalities and intersection theorems in non-linear functional analysis:

Theorem 5.22 *Let* X *be a non-empty compact convex subset of a Hausdorff topological vector space and let* $T \colon X \to 2^X$ *be a multi-map satisfying the following:*

(1) *for each* $x \in X$, $T(x)$ *is convex,*

(2) *for each* $y \in X$, $T^{-1}(y)$ *is open.*

Then T *has a fixed point* $x \in X$, *i.e.* $\hat{x} \in T(\hat{x})$.

The Fan-Browder theorem can be proved by using Brouwer's fixed point theorem or the KKM-theorem. Till now, there have been numerous generalization and applications of this Theorem by several authors; e.g. see (Ding et al. (1992), Horvath (1987)) and the reference there.

In the paper (Ding et al. (1992)), Ding-Kim-Tan further generalized the above result to non-compact sets in locally convex spaces and the following is a special case of the fixed point version of their Theorem 1 there:

Lemma 5.12 *Let X be a non-empty convex subset of a locally convex Hausdorff topological vector space and D be a non-empty compact subset of X. Let $T : X \to 2^D$ be multi-map satisfying the following:*

(1) *for each $x \in X$, $coT(x) \subset D$*

(2) *for each $y \in X$, $T-1(y)$ is open in X*

Then there exits a point $\hat{x} \in X$ such that $\hat{x} \in \mathrm{co}\, T(\hat{x})$.

The following Lemma, which is due to Kim and Tan, will be required in proving their first theorem of this section:

Lemma 5.13 *Let E be a topological vector space and E^* be the dual space of E equipped with the strong topology. Let X be a non-empty bounded subset of E and $T : X \to 2^{E^*}$ be an upper semi-continuous multi-map such that each $T(x)$ is (strongly) compact. Then for each $y \in E$, the real-valued function $g_y : X \to R$ defined by*

$$g_y(x) = \inf_{w \in T(x)} Re\langle w, x - y\rangle \text{ for each } x \in X,$$

is lower semi-continuous.

Proof. Let $x_0 \in X$ be given. For any $\epsilon > 0$, we shall show that there exits an open neighborhood $N(x_0)$ of x_0 such that

$$g_y(x) \geq g_y(x_0) - \epsilon \text{ for each } x \in N(x_0)$$

Indeed, let $V := p \in E* : \sup_{t \in X-y} |p(t)| < \frac{\epsilon}{3}$ where$X - y = x - y : x \in X$.Then V is a strongly open neighborhood of 0 in E^*. Since $X - y$ is a bounded set E. Since T is upper semi-continuous at x_0 and $T(x_0) + V$ is a strongly open set containing $T(x_0)$ there exists an open neighborhood N_0 of x_0 in X such that $T(x) \subset T(x_0)+V$ for each $x \in N_0$.

Next, for each $u \in T(x_0)$,we let

$$V_u := \{p \in E^* : \sup_{t \in X-X} |p(t) - u(t)| < \frac{\epsilon}{3}\},$$

where $X - X = \{x - z : x, z \in X\}$; then V_u is also a strongly open neighborhood of u in E since $X - X$ is bounded set in E. Since $T(x_0)$ is strongly compact and $T(x_0) \subset \cup_{u \in T(x_0)} V_u$, there exists a finite subset $u_1, \ldots, u_n$ of $T(x_0)$ with $T(x_0) \subset$

$\cap_{i=1}^n V_u$. For each $i = 1, \cdots n$, since u_i is a continuous linear functional, there exists an open neighborhood N_i of x_0 in X such that $|u_i(x) - u_i(x_0)| < \frac{\epsilon}{3}$ for each $x \in N_i$.

Now let $N(x_0) := \cap_{i-0}^n N_i$; then $N(x_0)$ is an open neighborhood of x_0 in X. We shall show that this open neighborhood $N(x_0)$ of x_0 is the desired one. For each $x \in N(x_0)$ and each $w \in T(x)$ since $x \in N_0$, there exists $u \in T(x_0)$ such that $w - u \in V$. Also, since $u \in T(x_0) \subset U_{i=1}^n$ there exists $i_0 \in \{1, \ldots, n\}$ such that $u \in V_0$. Therefore we have

$$|Re\langle w, x-y\rangle Re\langle u, x-y\rangle| \leq |\langle w-u, x-y\rangle| < \frac{\epsilon}{3},$$

so that

$$\begin{aligned} Re\langle w, x-y\rangle > Re\langle u, x-y\rangle - \frac{\epsilon}{3} &= Re\langle u, x_0 - y\rangle + Re\langle u, x - x_0\rangle - \frac{\epsilon}{3} \\ &= Re\langle u, x_0 - y\rangle + Re\langle u - u_{i_0}, x - x_0\rangle \\ &\quad + Re\langle u_{i_0}, x - x_0 > -\frac{\epsilon}{3} \\ &\geq \inf_{v \in T(x_0)} Re\langle v, x_0 - y\rangle - \epsilon \\ &= g_y(x_0) - \epsilon. \end{aligned}$$

Since $w \in T(x)$ is arbitrary, we have $g_y(x) = \inf_{w \in T(x)} Re\langle w, x-y\rangle \geq g_y(x_0) - \epsilon$, which completes the proof. □

Lemma 5.13 is a multi-valued generalization of Lemma 5.12 in [Browder (1967)] (see also (Shih and Tan (1986), Lemma 1) where it was observed that the result holds for X being bounded instead of compact).

The following is a variational inequality of Kim and Tan in non-compact sets.

Theorem 5.23 *Let X be a bounded convex subset of a locally convex Hausdorff topological vector space E and D be a non-empty compact subset of X. Let $T : X \to 2^{E^*}$ be an upper semi-continuous multi-map from the relative topology of X to the strong topology of E^* such that each $T(x)$ is (strongly) compact. Suppose further that for each $x \in X \setminus D$*

$$\inf_{w \in T(y)} Re\langle w, y - x\rangle \leq 0 \text{ for all } y \in X \tag{5.68}$$

Then there exists a point $\hat{x} \in X$ such that

$$\inf_{w \in T(\hat{x})} Re\langle w, \hat{x} - x\rangle \leq 0 \text{ for all } x \in X.$$

Furthermore, if $T(\hat{x})$ is also convex, then there exists a point $\hat{w} \in T(\hat{x})$ such that

$$Re\langle \hat{w}, \hat{x} - x\rangle \leq 0 \text{ for all } x \in X$$

Proof. Suppose that for each $x \in X$ there exists a point $\hat{x} \in X$ such that $\inf_{w \in T(x)} Re\langle w, x - \tilde{x}\rangle > 0$. Then by the assumption (5.68) $\tilde{x} \in D$ Now we define a multi-map $P : X \to 2^D$ by

$$P(x) : \{y \in D : \inf_{w \in T(x)} Re\langle w, x - y\rangle > 0\} \text{ for all } x \in X.$$

Then for each $x \in X$, $P(x)$ is non-empty. For each $x \in X$, we shall show that $coP(x) \subset D$. Indeed, let $n \in N, y_1, \ldots, y_n \in P(x)$ and $t_1, \ldots, t_n \in [0, 1]$ with $\sum_{i=1}^{n} t_i = 1$ then for each $i = 1, \ldots, n$,

$$\inf_{w \in T(x)} Re\langle w, x - y_I\rangle > 0,$$

it follows that

$$\inf_{w \in T(x)} Re\langle w, x - \sum_{t=1}^{n} t_i y_i\rangle \geq \sum_{i=1}^{n} t_i \inf_{w \in T(x)} Re\langle w, x - y_i\rangle > 0.$$

Since $\sum_{i=1}^{n} t_i y_i \in X$, by the assumption (5.68) again $\sum_{i=1}^{n} t_i y_i \in D$. Hence $coP(x) \subset D$.

Next for each $y \in D$, we shall show that $P^{-1}(y)$is open in X. Let $(x_\alpha)_{\alpha \in \Gamma}$ be a net in $X \setminus P^{-1}(y)$, which converges to some $x_0 \in X$. Then we have

$$\inf_{w \in T(x_\alpha)} Re\langle w, x_\alpha - y\rangle \leq 0 \text{ for all } \alpha \in \Gamma.$$

By Lemma 5.13, the real-valued function

$$x \longrightarrow \inf_{w \in T(x_\alpha)} Re\langle w, x - y\rangle$$

is lower semi-continuous it follows that

$$\inf_{w \in T(x_0} Re\langle w, x_0 - y\rangle \leq 0.$$

Therefore $X \setminus P^{-1}(y)$ is closed, and hence $P^{-1}(y)$ is open in X. Thus all the hypotheses of Lemma 5.12 are satisfied, so by Lemma 5.12 there exists a point $\hat{x} \in X$ such that $\hat{x} \in coP(\hat{x})$. But then there exist $y_1, \ldots, y_m \in P(\hat{x})$ and $\lambda_1, \ldots, \lambda_m \geq 0$ with $\sum_{i=1}^{m} \lambda_i = 1$ such that $\hat{x} = \sum_{i=1}^{m} \lambda_i y_i$. Therefore we have

$$\begin{aligned} 0 &= \inf_{w \in T(\hat{x})} Re\langle w, \hat{x} - \hat{x}\rangle \\ &= \inf_{w \in T(\hat{x})} Re\langle w, \hat{x} - \sum_{i=1}^{m} \lambda_i y_i\rangle \\ &= \inf_{w \in T(\hat{x})} \sum_{i=1}^{m} \lambda_i Re\langle w, \hat{x} - y_i\rangle \\ &\geq \sum_{i=1}^{m} \lambda_i \inf_{w \in T(\hat{x})} Re\langle w, \hat{x} - y_i\rangle > 0, \end{aligned}$$

which is a contradiction. Hence there must exist a point $\hat{x} \in X$ such that

$$\inf_{w \in T(\hat{x})} Re\langle w, \hat{x} - \hat{x}\rangle \leq 0 \text{ for all } x \in X \tag{5.69}$$

To prove the second assertion, suppose that $T(\hat{x})$ is convex. Then we define $f : X \times T(\hat{x}) \to R$ by

$$f(x, w) := Re\langle w, \hat{x} - x\rangle \text{ for each } (x, w) \in X \times T(\hat{x})$$

Note that for each fixed $x \in X$, $w \to Re\langle w, \hat{x} - x\rangle$ is continuous affine, and for each $w \in T(\hat{x})$, $x \to Re\langle w, \hat{x} - x\rangle$ is affine. Thus, by Kneser's minimax theorem [Kneser (1952)], we have

$$\min_{w \in T(\hat{x})} \sup_{x \in X} f(x, w) = \sup_{x \in X} \min_{w \in T(\hat{x})} f(x, w)$$

Thus

$$\min_{w \in T(\hat{x})} \sup_{x \in X} Re\langle w, \hat{x} - x\rangle \leq 0 \text{ by (5.68)}$$

Since $T(\hat{x})$ is compact, there exists $\hat{w} \in T(\hat{x})$ such that

$$\sup_{x \in X} Re\langle \hat{w}, \hat{x} - x\rangle = \min_{w \in T(\hat{x})} \sup_{x \in X} Re\langle w, \hat{x} - x\rangle.$$

Therefore $Re\langle \hat{w}, \hat{x} - x\rangle \leq 0$ for all $x \in X$.

This completes the proof. □

When $X = D$ is compact convex, the following generalization of Hartman-Stampacchia's variational inequality (Hartman and Stampacchia (1966)) due to Browder (Browder (1968), Theorem 6) is obtained:

Corollary 5.23.1 *Let X be a non-empty compact convex subset of a locally convex Hausdorff topological vector space E and let $T : X \to 2^{E^*}$ be an upper semi-continuous multi-map from the relative topology of X to the strong topology of E^* such that each $T(x)$ is a (strongly) compact convex subset of E^*.*

Then there exists a point $\hat{x} \in X$ and $\hat{w} \in T(\hat{x})$ such that

$$Re\langle \hat{w}, \hat{x} - x\rangle \leq 0 \textit{ for all } x \in X$$

The following is a single-valued version of Theorem 5.22:

Corollary 5.23.2 *Let X be a bounded convex subset of s locally convex Hausdorff topological vector space E and D be a non-empty compact subset of X. Let $T : X \to E^*$ be a continuous mapping from the relative topology of X to the strong topology of E^* satisfying the following condition:*

for each $x \in X \setminus D$, $Re\langle T(y), y - x\rangle \leq 0$ for all $y \in X$

Then there exists a point $\hat{x} \in X$ such that

$$Re\langle T(\hat{x}), \hat{x} - x\rangle \leq 0 \textit{ for all } x \in X.$$

Let E be a topological vector space and M be a topological space. Recall that a multi-map $F : M \to 2^E$ is *upper hemi-continuous* (e.g. see Aubin and Ekeland (1984, p. 122)) if for each $P \in E^*$ and for each $\lambda \in R$, the set $\{x \in M : \sup_{u \in f(x)} Re\langle p, u\rangle < \lambda\}$ is open in M.

For relationships among upper semi-continuity, upper demi-continuity and upper hemi-continuity, we refer to Shih and Tan (1987, Propositions 1 and 2 and Examples 1 and 2).

The following fixed point theorem is obtained as an application of Corollary 5.23.2:

Theorem 5.24 *Let X be a non-empty para-compact bounded convex subset of a locally convex Hausdorff topological vector space E, D he a non-empty compact subset of X. Let $F : X \to 2^E$ be an upper hemi-continuous multi-map satisfying the following:*

(1) *for each $x \in X$, $F(x)$ is non-empty closed convex,*
(2) *for each $x \in X$, $F(x) \cap cl(x + \cup_{\lambda>0}\lambda(X - x)) \neq \emptyset$,*
(3) *for each $x \in X \setminus D$, $y \in X$ and $P \in E^*$, if $\inf\{Re\langle p, y - z\rangle : z \in f(y)\} > 0$, then $Re\langle p, y - x\rangle \leq 0$.*

Then there exists $\hat{x} \in X$ such that $\hat{x} \in F(\hat{x})$.

Proof. Since F is upper hemi-continuous, for each $P \in E^*$, the set

$$\begin{aligned} U(P) &= \{x \in X : \sup_{z \in F(x)} Re\langle p, z\rangle < Re\langle p, x\rangle\} \\ &= \cup_{\lambda \in R}[\{x \in X : \sup_{z \in F(x)} Re\langle p, z\rangle < \lambda\} \cap \{x \in X : Re\langle p, x\rangle > \lambda\}] \end{aligned}$$

is open in X. Suppose $x \notin F(x)$ for each $x \in X$. Then for each $x \in X$ there exists $P \in E^*$ such that $\sup_{z \in F(x)} Re\langle p, z\rangle < Re\langle p, x\rangle$ so that $x \in U(p)$. Thus $\{U(p) : P \in E^*\}$ is an open cover of the para-compact space X. Let $\{V(P) : P \in E^*\}$ be a locally finite open precise refinement of $\{U(p) : P \in E^*\}$ and $\{\beta_p : P \in E^*\}$ be the continuous partition of unity subordinated to this refinement $\{V(P) : P \in E^*\}$. Define a mapping $T : X \to E^*$ by

$$T(x) = \sum_{p \in E^*} \beta_p(x)p \text{ for all } x \in X.$$

Let $x \in X$ be given. If $P \in E^*$ and $\beta_p(x) > 0$,then $x \in V(P) \subset U(p)$ so that $\sup_{z \in F(x)} Re\langle p, z\rangle < Re\langle p, x\rangle$; it follows that

$$\inf_{z \in F(x)} Re\langle p, x - z\rangle > 0.$$

Therefore for each $x \in X$,

$$\inf_{z\in F(x)} Re\langle T(x), x-z\rangle = \inf_{z\in F(x)} \sum_{p\in E^*} \beta_p(x) Re\langle p, x-z\rangle$$

$$\geq \sum_{p\in E^*} \beta_p(x) \inf_{z\in F(x)} Re\langle p, x-z\rangle \tag{5.70}$$

$$> 0. \tag{5.71}$$

Now we shall show that T satisfies all hypotheses of Corollary 5.23.2. To show that T is continuous from the relative topology of X to the strong topology of E^*, let $(x_\alpha)_{\alpha\in\Gamma}$ be a net in X which converges to some $x_0 \in X$. Since $\{V(p) : p \in E^*\}$is locally finite, there is an open neighborhood U_0 of x_0 in X such that $\{p \in E^* : v(p) \cap U_0 \neq \phi\}$ is finite, so we let $\{p \in E^* : v(p) \cap U_0 \neq \emptyset\} = p_1, \ldots, p_n$. Let B be any non-empty bounded subset of E, then by Theorem 1.18 [Rudin (1973)], $M = \max_{1\leq i\leq n} \sup\{|p_i(x)| : x \in B\} < \infty$. Since each β_p is continuous, there exists $\alpha_1 \in \Gamma$ such that for each $\alpha \geq \alpha_1$,

$$|\beta_p(x_\alpha) - \beta_p(x_0)| < \frac{\epsilon}{M_n} \text{ for all } i = 1, \ldots, n.$$

Also since (x_α) converges to x_0 and U_0 is an open neighborhood of x_0, there exists $\alpha_2 \in \Gamma$ such that for each $\alpha \geq \alpha_2$, $x_\alpha \in U_0$. Let $\alpha_0 \geq \max\{\alpha_1, \alpha_2\}$. Then for each $\alpha \geq \alpha_0$, we have

$$\sup_{z\in B} | < T(x_\alpha) - T(x_0), z > | = \sup_{z\in B} | \sum_{p\in E^*} (\beta_p(x_\alpha) - \beta_p(x_0))p(z)|$$

$$= \sup_{z\in B} | \sum_{i=1}^{n} (\beta_{p_i}(x_\alpha) - \beta_{p_i}(x_0))p_i(z)|$$

$$\leq \sum_{i=1}^{n} |(\beta_{p_i}(x_\alpha) - \beta_{p_i}(x_0))| \sup_{z\in B} p_i(z)|$$

$$< \sum_{i=1}^{n} \frac{\epsilon}{M_n} M = \epsilon,$$

and hence $T(x_\alpha)$ converges to $T(x_0)$ in the strong topology of E^*.

Next suppose there exists $x_1 \in X \setminus D$ such that for some $y \in X$,

$$Re\langle T(y), y - x_1\rangle = \sum_{p\in E^*} \beta_p(y) Re\langle p, y - x_1\rangle > 0. \tag{5.72}$$

If $\beta_p(y) > 0$, then $\inf_{z\in F(y)} Re\langle p, y - z\rangle > 0$ so that by assumption (3), $Re\langle p, y - x_1\rangle \leq 0$, which contradicts (5.72).

Therefore by Corollary 5.23.2, there exists $\hat{x} \in X$ such that

$$Re\langle T(\hat{x}), \hat{x} - y\rangle \leq 0 \text{ for all } y \in X. \tag{5.73}$$

By the assumption (2), $F(\hat{x}) \cap cl(\hat{x} + \cup_{\lambda>0}\lambda(X - \hat{x}) \neq \emptyset$. Let $\hat{y} \in F(\hat{x})$, $(\lambda_\alpha)_{\alpha\in\Gamma}$ be a net in $(0, \infty)$ and $(u_\alpha)_{\alpha\in\Gamma}$ be a net in X such that $(\hat{x} + \lambda_\alpha(u_\alpha - \hat{x})) \to \hat{y}$. Then we have

$$\begin{aligned} Re\langle T(\hat{x}, \hat{x} - \hat{y}\rangle &= \lim_\alpha Re\langle T(\hat{x}, \hat{x} - (\hat{x} + \lambda_\alpha(u_\alpha - \hat{x}))\rangle \\ &= \lim_\alpha \lambda_\alpha Re\langle T(\hat{x}, \hat{x} - u_\alpha\rangle \\ &= \leq 0 \quad \text{by 5.73}). \end{aligned}$$

Hence $\inf_{z\in F(\hat{x})} Re\langle T(\hat{x}), \hat{x} - z\rangle \leq 0$ which contradicts (5.70). This completes the proof. □

Theorem 5.24 generalizes Theorem 2 of Halpern (1970), p. 88, in the following ways:

(i) X need not be compact and (ii) F is upper hemi-continuous instead of upper semi-continuous.

The following is a re-formulation of Proposition 3.1.21 of Aubin-Ekeland (Aubin and Ekeland (1984)):

Lemma 5.14 *Let X and Y be topological spaces and $\Phi : X \times Y \times R$ be a real valued lower semi-continuous function on $X \times Y$ and $T : Y \to 2^X$ be an upper semi-continuous multi-map at $y_0 \in Y$ and $T(y_0)$ is non- empty compact. Then a real-valued function $g : Y \to R$ defined by*

$$g(y) := \inf_{x\in T(y)} \Phi(x, y), \text{ for all } y \in Y,$$

is lower semi-continuous at y_0.

Lemma 5.15 *Let E be a normed space, X be a non-empty subset of E and $T : X \to 2^{E^*}$ be an upper semi-continuous multi-map, such that each $T(x)$ is (norm-) compact. Then for each $y \in E$, the real-valued function $g_y : X \to R$ defined by*

$$g_y(x) := \inf_{w\in T(x)} Re\langle w, x - y\rangle, \text{ for each } x \in X,$$

is lower semi-continuous.

Proof. Define $\Phi : X \times E^* \to R$ by

$$\Phi(x, w) = Re\langle w, x - y\rangle \text{ for each } (x, w) \in X \times E^*.$$

Let (x_n) be a sequence in X which converges to $x \in X$ and (w_n) be a sequence in E^* which converges to $w \in E^*$. Then we have

$$\begin{aligned}|\Phi(x_n, w_n) - \Phi(x, w)| &= |Re\langle w_n, x_n - y\rangle - Re\langle w, x - y\rangle| \\ &\leq |\langle w_n - w, x - y\rangle| + |w_n, x_n - x\rangle| \\ &\leq \|w_n - w\|\|x - y\| + \|w_n\|\|x_n - x\| \to 0,\end{aligned}$$

since $\{\|w_n\| : n \geq 1\}$ is bounded.

Thus Φ is continuous. By Lemma 5.14, g_y is lower semi- continuous. This completes the proof. □

We remark that in the proof of Theorem 5.23, the condition "X is bounded" was never needed until Lemma 5.13 was quoted. In view of Lemma 5.15, the same proof of Theorem 5.23 gives the following:

Theorem 5.25 *Let X be a convex subset of a normed linear space E and D be a non-empty compact subset of X. Let $T : X \to 2^{E^*}$ be an upper semi-continuous multi-map from the relative topology of X to the norm topology of E^* such that each $T(x)$ is (norm-) compact in E^*. Suppose further that for each $x \in X \setminus D$,*

$$\inf_{w \in T(y)} Re\langle w, y - x\rangle \leq 0 \quad \textit{for all} \quad y \in X. \tag{5.74}$$

Then there exists a point $\hat{x} \in X$ such that

$$\inf_{w \in T(\hat{x})} Re\langle w, \hat{x} - x\rangle \leq 0 \quad \textit{for all} \quad x \in X.$$

Furthermore, if $T(\hat{x})$ is also convex, then there exists a point $\hat{w} \in T(\hat{x}))$ such that

$$Re\langle \hat{w}, \hat{x} - x\rangle \leq 0 \quad \textit{for all} \quad x \in X.$$

When E is a normed space, we can delete the condition in Corollary 5.23.2 (for the same reason) that X is bounded. Thus, recalling that every metric space is paracompact, we obtain the following norm-version of Theorem 5.24:

Theorem 5.26 *Let X be a non-empty convex subset of a normed linear space E and D be a non-empty compact subset of X. Let $F : X \to 2^E$ be an upper hemi-continuous multi-map satisfying the following:*

(1) *for each $x \in X$, $F(x)$ is non-empty closed convex,*
(2) *for each $x \in X$, $F(x) \cap cl(x + \cup_{\lambda > 0}\lambda(X - x)) \neq \emptyset$,*
(3) *for each $x \in X \setminus D$, $y \in X$, and $p \in E^*$, if $\inf\{Re\langle p, y - z\rangle : z \in F(y)\} > 0$, then $Re\langle p, y - x\rangle \leq 0$.*

Then there exists $\hat{x} \in X$ such that $\hat{x} \in F(\hat{x})$.

Note that the main results of this section have been written and presented after a thorough revision of some results of an article in Kim and Tan (1992)

5.6 Browder-Hartman-Stampacchia Variational Inequalities for Set-Valued Monotone Operators

If X, Y are topological spaces, then a map $g :\to 2^Y$ is said to be lower semi-continuous on X (Berge (1963), p. 109) if for every $x_0 \in X$ and for every open set G in Y with $g(x_0) \cap G \neq \phi$, there exists an open neighborhood $N(x_0)$ of x_0 in X such that $g(x) \cap G \neq \phi$, for all $x \in N(x_0)$.

If $X \subset E$, a map $T : X \to 2^{E'}$ is said to be *monotone* on X (Browder (1976), p. 79) if for any $x, y \in X$, $u \in T(x)$ and $w \in T(y)$, $Re\langle w - u, y - x\rangle \geq 0$. If $X \subset E$ is convex , a real-valued function Ψ on X is said to be (*quasi-concave*) if for every real number t, the set $\{x \in X : \Psi(x) > t\}$ is convex.

In this section we shall present first a minimax inequality of Shih and Tan which is a generalization of the celebrated minimax inequality of Ky Fan (Fan (1972)). A fixed point version of the minimax inequality will be given. As an application, some generalizations of Browder-Hartman-Stampacchia's variational inequalities to multi-valued monotone operators will be presented which are lower semi-continuous along the line segments.

5.6.1 *A Minimax Inequality*

We shall next present the following generalization of the celebrated minimax inequality of Ky Fan (Fan (1972)). We observe that next result is an useful tool for the study of existence theorems of variational inequalities for multivalued monotone operators:

Theorem 5.27 *Let X be a non-empty convex subset of a Hausdorff topological Vector space and let ϕ, Ψ be two real-valued functions on the product space $X \times X$. Suppose that*

(a) *$\phi(x, y) \leq \Psi(x, y)$ for all $(x, y) \in X \times X$ and $\Psi(x, x) \leq 0$ for all $x \in X$*
(b) *For each fixed $x \in X$, $\psi(x, y)$ is a lower semi-continuous function of y on X.*
(c) *For each fixed $y \in X$, $Psi(x, y)$ is a quasi-concave function of x on X.*
(d) *There exists a non-empty compact subset K of X and $x_0 \in X$ such that $\Psi(x_0, y) > 0$ for all $y \in X \setminus K$.*

Then there exists $\hat{y} \in X$ such that $\phi(x, \hat{y}) \leq 0$ for all $x \in X$.

Proof. For each $x \in X$, let

$$F(x) := y \in X : \Psi(x, y) \leq 0$$
$$G(x) := y \in X : \phi(x, y) \leq 0.$$

Then

(i) by (a), $F(x) \subset G(x)$ for each $x \in X$;
(ii) by (b), $G(x)$ is closed in X for each $x \in X$;
(iii) by (a) and (d), $F(x_0) \subset K$ so that $\overline{F}(x_0)$is compact;

(iv) by (a) and (c), given any finite subset $\{x_1, \cdots, x_n\}$ of X,

$$C_0\{x_1, \cdots, x_n\} \subset \bigcup_{i=1}^{n} F(x_i).$$

Thus by Ky Fan's infinite dimensional generalization (see Fan (1961)) of the covering theorem of Knaster-Kuratowski-Mazurkiewicz (see ? (?)), we have $\cap_{x \in X} \overline{F(x)} \neq \emptyset$. By (i) and (ii), $\cap_{x \in X} G(x) \neq \emptyset$. Take any $\hat{y} \in \cap_{x \in X}$; then $\phi(x, \hat{y}) \leq 0$ for all $x \in X$.

When X is a compact convex set, by taking $K = X$, Theorem 5.27 reduces to Yen's generalization [Yen (1981)] of Ky Fan's minimax inequality and by further taking $\phi = \Psi$, Theorem 5.27 reduces to Ky Fan's minimax inequality. □

The following fixed point theorem, represents a fixed point version of Theorem 5.27. This fixed point version is a generalization of an earlier basic fixed point theorem of Fan and Browder (see Fan (1961, Lemma 4), Fan (1972, Theorem 2) and Browder (1968, Theorem 1)).

Theorem 5.28 *Let X be a non-empty convex subset of a Hausdorff topological vector space and F ,$G : X \to 2^X$. Suppose that*

(a) *$F(x) \subset G(x)$ for each $x \in X$;*
(b) *For each $x \in X$, $F(x)$ is open in X;*
(c) *For each $y \in X$, $F^{-1}(y) \neq \emptyset$ and $G^{-1}(y)$ is convex;*
(d) *There exist a non-empty compact subset K of X and $X_0 \in X$ such that $X \backslash K \subset G(x_0)$.*

Then there exists $\hat{x} \in X$ such that $\hat{x} \in G(x)$

5.6.2 *An Existence Theorem of Variational Inequalities*

We first state the following lemma, which is a generalization of Minty's lemma (see Minty (1962)) and is a special case of Lemma 4 in Shih and Tan (1989):

Lemma 5.16 *Let E be a Hausdorff topological vector space, X be a non-empty subset of E and $T : X \to 2^{E'}$ be lower semi-continuous along the line segments in X to the weak-topology of E' . Then for $\hat{y} \in X$ the inequality*

$$\sup_{u \in T(x)} Re\langle u, \hat{y} - x\rangle \leq 0 \ \ for\ all x \in X$$

implies the inequality

$$\sup_{u \in T(\hat{y})} Re\langle w, \hat{y} - x\rangle \leq 0 \ \ for\ all x \in X.$$

We shall present the following existence theorem of variational inequalities as an application of Theorem 5.27:

Theorem 5.29 *Let E be a reflexive Banach space equipped with norm $\|\cdot\|$ and X be a non-empty closed convex subset of E. Suppose that $T : X \to 2^{E'}$ is monotone* *on X and is lower semi-continuous from the line segments in X to the weak topology of E′. Assume there exists $x_0 \in X$ such that*

$$\lim_{\substack{\|y\|\to\infty \\ y\in X}} \inf_{w\in T(x)} Re\langle w, y - x_0\rangle > 0 \tag{5.75}$$

Then there exists $\hat{y} \in X$ such that

$$\sup_{w\in T(\hat{y})} Re\langle w, \hat{y} - x\rangle \le 0 \ \textit{for all}\, x \in X.$$

Proof. Define $\phi,\ \Psi : X \times X \to \mathbb{R}$

$$\phi(x, y) = \sup_{u\in T(x)} Re\langle u, y - x\rangle, \tag{5.76}$$

$$\Psi(x, y) = \inf_{w\in T(y)} Re\langle w, y - x\rangle. \tag{5.77}$$

As T is monotone, for each $x, y \in X$, we have

$$\inf_{w\in T(y)} Re\langle w, y - x\rangle \ge \sup_{u\in T(x)} Re\langle u, y - x\rangle,$$

so that $\phi(x, y) \le \Psi(x, y)$ for all $x, y \in X$. By coercive condition (1), there exists a sufficiently large $R > 0$ such that

$$\inf_{w\subset T(y)} Re\langle w, y - x_0\rangle > 0$$

for all $y \in X$ with $\|y\| > R$.

Take $K := \{y \in X : \|y\| \le R\}$; then K is a weakly compact convex subset of X as E is reflexive and X is a closed convex subset of E. Thus $\Psi(x_0, y) > 0$ for all $y \in X\backslash K$. It is clear that $\Psi(x, x) \le 0$ for all $x \in X$ and for each $x \in X$, $\phi(x, y)$ is a weakly lower semi-continuous function of y on X. It is easy to see that for each $y \in X$, $\Psi(x, y)$ is a quasi-concave function of x on X. Thus when E is equipped with weak topology, all conditions in Theorem 5.27 are satisfied so that by Theorem 5.27, there exists $\hat{y} \in X$ such that $\phi(x, \hat{y}) \le 0$ for all $x \in X$, i.e.

$$\sup_{u\in T(x)} Re\langle u, \hat{y} - x\rangle \le 0 \ \text{for all}\, x \in X.$$

By Lemma above, we have

$$\sup_{w\in T(\hat{y})} Re < w, \hat{y} - x\rangle \le 0 \ \text{for all}\, x \in X. \qquad \square$$

We observe that an immediate consequence of Theorem 5.29 can be obtained (as given below) which is a generalization of Browder-Hartman-Stampacchia's variational inequalities for multi-valued monotone operators. We will note that these variational inequalities are lower semi-continuous along the line segments.

Theorem 5.30 *Let E be reflexive Banach space equipped with norm $\|\cdot\|$ and X be a non-empty closed convex subset of E. Suppose that $T : X \to 2^{E'}$ is monotone on X and is lower semi-continuous from line segments in X to the weak topology of E'. Assume there exists $x_0 \in X$ such that*

$$\lim_{\substack{\|y\|\to\infty \\ y\in X}} \inf_{w\in T(y)} Re\langle w, y - y_0\rangle / \|y\| = +\infty.$$

Then for each given $w_0 \in E'$ there exists $\hat{y} \in X$ such that

$$\sup_{w\in T(\hat{y})} Re\langle w - w_0, \hat{y} - x\rangle \leq 0 \text{ for all } x \in X.$$

When T is single-valued, Theorem 5.30 reduces to Browder-Hartman-Stampacchia's variational inequalities (Browder (1968), Hartman and Stampacchia (1966)).

Remark 5.19 We first recall that if X, Y are topological spaces, a map $g : X \to 2^Y$ is said to be upper semi-continuous on X [Berge (1963), p. 109] if for every $x_0 \in X$ and for every open set G in Y with $g(x_0) \subset G$ there exists an open neighborhood $N(x_0)$ in X such that $g(x) \subset G$ for all $x \in N(x_0)$. The following different generalization of Browder-Hartman-Stampacchia variational inequalities for multi-valued monotone operators which are upper semi-continuous along the line segments was obtained in Shih and Tan (1988a).

Theorem 5.31 *Let E be reflexive Banach space equipped with norm $\|\cdot\|$ and X be a non-empty closed convex subset of E. Suppose that $T : X \to 2^{E'}$ is monotone on X such that each $T(x)$ is weakly compact convex subset of E' and T is upper semi-continuous from line segments in X to the weak topology of E'. Assume there exists $x_0 \in X$ such that*

$$\lim_{\substack{\|y\|\to\infty \\ y\in X}} \inf_{w\in T(y)} Re\langle w, y - y_0\rangle / \|y\| = +\infty.$$

Then for each given $w_0 \in E'$ there exists $\hat{y} \in X$ and $\hat{w} \in T(\hat{y})$ such that

$$Re\langle \hat{w} - w_0, \hat{y} - x\rangle \leq 0 \text{ for all } x \in X.$$

We note here that Theorem 5.31 can also be proved by Theorem 5.27. We shall see that Theorem 5.31 has an application in convex optimization as follows:

Let E be a Banach space, X be a non-empty convex subset of E and f be a convex function from X into $\mathbb{R}$. If $w_0 \in E'$ and $u_0 \in X$ then w_0 is said to be a *subgradient* of f at u_0 if for all $u \in X$

$$f(u) \geq f(u_0) + \langle w_0, u - u_0\rangle.$$

The subdifferential ∂f of f is the map of X into $2^{E'}$ given by

$$(\partial f)(x) := \{w \in E' : w \text{ is a subgradient of } f \text{ at } x.$$

Applying Theorem 5.31 together with the fact that the subdifferential ∂f is a (multi-valued) monotone operator (Browder (1976), p. 97) we have the following:

Theorem 5.32 *Let E be a reflexive Banach space equipped with norm $\|\cdot\|$ and $f : E \to \mathbb{R}$ be a lower semi-continuous convex function. Suppose $\partial f : E \to 2^{E'}$ is upper semi-continuous from line segments in E to the weak topology of E' such that each $\partial f(x)$ is bounded and there exists $x_0 \in E$ such that*

$$\lim_{\|y\|\to\infty} \inf_{w\in\partial f(y)} Re\langle w, y - x_0\rangle / \|y\| = +\infty.$$

Then there exists $\hat{y} \in E$ such that $0 \in \partial f(\hat{y})$, i.e. $\hat{y}$ is a critical point of f.

Finally we remark that in Theorem 5.30, if $X = E$, T is necessarily single-valued (and maximal monotone; for definition of maximal monotone, we refer to (Browder (1976), p. 791). Thus, in Theorem 5.32 if ∂f is assumed to be lower semi-continuous along line segments instead, then the conclusion becomes "there exists $\hat{y} \in E$ such that $\partial f(\hat{y}) = \{0\}$".

Note that the results of this section have been presented after a thorough revision of some results of an article in Shih and Tan (1988c).

5.7 Some Generalized Variational Inequalities with Their Applications

As an application of Fan-Browder type fixed point theorem which was given by Tarafdar (Tarafdar (1987)), first a generalized version of Ky Fan minimax principle is derived. This generalized version of Ky Fan minimax principle is then applied to establish existence theorems of solutions for non-compact variational inequalities in which mappings are either monotone or upper semi-continuous in locally convex topological vector spaces. Using these results some results on the existence of minimizers for minimization problems (consisting of the sum of convex lower semi-continuous functions) are obtained.

In this section we shall denote by 2^X the family of all subsets of X. Moreover, all vector spaces E will be assumed over the complex field Φ and all topological spaces will be assumed to be Hausdorff.

If X is a non-empty subset of E, then a mapping $T : X \to 2^{E^*}$ is said to be *monotone* on X if for each $x, y \in X$, $u \in T(x)$ and $w \in T(y)$, $Re\langle w - u, y - x\rangle \geq 0$.

5.7.1 *Some Generalized Variational Inequalities*

In this subsection, we shall first present a generalization of Ky Fan minimax principle which is then applied to study the existence of solutions for non-compact generalized variational inequalities in which the mappings are either monotone or upper semi-continuous in locally convex topological vector spaces.

The following fixed point theorem proved by Tarafdar in (Tarafdar (1987)) will be very useful in deriving other results of this section.

Theorem 5.33 (Fixed point theorem of Fan-Browder Type). *Let X be a non-empty convex subset of a real Hausdorff topological vector space. Let $F : X \to 2^X$ be a set-valued mapping such that*

(1) *for each $x \in X$, $F(x)$ is non-empty convex subset of X;*
(2) *for each $y \in X$, $F^{-l}(y) = \{x \in X : y \in F(x)\}$ contains relatively open subset O_y of X (O_y, may be empty for some $y \in X$ such that $U_{x\in X} O_x = X$; and*
(3) *X contains a non-empty subset X_0 which is contained in a compact convex subset X_1 of X such that the set $D = \cap_{x\in X_0} O_x^c$ is either empty or compact, where O_x^c denotes the complement of O_x in X.*

Then there exists a point $x_0 \in X$ such that $x_0 \in F(x_0)$.

The following lemma is a variant of Ky Fan inequality proved by Tarafdar in (Tarafdar (1986)). A different proof is given below which was obtained by Tarafdar and Yuan using the Fan-Browder type fixed point theorem mentioned above.

Lemma 5.17 *Let X be a non-empty convex subset of a Hausdorff topological vector space and $f : X \times X \to \mathbb{R} \cup \{-\infty, +\infty\}$ be an extended valued function such that*

(*i*) *for each $x \in X$, $f(x, \cdot)$ is lower semi-continuous on X;*
(*ii*) *for each $y \in X$,$g(\cdot, y)$ is quasi-concave on X (i.e., the set $\{y \in X : g(x, y) > \lambda\}$ is convex for each fixed $x \in X$ and $\lambda \in \mathbb{R}$);*
(*iii*) *for each $(x, y) \in X \times X$, $f(x, y) > 0 \Rightarrow g(x, y) \le 0$ and $g(x, x) \le 0$;*
(*iv*) *there exists a non-empty subset X_0 contained in a non-empty compact convex subset X_1, of X such that for each $y \in X \backslash X_1$, there exists $x \in X_0$ with $f(x, y) > 0$.*

Then the set $S := \{y \in X : f(x, y) \le 0 \text{ for all} x \in X\}$ is a non-empty compact subset of X_1.

Proof. For each $x \in X$, we define $H(x) = \{y \in X : f(x, y) \le 0\}$. Then by (i), $H(x)$ is a closed subset of X for each $x \in X$. It is trivial to see that $S = \cap_{x\in X} H(x)$. We first prove that $S \neq \emptyset$. We assume, if possible, that $S \neq \emptyset$. Then for each $y \in X$, the set

$$F(Y) = \{x \in X : y \notin H(x)\} = \{x \in X : f(x, y) > 0\}$$

is non-empty. Hence by (iv) for each $y \in X$, the set $G(y) = \{x \in X : g(x, y) > 0\} \supset F(y)$. Also it follows by (ii) $G(y)$ is a convex set for each $y \in X$. Thus $G : X \to 2^X$ defines a set-valued mapping such that for each $y \in X$, $G(y)$ is non-empty and convex. Now for each $x \in X$, the set

$$\begin{aligned} G^{-1}(x) &= \{y \in X : x \in G(y)\} = \{y \in X : g(x,y) > 0\} \\ &\subset \{y \in X : f(x,y) > 0\} \\ &= \{y \in X : f(x,y) \geq 0\}^c \\ &= [H(x)]^c \\ &= O_x \end{aligned}$$

which is a relatively open set in X. Also by (iii) for each $y \in X \backslash X_1$, there exists $x \in X_0$ such that $f(x,y) > 0$, i.e., $y \notin H(x)$. This implies that

$$D = \cap_{x \in X_0} O_x^c = \cap_{x \in X_0} H(x) \subset X_1.$$

Finally since $O_x = \{y \in X : f(x,y) > 0\} = F^{-1}(x)$, we immediately obtain

$$\cup_{x \in X} O_x = \cup_{x \in X} F^{-1}(x) = X.$$

To see this, let $x \in X$. As $F(x) \neq \emptyset$,we can choose $y \in F(x)$. Hence $x = F^{-1}(y)$. Hence by the fixed point Theorem 5.33, there is a fixed point $x_0 \in X$ such that $x_0 \in G(x_0)$ i.e., $g(x_0, x_0) > 0$ which contradicts (iv). Therefore $S \neq \emptyset$. We conclude the proof by noting that $S = \cap_{x \in X} H(x)$ being a closed subset of the compact subset $D = \cap_{x \in X_0} H(x)$ is compact and the proof is complete. □

Please note that Lemma 5.17 generalizes the celebrated Ky Fan minimax principle (Fan (1972)) in several aspects (see also Ding and Tan [Ding and Tan (1992b)], Yen [Yen (1981)] and references therein).

The proof of the following lemma can be found in Lemma 1 of Shih and Tan (Shih and Tan (1985), p. 334):

Lemma 5.18 *Let X be a non-empty subset of a Hausdorff topological vector space E and $S : X \to 2^E$ be upper semi-continuous and bounded. Then for each $P \in E^*$, the mappings $f_p : X \to \mathbb{R}$ defined by $f_p(y) := \sup_{x \in S(y)} Re\langle P, x\rangle$ for each $y \in X$ is upper semi-continuous.*

The following lemma is a set-valued version of well-known lemma due to Minty and Browder in (Minty (1962)) and (Browder (1968)) (see also Tarafdar (Tarafdar (1987)) and Tan and Yuan (Tan and Yuan (1994))).

Lemma 5.19 *Let X be a non-empty convex subset of a topological vector space E. Let $S : X \to 2^X$ be upper semi-continuous such that for each $x \in X$, $S(x)$ is a non-empty convex subset of X and $T : X \to 2^{E^*}$ be such that $T(x)$ is non-empty for each $x \in X$ and T is lower semi-continuous from the line segment to the weak topology of E^*. Then following statement holds:*

If there exists $\hat{y} \in X$ with $\hat{y} \in S(\hat{y})$ such that

$$\sup_{u \in T(x)} Re\langle u, \hat{y} - x\rangle \leq 0$$

for all $x \in S(\hat{y})$, *then we have*

$$\sup_{u \in T(\hat{y})} Re\langle u, \hat{y} - x\rangle \leq 0$$

for all $x \in S(\hat{y})$.

Proof. The proof is contained in the proof of Theorem 1 of Shih and Tan (Shih and Tan (1985)). We include the proof for the sake of completeness. Let $x \in S(\hat{y})$ be arbitrary and let $z_t = tx + (1-t)\hat{y} = \hat{y} + t(x - \hat{y})$, where $t \in [0,1]$. As $S(\hat{y})$ is convex, $x \in S(\hat{y})$ and $\hat{y} \in S(\hat{y})$, $z_t \in S(\hat{y})$for each $t \in [0,1]$. Hence

$$\sup_{u \in T(z_t)} Re\langle u, \hat{y} - x\rangle \leq 0$$

for all $t \in [0,1]$. Thus

$$\sup_{u \in T(z_t)} Re\langle u, \hat{y} - x\rangle \leq 0 \tag{5.78}$$

for all all $t \in [0,1]$. Now let $w_0 \in T(\hat{y})$ be arbitrary but fixed. For each $\epsilon > 0$, let

$$U_{w_0} = \{w \in E^* : |\langle w_0 - w, \hat{y} - x\rangle| < \epsilon\}.$$

Then U_{w_0} is a weakly open neighborhood of w_0. Since $U_{w_0} \cap T(\hat{y}) \neq \emptyset$ and T is lower semi-continuous from the line segment $L = \{z_t : t \in [0,1]\}$ to the weak topology of E^*, there is an open neighborhood $N(\hat{y})$ of $\hat{y}$ in L such that $T(z) \cap U_{w_0} \neq \emptyset$ whenever $z \in N(\hat{y})$. Now there exists $\delta \in (0,1)$ such that $z_t \in N(\hat{y})$ for all $t \in (0,\delta)$. Thus for each arbitrarily fixed $t \in (0,\delta)$ there exists $u \in T(z_t) \cap U_{w_0}$ i.e.,$|\langle w_0 - u, \hat{y} - x\rangle| < \epsilon$. This implies that $Re\langle w_0, \hat{y} - x\rangle < Re\langle u, \hat{y} - x\rangle + \epsilon$, which together with (5.78) implies that $Re\langle w_0, \hat{y} - x\rangle < \epsilon$. Now as $\epsilon > 0$ is arbitrary we obtain $Re\langle w_0, \hat{y} - x\rangle < 0$ and as $w_0 \in T(\hat{y})$ is arbitrary, we have

$$\sup_{w \in \hat{y}} Re\langle w, \hat{y} - x\rangle \leq 0$$

for all $x \in S(\hat{y})$. Thus the proof is complete. □

Theorem 5.34 *Let X be a non-empty closed convex subset of a locally convex Hausdorff vector topological space E. Let $S : X \to 2^X$ be upper semi-continuous such that for each $x \in X$, $S(x)$ is a non-empty closed convex bounded subset of X and $T : X \to 2^{E^*}$ be monotone and lower semi-continuous from the line segment to the weak topology $\sigma(E^*, E)$ of E^* and $T(x)$ is non-empty for each $x \in X$. Further assume that the set*

1. *the set $\sum_1 = \{y \in X : \sup_{x \in S(y)} \sup_{u \in T(x)} Re\langle u, y - x\rangle > 0\}$ is open;*
2. *there exists a non-empty subset X_0 contained in a compact convex subset X_1 of X such that for each $y \in X \backslash X_1$, there exists an $x \in X_0$ such that $\sup_{u \in T(x)} Re\langle u, y - x\rangle > 0$.*

Then there exists $\hat{y} \in X$ such that

(*i*) $\hat{y} \in S(\hat{y})$;
(*ii*) $\sup_{w \in T(\hat{y})} Re(w, \hat{y} - x) \leq 0$ *for all* $x \in S(\hat{y})$.

Proof. In view of our Lemma 5.19, it would suffice to prove that there exists a point $\hat{y} \in X$ such that $\hat{y} \in S(\hat{y})$ and

$$\sup_{u \in T(x)} Re\langle u, \hat{y} - x\rangle \leq 0$$

for all $x \in S(\hat{y})$.

If possible, suppose the above is false. Then for each $y \in X$, either $(\alpha) : y \notin S(y)$ or (β) : there exists a point $x \in S(y)$ such that

$$\sup_{u \in T(x)} (u, y - x) > 0.$$

Now whenever $y \notin S(y)$,there exists by Hahn-Banach separation theorem $p \in E^*$ such that

$$Re\langle p, y\rangle - \sup_{x \in S(y)} Re\langle p, x\rangle > 0.$$

For each $y \in X$, let $\alpha(y) := \sup_{x \in S(y)} \sup_{u \in T(x)} Re\langle u, y - x\rangle$. Also let $V_0 = \{y \in X : \alpha(y) > 0\}$ which is open in X by (i). Now $X \setminus X_1 \subset V_0$.To see this let $y \in X \setminus X_1$. Then either $y \in S(y)$ or $y \notin S(y)$. If $y \in S(y)$, then by (ii), we have $\alpha(y) > 0$). Hence $y \in V_0$. If $y \notin S(y)$, then by (ii) and (β), it follows that $\alpha(y) > 0$, i.e., $y \in V_0$. Thus $y \in V_0$ in either case. For each $p \in E^*$, let $V(P) := \{y \in X : Re\langle p, y\rangle - \sup_{x \in S(y)} Re\langle p, x\rangle > 0\}$. By Lemma 5.18, $V(p)$ is open for each $p \in E^*$. It is clear that $X = V_0 \cup \{\cup_{p \in E^*} V(p)\}$. Since X_1, is compact, there exists $p_l, p_2, \ldots, p_n \in E^*$ such that $X_1 \subset \cup_{i=1}^n V(p_i)$. Thus $X = V_0 \cup \cup_{i=1}^n V(p_i)$. Now we repeat the argument as given in Shih and Tan (Shih and Tan (1985, pp. 336–337)). Let $\{\beta_0, \beta_1, \ldots, \beta_n\}$ be a partition of unity corresponding to the covering $\{V_0, V(p_1), \ldots, V(p_n)\}$, i.e ,$\beta_0, \beta_1, \ldots, \beta_n$ are continuous nonnegative real valued functions defined on X such that β_i vanishes out $V(p_i)$ for $i = 1, 2, \ldots, n$ and β_0 vanishes outside V_0 and $\sum_{i=0}^n \beta_i(x) = 1$ for all $x \in X$. We define $f, g : X \times X \to \mathbb{R}$ by setting

$$f(x, y) = \beta_0(y) \sup_{u \in T(x)} Re\langle u, y - x\rangle + \sum_{i=1}^n \beta_i(y) Re\langle p_i, y - x\rangle,$$

and

$$g(x, y) = \beta_0(y) \inf_{u \in T(y)} Re\langle u, y - x\rangle + \sum_{i=1}^n \beta_i(y) Re\langle p_i, y - x\rangle,$$

for each $(x, y) \in X \times X$. Now since T is monotone, it follows that $g(x, y) > 0$ whenever $f(x, y) > 0$ for all $(x, y) \in X \times X$ (note that,$\beta_0(y) = 0$, then $f(x, y) = g(x, y)$). Obviously, $g(x, x) = 0$ for each $x \in X$. For each fixed $x \in X$, since β_i, where $i = 1, 2, \ldots, n$ are continuous functions of y on X and $\sup_{u \in T(x)} Re\langle u, y - x\rangle$ and $Re\langle p_i, y - x\rangle$ $(i = 1, 2, \ldots, n)$ are lower semi-continuous functions of y on X, by Lemma 3 of Takahashi (1976, p. 177), $y \to f(x, y)$ is lower semi-continuous on X. Furthermore, for each fixed $y \in X$, $x \in g(x, y)$ is quasi-concave. Hence all the conditions of Lemma 5.17 are satisfied. Thus there exists by Lemma 5.17 a point $\hat{y} \in X$ such that $f(x, \hat{y}) \leq 0$ for all $x \in X$, i.e.,

$$\beta_0(\hat{y}) \sup_{u \in T(x)} Re\langle u, \hat{y} - x\rangle + \sum_{i=1}^{n} \beta_i(\hat{y}) Re\langle p_i, \hat{y} - x\rangle \leq 0 \tag{5.79}$$

for all $x \in X$. Since $\sum_{i=0}^{n} \beta_i(\hat{y}) \geq 0$ for each $i = 0, 1, \ldots, n$ and $\beta_i(y) > 0$ for at least one $i = 0, 1, \ldots, n$. We choose $\hat{x} \in S(\hat{y})$ such that $sup_{u \in T(\hat{x})} Re\langle u, \hat{y} - \hat{x}\rangle \geq \frac{\alpha(\hat{y})}{2}$ whenever $\alpha(\hat{y}) > 0$. If $\beta_0(\hat{y}) > 0$, then $\hat{y} \in V_0$. Thus $\alpha(\hat{y}) > 0$. If $\beta_k(\hat{y}) > 0$ for some $k = 1, 2, \ldots, n$ then $\hat{y} \in V(p_k)$ and hence $Re\langle p_k, \hat{y} - \hat{x}\rangle > \sup_{x \in S(\hat{y})} Re\langle p_k, x\rangle \geq (p_k, \hat{x})$. Hence $Re\langle p_k, \hat{y} - \hat{x}\rangle > 0$. Thus it follows that

$$\beta_0(\hat{y}) \sup_{u \in T(\hat{x})} Re(u, \hat{y} - \hat{x}) + \sum_{i=1}^{n} \beta_i(\hat{y}) Re\langle p_i, \hat{y} - \hat{x}\rangle \leq 0$$

which contradicts (5.79). This proves the theorem and the proof is complete. □

Remark 5.20 Theorem 5.34 contains Theorem 1 of Shih and Tan (Shih and Tan (1985)) as a special case.

Lemma 5.20 *Let X be a non-empty bounded subset of a locally convex Hausdorff topological vector space E. Let $S : X \to 2^X$ be lower semi-continuous such that for each $x \in X$, $S(x)$ is a non-empty subset of X and $T : X \to 2^{E^*}$ be lower semi-continuous from the relative topology of X to the strongly topology of E^* such that for each $x \in X$, $T(x)$ is non-empty. Then the set*

$$\sum_1 = \{y \in X : \sup_{x \in S(y)} \sup_{u \in T(x)} Re\langle u, y - x\rangle > 0\}$$

is an open set in X.

Proof. The proof is contained in the proof of Theorem 2 of Shih and Tan in (Shih and Tan (1985)) and thus we omit it. □

Combining Theorem 5.34 and Lemma 5.20, the following theorem is obtained:

Theorem 5.35 *Let X be a non-empty bounded closed convex subset of a locally convex Hausdorff topological vector space E. Let $S : X \to 2^X$ be continuous such that for each $x \in X$, $S(x)$ is a non-empty closed convex bounded subset of X and $T : X \to 2^X$ be a monotone mapping such that $T(x) \neq \emptyset$ for each $x \in X$ and T is lower semi-continuous from the line segment to the weak topology $\sigma(E^*, E)$ of E^*. Further assume that there is a non-empty set X_0 contained in a compact convex* *subset X_1 of X such that for each $y \in X \setminus X_1$ there exists an $x \in X_0$ such that*

$$\sup_{u \in T(x)} \langle u, y - x \rangle > 0.$$

Then there exists $\hat{y} \in X$ such that $\hat{y} \in S(\hat{y})$ and

$$\sup_{w \in T(\hat{y})} Re\langle w, \hat{y} - x \rangle \leq 0$$

for all $x \in S(\hat{y})$.

Corollary 5.35.1 *Let X be a non-empty closed convex subset of a locally convex Hausdorff space E. let $S : X \to 2^X$ be continuous such that for each $x \in X$, $S(x)$ is non-empty closed bounded convex subset of X and $T : X \to 2^{E^*}$ be a monotone mapping such that for each $x \in X$, $T(x) \neq \emptyset$ and T is lower semi-continuous from the relative topology of X to the strong topology of E^*. Further assume that there is a point $x_0 \in X$ such that $\sup_{u \in T(x_0)} Re\langle u, y - x_0 \rangle > 0$ for all $y \in X$ and $y \neq x_0$. Then there exists $\hat{y} \in X$ such that $\hat{y} \in S(\hat{y})$ and*

$$\sup_{w \in T(\hat{y})} Re\langle w, \hat{y} - x \rangle \leq 0$$

for all $x \in S(\hat{y})$.

Proof. We take $X_1 = X_0 = \{x_0\}$ in Theorem 5.35 to obtain the Corollary. □

Corollary 5.35.2 *Let X be a non-empty closed convex subset of a locally convex Hausdorff space E. Let $T : X \to 2^{E^*}$ be a monotone mapping such that for each $x \in X$, $T(x) \neq \emptyset$ and T is lower semi-continuous from the relative topology of X to the strong topology of E^*. Further assume that there is a non-empty subset X_0 containing in a compact convex subset X_1, of X such that for each $y \in X \setminus X_1$, there exists $x \in X_0$ such that $\sup_{u \in T(x)} Re\langle u, y - x \rangle > 0$. Then there exists $\hat{y} \in X$ such that*

$$\sup_{w \in T(\hat{y})} Re(w, \hat{y} - x) \leq 0$$

for all $x \in S(\hat{y})$.

Proof. We define $S : X \to 2^X$ by $S(x) = X$ for each $x \in X$. Then the conclusion follows by Theorem 5.35. □

When T is a constant monotone mapping in Theorem 5.34, we have the following non-compact version of the well-known Fan and Glicksberg fixed point theorem (see Fan-Glicksberg (1952)).

Theorem 5.36 *Let X be a non-empty convex subset of a locally convex Hausdorff topological space E. Let $S : X \to 2^X$ be upper semi-continuous such that for each $x \in X$, $S(x)$ is a non-empty closed convex bounded subset of X and suppose there exists $p \in E^*$ such that the set $\sum_1 = \{y \in X : \sup_{x \in S(y)} Re\langle p, u - x \rangle > 0\}$ is open* *and there exists a non-empty subset x_0 contained in a compact convex subset X_1, of X such that for each $y \in X \setminus X_1$ there exists an $x \in X_0$ such that $Re\langle p, y - x\rangle > 0$. Then there exists $\hat{y} \in X$ such that $\hat{y} \in s(\hat{y})$.*

Proof. Let $T : X \to E^*$ be defined by $T(x) = p$ for each $x \in X$. Then T is a monotone mapping and all hypotheses of Theorem 5.34 are satisfied. Thus the conclusion follows from Theorem 5.34 and proof is complete. □

In what follows, we shall study the existence of solutions for non-compact generalized variational inequalities in which T is upper semi-continuous instead of being monotone in locally convex spaces.

Theorem 5.37 *Let X be a non-empty closed convex subset of a locally convex Hausdorff topological vector space E. Let $S : X \to 2^X$ be upper semi-continuous such that for each $x \in X$, $S(x)$ is a non-empty closed convex bounded subset of X and $T : X \to 2^{E^*}$ be upper semi-continuous from the relative topology of X to the strong topology of E^* such that for each $x \in X$, $T(x)$ is a non-empty compact convex subset of E^*. Further assume that (1) the set $\sum_2 = \{y \in X : \sup_{x\in S(y)} \inf_{u\in T(y)} Re\langle u, y - x\rangle > 0\}$ is open; (2) there exists a non-empty subset X_0 contained in a compact convex subset X_1 of X such that for each $y \in X \setminus X_1$ there exists an $x \in X_0$ such that $\inf_{u\in T(y)} Re\langle u, y - x\rangle > 0$.*

Then there exists $\hat{y} \in X$ such that (i) $\hat{y} \in S(\hat{y})$; (ii) there exists a point $\hat{z} \in T(\hat{y})$ such that $Re(\hat{z}, \hat{y} - x) \leq 0$ for all $x \in S(\hat{y})$.

Proof. We shall prove the result essentially by following the idea used in Theorem 5.34 (see also Theorem 3 of Shih and Tan (Shih and Tan (1985, p. 340)).

First we show that there exists a point $\hat{y} \in X$ such that $\hat{y} \in S(\hat{y}$ and

$$\sup_{x\in S(\hat{y})} \inf_{z\in T(\hat{y})} Re\langle z, \hat{y} - x\rangle \leq 0.$$

Suppose the assertion were false. Then for each $y \in X$, either $y \notin S(y)$ or there exists $x \in S(y)$ such that $\inf_{z\in T(y)} Re\langle z, y - x\rangle > 0$. Observe that whenever $y \notin S(y)$, there exists $p \in E^*$ such that $Re\langle p, y\rangle - \sup_{x\in S(y)} Re\langle p, x\rangle > 0$. For each $y \in X$, we set $\alpha(y) := \sup_{x\in S(y)} \inf_{z\in T(y)} Re\langle z, y - x\rangle$, $V_0 = \{y \in X : \alpha(y)\}$ and $V(p) := \{y \in X : Re\langle p, y\rangle - \sup_{x\in S(y)} Re\langle p, x\rangle > 0\}$ for each $P \in E^*$. By the condition (2) and the same argument used in the Proof of Theorem 5.34, there exists $p_1, p_2, \ldots, p_n \in E^*$ such that $X_1 \subset \cup_{n=1}^n V(p_i)$. Thus $X = V_0 \cup \cup_{i=1}^n V(p_i)$.Let $\beta_0, \beta_1, \ldots, \beta_n$ be a partition of unity corresponding to the covering $\{V_0, V(p_1), V(p_2), \ldots, V(p_n)\}$ i.e., $\beta_0, \beta_1, \ldots, \beta_n$ are continuous nonnegative real valued functions defined on X such that β_i vanishes out $V(p_i)$ for $i = 1, 2, \ldots, n$ and β_0 vanishes outside V_0 and $\sum_{i=0}^n \beta_i(x) = 1$ for all $x \in X$. We define $f : X \times X \to \mathbb{R}$ by

$$f(x, y) = \beta_0(y) \inf_{w\in T(y)} Re\langle w, y - x\rangle + \sum_{i=1}^n \beta_i(y) Re\langle p_i, y - x\rangle,$$

for each $(x, y) \in X \times X$. Clear ϕ satisfies all hypotheses of Lemma 5.17, (with $f = g$). By Lemma 5.17 there exists $\hat{y} \in X$ such that $f(x, \hat{y})$ for all $x \in X$. This contradicts that there is a point $\hat{x} \in X$ with $f(\hat{x}, \hat{y}) > 0$ can be achieved by using the corresponding proof of Theorem 5.34 above.

Next, we shall show that there exists $\hat{z} \in T(\hat{y})$ such that $Re\langle \hat{z}, \hat{y}M -^V z\rangle \leq 0$ for all $x \in S(\hat{y})$. We define a mapping $\psi : S(\hat{y}) \times T(\hat{y}) \to \mathbb{R}$ by $\psi(x, y) = Re\langle z, \hat{y} - x\rangle$ for each $(x, y) \in S(\hat{y}) \times T(\hat{y})$. Note that for each fixed $x \in T(\hat{y})$, $z \mapsto \psi(x, z)$ is continuous and affine; and for each fixed $z \in T(\hat{y})$, $x \mapsto \psi(x, z)$ is affine. By Kneser's minimax theorem (e.g., see Kneser (1952)), it follows that

$$\min_{z \in T(\hat{y})} \max_{x \in S(\hat{y})} \psi(x, z) = \max_{x \in S(\hat{y})} \min_{z \in T(\hat{y})} \psi(x, z).$$

Thus we have $\min_{z \in T(\hat{y})} \sup_{x \in S(\hat{y})} Re\langle z, \hat{y} - x\rangle \leq 0$. As $T(\hat{y})$ is compact, there exists $\hat{z} \in T(\hat{y})$ such that $Re\langle \hat{z}, \hat{y} - x\rangle \leq 0$ for all $x \in S(\hat{y})$ and we complete the proof. □

Assuming additional lower semi-continuity on S, the following theorem is obtained:

Theorem 5.38 *Let X be a non-empty closed bounded convex subset of a locally convex Hausdorff topological space E. Let $S : X \to 2^X$ be continuous such that for each $x \in X$, $S(x)$ is a non-empty closed convex bounded subset of X and $T : X \to 2^{E^*}$ be upper semi-continuous from the relative topology of X to the strong topology of E^* such that for each $x \in X$, $T(x)$ is a non-empty compact convex subset of E^*. Further assume that there exists a non-empty subset X_0 contained in a compact convex subset X_1, of X such that for each $y \in X \setminus X_1$, there exists an $x \in X_0$ such that $\inf_{u \in T(y)} Re\langle u, y - x\rangle > 0$.*

Then there exists $\hat{y} \in X$ such that

(*i*) $\hat{y} \in S(\hat{y})$;

(*ii*) *there exists a point $\hat{z} \in T(y)$ such that $Re\langle \hat{z}, \hat{y} - x\rangle \leq 0$ for all $x \in S(\hat{y})$.*

Proof. Note that X is bounded. By the upper semi-continuity of T and lower semi-continuity of S, it follows that $y \mapsto \sup_{x \in s(y)} \inf_{u \in T(y)} Re\langle u, y - x\rangle$ is lower semi-continuous by III-Propositions 19 and 21 of Aubin and Ekeland (1984, pp. 118, 119). Thus the set $\sum_2$ of Theorem 5.37 is open in X and the conclusion follows by Theorem 5.37. □

Remark 5.21 For the study of variational inequalities and its various applications in Banach spaces, the interested readers are referred to Browder (1963), Browder (1968), Ding and Tarafdar (1996), Tan and Yuan (1994) and references therein.

5.7.2 *Applications to Minimization Problems*

As applications of variational inequalities which have been established by Tarafdar and Yuan in the previous sections, we shall now present another result of Tarafdar and Yuan for the existence of solutions for a minimization problem.

$$\inf_{x \in E} f(x) \tag{5.80}$$

where f is the sum of two extended real-valued functions $g, h : E \to (-\infty, +\infty)$ and E is a normed space. Before we prove the existence of solutions for (5.80), we recall the following definition of subdifferential (e.g., see Aubin and Ekeland (1984, p. 187):

Definition 5.9 Let X be a non-empty convex subset of a topological vector space E. Suppose $f : X \to (-\infty, +\infty)$ is a function with non-empty domain. If $x_0 \in \text{Dom } f$, the "*subdifferential* $\partial f(x_0)$ of f at x_0," is the subset (which may be empty) of E^* defined by

$$\partial f(x_0) = \{p \in E^* : f(x_0) - f(x) \leq \langle p, x_0 - x \rangle \text{ for all } x \in X\}. \tag{5.81}$$

The elements $p \in \partial f(x_0)$ are also called subgradients. Also $p_0 \in \partial f(x_0)$ if and only if $f^{**}(x_0) = f(x_0)$ (see, e.g., Aubin and Ekeland Aubin and Ekeland (1984, pp. 216, 217)).

The following simple Proposition shows that the existence of solutions of variational inequalities guarantee sufficiently the existence of the minimizers for the minimization problem (5.80).

Proposition 5.3 *Let X be a non-empty convex subset of a Hausdorff topological vector space E. Suppose $f = g + h$ is the sum of a convex function g and a subdifferential function h defined on a non-empty convex subset X, i.e., $g, h : X \to (-\infty, +\infty]$. Then a point $\hat{x} \in X$ minimizes f if there exists $p \in \partial h(\hat{x})$ such that*

$$\sup_{x \in X} [\langle p, \hat{x} - x \rangle + g(\hat{x}) - g(x)] \leq 0. \tag{5.82}$$

Proof. Suppose there exists $p \in \partial h(\hat{x})$ such that $(p, \hat{x} - x) + g(\hat{x}) - g(x) \leq 0$ for all $x \in X$. Then we have that

$$f(\hat{x}) - f(x) = h(\hat{x}) - h(x) + g(\hat{x}) - g(x) \leq (p, \hat{x} - x) + g(\hat{x}) - g(x) \leq 0$$

for all $x \in X$. Thus $\hat{x}$ minimizes f and the proof is complete. □

Now we have the following general existence theorem which guarantee the existence of minimizers for the minimization problem (5.80).

Theorem 5.39 *Let $f : E \to \mathbb{R}$ be a subdifferentiable function such that the mapping $T : E \to 2^{E^*}$ defined by $T(x) = \partial f(x)$ for each $x \in X$ is lower semicontinuous from the topology of E to the strong topology of E^*. Further assume that there exists a non-empty subset X_0 contained in a compact convex subset X_1, of X*

such that for each $y \in E \setminus X_1$ *there exists an* $x \in X_0$ *such that* $\sup_{u \in T(x)} \langle u, y - x \rangle > 0$. *Then there exists a point* $\hat{y} \in E$ *such that*

$$\sup_{w \in T(\hat{y})} Re\langle w, \hat{y} - x \rangle \leq 0$$

for all $y \in E$, *i.e.,* $\hat{y}$ *minimizes* f *on* E. *Furthermore, for each* $u \in T(\hat{y})$, $\hat{y} \in T^{-1}(u) = \partial f^*(u)$, *and* $f(\hat{y}) = f^{**}(\hat{y})$, *where* $f^* : E^* \to (-\infty, +\infty)$ *is defined by*

$$f^*(u) = \sup_{u \in E} [(u, x) - f(x)]$$

for each $x \in X$.

Proof. By Corollary 5.35.2, there exists $\hat{y} \in X$ such that

$$\sup_{w \in T(\hat{y})} Re\langle w, \hat{y} - x \rangle \leq 0$$

for all $x \in X$, i.e., for each $u \in T(\hat{y})$, we have $(u, \hat{y} - x) \leq 0$ for all $u \in T(\hat{y}) = \partial f(\hat{y})$. Hence $\hat{y}$ minimizes f on E. The result follows easily from our discussion above. □

Note that the results of this section have been written and presented after a thorough revision of some results of an article in Tarafdar and Yuan (1997a).

5.8 Some Results of Tarafdar and Yuan on Generalized Variational Inequalities in Locally Convex Topological Vector Spaces

The results of this section will be established on topological vector spaces E (respectively, locally convex topological vector spaces E) which are Hausdorff. If X is a nonempty set, we shall denote by 2^X the family of all subsets of X.

Ding (Ding (1991, Theorem 2.2)) proved the following existence theorem of solutions for generalized variational inequalities (the proof in an English version can also be found in the Appendix of Ding and Tarafdar (1996)). He used this result in studying the existence of solutions of *generalized complementarity problems* ($GCP(f, K)$) for quasi-monotone mappings (definition follows) in general settings.

Theorem 5.40 *Let* K *be a nonempty convex subset of a Hausdorff locally convex topological vector space* E *which is of second category . Let* $f : K \to 2^{E^*}$ *be a quasi-monotone mapping such that* f *is upper semi-continuous from the line segments in* K *to the weak* topology of* E^* *and each* $f(x)$ *is nonempty compact convex in the strong topology on* E^*. *Suppose that there exist nonempty weakly compact convex subset* K_0 *of* K *and nonempty weakly compact subset* D *of* K *such that for each* $y \in K \setminus D$, *there exists* $x \in co(K_0 \cup \{y\})$ *with*

$$Re\langle y - x, u \rangle > 0, \quad \textit{for all } u \in f(x).$$

Then there exists a point $\hat{y} \in K$ such that

$$\inf_{v \in f(\hat{y})} Re(\hat{y} - x, v) \le 0, \text{ for all } x \in K.$$

If in addition, $f(\hat{y})$ is also convex, then there exists $\hat{v} \in f(\hat{y})$ such that

$$Re\langle x - \hat{y}, \hat{v}\rangle \ge 0 \text{ for all } x \in K$$

that is, $(\hat{y}, \hat{v})$ is a solution of GVIP(f, K).

We recall that a topological space X is said to be of second category if X cannot be expressed as the union of a sequence of nowhere dense sets [see Rudin (1973), p. 41]). Of course, each Banach space is of second category. However, a topological vector space (respectively, a locally convex topological vector space) need not be of second category in general.

As Theorem 5.40 above has found many applications in the study of variational theory itself, mathematical programming and operations research such as complementarity problems, and so on, it is our purpose in this note to generalize Theorem 5.40 to locally convex topological vector spaces (hence, which need not be of second category). Our generalization of GVIP(f, K) includes corresponding results of Cottle and Yao (1992), Saigal (1976), Fang and Peterson (1982), Harker and Pang (1990), Siddiqi and Ansari (1989), Shih and Tan (1988b), Tan and Yuan (1994), and Zhou and Chen (1988) as special cases.

Let X and Y be topological spaces and $f : X \to 2^Y$. Then f is said to be upper semi-continuous (in short, u.s.c.) on X if for each $x_0 \in X$ and any open set V in Y containing $f(x_0)$, there exists an open neighborhood U of $x_0 \in X$ such that $f(x) \subset V$ for all $x \in U$. If X is a topological space and $f : X \to \mathbb{R} \cup \{-\infty, +\infty\}$ is an extended real valued function. Then f is said to be compactly lower semi-continuous if f is lower semi-continuous on each nonempty compact subset of its domain X.

Definition 5.10 Let K be a nonempty subset of a topological vector space E and $f : K \to 2^{E^*}$ a set-valued mapping with nonempty values. Then f is said to be (1) *monotone* if for each $x, y \in K$, $u \in f(x)$, and $v \in f(y)$, $Re\langle x - y, u - v\rangle \ge 0$; (2) *pseudo-monotone* (see Saigal (1976), p. 263, Definition 3.4]) if for each $x, y \in K$, $u \in f(x)$, and $v \in f(y)$, $Re\langle x - y, v\rangle \ge 0$ implies $Re\langle x - y, u\rangle \ge 0$; (3) *quasi-monotone* (compare with Ding (1991)) if for each $x, y \in K$,

$$\inf_{v \in f(y)} Re\langle x - y, v\rangle \ge 0 \,\text{implies} \inf_{u \in f(y)} Re\langle x - y, u\rangle \ge 0$$

Clearly, each monotone mapping is pseudo-monotone, and Proposition 3.2 of [Karamardian and Schaible (1990), p. 39] shows that the definition (3) above is a set-valued generalization of the definition (2) for a single-valued mapping.

Let X be a nonempty convex subset of a topological vector space and $\psi : X \times X \to \mathbb{R} \cup \{-\infty, +\infty\}$ an extended real-valued function. We also recall that (see

[Zhou and Chen (1988)]) ψ is said to be 0-*diagonal quasi-concave in* y if for each $A \in \mathcal{F}(X)$ and $x_0 \in co(A)$, we have $min_{y\in A}\psi(x_0, y) \leq 0$.

Let $\psi(x,x) \leq 0$ for each $x \in X$. It is clear that if $\psi(x,y)$ is quasi-concave in y for each $x \in X$, $\psi(x,y)$ is 0-diagonal concave in y; but the converse does not hold by Remark 2.2 of Zhou and Chen (1988, p. 215).

5.8.1 *Some Generalized Variational Inequalities*

We shall now present some results of Yuan and Tarafdar on existence theorem of GVIP(f, K) for quasi-monotone set-valued mappings in locally convex Hausdorff topological vector spaces.

Theorem 5.41 *Let K be nonempty convex subset of a locally convex Hausdorff topological vector space E with $E^* \neq \emptyset$ (hence, E need not be of second category). Let $f : K - 2^{E^*}$ be a quasi-monotone mapping such that f is upper semi-continuous from the line segments in K to the weak* topology of E^* and each $f(x)$ is nonempty compact in the strong topology on E^*. Suppose that there exist nonempty weakly compact convex subset K_0 of K and nonempty weakly compact subset D of K such that for each $y \in K\backslash D$, there exists $x \in co(K_0 \cup \{y\})$ with*

$$Re\langle y - x, u\rangle \geq 0, \text{ for all } u \in f(x).$$

Then there exists a point $\hat{y} \in K$ such that

$$\inf_{v\in f(\hat{y})} Re\langle \hat{y} - x, v\rangle \leq 0, \text{ for all } x \subset K.$$

If, in addition, $f(\hat{y})$ is also convex, then there exists $\hat{v} \in f(\hat{y})$ such that

$$Re\langle x - \hat{y}, \hat{v}\rangle \geq 0, \text{ for all } x \in K,$$

that is, $(\hat{y}, \hat{v})$ is a solution of GVIP(f, K).

Remark 5.22 Theorem 5.41 shows that Theorem 5.40 above still holds without the assumption that E is of secondary category, thus Theorem 5.41 includes Theorem A of Ding (1991) as a special case. To prove Theorem 5.41, Tarafdar and Yuan used the following generalization of the Ky Fan minimax principle which is due to Ding and Tan (1992b).

Lemma 5.21 *Let K be a nonempty convex subset of a Hausdorff topological vector space E and $\psi : X \times X \to \mathbb{R} \cup \{-\infty, +\infty\}$ be an extended real-valued function such that*

1. *for each fixed $x \in K$, $y \mapsto \psi(x,y)$ is compactly lower semi-continuous;*
2. *for each $A \in \mathcal{F}(\mathcal{X})$ and $y_0 \in co(A)$, $\min_{x\in A} \psi(x, y_0) \leq 0$;*
3. *there exist a nonempty compact and convex subset K_0 of K and a nonempty compact subset D of K such that for each $y \in K\backslash D$, there exists $x \in co(K_0 \cup \{x\})$ with $\psi(x,y) > 0$.*

Then there exists $\hat{y} \in D$ *such that* $\sup_{x \in X} \psi(x, \hat{y}) \leq 0$.

Tarafdar and Yuan also used the following result which was originally proved by [Tan and Yuan (1994), Lemma 8].

Lemma 5.22 *Let K be a nonempty convex subset of a topological vector space E and $T : K \to 2^{E^*}$ be upper semi-continuous from the line segments in K to the weak* topology E^* such that each $T(x)$ is nonempty weak*-compact. If $\hat{y} \in K$, then the inequality*

$$\inf_{u \in T(x)} Re\langle u, \hat{y} - x\rangle \leq 0, \text{ for all } x \in K$$

implies the inequality

$$\inf_{w \in T(\hat{y})} Re\langle w, \hat{y} - x\rangle \leq 0, \text{ for all } x \in K$$

Lemma 5.23 *Let K be a nonempty convex subset of a topological vector space E and $f : K \to 2^{E^*}$ a set-valued quasi-monotone mapping with nonempty values. Then the mapping $\psi : K \times K \to \mathbb{R} \cup \{-\infty, +\infty\}$ defined by*

$$\psi(x, y) := \inf_{u \in f(y)} Re\langle u, x - y\rangle,$$

for each $(x, y) \in K \times K$ is 0-diagonal concave in y for each fixed $x \in K$.

By combining Lemmas 5.21, 5.22 and 5.23 Tarafdar and Yuan proved Theorem 5.41 as follows:

Proof of Theorem 5.41. Define a mapping $\varphi : K \times K \to \mathbb{R}$ by $\varphi(x, y) := \inf_{u \in f(x)} Re\langle u, y - x\rangle$, for each $(x, y) \in K \times K$. Then we have

(a) for each fixed $y \in K$, $\varphi(x, y)$ is 0-diagonal concave in x by Lemma 5.14 as f is quasi-monotone;
(b) for each $x \in K$, the function $y \mapsto \varphi(x, y)$ is weakly lower semi-continuous on each weakly compact subset A of K.

Note that E is locally convex, and for each $x \in K$ and for any nonempty weakly compact subset A of K, $A - x$ is weakly bounded and hence, by Rudin (1973, Theorem 3.18), $A - x$ is strongly bounded in E. Now we only need to prove that for each fixed $x \in K$ and for any real number $\lambda \in R$ the set $A_\lambda = \{y \in A : \inf_{u \in f(x)} Re\langle y - x, u\rangle \leq \lambda\}$ is weakly closed. Let $\{y_\alpha\}_{\alpha \in \Gamma} \subset A_\lambda$ be a net, $y_0 \in A$ and $y_\alpha \rightharpoonup y_0$. Since $f(x)$ is strong-compact, for each $\alpha \in \Gamma$, there exists an $u_\alpha \in f(x)$ such that

$$Re\langle y_\alpha - x, u_\alpha\rangle = \inf_{u \in f(x)} Re\langle y_\alpha - x, u\rangle \leq \lambda,$$

and there exists a subnet $\{u_\beta\}$ of $\{u_\alpha\}_{\alpha \in \Gamma}$, such that $u_\beta \to u_0 \in f(x)$. Let $B = \{y_\alpha - y : \alpha \in \Gamma\} \cup \{y_0\}$. Then B is strongly bounded. For any given $\varepsilon > 0$, let

$W(B,\varepsilon) = \{u \in E^* : \sup_{y\in B} |\langle y, u - u_0\rangle| < \varepsilon/3\}$, then $W = W(B,\varepsilon)$ is a strongly open neighborhood of u_0 in E^*. Let $U = \{y \in A : |\langle y - y_0, u\rangle| < \varepsilon/3\}$, It follows that U is a weakly open neighborhood of y_0 in A. Choose $\beta_0 \in \Gamma$ such that for all $\beta \geq \beta_0$,$u_\beta \in W$ and $y_\beta \in U$. It follows that

$$|Re\langle y_\beta - x, u_\beta\rangle - Re\langle y_0 - x, u_0\rangle| \leq |\langle y_\beta - y_0, u_\beta - u_0\rangle| + |\langle y_0 - x, u_\beta - u_0\rangle| \tag{5.83}$$

$$+ |\langle y_\beta - y_0, u_0\rangle| < \frac{\varepsilon}{3} + \frac{\varepsilon}{3} + \frac{\varepsilon}{3} = \varepsilon, \tag{5.84}$$

and so $\lim_{\beta\in\Gamma} Re\langle y_\beta - x, u_\beta\rangle = Re\langle y_0 - x, u_0\rangle$.

Hence, we have

$$\inf_{u\in f(x)} Re\langle y_0 - x, u\rangle \leq Re\langle y_0 - x, u_0\rangle = \lim_{\beta\in\Gamma} Re\langle y_\beta - x, u_\beta\rangle \leq \lambda,$$

i.e., $y_0 \in A_\lambda$ and A_λ is weakly closed in A. This shows that for each $x \in K$,$y \mapsto \varphi(x,y)$ is weakly lower semi-continuous on each weakly compact subset A of K.

(c) There exists a nonempty weakly compact convex subset K_0 of K and a nonempty weakly compact subset D of K such that for each $y \in K\backslash D$, there exists $x \in co(K_0 \cup \{y\})$ with

$$Re\langle y - x, u\rangle > 0, \text{ for all}\, u \in f(x).$$

Since $f(x)$ is strongly compact, therefore $\varphi(x,y) = \inf_{u\in f(x)} Re\langle y - x, u\rangle > 0$.

Now equip E with the weak topology, and we see that all hypotheses of Lemma 5.21 hold. Hence, it follows that there exists an $\hat{y} \in K$ such that

$$\varphi(x,\hat{y}) = \inf_{u\in f(x)} Re\langle \hat{y} - x, u\rangle \leq 0, \text{for all} x \in K. \tag{5.85}$$

By Lemma 5.22, we have that

$$\varphi(x,\hat{y} = \inf_{u\in f(\hat{y})} Re\langle \hat{y} - x, u\rangle \leq 0, \text{ for all}\, x \in K.$$

Finally, if $f(\hat{y})$ is convex, define $g : K \times f(\hat{y}) \to \mathbb{R}$ by

$$g(x,v) = Re(\hat{y} - x, v).$$

Then for each $x \in K$, $v \mapsto g(x,v)$ is weak* continuous and affine and for each $v \in f(\hat{y})$, $x \mapsto g(x,y)$ is also affine. By the minimax theorem of Kneser (1952) (see also Aubin (1979)),

$$\min_{v\in f(\hat{y})} \sup_{x\in K} Re\langle \hat{y} - x, v\rangle = \sup_{x\in K} \min_{v\in f(\hat{y})} Re\langle \hat{y} - x, v\rangle \leq 0.$$

Since $f(\hat{y})$ is norm compact, there exists $\hat{v} \in f(\hat{y})$ such that $Re\langle \hat{y} - x, \hat{v}\rangle \leq 0$, for all $x \in K$, and so $Re\langle x - \hat{y}, \hat{v}\rangle \geq 0$, for all $x \in K$, i.e., $(\hat{y}, \hat{v})$ is a solution of GIVP(f, K) □

Remark 5.23 Tarafdar and Yuan noted that Theorem 5.41 not only showed that Theorem 5.40 of Ding Ding (1991) was true without the assumption that E is of secondary category, but it also showed that the conclusion of Theorem 5.40 can hold in topological vector spaces instead of locally convex topological vector spaces. By following the same ideas used in literature such as Ding (1991), Ding and Tarafdar (1996), Cottle and Yao (1992), Saigal (1976), Harker and Pang (1990), Siddiqi and Ansari (1989), Tan and Yuan (1994), Aubin (1979), Karamardian (1971) and so on, a number of existence theorems of GVIP(f, K) and GCP(f, K) can be proved in which the underlying spaces are not of secondary category.

Tarafdar and Yuan also discussed the difference between their Theorem 5.41 and Theorem 1 of Cubiotti (1993) (which was an extension of Cubiotti (1992, Theorem 2.1), which also concerned the existence of solutions for GQVI in finite dimensional spaces). First, it was noted that Cublotti's coercive condition (iii) of Theorem 1 in Cubiotti (1993) was different from ours in Theorem 5.41. Secondly, the set-valued mapping f in Theorem 5.41 was quasi-monotone and upper semi-continuous (thus it may not be lower semi-continuous); however, the corresponding set-valued mapping Γ of Theorem 1 in Cubiotti (1993) was lower semi-continuous and its graph was closed (hence, it may not be quasi-monotone nor upper semi-continuous). Furthermore, the underlying space E of Theorem 5.41 was a locally convex topological vector space which may be finite or infinite-dimensional, but the underlying space of Theorem 1 (see also Cubiotti (1992, Theorem 2.41)), was finite dimensional. Thus, Theorem 5.41 and Theorem 1 of Cubiotti (1993) (see also Cubiotti (1993, Theorem 2.4)), were independent of each other.

Note that the results of this section have been written and presented after a thorough revision of some results of an article in Tarafdar and Yuan (1997c).

5.9 Generalized Variational Inequalities for Quasi-Monotone and Quasi-Semi-Monotone Operators

Chowdhury and Tan obtained some existence theorems of generalized variational inequalities with applications to existence theorems of generalized complementarity problems and fixed point theorems in Hilbert spaces. Their main results are listed as Theorems 5.45, 5.50, 5.54, 5.58, 5.66, 5.67 and 5.68.

We note that all topological vector spaces are not assumed to be Hausdorff unless explicitly stated.

In obtaining the results of this section, Chowdhury and Tan mainly used the following Ky Fan's infinite dimensional generalization of the classical Knaster–Kuratowski–Mazurkiewicz Theorem [Knaster et al. (1929)]:

Theorem 5.42 (Fan (1961, Lemma 1)) *Let E be a topological vector space, X and Y be non-empty subsets of E such that $X \subset Y$ and Y is convex. Suppose*

$F : X \to 2^Y$ is such that

(*a*) *F is a KKM-map;*

(*b*) *for each $x \in X$, $F(x)$ is closed in Y;*

(*c*) *there exists $x_0 \in X$ such that $F(x_0)$ is compact.*

Then $\cap_{x \in X} F(x) \neq \emptyset$.

The following result is equivalent to the celebrated 1972 Ky Fan's minimax inequality [Fan (1972), Theorem 1]:

Theorem 5.43 *Let E be a topological vector space and X be a non-empty compact convex subset of E. Let f be a real-valued function defined on $X \times X$ such that*

(*a*) *for each $x \in X, f(x,x) \leq 0$;*

(*b*) *for each fixed $x \in X$, $f(x,y)$ is a lower semicontinuous function of y on X;*

(*c*) *for each fixed $y \in X$, $f(x,y)$ is a quasi-concave function of x on X.*

Then there exists $\hat{y} \in X$ such that $f(x, \hat{y}) \leq 0$ for all $x \in X$.

Ky Fan's minimax inequality has become a versatile tool in nonlinear functional analysis [Fan (1972)], convex analysis, game theory and economic theory [Aubin (1982)]. There have been numerous generalizations of Ky Fan's minimax inequality by weakening the compactness assumption or the convexity assumption; e.g., due to [Allen (1977), Bae et al. (1993), H. Brézis and Stampacchia (1972), Ding and Tan (1992b), Shih and Tan (1984), Tan [27], Tan and Yuan [30], Yen [31] and Fan himself [17].

Kneser's minimax theorem (Kneser (1952, pp. 2418–2420), see also Aubin (1979, pp. 40–41)) will also be used in obtaining some results of this section:

Theorem 5.44 *Let X be a non-empty convex subset of a vector space and Y be a non-empty compact convex subset of a Hausdorff topological vector space. Suppose that f is a real-valued function on $X \times Y$ such that for each fixed $x \in X$, $f(x,y)$ is lower semicontinuous and convex on Y and for each fixed $y \in Y$, $f(x,y)$ is concave on X. Then*

$$\min_{y \in Y} \sup_{x \in X} f(x,y) = \sup_{x \in X} \min_{y \in Y} f(x,y).$$

The following definitions were introduced by Chowdhury and Tan in their paper Chowdhury and Tan (1997b):

Definition 5.11 Let E be a topological vector space, X be a non-empty convex subset of E and $T : X \to 2^{E^*}$. Then T is said to be

(*a*) *lower hemicontinuous* on X if and only if for each $p \in E$, the function $f_p : X \to \mathbb{R} \cup \{+\infty\}$, defined by

$$f_p(z) = \sup_{u \in T(z)} Re\langle u, p \rangle \text{ for each } z \in X,$$

is lower semicontinuous on X (if and only if for each $p \in E$, the function $g_p : X \to \mathbb{R} \cup \{-\infty\}$, defined by

$$g_p(z) = \inf_{u \in T(z)} Re\langle u, p\rangle \text{ for each } z \in X,$$

is upper semicontinuous on X);

(*b*) *upper hemicontinuous* on X if and only if for each $p \in E$, the function $f_p : X \to \mathbb{R} \cup \{+\infty\}$, defined by

$$f_p(z) = \sup_{u \in T(z)} Re\langle u, p\rangle \text{ for each } z \in X,$$

is upper semicontinuous on X (if and only if for each $p \in E$, the function $g_p : X \to \mathbb{R} \cup \{-\infty\}$, defined by

$$g_p(z) = \inf_{u \in T(z)} Re\langle u, p\rangle \text{ for each } z \in X,$$

is lower semicontinuous on X).

It is observed that if X is convex, then the notions of lower hemicontinuity along line segments in X and upper hemicontinuity along line segments in X are independent of the vector topology τ on E as long as τ is Hausdorff and the continuous dual E^* remains unchanged. Note also that if T, $S : X \to 2^{E^*}$ are lower (respectively, upper) hemicontinuous and $\alpha \in \mathbb{R}$, then $T + S$ and αT are also lower (respectively, upper) hemicontinuous.

Proposition 5.4 *Let E be a topological vector space and X be a non-empty convex subset of E. Let $T : X \to 2^{E^*}$ be lower semicontinuous from relative topology on X to the weak* topology $\sigma\langle E^*, E\rangle$ on E^*. Then T is lower hemicontinuous on X.*

Proof. For each fixed $p \in E$, define $f_p : X \to \mathbb{R} \cup \{+\infty\}$ by

$$f_p(z) = \sup_{u \in T(z)} Re\langle u, p\rangle \text{ for each } z \in X.$$

Fix any $p \in E$. Let $\lambda \in \mathbb{R}$ be given and let $A = \{z \in X : f_p(z) > \lambda\}$. Take any $z_0 \in A$. Then $f_p(z_0) = \sup_{u \in T(z_0)} Re\langle u, p\rangle = \sup_{u \in T(z_0)} Re\ \hat{p}(u) > \lambda$. Choose any $u_0 \in T(z_0)$ such that $Re\ \hat{p}(u_0) > \lambda$. Thus $(Re\ \hat{p})^{-1}(\lambda, \infty) \cap T(z_0) \neq \emptyset$ where $(Re\ \hat{p})^{-1}(\lambda, \infty)$ is a weak* open set in E^*. Since T is lower semicontinuous at z_0, there exists an open neighborhood N_{z_0} of z_0 in X such that $T(z) \cap (Re\ \hat{p})^{-1}(\lambda, \infty) \neq \emptyset$ for all $z \in N_{z_0}$. Hence $f_p(z) = \sup_{u \in T(z)} Re\ \hat{p}(u) = \sup_{u \in T(z)} Re\langle u, p\rangle > \lambda$ for all $z \in N_{z_0}$. Thus $N_{z_0} \subset A$. Consequently, f_p is lower semicontinuous on X. Hence T is lower hemicontinuous on X. □

The converse of Proposition 5.4 is not true in general as was observed by Chowdhury and Tan in their paper Chowdhury and Tan (1997b):

Example 5.7 Let $X = [0, 1]$ and $E = \mathbb{R}$. Then $E^* = \mathbb{R}$. Let $T : X \to 2^{E^*}$ be defined by

$$T(x) = \begin{cases} \{1,3\}, & \text{if } x < 1, \\ \{1,2,3\}, & \text{if } x = 1. \end{cases}$$

If $p \in E$ and $p \geq 0$, the function $f_p : X \to \mathbb{R} \cup \{+\infty\}$, defined by $f_p(z) = \sup_{u \in T(z)} Re\langle u, p\rangle = 3p$ for each $z \in X$, is continuous. If $p \in E$ and $p < 0$, the function $f_p : X \to \mathbb{R} \cup \{+\infty\}$, defined by $f_p(z) = \sup_{u \in T(z)} Re\langle u, p\rangle = p$ for each $z \in X$, is also continuous. Thus T is lower (and upper) hemicontinuous on X.

But T is not lower semicontinuous (along line segments) in X. Indeed, let $x_0 = 1$ and $U = (\frac{3}{2}, \frac{5}{2})$, then U is an open set in $\mathbb{R}$ such that $U \cap T(x_0) = \{2\} \neq \emptyset$. But for any open neighborhood V of x_0 in X and for any $x \in V$ with $x \neq x_0$, $U \cap T(x) = \emptyset$. This shows that T is not lower semicontinuous (along line segments) in X.

Proposition 5.5 *Let E be a topological vector space and X be a non-empty convex subset of E. Let $T : X \to 2^{E^*}$ be upper semicontinuous from relative topology on X to the weak* topology $\sigma\langle E^*, E\rangle$ on E^*. Then T is upper hemicontinuous on X.*

Proof. For each fixed $p \in E$, define $f_p : X \to \mathbb{R} \cup \{+\infty\}$ by

$$f_p(z) = \sup_{u \in T(z)} Re\langle u, p\rangle, \text{ for each } z \in X.$$

Fix any $p \in E$. Let $\lambda \in \mathbb{R}$ be given and let $A = \{z \in X : f_p(z) < \lambda\}$. Take any $z_0 \in A$. Then $f_p(z_0) = \sup_{u \in T(z_0)} Re\langle u, p\rangle = \sup_{u \in T(z_0)} Re\, \hat{p}(u) < \lambda$. Thus there exists $\epsilon > 0$ such that $f_p(z_0) < \lambda - \epsilon < \lambda$. Therefore $Re\, \hat{p}(u) < \lambda - \epsilon < \lambda$ for all $u \in T(z_0)$. Hence $T(z_0) \subset (Re\, \hat{p})^{-1}(-\infty, \lambda - \epsilon)$ which is weak* open in E^*. Since T is upper semicontinuous at z_0, there exists an open neighborhood N_{z_0} of z_0 in X such that $T(z) \subset (Re\, \hat{p})^{-1}(-\infty, \lambda - \epsilon)$ for all $z \in N_{z_0}$. Thus $Re\, \hat{p}(u) < \lambda - \epsilon$ for all $u \in T(z)$ and for all $z \in N_{z_0}$. Hence $\sup_{u \in T(z)} Re\, \hat{p}(u) \leq \lambda - \epsilon < \lambda$ for all $z \in N_{z_0}$; i.e., $f_p(z) = \sup_{u \in T(z)} Re\langle u, p\rangle \leq \lambda - \epsilon < \lambda$ for all $z \in N_{z_0}$. Therefore $N_{z_0} \subset A$ so that A is open in X. Consequently, f_p is upper semicontinuous on X. Hence T is upper hemicontinuous on X. □

The converse of Proposition 5.5 is not true in general as was observed by Chowdhury and Tan in their paper Chowdhury and Tan (1997b):

Example 5.8 Let $E = \mathbb{R}^2$ and $X = \{(x, y) \in \mathbb{R}^2 : x^2 + y^2 \leq 1 \text{ and } x, y > 0\}$. Define $f, g : X \to 2^{E^*}$ by

$$f(rcos\theta, rsin\theta) = \{(tcos\theta, tsin\theta) : r \leq t \leq 2\} \text{ for all } r \in (0,1),\ \ \theta \in (0, \frac{\pi}{2}),$$

and

$$g(x, y) = \{(z, 0) : z \geq x\} \text{ for all } (x, y) \in X.$$

Then f and g are upper semicontinuous on X so that f and g are upper hemicontinuous on X by Proposition 2.4 and hence $f+g$ is also upper hemicontinuous. However it is easy to see that $f+g$ is not upper semicontinuous along line segments in X.

Chowdhury and Tan introduced the following definitions in Chowdhury and Tan (1997b):

Definition 5.12 Let E be a topological vector space, X be a non-empty subset of E and $T: X \to 2^{E^*}$. If $h: X \to \mathbb{R}$, then T is said to be

(1) *h-quasi-monotone* if for each $x, y \in X$, $\inf_{w\in T(y)} Re\langle w, y-x\rangle + h(y) - h(x) > 0$ whenever $\sup_{u\in T(x)} Re\langle u, y-x\rangle + h(y) - h(x) > 0$;

(2) *quasi-monotone* if T is h-quasi-monotone with $h \equiv 0$;

(3) *h-quasi-semi-monotone* if for each $x, y \in X$, $\inf_{w\in T(y)} Re\langle w, y-x\rangle + h(y) - h(x) > 0$ whenever $\inf_{u\in T(x)} Re\langle u, y-x\rangle + h(y) - h(x) > 0$;

(4) *quasi-semi-monotone* if T is h-quasi-semi-monotone with $h \equiv 0$.

Clearly, monotonicity implies quasi-monotonicity, but the converse is not true. Moreover, quasi-monotonicity does not imply semi-monotonicity and semi-monotonicity does not imply quasi-monotonicity. For examples please see Chowdhury and Tan's paper (Chowdhury and Tan (1997a))

Clearly, semi-monotonicity implies quasi-semi-monotonicity and quasi-monotonicity implies quasi-semi-monotonicity; but the converses are not true in general as shown in Chowdhury and Tan (1997a).

Definition 5.13 Let $(E, \|\cdot\|)$ be a normed space and X be a non-empty subset of E. Then $T: X \to 2^{E^*}$ is quasi-non-expansive if for each $x, y \in X$, each $u \in T(x)$ and each $w \in T(y)$, $Re\langle w-u, y-x\rangle \le \|y-x\|^2$.

It is clear that if T is single-valued and non-expansive (i.e., $\|T(x)-T(y)\| \le \|x-y\|$ for all $x, y \in X$), then T is quasi-non-expansive. The converse is false in general as shown in Chowdhury and Tan (1997a).

In what follows, if H is a Hilbert space, we shall denote by I the identity operator on H; i.e., $I(x) = x$ for all $x \in H$.

The following is a proposition in Chowdhury and Tan (1997a):

Proposition 5.6 *If X is a non-empty subset of a Hilbert space H and $T: X \to 2^H$, then T is quasi-non-expansive if and only if $I-T$ is monotone.*

Proof. Suppose T is quasi-non-expansive. Let $x, y \in X$ be given and choose any $u_0 \in T(x)$. Then for each $w \in T(y)$,

$$Re\langle y-w, y-x\rangle = Re\langle y-x+x-u_0+u_0-w, y-x\rangle \tag{5.86}$$

$$= \|y-x\|^2 + Re\langle x-u_0, y-x\rangle + Re\langle u_0-w, y-x\rangle \tag{5.87}$$

$$\ge Re\langle x-u_0, y-x\rangle \tag{5.88}$$

since T is quasi-non-expansive. Thus $\inf_{w\in T(y)} Re\langle y-w, y-x\rangle \geq Re\langle x-u_0, y-x\rangle$. As $u_0 \in T(x)$ is arbitrary, $\inf_{w\in T(y)} Re\langle y-w, y-x\rangle \geq \sup_{u\in T(x)} Re\langle x-u, y-x\rangle$; i.e.,

$$\inf_{w\in(I-T)(y)} Re\langle w, y-x\rangle \geq \sup_{u\in(I-T)(x)} Re\langle u, y-x\rangle.$$

Thus $I-T$ is monotone.

Conversely, suppose $I-T$ is monotone. Then for each $x, y \in X, u \in T(x)$ and $w \in T(y), Re\langle (y-w)-(x-u), y-x\rangle \geq 0$ so that $Re\langle w-u, y-x\rangle = Re\langle w-y+y-x+x-u, y-x\rangle = Re\langle w-y, y-x\rangle + \|y-x\|^2 + Re\langle x-u, y-x\rangle \leq \|y-x\|^2$. Thus T is quasi-non-expansive. □

Proposition 5.6 is a generalization of Proposition 1 in Browder (1967).

Definition 5.14 Let $(E, \|\cdot\|)$ be a normed space and X be a non-empty subset of E. Then $T : X \to 2^{E^*}$ is semi-non-expansive if for each $x, y \in X$, $\inf_{u\in T(x)} \sup_{w\in T(y)} Re\langle w-u, y-x\rangle \leq \|y-x\|^2$.

The following is another proposition in Chowdhury and Tan (1997a):

Proposition 5.7 *If X is a non-empty subset of a Hilbert space H and $T : X \to 2^H$, then T is semi-non-expansive if and only if $I-T$ is semi-monotone.*

It is clear from definitions that a quasi-non-expansive operator is semi-non-expansive. The converse does not hold in general (for an example please see Chowdhury and Tan (1997a)).

Then by Propositions 5.6 and 5.7, the operator $S = I-T$ is semi- non-expansive but not quasi-non-expansive.

The proof of Lemma 2 in Shih and Tan (1986) (see also Tan (1994, Lemma 2.4.1), can be easily modified to give the following simple but useful result which was first given by Karamardian for single-valued operators in Karamardian (1971, Lemma 3.1):

Lemma 5.24 *Let X be a cone in a topological vector space E and $T : X \to 2^{E^*}$. Then the following statements are equivalent:*

(*a*) *There exists $\hat{y} \in X$ such that $\sup_{w\in T(\hat{y})} Re\langle w, \hat{y}-x\rangle \leq 0$ for all $x \in X$.*

(*b*) *There exists $\hat{y} \in X$ such that $Re\langle w, \hat{y}\rangle = 0$ for all $w \in T(\hat{y})$ and $T(\hat{y}) \subset \widehat{X}$.*

A result of S. C. Fang (e.g. see Chan and Pang (1982) and Shih and Tan (1986, p. 59), can be modified as follows (see also Tan (1994, Lemma 2.4.2)):

Lemma 5.25 *Let X be a cone in a topological vector space E and $T : X \to 2^{E^*}$. Then the following statements are equivalent:*

(*a*) *There exist $\hat{y} \in X$ and $\hat{w} \in T(\hat{y})$ such that $Re\langle \hat{w}, \hat{y}-x\rangle \leq 0$ for all $x \in X$.*

(*b*) *There exist $\hat{y} \in X$ and $\hat{w} \in T(\hat{y})$ such that $Re\langle \hat{w}, \hat{y}\rangle = 0$ and $\hat{w} \in \widehat{X}$.*

The following simple result is Lemma 2.1.6 in [Tan (1994)]:

Lemma 5.26 *Let E be a topological vector space and A be a non-empty bounded subset of E. Let C be a non-empty strongly compact subset of E^*. Define $f : A \to \mathbb{R}$ by $f(x) = \min_{u \in C} Re\langle u, x\rangle$ for all $x \in A$. Then f is weakly continuous on A.*

5.9.1 *Generalization of Ky Fan's Minimax Inequality*

We shall begin with the following result which generalizes Theorem 5.43 in several aspects:

Theorem 5.45 *Let E be a topological vector space, X be a non-empty convex subset of E and $f, g : X \times X \to \mathbb{R} \cup \{-\infty, +\infty\}$ be such that*

(a_1) *for each $x, y \in X$, $f(x, y) > 0$ implies $g(x, y) > 0$;*

(b_1) *for each fixed $x \in X$, $y \longmapsto f(x, y)$ is lower semicontinuous on non-empty compact subsets of X;*

(c_1) *for each $A \in \mathcal{F}(\mathcal{X})$ and for each $y \in co(A)$, $\min_{x \in A} g(x, y) \leq 0$;*

(d_1) *there exist a non-empty closed and compact subset K of X and $x_0 \in K$ such that $g(x_0, y) > 0$ for all $y \in X \setminus K$.*

Then there exists $\hat{y} \in K$ such that $f(x, \hat{y}) \leq 0$ for all $x \in X$.

Proof. Define $F : X \to 2^K$ by

$$F(x) = \{y \in K : f(x, y) \leq 0\} \text{ for all } x \in X.$$

Note that by (b_1), each $F(x)$ is closed in K. We shall first show that the family $\{F(x) : x \in X\}$ has the finite intersection property. Indeed, let $\{x_1, \ldots, x_n\}$ be any finite subset of X. Set $C = co(\{x_0, x_1, \ldots, x_n\})$, then C is non-empty compact convex. Note that by (c_1), $g(x, x) \leq 0$ for all $x \in X$. Define $G : C \to 2^C$ by $G(x) = \{y \in C : g(x, y) \leq 0\}$ for all $x \in C$. We observe that: (i) if A is any finite subset of C, then $co(A) \subset \cup_{x \in A} G(x)$; for if this were false, then there exist a finite subset $\{z_1, \ldots, z_m\}$ of C and $z \in co(\{z_1, \ldots, z_m\})$ with $z \notin \cup_{j=1}^m G(z_j)$ so that $g(z_j, z) > 0$ for all $j = 1, \ldots, m$ which contradicts (c_1); (ii) for each $x \in C$, $c\ell_C(G(x))$ is closed in C and is therefore also compact. By Theorem 5.42, $\cap_{x \in C} c\ell_C(G(x)) \neq \emptyset$. Take any $\bar{y} \in \cap_{x \in C} c\ell_C(G(x))$. Note that $x_0 \in C$ and $G(x_0) \subset K$ by (d_1); thus $\bar{y} \in c\ell_C(G(x_0)) \subset c\ell_X(G(x_0)) = c\ell_K(G(x_0)) \subset K$. Since we also have $\bar{y} \in \cap_{j=1}^n c\ell_C(G(x_j))$ and for each $j = 1, \ldots, n$, $c\ell_C(G(x_j)) = c\ell_C(\{y \in C : g(x_j, y) \leq 0\}) \subset c\ell_C(\{y \in C : f(x_j, y) \leq 0\}) = \{y \in C : f(x_j, y) \leq 0\}$ by (a_1) and (b_1), we have $f(x_j, \bar{y}) \leq 0$ for all $j = 1, \ldots, n$ and hence $\bar{y} \in \cap_{j=1}^n F(x_j)$. Therefore $\{F(x) : x \in X\}$ has the finite intersection property.

By compactness of K, $\cap_{x \in X} F(x) \neq \emptyset$. Take any $\hat{y} \in \cap_{x \in X} F(x)$, then $\hat{y} \in K$ and $f(x, \hat{y}) \leq 0$ for all $x \in X$. □

The following fixed point theorem is equivalent to Theorem 5.45:

Theorem 5.46 *Let E be a topological vector space, X be a non-empty convex subset of E and $F, G : X \to 2^X \cup \{\emptyset\}$ be such that*

(a_2) *for each $x \in X$, $F(x) \subset G(x)$;*
(b_2) *for each $x \in X$, $F^{-1}(x)$ is compactly open (i.e., $F^{-1}(x) \cap C$ is open in C for each non-empty compact subset C of X);*
(c_2) *there exist a non-empty closed and compact subset K of X and $x_0 \in K$ such that $X \setminus K \subset G^{-1}(x_0)$;*
(d_2) *for each $x \in K$, $F(x) \neq \emptyset$,*
(e_2) *for each $x \in X$, $G(x)$ is convex.*

Then there exists $\bar{y} \in X$ such that $\bar{y} \in G(\bar{y})$.

To show Theorem 5.45 implies Theorem 5.46:

Define $f, g : X \times X \to \mathbb{R}$ by

$$f(x,y) = \begin{cases} 1, & \text{if } x \in F(y), \\ 0, & \text{if } x \notin F(y), \end{cases}$$

$$g(x,y) = \begin{cases} 1, & \text{if } x \in G(y), \\ 0, & \text{if } x \notin G(y) \end{cases}$$

for all $x, y \in X$. It is easy to see that the conditions (a_1), (b_1) and (d_1) of Theorem 5.45 are satisfied. If the hypothesis (c_1) of Theorem 5.45 is also satisfied, then by Theorem 5.45, there exists $\hat{y} \in K$ such that $f(x, \hat{y}) \leq 0$ for all $x \in X$. It follows that $F(\hat{y}) = \emptyset$ which is impossible. Thus the hypothesis (c_1) of Theorem 5.45 does not hold. Hence there exist $A \in \mathcal{F}(\mathcal{X})$ and $\bar{y} \in co(A)$ such that $\min_{x \in A} g(x, \bar{y}) > 0$ so that $x \in G(\bar{y})$ for all $x \in A$. Therefore $\bar{y} \in co(A) \subset G(\bar{y})$ by (e_2). □

To show Theorem 5.46 implies Theorem 5.45:

Define $F, G : X \to 2^X \cup \{\emptyset\}$ by $F(y) = \{x \in X : f(x,y) > 0\}$ and $G(y) = co(\{x \in X : g(x,y) > 0\})$ for all $y \in X$. It is easy to see that the conditions (a_2), (b_2), (c_2) and (e_2) of Theorem 5.46 are satisfied. If the hypothesis (d_2) of Theorem 5.46 is also satisfied, then by Theorem 3.2, here exists $\bar{y} \in X$ such that $\bar{y} \in G(\bar{y})$. But then there exist $x_1, \ldots, x_n \in X$, $\lambda_1, \ldots, \lambda_n \in [0,1]$ such that $g(x_i, \bar{y}) > 0$ for all $i = 1, \ldots, n$, $\sum_{i=1}^n \lambda_i = 1$ and $\bar{y} = \sum_{i=1}^n \lambda_i x_i$. This contradicts ($c_1$) because $\bar{y} \in co(A)$, where $A = \{x_1, \ldots, x_n\}$. Hence the hypothesis (d_2) of Theorem 5.46 does not hold. Thus there exists $\hat{y} \in K$ such that $F(\hat{y}) = \emptyset$. It follows that $f(x, \hat{y}) \leq 0$ for all $x \in X$. □

Note that Theorem 5.46 is Theorem 2.4′ in [Tan and Yuan (1993)].

Clearly, Theorem 5.45 implies the following result which is Theorem 2.2 in [Tan and Yuan (1993)]:

Theorem 5.47 *Let X be a non-empty convex subset of a topological vector space and $\phi, \psi : X \times X \to \mathbb{R} \cup \{-\infty, \infty\}$ be such that*

(*a*) $\phi(x,y) \leq \psi(x,y)$ *for each* $(x,y) \in X \times X$;
(*b*) *for each fixed* $x \in X$, $y \longmapsto \phi(x,y)$ *is lower semicontinuous on non-empty compact subsets of* X;
(*c*) *for each* $A \in \mathcal{F}(\mathcal{X})$ *and for each* $y \in co(A)$, $\min_{x\in A} \psi(x,y) \leq 0$;
(*d*) *there exist a non-empty closed and compact subset* K *of* X *and* $x_0 \in X$ *such that* $\psi(x_0, y) > 0$ *for all* $y \in X \setminus K$.

Then there exists $y \in K$ *such that* $\phi(x,y) \leq 0$ *for all* $x \in X$.

It is shown in Tan and Yuan (1993) that Theorem 5.47 implies the following result which is Theorem 2.4 in Tan and Yuan (1993):

Theorem 5.48 *Let* X *be a non-empty convex subset of a topological vector space and* $\phi, \psi : X \times X \to \mathbb{R} \cup \{-\infty, +\infty\}$ *be such that*

(*a*) $\phi(x,y) \leq \psi(x,y)$ *for each* $(x,y) \in X \times X$ *and* $\psi(x,x) \leq 0$ *for each* $x \in X$;
(*b*) *for each fixed* $x \in X$, $y \longmapsto \phi(x,y)$ *is lower semicontinuous on non-empty compact subsets of* X;
(*c*) *for each fixed* $y \in X$, *the set* $\{x \in X : \psi(x,y) > 0\}$ *is convex;*
(*d*) *there exist a non-empty closed and compact subset* K *of* X *and a point* $x_0 \in X$ *such that* $\psi(x_0, y) > 0$ *for all* $y \in X \setminus K$.

Then there exists $\hat{y} \in K$ *such that* $\phi(x,\hat{y}) \leq 0$ *for all* $x \in X$.

It is shown in [Tan and Yuan (1993)] that Theorem 5.48 is equivalent to Theorem 5.46. Thus Theorems 5.45, 5.46 (Theorem 2.4′ in [Tan and Yuan (1993)]), 5.47 (Theorem 2.2 in Tan and Yuan (1993)) and 5.48 (Theorem 2.4 in [Tan and Yuan (1993)]) above are all equivalent and are also equivalent to Theorems 2.2′, 2.2″, 2.3‴, 2.4″, 2.4‴, 2.4⁗ and 2.4′′′′′ in [Tan and Yuan (1993)]. Note however that the equivalence of Theorem 2.2 in [Tan and Yuan (1993)] and Theorem 2.4 in [Tan and Yuan (1993)] was not established in [Tan and Yuan (1993)]. Note also that Theorem 5.47 does not imply Theorem 5.45 directly and Theorem 5.48 does not imply Theorems 5.45 and 5.47 directly. For applications to existence of equilibrium points of generalized games, we refer to Tan and Yuan (1993).

5.9.2 *Generalized Variational Inequalities*

Applying Theorem 5.45 Chowdhury and Tan obtained some existence theorems of generalized variational inequalities(GVI). As applications of these GVI some existence theorems of generalized complementarity problems(GCP) were obtained. Maximality of monotone operators, and surjectivity of monotone or semi-monotone operators will be dealt with at the end of this section.

The following is a result in Chowdhury and Tan (1997a):

Lemma 5.27 *Let* E *be a topological vector space,* X *be a non-empty convex subset of* E, $h : X \to \mathbb{R}$ *be convex and* $T : X \to 2^{E^*}$ *be lower hemicontinuous* *along line segments in X. Suppose $\hat{y} \in X$ is such that $\sup_{u\in T(x)} Re\langle u, \hat{y} - x\rangle \leq h(x) - h(\hat{y})$ for all $x \in X$. Then*

$$\sup_{w\in T(\hat{y})} Re\langle w, \hat{y} - x\rangle \leq h(x) - h(\hat{y}) \text{ for all } x \in X.$$

Proof. Suppose that $\sup_{u\in T(x)} Re\langle u, \hat{y} - x\rangle \leq h(x) - h(\hat{y})$ for all $x \in X$. Let $x \in X$ be arbitrarily fixed. Let $z_t = tx + (1-t)\hat{y} = \hat{y} - t(\hat{y} - x)$ for all $t \in [0,1]$. Then $z_t \in X$ as X is convex. Let $L = \{z_t : t \in [0,1]\}$. Thus $\sup_{u\in T(z_t)} Re\langle u, \hat{y} - z_t\rangle \leq h(z_t) - h(\hat{y})$ for all $t \in [0,1]$. Therefore $\sup_{u\in T(z_t)} Re\langle u, \hat{y} - x\rangle \leq h(x) - h(\hat{y})$ for all $t \in (0,1]$.

Since T is lower hemicontinuous on L, the function $f_{\hat{y}-x} : L \to \mathbb{R} \cup \{+\infty\}$, defined by

$$f_{\hat{y}-x}(z_t) = \sup_{u\in T(z_t)} Re\langle u, \hat{y} - x\rangle \text{ for each } z_t \in L,$$

is lower semicontinuous on L. Thus the set $A = \{z_t \in L : f_{\hat{y}-x}(z_t) \leq h(x) - h(\hat{y})\}$ is closed in L. Now $z_t \to \hat{y}$ in L as $t \to 0^+$. Since $z_t \in A$ for all $t \in (0,1]$ we have $\hat{y} \in A$. Hence $f_{\hat{y}-x}(\hat{y}) = \sup_{u\in T(\hat{y})} Re\langle u, \hat{y} - x\rangle \leq h(x) - h(\hat{y})$. Since $x \in X$ is arbitrary, we have $\sup_{w\in T(\hat{y})} Re\langle w, \hat{y} - x\rangle \leq h(x) - h(\hat{y})$ for all $x \in X$. □

The following is another result in Chowdhury and Tan (1997a):

Lemma 5.28 *Let E be a topological vector space, X be a non-empty convex subset of E, $h : X \to \mathbb{R}$ be convex and $T : X \to 2^{E^*}$ be upper hemicontinuous along line segments in X. Suppose $\hat{y} \in X$ is such that $\inf_{u\in T(x)} Re\langle u, \hat{y} - x\rangle \leq h(x) - h(\hat{y})$ for all $x \in X$. Then*

$$\inf_{w\in T(\hat{y})} Re\langle w, \hat{y} - x\rangle \leq h(x) - h(\hat{y}) \text{ for all } x \in X.$$

Note that if E is a locally convex space, X is a non-empty convex subset of E and $h : X \to \mathbb{R}$ is convex, then h is lower semicontinuous on X if and only if h is weakly lower semicontinuous on X.

Lemma 5.29 *Let E be a topological vector space, X be a non-empty convex subset of E and $h : E \to \mathbb{R}$ be convex. Suppose $\hat{y} \in X$ and $\hat{w} \in E^*$ are such that $Re\langle \hat{w}, \hat{y} - x\rangle \leq h(x) - h(\hat{y})$ for all $x \in X$, then $Re\langle \hat{w}, \hat{y} - x\rangle \leq h(x) - h(\hat{y})$ for all $x \in I_X(\hat{y})$.*

Proof. Let $x \in I_X(\hat{y})$ be arbitrarily fixed; then $x = \hat{y} + r(u - \hat{y})$ for some $u \in X$ and $r > 0$.

Case 1. Suppose $0 < r \leq 1$, then $x = ru + (1-r)\hat{y} \in X$ as X is convex and $u, \hat{y} \in X$. By assumption, we have $Re\langle \hat{w}, \hat{y} - x\rangle \leq h(x) - h(\hat{y})$.

Case 2. Suppose $r > 1$, then $u = (1 - \frac{1}{r})\hat{y} + \frac{1}{r}x$. Since $u \in X$, by assumption again, we have

$$\frac{1}{r}Re\langle\hat{w},\hat{y}-x\rangle = Re\langle\hat{w},\hat{y}-u\rangle \tag{5.89}$$

$$\leq h(u)-h(\hat{y}) \leq (1-\frac{1}{r})h(\hat{y})+\frac{1}{r}h(x)-h(\hat{y}) \tag{5.90}$$

$$= \frac{1}{r}(h(x)-h(\hat{y})) \tag{5.91}$$

so that $Re\langle\hat{w},\hat{y}-x\rangle \leq h(x)-h(\hat{y})$.

Thus in either case, $Re\langle\hat{w},\hat{y}-x\rangle \leq h(x)-h(\hat{y})$ for all $x \in I_X(\hat{y})$. □

Theorem 5.49 *Let E be a topological vector space, X be a non-empty convex subset of E, $h : X \to \mathbb{R}$ be convex and weakly lower semicontinuous on weakly compact subsets of X and $T : X \to 2^{E^*}$ be h-quasi-monotone. Suppose there exist a non-empty weakly compact and weakly closed subset K of X and $x_0 \in K$ such that $\inf_{w\in T(y)} Re\langle w, y-x_0\rangle + h(y) - h(x_0) > 0$ for all $y \in X \setminus K$. Then there exists $\hat{y} \in K$ such that $\sup_{u\in T(x)} Re\langle u,\hat{y}-x\rangle \leq h(x)-h(\hat{y})$ for all $x \in X$.*

Proof. Define $f, g : X \times X \to \mathbb{R}$ by

$$f(x,y) = \sup_{u\in T(x)} Re\langle u, y-x\rangle + h(y) - h(x),$$

$$g(x,y) = \inf_{w\in T(y)} Re\langle w, y-x\rangle + h(y) - h(x)$$

for all $x, y \in X$. Then we have the following:

(1) For each $x, y \in X$, since T is h-quasi-monotone, $f(x,y) > 0$ implies $g(x,y) > 0$.

(2) For each fixed $x \in X$, $y \longmapsto f(x,y)$ is weakly lower semicontinuous on non-empty weakly compact subsets of X.

(3) For each $A \in \mathcal{F}(\mathcal{X})$ and $y \in co(A)$, $\min_{x\in A} g(x,y) \leq 0$. Indeed, if this were false, then for some $A = \{x_1, \ldots, x_n\} \in \mathcal{F}(\mathcal{X})$ and some $y \in co(A)$, say $y = \sum_{i=1}^n \lambda_i x_i$ where $\lambda_1, \ldots, \lambda_n \geq 0$ with $\sum_{i=1}^n \lambda_i = 1$, such that $\min_{1\leq i\leq n} g(x_i, y) > 0$. Then for each $i = 1, \ldots, n$, $\inf_{w\in T(y)} Re\langle w, y-x_i\rangle + h(y) - h(x_i) > 0$ so that $0 = g(y,y) = \inf_{w\in T(y)} Re\langle w, y - \sum_{i=1}^n \lambda_i x_i\rangle + h(y) - h(\sum_{i=1}^n \lambda_i x_i) \geq \sum_{i=1}^n \lambda_i (\inf_{w\in T(y)} Re\langle w, y-x_i\rangle + h(y) - h(x_i)) > 0$, which is a contradiction.

(4) K is a weakly compact and weakly closed subset of X and $x_0 \in K$ such that for all $y \in X \setminus K$, $g(x_0, y) > 0$.

Equip E with the weak topology. Then f and g satisfy all the hypotheses of Theorem 5.45 so that by Theorem 5.45, there exists $\hat{y} \in K$ such that $f(x,\hat{y}) \leq 0$ for all $x \in X$; i.e., $\sup_{u\in T(x)} Re\langle u,\hat{y}-x\rangle + h(\hat{y}) - h(x) \leq 0$ for all $x \in X$. □

Theorem 5.50 *Let E be a topological vector space, X be a non-empty convex subset of E, $h : X \to \mathbb{R}$ be convex and weakly lower semicontinuous on weakly* *compact subsets of* X *and* $T : X \to 2^{E^*}$ *be h-quasi-monotone and lower hemicontinuous along line segments in* X *to the weak*-topology on* E^*. *Suppose there exist a non-empty weakly compact and weakly closed subset* K *of* X *and* $x_0 \in K$ *such that* $\inf_{w\in T(y)} Re\langle w, y - x_0\rangle + h(y) - h(x_0) > 0$ *for all* $y \in X \setminus K$. *Then there exists* $\hat{y} \in K$ *such that* $\sup_{w\in T(\hat{y})} Re\langle w, \hat{y} - x\rangle \le h(x) - h(\hat{y})$ *for all* $x \in X$. *Moreover, if* h *is defined on all of* E *and is convex, then* $\sup_{w\in T(\hat{y})} Re\langle w, \hat{y} - x\rangle \le h(x) - h(\hat{y})$ *for all* $x \in I_X(\hat{y})$.

Proof. By Theorem 5.49, there exists $\hat{y} \in K$ such that $\sup_{u\in T(x)} Re\langle u, \hat{y} - x\rangle \le h(x) - h(\hat{y})$ for all $x \in X$.

Since h is convex and T is lower hemicontinuous along line segments in X, by Lemma 5.27, we have

$$\sup_{w\in T(\hat{y})} Re\langle w, \hat{y} - x\rangle \le h(x) - h(\hat{y}) \quad \text{for all} \quad x \in X. \tag{5.92}$$

Now if h is defined on all of E and is convex, then by (5.92) and Lemma 5.29, we have

$$\sup_{w\in T(\hat{y})} Re\langle w, \hat{y} - x\rangle \le h(x) - h(\hat{y}) \text{ for all } x \in I_X(\hat{y}).$$ □

It was noted that Theorem 5.47 (i.e., Theorem 2.2 in [Tan (1994)]) cannot be applied directly to prove Theorem 5.50.

Remark 5.24 Theorem 5.50. improves Theorem 3 of Shih and Tan in Shih and Tan (1984, pp. 283–285), in the following ways:

(1) f is h-quasi-monotone instead of monotone;
(2) f is lower hemicontinuous along line segments instead of lower semicontinuous along line segments in X.

Theorem 5.51 *Let* $(E, \|\cdot\|)$ *be a reflexive Banach space,* X *be a non-empty closed convex subset of* E, $h : X \to \mathbb{R}$ *be convex and lower semicontinuous on weakly compact subsets of* X *and* $T : X \to 2^{E^*}$ *be h-quasi-monotone and lower hemicontinuous along line segments in* X *to the weak topology on* E^*. *Suppose there is* $x_0 \in X$ *such that*

$$\lim_{\substack{\|y\|\to\infty \\ y\in X}} \inf_{w\in T(y)} Re\langle w, y - x_0\rangle + h(y) - h(x_0) > 0. \tag{5.93}$$

Then there exists $\hat{y} \in X$ *such that* $\sup_{w\in T(\hat{y})} Re\langle w, \hat{y}-x\rangle \le h(x)-h(\hat{y})$ *for all* $x \in X$. *Moreover, if* h *is defined on all of* E *and is convex, then* $\sup_{w\in T(\hat{y})} Re\langle w, \hat{y} - x\rangle \le h(x) - h(\hat{y})$ *for all* $x \in I_X(\hat{y})$.

Proof. Let $\alpha = \lim_{\substack{\|y\|\to\infty \\ y\in X}} \inf_{w\in T(y)} Re\langle w, y - x_0\rangle + h(y) - h(x_0)$. Then by (4.2), $\alpha > 0$. Let $M > 0$ be such that $\|x_0\| \le M$ and $\inf_{w\in T(y)} Re\langle w, y - x_0\rangle + h(y) - h(x_0) > \frac{\alpha}{2}$

for all $y \in X$ with $\|y\| > M$. Let $K = \{x \in X : \|x\| \leq M\}$; then K is a non-empty weakly compact subset of X. Note that for any $y \in X \setminus K$, $\inf_{w \in T(y)} Re\langle w, y - x_0\rangle + h(y) - h(x_0) > \frac{\alpha}{2} > 0$. The conclusion now follows from Theorem 5.50. □

By taking $h \equiv 0$ in Theorem 5.50 and applying Lemma 5.24, Chowdhury and Tan obtained the following existence theorem of a generalized complementarity problem in Chowdhury and Tan (1997a):

Theorem 5.52 *Let X be a cone in a topological vector space E. Let $T : X \to 2^{E^*}$ be quasi-monotone and lower hemicontinuous along line segments in X to the weak*-topology on E^*. Suppose there exist a non-empty weakly compact and weakly closed subset K of X and $x_0 \in K$ such that $\inf_{w \in T(y)} Re\langle w, y - x_0\rangle > 0$ for all $y \in X \setminus K$. Then there exists $\hat{y} \in K$ such that $Re\langle w, \hat{y}\rangle = 0$ for all $w \in T(\hat{y})$ and $T(\hat{y}) \subset \widehat{X}$.*

By taking $h \equiv 0$ in Theorem 5.51 and applying Lemma 5.24 (or by the same argument as in the proof of Theorem 5.51 and by Theorem 5.52), Chowdhury and Tan obtained the following existence theorem of a generalized complementarity problem in Chowdhury and Tan (1997a):

Theorem 5.53 *Let $(E, \|\cdot\|)$ be a reflexive Banach space, X be a closed cone in E and $T : X \to 2^{E^*}$ be quasi-monotone and lower hemicontinuous along line segments in X to the weak topology on E^*. Suppose there is $x_0 \in X$ such that*

$$\lim_{\substack{\|y\| \to \infty \\ y \in X}} \inf_{w \in T(y)} Re\langle w, y - x_0\rangle > 0.$$

Then there exists $\hat{y} \in X$ such that $Re\langle w, \hat{y}\rangle = 0$ for all $w \in T(\hat{y})$ and $T(\hat{y}) \subset \widehat{X}$.

Theorem 5.54 *Let E be a Hausdorff topological vector space, X be a non-empty convex subset of E, $h : X \to \mathbb{R}$ be convex and weakly lower semicontinuous on weakly compact subsets of X and $T : X \to 2^{E^*}$ be h-quasi-monotone and upper hemicontinuous along line segments in X to the weak* topology on E^* such that each $T(x)$ is weak* compact convex. Suppose there exist a non-empty weakly compact and weakly closed subset K of X and $x_0 \in K$ such that $\inf_{w \in T(y)} Re\langle w, y - x_0\rangle + h(y) - h(x_0) > 0$ for all $y \in X \setminus K$. Then there exist $\hat{y} \in K$ and $\hat{w} \in T(\hat{y})$ such that $Re\langle \hat{w}, \hat{y} - x\rangle \leq h(x) - h(\hat{y})$ for all $x \in X$. Moreover, if h is defined on all of E and is convex, then $Re\langle \hat{w}, \hat{y} - x\rangle \leq h(x) - h(\hat{y})$ for all $x \in I_X(\hat{y})$.*

Proof. By Theorem 5.49, there exists $\hat{y} \in K$ such that $\sup_{u \in T(x)} Re\langle u, \hat{y} - x\rangle + h(\hat{y}) - h(x) \leq 0$ for all $x \in X$. It follows that $\inf_{u \in T(x)} Re\langle u, \hat{y} - x\rangle + h(\hat{y}) - h(x) \leq 0$ for all $x \in X$. Since h is convex and T is upper hemicontinuous, by Lemma 5.28, $\inf_{w \in T(\hat{y})} Re\langle w, \hat{y} - x\rangle + h(\hat{y}) - h(x) \leq 0$ for all $x \in X$.

Define $\phi : X \times T(\hat{y}) \to \mathbb{R}$ by $\phi(x, w) = Re\langle w, \hat{y} - x\rangle + h(\hat{y}) - h(x)$ for all $(x, w) \in X \times T(\hat{y})$. Then for each fixed $x \in X$, $w \longmapsto \phi(x, w)$ is weak* lower semicontinuous and convex and for each fixed $w \in T(\hat{y})$, $x \longmapsto \phi(x, w)$ is concave.

By Theorem 5.44,

$$\min_{w\in T(\hat{y})}\sup_{x\in X}\phi(x,w)=\sup_{x\in X}\min_{w\in T(\hat{y})}\phi(x,w)\leq 0.$$

Since $T(\hat{y})$ is *weak**-compact, there exists $\hat{w}\in T(\hat{y})$ such that

$$\sup_{x\in X}\phi(x,\hat{w})=\min_{w\in T(\hat{y})}\sup_{x\in X}\phi(x,w)\leq 0.$$

Therefore

$$Re\langle\hat{w},\hat{y}-x\rangle\leq h(x)-h(\hat{y})\ \text{ for all }\ x\in X. \tag{5.94}$$

Now suppose h is defined on all of E and is convex. Then by (5.94) and Lemma 5.29,

$$Re\langle\hat{w},\hat{y}-x\rangle\leq h(x)-h(\hat{y})\ \text{ for all }\ x\in I_X(\hat{y}).$$ □

Remark 5.25 Theorem 5.54 extends Theorem 5 of Bae-Kim-Tan in Bae et al. (1993, pp. 238–240), in the following ways:

(1) E^* is not equipped with strong topology,
(2) T is h-quasi-monotone instead of semi-monotone,
(3) each $T(x)$ is weak*-compact instead of strongly compact,
(4) T is upper hemicontinuous along line segments instead of upper semi-continuous along line segments in X.

Note however that the coercive conditions in Theorem 5.54 here and in Theorem 5 of [Bae et al. (1993)] are not comparable.

Theorem 5.55 *Let $(E,\|\cdot\|)$ be a reflexive Banach space, X be a non-empty closed convex subset of E, $h:X\to\mathbb{R}$ be convex and lower semicontinuous on weakly compact subsets of X and $T:X\to 2^{E^*}$ be h-quasi-monotone and upper hemicontinuous along line segments in X to the weak topology on E^* such that each $T(x)$ is weakly compact convex. Suppose there is $x_0\in X$ such that*

$$\lim_{\substack{\|y\|\to\infty\\ y\in X}}\ \inf_{w\in T(y)} Re\langle w,y-x_0\rangle+h(y)-h(x_0)>0.$$

Then there exist $\hat{y}\in X$ and $\hat{w}\in T(\hat{y})$ such that $Re\langle\hat{w},\hat{y}-x\rangle\leq h(x)-h(\hat{y})$ for all $x\in X$. Moreover, if h is defined on all of E and is convex, then $Re\langle\hat{w},\hat{y}-x\rangle\leq h(x)-h(\hat{y})$ for all $x\in I_X(\hat{y})$.

Proof. By using the same argument in the proof of Theorem 5.51 and by Theorem 5.54, the conclusion follows. □

By taking $h\equiv 0$ in Theorem 5.54 and applying Lemma 5.25 Chowdhury and Tan obtained the following existence theorem of a generalized complementarity problem in Chowdhury and Tan (1997a):

Theorem 5.56 *Let X be a cone in a Hausdorff topological vector space E. Let $T : X \to 2^{E^*}$ be quasi-monotone and upper hemicontinuous along line segments in X to the weak*-topology on E^* such that each $T(x)$ is weak*-compact convex. Suppose there exist a non-empty weakly compact and weakly closed subset K of X and $x_0 \in K$ such that $\inf_{w\in T(y)} Re\langle w, y - x_0\rangle > 0$ for all $y \in X \setminus K$. Then there exist $\hat{y} \in K$ and $\hat{w} \in T(\hat{y})$ such that $Re\langle \hat{w}, \hat{y}\rangle = 0$ and $\hat{w} \in \widehat{X}$.*

By taking $h \equiv 0$ in Theorem 5.55 and applying Lemma 5.25 (or by a similar argument in proving Theorem 5.51 and by Theorem 5.56), Chowdhury and Tan obtained the following existence theorem of a generalized complementarity problem in Chowdhury and Tan (1997a):

Theorem 5.57 *Let $(E, \|\cdot\|)$ be a reflexive Banach space, X be a closed cone in E and $T : X \to 2^{E^*}$ be quasi-monotone and upper hemicontinuous along line segments in X to the weak topology on E^* such that each $T(x)$ is weakly compact convex. Suppose there is $x_0 \in X$ such that*

$$\lim_{\substack{\|y\|\to\infty \\ y\in X}} \inf_{w\in T(y)} Re\langle w, y - x_0\rangle > 0.$$

Then there exist $\hat{y} \in X$ and $\hat{w} \in T(\hat{y})$ such that $Re\langle \hat{w}, \hat{y}\rangle = 0$ and $\hat{w} \in \widehat{X}$.

Theorem 5.58 *Let E be a Hausdorff locally convex topological vector space, X be a non-empty convex subset of E, $h : X \to \mathbb{R}$ be convex and weakly lower semicontinuous on weakly compact subsets of X and $T : X \to 2^{E^*}$ be h-quasi-semi-monotone and upper hemicontinuous along line segments in X to the weak*-topology on E^* such that each $T(x)$ is strongly compact convex. Suppose there exist a non-empty weakly compact subset K of X and $x_0 \in K$ such that for each $y \in X \setminus K$, $\min_{w\in T(y)} Re\langle w, y - x_0\rangle + h(y) - h(x_0) > 0$. Then there exist $\hat{y} \in K$ and $\hat{w} \in T(\hat{y})$ such that $Re\langle \hat{w}, \hat{y} - x\rangle \le h(x) - h(\hat{y})$ for all $x \in X$. Moreover, if h is defined on all of E and is convex, then $Re\langle \hat{w}, \hat{y} - x\rangle \le h(x) - h(\hat{y})$ for all $x \in I_X(\hat{y})$.*

Proof. Define $f, g : X \times X :\to \mathbb{R} \cup \{-\infty, +\infty\}$ by

$$f(x, y) = \min_{u\in T(x)} Re\langle u, y - x\rangle + h(y) - h(x),$$

$$g(x, y) = \min_{w\in T(y)} Re\langle w, y - x\rangle + h(y) - h(x)$$

for all $x, y \in X$. Then we have the following:

(1) For each $x, y \in X$, since T is h-quasi-semi-monotone, $f(x, y) > 0$ implies $g(x, y) > 0$.

(2) For each fixed $x \in X$, since $T(x)$ is strongly compact, by Lemma 5.26, $y \longmapsto f(x, y)$ is weakly lower semicontinuous on non-empty bounded subsets of X and hence also weakly lower semicontinuous on weakly compact subsets of X.

(3) For each $A \in \mathcal{F}(\mathcal{X})$ and $y \in co(A)$, $\min_{x \in A} g(x, y) \leq 0$ by using the same argument in the proof of (3) in proving Theorem 5.49.

(4) By assumption, K is a weakly compact and weakly closed subset of X and $x_0 \in K$ such that for each $y \in X \setminus K$, $\min_{w \in T(y)} Re\langle w, y - x_0\rangle + h(y) - h(x_0) > 0$, i.e., $g(x_0, y) > 0$.

Equip E with the weak topology. Then f and g satisfy all the hypotheses of Theorem 3.1 so that by Theorem 5.45, there exists $\hat{y} \in K$ such that $f(x, \hat{y}) \leq 0$ for all $x \in X$; i.e., $\min_{u \in T(x)} Re\langle u, y - x\rangle + h(y) - h(x) \leq 0$ for all $x \in X$. Since h is convex and T is upper hemicontinuous, by Lemma 5.28, we have

$$\min_{w \in T(\hat{y})} Re\langle w, \hat{y} - x\rangle \leq h(x) - h(\hat{y}) \quad \text{for all} \quad x \in X.$$

By following the same argument as in proving Theorem 5.54, the conclusion follows. □

Remark 5.26 Theorem 5.58 extends Theorem 5 of Bae-Kim-Tan in [Bae et al. (1993), pp. 238–240] in the following ways:

(1) T is upper hemicontinuous along line segments instead of upper semicontinuous along line segments in X,

(2) T is h-quasi-semi-monotone instead of semi-monotone.

Note however that the coercive conditions in Theorem 4.15 here and in Theorem 5 of [Bae et al. (1993)] are not comparable.

Theorem 5.59 *Let $(E, \|\cdot\|)$ be a reflexive Banach space, X be a non-empty closed convex subset of E, $h : X \to \mathbb{R}$ be convex and lower semicontinuous on weakly compact subsets of X and $T : X \to 2^{E^*}$ be h-quasi-semi-monotone and upper hemicontinuous along line segments in X to the weak topology on E^* such that each $T(x)$ is compact convex. Suppose there is $x_0 \in X$ such that*

$$\lim_{\substack{\|y\| \to \infty \\ y \in X}} \inf_{w \in T(y)} Re\langle w, y - x_0\rangle + h(y) - h(x_0) > 0.$$

Then there exist $\hat{y} \in X$ and $\hat{w} \in T(\hat{y})$ such that $Re\langle \hat{w}, \hat{y} - x\rangle \leq h(x) - h(\hat{y})$ for all $x \in X$. Moreover, if h is defined on all of E and is convex, then $Re\langle \hat{w}, \hat{y} - x\rangle \leq h(x) - h(\hat{y})$ for all $x \in I_X(\hat{y})$.

Proof. By using the same argument in the proof of Theorem 4.7 and by Theorem 4.15, the conclusion follows. □

By taking $h \equiv 0$ in Theorem 5.58 and applying Lemma 5.25, Chowdhury and Tan obtained the following existence theorem of a generalized complementarity problem in Chowdhury and Tan (1997a):

Theorem 5.60 *Let X be a cone in a Hausdorff locally convex topological vector space E. Let $T : X \to 2^{E^*}$ be quasi-semi-monotone and upper hemicontinuous along line segments in X to the weak*-topology on E^* such that each $T(x)$ is strongly* *compact convex. Suppose there exist a non-empty weakly compact subset K of X and $x_0 \in K$ such that for each $y \in X \setminus K$, $\min_{w\in T(y)} Re\langle w, y - x_0\rangle > 0$. Then there exist $\hat{y} \in K$ and $\hat{w} \in T(\hat{y})$ such that $Re\langle \hat{w}, \hat{y}\rangle = 0$ and $\hat{w} \in \widehat{X}$.*

Theorem 5.61 *Let $(E, \|\cdot\|)$ be a reflexive Banach space, X be a closed cone in E and $T : X \to 2^{E^*}$ be quasi-semi-monotone and upper hemicontinuous along line segments in X to the weak topology on E^* such that each $T(x)$ is compact convex. Suppose there is $x_0 \in X$ such that*

$$\lim_{\substack{\|y\|\to\infty \\ y\in X}} \inf_{w\in T(y)} Re\langle w, y - x_0\rangle > 0.$$

Then there exist $\hat{y} \in X$ and $\hat{w} \in T(\hat{y})$ such that $Re\langle \hat{w}, \hat{y}\rangle = 0$ and $\hat{w} \in \widehat{X}$.

It is observed that in all generalized variational inequalities and generalized complementarity problems stated above, (1) when T is lower hemicontinuous along line segments, T is only required to have non-empty values, (2) when T is upper hemicontinuous along line segments and quasi-monotone, T is required to have weak*-compact-convex values and (3) when T is upper hemicontinuous along line segments and quasi-semi-monotone, T is required to have strongly-compact-convex values.

Next maximality of monotone operators will be discussed briefly:

Let X be a non-empty subset of a topological vector space E; then $T : X \to 2^{E^*}$ is *maximal monotone* if T is monotone and if $T^* : X \to 2^{E^*}$ is monotone such that $T(x) \subset T^*(x)$ for all $x \in X$, then $T = T^*$.

Theorem 5.62 *Let E be a topological vector space and $T : E \to 2^{E^*}$ be monotone and lower (respectively, upper) hemicontinuous along line segments in E such that each $T(x)$ is weak* compact convex. Then T is maximal monotone.*

Proof. Let $T^* : E \to 2^{E^*}$ be monotone such that $T(x) \subset T^*(x)$ for all $x \in X$. Let $y_0 \in E$ be arbitrarily fixed and let $w_0 \in T^*(y_0)$. Since T^* is monotone, for each $x \in E$ and each $u \in T^*(x)$, $Re\langle u - w_0, y_0 - x\rangle \le 0$. It follows that $\sup_{u\in T(x)} Re\langle u - w_0, y_0 - x\rangle \le 0$ for all $x \in E$. By Lemma 5.27 (respectively, Lemma 5.28), $\sup_{w\in T(y_0)} Re\langle w - w_0, y_0 - x\rangle \le 0$ (respectively, $\inf_{w\in T(y_0)} Re\langle w - w_0, y_0 - x\rangle \le 0$) for all $x \in E$. Thus $\sup_{x\in E} \inf_{w\in T(y_0)} Re\langle w - w_0, y_0 - x\rangle \le 0$. By Theorem 5.44, $\inf_{w\in T(y_0)} \sup_{x\in E} Re\langle w - w_0, y_0 - x\rangle \le 0$. Since $T(y_0)$ is weak* compact, there exists $\hat{w} \in T(y_0)$ such that $\sup_{x\in E} Re\langle \hat{w} - w_0, y_0 - x\rangle = \inf_{w\in T(y_0)} \sup_{x\in E} Re\langle w - w_0, y_0 - x\rangle \le 0$. Therefore $w_0 = \hat{w} \in T(y_0)$. Since $w_0 \in T^*(y_0)$ is arbitrary, $T(y_0) = T^*(y_0)$. Since $y_0 \in E$ is also arbitrary, we conclude that $T = T^*$. Hence T is maximal monotone. □

Theorem 5.62 improves Lemma 3 of [Shih and Tan (1988a)] in several aspects.

Finally, Chowdhury and Tan introduced some results on the surjectivity of monotone or semi-monotone operators in Chowdhury and Tan (1997a).

Theorem 5.63 *Let $(E, \|\cdot\|)$ be a reflexive Banach space, X be a non-empty closed convex subset of E and $T : X \to 2^{E^*}$ be monotone and lower hemicontinuous along line segments in X to the weak topology on E^*. Suppose there is $x_0 \in X$ such that*

$$\lim_{\substack{\|y\|\to\infty \\ y\in X}} \inf_{w\in T(y)} Re\langle w, y - x_0\rangle/\|y\| = \infty.$$

Then for each given $w_0 \in E^*$, there exist $\hat{y} \in X$ such that $\sup_{w\in T(\hat{y})} Re\langle w - w_0, \hat{y} - x\rangle \le 0$ for all $x \in X$. In particular, when $X = E$, then T is surjective; in fact, for each $w \in E^*$, there is $y \in E$ such that $T(y) = \{w\}$.

Proof. Let $w_0 \in E^*$ be given. Then

$$\lim_{\substack{\|y\|\to\infty \\ y\in X}} \left(\inf_{w\in T(y)} Re\langle w - w_0, y - x_0\rangle/\|y\|\right) \tag{5.95}$$

$$= \lim_{\substack{\|y\|\to\infty \\ y\in X}} \left(\left(\inf_{w\in T(y)} Re\langle w, y - x_0\rangle/\|y\|\right) - \|w_0\|\right) = \infty. \tag{5.96}$$

Define $T^* : X \to 2^{E^*}$ by $T^*(x) = T(x) - w_0$ for all $x \in X$. Then T^* is monotone and lower hemicontinuous along line segments in X to the weak topology on E^* and

$$\lim_{\substack{\|y\|\to\infty \\ y\in X}} \inf_{w\in T^*(y)} Re\langle w, y - x_0\rangle/\|y\| = \infty.$$

Therefore by Theorem 5.51, there exist $\hat{y} \in X$ such that $\sup_{w\in T^*(\hat{y})} Re\langle w, \hat{y} - x\rangle \le 0$ for all $x \in X$. That is, $\sup_{w\in T(\hat{y})} Re\langle w - w_0, \hat{y} - x\rangle \le 0$ for all $x \in X$. Now if $X = E$, then $w - w_0 = 0$ so that $w_0 = w$ for all $w \in T(\hat{y})$ and hence $T(\hat{y}) = \{w_0\}$. This shows that T is surjective such that for each $w \in E^*$, there is $y \in E$ with $T(y) = \{w\}$. □

Theorem 5.64 *Let $(E, \|\cdot\|)$ be a reflexive Banach space, X be a non-empty closed convex subset of E and $T : X \to 2^{E^*}$ be monotone and upper hemicontinuous along line segments in X to the weak topology on E^* such that each $T(x)$ is weakly compact convex. Suppose there is $x_0 \in X$ such that*

$$\lim_{\substack{\|y\|\to\infty \\ y\in X}} \inf_{w\in T(y)} Re\langle w, y - x_0\rangle/\|y\| = \infty.$$

Then for each given $w_0 \in E^*$, there exist $\hat{y} \in X$ and $\hat{w} \in T(\hat{y})$ such that $Re\langle \hat{w} - w_0, \hat{y} - x\rangle \le 0$ for all $x \in X$. In particular, if $X = E$, then T is surjective.

Proof. Let $w_0 \in E^*$ be given. Then

$$\lim_{\substack{\|y\|\to\infty \\ y\in X}} \big(\inf_{w\in T(y)} Re\langle w - w_0, y - x_0\rangle / \|y\| \big) \tag{5.97}$$

$$= \lim_{\substack{\|y\|\to\infty \\ y\in X}} \big(\big(\inf_{w\in T(y)} Re\langle w, y - x_0\rangle / \|y\| \big) - \|w_0\| \big) = \infty. \tag{5.98}$$

Define $T^* : X \to 2^{E^*}$ by $T^*(x) = T(x) - w_0$ for all $x \in X$. Then T^* is monotone and upper hemicontinuous along line segments in X to the weak topology on E^* such that each $T^*(x)$ is weakly compact convex and

$$\lim_{\substack{\|y\|\to\infty \\ y\in X}} \inf_{w\in T^*(y)} Re\langle w, y - x_0\rangle / \|y\| = \infty.$$

Therefore by Theorem 5.55, there exist $\hat{y} \in X$ and $\bar{w} \in T^*(\hat{y})$ such that $Re\langle \bar{w}, \hat{y} - x\rangle \leq 0$ for all $x \in X$. But then there exists $\hat{w} \in T(\hat{y})$ with $\bar{w} = \hat{w} - w_0$ so that $Re\langle \hat{w} - w_0, \hat{y} - x\rangle \leq 0$ for all $x \in X$. Now if $X = E$, then $\hat{w} - w_0 = 0$ so that $w_0 = \hat{w} \in T(\hat{y})$. This shows that T is surjective. □

By using an argument similar to the proof of Theorem 5.64 and by applying Theorem 5.59 (instead of Theorem 5.55), the following surjectivity of semi-monotone operators was obtained in Chowdhury and Tan (1997a):

Theorem 5.65 *Let $(E, \|\cdot\|)$ be a reflexive Banach space, X be a non-empty closed convex subset of E and $T : X \to 2^{E^*}$ be semi-monotone and upper hemicontinuous along line segments in X to the weak topology on E^* such that each $T(x)$ is compact convex. Suppose there is $x_0 \in X$ such that*

$$\lim_{\substack{\|y\|\to\infty \\ y\in X}} \inf_{w\in T(y)} Re\langle w, y - x_0\rangle / \|y\| = \infty.$$

Then for each given $w_0 \in E^*$, there exist $\hat{y} \in X$ and $\hat{w} \in T(\hat{y})$ such that $Re\langle \hat{w} - w_0, \hat{y} - x\rangle \leq 0$ for all $x \in X$. In particular, if $X = E$, then T is surjective.

It is observed that the proofs of Theorems 5.63 and 5.64 are slight modification of the proof of Theorem 2 in [Shih and Tan (1988a)] and improve Theorem 2 in [Shih and Tan (1988a)] from upper semicontinuous along line segments to lower or upper hemicontinuous along line segments.

5.9.3 *Fixed Point Theorems*

Applying the main results from Subsection 5.9.2, namely, Theorems 5.50, 5.54 and 5.58, Chowdhury and Tan obtained some fixed point theorems in Chowdhury and Tan (1997a) for operators which are either quasi-monotone, quasi-semi-monotone or quasi-non-expansive.

In all the ramining results, H will denote a Hilbert space with inner product $\langle \cdot, \cdot \rangle$ and the corresponding induced norm $\| \cdot \|$, and $bc(H)$ will denote the family of all non-empty bounded closed subsets of H. If X is a non-empty subset of H, we shall denote by $\partial_H(X)$ the boundary of X in H. If $x \in H$ and $r > 0$, let $B_r(x) = \{y \in H : \|x - y\| < r\}$. If d is the metric on H induced by the norm $\| \cdot \|$, let D be the Hausdorff metric on $bc(H)$ induced by d. Recall that $T : X \to bc(H)$ is non-expansive if $D(T(x), T(y)) \leq \|x - y\|$ for all $x, y \in X$.

Let K be a non-empty closed convex subset of H. For each $x \in H$, there is a unique point $\pi_K(x)$ in K such that $\|x - \pi_K(x)\| = \inf_{z \in K} \|x - z\|$. $\pi_K(x)$ is called the projection of x on K and is characterized as follows Kinderlehrer and Stampacchia (1980, Theorem 1.2.3, p. 9):

Proposition 5.8 *Let K be a non-empty closed convex subset of H. Then for each $x \in H$ and $y \in K$, $y = \pi_K(x)$ if and only if $Re\langle x - y, z - y \rangle \leq 0$ for all $z \in K$.*

As an application of Theorem 5.50, Chowdhury and Tan obtained the following fixed point theorem in Chowdhury and Tan (1997a):

Theorem 5.66 *Let X be a non-empty convex subset of H and $T : X \to 2^H$ be lower hemicontinuous along line segments in X such that each $T(x)$ is closed convex and $I - T$ is quasi-monotone. Suppose there exist a non-empty weakly compact subset K of X and $x_0 \in K$ such that for each $y \in X \setminus K$, $\inf_{w \subset T(y)} Re\langle y - w, y - x_0 \rangle > 0$. Then there exists $\hat{y} \in K$ such that $Re\langle \hat{y} - w, \hat{y} - x \rangle \leq 0$ for all $x \in \overline{I_X(\hat{y})}$ and for all $w \in T(\hat{y})$. Moreover, if either $\hat{y} \in int_H(X)$ or $\pi_{T(\hat{y})}(\hat{y}) \in \overline{I_X(\hat{y})}$, then $\hat{y}$ is a fixed point of T.*

Proof. Equip H with the weak topology. Since T is lower hemicontinuous along line segments in X, $I - T : X \to 2^H$ is also lower hemicontinuous along line segments in X and satisfies all the hypotheses of Theorem 5.50 with $h \equiv 0$. Hence by Theorem 5.50, there exists $\hat{y} \in K$ such that $\sup_{w \in T(\hat{y})} Re\langle \hat{y} - w, \hat{y} - x \rangle \leq 0$ for all $x \in I_X(\hat{y})$. By continuity, we have

$$Re\langle \hat{y} - w, \hat{y} - x \rangle \leq 0 \quad \text{for all} \quad x \in \overline{I_X(\hat{y})} \quad \text{and for all} \quad w \in T(\hat{y}). \tag{5.99}$$

Case 1. Suppose $\hat{y} \in int_H(X)$. Fix an arbitrary $\hat{w} \in T(\hat{y})$. Take any $r > 0$ such that $B_r(\hat{y}) \subset X$. Then for each $z \in H$ with $z \neq \hat{y}$, let $u = \hat{y} + \frac{r}{2} \cdot \frac{\hat{y} - z}{\|\hat{y} - z\|}$, then $u \in B_r(\hat{y}) \subset X \subset I_X(\hat{y})$. By (5.1), $Re\langle \hat{y} - \hat{w}, \frac{r}{2} \cdot \frac{z - \hat{y}}{\|\hat{y} - z\|} \rangle \leq 0$ so that $\frac{r}{2\|\hat{y} - z\|} Re\langle \hat{y} - \hat{w}, z - \hat{y} \rangle \leq 0$ and hence $Re\langle \hat{y} - \hat{w}, z - \hat{y} \rangle \leq 0$ for all $z \in H$.

It follows that $Re\langle \hat{y} - \hat{w}, z \rangle = 0$ for all $z \in H$ so that $\hat{y} = \hat{w} \in T(\hat{y})$. As $\hat{w} \in T(\hat{y})$ is arbitrary, we conclude that in fact $T(\hat{y}) = \{\hat{y}\}$.

Case 2. Suppose $p := \pi_{T(\hat{y})}(\hat{y}) \in \overline{I_X(\hat{y})}$. Fix an arbitrary $\hat{w} \in T(\hat{y})$. By Proposition

5.1, we have $p \in T(\hat{y})$ and $Re\langle \hat{y} - p, \hat{w} - p\rangle \leq 0$. Thus

$$0 \leq Re\langle p - \hat{y}, \hat{w} - p\rangle \tag{5.100}$$
$$= Re\langle p - \hat{y}, \hat{w} - \hat{y} + \hat{y} - p\rangle \tag{5.101}$$
$$= Re\langle p - \hat{y}, \hat{w} - \hat{y}\rangle - \|\hat{y} - p\|^2. \tag{5.102}$$

Therefore $\|\hat{y} - p\|^2 \leq Re\langle \hat{y} - \hat{w}, \hat{y} - p\rangle \leq 0$ by (5.1) as $p \in \overline{I_X(\hat{y})}$. Thus $\hat{y} = p = \pi_{T(\hat{y})}(\hat{y}) \in T(\hat{y})$. □

As it can be seen in an Example in Chowdhury and Tan (1997a), if we define $T : \mathbb{R}^+ \to bc(\mathbb{R})$ by $T(x) = [-x, 0]$ for all $x \in \mathbb{R}^+$, then $I - T$ is quasi-monotone but not monotone.

The following result follows from Theorem 5.46 and Proposition 5.6:

Corollary 5.66.1 *Let X be a non-empty convex subset of H and $T : X \to 2^H$ be quasi-non-expansive and lower hemicontinuous along line segments in X such that each $T(x)$ is closed and convex. Suppose there exist a non-empty weakly compact subset K of X and $x_0 \in K$ such that (i) for each $y \in K \cap \partial_H(X), \pi_{T(y)}(y) \in \overline{I_X(y)}$ and (ii) for each $y \in X \setminus K$, $\inf_{w \in T(y)} Re\langle y - w, y - x_0\rangle > 0$. Then T has a fixed point in K.*

Corollary 5.66.2 *Let X be a non-empty bounded closed convex subset of H and $T : X \to 2^H$ be quasi-non-expansive and lower hemicontinuous along line segments in X such that each $T(x)$ is closed and convex. If $\pi_{T(y)}(y) \in \overline{I_X(y)}$ for each $y \in \partial_H(X)$, then T has a fixed point in X.*

Corollary 5.66.3 *Let X be a non-empty bounded closed convex subset of H and $T : X \to 2^X$ be quasi-non-expansive and lower hemicontinuous along line segments in X such that each $T(x)$ is closed and convex. Then T has a fixed point in X.*

As an application of Theorem 5.54, Chowdhury and Tan obtained the following fixed point theorem in Chowdhury and Tan (1997a):

Theorem 5.67 *Let X be a non-empty convex subset of H and $T : X \to 2^H$ be upper hemicontinuous along line segments in X such that each $T(x)$ is weakly compact convex and $I - T$ is quasi-monotone. Suppose there exist a non-empty weakly compact subset K of X and $x_0 \in K$ such that for each $y \in X \setminus K$, $\inf_{w \in T(y)} Re\langle y - w, y - x_0\rangle > 0$. Then there exist $\hat{y} \in K$ and $\hat{w} \in T(\hat{y})$ such that $Re\langle \hat{y} - \hat{w}, \hat{y} - x\rangle \leq 0$ for all $x \in \overline{I_X(\hat{y})}$. Moreover, if either $\hat{y} \in int_H(X)$ or $\pi_{T(\hat{y})}(\hat{y}) \in \overline{I_X(\hat{y})}$, then $\hat{y}$ is a fixed point of T, i.e., $\hat{y} \in T(\hat{y})$.*

Proof. Equip H with the weak topology. Since T is upper hemicontinuous along line segments in X, $I - T : X \to 2^H$ is also upper hemicontinuous along line segments in X. Note that $I - T$ satisfies all the hypotheses of Theorem 5.54 with $h \equiv 0$. Thus by Theorem 5.54, there exist $\hat{y} \in K$ and $\hat{w} \in T(\hat{y})$ such that $Re\langle \hat{y} - \hat{w}, \hat{y} - x\rangle \leq 0$

for all $x \in I_X(\hat{y})$. By continuity,

$$Re\langle \hat{y} - \hat{w}, \hat{y} - x\rangle \leq 0 \quad \text{for all} \quad x \in \overline{I_X(\hat{y})}. \tag{5.103}$$

Case 1. Suppose $\hat{y} \in int_H(X)$, then there exists $r > 0$ such that $B_r(\hat{y}) \subset X$. Then for each $z \in H$ with $z \neq \hat{y}$, let $u = \hat{y} + \frac{r}{2} \cdot \frac{\hat{y}-z}{\|\hat{y}-z\|}$, then $u \in B_r(\hat{y}) \subset X \subset I_X(\hat{y})$. Thus $Re\langle \hat{y} - \hat{w}, \frac{r}{2} \cdot \frac{z-\hat{y}}{\|\hat{y}-z\|}\rangle \leq 0$ so that $\frac{r}{2\|\hat{y}-z\|} Re\langle \hat{y} - \hat{w}, z - \hat{y}\rangle \leq 0$ and hence $Re\langle \hat{y} - \hat{w}, z - \hat{y}\rangle \leq 0$ for all $z \in H$.

It follows that $Re\langle \hat{y} - \hat{w}, z\rangle = 0$ for all $z \in H$ so that $\hat{y} = \hat{w} \in T(\hat{y})$.

Case 2. Suppose $p := \pi_{T(\hat{y})}(\hat{y}) \in \overline{I_X(\hat{y})}$. By Proposition 5.8, we have

$$p \in T(\hat{y}) \text{ and } Re\langle \hat{y} - p, w - p\rangle \leq 0 \quad \text{for all} \quad w \in T(\hat{y}). \tag{5.104}$$

Since $\hat{w} \in T(\hat{y})$, by (5.104) we have

$$0 \leq Re\langle p - \hat{y}, \hat{w} - p\rangle \tag{5.105}$$

$$= Re\langle p - \hat{y}, \hat{w} - \hat{y} + \hat{y} - p\rangle \tag{5.106}$$

$$= Re\langle p - \hat{y}, \hat{w} - \hat{y}\rangle - \|\hat{y} - p\|^2. \tag{5.107}$$

Therefore $\|\hat{y} - p\|^2 \leq Re\langle \hat{y} - \hat{w}, \hat{y} - p\rangle \leq 0$ by (5.2). Thus $\hat{y} = p = \pi_{T(\hat{y})}(\hat{y}) \in T(\hat{y})$. □

The following result follows from Theorem 5.67 and Proposition 5.6:

Corollary 5.67.1 *Let X be a non-empty convex subset of H and $T : X \to 2^H$ be quasi-non-expansive and upper hemicontinuous along line segments in X such that each $T(x)$ is weakly compact convex. Suppose there exist a non-empty weakly compact subset K of X and $x_0 \in K$ such that (i) for each $y \in K \cap \partial_H(X)$, $\pi_{T(y)}(y) \in \overline{I_X(y)}$ and (ii) for each $y \in X \setminus K$, $\inf_{w \in T(y)} Re\langle y - w, y - x_0\rangle > 0$. Then T has a fixed point in K.*

Corollary 5.67.2 *Let X be a non-empty bounded closed convex subset of H and $T : X \to 2^H$ be quasi-non-expansive and upper hemicontinuous along line segments in X such that each $T(x)$ is weakly compact convex. If $\pi_{T(y)}(y) \in \overline{I_X(y)}$ for each $y \in \partial_H(X)$, then T has a fixed point in X.*

Corollary 5.67.3 *Let X be a non-empty bounded closed convex subset of H and $T : X \to 2^X$ be quasi-non-expansive and upper hemicontinuous along line segments in X such that each $T(x)$ is weakly compact convex. Then T has a fixed point in X.*

As an application of Theorem 5.58 with $h \equiv 0$ Chowdhury and Tan obtained the following fixed point theorem in Chowdhury and Tan (1997a):

Theorem 5.68 *Let X be a non-empty convex subset of H, $T : X \to 2^H$ be upper hemicontinuous along line segments in X such that each $T(x)$ is compact convex and $I - T$ is quasi-semi-monotone. Suppose there exist a non-empty weakly compact* *subset K of X and $x_0 \in K$ such that for each $y \in X\backslash K$, $\inf_{w\in T(y)} Re\langle y-w, y-x_0\rangle > 0$. Then there exist $\hat{y} \in K$ and $\hat{w} \in T(\hat{y})$ such that $Re\langle \hat{y}-\hat{w}, \hat{y}-x\rangle \le 0$ for all $x \in \overline{I_X(\hat{y})}$. Moreover, if either $\hat{y} \in int_H(X)$ or $\pi_{T(\hat{y})}(\hat{y}) \in \overline{I_X(\hat{y})}$, then $\hat{y}$ is a fixed point of T.*

Proof. Equip H with the weak topology. Since T is upper hemicontinuous along line segments in X, $I - T : X \to 2^H$ is also upper hemicontinuous along line segments in X and satisfies all the hypotheses of Theorem 4.15 with $h \equiv 0$. Hence by Theorem 5.58, there exist $\hat{y} \in K$ and $\hat{w} \in T(\hat{y})$ such that $Re\langle \hat{y}-\hat{w}, \hat{y}-x\rangle \le 0$ for all $x \in I_X(\hat{y})$. By continuity of $\hat{w}$, $Re\langle \hat{y}-\hat{w}, \hat{y}-x\rangle \le 0$ for all $x \in \overline{I_X(\hat{y})}$. Now the rest of the proof is similar to that of Theorem 5.6 and the conclusion follows. □

Remark 5.27 Theorem 5.68 extends Theorem 6 of Bae-Kim-Tan in Bae et al. (1993, pp. 242, 243), in the following ways:

(1) $I - T$ is quasi-semi-monotone instead of T is pseudo-contractive [Bae et al. (1993), p. 240];
(2) T is upper hemicontinuous along line segments instead of upper semicontinuous along line segments in X.

Note however that the coercive conditions in our Theorem 5.68 here and in Theorem 6 of [Bae et al. (1993)] are not comparable.

The following result is an immediate consequence of Theorem 5.68:

Theorem 5.69 *Let X be a non-empty convex subset of H and $T : X \to 2^H$ be upper hemicontinuous along line segments in X such that each $T(x)$ is compact convex and $I - T$ is quasi-semi-monotone. Suppose there exist a non-empty weakly compact subset K of X and $x_0 \in K$ such that (i) for each $y \in K \cap \partial_H(X)$, $\pi_{T(y)}(y) \in \overline{I_X(y)}$ and (ii) for each $y \in X \setminus K$, $\inf_{w\in T(y)} Re\langle y-w, y-x_0\rangle > 0$. Then T has a fixed point in K.*

As it can be seen in an Example in Chowdhury and Tan (1997a), if we define $T : \mathbb{R}^+ \to bc(\mathbb{R})$ by

$$T(x) = \begin{cases} [\frac{x^2-1}{x}, 0], & \text{if } 0 < x < 1, \\ [0, \frac{x^2-1}{x}], & \text{if } x \ge 1, \end{cases}$$

then $I - T$ is quasi-semi-monotone but not semi-monotone.

Corollary 5.69.1 *Let X be a non-empty bounded closed convex subset of H and $T : X \to 2^H$ be semi-non-expansive and upper hemicontinuous along line segments in X such that each $T(x)$ is compact convex. If $\pi_{T(y)}(y) \in \overline{I_X(y)}$ for each $y \in \partial_H(X)$, then T has a fixed point in X.*

Corollary 5.69.2 *Let X be a non-empty bounded closed convex subset of H and $T : X \to 2^X$ be semi-non-expansive and upper hemicontinuous along line segments in X such that each $T(x)$ is compact convex. Then T has a fixed point in X.*

By Proposition in Bae et al. (1993), if T is non-expansive, then $I - T$ is semi-monotone; it follows from Proposition 5.7 that T is semi-non-expansive. Also observe that a set-valued non-expansive operator is necessarily upper semicontinuous (and also lower semicontinuous) so that it is upper hemicontinuous by Proposition 5.5. Thus we have:

Corollary 5.69.3 *Let X be a non-empty convex subset of H and $T : X \to bc(H)$ be non-expansive such that each $T(x)$ is compact convex. Suppose there exist a non-empty weakly compact subset K of X and $x_0 \in K$ such that (i) for each $y \in K \cap \partial_H(X)$, $\pi_{T(y)}(y) \in \overline{I_X(y)}$ and (ii) for each $y \in X \backslash K$, $\inf_{w \in T(y)} Re\langle y-w, y-x_0\rangle > 0$. Then T has a fixed point in K.*

Corollary 5.69.4 *Let X be a non-empty bounded closed convex subset of H and $T : X \to 2^H$ be non-expansive such that each $T(x)$ is compact convex. If $\pi_{T(y)}(y) \in \overline{I_X(y)}$ for each $y \in \partial_H(X)$, then T has a fixed point in X.*

Corollary 5.69.5 *Let X be a non-empty bounded closed convex subset of H and $T : X \to 2^X$ be non-expansive such that each $T(x)$ is compact convex. Then T has a fixed point in X.*

Remark 5.28 Theorem 1 of Browder (1965b) states that if X is a non-empty bounded closed convex subset of H and $f : X \to X$ is non-expansive, then f has a fixed point in X. Thus Corollaries 5.66.3 and 5.67.3 (respectively, Corollary 5.69.5) generalize Browder's fixed point theorem [Browder (1965b), Theorem 1] to set-valued quasi-non-expansive (respectively, non-expansive) operators while Corollaries 5.66.2 and 5.67.2 (respectively, Corollary 5.69.4) generalize Browder's fixed point theorem [Browder (1965b)], Theorem 1] to set-valued quasi-non-expansive (respectively, non-expansive) operators which need not be self-maps.

Chowdhury and Tan observed in Chowdhury and Tan (1997a) that in all fixed point theorems stated above, (1) when T is lower hemicontinuous along line segments, T is required to have closed-convex values, (2) when T is upper hemicontinuous along line segments and quasi-monotone, T is required to have weakly-compact-convex values and (3) when T is upper hemicontinuous along line segments and quasi-semi-monotone, T is required to have compact-convex values.

Note that the results of this section have been written and presented after a thorough revision of some results of an article in Chowdhury and Tan (1997b).

5.10 Generalization of Ky Fan's Minimax Inequality with Applications to Generalized Variational Inequalities for Pseudo-Monotone Type I Operators and Fixed Point Theorems

In obtaining the results of this section, Chowdhury and Tan mainly used the following celebrated 1972 Ky Fan's minimax inequality (Fan (1972), Theorem 1).

Theorem 5.70 *Let E be a Hausdorff topological vector space and X a non-empty compact convex subset of E. Let f be a real-valued function defined on $X \times X$ such that*

(a) for each fixed $x \in X$, $f(x, y)$ is a lower semicontinuous function of y on X;

(b) for each fixed $y \in X$, $f(x, y)$ is a quasi-concave function of x on X.

Then the minimax inequality

$$\min_{y \in X} \sup_{x \in X} f(x, y) \leq \sup_{x \in X} f(x, x)$$

holds.

Ky Fan's minimax inequality has become a versatile tool in non-linear functional analysis [Fan (1972), pp. 103–113], convex analysis, game theory and economic theory [Aubin (1982)]. There have been numerous generalizations of Ky Fan's minimax inequality by weakening the compactness assumption or the convexity assumption; e.g., see Allen [Allen (1977)], [Bae et al. (1993)], H. Brézis and Stampacchia (1972), Ding and Tan (1992b), [Shih and Tan (1984)], [Tan (1983)], [Tan and Yuan (1993)], [Yen (1981)] and [Fan (1984)].

In this section, we shall present Chowdhury and Tan's generalization of Ky Fan's minimax inequality with applications to generalized variational inequalities for pseudo-monotone type I operators. Later, we shall also give some of their results of fixed point theorems in Hilbert spaces for set-valued operators which are upper semi-continuous such that $I - T$ are of pseudo-monotone type I.

We shall now define pseudo-monotone type I operators which were first introduced by Chowdhury and Tan as set-valued pseudo-monotone operators in Chowdhury and Tan (1996). Later, these operators were re-named as pseudo-monotone type I operators in Chowdhury (2000).

Definition 5.15 Let E be a topological vector space, X be a non-empty subset of E and $T : X \to 2^{E^*}$. If $h : X \to \mathbb{R}$, then T is said to be an *h-pseudo-monotone* (respectively, *h-demi-monotone*) operator if for each $y \in X$ and every net $\{y_\alpha\}_{\alpha \in \Gamma}$ in X converging to y (respectively, weakly to y) with

$$\limsup_{\alpha} [\inf_{u \in T(y_\alpha)} \mathrm{Re}\langle u, y_\alpha - y\rangle + h(y_\alpha) - h(y)] \leq 0$$

we have

$$\limsup_{\alpha} [\inf_{u \in T(y_\alpha)} Re\langle u, y_\alpha - x\rangle + h(y_\alpha) - h(x)] \tag{5.108}$$

$$\geq \inf_{w \in T(y)} Re\langle w, y - x\rangle + h(y) - h(x) \text{ for all } x \in X; \tag{5.109}$$

T is said to be *pseudo-monotone* (respectively, *demi-monotone*) if T is h-pseudo-monotone (respectively, h-demi-monotone) with $h \equiv 0$. In general, we shall call all such operators in this definition as h-pseudo-monotone type I operators or pseudo-monotone type I operators.

Note that if T is single-valued and is pseudo-monotone in the sense of H. Brézis and Stampacchia (1972, p. 297), then T is pseudo-monotone type I in the sense of Definition 5.15 above.

We observe that monotonicity implies semi-monotonicity. But these two operators coincide and become one operator when they are single-valued operators.

Proposition 5.9 *Let X be a non-empty subset of a topological vector space E. If $T : X \to E^*$ is monotone and continuous from the relative weak topology on X to the weak* topology on E^*, then T is pseudo-monotone and demi-monotone.*

Proof. Suppose $\{y_\alpha\}_{\alpha\in\Gamma}$ is a net in X and $y \in X$ with $y_\alpha \to y$ (respectively, $y_\alpha \to y$ weakly) (and $\limsup_\alpha Re\langle Ty_\alpha,\, y_\alpha - y\rangle \leq 0$). Then for any $x \in X$ and $\epsilon > 0$, there are $\beta_1, \beta_2 \in \Gamma$ with $|Re\langle Ty, y_\alpha - y\rangle| < \frac{\epsilon}{2}$ for all $\alpha \geq \beta_1$ and $|Re\langle Ty_\alpha - Ty, y - x\rangle| < \frac{\epsilon}{2}$ for all $\alpha \geq \beta_2$. Choose $\beta_0 \in \Gamma$ with $\beta_0 \geq \beta_1, \beta_2$. Thus

$$
\begin{aligned}
Re\langle Ty_\alpha, y_\alpha - x\rangle &= Re\langle Ty_\alpha, y_\alpha - y\rangle + Re\langle Ty_\alpha, y - x\rangle && (5.110)\\
&\geq Re\langle Ty, y_\alpha - y\rangle + Re\langle Ty_\alpha, y - x\rangle && (5.111)\\
&= Re\langle Ty, y_\alpha - y\rangle + Re\langle Ty_\alpha - Ty, y - x\rangle + Re\langle Ty, y - x\rangle && (5.112)\\
&> -\frac{\epsilon}{2} - \frac{\epsilon}{2} + Re\langle Ty, y - x\rangle \text{ for all } \alpha \geq \beta_0. && (5.113)
\end{aligned}
$$

It follows that

$$\limsup_\alpha Re\langle Ty_\alpha, y_\alpha - x\rangle \geq -\epsilon + Re\langle Ty, y - x\rangle.$$

As $\epsilon > 0$ is arbitrary,

$$\limsup_\alpha Re\langle Ty_\alpha, y_\alpha - x\rangle \geq Re\langle Ty, y - x\rangle.$$

Hence T is pseudo-monotone (respectively, demi-monotone). □

5.10.1 *Generalization of Ky Fan's Minimax Inequality*

We shall begin with a lemma of Chowdhury and Tan in (Chowdhury and Tan (1996)):

Lemma 5.30 *Let E be a topological vector space, X be a non-empty convex subset of E. Let $F : X \to 2^X$ be a KKM-map such that*

- (a) *$cl_X F(x_0)$ is compact for some $x_0 \in X$;*
- (b) *for each $A \in \mathcal{F}(X)$ with $x_0 \in A$ and each $x \in co(A)$, $F(x) \cap co(A)$ is closed in $co(A)$ and*
- (c) *for each $A \in \mathcal{F}(X)$ with $x_0 \in A$,*

$$(cl_X(\bigcap_{x\in co(A)} F(x))) \cap co(A) = (\bigcap_{x\in co(A)} F(x)) \cap co(A).$$

Then $\bigcap_{x \in X} F(x) \neq \emptyset$.

Proof. Fix any $A \in \mathcal{F}(\mathcal{X})$ with $x_0 \in A$. Define $G_A : co(A) \to 2^{co(A)}$ by $G_A(x) = F(x) \cap co(A)$ for each $x \in co(A)$.

Now, for each $x \in co(A)$, $G_A(x)$ is non-empty since F is a KKM map and closed in $co(A)$ by (b). Note that $co(A)$ is compact. Thus each $G_A(x)$ is also compact. For each $B \in \mathcal{F}(\rfloor\wr(\mathcal{A}))$ we have $B \in \mathcal{F}(\mathcal{X})$ as $co(A) \subset X$ and so $co(B) \subset \cup_{x \in B} F(x)$. But $co(B) \subset co(A)$; it follows that

$$co(B) \subset (\bigcup_{x \in B} F(x)) \cap co(A) = \bigcup_{x \in B} (F(x) \cap co(A)) = \bigcup_{x \in B} G_A(x).$$

Thus G_A is a KKM-map on $co(A)$. Hence by Ky Fan's lemma in Fan (1961, Lemma 1), we have

$$\bigcap_{x \in co(A)} G_A(x) \neq \emptyset, \quad \text{i.e.,} \quad \bigcap_{x \in co(A)} F(x) \cap co(A) \neq \emptyset. \tag{5.114}$$

Note that we do not require Hausdorff condition in Ky Fan's Lemma. This was observed by Ding and Tan in [Ding and Tan (1992b), Lemma 3].

Let $\{E_i\}_{i \in I}$ be the family of all convex hulls of finite subsets of X containing the point x_0, partially ordered by $\subset$.

Now, for each $i \in I$, let $E_i = co(A_i)$, where $A_i \in \mathcal{F}_\prime(\mathcal{X})$ = the family of all non-empty finite subsets of X containing the point x_0.

By (5.114), for each $i \in I$, $\bigcap_{x \in E_i} F(x) \cap E_i \neq \emptyset$. Fix any $u_i \in \bigcap_{x \in E_i} F(x) \cap E_i$. For each $i \in I$, let

$$\Phi_i = \{u_j | j \geq i, j \in I\}.$$

Clearly, (i) $\{\Phi_i | i \in I\}$ has the finite intersection property and (ii) $\Phi_i \subset F(x_0)$ for all $i \in I$. Then $cl_X \Phi_i \subset cl_X F(x_0)$ for all $i \in I$. By compactness of $cl_X F(x_0)$, $\bigcap_{i \in I} cl_X \Phi_i \neq \emptyset$. Choose any $\hat{x} \in \bigcap_{i \in I} cl_X \Phi_i$. Note that for any $i \in I$ and for all $j \in I$ with $j \geq i$,

$$u_j \in \bigcap_{x \in E_j} F(x) \cap E_j \subset \bigcap_{x \in E_i} F(x) \cap E_j \subset \bigcap_{x \in E_i} F(x).$$

Therefore

$$\Phi_i \subset \bigcap_{x \in E_i} F(x). \tag{5.115}$$

Now, for any $x \in X$, there exists $i_0 \in I$ such that $x, \hat{x} \in E_{i_0}$. Therefore for all $i \geq i_0$, we have $x, \hat{x} \in E_{i_0} \subset E_i$. It follows that

$$\hat{x} \in E_i \cap cl_X \Phi_i \subset E_i \cap (cl_X \bigcap_{z \in E_i} F(z)) \tag{5.116}$$

$$= (\bigcap_{z \in E_i} F(z)) \cap E_i \quad \text{(by (c))} \tag{5.117}$$

$$= (\bigcap_{z \in E_i} F(z) \cap E_i) \subset F(x). \tag{5.118}$$

Thus $\hat{x} \in F(x)$ for all $x \in X$. Hence $\bigcap_{x \in X} F(x) \neq \emptyset$. □

Under the hypotheses of Lemma 5.30, we see that if for each $x \in X$ and each finite dimensional subspace L of E, $F(x) \cap L$ is closed in L, then for each $A \in \mathcal{F}(\mathcal{X})$ with $x_0 \in A$ and each $x \in co(A)$, $F(x) \cap co(A)$ is also closed in $co(A)$.

The following example of Chowdhury and Tan in (Chowdhury and Tan (1996)) shows that the converse is not true in general.

Example 5.9 Let $E = \mathbb{R}^2$. Consider the following non-empty convex subset X of E:

$$X = \{(u, v) \in \mathbb{R}^2 | 0 < u < 1 \text{ and } 0 < v \leq 1-u\} \cup \{(u, v) \in \mathbb{R}^2 | u = 0 \text{ and } 0 \leq v \leq 1\}.$$

Fix $x_0 = (\frac{1}{2}, \frac{1}{2}) \in X$. For each $x \in X$ with $x \neq (0,0)$ and $x \neq x_0$, let A_x denote the following set.

A_x = the closed region in X bounded by the line $v = 1 - u$ and the line passing through the point x and parallel to the line $v = 1 - u$.

Now, we define $F : X \to 2^X$ by

$$F(x) = \begin{cases} A_x \cup \{(0,0)\} \cup \left\{\left(\frac{1}{n+2}, \frac{1}{n+2}\right) : n = 1, 2, 3, \ldots\right\}, \\ \quad \text{if } x \notin (0,0) \text{ and } x \notin x_0; \\ X, \text{if } x = (0,0); \\ \{(0,0)\} \cup \left\{\left(\frac{1}{n+2} \frac{1}{n+2}\right) : n \neq 1, 2, \ldots\right\}, \\ \quad \text{if } x = x_0 \end{cases}$$

Then for each $A \in \mathcal{F}(\mathcal{X})$ with $x_0 \in A$ and for each $x \in co(A)$, $F(x) \cap co(A)$ is closed in $co(A)$. However, consider $L = \mathbb{R}^2$ and $x = (0,0)$; then $F(x) \cap L = F((0,0)) \cap \mathbb{R}^2 = X$ is not closed in L. Note that F is a KKM-map such that $cl_X F(x_0) = F(x_0)$ is compact and the condition (c) of Lemma 5.30 is also satisfied. Thus Lemma 5.30 is applicable but Lemma 1 of [H. Brézis and Stampacchia (1972)] is not.

We remark here that the hypotheses (b) and (c) of Lemma 5.30 are more general than the hypotheses (8) and (9) of Lemma 1 in [H. Brézis and Stampacchia (1972), pp. 294–295]. But Lemma 5.30 requires that X be convex while Lemma 1 of Fan in [Fan (1961)] and Lemma 1 of [H. Brézis and Stampacchia (1972)] do not.

However, in all applications of Lemma 1 in [Fan (1961)] or Lemma 1 in [H. Brézis and Stampacchia (1972)], the set X is always assumed to be convex.

We shall now present the following minimax inequality in [Chowdhury and Tan (1996)]:

Theorem 5.71 *Let E be a topological vector space, X be a non-empty convex subset of E, $h : X \to \mathbb{R}$ be lower semicontinuous on $co(A)$ for each $A \in \mathcal{F}(X)$ and $f : X \times X \to \mathbb{R} \cup \{-\infty, +\infty\}$ be such that*

(*a*) *for each $A \in \mathcal{F}(X)$ and each fixed $x \in co(A)$, $y \longmapsto f(x, y)$ is lower semicontinuous on $co(A)$;*

(*b*) *for each $A \in \mathcal{F}(\mathcal{X})$ and each $y \in co(A)$, $\min_{x\in A}[f(x, y) + h(y) - h(x)] \le 0$;*

(*c*) *for each $A \in \mathcal{F}(X)$ and each $x, y \in co(A)$ and every net $\{y_\alpha\}_{\alpha\in\Gamma}$ in X converging to y with*

$$f(tx+(1-t)y, y_\alpha)+h(y_\alpha)-h(tx+(1-t)y) \le 0 \textit{ for all } \alpha \in \Gamma \textit{ and all } t \in [0, 1],$$

we have $f(x, y) + h(y) - h(x) \le 0$;

(*d*) *there exist a non-empty closed and compact subset K of X and $x_0 \in K$ such that $f(x_0, y) + h(y) - h(x_0) > 0$ for all $y \in X\backslash K$.*

Then there exists $\hat{y} \in K$ such that $f(x, \hat{y}) \le h(x) - h(\hat{y})$ for all $x \in X$.

Proof. Define $F : X \to 2^X$ by

$$F(x) = \{y \in X : f(x, y) + h(y) - h(x) \le 0\} \text{ for each } x \in X.$$

If F is not a KKM-map, then for some finite subset $\{x_1, \dots, x_n\}$ of X and $\alpha_i \ge 0$ for $i = 1, \dots, n$ with $\sum_{i=1}^n \alpha_i = 1$, we have $\bar{y} = \sum_{i=1}^n \alpha_i x_i \notin \bigcup_{i=1}^n F(x_i)$. Thus $f(x_i, \bar{y}) + h(\bar{y}) - h(x_i) > 0$ for $i = 1, \dots, n$ so that

$$\min_{1\le i\le n} [f(x_i, \bar{y}) + h(\bar{y}) - h(x_i)] > 0,$$

which contradicts the assumption (*b*). Hence $F : X \to 2^X$ is a KKM-map. Moreover we have,

(*i*) $F(x_0) \subset K$ by (*d*), so that $cl_X F(x_0) \subset cl_X K = K$ and hence $cl_X F(x_0)$ is compact in X;

(*ii*) for each $A \in \mathcal{F}(\mathcal{X})$ with $x_0 \in A$ and each $x \in co(A)$,

$$\begin{aligned} F(x) \cap co(A) &= \{y \in co(A) : f(x, y) + h(y) - h(x) \le 0\} \\ &= \{y \in co(A) : f(x, y) + h(y) \le h(x)\} \end{aligned}$$

is closed in $co(A)$ by (*a*) and the fact that h is lower semi-continuous on $co(A)$;

(*iii*) for each $A \in \mathcal{F}(X)$ with $x_0 \in A$, if $y \in (cl_X(\bigcap_{x\in co(A)} F(x))) \cap co(A)$, then $y \in co(A)$ and there is a net $\{y_\alpha\}_{\alpha\in\Gamma}$ in $\bigcap_{x\in co(A)} F(x)$ such that $y_\alpha \to y$. For each $x \in co(A)$, since $tx + (1-t)y \in co(A)$ for all $t \in [0, 1]$, we have $y_\alpha \in F(tx + (1-t)y)$

for all $\alpha \in \Gamma$ and all $t \in [0,1]$. This implies that $f(tx+(1-t)y, y_\alpha)+h(y_\alpha)-h(tx+(1-t)y) \leq 0$ for all $\alpha \in \Gamma$ and all $t \in [0,1]$ so that by (c), $f(x,y)+h(y)-h(x) \leq 0$; it follows that $y \in (\bigcap_{x\in co(A)} F(x)) \cap co(A)$. Hence, $(cl_X(\bigcap_{x\in co(A)} F(x))) \cap co(A) = (\bigcap_{x\in co(A)} F(x)) \cap co(A)$.

Hence by Lemma 5.30 we have $\bigcap_{x\in X} F(x) \neq \emptyset$. Then there exists $\hat{y} \in \bigcap_{x\in X} F(x)$, so that $f(x,\hat{y})+h(\hat{y})-h(x) \leq 0$ for all $x \in X$, *i.e.*, $f(x,\hat{y}) \leq h(x)-h(\hat{y})$, for all $x \in X$. □

When $h \equiv 0$, Theorem 5.71 reduces to the following:

Theorem 5.72 *Let E be a topological vector space, X be a non-empty convex subset of E and $f : X \times X \to \mathbb{R} \cup \{-\infty, +\infty\}$ be such that*

(a) *for each $A \in \mathcal{F}(X)$ and each fixed $x \in co(A)$, $y \longmapsto f(x,y)$ is lower semi-continuous on $co(A)$;*
(b) *for each $A \in \mathcal{F}(\mathcal{X})$ and each $y \in co(A)$, $\min_{x\in A} f(x,y) \leq 0$;*
(c) *for each $A \in \mathcal{F}(X)$ and each $x, y \in co(A)$ and every net $\{y_\alpha\}_{\alpha\in\Gamma}$ in X converging to y with $f(tx+(1-t)y, y_\alpha) \leq 0$ for all $\alpha \in \Gamma$ and all $t \in [0,1]$, we have $f(x,y) \leq 0$;*
(d) *there exist a non-empty closed and compact subset K of X and $x_0 \in K$ such that $f(x_0,y) > 0$ for all $y \in X\backslash K$.*

Then there exists $\hat{y} \in K$ such that $f(x,\hat{y}) \leq 0$ for all $x \in X$.

Note that Theorem 5.72 implicitly implies the following minimax inequality:

Theorem 5.73 *Let E be a topological vector space, X be a non-empty convex subset of E and $f : X \times X \to \mathbb{R} \cup \{-\infty, +\infty\}$ be such that*

(a) *for each $A \in \mathcal{F}(X)$ and each fixed $x \in co(A)$, $y \longmapsto f(x,y)$ is lower semi-continuous on $co(A)$;*
(b) *for each $A \in \mathcal{F}(\mathcal{X})$ and each $y \in co(A)$, $\min_{x\in A} f(x,y) \leq 0$;*
(c) *for each $A \in \mathcal{F}(X)$ and each $x, y \in co(A)$ and every net $\{y_\alpha\}_{\alpha\in\Gamma}$ in X converging to y with $f(tx+(1-t)y, y_\alpha) \leq 0$ for all $\alpha \in \Gamma$ and all $t \in [0,1]$, we have $f(x,y) \leq 0$;*
(d) *there exist a non-empty closed and compact subset K of X and $x_0 \in K$ such that whenever $\sup_{x\in X} f(x,x) < \infty$, $f(x_0,y) > \sup_{x\in X} f(x,x)$ for all $y \in X\backslash K$.*

Then the minimax inequality

$$\min_{y\in K} \sup_{x\in X} f(x,y) \leq \sup_{x\in X} f(x,x)$$

holds.

Proof. Let $t = \sup_{x\in X} f(x,x)$. Clearly, we may assume that $t < +\infty$. Define for any $x, y \in X, g(x,y) = f(x,y) - t$. Then g satisfies all the hypotheses of Theorem 5.72 when f is replaced by g. Hence by Theorem 5.72, there exists an $\hat{y} \in K$ such that $g(x, \hat{y}) \leq 0$ for all $x \in X$. This implies $f(x, \hat{y}) \leq t$ for all $x \in X$, so that $\sup_{x \in X} f(x, \hat{y}) \leq t$ and therefore

$$\min_{y \in K} \sup_{x \in X} f(x, y) \leq \sup_{x \in X} f(x, \hat{y}) \leq t = \sup_{x \in X} f(x, x),$$

i.e.,

$$\min_{y \in K} \sup_{x \in X} f(x, y) \leq \sup_{x \in X} f(x, x).$$ □

Theorem 5.73 generalizes Theorem 5.70 in several ways.

Theorem 5.74 *Let E be a topological vector space, X be a non-empty convex subset of E. Let $f, g : X \times X \longrightarrow \mathbb{R} \cup \{-\infty, +\infty\}$ be such that*

(a) *$f(x, y) \leq g(x, y)$ for all $x, y \in X$ and $g(x, x) \leq 0$ for all $x \in X$;*
(b) *for each $A \in \mathcal{F}(X)$ and each fixed $x \in co(A)$, $y \longmapsto f(x, y)$ is lower semi-continuous on $co(A)$;*
(c) *for each $y \in X$, the set $\{x \in X : g(x, y) > 0\}$ is convex;*
(d) *for each $A \in \mathcal{F}(X)$ and each $x, y \in co(A)$ and every net $\{y_\alpha\}_{\alpha \in \Gamma}$ in X converging to y with $f(tx + (1-t)y, y_\alpha) \leq 0$ for all $\alpha \in \Gamma$ and all $t \in [0, 1]$, we have $f(x, y) \leq 0$;*
(e) *there exist a non-empty closed and compact subset K of X and $x_0 \in K$ such that $f(x_0, y) > 0$ for all $y \in X \backslash K$.*

Then there exists $\hat{y} \in K$ such that $f(x, \hat{y}) \leq 0$ for all $x \in X$.

Proof. It is easy to see that the conditions (a) and (c) here imply the condition (b) of Theorem 5.72 so that the conclusion follows. □

Note that Theorem 5.74 generalizes Theorem 1 of Shih and Tan in [Shih and Tan (1984), pp. 280–282].

Theorem 5.75 *Let E be topological vector space, C be a non-empty closed convex subset of E and $f : C \times C \to \mathbb{R}$ be such that*

(a) *$f(x, x) \leq 0$ for all $x \in C$;*
(b) *for each $A \in \mathcal{F}(C)$ and each fixed $x \in co(A)$, $y \longmapsto f(x, y)$ is lower semi-continuous on $co(A)$;*
(c) *for each $y \in C$, the set $\{x \in C : f(x, y) > 0\}$ is convex;*
(d) *for each $A \in \mathcal{F}(C)$ and each $x, y \in co(A)$ and every net $\{y_\alpha\}_{\alpha \in \Gamma}$ in C converging to y with $f(tx + (1-t)y, y_\alpha) \leq 0$ for all $\alpha \in \Gamma$ and all $t \in [0, 1]$, we have $f(x, y) \leq 0$;*
(e) *there exist a non-empty closed and compact subset L of E and $x_0 \in C \cap L$ such that $f(x_0, y) > 0$ for all $y \in C \setminus L$.*

Then there exists $\hat{y} \in C \cap L$ such that $f(x, \hat{y}) \leq 0$ for all $x \in C$.

Proof. Let $f = g$, $K = C \cap L$ and $X = C$ in Theorem 5.74, the conclusion follows. □

Theorem 5.75 improves Theorem 1 of Brézis et al. in [H. Brézis and Stampacchia (1972)]. Note that if the compact set L is a subset of C, C is not required to be closed in E in Theorem 5.75. Note also that in Theorem 1 of [H. Brézis and Stampacchia (1972)], the set C was not assumed to be closed in E. However this is false in general as is observed by Chowdhury and Tan by the following example in [Tan (1994), Example 1.3.14].

Example 5.10 Let $E = \mathbb{R}^2$, $C = \{(u,v) \in \mathbb{R}^2 : u^2 + v^2 \le 1, u, v > 0\}$, $L = \{(u,v) \in \mathbb{R}^2 : u^2 + v^2 \le \frac{1}{4}\}$, $x_0 = (\frac{1}{2\sqrt{2}}, \frac{1}{2\sqrt{2}})$, and $f : C \times C \to \mathbb{R}$ be defined by $f(x,y) = \|y\| - \|x\|$ for all $x, y \in C$. Then all the hypotheses of Theorem 1 in [H. Brézis and Stampacchia (1972)] are satisfied. However there does not exist $\hat{y} \in C \cap L$ such that $f(x, \hat{y}) \le 0$ for all $x \in C$.

The following is equivalent to Theorem 5.72:

Theorem 5.76 *Let E be a topological vector space, X be a non-empty convex subset of E, and $G : X \to 2^X \cup \{\emptyset\}$ be a set-valued map such that*

- (a_1) *for each $A \in \mathcal{F}(X)$ and each fixed $x \in co(A)$, $G^{-1}(x) \cap co(A) = \{y \in co(A) : x \in G(y)\}$ is open in $co(A)$;*
- (b_1) *for each $A \in \mathcal{F}(X)$ and each $y \in co(A)$, there exists $x \in A$ such that $x \notin G(y)$;*
- (c_1) *for each $A \in \mathcal{F}(X)$ and each $x, y \in co(A)$ and every net $\{y_\alpha\}_{\alpha\in\Gamma}$ in X converging to y such that $tx + (1-t)y \notin G(y_\alpha)$ for all $\alpha \in \Gamma$ and all $t \in [0,1]$, we have $x \notin G(y)$;*
- (d_1) *there exist a non-empty closed and compact subset K of X and $x_0 \in K$ such that $x_0 \in G(y)$ for all $y \in X \backslash K$.*

Then there exists a point $\hat{y} \in K$ such that $G(\hat{y}) = \emptyset$.

Sketch of Proof. Theorem 5.72 implies Theorem 5.76:

Define $f : X \times X \to \mathbb{R}$ by

$$f(x,y) = \begin{cases} 1, \text{ if } x \in G(y), \\ 0, \text{ if } x \notin G(y), \end{cases}$$

for all $x, y \in X$ and apply Theorem 5.72. The conclusion of Theorem 5.76 follows. □

Sketch of Proof. Theorem 5.76 implies Theorem 5.72:

Define $G : X \to 2^X$ by $G(x) = \{y \in X : f(x,y) > 0\}$ for all $x \in X$ and apply Theorem 5.76. The conclusion of Theorem 5.72 follows. □

The following fixed point theorem is an immediate consequence of Theorem 5.76:

Theorem 5.77 *Let X be a non-empty convex subset of a topological vector space E and let $G : X \to 2^X \cup \{\emptyset\}$ be a set-valued map such that*

- (a_2) *for each $A \in \mathcal{F}(X)$ and each fixed $x \in co(A)$, $G^{-1}(x) \cap co(A)$ is open in $co(A)$;*

(b_2) *for each* $y \in X$, $G(y)$ *is convex;*

(c_2) *for each* $A \in \mathcal{F}(X)$ *and each* $x, y \in co(A)$ *and every net* $\{y_\alpha\}_{\alpha\in\Gamma}$ *in* X *converging to* y *such that* $tx + (1-t)y \notin G(y_\alpha)$ *for all* $\alpha \in \Gamma$ *and for all* $t \in [0,1]$, *we have* $x \notin G(y)$;

(d_2) *there exists a non-empty closed and compact subset* K *of* X *and* $x_0 \in K$ *such that* $x_0 \in G(y)$ *for all* $y \in X\backslash K$;

(e_2) *for each* $y \in K$, $G(y) \neq \emptyset$.

Then there exists a point $y_0 \in X$ *such that* $y_0 \in G(y_0)$.

Proof. By Theorem 5.76, there exist $A \in \mathcal{F}(X)$ and $y \in co(A)$ such that $x \in G(y)$ for all $x \in A$. Thus $y \in co(A) \subset G(y)$ as $G(y)$ is convex. □

5.10.2 *Generalized Variational Inequalities*

In this section we shall present some results of Chowdhury and Tan on generalized variational inequalities in Chowdhury-Tan (1996). We shall first state a lemma in Chowdhury-Tan (1996) with its proof:

Lemma 5.31 *Let* E *be a Hausdorff topological vector space,* $A \in \mathcal{F}(E)$, $X = co(A)$ *and* $T : X \to 2^{E^*}$ *be upper semi-continuous from* X *to the weak*-topology on* E^* *such that each* $T(x)$ *is weak*-compact. Let* $f : X \times X \to \mathbb{R}$ *be defined by* $f(x,y) = \inf_{w\in T(y)} Re\langle w, y-x\rangle$ *for all* $x, y \in X$. *Then for each fixed* $x \in X$, $y \longmapsto f(x,y)$ *is lower semi-continuous on* X.

Proof. Let $\lambda \in \mathbb{R}$ be given and let $x \in X$ be arbitrarily fixed. Let $C_\lambda = \{y \in X : f(x,y) \le \lambda\}$. Suppose $\{y_\alpha\}_{\alpha\in\Gamma}$ is a net in C_λ and $y_0 \in X$ such that $y_\alpha \to y_0$. Then for each $\alpha \in \Gamma$, $\lambda \ge f(x, y_\alpha) = \inf_{w\in T(y_\alpha)} Re\langle w, y_\alpha - x\rangle$ so that by weak*-compactness of $T(y_\alpha)$, there exists $w_\alpha \in T(y_\alpha)$ such that $\lambda \ge \inf_{w\in T(y_\alpha)} Re\langle w, y_\alpha - x\rangle = Re\langle w_\alpha, y_\alpha - x\rangle$. Since T is upper semi-continuous from X to the weak*-topology on E^*, X is compact and each $T(z)$ is weak*-compact, $\cup_{z\in X} T(z)$ is also weak*-compact. Thus there is a subnet $\{w_{\alpha'}\}_{\alpha'\in\Gamma'}$ of $\{w_\alpha\}_{\alpha\in\Gamma}$ and $w_0 \in E^*$ with $w_{\alpha'} \to w_0$ in the weak*-topology. Again, as T is upper semi-continuous with weak$*$-closed values, $w_0 \in T(y_0)$.

Suppose $A = \{a_1, \ldots, a_n\}$ and let $t_1, \ldots, t_n \ge 0$ with $\sum_{i=1}^n t_i = 1$ such that $y_0 = \sum_{i=1}^n t_i a_i$. For each $\alpha' \in \Gamma$, let $t_1^{\alpha'}, \ldots, t_n^{\alpha'} \ge 0$ with $\sum_{i=1}^n t_i^{\alpha'} = 1$ such that $y_{\alpha'} = \sum_{i=1}^n t_i^{\alpha'} a_i$. Since E is Hausdorff and $y_{\alpha'} \to y_0$, we must have $t_i^{\alpha'} \to t_i$ for each $i = 1, \ldots, n$. Thus

$$\lambda \ge Re\langle w_{\alpha'}, y_{\alpha'} - x\rangle = \sum_{i=1}^n t_i^{\alpha'} Re\langle w_{\alpha'}, a_i - x\rangle \longrightarrow \sum_{i=1}^n t_i Re\langle w_0, a_i - x\rangle$$

$$= Re\langle w_0, \sum_{i=1}^n t_i(a_i - x)\rangle = Re\langle w_0, y_0 - x\rangle$$

so that $\lambda \geq \inf_{w\in T(y_0)} Re\langle w, y_0 - x\rangle$ and hence $y_0 \in C_\lambda$. Thus C_λ is closed in X for each $\lambda \in \mathbb{R}$. Therefore $y \longmapsto f(x,y)$ is lower semi-continuous on X. □

We remark here that in Lemma 5.31, T is only assumed to be upper semi- continuous from $X = co(A)$ to the weak*-topology on E^* and T is weak*-compact valued. If X is any non-empty compact subset of E, strong topology on E^*, strongly-compact-valued are generally required, see e.g. [Shih and Tan (1989), Lemma 2].

The following result is Lemma 4.3 in [Chowdhury and Tan (1997b)]:

Lemma 5.32 *Let E be a topological vector space, X be a non-empty convex subset of E and $h : E \to \mathbb{R}$ be convex. Suppose $\hat{y} \in X$ and $\hat{w} \in E^*$ are such that $Re\langle \hat{w}, \hat{y} - x\rangle \leq h(x) - h(\hat{y})$ for all $x \in X$, then $Re\langle \hat{w}, \hat{y} - x\rangle \leq h(x) - h(\hat{y})$ for all $x \in I_X(\hat{y})$.*

We shall now present the following result in Chowdhury and Tan (1996):

Theorem 5.78 *Let X be a non-empty convex subset of a Hausdorff topological vector space E and $h : E \to \mathbb{R}$ be convex. Let $T : X \to 2^{E^*}$ be h-pseudo-monotone (respectively, h-demi-monotone) and be upper semi-continuous from $co(A)$ to the weak*-topology on E^* for each $A \in \mathcal{F}(X)$ such that each $T(x)$ is weak*-compact convex. Suppose there exist a non-empty compact (respectively, weakly closed and weakly compact) subset K of X and $x_0 \in K$ such that for each $y \in X \setminus K$, $\min_{w\in T(y)} Re\langle w, y - x_0\rangle + h(y) - h(x_0) > 0$. Then there exist $\hat{y} \in K$ and $\hat{w} \in T(\hat{y})$ such that $Re\langle \hat{w}, \hat{y} - x\rangle \leq h(x) - h(\hat{y})$ for all $x \in I_X(\hat{y})$.*

Proof. We first note that for each $A \in \mathcal{F}(X)$, h is continuous on $co(A)$ (see e.g. [14, Corollary 10.1.1, p. 83]). Define $\phi : X \times X \to \mathbb{R}$ by $\phi(x,y) = \min_{w\in T(y)} Re\langle w, y-x\rangle$, for each $x, y \in X$. Then we have the following:

(*a*) For each $A \in \mathcal{F}(X)$ and each fixed $x \in co(A)$, since E is Hausdorff and $co(A)$ is compact, and the relative weak topology on $co(A)$ coincide with its relative topology, it follows that $y \longmapsto \phi(x,y)$ is lower semi-continuous (respectively, weakly lower semi-continuous) on $co(A)$, by Lemma 5.31.

(*b*) Clearly, for each $x \in X$, $\phi(x,x) = 0$ and for each fixed $y \in X$, $x \longmapsto \phi(x,y)$ is quasi-concave. It follows that for each $A \in \mathcal{F}(X)$ and each $y \in co(A)$, $\min_{x\in A}[\phi(x,y) + h(y) - h(x)] \leq 0$.

(*c*) Suppose $A \in \mathcal{F}(X)$, $x, y \in co(A)$ and $\{y_\alpha\}_{\alpha\in\Gamma}$ is a net in X converging to y (respectively, weakly to y) with

$$\phi(tx+(1-t)y, y_\alpha) + h(y_\alpha) - h(tx+(1-t)y) \leq 0 \quad \text{for all} \quad \alpha \in \Gamma \quad \text{and all} \quad t \in [0,1].$$

Then for $t = 0$ we have $\phi(y, y_\alpha) + h(y_\alpha) - h(y) \leq 0$ for all $\alpha \in \Gamma$, i.e., $\min_{w\in T(y_\alpha)} Re\langle w, y_\alpha - y\rangle + h(y_\alpha) - h(y) \leq 0$ for all $\alpha \in \Gamma$. Hence

$$\limsup_{\alpha} \left[\min_{w\in T(y_\alpha)} Re\langle w, y_\alpha - y\rangle + h(y_\alpha) - h(y) \right] \leq 0.$$

Since T is h-pseudo-monotone (respectively, h-demi-monotone), we have

$$\limsup_{\alpha}\left[\min_{w\in T(y_\alpha)} Re\langle w, y_\alpha - x\rangle + h(y_\alpha) - h(x)\right] \geq \inf_{w\in T(y)} Re\langle w, y-x\rangle + h(y) - h(x). \tag{5.119}$$

For $t = 1$ we also have $\phi(x, y_\alpha) + h(y_\alpha) - h(x) \leq 0$ for all $\alpha \in \Gamma$, i.e., $\min_{w\in T(y_\alpha)} Re\langle w, y_\alpha - x\rangle + h(y_\alpha) - h(x) \leq 0$ for all $\alpha \in \Gamma$. It follows that

$$\limsup_{\alpha}\left[\min_{w\in T(y_\alpha)} Re\langle w, y_\alpha - x\rangle + h(y_\alpha) - h(x)\right] \leq 0. \tag{5.120}$$

By (5.119) and (5.120), $\phi(x, y) + h(y) - h(x) \leq 0$.

(*d*) By assumption, K is a compact (respectively, weakly closed and weakly compact) subset of X and $x_0 \in K$ such that for each $y \in X\backslash K$, $\min_{w\in T(y)} Re\langle w, y - x_0\rangle + h(y) - h(x_0) > 0$, i.e., $\phi(x_0, y) + h(y) - h(x_0) > 0$.

(If T is h-demi-monotone, we equip E with the weak topology.) Then ϕ satisfies all the hypotheses of Theorem 5.71. Hence by Theorem 5.71, there exists a point $\hat{y} \in K$ with $\phi(x, \hat{y}) \leq h(x) - h(\hat{y})$ for all $x \in X$; in other words, $\min_{w\in T(\hat{y})} Re\langle w, \hat{y} - x\rangle \leq h(x) - h(\hat{y})$ for all $x \in X$.

Define $f : X \times T(\hat{y}) \longrightarrow \mathbb{R}$ by

$$f(x, w) = Re\langle w, \hat{y} - x\rangle + h(\hat{y}) - h(x) \text{ for all } x \in X \text{ and for all } w \in T(\hat{y}).$$

Note that for each fixed $x \in X$, $w \longmapsto f(x, w)$ is weak* continuous and convex and for each fixed $w \in T(\hat{y})$, $x \longmapsto f(x, w)$ is concave. Thus by Theorem 5.44 we have

$$\min_{w\in T(\hat{y})}\sup_{x\in X}(Re\langle w, \hat{y}-x\rangle + h(\hat{y}) - h(x)) = \sup_{x\in X}\min_{w\in T(\hat{y})}(Re\langle w, \hat{y}-x\rangle + h(\hat{y}) - h(x)) \leq 0.$$

Hence there exists a point $\hat{w} \in T(\hat{y})$ such that

$$Re\langle \hat{w}, \hat{y} - x\rangle \leq h(x) - h(\hat{y}) \quad \text{for all} \quad x \in X. \tag{5.121}$$

Since h is defined on all of E and is convex, by (4.3) and Lemma 5.32, we have $Re\langle \hat{w}, \hat{y} - x\rangle \leq h(x) - h(\hat{y})$ for all $x \in I_X(\hat{y})$. □

Remark 5.29 If T is h-pseudo-monotone (respectively, h-demi-monotone), Theorem 5.78 generalizes (respectively, extends or improves) Application 3 in [H. Brézis and Stampacchia (1972), p. 297] in the following ways: (1), (2) and (3) (respectively, following ways (1) and (2)):

(1) T is set-valued and upper semi-continuous from $co(A)$ to the weak* topology on E^* for each $A \in \mathcal{F}(X)$ instead of single-valued and continuous on any finite dimensional subspace;

(2) h need not be lower semi-continuous on X;

(3) As noted earlier, the definition of pseudo-monotonicity, even in the single-valued case, is more general.

The results from 5.79 to 5.82 have been taken from Chowdhury and Tan (1996).

Theorem 5.79 *Let $(E, \|\cdot\|)$ be a reflexive Banach space, X be a non-empty closed convex subset of E and $h : E \to \mathbb{R}$ be convex. Let $T : X \to 2^{E^*}$ be h-demi-monotone and be upper semi-continuous from $co(A)$ to the weak topology on E^* for each $A \in \mathcal{F}(X)$ such that each $T(x)$ is weakly compact convex. Suppose there is $x_0 \in X$ such that*

$$\lim_{\substack{\|y\|\to\infty \\ y\in X}} \inf_{w\in T(y)} Re\langle w, y - x_0\rangle + h(y) - h(x_0) > 0. \tag{5.122}$$

Then there exist $\hat{y} \in X$ and $\hat{w} \in T(\hat{y})$ such that $Re\langle \hat{w}, \hat{y} - x\rangle \leq h(x) - h(\hat{y})$ for all $x \in I_X(\hat{y})$.

Proof. Let $\alpha = \lim_{\substack{\|y\|\to\infty \\ y\in X}} \inf_{w\in T(y)} Re\langle w, y - x_0\rangle + h(y) - h(x_0)$. Then by (5.122), $\alpha > 0$. Let $M > 0$ be such that $\|x_0\| \leq M$ and $\inf_{w\in T(y)} Re\langle w, y - x_0\rangle + h(y) - h(x_0) > \frac{\alpha}{2}$ for all $y \in X$ with $\|y\| > M$. Let $K = \{x \in X : \|x\| \leq M\}$; then K is a non-empty weakly compact subset of X. Note that for any $y \in X \setminus K$, $\inf_{w\in T(y)} Re\langle w, y - x_0\rangle + h(y) - h(x_0) > \frac{\alpha}{2} > 0$. The conclusion now follows from Theorem 5.78. □

By taking $h \equiv 0$ in Theorem 5.78 and applying Lemma 1.1, the following theorem is obtained on existence theorem of a generalized complementarity problem:

Theorem 5.80 *Let X be a cone in a Hausdorff topological vector space E. Let $T : X \to 2^{E^*}$ be pseudo-monotone (respectively, demi-monotone) and be upper semi-continuous from $co(A)$ to the weak*-topology on E^* for each $A \in \mathcal{F}(X)$ such that each $T(x)$ is weak*-compact convex. Suppose there exist a non-empty compact (respectively, weakly closed and weakly compact) subset K of X and $x_0 \in K$ such that for each $y \in X\backslash K$, $\inf_{w\in T(y)} Re\langle w, y - x_0\rangle > 0$. Then there exist $\hat{y} \in K$ and $\hat{w} \in T(\hat{y})$ such that $Re\langle \hat{w}, \hat{y}\rangle = 0$ and $\hat{w} \in \widehat{X}$.*

By taking $h \equiv 0$ in Theorem 5.79 and applying Lemma 1.1, the following theorem is obtained on existence theorem of a generalized complementarity problem:

Theorem 5.81 *Let $(E, \|\cdot\|)$ be a reflexive Banach space, X be a closed cone in E and $T : X \to 2^{E^*}$ be demi-monotone and be upper semi-continuous from $co(A)$ to the weak topology on E^* for each $A \in \mathcal{F}(X)$ such that each $T(x)$ is weakly compact convex. Suppose there is $x_0 \in X$ such that*

$$\lim_{\substack{\|y\|\to\infty \\ y\in X}} \inf_{w\in T(y)} Re\langle w, y - x_0\rangle > 0.$$

Then there exist $\hat{y} \in X$ and $\hat{w} \in T(\hat{y})$ such that $Re\langle \hat{w}, \hat{y}\rangle = 0$ and $\hat{w} \in \widehat{X}$.

Finally, a result in Chowdhury and Tan (1996) on the surjectivity of demi-monotone operators is presented:

Theorem 5.82 *Let $(E, \|\cdot\|)$ be a reflexive Banach space, X be a non-empty closed convex subset of E and $T : X \to 2^{E^*}$ be demi-monotone and be upper semi-continuous from $co(A)$ to the weak topology on E^* for each $A \in \mathcal{F}(X)$ such that each $T(x)$ is weakly compact convex. Suppose there is $x_0 \in X$ such that*

$$\lim_{\substack{\|y\|\to\infty \\ y\in X}} \inf_{w\in T(y)} Re\langle w, y - x_0\rangle/\|y\| = \infty.$$

Then for each given $w_0 \in E^*$, there exist $\hat{y} \in X$ and $\hat{w} \in T(\hat{y})$ such that $Re\langle \hat{w} - w_0, \hat{y} - x\rangle \leq 0$ for all $x \in X$. In particular, if $X = E$, then T is surjective.

Proof. Let $w_0 \in E^*$ be given. Then

$$\lim_{\substack{\|y\|\to\infty \\ y\in X}} \left(\inf_{w\in T(y)} Re\langle w - w_0, y - x_0\rangle/\|y\|\right)$$

$$= \lim_{\substack{\|y\|\to\infty \\ y\in X}} \left(\left(\inf_{w\in T(y)} Re\langle w, y - x_0\rangle/\|y\|\right) - \|w_0\|\right) = \infty.$$

Define $T^* : X \to 2^{E^*}$ by $T^*(x) = T(x) - w_0$ for all $x \in X$. Then T^* is upper semi-continuous from $co(A)$ to the weak topology on E^* for each $A \in \mathcal{F}(X)$ such that each $T^*(x)$ is weakly compact convex and

$$\lim_{\substack{\|y\|\to\infty \\ y\in X}} \inf_{w\in T^*(y)} Re\langle w, y - x_0\rangle/\|y\| = \infty.$$

Suppose $y \in X$ and $\{y_\alpha\}_{\alpha\in\Gamma}$ is a net in X converging weakly to y with

$$\limsup_{\alpha} \left[\inf_{u\in T^*(y_\alpha)} Re\langle u, y_\alpha - y\rangle\right] \leq 0.$$

It follows that

$$\limsup_{\alpha} \left[\inf_{u\in T(y_\alpha)} Re\langle u, y_\alpha - y\rangle\right]$$

$$\leq \limsup_{\alpha} \left[\inf_{u\in T(y_\alpha)} Re\langle u - w_0, y_\alpha - y\rangle\right] + \limsup_{\alpha} Re\langle w_0, y_\alpha - y\rangle$$

$$= \limsup_{\alpha} \left[\inf_{u\in T^*(y_\alpha)} Re\langle u, y_\alpha - y\rangle\right] \leq 0.$$

Since T is demi-monotone,

$$\limsup_{\alpha} \left[\inf_{u\in T(y_\alpha)} Re\langle u, y_\alpha - x\rangle\right] \geq \inf_{w\in T(y)} Re\langle w, y - x\rangle \quad \text{for all} \quad x \in X.$$

Hence for each $x \in X$,

$$\begin{aligned}
&\inf_{w\in T^*(y)} Re\langle w, y-x\rangle \\
&= \inf_{w\in T(y)} Re\langle w-w_0, y-x\rangle \\
&= \inf_{w\in T(y)} Re\langle w, y-x\rangle - Re\langle w_0, y-x\rangle \\
&\le \limsup_{\alpha}\left[\inf_{u\in T(y_\alpha)} Re\langle u, y_\alpha - x\rangle\right] - Re\langle w_0, y-x\rangle \\
&\le \limsup_{\alpha}\left[\inf_{u\in T(y_\alpha)} Re\langle u-w_0, y_\alpha - x\rangle + Re\langle w_0, y_\alpha - x\rangle\right] - Re\langle w_0, y-x\rangle \\
&\le \limsup_{\alpha}\left[\inf_{u\in T^*(y_\alpha)} Re\langle u, y_\alpha - x\rangle\right] + \limsup_{\alpha} Re\langle w_0, y_\alpha - x\rangle - Re\langle w_0, y-x\rangle \\
&= \limsup_{\alpha}\left[\inf_{u\in T^*(y_\alpha)} Re\langle u, y_\alpha - x\rangle\right].
\end{aligned}$$

Therefore T^* is also demi-monotone.

Hence by Theorem 5.79 with $h \equiv 0$, there exist $\hat{y} \in X$ and $\bar{w} \in T^*(\hat{y})$ such that $Re\langle \bar{w}, \hat{y} - x\rangle \le 0$ for all $x \in X$. But then there exists $\hat{w} \in T(\hat{y})$ with $\bar{w} = \hat{w} - w_0$ so that $Re\langle \hat{w} - w_0, \hat{y} - x\rangle \le 0$ for all $x \in X$. Now if $X = E$, then $\hat{w} - w_0 = 0$ so that $w_0 = \hat{w} \in T(\hat{y})$. This shows that T is surjective. □

5.10.3 *Applications to Fixed Point Theorems*

In the following results in Chowdhury and Tan (1996), H will denote a Hilbert space with inner product $\langle\ ,\ \rangle$ and its corresponding induced norm $\|\cdot\|$.

The following fixed point theorem is obtained in Chowdhury and Tan (1996) as an application of Theorem 5.78:

Theorem 5.83 *Let X be a non-empty convex subset of H and $T : X \to 2^H$ be upper semi-continuous from $co(A)$ to the weak topology on H for each $A \in \mathcal{F}(X)$ such that each $T(x)$ is weakly compact convex and $I - T$ is pseudo-monotone (respectively, demi-monotone). Suppose there exist a non-empty compact (respectively, weakly compact) subset K of X and $x_0 \in K$ such that for each $y \in X \setminus K$, $\inf_{w\in T(y)} Re\langle y-w, y-x_0\rangle > 0$. Then there exist $\hat{y} \in K$ and $\hat{w} \in T(\hat{y})$ such that*

$$Re\langle \hat{y} - \hat{w}, \hat{y} - x\rangle \le 0 \quad \textit{for all} \quad x \in \overline{I_X(\hat{y})}.$$

Moreover, if either $\hat{y}$ is an interior point of X in H or $p(\hat{y}) \in \overline{I_X(\hat{y})}$, where $p(\hat{y})$ is the projection of $\hat{y}$ on $T(\hat{y})$, then $\hat{y}$ is a fixed point of T, i.e., $\hat{y} \in T(\hat{y})$.

Proof. (If $I - T$ is demi-monotone, we equip H with the weak topology.) Since T is upper semi-continuous from $co(A)$ to the weak topology on H for each $A \in \mathcal{F}(X)$,

$I - T : X \to 2^H$ is also upper semi-continuous from $co(A)$ to the weak topology on H for each $A \in \mathcal{F}(X)$ and satisfies all the hypotheses of Theorem 5.78 with $h \equiv 0$. By Theorem 5.78, there exist $\hat{y} \in K$ and $\hat{w} \in T(\hat{y})$ such that $Re\langle \hat{y} - \hat{w}, \hat{y} - x\rangle \le 0$ for all $x \in I_X(\hat{y})$. By continuity,

$$Re\langle \hat{y} - \hat{w}, \hat{y} - x\rangle \le 0 \text{ for all } x \in \overline{I_X(\hat{y})}. \tag{5.123}$$

Case 1. Suppose $\hat{y}$ is an interior point of X in H, *i.e.*, $\hat{y} \in int_H X$, then there exists $r > 0$ such that $B_r(\hat{y}) \subset X$. Then for each $z \in H$ with $z \neq \hat{y}$, let $u = \hat{y} + \frac{r}{2} \cdot \frac{\hat{y}-z}{\|\hat{y}-z\|}$, then $u \in B_r(\hat{y}) \subset X \subset I_X(\hat{y})$. Thus $Re\langle \hat{y} - \hat{w}, \frac{r}{2} \cdot \frac{z-\hat{y}}{\|\hat{y}-z\|}\rangle \le 0$ so that $\frac{r}{2\|\hat{y}-z\|} Re\langle \hat{y} - \hat{w}, z - \hat{y}\rangle \le 0$ and hence $\mathrm{Re}\langle \hat{y} - \hat{w}, z - \hat{y}\rangle \le 0$ for all $z \in H$.

It follows that $Re\langle \hat{y} - \hat{w}, z\rangle = 0$ for all $z \in H$ so that $\hat{y} = \hat{w} \in T(\hat{y})$.

Case 2. Suppose $p(\hat{y}) \in \overline{I_X(\hat{y})}$. By Proposition 1.1, the projection $p(\hat{y})$ of $\hat{y}$ on $T(\hat{y})$ has the following property:

$$p(\hat{y}) \in T(\hat{y}) \text{ and } Re\langle \hat{y} - p(\hat{y}), w - p(\hat{y})\rangle \le 0 \text{ for all } w \in T(\hat{y}). \tag{5.124}$$

Since $\hat{w} \in T(\hat{y})$, by (5.124) we have

$$\begin{aligned} 0 &\le Re\langle p(\hat{y}) - \hat{y}, \hat{w} - p(\hat{y})\rangle \\ &= Re\langle p(\hat{y}) - \hat{y}, \hat{w} - \hat{y} + \hat{y} - p(\hat{y})\rangle \\ &= Re\langle p(\hat{y}) - \hat{y}, \hat{w} - \hat{y}\rangle - \|\hat{y} - p(\hat{y})\|^2. \end{aligned}$$

Therefore

$$\|\hat{y} - p(\hat{y})\|^2 \le Re\langle \hat{y} - \hat{w}, \hat{y} - p(\hat{y})\rangle \le 0 \ \text{ by(5.123)}$$

Thus $\hat{y} = p(\hat{y}) \in T(\hat{y})$. □

The following fixed point theorem obtained in Chowdhury and Tan (1996) is an immediate consequence of Theorem 5.83:

Theorem 5.84 *Let X be a non-empty convex subset of H and $T : X \to bc(H)$ be upper semi-continuous from $co(A)$ to the weak topology on H for each $A \in \mathcal{F}(X)$ such that each $T(x)$ is weakly compact convex and $I - T$ is pseudo-monotone (respectively, demi-monotone). Suppose there exist a non-empty compact (respectively, weakly compact) subset K of X and $x_0 \in K$ such that (i) for each $y \in K \cap \partial_H(X)$, $\pi_{T(y)}(y) \in \overline{I_X(y)}$ and (ii) for each $y \in X \setminus K$, $\inf_{w \in T(y)} Re\langle y - w, y - x_0\rangle > 0$. Then T has a fixed point in K.*

Corollary 5.84.1 *Let X be a non-empty compact (respectively, bounded closed) convex subset of H and $T : X \to bc(H)$ be upper semi-continuous from $co(A)$ to the weak topology on H for each $A \in \mathcal{F}(X)$ such that each $T(x)$ is weakly compact convex and $I - T$ is pseudo-monotone (respectively, demi-monotone). Suppose that for each $y \in \partial_H(X)$, $\pi_{T(y)}(y) \in \overline{I_X(y)}$. Then T has a fixed point in X.*

Corollary 5.84.2 *Let X be a non-empty compact (respectively, bounded closed) convex subset of H and $T : X \to bc(X)$ be upper semi-continuous from $co(A)$ to the relative weak topology on X for each $A \in \mathcal{F}(X)$ such that each $T(x)$ is weakly compact convex and $I - T$ is pseudo-monotone (respectively, demi-monotone). Then T has a fixed point in X.*

Note that the results of this section have been written and presented after a thorough revision of some results of an article in Chowdhury and Tan (1996).

5.11 Generalized Variational-Like Inequalities for Pseudo-Monotone Type I Operators

The generalized variational-like inequalities were used by Chowdhury and Tan in the simplest form in Chowdhury and Tan (1996) using the name of the operators as generalized variational inequalities (GVI). Later, Tarafdar and Ding gave a generalization of these GVI and called these operators as generalized variational-like inequalities (GVLI) in their paper in Archiv der Mathematik mentioned above. Ding and Tarafdar's generalized variational-like inequality was introduced in problem (1.1).

In this section, E and F will denote vector spaces over a scalar field Φ (either the real field or the complex field). We shall denote by 2^F the family of all subsets of F.

Let X be a nonempty subset of E and $\langle\ ,\ \rangle : F \times E \to \Phi$ be a bilinear functional satisfying the following property:

(P) the family of $\{\langle \cdot, x\rangle\}_{x \in E}$ separates the points of F.

Unless otherwise stated, E, F and $\langle \cdot, \cdot\rangle$ will be assumed to satisfy the property (P) in all results of this section.

In all results of this section related to the problem (1.1), the underlying space F is equipped with the $\sigma(F, E)$-topology.

In Ding and Tarafdar (2000), a slight generalization of Chowdhury and Tan's definition of h-pseudo-monotone type I operators or pseudo-monotone type I operators was given. We shall call all such operators (η, h)-pseudo-monotone type I (resectively, a strong (η, h)-pseudo-monotone type I) operators. We introduce these operators as follows:

Definition 5.16 Let X be a nonempty subset of a topological vector space E over Φ and F be a vector space over Φ which is equipped with the $\sigma(F, E)$-topology. Let $T : X \to 2^F \setminus \{\phi\}$, $\eta : X \times X \to E$ and $h : X \times X \to \mathbb{R}$. Then T is said to be an (η, h)-pseudo-monotone type I (respectively, a strong (η, h)-pseudo-monotone type I) operators, if for each $y \in X$ and every net $\{y_\alpha\}_{\alpha \in \Gamma}$ in X converging to y

(respectively, weakly to y) with

$$\limsup_{\alpha}[\inf_{w\in T(y_\alpha)} Re\langle w,\, \eta(y_\alpha,\, y)\rangle + h(y_\alpha,\, y)] \leq 0,$$

we have

$$\limsup_{\alpha}[\inf_{w\in T(y_\alpha)} Re\langle w,\, \eta(y_\alpha,\, x)\rangle + h(y_\alpha,\, x)] \tag{5.125}$$

$$\geq \inf_{w\in T(y)} Re\langle w,\, \eta(y,\, x)\rangle + h(y,\, x) \quad \forall x \in X. \tag{5.126}$$

T is said to be an h-pseudo-monotone type I (respectively, a strong h-pseudo-monotone type I) operators, if T is an $(\eta,\, h)$-pseudo-monotone type I (respectively, a strong $(\eta,\, h)$-pseudo-monotone type I) operators, with $\eta(x,\, y) = x - y$ and for some $h' : X \to \mathbb{R}$, $h(x, y) = h'(x) - h'(y)$ for all $x,\, y \in X$. Note that if $F = E^*$, the topological dual space of E, then the notions of the h-pseudo-monotonicity of type I (respectively, strong h-pseudo-monotonicity of type I) operators coincide with those in Chowdhury and Tan (1996).

Definition 5.17 Let $T : X \to 2^F \setminus \{\phi\}$, $\eta : X \times X \to E$ and $g : X \to E$. The mappings T and η are said to have 0-diagonally concave relation (in short, 0-DCVR) if the function $\phi : X \times X \to \mathbb{R} \cup \{\pm\infty\}$ defined by

$$\phi(x,\, y) = \inf_{w\in T(x)} Re\langle w,\, \eta(x,\, y)\rangle \tag{5.127}$$

is 0-DCV in y. The mappings T and g are said to have 0-diagonally concave relation if T and $\eta(x,\, y) = g(x) - g(y)$ have the 0-DCVR.

We shall now present some lemmas from Ding and Tarafdar (2000):

Lemma 5.33 *Let $T : X \to 2^F \setminus \{\phi\}$ and $\eta : X \times X \to E$ be such that for each fixed $x \in X$, $\inf_{u\in T(x)} \langle u,\, \eta(x,\, x)\rangle = 0$ and $\eta(x,\, \cdot)$ is an affine mapping. Then T and η have 0-DCVR.*

Proof. By the assumptions on T and η, for any finite set $\{y_1, \ldots, y_m\} \subset X$ and for any $y_0 = \sum_{i=1}^{m} \lambda_i y_i$ ($\lambda_i \geq 0$, $\sum_{i=1}^{m} \lambda_i = 1$), we have

$$\begin{aligned}\sum_{i=1}^{m} \lambda_i \phi(y_0,\, y_i) &= \sum_{i=1}^{m} \lambda_i [\inf_{w\in T(y_0)} \langle w,\, \eta(y_0,\, y_i)\rangle] \\ &\leq \inf_{w\in T(y_0)} [\langle w,\, \sum_{i=1}^{m} \lambda_i \eta(y_0,\, y_i)\rangle] \\ &\leq \inf_{w\in T(y_0)} \langle w,\, \eta(y_0,\, y_0)\rangle = 0.\end{aligned}$$

This shows that T and η have 0-DCVR.

The following is another lemma in (Ding and Tarafdar (2000)). We shall also give the proof of this lemma exactly given in (Ding and Tarafdar (2000)). However, there is a lacuna in its proof in (Ding and Tarafdar (2000)): the point-wise boundedness was not proved which is an essential part in applying the Banach-Steinhauss' Theorem.

Lemma 5.34 *Let E be a topological vector space over Φ, $A \in \mathcal{F}(E)$ and $X = \mathrm{co}(A)$. Let F be a vector space over Φ which is equipped with $\sigma(F, E)$-topology such that for each $w \in F$, $x \mapsto \langle w, x\rangle$ is continuous. Let $\eta : X \times X \to E$ be continuous in first argument and $T : X \to 2^F \setminus \{\phi\}$ be u.s.c. from X to the $\sigma(F, E)$-topology on F such that each $T(x)$ is $\sigma(F, E)$-compact. Let $\phi : X \times X \to \mathbb{R}$ be defined by*

$$\phi(x, y) = \inf_{w \in T(y)} \langle w, \eta(y, x)\rangle.$$

Then for each fixed $x \in X$, the function $y \mapsto \phi(x, y)$ is lower semicontinuous on X.

Proof. For given $\lambda \in \mathbb{R}$ and arbitrary fixed $x \in X$, let

$$C_\lambda = \{y \in X : \phi(x, y) \leq \lambda\}.$$

Suppose that $\{y_\alpha\}_{\alpha\in\Gamma} \subset C_\lambda$ is a net such that $y_\alpha \to y_0 \in X$, then for each $\alpha \in \Gamma$,

$$\inf_{w \in T(y_\alpha)} \langle w, \eta(y_\alpha, x)\rangle = \phi(x, y_\alpha) \leq \lambda.$$

By the definition of the $\sigma(F, E)$-topology of F, it is known that for each $x \in E$, the function $w \mapsto \langle w, x\rangle$ is continuous. By the continuity of $\eta(\cdot, x)$, we obtain $\eta(y_\alpha, x) \to \eta(y_0, x)$. Since each $T(y_\alpha)$ is $\sigma(F, E)$-compact, there exists $w_\alpha \in T(y_\alpha)$ such that

$$\langle w_\alpha, \eta(y_\alpha, x)\rangle = \inf_{w \in T(y_\alpha)} \langle w, \eta(y_\alpha, x)\rangle \leq \lambda.$$

Since T is u.s.c. from X to the $\sigma(F, E)$-topology on F, X is compact and each $T(x)$ is $\sigma(F, E)$-compact, it follows from Proposition 3.1.11 of Aubin and Ekeland Aubin and Ekeland (1984) that the set $\bigcap_{y\in X} T(y)$ is also $\sigma(F, E)$-compact. Hence there exists a subnet $\{w_\beta\}$ of $\{w_\alpha\}$ such that $\{w_\beta\}$ converges to a point $w_0 \in \bigcap_{y\in X} T(y)$. By the upper semi-continuity of T, we have $w_0 \in T(y_0)$. Since $w_\beta \to w_0$ in the $\sigma(F, E)$-topology, we have

$$\langle w_\beta - w_0, \eta(y_0, x)\rangle \to 0.$$

Since $\eta(y_\beta, x) \to \eta(y_0, x)$ and for each $w \in F$, the function $x \mapsto \langle w, x\rangle$ is continuous, we have

$$\langle w_0, \eta(y_\beta, x) - \eta(y_0, x)\rangle \to 0.$$

It follows from the compactness of X that it is second category, by Banach-Steinhaus theorem (see, Theorem 2.5 in Rudin (1973)), the family of functions $\{(w_\beta - w_0, \cdot\rangle\}$ is equi-continuous on X and hence

$$\langle w_\beta - w_0, \eta(y_\beta, x) - \eta(y_0, x)\rangle \to 0.$$

It follows that

$$\begin{aligned}
&|\langle w_\beta, \eta(y_\beta, x)\rangle - \langle w_0, \eta(y_0, x)\rangle| \\
&\quad = |\langle w_\beta - w_0, \eta(y_0, x)\rangle + \langle w_\beta - w_0, \eta(y_\beta, x) - \eta(y_0, x)\rangle \\
&\qquad + \langle w_0, \eta(y_\beta, x) - \eta(y_0, x)\rangle| \\
&\quad \leq |\langle w_\beta - w_0, \eta(y_0, x)\rangle| + |\langle w_\beta - w_0, \eta(y_\beta, x) - \eta(y_0, x)\rangle| \\
&\qquad + |\langle w_0, \eta(y_\beta, x) - \eta(y_0, x)\rangle| \to 0.
\end{aligned}$$

Thus we obtain that for each $x \in X$,

$$\phi(x, y_0) = \inf_{w \in T(y_0)} \langle w, \eta(y_0, x)\rangle \leq \langle w_0, \eta(y_0, x)\rangle = \lim_\beta \langle w_\beta, \eta(y_\beta, x)\rangle \leq \lambda.$$

Therefore $y_0 \in C_\lambda$ and C_λ is closed in X. Hence for each fixed $x \in X$, the function $y \mapsto \phi(x, y)$ is lower semi-continuous on X. □

Remark 5.30 If $F = E^*$, the topological dual space of E, and $\eta(x, y) = x - y$ for all $x, y \in X$, then Lemma 2.2 reduces to Lemma 3 in [Chowdhury and Tan (1996)].

The following result is a slightly improved form of Lemma 2 in Chowdhury and Tan (1996)].

Lemma 5.35 *Let X be a nonempty convex subset of a topological vector space E and $G : X \to 2^X \setminus \{\phi\}$ be such that*

(i) for some $M \in \mathcal{F}(X)$, $\mathrm{cl}_X(\bigcap_{x \in M} G(x))$ is compact,

(ii) for each $A \in \mathcal{F}(X)$ with $M \subset A$ and for $x \in \mathrm{co}(A)$, $G(x) \cap \mathrm{co}(A)$ is closed in $\mathrm{co}(A)$,

(iii) G is a KKM mapping, i.e., $\mathrm{co}(A) \subset \bigcap_{x \in A} G(x)$ for all $A \subset \mathcal{F}(X)$,

(iv) for each $A \in \mathcal{F}(X)$ with $M \subset A$,

$$(\mathrm{cl}_X(\bigcap_{x \in \mathrm{co}(A)} G(x))) \bigcap \mathrm{co}(A) = (\bigcap_{x \in \mathrm{co}(A)} G(x)) \bigcap \mathrm{co}(A).$$

Then $\bigcap_{x \in X} G(x) \neq \emptyset$.

Proof. If x_0 and $\mathrm{cl}_X G(x_0)$ are replaced by M and $\mathrm{cl}_X(\bigcap_{x \in M} G(x))$ respectively in the proof of Lemma 2 of Chowdhury and Tan Chowdhury and Tan (1996)], then by using similar argument, we can prove that the conclusion of Lemma 5.35 holds. □

Example 5.11 Let $X = E = \mathbb{R}$ and $G : X \to 2^X$ be defined as follows:

$$G(x) = \begin{cases} \mathbb{R}, & \text{if } x \in \mathbb{R} \setminus \{1, 2\} \\ [1, \infty) & \text{if } x = 1 \\ (-\infty, 2], & \text{if } x = 2. \end{cases}$$

By taking $M = \{1, 2\}$, it is easy to check that all conditions of Lemma 5.35 are satisfied and $\bigcap_{x \in X} G(x) = [1, 2] \neq \emptyset$. But for any $x \in X$, $\mathrm{cl}_X\, G(x)$ is not compact and hence Lemma 2 of Chowdhury and Tan [Chowdhury and Tan (1996)] cannot be applied.

5.11.1 *Existence Theorems for GV LI(T, η, h, X, F)*

In this section, we shall present some results of Ding and Tarafdar on existence theorems for the solutions to the $GV\,LI(T, \eta, h, X, F)$ involving an (η, h)-pseudo-monotone type I operator T with non-compact domain in Hausdorff topological vector spaces.

Theorem 5.85 *Let X be a nonempty convex subset of a Hausdorff topological vector space E over Φ and F be a vector space over Φ with the $\sigma(F, E)$-topology such that for each $w \in F$, the function $x \mapsto Re\langle w, x\rangle$ is continuous. Let $T : X \to 2^F \setminus \{\phi\}$, $\eta : X \times X \to E$ and $h : X \times X \to \mathbb{R}$ be mappings such that*

- *(i) for each $A \in \mathcal{F}(X)$, T is u.s.c. from $\mathrm{co}(A)$ to the $\sigma(F, E)$-topology on F such that for each $x \in X$, $T(x)$ is $\sigma(F, E)$-compact,*
- *(ii) T and η have the 0-DCVR,*
- *(iii) for each fixed $y \in X$, $\eta(\cdot, y)$ is continuous and $h(\cdot, y)$ is lower semi-continuous on $\mathrm{co}(A)$ for each $A \in \mathcal{F}(X)$, and for each fixed $x \in X$, $h(x, \cdot)$ is concave and $h(x, x) = 0$,*
- *(iv) T is an (η, h)-pseudo-monotone type I (respectively, a strong (η, h)-pseudo-monotone type I) operator,*

Suppose that there exist a nonempty compact (respectively, weakly compact and weakly closed) subset K of X and a finite set $M \in \mathcal{F}(X)$ such that for each $y \in X \setminus K$, there is an $x \in M$ satisfying

$$\inf_{w \in T(y)} Re\langle w, \eta(y, x)\rangle + h(y, x) > 0.$$

Then there exists a point $\hat{y} \in K$ such that

$$\inf_{w \in T(\hat{y})} Re\langle w, \eta(\hat{y}, x)\rangle + h(\hat{y}, x) \leq 0, \quad \forall x \in X. \tag{5.128}$$

If, in addition, $T(\hat{y})$ is convex and for each $x \in X$ and $\eta(x, \cdot)$ is affine then there exists a point $\hat{w} \in T(\hat{y})$ such that

$$Re\langle \hat{w}, \eta(\hat{y}, x)\rangle + h(\hat{y}, x) \leq 0, \quad \forall x \in X.$$

Proof. Define a set-valued mapping $G : X \to 2^X \setminus \{\phi\}$ by

$$G(x) = \{y \in X : \inf_{w \in T(y)} Re\langle w, \eta(y, x)\rangle + h(y, x) \leq 0\}.$$

For each $A \in \mathcal{F}(X)$ with $M \subset A$, since E is Hausdorff and co(A) is compact, and the relative weak topology on co(A) coincides with the relative topology on co(A), it follows from Lemma 5.34 and the condition (iii) that for each fixed $x \in X$, the function

$$y \mapsto \inf_{w \in T(y)} Re\langle w, \eta(y, x)\rangle + h(y, x)$$

is lower semi-continuous (respectively, weakly lower semicontinuous) on co(A) and so $G(x) \cap \text{co}(A)$ is closed (respectively, weakly closed) in co(A). Hence condition (ii) of Lemma 5.35 is satisfied.

We claim that G is an KKM mapping. If it is false, then there exist a finite set $A = \{x_1, \ldots, x_n\} \in \mathcal{F}(X)$ and $y = \sum_{i=1}^n \lambda_i x_i$ ($\lambda_i \geq 0$, $\sum_{i=1}^n \lambda = 1$) such that $y \notin \bigcap_{i=1}^n G(x_i)$. Then we have

$$\inf_{w \in T(y)} Re\langle w, \eta(y, x_i)\rangle + h(y, x_i) > 0, \forall i = 1, \ldots, n.$$

It follows from (iii) that

$$\begin{aligned} 0 &< \sum_{i=1}^n \lambda_i \inf_{w \in T(y)} Re\langle w, \eta(y, x_i)\rangle + \sum_{i=1}^n \lambda_i h(y, x_i) \\ &\leq \sum_{i=1}^n \lambda_i \inf_{w \in T(y)} Re\langle w, \eta(y, x_i)\rangle + h(y, y) \\ &= \sum_{i=1}^n \lambda_i \inf_{w \in T(y)} \langle w, \eta(y, x_i)\rangle \end{aligned}$$

which contradicts the condition (ii). Hence G is an KKM mapping and the condition (iii) of Lemma 5.35 is satisfied.

By the condition (v), it is easy to see that $\bigcap_{x \in M} G(x) \subset K$ and hence $\text{cl}_X(\bigcap_{x \in M} G(x))$ is compact (respectively, weakly compact) and the condition (i) of Lemma 5.35 is satisfied. For each $A \in \mathcal{F}(X)$ with $M \subset A$, if $y \in (\text{cl}_X(\bigcap_{x \in \text{co}(A)} G(x))) \bigcap \text{co}(A)$, then $y \in \text{co}(A)$ and there exists a net $\{y_\alpha\}_{\alpha \in \Gamma}$ in $\bigcap_{x \in \text{co}(A)} G(X)$ such that $\{y_\alpha\}_{\alpha \in \Gamma}$ converges to y (respectively, weakly to y). Since $tx + (1 - t)y \in \text{co}(A)$ for all $t \in [0, 1]$ and $x \in \text{co}(A)$, we have $\inf_{w \in T(y_\alpha)} Re\langle w, \eta(y_\alpha, tx + (1 - t)y)\rangle + h(y_\alpha, tx + (1 - t)y) \leq 0$, $\forall \alpha \in \Gamma$, and $t \in [0, 1]$.

Then for $t = 0$ we have

$$\inf_{w \in T(y_\alpha)} Re\langle w, \eta(y_\alpha, y)\rangle + h(y_\alpha, y) \leq 0, \quad \forall \alpha \in \Gamma.$$

It follows that

$$\limsup_\alpha [\inf_{w \in T(y_\alpha)} Re\langle w, \eta(y_\alpha, y) + h(y_\alpha, y)] \leq 0.$$

Since T is an (η, h)-pseudo-monotone type I (respectively, a strong (η, h)-pseudo-monotone type I) operator, we have

$$\limsup_\alpha [\inf_{w \in T(y_\alpha)} Re\langle w, \eta(y_\alpha, x)\rangle + h(y_\alpha, x)] \geq \inf_{w \in T(y)} Re\langle w, \eta(y, x)\rangle + h(y, x). \tag{5.129}$$

For $t = 1$, we have

$$\inf_{w \in T(y_\alpha)} Re\langle w, \eta(y_\alpha, x)\rangle + h(y_\alpha, x) \leq 0, \quad \forall \alpha \in \Gamma.$$

It follows that

$$\limsup_\alpha [\inf_{w \in T(y_\alpha)} Re\langle w, \eta(y_\alpha, x)\rangle + h(y_\alpha, x)] \leqq 0. \tag{5.130}$$

By (5.129) and (5.130), we obtain

$$\inf_{w \in T(y)} Re\langle w, \eta(y, x)\rangle + h(y, x) \leq 0$$

and so $y \in G(x)$ for all $x \in \text{co}(A)$. Therefore we have $y \in (\bigcap_{x \in \text{co}(A)} G(x)) \bigcap \text{co}(A)$.

Hence we must have

$$\left(\text{cl}_X \left(\bigcap_{x \in \text{co}(A)} G(x)\right)\right) \bigcap \text{co}(A) = \left(\bigcap_{x \in \text{co}(A)} G(x)\right) \bigcap \text{co}(A)$$

and the condition (iv) of Lemma 5.35 is satisfied. By Lemma 5.35, we have $\bigcap_{x \in X} G(x) \neq \emptyset$. Hence there exists $\hat{y} \in \bigcap_{x \in X} G(x)$. This means that

$$\inf_{w \in T(\hat{y})} Re\langle w, \eta(\hat{y}, x)\rangle + h(\hat{y}, x) \leq 0, \quad \forall x \in X.$$

Also from this and condition (v) it follows that $\hat{y} \in K$.

This proves that the conclusion (5.128) holds.

Now suppose that $T(\hat{y})$ is convex and for each $x \in X$ and $\eta(x, \cdot)$ is affine. Define a functional $f : X \times T(\hat{y}) \to \mathbb{R}$ by

$$f(x, w) = Re\langle w, \eta(\hat{y}, x)\rangle + h(\hat{y}, x).$$

By the definition of the $\sigma(F, E)$-topology on F, for each $x \in E$, the function $w \mapsto \langle w, x\rangle$ is continuous, it follows that for each $x \in X$, the functional $w \mapsto f(x, w)$ is continuous and affine and for each $w \in T(\hat{y})$, the functional $x \mapsto f(x, w)$ is concave since $\eta(\hat{y}, \cdot)$ is affine and $h(\hat{y}, \cdot)$ is concave. By applying Kneser's minimax Theorem 5.44, we obtain

$$\begin{aligned}
&\min_{w\in T(\hat{y})} \sup_{x\in X}[Re\langle w, \eta(\hat{y}, x)\rangle + h(\hat{y}, x)] \\
&\quad = \sup_{x\in X} \min_{w\in T(\hat{y})}[Re\langle w, \eta(\hat{y}, x)\rangle + h(\hat{y}, x)] \\
&\quad = \sup_{x\in X}\left[\inf_{w\in T(y)} Re\langle w, \eta(\hat{y}, x)\rangle + h(\hat{y}, x)\right] \leq 0.
\end{aligned}$$

Since $T(\hat{y})$ is $\sigma(F, E)$-compact, there exists a point $\hat{w} \in T(\hat{y})$ such that

$$\sup_{x\in X}[Re\langle \hat{w}, \eta(\hat{y}, x)\rangle + h(\hat{y}, x) \leq 0.$$

and hence

$$Re\langle \hat{w}, \eta(\hat{y}, x)\rangle + h(\hat{y}, x) \leq 0, \quad \forall x \in X,$$

that is, the pair $\hat{y} \in K$ and $\hat{w} \in T(\hat{y})$ is a solution of the $GV\,LI(F, \eta, h, X, F)$ (2). □

By taking $\eta(x, y) = g(x) - g(y)$ for all $x, y \in X$, we obtain the following result of Ding and Tarafdar from Theorem 5.85.

Corollary 5.85.1 *Let X, E and F be same as in Theorem 5.85. Let $T : X \to 2^F \setminus \{\phi\}$, $g : X \to E$ and $b : X \times X \to \mathbb{R}$ be such that*

(i) *T satisfies the condition (i) of Theorem 5.85,*
(ii) *T and g have the 0-DCVR,*
(iii) *g is continuous on $\mathrm{co}(A)$ for each $A \in \mathcal{F}(X)$, and b is lower semicontinuous in $\{(x, x) : x \in X\}$ such that for each $x \in X$, $b(x, \cdot)$ is a convex function and b is upper semi-continuous in its first argument,*
(iv) *T is an (η, h)-pseudo-monotone type I operator where $\eta(x, y) = g(x) - g(y)$ and $h(x, y) = b(x, x) - b(x, y)$ for all $x, y \in X$,*
(v) *there exist a nonempty compact subset K of X and a finite set $M \in \mathcal{F}(X)$ such that for each $y \in X \setminus K$, there is an $x \in M$ satisfying*

$$\inf_{w\in T(y)} Re\langle w, g(y) - g(x)\rangle + b(y, y) - b(y, x) > 0.$$

Then there exists a point $\hat{y} \in K$ such that

$$\inf_{w\in T(\hat{y})} Re\langle w, g(\hat{y}) - g(x)\rangle + b(\hat{y}, \hat{y}) - b(\hat{y}, x)\rangle \leq 0, \quad \forall x \in X.$$

If, in addition, $T(\hat{y})$ is convex and g is affine, then there exists a point $\hat{w} \in T(\hat{y})$ such that

$$Re\langle \hat{w}, g(\hat{y}) - g(x)\rangle + b(\hat{y}, \hat{y}) - b(\hat{y}, x) \leq 0, \quad \forall x \in X.$$

Remark 5.31 If X is also compact in E, then the condition (v) of Theorem 5.85 and Corollary 5.85.1 is satisfied trivially. We note here that in Theorem 5.85 and Corollary 5.85.1, the space E may not be locally convex, T is only assumed to be upper semi-continuous from co(A) to the $\sigma(F, E)$-topology on F and g is only assumed to be continuous in co(A) for each $A \in \mathcal{F}(X)$. Since E is Hausdorff and co(A) is compact, the relative weak topology on co(A) coincides with its relative topology. Therefore Theorem 5.85 and Corollary 5.85.1 are different from Theorem 3.1 in Ding and Tarafdar (1994), Theorem 3.1 of Yu and Yao Yu and Yao (1996) and Theorem 1 of Yu, Lai and Yao Yu, Lai, and Yao (1966). Theorem 5.85 and Corollary 5.85.1 require that T is an (η, h)-pseudo-monotone type I operator but the results in [Ding and Tarafdar (1994), Yu et al. (1966), Yu and Yao (1996)], though do not require the pseudo-monotonicity, do require some other strong conditions. Hence Theorem 5.85 and Corollary 5.85.1 are new results which are different from those in [Ding and Tarafdar (1994), Yu et al. (1966), Yu and Yao (1996)].

When $\eta(x, y) = x - y$ and $h(x, y) = h'(x) - h'(y)$ for some convex function $h' : X \to \mathbb{R}$ and for all $x, y \in X$, we obtain the following result of Ding and Tarafdar from Theorem 5.85.

Theorem 5.86 *Let X, E and F be same as in Theorem 5.85. Let $T : X \to 2^F \setminus \{\emptyset\}$ and $h' : X \to \mathbb{R}$ be such that*

(*i*) *T satisfies the condition (i) of Theorem 5.85 and each $T(x)$ is also convex,*
(*ii*) *T is an h'-pseudo-monotone type I (respectively, a strong h'-pseudo-monotone type I) operator,*
(*iii*) *h' is a convex function,*
(*iv*) *there exist a nonempty compact (respectively, weakly closed and weakly compact) subset K of X and a finite set $M \in \mathcal{F}(X)$ such that for each $y \in X \setminus K$, there is an $x \in M$ satisfying*

$$\inf_{w \in T(y)} Re\langle w, y - x\rangle + h'(y) - h'(x) > 0.$$

Then there exist $\hat{y} \in K$ and $\hat{w} \in T(\hat{y})$ such that

$$Re\langle \hat{w}, \hat{y} - x\rangle \leq h'(x) - h'(\hat{y}), \quad \forall x \in I_X(\hat{y}),$$

where $I_X(\hat{y}) = \{\hat{y} + r(x - \hat{y}) : x \in X \text{ and } r \geq 0\}$.

Proof. We first note that for each $A \in \mathcal{F}(X)$, h' is continuous on co(A) (see, Rockafeller (1970, Corollary 10.1.1, p. 83). Let $\eta(x, y) = x - y$ and $h(x, y) = h'(x) - h'(y)$ for all $x, y \in X$ in Theorem 5.85. It is easy to check that all conditions of Theorem 5.85 are satisfied. Hence there exist $\hat{y} \in K$ and $\hat{w} \in T(\hat{y})$ such that

$$Re\langle \hat{w}, \hat{y} - x\rangle + h'(\hat{y}) - h'(x) \leq 0, \quad \forall x \in X.$$

Let $z \in I_X(\hat{y}) \setminus X$. Since X is convex, there exist $x \in X$ and $r > 1$ such that $z = \hat{y} + r(x - \hat{y})$. Suppose that

$$Re\langle \hat{w}, \hat{y} - z\rangle + h'(\hat{y}) - h'(z) > 0.$$

Then we have $x = (1 - 1/r)\hat{y} + (1/r)z \in X$ and

$$\begin{aligned} 0 &\geq \Re\langle \hat{w}, \hat{y} - x\rangle + h'(\hat{y}) - h'(x) \\ &\geq (1/r)Re\langle \hat{w}, \hat{y} - z\rangle + h'(\hat{y}) - [(1 - 1/r)h'(\hat{y}) + (1/r)h'(z)] \\ &= (1/r)[Re\langle \hat{w}, \hat{y} - z\rangle + h'(\hat{y}) - h'(z)] > 0 \end{aligned}$$

which is impossible. Hence we must have

$$Re\langle \hat{w}, \hat{y} - x\rangle \leq h'(x) - h'(\hat{y}), \text{for all} x \in I_X(\hat{y}). \qquad \square$$

Note that the results of this section have been written and presented after a thorough revision of some results of an article in Ding and Tarafdar (2000).

5.12 Generalized Quasi-Variational Inequalities

In this section, E will denote a Hausdorff topological vector space, and if X is an arbitrary non-empty subset of E then 2^X will denote the family of all subsets of X.

In this section we shall present some results of Shih and Tan on general theorems on solutions of the generalized quasi-variational inequalities (GQVI). In obtaining these results, Shih and Tan used the Ky Fan minimax principle [Fan (1972)] or the following generalized version due to Yen [Yen (1981)] as their basic tool:

Theorem 5.87 *Let X be a nonempty compact convex set in a Hausdorff topological vector space E. Let ϕ and ψ be two real-valued functions on $X \to X$ having the following properties:*

(a) *$\phi \leq \psi$ on $X \times X$ and $\psi(x, x)) \leq 0$ for all $x \in X$;*
(b) *For each fixed $x \in X, \phi(x, y)$ is a lower semi-continuous function of y on X;*
(c) *For each fixed $y \in X, \psi(x, y)$ is a quasi-concave function of x on X.*

Then there exists a point $\hat{y} \in X$ such that $\phi(x, \hat{y}) \leq 0$ for all $x \in X$.

5.12.1 *Generalized Quasi-Variational Inequalities for Monotone and Lower Semi-Continuous Mappings*

We shall present some results from Shih and Tan (1985) on generalized quasi-variational inequalities (GQVI) for monotone and lower semicontinuous operators on compact sets. Before doing that we shall start with a lemma in Shih and Tan (1985).

Lemma 5.36 *Let E be a Hausdorff topological vector space, $X \subset E$ be nonempty and $S : X \to 2^E$ be upper semi-continuous such that for each $x \in X, S(x)$ is nonempty and bounded. Then for $p \in E'$ the map $f_p : X \to \mathbb{R}$ defined by $f_p(y) := \sup_{x \in S(y)} Re\langle p, x\rangle$ is upper semi-continuous.*

Proof. Let $y_0 \in X$ and $\varepsilon > 0$ be given. Let

$$U_\varepsilon := \{x \in E : |\langle p, x\rangle| < \varepsilon/2\};$$

then U_ε is an open neighborhood of 0. As $S(y_0) + U_\varepsilon$ is an open set containing $S(y_0)$, by upper semicontinuity of S at y_0, there exists a neighborhood $N(y_0)$ of y_0 in X such that if $y \in N(y_0)$ then $S(y) \subset S(y_0) + U_\varepsilon$. Thus, for each $y \in N(y_0)$,

$$f_p(y_0) = \sup_{x \in S(y)} Re\langle p, x\rangle \tag{5.131}$$

$$\leq \sup_{x \in S(y_0)+U_\varepsilon} Re\langle p, x\rangle \tag{5.132}$$

$$\leq \sup_{x \in S(y_0)} Re\langle p, x\rangle + \sup_{x \in U_\varepsilon} Re\langle p, x\rangle \tag{5.133}$$

$$< f_p(y_0) + \varepsilon. \tag{5.134}$$

Hence f_p is upper semi continuous and the proof is completed. □

We now present the first result in Shih and Tan (1985):

Theorem 5.88 *Let E be a locally convex Hausdorff topological vector space and X be a nonempty compact convex subset of E. Let $S : X \to 2^X$ be upper semi-continuous such that for each $x \in X, S(x)$ is a nonempty closed convex subset of X, and let $T : X \to 2^{E'}$ be monotone such that for all $x \in X, T(x)$ is a nonempty subset of E' and for each one-dimensional flat $L \subset E, T|L \cap X$ is lower semi-continuous from the topology of E to the weak*- topology $\sigma(E', E)$ of E'. Suppose further that the set $\Sigma_1 := \{y \in X : \sup_{x \in S(y)} \sup_{u \in T(x)} Re\langle u, y - x\rangle > 0\}$ is open in X. Then there exists a point $\hat{y} \in X$ such that*

(i) $\hat{y} \in S(\hat{y})$ and

(ii) $\sup_{w \in T(\hat{y})} Re\langle w, \hat{y} - x\rangle \leq 0$ for all xex $\in S(\hat{y})$.

Proof. The proof is divided into two steps as follows:

Step 1. There exists a point $\hat{y} \in X$ such that $\hat{y} \in S(\hat{y})$ and $\sup_{u \in T(x)} Re\langle u, y - x\rangle \leq 0$ for all $x \in S(\hat{y})$.

Suppose the assertion were false. Then for all $y \in X$, either $y \notin S(y)$ or there exists a point $x \in S(y)$ such that $\sup_{u \in S(y)} Re\langle u, y-x\rangle > 0$. Observe that whenever $y \notin S(y)$, there exists $p \in E'$ such that

$$Re\langle p, y\rangle - \sup_{x \in S(y)} Re\langle p, x\rangle > 0$$

by applying the Hahn-Banach separation theorem. For each $y \in X$, we set

$$\alpha(y) := \sup_{x \in S(y)} \sup_{u \in T(x)} Re\langle u, y - x \rangle.$$

Let

$$V_0 := \{y \in X : \alpha(y) > 0\}.$$

For each $p \in E'$, we set

$$V(p) := \{y \in X : Re\langle p, y \rangle - \sup_{x \in S(y)} Re\langle p, x \rangle 0\}.$$

Then $X = V_0 \cup \bigcup_{p \in E'} V(p)$. By hypothesis, V_0 is open in X. By Lemma 5.36, $V(p)$ is open in X for each $p \in E'$. Since X is compact, there exist $p_1, \dots, p_n \in E'$ such that $X = V_0 \cup \bigcup_{i=1}^{n} V(p_1)$ and a continuous partition of unity $\{\beta_0, \beta_1, \dots, \beta_n\}$ subordinated to the covering $\{V_0, V(p_1), \dots, V(p_n)\}$, that is, $\beta_0, \beta_1, \dots, \beta_n$ are continuous nonnegative real-valued functions on X such that β_0 vanishes on X V_0 and for each $1 \leq i \leq n, \beta_I$ vanishes on X $V(p_i)$ and $\Sigma_{i=0}^{n} \beta_i(x) = 1$ for all $x \in X$.

Define $\phi, \psi : X \times X \to \mathbb{R}$ by setting

$$\phi(x, y) := \beta_0(y) \sup_{u \in T(x)} Re\langle u, y - x \rangle + \sum_{i=1}^{n} \beta_i(y) Re\langle p_i, y - x \rangle,$$

$$\psi(x, y) := \beta_0(y) \inf_{w \in T(y)} Re\langle w, y - x \rangle + \sum_{i=1}^{n} \beta_i(y) Re\langle p_i, y - x \rangle.$$

By monotonicity of T, we have

$$\sup_{u \in T(x)} Re\langle u, y - x \rangle \leq \inf_{w \in T(y)} Re\langle w, y - x \rangle \quad \text{for all} \quad x, y \in X.$$

It follows that $\phi \leq \psi$ on $X \times X$. Clearly $\psi(x, x) = 0$ for all $x \in X$. For each fixed $x \in X$, since, $\beta_i (i = 0, 1, \dots n)$ are continuous nonnegative functions of y on X and $\sup_{u \in T(x)} Re\langle u, y - x \rangle (i = l, \dots, n)$ are lower semi-continuous functions of y on X, by Lemma 3 in [Takahashi (1976), p. 177], $y \to \phi(x, y)$ is lower semi-continuous on X. Furthermore, for each fixed $y \in X, x \to \psi(x, y)$ is quasi-concave. Hence, all the conditions of Theorem 5.87 are satisfied, so that there exists a point $\hat{y} \in X$ such that $\phi(x, \hat{y}) \leq 0$ for all $x \in X$; that is,

$$\beta_0(\hat{y}) \sup_{u \in T(x)} Re\langle u, \hat{y} - x \rangle + \sum_{i=1}^{n} \beta_i(\hat{y}_i) Re\langle p_i, \hat{y} - x \rangle \leq 0 \quad \text{for all} \quad x \in X. \qquad (*)$$

Since $\{\beta_0, \beta_1, \dots, \beta_n\}$ is a partition of unity, $|be_i(\hat{y}) > 0$ for at least one index $I \in \{0, 1, \dots, n\}$. Choose any $\hat{x} \in S(\hat{y})$ such that

$$\sup_{u \in T(\hat{x})} Re\langle u, \hat{y} - \hat{x} \rangle \geq \frac{\alpha(\hat{y})}{2} \text{ whenever } \alpha(\hat{y}) > 0.$$

If $\beta_0(\hat{y}) > 0$, then $\hat{y} \in V_0$ so that $\alpha(\hat{y}) > 0$. Hence,

$$\sup_{u\in T(\hat{x})} Re\langle u, \hat{y} - \hat{x}\rangle \geq \frac{\alpha(\hat{y})}{2} > 0.$$

If $\beta_i(\hat{y}) > 0$, for $i = 1, \ldots, n$, then $\hat{y} \in V(p_i)$ and hence

$$Re\langle p_i, \hat{y}\rangle > \sup_{x\in S(y)} Re\langle p_i, x\rangle \geq Re\langle p_i, \hat{x}\rangle$$

so that $Re\langle p_i, \hat{y} - \hat{x}\rangle > 0$. It follows that

$$\beta_0(\hat{y}) \sup_{u\in T(x)} Re\langle u, \hat{y} - \hat{x}\rangle + \sum_{i=1}^{n} \beta_i(\hat{y}_i) Re\langle p_i, \hat{y} - \hat{x}\rangle > 0$$

contradicting $(*)$. This contradiction proves Step 1.

Step 2.

$$\sup_{w\in T(\hat{y})} Re\langle w, \hat{y} - x\rangle \leq 0 \quad \text{for all} \quad x \in S(\hat{y}).$$

Let $x \in S(\hat{y})$ be arbitrarily fixed and let $z_t := tx + (1-t)\hat{y} \equiv \hat{y} - t(\hat{y} - x)$ for $t \in [0,1]$. As $S(\hat{y})$ is convex, we have $z_t \in S(\hat{y})$ for $t \in [0,1]$. Therefore by Step 1, we have

$$\sup_{u\in T(z_t)} Re\langle u, \hat{y} - z_t\rangle \leq 0 \quad \text{for all} \quad t \in [0,1],$$

and it follows that

$$\sup_{u\in T(z_t)} Re\langle u, \hat{y} - x\rangle \leq 0 \quad \text{for all} \quad t \in (0,1]. \tag{**}$$

Let $w_0 \in T(\hat{y})$ be arbitrarily fixed. For each $\varepsilon > 0$, let

$$U_{w0} := \{w \in E' : |\langle w_0 - w, \hat{y} - x\rangle| < \varepsilon\};$$

then U_{w0} is a $\sigma(E', E)$-neighborhood of w_0. Since $T|L\cap X$ is lower semi-continuous, where $L := \{z_t : t \in [0,1]\}$, and $U_{w0} \cap T(\hat{y}) \neq \emptyset$, there exists a neighborhood $N(\hat{y})$ of $\hat{y}$ in L such that if $z \in N(\hat{y})$ then $T(z)\cap U_{w0} \neq \emptyset$. But then there exists $\delta \in (0,1)$ such that $z_t \in N(\hat{y})$ for all $t \in (0,\delta)$. Fix any $t \in (0,\delta)$ and $u \in T(z_t) \cap U_{w0}$, we have

$$|\langle w_0 - u, \hat{y} - x\rangle| < \varepsilon.$$

This implies

$$Re\langle w_0, \hat{y} - x\rangle < Re\langle u, \hat{y} - x\rangle + \varepsilon.$$

By (**), we have $Re\langle w_0, \hat{y} - x\rangle < \varepsilon$. Since $\varepsilon > 0$ is arbitrary, $Re\langle w_0, \hat{y} - x\rangle \leq 0$. As $w_0 \in T(\hat{y})$ is arbitrary,

$$\sup_{w \in T(\hat{y})} Re\langle w, \hat{y} - x\rangle \leq 0 \quad \text{for all} \quad x \in S(\hat{y}).$$

This concludes the proof of our theorem. □

When $T \equiv 0$, Theorem 5.88 gives the well-known Fan-Glicksberg fixed point theorem [Fan (1952), Glicksberg (1952)].

Corollary 5.88.1 (Fan and Glicksberg) *Let E be a locally convex Hausdorff topological vector space and X a nonempty compact convex set in E. Let $S : X \to 2^X$ be upper semi-continuous such that for each $x \in X, S(x)$ is a nonempty closed convex subset of X. Then there exists a point $\hat{x} \in X$ such that $\hat{x} \in S(\hat{x})$.*

Shih and Tan observed that, if S is assumed to be continuous and T is assumed to be lower semi-continuous in Theorem 5.88, then the set Σ_1 becomes open in X. The following theorem presents the results with these additional continuity assumptions.

Theorem 5.89 *Let E be a locally convex Hausdorff topological vector space and X be a nonempty compact convex subset of E. Let $S : X \to 2^X$ be continuous such that for each $x \in X$, $S(x)$ is a nonempty closed convex subset of X, and $T : X \to 2^{E'}$ be monotone such that for each $x \in X$, $T(x)$ is a nonempty subset of E' and T is lower semi-continuous from the relative topology of X to the strong topology of E'. Then there exists a point $\hat{y} \in X$ such that*

(i) $\hat{y} \in S(\hat{y})$ and

(ii) $\sup_{w \in T(\hat{y})} Re\langle w, \hat{y} - x\rangle \leq 0$ for all $x \in S(\hat{y})$.

Proof. By virtue of Theorem 5.88, we need only show that

$$\Sigma_1 := \{y \in X : \sup_{x \in S(y)} \sup_{u \in T(x)} Re\langle u, y - x\rangle > 0\}$$

is open in X. Let $y_0 \in \Sigma_1$; then there exist $x_0 \in S(y_0)$ and $f_0 \in T(x_0)$ such that

$$\alpha := Re\langle f_0, y_0 - x_0\rangle > 0.$$

Since f_0 is continuous at x_0 and at y_0, there exist an open neighborhood N_1, of x_0 and an open neighborhood U_1 of y_0 such that

$$x \in N_1 \Rightarrow |\langle f_0, x_0\rangle - \langle f_0, x\rangle| < \alpha/6,$$

$$y \in U_1 \Rightarrow |\langle f_0, y_0\rangle - \langle f_0, y\rangle| < \alpha/6.$$

Let

$$W := \{f \in E' : \sup_{z_1, z_2 \in X} |\langle f - f_0, z_1 - z_2\rangle| < \alpha/6\};$$

then W is a strongly open neighborhood of f_0 and $W \cap T(x_0) \neq \emptyset$ so that by lower semi-continuity of T at x_0, there exists an open neighborhood N_2 of x_0 such that

$$x \in N_2 \Rightarrow T(x) \cap W \neq \emptyset.$$

Let $N := N_1 \cap N_2$; since N is a neighborhood of x_0 and $N \cap S(y_0) \neq \emptyset$, by lower semi-continuity of S at y_0, there exists an open neighborhood U_2 of y) such that

$$y \in U_2 \Rightarrow S(y) \cap N \neq \emptyset.$$

Let $U := U_1 \cap U_2$; then U is an open neighborhood of y_0. For each $y_1 \in U$, choose $x_1 \in S(y_1) \cap N$ and $f_1 \in T(x_1) \cap W$; it follows that

$$\begin{aligned} \alpha &= Re\langle f_0, y_0 - x_0\rangle && (5.135)\\ &= Re\langle f_1, y_1 - x_1\rangle + Re\langle f_0, y_0 - y_1\rangle + Re\langle f_0 - f_1, y_1 - x_1\rangle && (5.136)\\ &\quad + Re\langle f_0, x_1 - x_0\rangle && (5.137)\\ &\quad < Re\langle f_1, y_1 - x_1\rangle + \alpha/2. && (5.138)\end{aligned}$$

Thus, $Re\langle f_1, y_1 - x_1\rangle > \alpha/2 > 0$ so that $y_1 \in \Sigma_1$ for all $y_1 \in U_4$. Hence Σ_1 is open in X and the proof is completed. □

When $S(x) \equiv X$, Theorem 5.89 gives a multi-valued version of the Hartman-Stampacchia variational inequality [Hartman and Stampacchia (1966)] as follows.

Corollary 5.89.1 *Let E be a locally convex Hausdorff topological vector space and X be a nonempty compact convex subset of E. Let $T : X \to 2^{E'}$ be monotone such that for each $x \in X$, $T(x)$ is a nonempty subset of E' and T is lower semi-continuous from the relative topology of X to the strong topology of E'. Then there exists a point $\hat{y} \in X$ such that*

$$\sup_{w \in T(\hat{y})} Re\langle w, \hat{y} - x\rangle \leq 0 \quad \textit{for all } x \in X.$$

5.12.2 *Generalized Quasi-Variational Inequalities for Upper Semi-Continuous Mappings Without Monotonicity*

In Theorems 5.88 and 5.89, Shih and Tan used a monotone operator T which also have some kind of lower semi-continuity. In the next result from Shih and Tan (1985) the operator T is upper semi-continuous; but T is not a monotone operator.

Theorem 5.90 *Let E be a locally convex Hausdorff topological vector space and X be a nonempty compact convex subset of E. Let $S : X \to 2^X$ be upper semi- continuous such that for each $x \in X$, $S(x)$ is a nonempty closed convex subset of X, and let $T : X \to 2^{E'}$ be upper semi-continuous from the relative topology of X to the strong topology of E' such that for each x c- X, $T(x)$ is a nonempty compact convex subset of E'. Suppose further that the set $\Sigma_2 = \{y \in X :$*

$\sup_{x\in S(y)} \inf_{z\in T(y)} Re\langle z, y-x\rangle > 0\}$ is open in X. Then there exists a point $\hat{y} \in X$ such that

(i) *$\hat{y} \in S(\hat{y})$ and*

(ii) *there exists a point $\hat{z} \in T(\hat{y})$ with $Re\langle \hat{z}, \hat{y} - x\rangle \leq 0$ for all $x \in S(\hat{y})$.*

Proof. The proof is divided into two steps as follows:

Step 1. There exists a point $\hat{y} \in S$ such that $\hat{y} \in S(\hat{y})$ and

$$\sup_{x\in S(\hat{y})} \inf_{z\in T(\hat{y})} Re\langle z, \hat{y} - x\rangle \leq 0.$$

Suppose the assertion were false. Then for all $y \in X$, either $y \notin S(y)$ or there exists $x \in S(y)$ such that $\inf_{z\in T(y)} Re\langle z, y-x\rangle > 0$. Observe that whenever $y \notin S(y)$, there exists $p \in E'$ with

$$Re\langle p, y\rangle - \sup_{x\in S(y)} Re\langle p, x\rangle > 0.$$

For each $y \in X$, we set

$$\alpha(y) := \sup_{x\in S(y)} \inf_{z\in T(y)} Re\langle z, y - x\rangle.$$

Let

$$V_0 := \{y \in X : \alpha(y) > 0\},$$

and for each $p \in E'$, we set

$$V(p) := \{y \in X : Re\langle p, y\rangle - \sup_{x\in S(y)} Re\langle p, x\rangle > 0\}.$$

Then $x = V_0 \cup \bigcup_{p\in E'} V(p)$. By hypothesis, V_0 is open in X. By Lemma 5.36, $V(p)$ is open in X for each $p \in E'$. Since X is compact, there exist $p_1, \ldots, p_n \in E'$ such that

$$X = V_0 \cup \bigcup_{i=1}^{n} V(p_i)$$

and a continuous partition of unity $\{\beta_0, \beta_1, \ldots, \beta_n\}$ subordinated to the covering $\{V_0, V(p_1), \ldots, V(p_n)\}$.

Define $\phi : X \times X \to \mathbb{R}$ by setting

$$\phi(x, y) := \beta_0(y) \inf_{w\in T(y)} Re\langle w, y - x\rangle + \sum_{i=1}^{n} \beta_i(y) Re\langle p_i, y - x\rangle.$$

Clearly $\phi(x, x) = 0$ for each $x \in X$. Note that for each fixed $x \in X$, $y \to \inf_{w\in T(y)} Re\langle w, y - x\rangle$ is lower semi-continuous as can be seen within the proof of Theorem 21 in [Takahashi (1976)], so that $y \to \phi(x, y)$ is lower semi-continuous. Also it is clear that for each fixed $y \in X$, $x \to \phi(x, y)$ is quasi-concave. Hence by the Ky Fan minimax principle (i.e., Theorem 5.87 with $\phi \equiv \psi$), there exists a point $\hat{y} \in X$ such that $\phi(x, \hat{y}) \leq 0$ for all $x \in X$. The contradiction that there is a point

$\hat{x} \in X$ with $\phi(\hat{x}, \hat{y}) > 0$ can be achieved by using the corresponding proof of Step 1 of Theorem 5.88.

Step 2. There exists a point $\hat{z} \in T(\hat{y}$ such that $Re\langle \hat{z}, \hat{y} - x\rangle \leq 0$ for all $x \in S(y)$.

Indeed, define $f : S(\hat{y}) \times T(\hat{y}) \to \mathbb{R}$ by

$$f(x, z) := Re\langle z, \hat{y} - x\rangle.$$

Note that for each fixed $x \in S(\hat{y})$, $z \to f(x, z)$ is continuous and affine, and for each $z \in T(\hat{y})$, $x \to f(x, z)$ is affine. Thus by Kneser's minimax theorem [Kneser (1952)], we have

$$\min_{z \in T(\hat{y})} \max_{x \in S(\hat{y})} f(x, z) = \max_{x \in S(\hat{y})} \min_{z \in T(\hat{y})} f(x, z).$$

Thus

$$\min_{z \in T(\hat{y})} \max_{x \in S(\hat{y})} Re\langle z, \hat{y} - x\rangle \leq 0$$

by Step 1. Since $T(\hat{y})$ is compact, there exists $\hat{z} \in T(\hat{y})$ such that

$$Re\langle z, \hat{y} - x\rangle \leq 0 \text{ for all } x \in S(\hat{y}).$$

Shih and Tan further observed that, when E is a normed linear space, if S is assumed to be continuous in Theorem 5.90, then the set Σ_2 becomes open in X. The following theorem is a result with this additional assumption.

Theorem 5.91 *Let E be a normed linear space and X be a nonempty compact convex subset of E. Let $S : X \to 2^X$ be continuous such that for each $x \in X$, $S(x)$ is a nonempty closed convex subset of X, and let $T : X \to 2^{E'}$ be upper semi-continuous such that for each $x \in X$, $T(x)$ is a nonempty compact convex subset of E'. Then there exists a point $\hat{y} \in X$ such that*

(i) $\hat{y} \in S(\hat{y})$ and

(ii) there exists a point $\hat{z} \in T(\hat{y})$ with $Re\langle \hat{z}, \hat{y} - x\rangle \leq 0$ for all $x \in S(\hat{y})$.

Proof. By virtue of Theorem 5.90, we need only show that the set

$$\Sigma_2 := \{y \in X : \sup_{x \in S(y)} \inf_{z \in T(y)} Re\langle z, y - x\rangle > 0\}$$

is open in X. For this purpose, let $y_0 \in \Sigma_2$, then there exists $x_0 \in S(y_0)$ with

$$\alpha = \inf_{z \in T(y)} Re\langle z, y_0 - x_0\rangle > 0\}.$$

Let

$$M := \max\{(x), \sup_{z \in T(y_0)} \|z\|\} \text{ and } B := \{f \in E' : \|f\| < 1\}.$$

Since T is upper semi-continuous at y_0, for $\eta = \alpha/6(1+M) > 0$, there exists $\delta_1 \in (0, \min\{1, \alpha/6(1+M)\})$ such that for all $y \in X$, $\|y - y_0\| < \delta_1$ implies $T(y) \subset T(y_0) + \eta B$. As S is lower semi-continuous at y_0, there exists $\delta_2 \in (0, \min\{1, \alpha/6(1+M)\})$ such that for all $y \in X$, $\|y - y_0\| < \delta_2$ implies

$$S(y) \cap \{x \in X : \|x - x_0\|, \eta\} \neq \emptyset.$$

Let $\delta := \min\{\delta_1, \delta_2\}$. Let $y_1 \in X$ be such that $\|y_1 - y_0\| < \delta$. Then $T(y_1) \subset T(y_0) + \eta B$ and we can choose $x_1 \in S(y_1)$ with $\|x_1 - x_0\| < \eta$. It follows that

$$\inf_{z\in T(y_1)} Re\langle z, y_1 - x_1\rangle \geq \inf_{Z\in T(y_0)+\eta B} Re\langle z, y_1 - x_1\rangle \tag{5.139}$$

$$\geq \inf_{z\in T(y_0)} Re\langle z, y_1 - x_1\rangle + \inf_{z\in \eta B} Re\langle z, y_1 - x_1\rangle \tag{5.140}$$

$$\geq \inf_{z\in T(y_0)} Re\langle z, y_1 - y_0\rangle + \inf_{z\in T(y_0)} Re\langle z, y_0 - x_0\rangle \tag{5.141}$$

$$+ \inf_{z\in T(y_0)} Re\langle z, x_0 - x_1\rangle - \eta\|y_1 - x_1\| \tag{5.142}$$

$$\geq - \sup_{z\in T(y_0)} \|z\|\|y_1 - y_0\| + \alpha \tag{5.143}$$

$$- \sup_{z\in T(y_0)} \|z\|\|x_0 - x_1\| - \alpha/6 \tag{5.144}$$

$$> \alpha/2 > 0. \tag{5.145}$$

Thus,

$$\sup_{x\in S(y_1)} \inf_{z\in T(y_1)} Re\langle z, y_1 - x\rangle > 0$$

so that $y_1 \in \Sigma_2$ whenever $y_1 \in X$ with $\|y_1 - y_0\| < \delta$. This shows that Σ_2 is open in X and the proof is completed. □

When $S(x) \equiv X$, the above result of Shih and Tan (1985) gives the following multi-valued version of the Hartman-Stampacchia variational inequality:

Corollary 5.91.1 *Let E be a normed linear space and $X \subset E$ a nonempty compact convex subset of E. Let $T : X \to 2^{E'}$ be upper semicontinuous such that for each $x \in X$, $T(x)$ is a nonempty compact convex subset of E'. Then there exist a point $\hat{y} \in X$ and a point $\hat{z} \in T(\hat{y})$ such that*

$$Re\langle \hat{z}, \hat{y} - x\rangle \leq 0 \quad \textit{for all} \quad x \in X.$$

Note that the results of this section have been written and presented after a thorough revision of some results of an article in Shih and Tan (1985).

5.13 Generalized Quasi-Variational Inequalities for Lower and Upper Hemi-Continuous Operators on Non-Compact Sets

In obtaining some main results of this section Chowdhury and Tan used the following result which is Theorem 1 in [Bae et al. (1993), p. 231]:

Theorem 5.92 *Let E be a topological vector space, X be a non-empty convex subset of E and $f, g : X \times X \to \mathbb{R} \cup \{-\infty, +\infty\}$ be such that*

(a) *$g(x, x) \leq 0$ for all $x \in X$ and $f(x, y) \leq g(x, y)$ for all $x, y \in X$;*
(b) *for each fixed $x \in X, y \mapsto f(x, y)$ is lower semicontinuous on non-empty compact subsets of X;*
(c) *for each fixed $y \in X$, the set $\{x \in X : g(x, y) > 0\}$ is convex;*
(d) *there exist a non-empty compact convex subset X_0 of X and a non- empty compact subset K of X such that for each $y \in X \backslash K$, there is an $x \in co(X_0 \cup \{y\})$ with $f(x, y) > 0$.*

Then there exists $\hat{y} \in K$ such that $f(x, \hat{y}) \leq 0$ for all $x \in X$.

The following is a definition in [Chowdhury and Tan (1997b), pp. 28–29]:

Definition 5.18 Let E be a topological vector space, X be a non-empty subset of E and $T : X \to 2^{E^*}$. Then T is said to be lower hemi-continuous on X if and only if for each $p \in E$, the function $f_p : X \to \mathbb{R} \cup \{+\infty\}$, defined by

$$f_p(z) = \sup_{u \in T(z)} Re\langle u, p \rangle \quad \text{for each} \quad z \in X,$$

is lower semicontinuous on X (if and only for each $p \in E$, the function $g_p; X \to \mathbb{R} \cup \{-\infty\}$, defined by

$$g_p(z) = \inf_{u \in T(z)} Re\langle u, p \rangle \quad \text{for each} \quad z \in X,$$

is upper semicontinuous on X).

The following is another definition in [Chowdhury and Tan (1997b), pp. 28–29]:

Definition 5.19 Let E be a topological vector space, X be a non-empty subset of E and $T : X \to 2^{E^*}$. Then T is said to upper hemi-continuous on X if and only if for each $p \in E$, the function $f_p : X \to \mathbb{R} \cup \{+\infty\}$, defined by

$$f_p(z) = \sup_{u \in T(z)} Re\langle u, p \rangle \quad \text{for each} \quad z \in X,$$

is upper semicontinuous on X (if and only if for each $p \in E$, the function $g_p : X \to \mathbb{R} \cup \{-\alpha\}$, is defined by

$$g_p(z) = \inf_{u \in T(z)} Re\langle u, p \rangle \quad \text{for each} \quad z \in X,$$

is lower semicontinuous on X).

The following result is in [Chowdhury and Tan (1997b), p. 29]:

Proposition 5.10 *Let E be a topological vector space and X be a non-empty subset of E. Let $T : X \to 2^{E^*}$ be lower semicontinuous from relative topology on X to the weak* topology $\sigma\langle E^*, E\rangle$ on E^*. Then T is lower hemi-continuous on X.*

Next, we give another result in [Chowdhury and Tan (1997b), p. 30]:

Proposition 5.11 *Let E be a topological vector space and X be a non-empty subset of E. Let $T : X \to 2^{E^*}$ be upper semicontinuous from relative topology on X to the weak* topology $\sigma\langle E^*, E\rangle$ on E^*. Then T is upper hemi-continuous on X.*

The following simple result is Lemma 2.1.6 in [Tan (1994)]:

Lemma 5.37 *Let E be a topological vector space and A be a non-empty bounded subset of E. Let C be a non-empty strongly compact subset of E^*. Define $f : A \to \mathbb{R}$ by $f(x) = \min_{u\in C} Re\langle u, x\rangle$ for all $x \in A$. Then f is weakly continuous on A.*

The following result is Lemma 4.2 in [Chowdhury and Tan (1997b), p. 38]:

Lemma 5.38 *Let E be a topological vector space, X be a non-empty convex subset of $E, h : X \to \mathbb{R}$ be convex and $T : X \to 2^{E^*}$ be upper hemi-continuous along line segments in X. Suppose $\hat{y} \in X$ is such that $\inf_{u\in T(x)} Re\langle u, \hat{y} - x\rangle \leq h(x) - h(\hat{y})$ for all $x \in X$.Then*

$$\inf_{w\in T(\hat{y})} Re\langle w, \hat{y} - x\rangle \leq h(x) - h(y) \quad \textit{for all} \quad x \in X$$

We shall end this section with a result which is Lemma 4.1 in [Chowdhury and Tan (1997b), pp. 37–38]:

Lemma 5.39 *Let E be a topological vector space, X be a non-empty convex subset of $E, h : X \to \mathbb{R}$ be convex and $T : X \to 2^{E^*}$ be lower hemi-continuous along line segments in X. Then*

$$\sup_{w\in T(\hat{y})} Re\langle w, \hat{y} - x\rangle \leq h(x) - h(\hat{y}) \quad \textit{for all} \quad x \in X.$$

5.13.1 *Generalized Quasi-Variational Inequalities for Lower Hemi-Continuous Operators*

In this subsection we shall present some existence theorems of Chowdhury and Tan on generalized quasi-variational inequalities for monotone and lower hemi-continuous operators on paracompact sets.

Theorem 5.93 *Let E be a locally convex Hausdorff topological vector space and X be a non-empty paracompact convex subset of E. Let $S : X \to 2^X$ be upper semicontinuous such that each $S(x)$ is compact convex and $T : X \to 2^{E^*}$*

be monotone and lower hemi-continuous along line segments in X *to the weak***-topology on* E^*. *Let* $h : X \to \mathbb{R}$ *be convex and continuous. Suppose that the set*

$$\Sigma = \{y \in X : \sup_{x\in S(y)} \sup_{u\in T(x)} Re\langle u, y - x\rangle + h(y) - h(x) > 0\}$$

is open in X. *Suppose further that there exist a non-empty compact convex subset* X_0 *of* X *and a non-empty compact subset* K *of* X *such that for each* $y \in X\backslash K$, *there exists a point* $x \in co(X_0 \cup \{y\}) \cap S(y)$ *with* $\sup_{u\in T(x)} Re\langle u, y-x\rangle + h(y) - h(x) > 0$. *Then there exists a point* $\hat{y} \in K$ *such that*

(*i*) $\hat{y} \in S(\hat{y})$ *and*

(*ii*) $\sup_{w\in T(\hat{y})} Re\langle w, \hat{y} - x\rangle \leq h(x) - h(\hat{y})$ *for all* $x \in S(\hat{y})$.

Proof. The proof is divided into two steps as follows:

Step 1. There exists a point $\hat{y} \in X$ such that $\hat{y} \in S(\hat{y})$ and

$$\sup_{x\in S(\hat{y})} [\sup_{u\in T(x)} Re\langle u, \hat{y} - x\rangle + h(\hat{y}) - h(x)] \leq 0.$$

Suppose the contrary. Then for each $y \in X$, either $y \notin S(y)$ or there exist $x \in S(y)$ and $u \in T(x)$ such that $Re\langle u, y - x\rangle + h(y) - h(x) > 0$; that is, for each $y \in X$, either $y \notin S(y)$ or $y \in \Sigma$. If $y \notin S(y)$, then by Hahn-Banach separation theorem, there exists $p \in E^*$ such that $Re\langle p, y\rangle - \sup_{x\in S(y)} Re\langle p, x\rangle > 0$. For each $y \in X$, set

$$\gamma(y) := \sup_{x\in S(y)} [\sup_{u\in T(x)} Re\langle u, y - x\rangle + h(y) - h(x)].$$

Let $V_0 := \{y \in X | \gamma(y) > 0\} = \Sigma$ and for each $p \in E^*$, set

$$V_p := \{y \in X : Re\langle p, y\rangle - \sup_{x\in S(y)} Re\langle p, x\rangle > 0\}.$$

Then $X = V_0 \cup \bigcup_{p\in E^*} V_p$. Since each V_p is open in X by Lemma 5.36 and V_0 is open in X by hypothesis, $\{V_0, V_p : p \in E^*\}$ is an open covering for X. Since X is paracompact, there is a continuous partition of unity $\{\beta_0, \beta_p : p \in E^*\}$ for X subordinated to the open cover $\{V_0, V_p : p \in E^*\}$ (see, e.g., Theorem VIII.4.2 of Dugundji in Dugundji (1966)); that is for each $p \in E^*$, $\beta_p : X \to [0, 1]$ and $\beta_0 : X \to [0, 1]$ are continuous functions such that for each $p \in E^*$, $\beta_p(y) = 0$ for all $y \in X \setminus V_p$ and $\beta_0(y) = 0$ for all $y \in X \setminus V_0$ and {support β_0, support $\beta_p : p \in E^*$} is locally finite and $\beta_0(y) + \Sigma_{p\in E^*} \beta_p(y) = 1$ for each $y \in X$. Define $\phi, \psi : X \times X \to \mathbb{R}$ by

$$\phi(x, y) = \beta_0(y)[\sup_{u\in T(x)} Re\langle u, y - x\rangle + h(y) - h(x)] + \Sigma_{p\in E^*} \beta_p(y) Re\langle p, y - x\rangle,$$

and

$$\psi(x,y) = \beta_0(y)[\inf_{w\in T(y)} Re\langle w, y-x\rangle + h(y) - h(x)] + \Sigma_{p\in E^*}\beta_p(y)Re\langle p, y-x\rangle,$$

for each $x, y \in X$. Then we have the following.

(1) For each $x, y \in X$, since T is monotone, $\phi(x,y) \le \psi(x,y)$ and $\psi(x,x) = 0$ for all $x \in X$.

(2) For each fixed $x \in X$ and each fixed $u \in T(x)$, the map

$$y \longmapsto Re\langle u, y-x\rangle + h(y) - h(x)$$

is continuous on X and therefore the map

$$y \longmapsto \beta_0(y)[\sup_{u\in T(x)} Re\langle u, y-x\rangle + h(y) - h(x)]$$

is lower semicontinuous on X by Lemma 3 in [Takahashi (1976), p. 177]. Also for each fixed $x \in X$,

$$y \longmapsto \Sigma_{p\in E^*}\beta_p(y)Re\langle p, y-x\rangle$$

is continuous on X. Hence, for each fixed $x \in X$, the map $y \longmapsto \phi(x,y)$ is lower semicontinuous on X.

(3) Clearly, for each $y \in X$, the set $\{x \in X : \psi(x,y) > 0\}$ is convex.

(4) By hypothesis, there exists a non-empty compact convex subset X_0 of X and a non-empty compact subset K of X such that for each $y \in X \setminus K$, there exists a point $x \in co(X_0 \cup \{y\}) \cap S(y)$ such that $\sup_{u\in T(x)} Re\langle u, y-x\rangle + h(y) - h(x) > 0$. Thus $\beta_0(y)[\sup_{u\in T(x)} Re\langle u, y-x\rangle + h(y) - h(x)] > 0$ whenever $\beta_0(y) > 0$. Also $Re\langle p, y-x\rangle > 0$ whenever $\beta_p(y) > 0$ for $p \in E^*$. Consequently, $\phi(x,y) = \beta_0(y)[\sup_{u\in T(x)} Re\langle u, y-x\rangle + h(y) - h(x)] + \Sigma_{p\in E^*}\beta_p(y)Re\ \langle p,\ y-x\rangle > 0$.

Then ϕ and ψ satisfy all the hypotheses of Theorem 5.92. Thus by Theorem 5.92, there exists $\hat{y} \in K$ such that $\phi(x, \hat{y}) \le 0$ for all $x \in X$, i.e.,

$$\beta_0(\hat{y})[\sup_{u\in T(x)} Re\langle u, \hat{y}-x\rangle + h(\hat{y}) - h(x)] + \Sigma_{p\in E^*}\beta_p(\hat{y})Re\langle p, \hat{y}-x\rangle \le 0 \qquad (5.146)$$

for all $x \in X$.

If $\beta_0(\hat{y}) > 0$, then $\hat{y} \in V_0 = \Sigma$ so that $\gamma(\hat{y}) > 0$. Choose $\hat{x} \in S(\hat{y}) \subset X$ such that

$$\sup_{u\in T(\hat{x})} Re\langle u, \hat{y}-\hat{x}\rangle + h(\hat{y}) - h(\hat{x}) \ge \frac{\gamma(\hat{y})}{2} > 0;$$

it follows that

$$\beta_0(\hat{y})[\sup_{u\in T(\hat{x})} Re\langle u, \hat{y}-\hat{x}\rangle + h(\hat{y}) - h(\hat{x})] > 0.$$

If $\beta_p(\hat{y}) > 0$ for some $p \in E^*$, then $\hat{y} \in V_p$ and hence

$$Re\langle p, \hat{y}\rangle > \sup_{x \in S(\hat{y})} Re\langle p, x\rangle \geq Re\langle p, \hat{x}\rangle$$

so that $Re\langle p, \hat{y} - \hat{x}\rangle > 0$. Then note that

$$\beta_p(\hat{y}) Re\langle p, \hat{y} - \hat{x}\rangle > 0 \quad \text{whenever} \quad \beta_p(\hat{y}) > 0 \text{ for } \ p \in E^*.$$

Since $\beta_0(\hat{y}) > 0$ or $\beta_p(\hat{y}) > 0$ for some $p \in E^*$, it follows that

$$\phi(\hat{x}, \hat{y}) = \beta_0(\hat{y})[\sup_{u \in T(\hat{x})} Re\langle u, \hat{y} - \hat{x}\rangle + h(\hat{y}) - h(\hat{x})] + \Sigma_{p \in E^*} \beta_p(\hat{y}) Re\langle p, \hat{y} - \hat{x}\rangle > 0,$$

which contradicts (5.146). This contradiction proves Step 1.

Step 2.

$$\sup_{w \in T(\hat{y})} Re\langle w, \hat{y} - x\rangle \leq h(x) - h(\hat{y}) \quad \text{for all} \quad x \in S(\hat{y}).$$

Indeed, from Step 1, $\hat{y} \in S(\hat{y})$ which is a convex subset of X, and

$$\sup_{u \in T(x)} Re\langle u, \hat{y} - x\rangle \leq h(x) - h(\hat{y}) \quad \text{for all} \quad x \in S(\hat{y}).$$

Hence by Lemma 5.39, we have

$$\sup_{w \in T(\hat{y})} Re\langle w, \hat{y} - x\rangle \leq h(x) - h(\hat{y}) \quad \text{for all} \quad x \in S(\hat{y}). \qquad \square$$

If X is compact, Theorem 5.93 reduces to the following of Chowdhury and Tan:

Theorem 5.94 *Let E be a locally convex Hausdorff topological vector space and X be a non-empty compact convex subset of E. Let $S : X \to 2^X$ be upper semicontinuous such that each $S(x)$ is closed convex and $T : X \to 2^{E^*}$ be monotone and lower hemi-continuous along line segments in X to the weak*-topology on E^*. Let $h : X \to \mathbb{R}$ be convex and continuous. Suppose that the set*

$$\Sigma = \{y \in X : \sup_{x \in S(y)} \sup_{u \in T(x)} Re\langle u, y - x\rangle + h(y) - h(x) > 0\}$$

is open in X. Then there exists a point $\hat{y} \in X$ such that

(i) *$\hat{y} \in S(\hat{y})$ and*

(ii) *$\sup_{w \in T(\hat{y})} Re\langle w, \hat{y} - x\rangle \leq h(x) - h(\hat{y})$ for all $x \in S(\hat{y})$.*

Remark 5.32 Theorem 5.93 and Theorem 5.94 generalize Theorem 1 of Shih and Tan (1985, p. 335).

Chowdhury and Tan observed that if X is also bounded in Theorem 5.93 and the map $S : X \to 2^X$ is, in addition, lower semicontinuous and for each $y \in \Sigma = \{y \in X : \sup_{x\in S(y)}[\sup_{u\in T(x)} Re\langle u, y-x\rangle + h(y) - h(x)] > 0\}$, T is lower semicontinuous at some point x in $S(y)$ with $\sup_{u\in T(x)} Re\langle u, y-x\rangle + h(y) - h(x) > 0$, then the set Σ in Theorem 5.93 is always open in X. The following theorem is presented with these additional assumptions.

Theorem 5.95 *Let E be a locally convex Hausdorff topological vector space and X be a non-empty paracompact convex and bounded subset of E. Let $S : X \to 2^X$ be continuous such that each $S(x)$ is compact convex and $T : X \to 2^{E^*}$ be monotone and be lower hemi-continuous along line segments in X to the weak*-topology on E^*. Let $h : X \to \mathbb{R}$ be convex and continuous. Suppose that for each $y \in \Sigma = \{y \in X : \sup_{x\in S(y)}[\sup_{u\in T(x)} Re\langle u, y-x\rangle + h(y) - h(x)] > 0\}$, T is lower semicontinuous at some point x in $S(y)$ with $\sup_{u\in T(x)} Re\langle u, y-x\rangle + h(y) - h(x) > 0$. Suppose further that there exist a non-empty compact convex subset X_0 of X and a non-empty compact subset K of X such that for each $y \in X \setminus K$, there exists a point $x \in co(X_0 \cup \{y\}) \cap S(y)$ with $\sup_{u\in T(x)} Re\langle u, y-x\rangle + h(y) - h(x) > 0$. Then there exists a point $\hat{y} \in K$ such that*

(i) $\hat{y} \in S(\hat{y})$ and

(ii) $\sup_{w\in T(\hat{y})} Re\langle w, \hat{y} - x\rangle \le h(x) - h(\hat{y})$ for all $x \in S(\hat{y})$.

Proof. By virtue of Theorem 5.93, we need only show that the set

$$\Sigma := \{y \in X : \sup_{x\in S(y)} [\sup_{u\in T(x)} Re\langle u, y-x\rangle + h(y) - h(x)] > 0\}$$

is open in X. Indeed, let $y_0 \in \Sigma$; then by hypothesis, T is lower semicontinuous at some point x_0 in $S(y_0)$ with $\sup_{u\in T(x_0)} Re\langle u, y_0 - x_0\rangle + h(y_0) - h(x_0) > 0$. Hence there exists $u_0 \in T(x_0)$ such that $Re\langle u_0, y_0 - x_0\rangle + h(y_0) - h(x_0) > 0$. Let

$$\alpha := Re\langle u_0, y_0 - x_0\rangle + h(y_0) - h(x_0).$$

Then $\alpha > 0$. Also let

$$U_1 := \{u \in E^* : \sup_{z_1,z_2\in X} |\langle u - u_0, z_1 - z_2\rangle| < \frac{\alpha}{6}\}.$$

Then U_1 is a strongly open neighborhood of u_0 in E^*. Since T is lower semicontinuous at x_0 and $U_1 \cap T(x_0) \neq \emptyset$, there exists an open neighborhood V_1 of x_0 in X such that $T(x) \cap U_1 \neq \emptyset$ for all $x \in V_1$.

As the map $x \longmapsto Re\langle u_0, x_0 - x\rangle + h(x_0) - h(x)$ is continuous at x_0, there exists an open neighborhood V_2 of x_0 in X such that

$$|Re\langle u_0, x_0 - x\rangle + h(x_0) - h(x)| < \frac{\alpha}{6} \quad \text{for all} \quad x \in V_2.$$

Let $V_0 := V_1 \cap V_2$; then V_0 is an open neighborhood of x_0 in X. Since $x_0 \in V_0 \cap S(y_0) \neq \emptyset$ and S is lower semicontinuous at y_0, there exists an open neighborhood N_1 of y_0 in X such that $S(y) \cap V_0 \neq \emptyset$ for all $y \in N_1$.

Since the map $y \longmapsto Re\langle u_0, y - y_0\rangle + h(y) - h(y_0)$ is continuous at y_0, there exists an open neighborhood N_2 of y_0 in X such that

$$|Re\langle u_0, y - y_0\rangle + h(y) - h(y_0)| < \frac{\alpha}{6} \quad \text{for all} \quad y \in N_2.$$

Let $N_0 := N_1 \cap N_2$. Then N_0 is an open neighborhood of y_0 in X such that for each $y_1 \in N_0$, we have

(i) $S(y_1) \cap V_0 \neq \emptyset$ as $y_1 \in N_1$; so we can choose any $x_1 \in S(y_1) \cap V_0$;

(ii) $|Re\langle u_0, y_1 - y_0\rangle + h(y_1) - h(y_0)| < \frac{\alpha}{6}$ as $y_1 \in N_2$;

(iii) $T(x_1) \cap U_1 \neq \emptyset$ as $x_1 \in V_1$; choose any $u_1 \in T(x_1) \cap U_1$ so that

$$\sup_{z_1, z_2 \in X} |\langle u_1 - u_0, z_1 - z_2\rangle| < \frac{\alpha}{6};$$

(iv) $|Re\langle u_0, x_0 - x_1\rangle + h(x_0) - h(x_1)| < \frac{\alpha}{6}$ as $x_1 \in V_2$.

It follows that

$$\begin{aligned}
&Re\langle u_1, y_1 - x_1\rangle + h(y_1) - h(x_1) \\
&\quad = Re\langle u_1 - u_0, y_1 - x_1\rangle + Re\langle u_0, y_1 - x_1\rangle + h(y_1) - h(x_1) \\
&\quad \geq -\frac{\alpha}{6} + Re\langle u_0, y_1 - y_0\rangle + h(y_1) - h(y_0) \\
&\qquad + Re\langle u_0, y_0 - x_0\rangle + h(y_0) - h(x_0) \\
&\qquad + Re\langle u_0, x_0 - x_1\rangle + h(x_0) - h(x_1) \ \text{(by (iii))}, \\
&\quad \geq -\frac{\alpha}{6} - \frac{\alpha}{6} + \alpha - \frac{\alpha}{6} = \frac{\alpha}{2} > 0 \ \text{(by } (ii) \text{ and (iv))};
\end{aligned}$$

therefore

$$\sup_{x \in S(y_1)} [\sup_{u \in T(x)} Re\langle u, y_1 - x\rangle + h(y_1) - h(x)] > 0$$

as $x_1 \in S(y_1)$ and $u_1 \in T(x_1)$. This shows that $y_1 \in \Sigma$ for all $y_1 \in N_0$, so that Σ is open in X. This completes the proof. □

If X is compact, Theorem 5.95 reduces to the following result of Chowdhury and Tan:

Theorem 5.96 *Let E be a locally convex Hausdorff topological vector space and X be a non-empty compact convex subset of E. Let $S : X \to 2^X$ be continuous such that each $S(x)$ is closed convex and $T : X \to 2^{E^*}$ be monotone and be lower hemi-continuous along line segments in X to the weak*-topology on E^*. Let $h : X \to \mathbb{R}$ be convex and continuous. Suppose that for each $y \in \Sigma = \{y \in X : \sup_{x \in S(y)}[\sup_{u \in T(x)} Re\langle u, y - x\rangle + h(y) - h(x)] > 0\}$, T is lower semicontinuous at* *some point x in $S(y)$ with $\sup_{u\in T(x)} Re\langle u, y-x\rangle + h(y) - h(x) > 0$. Then there exists a point $\hat{y} \in X$ such that*

(i) *$\hat{y} \in S(\hat{y})$ and*

(ii) *$\sup_{w\in T(\hat{y})} Re\langle w, \hat{y}-x\rangle \leq h(x) - h(\hat{y})$ for all $x \in S(\hat{y})$.*

Remark 5.33 Theorem 5.95 and Theorem 5.96 generalize Theorem 2 of Shih-Tan in [Shih and Tan (1985), p. 338]

5.13.2 *Generalized Quasi-Variational Inequalities for Upper Hemi-Continuous Operators*

In this subsection we shall present some existence theorems of Chowdhury and Tan on generalized quasi-variational inequalities for semi-monotone and upper hemi-continuous operators on paracompact sets.

We present the following result of Chowdhury and Tan on these operators mentioned above:

Theorem 5.97 *Let E be a locally convex Hausdorff topological vector space and X be a non-empty paracompact convex and bounded subset of E. Let $S : X \to 2^X$ be upper semicontinuous such that each $S(x)$ is compact convex and $T : X \to 2^{E^*}$ be semi-monotone and be upper hemi-continuous along line segments in X to the weak*-topology on E^* such that each $T(x)$ is strongly compact convex. Let $h : X \to \mathbb{R}$ be convex and continuous. Suppose that the set*

$$\Sigma = \{y \in X : \sup_{x\in S(y)} [\inf_{u\in T(x)} Re\langle u, y-x\rangle + h(y) - h(x)] > 0\}$$

is open in X. Suppose further that there exist a non-empty compact convex subset X_0 of X and a non-empty compact subset K of X such that for each $y \in X\backslash K$, there exists a point $x \in co(X_0 \cup \{y\}) \cap S(y)$ with $\inf_{u\in T(x)} Re\langle u, y-x\rangle + h(y) - h(x) > 0$. Then there exists a point $\hat{y} \in K$ such that

(i) *$\hat{y} \in S(\hat{y})$ and*

(ii) *there exists a point $\hat{w} \in T(\hat{y})$ with $Re\langle \hat{w}, \hat{y}-x\rangle \leq h(x) - h(\hat{y})$ for all $x \in S(\hat{y})$.*

Proof. The proof is divided into three steps as follows:

Step 1. There exists a point $\hat{y} \in X$ such that $\hat{y} \in S(\hat{y})$ and

$$\sup_{x\in S(\hat{y})} [\inf_{u\in T(x)} Re\langle u, \hat{y}-x\rangle + h(\hat{y}) - h(x)] \leq 0.$$

Suppose the contrary. Then for each $y \in X$, either $y \notin S(y)$ or there exists $x \in S(y)$ such that $\inf_{u\in T(x)} Re\langle u, y-x\rangle + h(y) - h(x) > 0$; that is, for each $y \in X$, either $y \notin S(y)$ or $y \in \Sigma$. If $y \notin S(y)$, then by Hahn-Banach separation theorem, there exists $p \in E^*$ such that $Re\langle p, y\rangle - \sup_{x\in S(y)} Re\langle p, x\rangle > 0$. For each $y \in X$, set

$$\gamma(y) := \sup_{x\in S(y)} [\inf_{u\in T(x)} Re\langle u, y-x\rangle + h(y) - h(x)].$$

Let $V_0 := \{y \in X | \gamma(y) > 0\} = \Sigma$ and for each $p \in E^*$, set

$$V_p := \{y \in X : Re\langle p, y\rangle - \sup_{x\in S(y)} Re\langle p, x\rangle > 0\}.$$

Then $X = V_0 \cup \bigcup_{p\in E^*} V_p$. Since each V_p is open in X by Lemma 5.36 and V_0 is open in X by hypothesis, $\{V_0, V_p : p \in E^*\}$ is an open covering for X. Since X is paracompact, there is a continuous partition of unity $\{\beta_0, \beta_p : p \in E^*\}$ for X subordinated to the open cover $\{V_0, V_p : p \in E^*\}$.

Define $\phi, \psi : X \times X \to \mathbb{R}$ by

$$\phi(x, y) = \beta_0(y)[\inf_{u\in T(x)} Re\langle u, y-x\rangle + h(y) - h(x)] + \Sigma_{p\in E^*}\beta_p(y)Re\langle p, y-x\rangle,$$

and

$$\psi(x, y) = \beta_0(y)[\inf_{w\in T(y)} Re\langle w, y-x\rangle + h(y) - h(x)] + \Sigma_{p\in E^*}\beta_p(y)Re\langle p, y-x\rangle$$

for each $x, y \in X$. Then we have the following.

(1) For each $x, y \in X$, since T is semi-monotone, $\phi(x, y) \leq \psi(x, y)$ and $\psi(x, x) = 0$ for all $x \in X$.

(2) For each fixed $x \in X$, the map

$$y \longmapsto \inf_{u\in T(x)} Re\langle u, y-x\rangle + h(y) - h(x)$$

is weakly lower semicontinuous (and therefore lower semicontinuous) on X by Lemma 5.37 and the fact that h is continuous; therefore the map

$$y \longmapsto \beta_0(y)[\inf_{u\in T(x)} Re\langle u, y-x\rangle + h(y) - h(x)]$$

is lower semicontinuous on X by Lemma 3 in [Takahashi (1976), p. 177]. Also for each fixed $x \in X$,

$$y \longmapsto \Sigma_{p\in E^*}\beta_p(y)Re\langle p, y-x\rangle$$

is continuous on X. Hence, for each fixed $x \in X$, the map $y \longmapsto \phi(x, y)$ is lower semicontinuous on X.

(3) Clearly, for each $y \in X$, the set $\{x \in X : \psi(x, y) > 0\}$ is convex.

(4) By hypothesis, there exists a non-empty compact convex subset X_0 of X and a non-empty compact subset K of X such that for each $y \in X \setminus K$, there exists a point $x \in co(X_0 \cup \{y\}) \cap S(y)$ such that $\inf_{u\in T(x)} Re\langle u, y-x\rangle + h(y) - h(x) > 0$. Thus $\beta_0(y)[\inf_{u\in T(x)} Re\langle u, y-x\rangle + h(y) - h(x)] > 0$ whenever $\beta_0(y) > 0$. Also $Re\langle p, y-x\rangle > 0$ whenever $\beta_p(y) > 0$ for $p \in E^*$. Consequently, $\phi(x, y) = \beta_0(y)[\inf_{u\in T(x)} Re\langle u, y-x\rangle + h(y) - h(x)] + \Sigma_{p\in E^*}\beta_p(y)Re\langle p, y-x\rangle > 0$.

Then ϕ and ψ satisfy all the hypotheses of Theorem 5.92. Thus by Theorem 5.92, there exists $\hat{y} \in K$ such that $\phi(x, \hat{y}) \leq 0$ for all $x \in X$, i.e.,

$$\beta_0(\hat{y})[\inf_{u \in T(x)} Re\langle u, \hat{y} - x\rangle + h(\hat{y}) - h(x)] + \Sigma_{p \in E^*} \beta_p(\hat{y}) Re\langle p, \hat{y} - x\rangle \leq 0 \quad (5.147)$$

for all $x \in X$.

If $\beta_0(\hat{y}) > 0$, then $\hat{y} \in V_0 = \Sigma$ so that $\gamma(\hat{y}) > 0$. Choose $\hat{x} \in S(\hat{y}) \subset X$ such that

$$\inf_{u \in T(\hat{x})} Re\langle u, \hat{y} - \hat{x}\rangle + h(\hat{y}) - h(\hat{x}) \geq \frac{\gamma(\hat{y})}{2} > 0;$$

it follows that

$$\beta_0(\hat{y})[\inf_{u \in T(\hat{x})} Re\langle u, \hat{y} - \hat{x}\rangle + h(\hat{y}) - h(\hat{x})] > 0.$$

If $\beta_p(\hat{y}) > 0$ for some $p \in E^*$, then $\hat{y} \in V_p$ and hence

$$Re\langle p, \hat{y}\rangle > \sup_{x \in S(\hat{y})} Re\langle p, x\rangle \geq Re\langle p, \hat{x}\rangle$$

so that $Re\langle p, \hat{y} - \hat{x}\rangle > 0$. Then note that

$$\beta_p(\hat{y}) Re\langle p, \hat{y} - \hat{x}\rangle > 0 \quad \text{whenever} \quad \beta_p(\hat{y}) > 0 \text{ for } \ p \in E^*.$$

Since $\beta_0(\hat{y}) > 0$ or $\beta_p(\hat{y}) > 0$ for some $p \in E^*$, it follows that

$$\phi(\hat{x}, \hat{y}) = \beta_0(\hat{y})[\inf_{u \in T(\hat{x})} Re\langle u, \hat{y} - \hat{x}\rangle + h(\hat{y}) - h(\hat{x})] + \Sigma_{p \in E^*} \beta_p(\hat{y}) Re\langle p, \hat{y} - \hat{x}\rangle > 0,$$

which contradicts (5.147). This contradiction proves Step 1.

Step 2.

$$\inf_{w \in T(\hat{y})} Re\langle w, \hat{y} - x\rangle \leq h(x) - h(\hat{y}) \quad \text{for all} \quad x \in S(\hat{y}).$$

Indeed, from Step 1, $\hat{y} \in S(\hat{y})$ which is a convex subset of X, and

$$\inf_{u \in T(x)} Re\langle u, \hat{y} - x\rangle \leq h(x) - h(\hat{y}) \quad \text{for all} \quad x \in S(\hat{y}).$$

Hence by Lemma 5.38, we have

$$\inf_{w \in T(\hat{y})} Re\langle w, \hat{y} - x\rangle \leq h(x) - h(\hat{y}) \quad \text{for all} \quad x \in S(\hat{y}). \quad (5.148)$$

Step 3. There exist a point $\hat{w} \in T(\hat{y})$ with $Re\langle \hat{w}, \hat{y} - x\rangle \leq h(x) - h(\hat{y})$ for all $x \in S(\hat{y})$.

Indeed, from Step 2 we have

$$\sup_{x \in S(\hat{y})} [\inf_{w \in T(\hat{y})} Re\langle w, \hat{y} - x\rangle + h(\hat{y}) - h(x)] \leq 0, \quad (5.149)$$

where $T(\hat{y})$ is a strongly compact convex subset of the Hausdorff topological vector space E^* and $S(\hat{y})$ is a convex subset of X.

Now, define $f : S(\hat{y}) \times T(\hat{y}) \to \mathbb{R}$ by $f(x,w) = Re\langle w, \hat{y} - x\rangle + h(\hat{y}) - h(x)$ for each $x \in S(\hat{y})$ and each $w \in T(\hat{y})$. Note that for each fixed $x \in S(\hat{y})$, the map $w \longmapsto f(x,w)$ is convex and continuous on $T(\hat{y})$ and for each fixed $w \in T(\hat{y})$, the map $x \longmapsto f(x,w)$ is concave on $S(\hat{y})$. Thus by applying Kneser's minimax Theorem 5.44, we have

$$\min_{w\in T(\hat{y})} \sup_{x\in S(\hat{y})} [Re\langle w, \hat{y}-x\rangle + h(\hat{y}) - h(x)] = \sup_{x\in S(\hat{y})} \min_{w\in T(\hat{y})} [Re\langle w, \hat{y}-x\rangle + h(\hat{y}) - h(x)].$$

Hence

$$\min_{w\in T(\hat{y})} \sup_{x\in S(\hat{y})} [Re\langle w, \hat{y}-x\rangle + h(\hat{y}) - h(x)] \leq 0, \quad \text{by (5.149)}.$$

Since $T(\hat{y})$ is compact, there exists $\hat{w} \in T(\hat{y})$ such that

$$Re\langle \hat{w}, \hat{y}-x\rangle \leq h(x) - h(\hat{y}) \quad \text{for all} \quad x \in S(\hat{y}). \qquad \square$$

If X is compact, Theorem 5.97 reduces to the following of Chowdhury and Tan:

Theorem 5.98 *Let E be a locally convex Hausdorff topological vector space and X be a non-empty compact convex subset of E. Let $S : X \to 2^X$ be upper semicontinuous such that each $S(x)$ is closed convex and $T : X \to 2^{E^*}$ be semi-monotone and be upper hemi-continuous along line segments in X to the weak*-topology on E^* such that each $T(x)$ is strongly compact convex. Let $h : X \to \mathbb{R}$ be convex and continuous. Suppose that the set*

$$\Sigma = \{y \in X : \sup_{x\in S(y)} [\inf_{u\in T(x)} Re\langle u, y-x\rangle + h(y) - h(x)] > 0\}$$

is open in X. Then there exists a point $\hat{y} \in K$ such that

(i) $\hat{y} \in S(\hat{y})$ and

(ii) there exists a point $\hat{w} \in T(\hat{y})$ with $Re\langle \hat{w}, \hat{y}-x\rangle \leq h(x) - h(\hat{y})$ for all $x \in S(\hat{y})$.

Chowdhury and Tan observed that if the map $S : X \to 2^X$ is, in addition, lower semicontinuous and for each $y \in \Sigma$, T is upper semicontinuous at some point x in $S(y)$ with $\inf_{u\in T(x)} Re\langle u, y-x\rangle + h(y) - h(x)] > 0$, then the set Σ in Theorem 5.97 becomes an open set in X. The following theorem is presented with these additional conditions:

Theorem 5.99 *Let E be a locally convex Hausdorff topological vector space and X be a non-empty paracompact convex and bounded subset of E. Let $S : X \to 2^X$ be continuous such that each $S(x)$ is compact convex and $T : X \to 2^F$ be semi-monotone and be upper hemi-continuous along line segments in X to the weak*-topology on E^* such that each $T(x)$ is strongly compact convex. Let $h : X \to \mathbb{R}$ be convex and continuous. Suppose that for each $y \in \Sigma = \{y \in X :$*

$\sup_{x\in S(y)}[\inf_{u\in T(x)} Re\langle u, y-x\rangle + h(y) - h(x)] > 0\}$, T is upper semicontinuous at some point x in $S(y)$ with $\inf_{u\in T(x)} Re\langle u, y-x\rangle + h(y) - h(x) > 0$. Suppose further that there exist a non-empty compact convex subset X_0 of X and a non-empty compact subset K of X such that for each $y \in X \setminus K$, there exists a point $x \in co(X_0 \cup \{y\}) \cap S(y)$ with $\inf_{u\in T(x)} Re\langle u, y-x\rangle + h(y) - h(x) > 0$. Then there exists $\hat{y} \in K$ such that

(*i*) *$\hat{y} \in S(\hat{y})$ and*

(*ii*) *there exist a point $\hat{w} \in T(\hat{y})$ with $Re\langle \hat{w}, \hat{y}-x\rangle \leq h(x)-h(\hat{y})$ for all $x \in S(\hat{y})$.*

Proof. By virtue of Theorem 5.97, it suffices to show that the set

$$\Sigma = \{y \in X : \sup_{x\in S(y)} [\inf_{u\in T(x)} Re\langle u, y-x\rangle + h(y) - h(x)] > 0\}$$

is open in X. Indeed, let $y_0 \in \Sigma$; then by hypothesis, T is upper semicontinuous at some point x_0 in $S(y_0)$ with $\inf_{u\in T(x_0)} Re\langle u, y_0-x_0\rangle + h(y_0) - h(x_0) > 0$. Let

$$\alpha := \inf_{u\in T(x_0)} Re\langle u, y_0 - x_0\rangle + h(y_0) - h(x_0).$$

Then $\alpha > 0$. Also let

$$W := \{w \in E^* : \sup_{z_1,z_2\in X} |\langle w, z_1 - z_2\rangle| < \alpha/6\}.$$

Then W is a strongly open neighborhood of 0 in E^* so that $U_1 := T(x_0) + W$ is an open neighborhood of $T(x_0)$ in E^*. Since T is upper semicontinuous at x_0, there exists an open neighborhood V_1 of x_0 in X such that $T(x) \subset U_1$ for all $x \in V_1$.

As the map $x \longmapsto \inf_{u\in T(x_0)} Re\langle u, x_0 - x\rangle + h(x_0) - h(x)$ is continuous at x_0, there exists an open neighborhood V_2 of x_0 in X such that

$$|\inf_{u\in T(x_0)} Re\langle u, x_0 - x\rangle + h(x_0) - h(x)| < \alpha/6 \quad \text{for all} \quad x \in V_2.$$

Let $V_0 := V_1 \cap V_2$; then V_0 is an open neighborhood of x_0 in X. Since $x_0 \in V_0 \cap S(y_0) \neq \emptyset$ and S is lower semicontinuous at y_0, there exists an open neighborhood N_1 of y_0 in X such that $S(y) \cap V_0 \neq \emptyset$ for all $y \in N_1$.

Since the map $y \longmapsto \inf_{u\in T(x_0)} Re\langle u, y - y_0\rangle + h(y) - h(y_0)$ is continuous at y_0, there exists an open neighborhood N_2 of y_0 in X such that

$$|\inf_{u\in T(x_0)} Re\langle u, y - y_0\rangle + h(y) - h(y_0)| < \alpha/6 \quad \text{for all} \quad y \in N_2.$$

Let $N_0 := N_1 \cap N_2$. Then N_0 is an open neighborhood of y_0 in X such that for each $y_1 \in N_0$, we have

(*i*) $S(y_1) \cap V_0 \neq \emptyset$ as $y_1 \in N_1$; so we can choose any $x_1 \in S(y_1) \cap V_0$;
(*ii*) $|\inf_{u\in T(x_0)} Re\langle u, y_1 - y_0\rangle + h(y_1) - h(y_0)| < \alpha/6$ as $y_1 \in N_2$;
(*iii*) $T(x_1) \subset U_1 = T(x_0) + W$ as $x_1 \in V_1$;
(*iv*) $|\inf_{u\in T(x_0)} Re\langle u, x_0 - x_1\rangle + h(x_0) - h(x_1)| < \alpha/6$ as $x_1 \in V_2$.

It follows that

$$\begin{aligned}
&\inf_{u\in T(x_1)} Re\langle u, y_1 - x_1\rangle + h(y_1) - h(x_1)\\
&\geq \inf_{[u\in T(x_0)+W]} Re\langle u, y_1 - x_1\rangle + h(y_1) - h(x_1) \quad (\text{by } (iii)),\\
&\geq \inf_{u\in T(x_0)} Re\langle u, y_1 - x_1\rangle + h(y_1) - h(x_1) + \inf_{u\in W} Re\langle u, y_1 - x_1\rangle\\
&\geq \inf_{u\in T(x_0)} Re\langle u, y_1 - y_0\rangle + h(y_1) - h(y_0)\\
&\quad + \inf_{u\in T(x_0)} Re\langle u, y_0 - x_0\rangle + h(y_0) - h(x_0)\\
&\quad + \inf_{u\in T(x_0)} Re\langle u, x_0 - x_1\rangle + h(x_0) - h(x_1) + \inf_{u\in W} Re\langle u, y_1 - x_1\rangle\\
&\geq -\frac{\alpha}{6} + \alpha - \frac{\alpha}{6} - \frac{\alpha}{6} = \frac{\alpha}{2} > 0 \quad (\text{by } (ii) \text{ and } (iv));
\end{aligned}$$

therefore

$$\sup_{x\in S(y_1)} [\inf_{u\in T(x)} Re\langle u, y_1 - x\rangle + h(y_1) - h(x)] > 0$$

as $x_1 \in S(y_1)$. This shows that $y_1 \in \Sigma$ for all $y_1 \in N_0$, so that Σ is open in X. This completes the proof. □

If X is compact, Theorem 5.99 reduces to the following result of Chowdhury and Tan:

Theorem 5.100 *Let E be a locally convex Hausdorff topological vector space and X be a non-empty compact convex subset of E. Let $S : X \to 2^X$ be continuous such that each $S(x)$ is closed convex and $T : X \to 2^{E^*}$ be semi-monotone and be upper hemi-continuous along line segments in X to the weak*-topology on E^* such that each $T(x)$ is strongly compact convex. Let $h : X \to \mathbb{R}$ be convex and continuous. Suppose that for each $y \in \Sigma = \{y \in X : \sup_{x\in S(y)}[\inf_{u\in T(x)} Re\langle u, y - x\rangle + h(y) - h(x)] > 0\}$, T is upper semicontinuous at some point x in $S(y)$ with $\inf_{u\in T(x)} Re\langle u, y - x\rangle + h(y) - h(x) > 0$. Then there exists $\hat{y} \in X$ such that*

(i) $\hat{y} \in S(\hat{y})$ and

(ii) there exists a point $\hat{w} \in T(\hat{y})$ with $Re\langle \hat{w}, \hat{y} - x\rangle \leq h(x) - h(\hat{y})$ for all $x \in S(\hat{y})$.

Note that the results of this section have been written and presented after a thorough revision of some results of an article in Chowdhury and Tan (1999).

5.14 Generalized Quasi-Variational Inequalities for Upper Semi-Continuous Operators on Non-Compact Sets

In this section we shall present some existence theorems of Chowdhury and Tan on generalized quasi-variational inequalities for upper semi-continuous operators on paracompact convex sets. In obtaining these results Chowdhury and Tan mainly used the following generalized version of Ky Fan's minimax inequality (Fan (1972)) in [Ding and Tan (1992b, Theorem 1)].

Theorem 5.101 *Let X be a non-empty convex subset of a topological vector space E and let $f : X \times X \to \mathbb{R} \cup \{-\infty, +\infty\}$ be such that*

(a) *for each fixed $x \in X$, the map $y \mapsto f(x,y)$ is lower semicontinuous on each non-empty compact subset C of X;*
(b) *for each $A \in \mathcal{F}(X)$ and for each $y \in co(A)$, $\min_{x \in A} f(x,y) \leq 0$;*
(c) *there exists a non-empty compact convex subset X_0 of X and a non-empty compact subset K of X such that for each $y \in X \setminus K$, there exists an $x \in co(X_0 \cup \{y\})$ with $f(x,y) > 0$.*

Then there exists a point $\hat{y} \in K$ such that $f(x,\hat{y}) \leq 0$ for all $x \in X$.

Chowdhury and Tan used the following result which is Lemma 2.2.7 in Tan (1991) (see also the proof of Theorem 21 of Takahashi in Takahashi (1976)):

Lemma 5.40 *Let E be a topological vector space and E^* be the continuous dual of E equipped with the strong topology. Let X be a non-empty compact subset of E and $T : X \to 2^{E^*}$ be upper semi-continuous such that $T(x)$ is strongly compact for each $x \in X$. Define $f : X \times X \to \mathbb{R}$ by $f(x,y) = \inf_{w \in T(y)} Re\langle w, y - x\rangle$, for each $x, y \in X$. Then f is lower semi-continuous on $X \times X$.*

In obtaining the main results, Chowdhury and Tan also used the following result which is Lemma 3 of Takahshi in [Takahashi (1976)](see also Lemma 3 in [Shih and Tan (1989)]):

Lemma 5.41 *Let X and Y be topological spaces, $f : X \to \mathbb{R}$ be non-negative and continuous and $g : Y \to \mathbb{R}$ be lower semi-continuous. Then the map $F : X \times Y \to \mathbb{R}$, defined by $F(x,y) = f(x)g(y)$ for all $(x,y) \in X \times Y$, is lower semi-continuous.*

5.14.1 *Non-Compact Generalized Quasi-Variational Inequalities*

We shall first present a result of Chowdhury and Tan to a non-compact setting which generalizes Theorem 3 in Shih and Tan (1985):

Theorem 5.102 *Let E be a locally convex Hausdorff topological vector space and X be a non-empty para-compact convex subset of E. Let $S : X \to 2^X$ be upper semi-continuous such that each $S(x)$ is a non-empty compact convex subset of X and $T : X \to 2^{E^*}$ be upper semi-continuous from the relative topology of X to the strong topology of E^* such that each $T(x)$ is a strongly compact convex subset of E^*. Let $h : X \to \mathbb{R}$ be convex and continuous. Suppose that the set*

$$\Sigma = \{y \in X : \sup_{x \in S(y)} [\inf_{w \in T(y)} Re\langle w, y - x\rangle + h(y) - h(x)] > 0\}$$

is open in X. Suppose further that there exist a non-empty compact convex subset X_0 of X and a non-empty compact subset K of X such that for each $y \in X \backslash K$, there exists a point $x \in co(X_0 \cup \{y\}) \cap S(y)$ with $\inf_{w \in T(y)} Re\langle w, y-x \rangle + h(y) - h(x) > 0$. Then there exists $\hat{y} \in X$ such that

(i) *$\hat{y} \in S(\hat{y})$ and*

(ii) *there exists $\hat{w} \in T(\hat{y})$ with $Re\langle \hat{w}, \hat{y} - x \rangle \leq h(x) - h(\hat{y})$ for all $x \in S(\hat{y})$.*

Proof. The proof is divided into two steps as follows:

Step 1. There exists a point $\hat{y} \in X$ such that $\hat{y} \in S(\hat{y})$ and

$$\sup_{x \in S(\hat{y})} [\inf_{w \in T(\hat{y})} Re\langle w, \hat{y} - x \rangle + h(\hat{y}) - h(x)] \leq 0.$$

Suppose the contrary. Then for each $y \in X$, either $y \notin S(y)$ or there exists $x \in S(y)$ such that $\inf_{w \in T(y)} Re\langle w, y - x \rangle + h(y) - h(x) > 0$; that is, $y \notin S(y)$ or $y \in \Sigma$. If $y \notin S(y)$, then by Hahn-Banach separation theorem, there exists $p \in E^*$ such that $Re\langle p, y \rangle - \sup_{x \in S(y)} Re\langle p, x \rangle > 0$. For each $y \in X$, set

$$\gamma(y) := \sup_{x \in S(y)} [\inf_{w \in T(y)} Re\langle w, y - x \rangle + h(y) - h(x)].$$

Let $V_0 := \{y \in X | \gamma(y) > 0\} = \Sigma$ and for each $p \in E^*$, set

$$V_p := \{y \in X : Re\langle p, y \rangle - \sup_{x \in S(y)} Re\langle p, x \rangle > 0\}.$$

Then $X = V_0 \cup \bigcup_{p \in E^*} V_p$. Since each V_p is open in X by Lemma 5.36 and V_0 is open in X by hypothesis, $\{V_0, V_p : p \in E^*\}$ is an open covering for X. Since X is paracompact, there is a continuous partition of unity $\{\beta_0, \beta_p : p \in E^*\}$ for X subordinated to the open cover $\{V_0, V_p : p \in E^*\}$.

Define $\phi : X \times X \to \mathbb{R}$ by

$$\phi(x, y) = \beta_0(y)[\inf_{w \in T(y)} Re\langle w, y - x \rangle + h(y) - h(x)] + \Sigma_{p \in E^*} \beta_p(y) Re\langle p, y - x \rangle,$$

for each $x, y \in X$. Then we have the following.

(1) For each fixed $x \in X$, the map

$$y \mapsto \inf_{w \in T(y)} Re\langle w, y - x \rangle + h(y) - h(x)$$

is lower semi-continuous on each non-empty compact subset of X by Lemma 5.40 and therefore the map

$$y \mapsto \beta_0(y)[\inf_{w \in T(y)} Re\langle w, y - x \rangle + h(y) - h(x)]$$

is lower semi-continuous on each non-empty compact subset of X by Lemma 5.41. Also for each fixed $x \in X$,

$$y \mapsto \Sigma_{p \in E^*} \beta_p(y) Re\langle p, y - x \rangle$$

is continuous on X. Hence, for each fixed $x \in X$, the map $y \mapsto \phi(x,y)$ is lower semi-continuous on each non-empty compact subset of X.

(2) For each $A \in \mathcal{F}(X)$ and for each $y \in co(A)$, $\min_{x\in A}\phi(x,y) \leq 0$. Indeed, if this were false, then for some $A = \{x_1,\cdots,x_n\} \in \mathcal{F}(X)$ and some $y \in co(A)$, say $y = \sum_{i=1}^n \lambda_i x_i$ where $\lambda_1,\cdots,\lambda_n \geq 0$ with $\sum_{i=1}^n \lambda_i = 1$, such that $\min_{1\leq i\leq n}\phi(x_i,y) > 0$. Then for each $i = 1,\cdots,n$, $\beta_0(y)[\inf_{w\in T(y)} Re\langle w, y-x_i\rangle + h(y) - h(x_i)] + \Sigma_{p\in E^*}\beta_p(y)Re\langle p, y-x_i\rangle > 0$ so that $0 = \phi(y,y) = \beta_0(y)[\inf_{w\in T(y)} Re\langle w, y - \sum_{i=1}^n \lambda_i x_i\rangle + h(y) - h(\sum_{i=1}^n \lambda_i x_i)] + \Sigma_{p\in E^*}\beta_p(y)Re\langle p, y - \sum_{i=1}^n \lambda_i x_i\rangle \geq \sum_{i=1}^n \lambda_i(\beta_0(y)[\inf_{w\in T(y)} Re\langle w, y-x_i\rangle + h(y) - h(x_i)] + \Sigma_{p\in E^*}\beta_p(y)Re\langle p, y-x_i\rangle) > 0$, which is a contradiction.

(3) By hypothesis, there exist a non-empty compact convex subset X_0 of X and a non-empty compact subset K of X such that for each $y \in X\setminus K$, there exists a point $x \in co(X_0 \cup \{y\}) \cap S(y)$ such that $\inf_{w\in T(y)} Re\langle w, y-x\rangle + h(y) - h(x) > 0$.

Thus $\beta_0(y)[\inf_{w\in T(y)} Re\langle w, y-x\rangle + h(y) - h(x)] > 0$ whenever $\beta_0(y) > 0$. Also $Re\langle p, y-x\rangle > 0$ whenever $\beta_p(y) > 0$ for $p \in E^*$. Consequently, $\phi(x,y) = \beta_0(y)[\inf_{w\in T(y)} Re\langle w, y-x\rangle + h(y) - h(x)] + \Sigma_{p\in E^*}\beta_p(y)Re\langle p, y-x\rangle > 0$.

Then ϕ satisfies all hypotheses of Theorem 5.101. Hence by Theorem 5.101, there exists a point $\hat{y} \in K$ such that $\phi(x,\hat{y}) \leq 0$ for all $x \in X$; i.e.,

$$\beta_0(\hat{y})[\inf_{w\in T(\hat{y})} Re\langle w, \hat{y}-x\rangle + h(\hat{y}) - h(x)] + \Sigma_{p\in E^*}\beta_p(\hat{y})Re\langle p, \hat{y}-x\rangle \leq 0 \quad (5.150)$$

for all $x \in X$.

If $\beta_0(\hat{y}) > 0$, then $\hat{y} \in V_0 = \Sigma$ so that $\gamma(\hat{y}) > 0$. Choose $\hat{x} \in S(\hat{y}) \subset X$ such that

$$\inf_{w\in T(\hat{y})} Re\langle w, \hat{y}-\hat{x}\rangle + h(\hat{y}) - h(\hat{x}) \geq \frac{\gamma(\hat{y})}{2} > 0;$$

it follows that

$$\beta_0(\hat{y})[\inf_{w\in T(\hat{y})} Re\langle w, \hat{y}-\hat{x}\rangle + h(\hat{y}) - h(\hat{x})] > 0.$$

If $\beta_p(\hat{y}) > 0$ for some $p \in E^*$, then $\hat{y} \in V_p$ and hence

$$Re\langle p, \hat{y}\rangle > \sup_{x\in S(\hat{y})} Re\langle p, x\rangle \geq Re\langle p, \hat{x}\rangle$$

so that $Re\langle p, \hat{y}-\hat{x}\rangle > 0$. Then note that

$$\beta_p(\hat{y})Re\langle p, \hat{y}-\hat{x}\rangle > 0 \quad \text{whenever} \quad \beta_p(\hat{y}) > 0 \text{ for } p \in E^*.$$

Since $\beta_0(\hat{y}) > 0$ or $\beta_p(\hat{y}) > 0$ for some $p \in E^*$, it follows that

$$\phi(\hat{x},\hat{y}) = \beta_0(\hat{y})[\inf_{w\in T(\hat{y})} Re\langle w, \hat{y}-\hat{x}\rangle + h(\hat{y}) - h(\hat{x})] + \Sigma_{p\in E^*}\beta_p(\hat{y})Re\langle p, \hat{y}-\hat{x}\rangle > 0,$$

which contradicts (5.150). This contradiction proves Step 1.

Step 2. There exists a point $\hat{w} \in T(\hat{y})$ such that

$$Re\langle \hat{w}, \hat{y} - x\rangle + h(\hat{y}) - h(x) \leq 0 \text{ for all } x \in S(\hat{y}).$$

Note that for each fixed $x \in S(\hat{y})$, $w \mapsto Re\langle w, \hat{y} - x\rangle + h(\hat{y}) - h(x)$ is convex and continuous on $T(\hat{y})$ and for each fixed $w \in T(\hat{y})$, $x \mapsto Re\langle w, \hat{y} - x\rangle + h(\hat{y}) - h(x)$ is concave on $S(\hat{y})$. Thus by Kneser's minimax theorem [Kneser (1952)], we have

$$\min_{w\in T(\hat{y})} \max_{x\in S(\hat{y})} [Re\langle w, \hat{y} - x\rangle + h(\hat{y}) - h(x)] = \max_{x\in S(\hat{y})} \min_{w\in T(\hat{y})} [Re\langle w, \hat{y} - x\rangle + h(\hat{y}) - h(x)].$$

Hence

$$\min_{w\in T(\hat{y})} \max_{x\in S(\hat{y})} [Re\langle w, \hat{y} - x\rangle + h(\hat{y}) - h(x)] \leq 0 \quad \text{by Step 1.}$$

Since $T(\hat{y})$ is compact, there exists $\hat{w} \in T(\hat{y})$ such that

$$Re\langle \hat{w}, \hat{y} - x\rangle + h(\hat{y}) - h(x) \leq 0 \quad \text{for all } x \in S(\hat{y}).$$ □

Chowdhury and Tan observed that if X is also bounded in Theorem 5.102 and the map $S : X \to 2^X$ is, in addition, lower semicontinuous, then the set Σ in Theorem 5.102 becomes an open set in X. The following theorem is presented with these additional conditions:

Theorem 5.103 *Let E be a locally convex Hausdorff topological vector space and X be a non-empty paracompact convex and bounded subset of E. Let $S : X \to 2^X$ be continuous such that each $S(x)$ is a non-empty compact convex subset of X and $T : X \to 2^{E^*}$ be upper semi-continuous from the relative topology of X to the strong topology of E^* such that each $T(x)$ is a strongly compact convex subset of E^*. Let $h : X \to \mathbb{R}$ be convex and continuous. Suppose further that there exists a non-empty compact convex subset X_0 of X and a non-empty compact subset K of X such that for each $y \in X \setminus K$, there exists a point $x \in co(X_0 \cup \{y\}) \cap S(y)$ with $\inf_{w\in T(y)} Re\langle w, y - x\rangle + h(y) - h(x) > 0$. Then there exists a point $\hat{y} \in X$ such that*

(i) $\hat{y} \in S(\hat{y})$ and

(ii) there exists a point $\hat{w} \in T(\hat{y})$ with $Re\langle \hat{w}, \hat{y} - x\rangle \leq h(x) - h(\hat{y})$ for all $x \in S(\hat{y})$.

Proof. Comparing with Theorem 5.102, we see that we shall only need to show that the set

$$\Sigma := \{y \in X : \sup_{x\in S(y)} [\inf_{w\in T(y)} Re\langle w, y - x\rangle + h(y) - h(x)] > 0\}$$

is an open set in X. For, let $y_0 \in \Sigma$; then there exists $x_0 \in S(y_0)$ such that $\alpha := \inf_{w\in T(y_0)} Re\,\langle w, y_0 - x_0\rangle + h(y_0) - h(x_0) > 0$. Let

$$W := \{w \in E^* : \sup_{z_1, z_2 \in X} |\langle w, z_1 - z_2\rangle| < \frac{\alpha}{6}\}.$$

Then W is a strongly open neighborhood of 0 in E^* so that $U_1 := T(y_0) + W$ is an open neighborhood of $T(y_0)$ in E^*. Since T is upper semi-continuous at y_0, there exists an open neighborhood N_1 of y_0 in X such that $T(y) \subset U_1$ for all $y \in N_1$.

As the map $x \longmapsto \inf_{w\in T(y_0)} Re\langle w, x_0 - x\rangle + h(x_0) - h(x)$ is continuous at x_0, there exists an open neighborhood V_1 of x_0 in X such that

$$|\inf_{w\in T(y_0)} Re\langle w, x_0 - x\rangle + h(x_0) - h(x)| < \frac{\alpha}{6} \quad \text{for all} \quad x \in V_1.$$

Since $x_0 \in V_1 \cap S(y_0) \neq \emptyset$ and S is lower semi-continuous at y_0, there exists an open neighborhood N_2 of y_0 in X such that $S(y) \cap V_1 \neq \emptyset$ for all $y \in N_2$.

Since the map $y \longmapsto \inf_{w\in T(y_0)} Re\langle w, y - y_0\rangle + h(y) - h(y_0)$ is continuous at y_0, there exists an open neighborhood N_3 of y_0 in X such that

$$|\inf_{w\in T(y_0)} Re\langle w, y - y_0\rangle + h(y) - h(y_0)| < \frac{\alpha}{6} \quad \text{for all} \quad y \in N_3.$$

Let $N_0 := N_1 \cap N_2 \cap N_3$. Then N_0 is an open neighborhood of y_0 in X such that for each $y_1 \in N_0$, we have

(*i*) $T(y_1) \subset U_1 = T(y_0) + W$ as $y_1 \in N_1$;

(*ii*) $S(y_1) \cap V_1 \neq \emptyset$ as $y_1 \in N_2$; so we can choose any $x_1 \in S(y_1) \cap V_1$;

(*iii*) $|\inf_{w\in T(y_0)} Re\langle w, y_1 - y_0\rangle + h(y_1) - h(y_0)| < \frac{\alpha}{6}$ as $y_1 \in N_3$;

(*iv*) $|\inf_{w\in T(y_0)} Re\langle w, x_0 - x_1\rangle + h(x_0) - h(x_1)| < \frac{\alpha}{6}$ as $x_1 \in V_1$.

It follows that

$$\begin{aligned}
&\inf_{w\in T(y_1)} Re\langle w, y_1 - x_1\rangle + h(y_1) - h(x_1)\\
&\geq \inf_{[w\in T(y_0)+W]} Re\langle w, y_1 - x_1\rangle + h(y_1) - h(x_1) \quad \text{(by (}i\text{))},\\
&\geq \inf_{w\in T(y_0)} Re\langle w, y_1 - x_1\rangle + h(y_1) - h(x_1) + \inf_{w\in W} Re\langle w, y_1 - x_1\rangle\\
&\geq \inf_{w\in T(y_0)} Re\langle w, y_1 - y_0\rangle + h(y_1) - h(y_0)\\
&\quad + \inf_{w\in T(y_0)} Re\langle w, y_0 - x_0\rangle + h(y_0) - h(x_0)\\
&\quad + \inf_{w\in T(y_0)} Re\langle w, x_0 - x_1\rangle + h(x_0) - h(x_1)\\
&\quad + \inf_{w\in W} Re\langle w, y_1 - x_1\rangle\\
&\geq -\frac{\alpha}{6} + \alpha - \frac{\alpha}{6} - \frac{\alpha}{6} = \frac{\alpha}{2} > 0 \quad \text{(by (}iii\text{) and (}iv\text{))};
\end{aligned}$$

therefore $\sup_{x\in S(y_1)}[\inf_{w\in T(y_1)} Re\langle w, y_1 - x\rangle + h(y_1) - h(x)] > 0$ as $x_1 \in S(y_1)$. This shows that $y_1 \in \Sigma$ for all $y_1 \in N_0$ so that Σ is open in X. This proves the theorem. □

Theorem 5.103 generalizes a result in [Shih and Tan (1985), Theorem 4] and in [Kim (1988), Theorem] which is a special case of Theorem 11 in [Shih and Tan (1989)] to non-compact setting.

Note that the results of this section have been written and presented after a thorough revision of some results of an article in Chowdhury and Tan (1997a).

5.15 Generalized Quasi-Variational Inequalities for Pseudo-Monotone Set-Valued Mappings

In this section, Chowdhury and Tan used a generalized version of Ky Fan's minimax inequality [Fan (1972), Theorem 1] in [Chowdhury and Tan (1996), Theorem 2] as a tool to obtain some general theorems on solutions of the generalized quasi-variational inequalities on paracompact sets X in locally convex Hausdorff topological vector spaces where the set-valued operators T are strong pseudo-monotone operators or pseudo-monotone operators and are upper semicontinuous from $co(A)$ to the weak*-topology on E^* for each $A \in \mathcal{F}(X)$.

In all results of this section, Chowdhury and Tan used the following set-valued generalization of the classical pseudo-monotone operator. The classical definition of a pseudo-monotone operator was introduced by Brézis, Nirenberg and Stampacchia in [H. Brézis and Stampacchia (1972)]. In [Chowdhury and Tan (1996), Definition 1], Chowdhury and Tan made a slightly general definition of a pseudo-monotone operator which we called pseudo-monotone type I operators.

Definition 5.20 Let E be a topological vector space, X be a non-empty subset of E and $T : X \to 2^{E^*}$. If $h : X \to \mathbb{R}$, then T is said to be

(1) *h-pseudo-monotone* if for each $y \in X$ and every net $\{y_\alpha\}_{\alpha\in\Gamma}$ in X converging to y with $\limsup_\alpha[\inf_{u\in T(y_\alpha)} \mathrm{Re}\langle u, y_\alpha - y\rangle + h(y_\alpha) - h(y)] \leq 0$, we have $\liminf_\alpha[\inf_{u\in T(y_\alpha)} Re\langle u, y_\alpha - x\rangle + h(y_\alpha) - h(x)] \geq \inf_{w\in T(y)} Re\langle w, y - x\rangle + h(y) - h(x)$ for all $x \in X$;

(2) *pseudo-monotone* if T is h-pseudo-monotone with $h \equiv 0$.

5.15.1 *Generalized Quasi-Variational Inequalities for Strong Pseudo-Monotone Operators*

We shall present Chowdhury and Tan's definition of strong pseudo-monotone operators and some general theorems on solutions of the generalized quasi-variational inequalities on paracompact sets in locally convex Hausdorff topological vector spaces.

Definition 5.21 Let E be a topological vector space, X be a non-empty subset of E and $T : X \to 2^{E^*}$. If $h : X \to \mathbb{R}$, then T is said to be (1) *strong h-pseudo-monotone* if for each continuous function $\theta : X \to [0, 1]$, for each $y \in X$ and every net $\{y_\alpha\}_{\alpha\in\Gamma}$ in X converging to y with $\limsup_\alpha[\theta(y_\alpha)(\inf_{u\in T(y_\alpha)} \mathrm{Re}\langle u, y_\alpha - y\rangle +$

$h(y_\alpha) - h(y))] \leq 0$ we have

$$\limsup_{\alpha}[\theta(y_\alpha)(\inf_{u\in T(y_\alpha)} Re\langle u, y_\alpha - x\rangle + h(y_\alpha) - h(x))] \tag{5.151}$$

$$\geq [\theta(y)(\inf_{w\in T(y)} Re\langle w, y - x\rangle + h(y) - h(x))] \tag{5.152}$$

for all $x \in X$; (2) *strong pseudo-monotone* if T is strong h-pseudo-monotone with $h \equiv 0$.

Clearly, every strong pseudo-monotone operator is also a pseudo-monotone operator as defined in [Chowdhury and Tan (1996)].

Proposition 5.12 *Let X be a non-empty subset of a topological vector space E. If $T : X \to E^*$ is monotone and continuous from the relative weak topology on X to the weak* topology on E^*, then T is strong pseudo-monotone.*

Proof. Let us consider any arbitrary continuous function $\theta : X \to [0,1]$. Suppose $\{y_\alpha\}_{\alpha\in\Gamma}$ is a net in X and $y \in X$ with $y_\alpha \to y$ (and $\limsup_\alpha[\theta(y_\alpha)(Re\langle Ty_\alpha,\, y_\alpha - y\rangle)] \leq 0$). Then for any $x \in X$ and $\epsilon > 0$, there are $\beta_1, \beta_2 \in \Gamma$ with $|\theta(y_\alpha)Re\langle Ty, y_\alpha - y\rangle| < \frac{\epsilon}{2}$ for all $\alpha \geq \beta_1$ and $|\theta(y_\alpha)Re\langle Ty_\alpha - Ty, y - x\rangle| < \frac{\epsilon}{2}$ for all $\alpha \geq \beta_2$. Choose $\beta_0 \in \Gamma$ with $\beta_0 \geq \beta_1, \beta_2$. Thus

$$\theta(y_\alpha)Re\langle Ty_\alpha, y_\alpha - x\rangle = \theta(y_\alpha)Re\langle Ty_\alpha, y_\alpha - y\rangle + \theta(y_\alpha)Re\langle Ty_\alpha, y - x\rangle \tag{5.153}$$

$$\geq \theta(y_\alpha)Re\langle Ty, y_\alpha - y\rangle + \theta(y_\alpha)Re\langle Ty_\alpha, y - x\rangle \tag{5.154}$$

$$= \theta(y_\alpha)Re\langle Ty, y_\alpha - y\rangle + \theta(y_\alpha)Re\langle Ty_\alpha - Ty, y - x\rangle + \theta(y_\alpha)Re\langle Ty, y - x\rangle \tag{5.155}$$

$$> -\frac{\epsilon}{2} - \frac{\epsilon}{2} + \theta(y_\alpha)Re\langle Ty, y - x\rangle \text{ for all } \alpha \geq \beta_0 \tag{5.156}$$

so that $\inf_{\alpha\geq\beta_0}\theta(y_\alpha)Re\langle Ty_\alpha, y_\alpha - x\rangle \geq -\epsilon + \inf_{\alpha\geq\beta_0}\theta(y_\alpha)Re\langle Ty, y - x\rangle$. It follows that $\limsup_\beta \theta(y_\beta)Re\langle Ty_\beta, y_\beta - x\rangle \geq \liminf_\beta \theta(y_\beta)Re\langle Ty_\beta, y_\beta - x\rangle \geq -\epsilon + \theta(y)Re\langle Ty, y - x\rangle$. As $\epsilon > 0$ is arbitrary, $\limsup_\beta \theta(y_\beta)Re\langle Ty_\beta, y_\beta - x \geq \theta(y)Re\langle Ty, y - x\rangle$. Hence T is a strong pseudo-monotone operator. □

We shall now present the following result of Chowdhury and Tan:

Theorem 5.104 *Let E be a locally convex Hausdorff topological vector space, X be a non-empty paracompact convex subset of E and $h : E \to \mathbb{R}$ be convex. Let $S : X \to 2^X$ be upper semicontinuous such that each $S(x)$ is compact convex and $T : X \to 2^{E^*}$ be strong h-pseudo-monotone and be upper semicontinuous from $co(A)$ to the weak*-topology on E^* for each $A \in \mathcal{F}(X)$ such that each $T(x)$ is weak*-compact convex. Suppose that the set $\Sigma = \{y \in X : \sup_{x\in S(y)}[\inf_{w\in T(y)} Re\langle w, y - x\rangle + h(y) - h(x)] > 0\}$ is open in X.*

Suppose further that there exist a non-empty compact subset K of X and a point $x_0 \in X$ such that $x_0 \in K \cap S(y)$ and $\inf_{w\in T(y)} Re\langle w, y - x_0\rangle + h(y) - h(x_0) > 0$ for all $y \in X \setminus K$. Then there exists $\hat{y} \in K$ such that

(i) $\hat{y} \in S(\hat{y})$ and

(ii) there exists $\hat{w} \in T(\hat{y})$ with $Re\langle \hat{w}, \hat{y} - x\rangle \leq h(x) - h(\hat{y})$ for all $x \in S(\hat{y})$.

Proof. The proof is divided into two steps as follows:

Step 1. There exists a point $\hat{y} \in X$ such that $\hat{y} \in S(\hat{y})$ and

$$\sup_{x\in S(\hat{y})}[\inf_{w\in T(\hat{y})} Re\langle w, \hat{y} - x\rangle + h(\hat{y}) - h(x)] \leq 0.$$

Suppose the contrary. Then for each $y \in X$, either $y \notin S(y)$ or there exists $x \in S(y)$ such that $\inf_{w\in T(y)} Re\langle w, y - x\rangle + h(y) - h(x) > 0$; that is, $y \notin S(y)$ or $y \in \Sigma$. If $y \notin S(y)$, then by Hahn-Banach separation theorem, there exists $p \in E^*$ such that $Re\langle p, y\rangle - \sup_{x\in S(y)} Re\langle p, x\rangle > 0$. For each $y \in X$, set $\gamma(y) := \sup_{x\in S(y)}[\inf_{w\in T(y)} Re\langle w, y - x\rangle + h(y) - h(x)]$. Let $V_0 := \{y \in X | \gamma(y) > 0\} = \Sigma$ and for each $p \in E^*$, set $V_p := \{y \in X : Re\langle p, y\rangle - \sup_{x\in S(y)} Re\langle p, x\rangle > 0\}$.

Then $X = V_0 \cup \bigcup_{p\in E^*} V_p$. Since each V_p is open in X by Lemma 1 in [Shih and Tan (1985)] and V_0 is open in X by hypothesis, $\{V_0, V_p : p \in E^*\}$ is an open covering for X. Since X is paracompact, there is a continuous partition of unity $\{\beta_0, \beta_p : p \in E^*\}$ for X subordinated to the open cover $\{V_0, V_p : p \in E^*\}$ (see, e.g., Theorem VIII.4.2 of Dugundji in [Dugundji (1966)]); that is, for each $p \in E^*$, $\beta_p : X \to [0, 1]$ and $\beta_0 : X \to [0, 1]$ are continuous functions such that for each $p \in E^*$, $\beta_p(y) = 0$ for all $y \in X \setminus V_p$ and $\beta_0(y) = 0$ for all $y \in X \setminus V_0$ and $\{$ support β_0, support $\beta_p : p \in E^*\}$ is locally finite and $\beta_0(y) + \Sigma_{p\in E^*}\beta_p(y) = 1$ for each $y \in X$. Note that for each $A \in \mathcal{F}(X)$, h is continuous on $co(A)$ (see e.g. [Rockafeller (1970), Corollary 10.1.1, p. 83]). Define $\phi : X \times X \to \mathbb{R}$ by $\phi(x, y) = \beta_0(y)[\min_{w\in T(y)} Re\langle w, y - x\rangle + h(y) - h(x)] + \Sigma_{p\in E^*}\beta_p(y)Re\langle p, y - x\rangle$ for each $x, y \in X$. Then we have the following.

(1) Since E is Hausdorff, for each $A \in \mathcal{F}(X)$ and each fixed $x \in co(A)$, the map $y \mapsto \min_{w\in T(y)} Re\langle w, y - x\rangle + h(y) - h(x)$ is lower semicontinuous on $co(A)$ by Lemma 3 in [Chowdhury and Tan (1996)] and the fact that h is continuous on $co(A)$ and therefore the map $y \mapsto \beta_0(y)[\min_{w\in T(y)} Re\langle w, y - x\rangle + h(y) - h(x)]$ is lower semicontinuous on $co(A)$ by Lemma 3 in [Takahashi (1976)]. Also for each fixed $x \in X$, $y \mapsto \Sigma_{p\in E^*}\beta_p(y)Re\langle p, y - x\rangle$ is continuous on X. Hence, for each $A \in \mathcal{F}(X)$ and each fixed $x \in co(A)$, the map $y \longmapsto \phi(x, y)$ is lower semicontinuous on $co(A)$.

(2) For each $A \in \mathcal{F}(X)$ and for each $y \in co(A)$, $\min_{x\in A} \phi(x, y) \leq 0$. Indeed, if this were false, then for some $A = \{x_1, \ldots, x_n\} \in \mathcal{F}(X)$ and some $y \in co(A)$, say $y = \sum_{i=1}^n \lambda_i x_i$ where $\lambda_1, \ldots, \lambda_n \geq 0$ with $\sum_{i=1}^n \lambda_i = 1$, such that $\min_{1\leq i\leq n} \phi(x_i, y) > 0$. Then for each $i = 1, \ldots, n$,

$$\beta_0(y)[\min_{w\in T(y)} Re\langle w, y - x_i\rangle + h(y) - h(x_i)] + \Sigma_{p\in E^*}\beta_p(y)Re\langle p, y - x_i\rangle > 0$$

so that

$$0 = \phi(y, y) = \beta_0(y)[\min_{w\in T(y)} Re\langle w, y - \sum_{i=1}^{n} \lambda_i x_i\rangle + h(y) - h(\sum_{i=1}^{n} \lambda_i x_i)] \quad (5.157)$$

$$+ \Sigma_{p\in E^*}\beta_p(y)Re\langle p, y - \sum_{i=1}^{n} \lambda_i x_i\rangle \quad (5.158)$$

$$\geq \sum_{i=1}^{n} \lambda_i(\beta_0(y)[\min_{w\in T(y)} Re\langle w, y - x_i\rangle + h(y) - h(x_i)] \quad (5.159)$$

$$+ \Sigma_{p\in E^*}\beta_p(y)Re\langle p, y - x_i\rangle) > 0, \quad (5.160)$$

which is a contradiction.

(3) Suppose $A \in \mathcal{F}(X)$, $x, y \in co(A)$ and $\{y_\alpha\}_{\alpha\in\Gamma}$ is a net in X converging to y with $\phi(tx + (1-t)y, y_\alpha) \leq 0$ for all $\alpha \in \Gamma$ and all $t \in [0, 1]$.

Then for $t = 0$ we have $\phi(y, y_\alpha) \leq 0$ for all $\alpha \in \Gamma$, i.e., $\beta_0(y_\alpha) \times [\min_{w\in T(y_\alpha)} Re\langle w, y_\alpha - y\rangle + h(y_\alpha) - h(y)] + \Sigma_{p\in E^*}\beta_p(y_\alpha)Re\langle p, y_\alpha - y\rangle \leq 0$ for all $\alpha \in \Gamma$. Hence

$$\limsup_{\alpha}[\beta_0(y_\alpha)(\min_{w\in T(y_\alpha)} Re\langle w, y_\alpha - y\rangle + h(y_\alpha) - h(y))] \quad (5.161)$$

$$+ \liminf_{\alpha}(\Sigma_{p\in E^*}\beta_p(y_\alpha)Re\langle p, y_\alpha - y\rangle) \quad (5.162)$$

$$\leq \limsup_{\alpha}[\beta_0(y_\alpha)(\min_{w\in T(y_\alpha)} Re\langle w, y_\alpha - y\rangle + h(y_\alpha) - h(y)) \quad (5.163)$$

$$+ \Sigma_{p\in E^*}\beta_p(y_\alpha)Re\langle p, y_\alpha - y\rangle] \leq 0. \quad (5.164)$$

Therefore $\limsup_\alpha[\beta_0(y_\alpha)(\min_{w\in T(y_\alpha)} Re\langle w, y_\alpha - y\rangle + h(y_\alpha) - h(y))] \leq 0$. Since T is a strong h-pseudo-monotone operator, we have

$$\limsup_{\alpha}[\beta_0(y_\alpha)(\min_{w\in T(y_\alpha)} Re\langle w, y_\alpha - x\rangle + h(y_\alpha) - h(x))] \quad (5.165)$$

$$\geq \beta_0(y)(\min_{w\in T(y)} Re\langle w, y - x\rangle + h(y) - h(x)). \quad (5.166)$$

Thus

$$\limsup_{\alpha}[\beta_0(y_\alpha)(\min_{w\in T(y_\alpha)} Re\langle w, y_\alpha - x\rangle + h(y_\alpha) - h(x))] \quad (5.167)$$

$$+ \Sigma_{p\in E^*}\beta_p(y)Re\langle p, y - x\rangle \quad (5.168)$$

$$\geq \beta_0(y)(\min_{w\in T(y)} Re\langle w, y - x\rangle + h(y) - h(x)) \quad (5.169)$$

$$+ \Sigma_{p\in E^*}\beta_p(y)Re\langle p, y - x\rangle. \quad (5.170)$$

For $t = 1$ we have $\phi(x, y_\alpha) \le 0$ for all $\alpha \in \Gamma$, i.e., $\beta_0(y_\alpha)[\min_{w\in T(y_\alpha)} Re\langle w, y_\alpha - x\rangle + h(y_\alpha) - h(x)] + \Sigma_{p\in E^*}\beta_p(y_\alpha)Re\langle p, y_\alpha - x\rangle \le 0$ for all $\alpha \in \Gamma$. Therefore

$$\limsup_\alpha[\beta_0(y_\alpha)(\min_{w\in T(y_\alpha)} Re\langle w, y_\alpha - x\rangle + h(y_\alpha) - h(x))] \tag{5.171}$$

$$+ \liminf_\alpha[\Sigma_{p\in E^*}\beta_p(y_\alpha)Re\langle p, y_\alpha - x\rangle] \tag{5.172}$$

$$\le \limsup_\alpha[\beta_0(y_\alpha)(\min_{w\in T(y_\alpha)} Re\langle w, y_\alpha - x\rangle + h(y_\alpha) - h(x)) \tag{5.173}$$

$$+ \Sigma_{p\in E^*}\beta_p(y_\alpha)Re\langle p, y_\alpha - x\rangle] \le 0. \tag{5.174}$$

Thus

$$\limsup_\alpha[\beta_0(y_\alpha)(\min_{w\in T(y_\alpha)} Re\langle w, y_\alpha - x\rangle + h(y_\alpha) - h(x))] \tag{5.175}$$

$$+ \Sigma_{p\in E^*}\beta_p(y)Re\langle p, y - x\rangle \le 0. \tag{5.176}$$

Hence by (5.167) and (5.175), we have $\phi(x, y) \le 0$.

(4) By hypothesis, there exist a non-empty compact (and therefore closed) subset K of X and a point $x_0 \in X$ such that $x_0 \in K \cap S(y)$ and $\inf_{w\in T(y)} Re\langle w, y - x_0\rangle + h(y) - h(x_0) > 0$ for each $y \in X \setminus K$. Thus for each $y \in X \setminus K$, $\beta_0(y)[\inf_{w\in T(y)} Re\langle w, y - x_0\rangle + h(y) - h(x_0)] > 0$ whenever $\beta_0(y) > 0$ and $Re\langle p, y - x_0\rangle > 0$ whenever $\beta_p(y) > 0$ for $p \in E^*$. Consequently, $\phi(x_0, y) = \beta_0(y)[\inf_{w\in T(y)} Re\langle w, y - x_0\rangle + h(y) - h(x_0)] + \Sigma_{p\in E^*}\beta_p(y)Re\langle p, y - x_0\rangle > 0$ for all $y \in X \setminus K$.

Then ϕ satisfies all hypotheses of Theorem 2 in [Chowdhury and Tan (1996)]. Hence by Theorem 2 in [Chowdhury and Tan (1996)], there exists a point $\hat{y} \in K$ such that $\phi(x, \hat{y}) \le 0$ for all $x \in X$; i.e.,

$$\beta_0(\hat{y})[\inf_{w\in T(\hat{y})} Re\langle w, \hat{y} - x\rangle + h(\hat{y}) - h(x)] + \Sigma_{p\in E^*}\beta_p(\hat{y})Re\langle p, \hat{y} - x\rangle \le 0 \tag{5.177}$$

for all $x \in X$. If $\gamma(\hat{y}) = 0$, choose any $\hat{x} \in S(\hat{y})$; if $\gamma(\hat{y}) > 0$, choose any $\hat{x} \in S(\hat{y})$ such that $\inf_{w\in T(\hat{y})} Re\langle w, \hat{y} - \hat{x}\rangle + h(\hat{y}) - h(\hat{x}) \ge \frac{\gamma(\hat{y})}{2} > 0$.

If $\beta_0(\hat{y}) > 0$, then $\hat{y} \in V_0 = \Sigma$ so that $\gamma(\hat{y}) > 0$; it follows that $\beta_0(\hat{y})[\inf_{w\in T(\hat{y})} Re\langle w, \hat{y} - \hat{x}\rangle + h(\hat{y}) - h(\hat{x})] > 0$.

If $\beta_p(\hat{y}) > 0$ for some $p \in E^*$, then $\hat{y} \in V_p$ and hence $Re\langle p, \hat{y}\rangle > \sup_{x\in S(\hat{y})} Re\langle p, x\rangle \ge Re\langle p, \hat{x}\rangle$ so that $Re\langle p, \hat{y} - \hat{x}\rangle > 0$. Then note that $\beta_p(\hat{y})Re\langle p, \hat{y} - \hat{x}\rangle > 0$ whenever $\beta_p(\hat{y}) > 0$ for $p \in E^*$.

Since $\beta_0(\hat{y}) > 0$ or $\beta_p(\hat{y}) > 0$ for some $p \in E^*$, it follows that

$$\phi(\hat{x}, \hat{y}) = \beta_0(\hat{y})[\inf_{w\in T(\hat{y})} Re\langle w, \hat{y} - \hat{x}\rangle + h(\hat{y}) - h(\hat{x})] + \Sigma_{p\in E^*}\beta_p(\hat{y})Re\langle p, \hat{y} - \hat{x}\rangle > 0,$$

which contradicts (5.177). This contradiction proves Step 1.

Step 2. There exists a point $\hat{w} \in T(\hat{y})$ such that $Re\langle \hat{w}, \hat{y} - x\rangle + h(\hat{y}) - h(x) \leq 0$ for all $x \in S(\hat{y})$.

Note that for each fixed $x \in S(\hat{y})$, $w \mapsto Re\langle w, \hat{y} - x\rangle + h(\hat{y}) - h(x)$ is convex and continuous on $T(\hat{y})$ and for each fixed $w \in T(\hat{y})$, $x \mapsto Re\langle w, \hat{y} - x\rangle + h(\hat{y}) - h(x)$ is concave on $S(\hat{y})$. Thus by Kneser's Minimax Theorem in [Kneser (1952)] (see also Aubin (1979, pp. 40, 41)), we have

$$\min_{w\in T(\hat{y})} \max_{x\in S(\hat{y})} [Re\langle w, \hat{y} - x\rangle + h(\hat{y}) - h(x)] = \max_{x\in S(\hat{y})} \min_{w\in T(\hat{y})} [Re\langle w, \hat{y} - x\rangle + h(\hat{y}) - h(x)].$$

Hence $\min_{w\in T(\hat{y})} \max_{x\in S(\hat{y})}[Re\langle w, \hat{y} - x\rangle + h(\hat{y}) - h(x)] \leq 0$ by Step 1. Since $T(\hat{y})$ is compact, there exists $\hat{w} \in T(\hat{y})$ such that $Re\langle \hat{w}, \hat{y} - x\rangle + h(\hat{y}) - h(x) \leq 0$ for all $x \in S(\hat{y})$. □

When X is compact, Chowdhury and Tan obtained the following result from the above Theorem 5.104:

Theorem 5.105 *Let E be a locally convex Hausdorff topological vector space, X be a non-empty compact convex subset of E and $h : E \to \mathbb{R}$ be convex. Let $S : X \to 2^X$ be upper semicontinuous such that each $S(x)$ is closed convex and $T : X \to 2^{E^*}$ be a strong h-pseudo-monotone operator and be upper semicontinuous from $co(A)$ to the weak*-topology on E^* for each $A \in \mathcal{F}(X)$ such that each $T(x)$ is weak*-compact convex. Suppose that the set $\Sigma = \{y \in X : \sup_{x\in S(y)}[\inf_{w\in T(y)} Re\langle w, y - x\rangle + h(y) - h(x)] > 0\}$ is open in X. Then there exists $\hat{y} \in X$ such that*

(i) $\hat{y} \in S(\hat{y})$ and

(ii) there exists $\hat{w} \in T(\hat{y})$ with $Re\langle \hat{w}, \hat{y} - x\rangle \leq h(x) - h(\hat{y})$ for all $x \in S(\hat{y})$.

Chowdhury and Tan observed that if X is also bounded in Theorem 5.104, the map $S : X \to 2^X$ is, in addition, lower semi-continuous and for each $y \in \Sigma$, T is upper semicontinuous at y in X, then the set Σ in Theorem 5.104 becomes an open set in X. The following theorem is presented with these additional assumptions:

Theorem 5.106 *Let E be a locally convex Hausdorff topological vector space, X be a non-empty paracompact convex and bounded subset of E and $h : E \to \mathbb{R}$ be convex. Let $S : X \to 2^X$ be continuous such that each $S(x)$ is compact convex and $T : X \to 2^{E^*}$ be a strong h-pseudo-monotone operator and be upper semicontinuous from $co(A)$ to the weak*-topology on E^* for each $A \in \mathcal{F}(X)$ such that each $T(x)$ is weak*-compact convex. Suppose that for each $y \in \Sigma = \{y \in X : \sup_{x\in S(y)}[\inf_{w\in T(y)} Re\langle w, y - x\rangle + h(y) - h(x)] > 0\}$, T is upper semicontinuous at y from the relative topology on X to the strong topology on E^*. Suppose further that there exist a non-empty compact subset K of X and a point $x_0 \in X$ such that $x_0 \in K \cap S(y)$ and $\inf_{w\in T(y)} Re\langle w, y - x_0\rangle + h(y) - h(x_0) > 0$ for all $y \in X \setminus K$. Then there exists $\hat{y} \in K$ such that*

(i) $\hat{y} \in S(\hat{y})$ and

(ii) there exists $\hat{w} \in T(\hat{y})$ with $Re\langle \hat{w}, \hat{y} - x\rangle \leq h(x) - h(\hat{y})$ for all $x \in S(\hat{y})$.

Proof. Comparing Theorem 5.104, we see that we need to show only that the set $\Sigma := \{y \in X : \sup_{x\in S(y)} [\inf_{w\in T(y)} Re\langle w, y-x\rangle + h(y) - h(x)] > 0\}$ is open in X. For, let $y_0 \in \Sigma$; then there exists $x_0 \in S(y_0)$ such that $\alpha := \inf_{w\in T(y_0)} Re\ \langle w, y_0 - x_0\rangle + h(y_0) - h(x_0) > 0$.

Let $W := \{w \in E^* : \sup_{z_1,z_2\in X} |\langle w, z_1 - z_2\rangle| < \frac{\alpha}{6}\}$. Then W is a strongly open neighborhood of 0 in E^* so that $U_1 := T(y_0) + W$ is an open neighborhood of $T(y_0)$ in E^*. Since T is upper semicontinuous at y_0 in X, there exists an open neighborhood N_1 of y_0 in X such that $T(y) \subset U_1$ for all $y \in N_1$.

Now, the rest of the proof is similar to the proof of Theorem 2.2 in [Chowdhury and Tan (1997b)]. Hence by the rest of the proof of Theorem 2.2 in [Chowdhury and Tan (1997b)], Σ is open in X. This proves the theorem. □

When X is compact, Chowdhury and Tan obtained the following result from Theorem 5.106:

Theorem 5.107 *Let E be a locally convex Hausdorff topological vector space, X be a non-empty compact convex subset of E and $h : E \to \mathbb{R}$ be convex. Let $S : X \to 2^X$ be continuous such that each $S(x)$ is closed convex and $T : X \to 2^{E^*}$ be strong h-pseudo-monotone and be upper semicontinuous from $co(A)$ to the weak*-topology on E^* for each $A \in \mathcal{F}(X)$ such that each $T(x)$ is weak*-compact convex. Suppose that for each $y \in \Sigma = \{y \in X : \sup_{x\in S(y)}[\inf_{w\in T(y)} Re\langle w, y-x\rangle + h(y) - h(x)] > 0\}$, T is upper semicontinuous at y from the relative topology on X to the strong topology on E^*. Then there exists $\hat{y} \in X$ such that*

(i) $\hat{y} \in S(\hat{y})$ and

(ii) there exists $\hat{w} \in T(\hat{y})$ with $Re\langle \hat{w}, \hat{y} - x\rangle \le h(x) - h(\hat{y})$ for all $x \in S(\hat{y})$.

Chowdhury and Tan observed that in Theorems 5.104–5.107, the condition "$h : E \to \mathbb{R}$ be convex" can be replaced by the condition "$h : X \to \mathbb{R}$ be convex such that $h|_{co(A)}$ is continuous for each $A \in \mathcal{F}(\mathcal{X})$".

5.15.2 *Generalized Quasi-Variational Inequalities for Pseudo-Monotone Set-Valued Mappings*

In this section we shall present some existence theorems of generalized quasi-variational inequalities for pseudo-monotone set-valued mappings (see Definition 5.20) on paracompact convex sets.

We shall first presnt the following result of Chowdhury and Tan:

Theorem 5.108 *Let E be a locally convex Hausdorff topological vector space, X be a non-empty paracompact convex and bounded subset of E and $h : E \to \mathbb{R}$ be convex such that $h(X)$ is bounded. Let $S : X \to 2^X$ be upper semicontinuous such that each $S(x)$ is compact convex and $T : X \to 2^{E^*}$ be h-pseudo-monotone and be upper semicontinuous from $co(A)$ to the weak*-topology on E^* for each $A \in \mathcal{F}(X)$*

such that each $T(x)$ is weak-compact convex and $T(X)$ is strongly bounded. Suppose that the set $\Sigma = \{y \in X : \sup_{x\in S(y)}[\inf_{w\in T(y)} Re\langle w, y-x\rangle + h(y) - h(x)] > 0\}$ is open in X. Suppose further that there exist a non-empty compact subset K of X and a point $x_0 \in X$ such that $x_0 \in K \cap S(y)$ and $\inf_{w\in T(y)} Re\langle w, y-x_0\rangle + h(y) - h(x_0) > 0$ for all $y \in X \setminus K$. Then there exists $\hat{y} \in K$ such that*

(i) $\hat{y} \in S(\hat{y})$ and

(ii) there exists $\hat{w} \in T(\hat{y})$ with $Re\langle \hat{w}, \hat{y} - x\rangle \le h(x) - h(\hat{y})$ for all $x \in S(\hat{y})$.

Proof. The proof is divided into two steps as follows:

Step 1. There exists a point $\hat{y} \in X$ such that $\hat{y} \in S(\hat{y})$ and $\sup_{x\in S(\hat{y})} \times [\inf_{w\in T(\hat{y})} Re\langle w, \hat{y} - x\rangle + h(\hat{y}) - h(x)] \le 0$.

Suppose the contrary. Then for each $y \in X$, either $y \notin S(y)$ or there exists $x \in S(y)$ such that $\inf_{w\in T(y)} Re\langle w, y - x\rangle + h(y) - h(x) > 0$; that is, $y \notin S(y)$ or $y \in \Sigma$. If $y \notin S(y)$, then by Hahn-Banach separation theorem, there exists $p \in E^*$ such that $Re\langle p, y\rangle - \sup_{x\in S(y)} Re\langle p, x\rangle > 0$. For each $y \in X$, set $\gamma(y) := \sup_{x\in S(y)}[\inf_{w\in T(y)} Re\langle w, y - x\rangle + h(y) - h(x)]$. Let $V_0 := \{y \in X | \gamma(y) > 0\} = \Sigma$ and for each $p \in E^*$, set $V_p := \{y \in X : Re\langle p, y\rangle - \sup_{x\in S(y)} Re\langle p, x\rangle > 0\}$.

Then $X = V_0 \cup \bigcup_{p\in E^*} V_p$. Since each V_p is open in X by Lemma 1 in [Shih and Tan (1985)] and V_0 is open in X by hypothesis, $\{V_0, V_p : p \in E^*\}$ is an open covering for X. Since X is paracompact, there is a continuous partition of unity $\{\beta_0, \beta_p : p \in E^*\}$ for X subordinated to the open cover $\{V_0, V_p : p \in E^*\}$.

Note that for each $A \in \mathcal{F}(X)$, h is continuous on $co(A)$ (see e.g. [Rockafeller (1970), Corollary 10.1.1, p. 83]). Define $\phi : X \times X \to \mathbb{R}$ by $\phi(x, y) = \beta_0(y)[\min_{w\in T(y)} Re\langle w, y - x\rangle + h(y) - h(x)] + \Sigma_{p\in E^*}\beta_p(y)Re\langle p, y - x\rangle$ for each $x, y \in X$. Then we have the following:

(1) The same argument in proving (1) in the proof of Theorem 5.104 shows that for each $A \in \mathcal{F}(X)$ and each fixed $x \in co(A)$, the map $y \longmapsto \phi(x, y)$ is lower semicontinuous on $co(A)$.

(2) The same argument in proving (2) in the proof of Theorem 5.104 shows that for each $A \in \mathcal{F}(X)$ and for each $y \in co(A)$, $\min_{x\in A} \phi(x, y) \le 0$.

(3) Suppose $A \in \mathcal{F}(X)$, $x, y \in co(A)$ and $\{y_\alpha\}_{\alpha\in\Gamma}$ is a net in X converging to y with $\phi(tx + (1-t)y, y_\alpha) \le 0$ for all $\alpha \in \Gamma$ and all $t \in [0, 1]$.

Case 1: $\beta_0(y) = 0$.

Note that $\beta_0(y_\alpha) \ge 0$ for each $\alpha \in \Gamma$ and $\beta_0(y_\alpha) \to 0$. Since $T(X)$ is strongly bounded and $\{y_\alpha\}_{\alpha\in\Gamma}$ is a bounded net, it follows that

$$\limsup_{\alpha}[\beta_0(y_\alpha)(\min_{w\in T(y_\alpha)} Re\langle w, y_\alpha - x\rangle + h(y_\alpha) - h(x))] = 0. \tag{5.178}$$

Also $\beta_0(y)[\min_{w\in T(y)} Re\langle w, y-x\rangle + h(y) - h(x)] = 0$. Thus

$$\limsup_{\alpha}[\beta_0(y_\alpha)(\min_{w\in T(y_\alpha)} Re\langle w, y_\alpha - x\rangle + h(y_\alpha) - h(x))] \tag{5.179}$$

$$+\Sigma_{p\in E^*}\beta_p(y)Re\langle p, y-x\rangle \tag{5.180}$$

$$= \Sigma_{p\in E^*}\beta_p(y)Re\langle p, y-x\rangle \text{ (by (3.1))} \tag{5.181}$$

$$= \beta_0(y)[\min_{w\in T(y)} Re\langle w, y-x\rangle + h(y) - h(x)] \tag{5.182}$$

$$+\Sigma_{p\in E^*}\beta_p(y)Re\langle p, y-x\rangle. \tag{5.183}$$

For $t = 1$ we have $\phi(x, y_\alpha) \leq 0$ for all $\alpha \in \Gamma$, i.e.,

$$\beta_0(y_\alpha)[\min_{w\in T(y_\alpha)} Re\langle w, y_\alpha - x\rangle + h(y_\alpha) - h(x)] + \Sigma_{p\in E^*}\beta_p(y_\alpha)Re\langle p, y_\alpha - x\rangle \leq 0 \tag{5.184}$$

for all $\alpha \in \Gamma$. Therefore

$$\limsup_{\alpha}[\beta_0(y_\alpha)(\min_{w\in T(y_\alpha)} Re\langle w, y_\alpha - x\rangle + h(y_\alpha) - h(x))] \tag{5.185}$$

$$+\liminf_{\alpha}[\Sigma_{p\in E^*}\beta_p(y_\alpha)Re\langle p, y_\alpha - x\rangle] \tag{5.186}$$

$$\leq \limsup_{\alpha}[\beta_0(y_\alpha)(\min_{w\in T(y_\alpha)} Re\langle w, y_\alpha - x\rangle + h(y_\alpha) - h(x)) \tag{5.187}$$

$$+\Sigma_{p\in E^*}\beta_p(y_\alpha)Re\langle p, y_\alpha - x\rangle] \tag{5.188}$$

$$\leq 0 \text{ (by (3.3))}. \tag{5.189}$$

Thus

$$\limsup_{\alpha}[\beta_0(y_\alpha)(\min_{w\in T(y_\alpha)} Re\langle w, y_\alpha - x\rangle + h(y_\alpha) - h(x))] + \Sigma_{p\in E^*}\beta_p(y)Re\langle p, y-x\rangle \leq 0. \tag{5.190}$$

Hence by (5.179) and (5.190), we have $\phi(x, y) \leq 0$.

Case 2: $\beta_0(y) > 0$.

Since $\beta_0(y_\alpha) \to \beta_0(y)$, there exists $\lambda \in \Gamma$ such that $\beta_0(y_\alpha) > 0$ for all $\alpha \geq \lambda$. Then for $t = 0$ we have $\phi(y, y_\alpha) \leq 0$ for all $\alpha \in \Gamma$, i.e., $\beta_0(y_\alpha)[\min_{w\in T(y_\alpha)} Re\langle w, y_\alpha - y\rangle + h(y_\alpha) - h(y)] + \Sigma_{p\in E^*}\beta_p(y_\alpha)Re\langle p, y_\alpha - y\rangle \leq 0$ for all $\alpha \in \Gamma$. Thus

$$\limsup_{\alpha}[\beta_0(y_\alpha)(\min_{w\in T(y_\alpha)} Re\langle w, y_\alpha - y\rangle + h(y_\alpha) - h(y)) \tag{5.191}$$

$$+\Sigma_{p\in E^*}\beta_p(y_\alpha)Re\langle p, y_\alpha - y\rangle] \leq 0. \tag{5.192}$$

Hence

$$\limsup_{\alpha}[\beta_0(y_\alpha)(\min_{w\in T(y_\alpha)} Re\langle w, y_\alpha - y\rangle + h(y_\alpha) - h(y))] \tag{5.193}$$

$$+ \liminf_{\alpha}[\Sigma_{p\in E^*}\beta_p(y_\alpha)Re\langle p, y_\alpha - y\rangle] \tag{5.194}$$

$$\leq \limsup_{\alpha}[\beta_0(y_\alpha)(\min_{w\in T(y_\alpha)} Re\langle w, y_\alpha - y\rangle + h(y_\alpha) - h(y)) \tag{5.195}$$

$$+ \Sigma_{p\in E^*}\beta_p(y_\alpha)Re\langle p, y_\alpha - y\rangle] \tag{5.196}$$

$$\leq 0 \text{ (by (5.191))}. \tag{5.197}$$

Since $\liminf_\alpha[\Sigma_{p\in E^*}\beta_p(y_\alpha)Re\langle p, y_\alpha - y\rangle] = 0$, we have

$$\limsup_{\alpha}[\beta_0(y_\alpha)(\min_{w\in T(y_\alpha)} Re\langle w, y_\alpha - y\rangle + h(y_\alpha) - h(y))] \leq 0. \tag{5.198}$$

Since $\beta_0(y_\alpha) > 0$ for all $\alpha \geq \lambda$, it follows that

$$\beta_0(y)\limsup_{\alpha}[\min_{w\in T(y_\alpha)} Re\langle w, y_\alpha - y\rangle + h(y_\alpha) - h(y))] \tag{5.199}$$

$$= \limsup_{\alpha}[\beta_0(y_\alpha)(\min_{w\in T(y_\alpha)} Re\langle w, y_\alpha - y\rangle + h(y_\alpha) - h(y))]. \tag{5.200}$$

Since $\beta_0(y) > 0$, by (5.198) and (5.199) we have $\limsup_\alpha[\min_{w\in T(y_\alpha)} Re\langle w, y_\alpha - y\rangle + h(y_\alpha) - h(y)] \leq 0$. Since T is an h-pseudo-monotone operator, we have $\liminf_\alpha[\min_{w\in T(y_\alpha)} Re\langle w, y_\alpha - x\rangle + h(y_\alpha) - h(x)] \geq \min_{w\in T(y)} Re\langle w, y - x\rangle + h(y) - h(x)$. Since $\beta_0(y) > 0$, we have

$$\beta_0(y)[\liminf_{\alpha}(\min_{w\in T(y_\alpha)} Re\langle w, y_\alpha - x\rangle + h(y_\alpha) - h(x))] \tag{5.201}$$

$$\geq \beta_0(y)[\min_{w\in T(y)} Re\langle w, y - x\rangle + h(y) - h(x)]. \tag{5.202}$$

Thus

$$\beta_0(y)[\liminf_{\alpha}(\min_{w\in T(y_\alpha)} Re\langle w, y_\alpha - x\rangle + h(y_\alpha) - h(x))] \tag{5.203}$$

$$+ \Sigma_{p\in E^*}\beta_p(y)Re\langle p, y - x\rangle \tag{5.204}$$

$$\geq \beta_0(y)[\min_{w\in T(y)} Re\langle w, y - x\rangle + h(y) - h(x)] \tag{5.205}$$

$$+ \Sigma_{p\in E^*}\beta_p(y)Re\langle p, y - x\rangle. \tag{5.206}$$

For $t = 1$ we also have $\phi(x, y_\alpha) \leq 0$ for all $\alpha \in \Gamma$, i.e.,

$$\beta_0(y_\alpha)[\min_{w\in T(y_\alpha)} Re\langle w, y_\alpha - x\rangle + h(y_\alpha) - h(x)] + \Sigma_{p\in E^*}\beta_p(y_\alpha)Re\langle p, y_\alpha - x\rangle \leq 0$$

for all $\alpha \in \Gamma$. Therefore

$$0 \geq \liminf_{\alpha}[\beta_0(y_\alpha)(\min_{w\in T(y_\alpha)} Re\langle w, y_\alpha - x\rangle + h(y_\alpha) - h(x)) \tag{5.207}$$

$$+\Sigma_{p\in E^*}\beta_p(y_\alpha)Re\langle p, y_\alpha - x\rangle] \tag{5.208}$$

$$\geq \liminf_{\alpha}[\beta_0(y_\alpha)(\min_{w\in T(y_\alpha)} Re\langle w, y_\alpha - x\rangle + h(y_\alpha) - h(x))] \tag{5.209}$$

$$+ \liminf_{\alpha}[\Sigma_{p\in E^*}\beta_p(y_\alpha)Re\langle p, y_\alpha - x\rangle] \tag{5.210}$$

$$= \beta_0(y)[\liminf_{\alpha}(\min_{w\in T(y_\alpha)} Re\langle w, y_\alpha - x\rangle + h(y_\alpha) - h(x))] \tag{5.211}$$

$$+\Sigma_{p\in E^*}\beta_p(y)Re\langle p, y - x\rangle. \tag{5.212}$$

Consequently, by (5.203) and (5.207), we have $\phi(x, y) \leq 0$.

Now, the rest of the proof of Step 1 is similar to the proofs in Step 1 of Theorem 5.104 and Theorem 3.1 in [Chowdhury and Tan (1997b)]. Thus Step 1 is proved.

Step 2. There exists a point $\hat{w} \in T(\hat{y})$ such that $Re\langle \hat{w}, \hat{y} - x\rangle + h(\hat{y}) - h(x) \leq 0$ for all $x \in S(\hat{y})$.

Also the same proof of Step 2 of Theorem 5.104 shows that there exists $\hat{w} \in T(\hat{y})$ such that $Re\langle \hat{w}, \hat{y} - x\rangle + h(\hat{y}) - h(x) \leq 0$ for all $x \in S(\hat{y})$. □

When X is compact, Chowdhury and Tan obtained the following result from Theorem 5.108:

Theorem 5.109 *Let E be a locally convex Hausdorff topological vector space, X be a non-empty compact convex subset of E and $h : E \to \mathbb{R}$ be convex such that $h(X)$ is bounded. Let $S : X \to 2^X$ be upper semicontinuous such that each $S(x)$ is closed convex and $T : X \to 2^{E^*}$ be h-pseudo-monotone and be upper semicontinuous from $co(A)$ to the weak*-topology on E^* for each $A \subset \mathcal{F}(X)$ such that each $T(x)$ is weak*-compact convex and $T(X)$ is strongly bounded. Suppose that the set $\Sigma = \{y \in X : \sup_{x\in S(y)} [\inf_{w\in T(y)} Re\langle w, y - x\rangle + h(y) - h(x)] > 0\}$ is open in X. Then there exists $\hat{y} \in X$ such that*

(i) $\hat{y} \in S(\hat{y})$ and

(ii) there exists $\hat{w} \in T(\hat{y})$ with $Re\langle \hat{w}, \hat{y} - x\rangle \leq h(x) - h(\hat{y})$ for all $x \in S(\hat{y})$.

Chowdhury and Tan observed that if the map $S : X \to 2^X$ is, in addition, lower semicontinuous and for each $y \in \Sigma$, T is upper semicontinuous at y in X, then the set Σ in Theorem 5.108 becomes an open set in X. The following theorem is presented with these additional assumptions:

Theorem 5.110 *Let E be a locally convex Hausdorff topological vector space, X be a non-empty paracompact convex and bounded subset of E and $h : E \to \mathbb{R}$ be convex such that $h(X)$ is bounded. Let $S : X \to 2^X$ be continuous such that* *each $S(x)$ is compact convex and $T : X \to 2^{E^*}$ be h-pseudo-monotone and be upper semicontinuous from $co(A)$ to the weak*-topology on E^* for each $A \in \mathcal{F}(X)$ such that each $T(x)$ is weak*-compact convex and $T(X)$ is strongly bounded. Suppose that for each $y \in \Sigma = \{y \in X : \sup_{x\in S(y)}[\inf_{w\in T(y)} Re\langle w, y - x\rangle + h(y) - h(x)] > 0\}$, T is upper semicontinuous at y from the relative topology on X to the strong topology on E^*. Suppose further that there exist a non-empty compact subset K of X and a point $x_0 \in X$ such that $x_0 \in K \cap S(y)$ and $\inf_{w\in T(y)} Re\langle w, y - x_0\rangle + h(y) - h(x_0) > 0$ for all $y \in X \setminus K$. Then there exists $\hat{y} \in K$ such that*

(i) *$\hat{y} \in S(\hat{y})$ and*

(ii) *there exists $\hat{w} \in T(\hat{y})$ with $Re\langle \hat{w}, \hat{y} - x\rangle \leq h(x) - h(\hat{y})$ for all $x \in S(\hat{y})$.*

Proof. Comparing Theorem 5.108, we see that we need to show only that the set $\Sigma := \{y \in X : \sup_{x\in S(y)} [\inf_{w\in T(y)} Re\langle w, y - x\rangle + h(y) - h(x)] > 0\}$ is open in X.

Now, following the same arguments as in the proofs of Theorem 3.2 in [Chowdhury and Tan (1997b)] and Theorem 5.106, we can similarly show that the set Σ is open in X. Hence by Theorem 5.108 the conclusion follows. □

When X is compact, Chowdhury and Tan obtained the following result from Theorem 5.110:

Theorem 5.111 *Let E be a locally convex Hausdorff topological vector space, X be a non-empty compact convex subset of E and $h : E \to \mathbb{R}$ be convex such that $h(X)$ is bounded. Let $S : X \to 2^X$ be continuous such that each $S(x)$ is closed convex and $T : X \to 2^{E^*}$ be h-pseudo-monotone and be upper semicontinuous from $co(A)$ to the weak*-topology on E^* for each $A \in \mathcal{F}(X)$ such that each $T(x)$ is weak*-compact convex and $T(X)$ is strongly bounded. Suppose that for each $y \in \Sigma = \{y \in X : \sup_{x\in S(y)}[\inf_{w\in T(y)} Re\langle w, y - x\rangle + h(y) - h(x)] > 0\}$, T is upper semicontinuous at y from the relative topology on X to the strong topology on E^*. Then there exists $\hat{y} \in X$ such that*

(i) *$\hat{y} \in S(\hat{y})$ and*

(ii) *there exists $\hat{w} \in T(\hat{y})$ with $Re\langle \hat{w}, \hat{y} - x\rangle \leq h(x) - h(\hat{y})$ for all $x \in S(\hat{y})$.*

We remark here that in Theorems 5.108–5.111, the condition "$h : E \to \mathbb{R}$ be convex" can be replaced by the condition "$h : X \to \mathbb{R}$ be convex such that $h|_{co(A)}$ is continuous for each $A \in \mathcal{F}(\mathcal{X})$".

Note that the results of this section have been written and presented after a thorough revision of some results of an article in Chowdhury and Tan (1998).

5.16 Non-Linear Variational Inequalities and the Existence of Equilibrium in Economics with a Riesz Space of Commodities

The object of this section is to present the results of Tarafdar and Mehta who investigated the existence of equilibrium in an economy with a Riesz space of commodities.

Riesz space methods have been used in economics by Aliprantis and Brown (1983). Their proof is based on a theorem by Ky Fan (1961) which generalizes the classic Knaster-Kuratowski-Mazurkiewicz Theorem. The main idea of the proof given by Aliprantis and Brown is to define a "revealed preference" relation on the space of prices and then to use Ky Fan's theorem to show the existence of a maximal element for this ordering. This maximal element is then proved to be an equilibrium point. The proof given in this section is not based on Ky Fan's theorem. Instead it is shown that the existence of an equilibrium point for the economy is equivalent to the solution of a non-linear variational inequality which was first proved by Hartman and Stampacchia (1966) and Browder (1965) independently (see also Tarafdar (1977), and Mosco (1976)). For applications of non-linear variational inequalities we refer the reader to Hartman and Stampacchia (1966) and Mosco (1976).

We shall now present the following result of Aliprantis and Burkinshaw (1981, pp. 189–190):

Theorem 5.112 (Riesz) *If L is a Riesz space, then its order dual $L^{\sim}$ is also a Riesz 5pace. If $f \in l$ and $u \in L^{+}$, then*

$$f^{+}(u) = \sup\{f(v)/0 \leq v \leq u\}$$
$$f^{-}(u) = \sup\{-f(v)/0 \leq v \leq u \ \ and \ \ |f|(u) = \sup\{f(v)/|v| \leq u\}.$$

Let $L^{\sim}_{+}$ denote the positive cone of $L^{\sim}$. Its members are called positive linear functionals *an L. f belongs to $L^{\sim}_{+}$ if and only if $f(u) \geq 0$ for all u in L^{+}. f is* strictly positive *($f >> 0$) if $u > 0$ implies $f(u) > 0$.*

The following is another result in Aliprantis and Burkinshaw (1981, pp. 190–191):

Theorem 5.113 *Let L be a Riesz space and let $f \in L^{\sim}$ such that $f \geq 0$. Then for every $x \in L$, $f(x^{+}) = \sup\{g(x)/g \in L^{\sim}, 0 \leq g \leq f), f(x^{-}) = \sup\{-g(x)/g \in L^{\sim}, 0 \leq g \leq f\}$ and $f(|x|) = \sup\{g(x)/|g| \leq f\}$.*

An *ideal* or *order-ideal* A of a Riesz space L is a vector sub-space of L such that $|f| \leq |g|$ and $g \in A$ imply $f \in A$.

If L is a Riesz space and L' an ideal of $L^{\sim}$ separating the points of L, then the dual pair (L, L') is called a *Riesz dual system.*

Let (L, L') be a Riesz dual system. A price-simplex D for (L, L') is a non-empty, w^*-compact and convex subset of $L^{\sim}_{+}$. Here, w^* is the weak-star topology on $L^{\sim}$, i.e. it is the $w(L^{\sim}, L)$ topology. We assume that D satisfies the following condition:

($*$) The cone generated by $S = \{p \in L'_{+} \cap D/p >> 0\}$ is w^*-dense in L'_{+}.

Let D be a price-simplex for a Riesz dual system (L, L') An excess demand function E is a mapping $E : D \to L$, satisfying the following condition (Walras' law):

$$pE(p) = 0 \text{ for all } p \in D.$$

By an *economy*, we mean a Riesz dual system (L, L'), a price-simplex D for (L, L') and an excess demand function E defined on D. An economy is said to have an *equilibrium price* p if $E(p) \leq 0$ where $\leq$ is the Riesz order on the space L.

Let K be a subset of Hausdorff linear topological space F over the reals and T a single-valued (non-linear) mapping of K into F', the topological dual of F. Recall that a point u_0 is said to satisfy the *variational inequality* if

$$(T(u_0), v - u_0) \leq 0 \text{ for all } v \text{ in } K.$$

Here, (,) denotes the pairing between F' and F. u_0 is also called a solution of the variational inequality. The mapping T is said to be *monotone* if $(T(u) - T(v), u - v) \leq 0$ for all u, v in K. T is said to be *hemicontinuous* if $T$is continuous from the line segments in K to the weak topology of F'.

5.16.1 *Existence of Equilibrium Lemma*

Lemma 5.42 *Let (L, L') be a Riesz dual system and let $u \in L$. Then $u \geq 0$ holds if and only if $f(u) \geq 0$ for all $f \geq 0$ in L'.*

Proof. The proof of the lemma is based on the two Riesz theorems cited in the preliminary remarks (see Aliprantis and Brown (1983, Theorem 2.2)). □

Next, we present the following result of Tarafdar and Mehta:

Theorem 5.114 *Any point p in D is an equilibrium price for $((L, L.), D, E)$ if and only if p is a solution of the variational inequality.*

Proof. Suppose that $E(p) \leq 0$ for some p. Then $qE(p) \leq 0$ for all q in D since q is a positive linear functional. This implies that $qE(p) \leq pE(p)$ since by Walras' law $pE(p) = 0$ for all p. Consequently, $(E(p), p - q) \leq 0$, or, equivalently, $(E(p), q - p) \leq 0$ for all q in D and p solves the variational inequality. Thus an equilibrium price p solves the variational inequality.

Conversely, suppose that p is a solution of the variational inequality. Then $(E(p), p - q) \geq 0$ which implies that $0 = E(p) \cdot p \geq E(p) \cdot q$ for all q in D, where the first equality holds by Walras' law. Hence, $E(p) \cdot \leq 0$ for all q in L'_+ by the density condition $(*)$. Now Lemma 5.42 implies that $E(p) \leq 0$ so that p is an equilibrium price. □

We shall now present the following result of Tarafdar and Mehta on the existence of equilibrium:

Theorem 5.115 *Let $((L, L'), D, E)$ be an economy. Then there exists an equilibrium price for this economy, if either one of the following conditions holds:*

(1) *$E : (D < W^*) \to (L, w(L, L')$ is continuous,*

(2) *E is hemicontinuous and monotone.*

Proof. Suppose first that E is continuous. Since $E : (D, w^*) \to (L, w(l, L')$ and D is w^*-compact and convex, E satisfies the conditions of Browder's Theorem 2 in [Browder (1968), p.286]. We conclude that there exists a p such that $(E(P), p-q) \leq 0$ for all q in D which implies that $(E(p), q-p) \leq 0$ for all q in D so that p solves the variational inequality. Theorem 5.114 now implies that p is an equilibrium price.

Suppose now that E is hemicontinuous and monotone. Again, since D is w^*-compact and convex, E satisfies the conditions of the corollary of Theorem 2 of Tarafdar (1977). We conclude as above that there exists an equilibrium price for this economy.

We now consider the existence of equilibrium prices for a more general class of economies. Suppose that (L, L') is a Riesz dual system and that D is a price-simplex for (L, L'). We now suppose that the domain D' of E is a subset of D. An *excess demand function* E is now defined to be a mapping $E : (D', w^*) \to (L, w(L, L')$ which satisfies the following properties:

(a) Density condition: D' is a w^*-dense convex subset of D.

(b) Walras' law: $pE(p) = 0$ for all p in D'.

(c) Boundary condition: It p_n is a net in D' which converges to q in $d \backslash D'$ then there exists a p in D' such that the upper limit $\lim p(E(p_n)) > 0$.

Tarafdar and Mehta proved the following theorem in Aliprantis and Brown (1983) without using the concept of a maximal element for the "revealed preference" relation on the space of prices.

Theorem 5.116 *Let $(L, L'), D, E)$ be an economy. Then there exists an equilibrium price for this economy if E is continuous.*

Proof. Let A denote the collection of all the finite subsets of D'. For each $a \in A$,let D_a be the convex hull of a. Each D_a is w^*-compact, and the restriction of E to D_a a is continuous so that Theorem 5.115 implies the existence of an equilibrium price p_a for the economy $(L, L'), D_a, E)$. Since $p_a$is an equilibrium price for D_a, p_a solves the variational inequality $(E(p_a), q - p_a) \leq 0$ by Theorem 5.114. This implies that $E(p) \leq 0$ for all q in D_a.

Although the rest of the argument is similar to that in [Aliprantis and Brown (1983), Theorem 3.6], we include it for the sake of completeness. Consider the net $\{p_a : a \in A\}$ where A is directed by inclusion. Since D is w^*-compact we may assume that $p_a \to q$ in the w^*-topology.

We show first that $q \in D'$. If $q \in D \backslash D'$, then by the boundary condition on the excess demand function there exists a $p \in D$ with $\overline{\lim} p(E(p_a)) > 0$. Since $\cup_{a \in D_a} = D', p \in D_a$ for some a, so that there exists $b \in A$ such that $p \in D_a$ for all $a \geq b$. But then for all $a \geq b$, $pE(p_a) \leq 0$ since p_a is an equilibrium price so that $\overline{\lim} pE(p_a) \leq 0$, a contradiction. Thus $q \in D'$.

We now show that q is an equilibrium price. To this end let $p \in D'$. The function $p \cdot E(r)$ from (D, w^*) to the reals is continuous as a composite of two continuous functions. It follows that

$$pE(q) = w^* - \lim pE(p_a) \text{ since } p_a \rightarrow^{w^*} q.$$

As above, there exists $b \in A$ satisfying $pE(p_a) \leq 0$ for all $a \geq b$, and so $pE(q) \leq 0$. This is true for all $p \in D'$.

Now the density condition on the excess demand function implies that $pE(q) \leq 0$ for all $p \in D$ since D' is w^*-dense in D. Since D is a price simpler, the density condition $*$ implies that $pE(q) \leq 0$ for all $p \in L'$. The lemma now implies that $E(q) \leq 0$ so that q is an equilibrium price. □

Remark 5.34 Theorem 5.116 can be obtained in the same manner under the assumption of monotonicity of the excess demand function E if the hemicontinuity condition on E is strengthened in the following way. Let $\{p_a : a \in A\}$ be a net in D' with $p_a \in D_a$ and D_a being as described in the above proof. Then we need to assume that if $p_a \rightarrow^{w^*} p$ in D' and for each a, $(E(p_a), q - p_a) \leq 0$ for all q in D_a, then it follows that $(E(p), q - p) \leq 0$ for all q in D'.

Note that the results of this section have been written and presented after a thorough revision of some results of an article in Tarafdar and Mehta (1986).

5.17 Equilibria of Non-Compact Generalized Games with $\mathcal{L}^*$ Majorized Preference Correspondences

Ding, Kim and Tan introduced the notions of correspondences of class $\mathcal{L}^*_\theta$, $\mathcal{L}^*_\theta$-majorant of ϕ at x and $\mathcal{L}^*_\theta$- majorized correspondences in Ding et al. (1992).

In this section we shall present an existence theorem of maximal elements in a non-compact set for $\mathcal{L}^*_\theta$-majorized correspondences (see Chapter 1 for details). Ding, Kim and Tan applied this maximal element theorem to prove an existence theorem of equilibrium in a qualitative game. Using this existence theorem of equilibrium, Ding et al. proved an existence theorem of equilibrium in a non-compact abstract economy with $\mathcal{L}^*$- majorized preference correspondences. Their result generalizes the result in (Yannelis and Prabhakar (1983, p. 242, Theorem 6.1)), in several ways and is closely related to the theorems of Toussaint (1984, pp. 102, 103; 6, p. 283).

5.17.1 *Existence of Maximal Elements*

We shall start with a result of Ding, Kim and Tan:

Lemma 5.43 *Let X be a regular topological space and Y be a non-empty subset of a vector space E. Let $\theta : X \to E$ and $P : X \to 2^Y$ be $\mathcal{L}_\theta{}^*$-majorized. If every open subset of X containing the set $\{x \in X : P(x) :\neq \emptyset\}$ is paracompact, then there* *exists a correspondence* $\phi : X \to 2^Y$ *of class* $\mathcal{L}_\theta{}^*$ *such that* $P(x) \subset \phi(x)$ *for all* $x \in X$.

Proof. Let $B = \{x \in X : P(x) \neq \emptyset\}$. Since P is $\mathcal{L}_\theta{}^*$-majorized, for each $x \in B$, let N_x be an open neighborhood of x in X and $\phi_x : X \to 2^Y$ be such that (i) $P(z) \subset \phi_x(z)$ and $\theta(z) \notin con\theta_x(z)$ for all $z \in N_x$, (ii) $con\phi_x(z) \subset Y$ for all $z \in X$ and (iii) $\phi_x^{-1}(y)$ is open in X for each $y \in Y$. Since the set $\{z \in X : \phi_x(z) \neq \emptyset\} = \bigcup_{y \in Y} \phi_x^{-1}(y)$ is an open neighborhood of x in X and X is regular, there exists an open neighborhood G_x of x in X such that $cl_x G_x \subset N_x$ and $\phi_x(z) \neq \emptyset$ for all $z \in cl_x G_x$. Let $g = \bigcup_{x \in B} G_x$; then G is an open subset of X which contains $B = \{x \in X : P(x) \neq \emptyset\}$ so that G is paracompact by assumption. By [Dugundji (1966), p. 162, Theorem VIII.1.4] the open covering $\{G_x\}$ of G has an open precise neighborhood-finite refinement $\{G'x\}$. For each $x \in B$, define $\phi'_x : G \to 2^Y$ by

$$\phi'x(z) = \begin{cases} \phi_x(z), & \text{if } z \in G \cap cl_x G'_x \\ Y, & \text{if } z \in G \backslash cl_x G'_x \end{cases}$$

We claim that for each $y \in Y, (\phi'_x)^{-1}(y) = \{z \in G : y \in \phi'_x(z)\}$ is opening X. Indeed, for each $y \in Y$, we have

$$\begin{aligned}(\phi'_x)^{-1}(y) &= \{z \in G \cap cl_x G'_x : y \in \phi'_x(z)\} \cup \{z \in G \backslash cl_x G'_x : y \in \phi'_x(z)\} \\ &= \{z \in G \cap cl_x G'_x : y \in \phi'_x(z)\} \cup \{z \in G \backslash cl_x G'_x : y \in Y\} \\ &= [G \cap cl_x G'_x) \cap \phi_x^{-1}(y)] \cup (G \backslash cl_x G'_x) \\ &= (G \cap \phi_x^{-1}(y)) \cup (G \backslash cl_x G'_x).\end{aligned}$$

Since $\phi_x^{-1}(y)$ is open in $X, (\phi'_x)^{-1}(y)$ is open in X.

Now we define $\phi : X \to 2^Y$ by

$$\phi(z) = \begin{cases} \bigcap_{x \in B} \phi'_x(z), & \text{if } z \in G \\ \emptyset, & \text{if } z \in X \backslash G \end{cases}$$

Let $z \in X$ be given. Clearly $con\phi(z) \subset Y$. If $z \in X \backslash G$, then $\phi(z) = \emptyset$ so that $\theta(z) \notin con\phi(z)$. If $z \in G$, then $z \in G \cap cl_x G'_x$ for some $x \in B$ so that $\phi'_x(z) = \phi_x(z)$, and hence $\phi(z) \subset \phi_x(z)$. As $\theta(z) \notin con\phi_x(z)$. We must also have $\theta(z) \notin con\phi(z)$. Therefore, $\theta(z) \notin con\phi(z)$ con 0(z) for all $z \in X$. Now we show that for each $y \in Y, \phi^{-1}(y)$ is open in X. Let $y \in X$ be such that $\phi^{-1}(y) \neq \emptyset$ and fix $u \in \phi^{-1}(y) = \{z \in X : y \in \phi(z)\} = \{z \in G : y \in \phi(z)\}$; then there exists an open neighborhood M_u of u in G such that $\{x \in B : M_u \cap G'_x \neq \emptyset\} = \{x(u,1), ..., x(u, n(u))\}$ since $\{G'_x\}$ is a neighborhood-finite refinement. Note that for each $x \in B$, if $x \notin \{x(u,1), ..., x(u, n(u))\}$, then $\emptyset = M_u \cap G'_x = M_u \cap cl_x G'_x$ so that $\phi'_x(z) = Y$ for all $z \in M_u$. Thus we have $\phi(z) = \bigcap_{x \in B} \phi'_x(z) = \bigcap_{i=1}^{n(u)} \phi'_{x(u,i)}(z)$

for all $z \in M_u$. Note that

$$\begin{aligned}\phi^{-1}(y) &= \{z \in X : y \in \phi(z)\} = \{z \in G : y \in \bigcap_{x\in B} \phi'_x(z)\} \\ &\subset \{z \in M_u : y \in \bigcap_{x \in B} \phi'_x(z)\} \\ &= \{z \in M_u : y \in \bigcap_{i=1}^{n(u)} \phi' x(u,i)(z)\} \\ &= M_u \cap \{z \in G : y \in \bigcap_{i=1}^{n(u)} \phi'_{x(u,1)}(z)\} \\ &\quad + M_u \cap \big[\bigcap_{i=1}^{n(u)} (\phi' x(u,i)^{-1}(y)\big].\end{aligned}$$

But then, $M'_u = M_u[\bigcap_{i=1}^{n(u)}(\phi'_{x(u,i)}(y)]$ is an open neighborhood of u in X such that $M'_u \subset \phi^{-1}(y)$. This shows that for each $y \in Y, \phi^{-1}(y)$ is open in X. Therefore, $\phi; X \to 2^Y$ is a correspondence of class $\mathcal{L}^*_\theta$. It remains to show that $P(w) \subset \phi(w)$ for all $w \in X$. Indeed, let $w \in X$ with $P(w) \neq \emptyset$. Note then $w \in G$. For each $x \in B$, if $w \in G \backslash cl_x G'^x$, then $\phi'_x(w) = Y \supset P(w)$ and if $w \in G \cap cl_x G'_x$, we have $w \in cl_x G'_x \subset cl_x G_x \subset N_x$ so that $P(w) \subset \phi(w) = \phi'_x(w)$. It follows that $P(w) \subset \phi'_x(w)$ for all $x \in B$ so that $P(w) \subset \bigcap_{x \in B}(w) = \phi(w)$. □

Note that Ding, Kim and Tan observed that Lemma 5.43 is exactly similar to the result in [Tulcea (1986), Proposition 1]. The only difference is in the definition of majorized correspondences,

In proving some of their main results, Ding et al. used the following simple result which is Lemma 1 in (Ding and Tan (1990)):

Lemma 5.44 *Let D be a non-empty compact subset of a topological vector space E. Then $conD$ is σ-compact and hence paracompact.*

We shall now present the following theorem of Ding, Kim and Tan on the existence of a maximal element:

Theorem 5.117 *Let X be a non-empty convex subset of a locally convex Hausdorff topological vector space and D be a non-empty compact subset of X. Let $P : X \to 2^D$ be $\mathcal{L}^*$-majorized (i.e., $\mathcal{L}^*_{1_x}$-majorized). Then there exists a point $\hat{x} \in conD \subset X$ such that $P(\hat{x}) = \emptyset$.*

Proof. Suppose the contrary, i.e., for all $x \in conD, P(x) \neq \emptyset$. Then the set $\{x \in conD : P(x) \neq \emptyset\} = conD$ is paracompact by Lemma 5.44. By Lemma 5.43, there exists a correspondence $\phi : conD \to 2^D$ of class $\mathcal{L}^*$ such that $P(x) \subset \phi(x)$ for each $x \in conD$. It is easy to see that ϕ satisfies all hypotheses of Yannelis-Prabhakar

(1984, Theorem 5.3), and hence there exists a point $\hat{x} \in conD$ such that $\phi(\hat{x}) = \emptyset$; it follows that $P(\hat{x}) = \emptyset$ which contradicts our assumption. Hence, the conclusion must hold. □

Theorem 5.117 generalizes Theorem 5.3 in [Yannelis-Prabhakar (1984)] in the following ways: (i) X need not be paracompact and (ii) P is $\mathcal{L}^*$-majorized but need not be of class $\mathcal{L}^*$.

As an application of Theorem 5.117, Ding, Kim and Tan proved the following equilibrium existence theorem for a 1-person game:

Theorem 5.118 *Let X be a non-empty convex subset of a locally convex Hausdorff topological vector space and D be a non-empty compact subset of X. Let $P : X \to 2^D$ be $\mathcal{L}^*$-majorized (i.e., $\mathcal{L}^*_{1_x}$-majorized) and $A, B : X \to 2^D$ be such that for each $x \in X, A(x)$ is non-empty and $conA(x) \subset B(x)$ and for each $y \in D, A^{-1}(y)$ is open in X. If the correspondence $clB : X \to 2^D$ is upper semicontinuous, then there exists an equilibrium choice $\hat{x} \in X$, i.e., $\hat{x} \in clB(\hat{x})$ and $A(\hat{x}) \cap P(\hat{x}) = \emptyset$.*

Proof. Let $F = \{x \in X : x \in clB(x)\}$. We first note that F is closed in X since clB is upper semicontinuous. Define $\psi : X \to 2^D$ by

$$\psi(x) = \begin{cases} A(x) \cap P(x), & \text{if } x \in F, \\ A(x) & \text{if } x \notin F. \end{cases}$$

If $x \notin F$, then $X \backslash F$ is an open neighborhood of x such that for each $z \in X \backslash F, z \notin clB(z)$ and $\psi(z) = A(z)$; since $conA(z) \subset B(z)$, we have $z \notin conA(z)$ and $conA(z) \subset D$. Since $A^{-1}(y)$ is open for all $y \in D$, A is an $\mathcal{L}^*$-majorant of ψ at x. Now suppose that $x \in F$ and $\psi(x) = A(x) \cap P(x) \neq \emptyset$; then by assumption there exist $\phi_x : X \to 2^D$ and an open neighborhood N_x of x in X such that (a) $P(z) \subset \phi_x(z)$ and $z \notin con\phi_x(z)$ for each $z \in N_x$, (b) $con\phi_x(z) \subset D$ for each $z \in X$ and (c) $\phi_x^{-1}(y)$ is open in X for each $y \in D$. Define $\phi'_x : X \to 2^D$ by

$$\phi'_x(z) = \begin{cases} A(z) \cap \phi_x(x), & \text{if } z \in F, \\ A(z), & \text{if } z \notin F; \end{cases}$$

then (i) for each $z \in N_x$ it is easy to see that $\psi(z) \subset \phi'_x(z)$ and $z \notin con\phi'_x(z)$, (ii) for each $z \in X, con\phi'_x(z) \subset conA(z) \subset B(z) \subset D$ and (iii) for each $y \in D, (\phi'_x)^{-1}(y) = [\phi_x^{-1}(y) \cap (X \backslash F) \cap A^{-1}$ is open in X. This shows that ϕ'_x is an $\mathcal{L}^*$-majorant of ψ at x. Therefore, ψ is $\mathcal{L}^*$-majorized. By Theorem 5.117, there exists a point $\hat{x} \in X$ such that $\psi(\hat{x}) = \emptyset$. Since $A(x) \neq \emptyset$ for each $x \in X$, we must have $\hat{x} \in clB(\hat{x})$ and $\psi(\hat{x}) = A(\hat{x}) \cap P(\hat{x}) = \emptyset$, □

5.17.2 *Existence of Equilibrium for Non-Compact Abstract Economies*

We shall present the following equilibrium existence theorem of Ding, Kim and Tan for a non-compact qualitative game (see Chapter 1 for definition). This follows from Theorem 5.117.

Theorem 5.119 *Let $\Gamma = (X_i, P_i)_{i\in I}$, be a qualitative game such that for each $i \in I$,*

(a) *X_i is a non-empty convex subset of a locally convex Hausdorff topological vector space,*
(b) *$P_i : X \in \Pi_{i\in i} X_i \to 2^{D_i}$ is $\mathcal{L}^*$-majorized (i.e., $\mathcal{L}^*_{\pi_i}$-majorized), where D_i is a non-empty compact subset of X_i*
(c) *the set $E^i = \{x \in X : P_i(x) \neq \emptyset\}$ is open in X,*
(d) *there exists a non-empty subset F_i of D_i such that $F_I \cap P_i(x) \neq \emptyset$ for each $x \in E^i$ and $conF_i \subset D_i$.*

Then Γ has an equilibrium point, i.e., there exists a point $\hat{x} \in X$ such that $P_i(\hat{x}) = \emptyset$ for all $i \in I$.

Proof. Let $D = \Pi_{i\in I} D_i$ then D is compact by Tychonoff's theorem (e.g., see [Tulcea (1986), p. 224]). For each $x \in X$, let $I(x) = \{i \in I : P_i(x) \neq \emptyset\}$. Define a correspondence $P : X \to 2^D$ by

$$P(x) = \begin{cases} \bigcap_{i\in I(x)} P_i'(x) & \text{if } I(x) \neq \emptyset, \\ \emptyset, & \text{if } I(x) = \emptyset, \end{cases}$$

where $P_i'(x) = \Pi_{j\neq, j\in I} F_j \oplus P_i(x)$ for each $x \in X$.

Then for each $x \in X$ with $I(x) \neq \emptyset, P(x) \neq \emptyset$. Let $x \in X$ be such that $P(x) \neq \emptyset$. Then $P_i'(x) \neq \emptyset$ for all $I \in I(x)$. Fix one $i \in I(x)$. By assumption (b), there exist an open neighborhood $N(x)$ of x in X and an $\mathcal{L}^*$-majorant ϕ_i of P_i at x such that

(i) for each $z \in N(x), P_i(z) \subset \phi_i(z)$and $z_I \notin con\phi_i(z)$,
(ii) for each $z \in X, con\phi_i(z) \subset D_i$, and
(iii) for each $y \in D_i, \phi_i{-}1(y)$ is open in X.

By assumption (c), we may assume $N(x) \subset E^i$, so that $P_i(z) \neq \emptyset$ for all $z \in N(x)$. Now we define $\Phi_x : X \to 2^D$ by

$$\Phi_x(z) = \Pi_{j\neq i, j\in I} F_j \oplus \phi_i(z) \text{for all} z \in X.$$

We now claim that Φ_x is an $\mathcal{L}^*$-majorant of P at x. Indeed, for each $z \in N(x)$, by (i),

$$P(z) = \bigcap_{j\in I(z)} P_j'(z) \subset P_i'(z) \subset \Phi_x(z)$$

and

$$z \notin con\Phi - x(z).$$

By assumption (d) and (ii), for each $z \in X$,

$$con\Phi_x(z) \subset \Pi_{j\neq i, j\in I} conF_j \oplus con\phi_i(z) \subset D.$$

Since for each $y \in D$,

$$\Phi_x^{-1}(y) = \begin{cases} \phi_i^{-1}(y_I), & \text{if } \ y_j \in F_j \ \text{ for all } \ j \neq i, \\ \emptyset, & \text{if } \ y_j \notin F_j \ \text{ for some } \ j \neq i \end{cases}$$

and ϕ_i^{-1} is open in $X, \Phi_x^{-1}(y)$ is also open in X.

Therefore, Φ_x is an $\mathcal{L}^*$-majorant of P at x. This shows that P is $\mathcal{L}^*$-majorized. By Theorem 5.117, there exists a point $\hat{x} \in X$ such that $P(\hat{x}) = \emptyset$ so that $I(\hat{x}) = \emptyset$ and hence $P_i(\hat{x}) = \emptyset$ for all $i \in I$. □

Remark 5.35 (1) If X_i is compact convex for each $i \in I$ then the condition (d) in Theorem 5.119 is satisfied if $D_i = X_i = F_i$ for each $i \in I$.

(2) Ding, Kim and Tan's Theorem 5.119 is an existence theorem of equilibrium for non-compact sets in locally convex spaces, while Toussaint (1984, p. 101, Theorem 2.4), is an existence theorem for compact sets in topological vector spaces.

(3) The following condition clearly implies the condition (d) in Theorem 5.119: $conP_i(X) = D_i$ for each $i \in I$.

As an application of Theorem 5.119, Ding, Kim and Tan proved the following existence theorem of equilibrium for a non-compact abstract economy (see definition in Chapter 1) with possibly an infinite number of agents.

Theorem 5.120 *Let $\Gamma = (X_i, A_i, B_i, P_i)_{i\in I}$ be an abstract economy such that for each $i \in I$,*

(a) *X_i a non-empty convex subset of a locally convex Hausdorff topological vector space,*

(b) *for each $x \in X = \Pi_{i\in I} X_i, A_i(x))$is non-empty and $conA_i(x) = B_i(x)$,*

(c) *for each $y \in X_i, A_i^{-1}(y)$ is open in X,*

(d) *the correspondence $clB_i : X \to 2^{D_i}$ is upper semicontinuous, where D_i is a non-empty compact subset of X_i,*

(e) *$A_i \cap P_i$ is $\mathcal{L}^*$-majorized,*

(f) *the set $E^i = \{x \in X : (A_i \cap P_i)(x) \neq \emptyset\}$ is open in X,*

(g) *there exists a non-empty subset K_i of D_i with $conK_I \subset D_i$ such that for each $x \in E^i \cap F_i, (A_i \cap P_i)(x) \cap K_i \neq \emptyset$ and for each $x \in X \backslash F_i, conA_i(x) \cap K_i \neq \emptyset$ where $F_i = \{x \in X : x_i \in clB_i(x)\}$.*

Then Γ has an equilibrium point, i.e., there exists a point $\hat{x} \in X$ such that $\hat{x}_i \in clB_i(\hat{x})$ and $A_i(\hat{x}) \cap P_i(\hat{x}) = \emptyset$ for each $I \in I$.

Proof. For each $I \in I$, as clB_i, is upper semicontinuous, the set F_i is closed in X. We now define $Q_i : X \to 2^{D_i}$ by

$$Q_i(x) = \begin{cases} (A_i \cap P_i)(x), & \text{if } x \in F_i, \\ conA_i(x), & \text{if } x \notin F_i. \end{cases}$$

We shall show that the qualitative game $\Gamma = (X_i, Q_i)_{i \in I}$, satisfies the hypotheses of Theorem 5.119. First we have that for each $I \in I$, the set

$$\begin{aligned} \{x \in X : Q_i(x) \neq \emptyset\} &= \{x \in X \backslash F_i : Q_i(x) \neq \emptyset\} \cup \{x \in F_i : Q_i(x) \neq \emptyset\} \\ &= (X \backslash F_i) \cup \{x \in F_i : A_i \cap P_i)(x) \neq \emptyset\} \\ &= (X \backslash F_i) \cup (F_i \cap E^i) \\ &= (X \backslash F_i) \cup E^i \end{aligned}$$

is open in X by (f).

Let $x \in X$ be such that $Q_i(x) \neq \emptyset$. We consider the following two cases:

Case 1. $x \notin F_i$.

Let $\Phi_x = conA_i$, and $N_x = X \backslash F_i$, N_x is an open neighborhood of x in X such that

(*) $Q_i(z) \subset \Phi_x(z)$ and by (b), $z \notin con\Phi_x(z)$ for each $z \in N_x$,

(**) $con\Phi_x(z) \subset D_i$ for each $z \in X$ by (b),

(***) $\Phi_x^{-1}(y)$ is open in X for all $y \in D_i$ by (c) and [Yannelis and Prabhakar (1983), p. 239, Lemma 5.1].

Therefore, Φ_x is an $\mathcal{L}^*$-majorant of Q_i at x.

Case 2. $x \in F_i$.

Since $Q_i(x) = (A_i \cap P_i)(x) \neq \emptyset$ and $A_i \cap P_i$ is $\mathcal{L}^*$-majorized, there exist an open neighborhood N_x of x in X and a correspondence $\phi_x : X \to 2^{D_i}$ such that $(A_i \cap P_i)(z) \subset \phi_x(z)$ and $z_i \notin con\phi_x(z)$ for each $z \in N_x, con\phi_x(z) \subset D_i$ for each $z \in X$ and $\phi_x^{-1}(y)$ is open in X for each $y \in D_i$. Define $\Phi_x : X \to 2^{D_i}$ by

$$\Phi_x(z) = \begin{cases} conA_i(z) \cap \phi_x(z), & \text{if } z \in F_i, \\ conA_i(z), & \text{if } z \notin F_i. \end{cases}$$

Note that as $(A_i \cap P_i)(z) \subset \phi_x(z)$ for each $z \in N_x$, we have $Q_i(z) \subset \Phi_x(z)$ and $con\Phi_x(z) \subset D_i$ for each $z \in N_x$. It is easy to see that $z_i \notin con\Phi_x(z)$ for all $z \in N_x$.

Moreover, for any $y \in D_i$, the set

$$\begin{aligned}(\Phi_x)^{-1}(y) &= \{z \in X : y \in \Phi_x(z)\} \\ &= \{z \in F_i : y \in \Phi_x(z)\} \cup \{z \in X \backslash F_i : y \in \Phi_x(z)\} \\ &= \{z \in F_i : y \in conA_i(z) \cap \phi_x(z)\} \cup \{z \in X \backslash F_i : y \in conA_i(z)\} \\ &= [F - i \cap (conA_i)^{-1}(y) \cap \phi_x^{-1}(y)] \cup [(X \backslash F_i) \cap (conA_i)^{-1}(y)] \\ &= [\phi_x^{-1}(y) \cup (X \backslash F_i)] \cap (conA_i)^{-1}(y)\end{aligned}$$

is open in X. Therefore, Φ_x is an $\mathcal{L}^*$-majorant of Q_i at x.

Therefore, Q_i, is an $\mathcal{L}^*$-majorized correspondence. Let $x \in X$ with $Q_i(z) \neq \emptyset$. Then $x \in (X \backslash F_i) \cup \{x' \in F_i : (A_i \cap P_i)(x') \neq \emptyset\}$.. If $x \in X \backslash F_i$, then by assumption (g) we have $\emptyset \neq conA_i(x) \cap K_i = Q_i(x) \cap K_i$. If $x \in \{x' \in F_i : (A_i \cap P_i)(x') \neq \emptyset\}$, then by assumption (g) again, $\emptyset \neq (A_i \cap P_i)(x) \cap K_i = Q_i(x) \cap K_i$. Therefore, we have $Q_i(x) \cap K_i \neq \emptyset$ in both cases. Also we have $conK_i \subset D_i$.

Therefore, all hypotheses of Theorem 5.119 are satisfied, so that there exists a point $\hat{x} \in X$ such that $Q_i(\hat{x}) = \emptyset$ for all $i \in I$. By (b), this implies that for each $i \in I, \hat{x}_i \in clB_i(\hat{x})$ and $A_i(\hat{x}) \cap P_i(\hat{x}) = \emptyset$. □

Remark 5.36 (1) If X_i is compact convex for each $i \in I$, then the condition (g) in Theorem 5.120is satisfied if $X_i = D_i = K_i$, for each $i \in I$.

(2) Thoorem 5.120 generalizes [Yannelis and Prabhakar (1983), p. 242, Theorem 6.1] in several ways:

(a) X_i need not be compact or metrizable,

(b) P_i need not be of class $\mathcal{L}$, and

(c) the set I of agents need not be countable.

In case X_i is compact convex and $A_i = B_i$ for each $i \in I$, then Ding, Kim and Tan obtained the following result from Theorem 5.120:

Corollary 5.120.1 *Let $\Gamma = (X_i, A_i, P_i)_{i \in I}$ be an abstract economy such that for each $i \in I$,*

(a) X_i is a non-empty compact convex subset of a locally convex Hausdorff topological vector space,

(b) for each $x \in X = \Pi_{i \in I} X_i, A_i(x)$ is non-empty convex,

(c) for each $y \in X_i, A_i^{-1}(y)$ is open in X,

(d) the correspondence $clA_iX \to 2^{X_i}$ is upper semicontinuous,

(e) $A_i \cap P_i$ is $\mathcal{L}^*$-majorized.

(f) the set $\{x \in X : (A_i \cap P_i)(x) \neq \emptyset\}$ is open in X.

Then Γ has an equilibrium.

Remark 5.37 The above corollary is closely related to [Tulcea (1988), p. 283, Theorem 4; Toussaint (1984) pp. 102–103, Theorem 2.5]; their results are existence theorems of equilibria for compact abstract economies under a stronger assumption (namely, $x_i \notin conP_i(x)$ for all $x \in X$) than Ding-Kim-Tan's corollary (and hence also Theorem 5.120) but their underlying spaces are only topological vector spaces (while Ding-Kim-Tan's are locally convex Hausdorff topological vector spaces).

Note that the results of this section have been written and presented after a thorough revision of some results of an article in Ding et al. (1992).

5.18 Equilibria of Non-Compact Generalized Games

In the last three decades, the classical Arrow-Debreu's existence theorem of Walrasian equilibria [Aubin and Ekeland (1984)] has been generalized in many directions. In finite dimensional spaces, Gale and Mas-Colell (1978) proved the existence of a competitive equilibrium without the assumptions of total or transitive preference correspondences. Shafer and Sonnenschein (1975) obtained results in the same direction and they proved the Arrow-Debreu Lemma for abstract economies for the case where preference correspondences may not be total or transitive. For the infinite dimensional strategy spaces and finite or infinite many players, the existence results of equilibria for generalized games were given by Aubin and Ekeland (1984), Bewley (1972), Border (1985), Chang (1990), Ding and Tan (1993), Ding *et al* (1992), Flam (1979), Florenzano (1983), Khan and Papageorgiou (1987), Kim and Richter (1986), Kim, Prikry, and Yannelis (1989), Lassonde and Schenkel (1992), Mehta and Tarafdar (1987), Tian (1992), Toussaint (1984), Tulcea (1986), Tulcea (1988), Yannelis and Prabhakar (1983), etc. All existence theorems mentioned above, however, are obtained by assuming that the constraint and preference correspondences have open graphs or have open lower (or upper) open sections. Besides, in most of these models, the strategy sets are assumed to be compact in topological vector spaces. These are restricted assumptions since it is well known that if a correspondence has an open graph, then it has open upper and lower sections and thus the correspondences with open lower sections are lower semi-continuous. However, a continuous correspondence does not hold open lower (or upper) sections properties in general. Moreover, we also know that in the infinite settings, the set of feasible allocations is generally not compact in any topology of commodity spaces. The motivations for economists interested in setting forth conditions for the existence of equilibria come from the importance of generalized games (also called abstract economy) in the study of markets and other general games and from the restrictions of the existing theorems.

In this section, by the approximate theorem for the upper semi-continuous correspondence of [Tulcea (1988)], Tarafdar and Yuan give the existence theorems of equilibria for non-compact generalized games in which constraint correspondences are upper semi-continuous instead of having lower (upper) open sections or open graph in locally convex topological vector spaces. Moreover in their framework, strategy spaces may be infinite-dimensional and non-compact; the number of players may be uncountable and preference correspondences may be non-total or non-transitive. Thus Tarafdar-Yuan's results generalize many of the existence theorems of equilibria in generalized games by relaxing the compactness of strategy spaces and the continuity of constraint correspondences. In particular, Tarafdar and Yuan answered the question raised by [Yannelis and Prabhakar (1983), p. 243] affirmatively with weaker assumptions. As applications, the Fan-Glicksberg fixed point theorem and an existence theorem for constrained game were derived by Tarafdar and Yuan. These results generalize the corresponding results due to Aubin and Ekeland (1984), Chang (1990), Shafer and Sonnenschein (1975), Toussaint (1984), Tulcea (1986),Tulcea (1988), Yannelis and Prabhakar (1983) and others.

Using Tulcea's approximate theorem for upper semi-continuous correspondences, Tarafdar and Yuan proved existence theorems of equilibria for generalized games in which the constraint correspondences are upper semi-continuous instead of having lower (upper) open sections or open graphs in infinite dimensional topological vector spaces. We shall also present an existence theorem for the constrained game derived by Tarafdar and Yuan.

Let A be a subset of a topological space X. A is said to be compactly open in X if for each non-empty compact subset C of X, $A \cap C$ is open in C. If X and Y are topological spaces and $T : X \to 2^Y$ is a correspondence, the Graph of T, denoted by $GraphT$, is the set $\{(x, y) \in X \times Y : y \in T(x)\}$ and the correspondence $\bar{T} : X \to 2^Y$ is defined by $\bar{T}(x) = \{y \in Y : (x, y) \in cl_{X\times Y}GraphT\}$ (the set $cl_{X\times Y}GraphT$ is called the adherence of the graph of T), and $clT : X \to 2^Y$ is defined by $clT(x) = cl_x(T(x))$ for each $x \in X$. It is easy to see that $clT(x) \subset \bar{T}(x))$ for each $x \in X$.

Let X be a topological space, Y a non-empty subset of a vector space $E, \theta : X \to E$ a map and let $\varphi : X \to 2^Y$ be a correspondence. Then (1) φ is said to be of class $L_{\theta,C}$ (e.g., see [Tan and Yuan (1993)]) if (a) for each $x \in X$, $conv\varphi(x) \subset Y$ and $\theta(x) \notin conv\varphi(x)$ W(x) and (b) there exists a correspondence $\psi : X \to 2^Y$ such that for each $x \in X, \varphi(x) \subset \varphi(x)$ and for each $y \in Y, \psi^{-1}(y)$ is compactly open in X and $\{x \in X : \varphi(x) \neq \theta\} = \{x \in X : \psi(x) \neq \theta\}$; $(2(\varphi_x, \psi_x; N_x)$ is an $l_{\theta,C}$-majorant of φ at x if $\varphi_x, \psi_x : X \to 2^Y$ and N_x is an open neighbourhood of x in X such that (a) for each $z \in N_x, \varphi(z) \subset \varphi(z)$ and $\theta)z) \notin conv\varphi_x(z)$, (b) for each $z \in X, \psi_x(z) \subset \varphi_x(z)$ and $conv\varphi_x(z) \subset Y$ and (c) for each $y \in Y, \psi_x^{-1}(y)$ is compactly open in X; (3) φ is said to be $l_{\theta,C}$-majorized if for each $x \in X$ with $\varphi(x) \neq \theta$, there exists an $L_{\theta,C}$-majorant (φ_x, ψ_x, N_x) of φ at x such that for any non-empty finite subset A of the set $\{x \in X : \varphi(x) \neq \emptyset\}$, we have

$$\left\{z \in \bigcap_{x\in A} N_x : \bigcap_{x\in A} conv\varphi_x(z) \neq \emptyset\right\} = \left\{z \in \bigcap_{x\in A} N_x : \bigcap_{x\in A} conv\psi_x(z) \neq \emptyset\right\}.$$

In this section, we shall deal mainly with either the case (I) where $X = Y$ and is a non-empty convex subset of the topological vector space E and $\theta = I_x$, the identity map on X, or the case (II) where X $= X = \Pi_{i\in I}X_i$ and $\theta = \pi_j : X \to X_j$ is the i-th projection of X onto X_j and $Y = X_j$ is a non-empty convex subset of a topological vector space. In both cases (I) and (II), we shall write L_C instead of of $L_{\theta,C}$.

Recall that if X and Y are topological spaces, (1) $F : X \to 2^Y$ is said to be lower semicontinuous (respectively, upper semicontinuous) if for each closed (respectively, open) subset C of Y, the set $\{x \in X : F(x) \subset C\}$ is closed (respectively, open) in X; (2) F is said to have compactly open lower (respectively, upper) sections if $F^{-1} := \{x \in X : y \in F(x)\}$ is compactly open for each $y \in Y$ (respectively, $F(x)$ is compactly open in Y for each $x \in X$) and (3) F is said to be compact if for each $x \in X$, there exists a neighbourhood V_x at x in X such that $F(V_x) = \bigcup_{z\in V_x} F(z)$ is relatively compact in Y. If X is a subset of a topological vector space E, X is said to have the property (K) (e.g. see [Tulcea (1988)]) if for every compact subset B of X, the convex hull of B is relatively compact in X.

Let X and Y be topological spaces. A correspondence $T : X \to 2^Y$ is said to he quasi-regular if:

(1) it has open lower sections, i.e. for each $y \in Y, T^{-1}(y)$ is open in X,

(2) $T(x)$ is non-empty and convex for each $x \in X$,

(3) $\bar{T}(x) = cl_Y T(x)$ for all $x \in X$.

The correspondence $T$is said to be regular if it is quasi-regular and has an open graph. Let I be a (possibly infinite) set of players. For each $i \in I$, let its choice or strategy set X_i be a non-empty subset of a topological vector space and $X = \Pi_{i\in I}X_i$. A generalized game (an abstract economy) is a family of quadruples $\Gamma = (X_i; A_i; B_i; P_i)_{i\in I}$ where I is a (finite or infinite) set of players (agents) such that for each $i \in I$, X_i is a non-empty subset of a topological vector space and $A_i, B_i : X = \Pi_{j\in I}X_j \to 2^{X_i}$ are constraint correspondences and $P_i : X \to 2^{X_i}$ is a preference correspondence. An equilibrium of Γ is a point $\hat{x} \in X$ such that for each $i \in I$, $\hat{x}_i = \pi_i(\hat{x}) \in \bar{B}_i(\hat{x})$ and $A_i(\hat{x}) \cap P_i(\hat{x}) = \emptyset$. We remark that when $\hat{B}_i(\hat{x}) = cl_{X_i}B_i(\hat{x})$ (which is the case when B_i has a closed graph in $X \times X_i$; in particular, when clB_i is upper semi-continuous with closed values), our definition of an equilibrium point coincides with that of Ding et al. [Ding et al. (1992)]; and if in addition, $A_i = B_i$ for each $i \in I$, our definition of an equilibrium point coincides with the standard definition; e.g. in Borglin and Keiding (1976), Tulcea (1986), and Yannelis and Prabhakar (1983).

Note that Tarafdar and Yuan's generalized game model is more general than the one given by Borglin and Keiding (1976) (see also [Tulcea (1986)] and [Yannelis and Prabhakar (1983)]) in the sense that the constraint mapping has been split into two parts A and B due to the fact that the "small" constraint mapping may have not enough fixed points but a "big" constraint mapping B does so.

The following example illustrates how Tarafdar and Yuan's generalized game model does work but the former one given by Borglin-Keiding et al. does not:

Example 5.12 Let $X = [0,1]$ and define the constraint mapping $A : X \to 2^X$ by

$$A(x) = \begin{cases} [0, 1-x), & \text{if } x \in (0,1]; \\ \{1\}, & \text{if } x = 0. \end{cases}$$

Define the preference mapping $P : X \to 2^X \cup \{\emptyset\}$ by

$$P(x) = \begin{cases} [0, 1-x), & \text{if } x \in [0,1); \\ \emptyset, & \text{if } x = 1. \end{cases}$$

Then the fixed point set of A is $(0, 1/2)$. It is also clear that $x \notin P(x)$ and P has open lower sections (due to the fact that $P^{-1}(y) = \emptyset$ if $y = 1$ and $P^{-1}(y) = (y, 1]$ if $y \in [0,1)$). Let $B : X \to 2^X$ be defined by $B(x) = A(x)$ for each $x \in X$. Note that the mapping $\bar{B} : 2^X$ is such that $\bar{B}(x) = [0, 1-x]$ for each $x \in X$ and $1/2$ is also a fixed point of B so that $1/2$ is an equilibrium point of the generalized game $\Gamma = ([0,1]; A, B; P))$ (in our sense); but $1/2$ is not an equilibrium point of the generalized game $([0,1]; A; P)$ (in the sense of Yannelis and Prabhakar (1983)) since $1/2 \notin A(1/2)$ even though $A(1/2) \cap P(1/2) = \emptyset$.

We shall now present the following existence theorem of equilibria for generalized games in Tarafdar and Yuan (1993) which follows from Theorem 4.3 in [Tan and Yuan (1993)]:

Theorem 5.121 *Let $\Gamma = (x_i, A_i, B_i, P_i)_{i \in I}$ be a generalized game such that $X = \Pi_{i \in I} X_i$ is paracompact. Suppose that the following conditions are satisfied:*

(a) *for each $i \in I$, X_i is a non-empty convex subset of a topological vector space,*
(b) *for each $i \in I$ and for each $x \in X A_i(x)$ is non-empty and $convA_i(x) \subset B_i(x)$,*
(c) *for each $i \in I$ and for each $y \in X_i, A_{-1}(y)$ and $P_i^{-1}(y)$ are open in X,*
(d) *for each $i \in I$ and for each $x \in X, x_i \notin convP_i(x)$;*
(e) *there exist a non-empty closed and compact subset K of X and $x^0 = (x_i^0)_{i \in I} \in X$ such that $x_i^0 \in conv(A_i(y) \cap P_i(y))$ for each $i \in I$ and for all $y \in X \backslash K$.*

Then Γ has an equilibrium in K.

Proof. Since $\{x \in X : (A_i \cap P_i)(x) \neq \emptyset\} = \bigcup_{y \in X_i} (A_i^{-1}(y) \cap P_i^{-1}(y))$ is open by (c), the conditions (c) and (e), all hypotheses of Theorem 4.3 of [Tan and Yuan (1993)] are satisfied. By Theorem 4.3 in [Tan and Yuan (1993)], the conclusion follows. □

Theorem 5.121 generalizes Theorem 2.5 of Toussaint in [Toussaint (1984)], Corollary 2 of Tulcea in [Tulcea (1986)] (also Corollary 2 in [Tulcea (1988)]) and Theorem 6.1 of Yannelis-Prabhakar in [Yannelis and Prabhakar (1983)] to non-compact case.

Let X be a non-empty set, Y a non-empty subset of topological vector space E and $F : X \to 2^Y$. A family $(f_j)_{j\in J}$ of correspondences between X and Y, indexed by a nonempty filtering set J (we denote by $\leq$ the order relation in J) is an upper approximating family for F [Tulcea (1988), p. 269] if

(A_I): $F(x) \subset f_j(x)$ for all $x \in X$ and all $j \in J$,

(A_{II}): for each $j \in J$ there is $j^* \in J$ such that for each $h \geq j^*$ and $h \in J, f_h(x) \subset f_j(x)$ for each $x \in X$,

(A_{III}): for each $x \in X$ and $V \in \mathcal{B}$, where $\mathcal{B}$ is a base for the zero neighbourhood in E, there is $j_{x,V} \in V$ such that $f_h(x) \subset F(x) + V$ if $h \in J$ and $j_{x,V} \leq h$.

From (A_I)–$(A_I II)$, it is easy to deduce that:

(A_{IV}): for each $x \in X$ and $k \in J, F(x) \subset \bigcap_{j\in J} f_j(x) = \bigcap_{\substack{k\leq j\\ k\in J}} f_j(x) \subset clF(x) \subset \bar{F}(x)$.

□

By Theorem 3 and its Remark in [Tulcea (1988), pp. 280–282], Tarafdar and Yuan gave the following:

Lemma 5.45 *Let $(X_i)_{i\in I}$ be a family of paracompact spaces and let $(Y_i)_{i\in I}$ be a family of set such that for each $i \in I, Y_i$ is a non-empty convex subset of a locally convex Hausdorff topological vector space E_i and Y_i has the property (K). For each $i \in I$, let $F_i : X_i \to 2^{Y_i}$ be such that F_i is compact and upper semi-continuous with compact convex values. Then there is a common filtering set J (independent of $i \in I$) such that for each $i \in I$, there is a family $(f_{ij})_{j\in J}$ of correspondences between X_i and F_i with the following properties:*

(a) *for each $j \in J$, f_{ij} is regular,*

(b) *$(f_{ij})_{j\in J}$ and $\overline{(f_{ij})}_{j\in J}$ are upper approximating families for F_i,*

(c) *for each $j \in J, f_{ij}$ is continuous if Y_i is compact.*

Remark 5.38 In the statement of above Lemma, Tulcea [Tulcea (1988), Theorem 3, p. 270]) assumed that Y_i is also closed. In his proof, the hypothesis "Y is closed" is not required.

5.18.1 *Equilibria of Generalized Games*

In this subsection, by the approximation Lemma 5.45 and Theorem 5.121, Tarafdar and Yuan gave the existence theorem of equilibria for the non-compact generalized game $\Gamma = (X_i; A_i, B_i; P_i)_{i\in I}$ in which the constraint correspondences are upper semi-continuous instead of having open lower or upper sections.

Theorem 5.122 *Let $\Gamma = (X_i, A_i, B_i; P_i)_{i\in I}$ be a generalized game such that $X = \Pi_{i\in I} X_i$ is paracompact. Suppose the following conditions are satisfied for each $i \in I$:*

(*a*) X_i *is a non-empty convex subset of locally convex Hausdorff topological vector space* E_i *and* X_i *has the property* (K),

(*b*) $A_i, B_i : X \to 2_i^X$ *such that* B_i *is compact and upper semi- continuous with nonempty compact convex values and* $A_i(x) \subset B_i(x)$ *for each* $x \in X$,

(*c*) $P_i : X \to 2^{X_i}$ *has compactly open lower sections and for each* $x \in X, x_i \notin$ $convP_i(x)$;

(*d*) *the set* $E^i = \{x \in X : (A_i \cap P_i)(x) \neq \emptyset\}$ *is open in* X,

(*e*) *there exist a non-empty compact subset* K *of* X *and* $x^0 \in X$ *for each* $y \in X \backslash K$ *with* $x_i^0 \in \{A_i(y) \cap P_i(y)\}$.

Then there exists $\bar{x} \in K$ *such that for each* $i \in I$, $A_i(\bar{x}) \cap P_i(\bar{x}) = \emptyset$ *and* $\bar{x}_i \in \bar{B}_i(\bar{x})$.

Proof. By Lemma 5.45, there is a common filtering set J such that for every $i \in I$, there exists a family $B_{ij})_{j \in J}$ of regular correspondences between X and X_i, such that both $((B_{ij})_{j \in J}$ and $(\overline{B_{ij}})_{j \in J}$ are the upper approximating of families for B_i.

Let $j \in J$ be arbitrarily fixed. The game $\Gamma_j = (X_i; B_{ij}, \overline{B_{ij}}; P_i)_{i \in I}$ satisfies all hypotheses of Theorem 5.121. Hence Γ_j has an equilibrium $\bar{x}^j \in K$ such that $Bij(\bar{x}^j) \cap P_i(\bar{x}^j) = \emptyset$, and $\pi_i(\bar{x}^j \in \overline{B_{ij}}(\bar{x}^j)$ for all $i \in I$.

Since $(\bar{x}^j)_{j \in J}$ is a net in the compact set K, without loss of generality we may assume that $((\bar{x}^j)_{j \in J}$, converges to $x^* \in K$. Then for each $i \in I, \pi_i(x^*) = \lim_{j \in J} \pi_i(\bar{x}^j)$. Noting that for every $j \in J$ and $x \in X, A_i(x) \subset B_i(x) \subset B_{ij}(x)$, we have $A_i(\bar{x}^j) \cap P_i(\bar{x}^j) = \emptyset$ for all $i \in I$. By condition (d), for every $i \in I, A_i(x^*) \cap P_i(x^*) = \emptyset$. As $\overline{B_{ij}}$ has closed graph, $(x^*, x_i^*) \in$ Graph $\overline{B_{ij}}$ for every $i \in I$. For each $i \in I$, since $(\overline{B_{ij}})_{j \in J}$ is also an upper approximation family for $B_i, \bigcap_{j \in J} \overline{B_{ij}}(x) \subset \overline{B_i}(x)$ for each $x \in X$ so that $(x^*, x_i^*) \in$ Graph$\overline{B_i}$. Therefore, for each $i \in I$, $A_i(x^*) \cap P_i(x^*) = \emptyset$ and $\pi_i(x^*) \in \overline{B_i}(x^*)$. □

Tarafdar and Yuan observed that Theorem 5.122 generalizes Theorem 2.5 of [Toussaint (1984), p. 103], Theorem 4.1 of [Chang (1990), p. 246], Theorem 6.1 of [Yannelis and Prabhakar (1983), p. 243] to non-compact generalized games and the constraint correspondences need not have open lower sections. In particular, Theorem 5.122 answers the question raised by [Yannelis and Prabhakar (1983), p. 243] in the affirmative with weaker conditions. If $A_i = B_i$ for each $i \in I$ in Theorem 5.122, Tarafdar and Yuan obtained the following:

Theorem 5.123 *Let* $\Gamma = (X_i, A_i, P_i)_{i \in J}$ *be a generalized game such that* $X = \Pi_{i \in I} X_i$ *is para-compact. Suppose the following conditions are satisfied for each* $i \in I$:

(*a*) X_i *is a non-empty convex subset of a locally convex Hausdorff topological vector space* E_i *and* X_i *has the property* (K),

(*b*) $A_i : X \to 2^{X_i}$ *is compact and upper semi-continuous with non-empty compact and convex values for each* $x \in X$,

(c) $P_i : X \to 2^{X_i}$ *has compactly open lower sections and for each* $x \in X, x_i \notin conv P_i(x)$

(d) *the set* $E^i = \{x \in X : (A_i \cap P_i)(x) \neq \emptyset\}$ *is open in* X,

(e) *there exist a non-empty compact subset* K *of* X *and* $x^0 \in X$ *for each* $y \in X \backslash K$ *with* $x_i^0 \in (A_i(y) \cap P_i(y))$.

Then there exists $\bar{x} \in K$ *such that for each* $i \in I, A_i(\bar{x}) \cap P_i(\bar{x}) = \emptyset$ *and* $\bar{x}_i \in A_i(\bar{x})$.

When X_i is compact, closed and convex in Theorem 5.123, Tarafdar and Yuan obtained the following:

Corollary 5.123.1 *Let* $\Gamma = (X_i; A_i; P_i)_{i \in I}$ *be a generalized game and* $X := \Pi_{i \in I} X_I$. *Suppose the following conditions are satisfied for each* $i \in I$:

(a) X_i *is a non-empty closed compact convex subset of locally convex Hausdorff topological vector space* E_i,

(b) $A_i : X \to 2^{X_i}$ *is upper semi-continuous with non-empty compact and convex values for each* $x \in X$,

(c) $Pi : X \to 2^{X_i}$ *has open lower sections and* $x_i \notin conv P_i(x)$ *for each* $x \in X$,

(d) *the set* $E^i = \{x \in X : (A_i \cap P_i)(x) \neq \emptyset\}$ *is open in* X,

Then there exists an $\bar{x} \in K$ *such that for each* $i \in I, A_i(\bar{x}) \cap P_i(\bar{x}) = \emptyset$ *and* $\bar{x} \in A_i(\bar{x})$.

Corollary 5.123.1 generalizes the Theorem of [Shafer and Sonnenschein (1975), p. 374] in the following ways: (1) I is uncountable or countable infinite instead of finite; (2) for each $i \in I$, the locally convex Hausdorff topological vector space is infinite dimensional instead of being finite dimensional; (3) for each $i \in I, A_i$ is upper semi-continuous instead of continuous and (4) P_i has open lower sections instead of an open graph.

As an application of Corollary 5.123.1, Tarafdar and Yuan obtained the well-known Fan-Glicksberg's fixed point theorem (see [Fan (1952)] or [Glicksberg (1952)]) for upper semi-continuous correspondence in locally convex topological vector spaces.

Corollary 5.123.2 *Let* X *be a convex and compact subset of a locally convex topological vector space and let* $A : X \to 2^X$ *be upper semi-continuous with non-empty closed and convex values for each* $x \in X$. *Then* A *has a fixed point.*

Proof. Let $I = \{1\}$ and $P_i(x) = \emptyset$ for each $x \in X$ in Corollary 5.123.1. The conclusion follows from Corollary 5.123.1. □

The following example shows that the condition (d) "for each $I \in I$, the set $E^I = \{x \in X : (A_i \cap P_i)(x) \neq \emptyset\}$ is open in X" in Theorem 5.122 is essential.

Example 5.13 Let $I = \{1\}$ and $X = [0,1]$. Define $A, P : X \to 2^X$ by

$$A(x) = \begin{cases} [1/2, 1], & \text{if } x \in [0, 1/2), \\ [0,1], & \text{if } x = 1/2, \\ [0, 1/2], & \text{if } x \in (1/2, 1], \end{cases}$$

and

$$P(x) = \begin{cases} \emptyset, & \text{if } x = 0, \\ [0, x) & \text{if } x \in (0, 1]. \end{cases}$$

Then A is upper semi-continuous with non-empty closed convex values and the fixed point set of A is the singleton set $1/2$. The correspondence P has convex values with open lower sections, since for each $y \in [0,1], P^{-1}(y) = (y, 1]$ which is open in X. Therefore A, P and X satisfy all conditions of Theorem 5.122 except that $E = \{x \in [0,1] : A(x) \cap P(x) \neq \emptyset\} = [1/2, 1]$ is closed but not open in $[0, 1]$. However, $A(1/2) \cap P(1/2) \neq \emptyset$, i.e. the generalized game $\Gamma = (0, 1]; A; P)$ has no equilibrium point.

5.18.2 *Tarafdar and Yuan's Application on Existence Theorem of Equilibria for Constrained Games*

In this subsection, as applications of Theorem 5.122, we shall present Tarafdar and Yuan's existence theorem of equilibria for constrained games in locally convex spaces. For simplicity, Tarafdar and Yuan only considered the compact constraint generalized games.

Let $I = \{1, 2, \ldots, N\}$. Each player i chooses a strategy x in a subset X of a locally convex topological vector space E_i. Denote by X the (Cartesian) product $\Pi_{j=1}^N X_j$ and X_{-i} the product $\Pi_{j\in I, j\neq i} X_j$. Denote by x and x_{-i} an element of X and X_{-i} respectively. Each player i has a payoff (utility) function $u_i : X \to \mathbb{R} \cup \{-\infty, +\infty\}$. Given x_{-i} (the strategies of others), the choice of the i-th player is restricted to a non-empty compact and convex set $A_i(x_{-i}) \subset X_i$, the *feasible strategy set*; the i-th player chooses $x_i \in A_i(x_{-i})$ so as to minimize $u_i(x_{-i}, x_i)$, over $A_i(x_{-i})$, where $x_{-i}, x_i)$ is the point $y = (y_j)_{j\in I}$ such that $y_I = x_{-i}$ and $y_i = x_i$. The family $\mathcal{G} = (X_i; A_i; u_i)_{i=1}^N$ is then called a constrained N-person game and an equilibrium for $\mathcal{G}$ is an $x^* \in X$ such that $x_i^* \in A_i(x_{-i}^*)$ and $u_i(x^*) \leq u_i(x_{-i}^*, x_i)$ for all $x_i \in A_i(x_{-i}^*)$ (e.g. $u_i(x^*) = inf_{x_i \in A_i(x_{-i}^*)} u_i(x_{-i}^*, x_i))$ for each $i = 1, 2, \ldots, N$.

Note that if $A_i(x_{-i}) = X_i$ for each $i = 1, 2, \ldots, N$, the constrained N-person game reduced to the conventional game $\mathcal{G} = (X_i; u_i)_{i\in I}$ and its equilibrium is called a Nash equilibrium.

Theorem 5.124 *Let $\mathcal{G} = (X_i; A_i; U_i)_{i=1}^N$ be a constrained game and $x = \Pi_{i=1}^N X_i$ where X_i is a non-compact convex subsets of a locally convex topological vector space E_i for each $i = 1, 2, \ldots, N$. Suppose the following conditions are satisfied:*

(a) *the correspondence* $A : X \to 2^X$ *defined by* $A(x) = \Pi_{i=1}^N A_i(x_{-i})$ *for each* $x = (x_{-i}, x_i) \in X$ *is upper semi-continuous with closed convex values,*

(b) *the function* $\psi : X \times X \to \mathbb{R} \cup \{-\infty, +\infty\}$ *defined by*

$$\psi(x,y) = \sum_{i=1}^{N} [u_i(x_{-i}, x_i) - u_i(x_{-i}, y_i)]$$

for each $(x,y) \in X \times X$ *is such that for each* $y \in X, x \mapsto \psi(x,y)$ *is lower semi-continuous on* X*, where* $x = (x_{-i}, x_i)$ *and* $y = (y_{-i}, y_i)$

(c) *for each* $x \in X, x \notin conv(\{y \in X : \psi(x,y) > 0\})$,

(d) *the set* $\{x \in X : \sup_{y \in A(x)} \psi(x,y) > 0\}$ *is open in* X*, Then there exists* $x^* \in X$ *such that for each* $i = 1, 2, \ldots, N$,

$$x_i^* \in A_i(x_{-i}^*) \text{ and } u_i(x^*) \leq \inf_{x_i \in A(x_{-i}^*)} u_i(x_{-i}^*, x_i).$$

Proof. Define $P : X \to 2^X$ by $P(x) = \{y \in X : \psi(x,y) > 0\}$ for each $x \in X$. Note that (b) implies that P has open lower sections in X. By (c), $x \notin convP_i(x)$ conv Pi (x) for each $x \in X$. The condition (d) implies that the set $\{x \in X : A(x) \cap P(x) \neq \emptyset\}$ is open in X. Therefore $\mathcal{G} = (X; A; P)$ (satisfies all the hypotheses of Theorem 5.122 with $A = B$. By Theorem 5.122, there exists an $x^* \in X$ such that $x^* \in A(x^*)$ and $A(x^*) \cap P(x^*) = \emptyset$. Since

$$\{x \in X : X(x) \cap P(x) \neq \emptyset\} = \{x \in X : \sup_{y \in A(x)} \psi(x,y) > 0\},$$

it follows $x^* \in A(x^*)$ and $\sup_{y \in A(x^*)} \psi(x^*, y_i) \leq 0$.

For each $i = 1, 2, \ldots, N$, and $y_i \in A_i(x_{-i}^*)$, let $y = (x_{-i}^*, y_i)$. Then $y \in A(x^*)$ so that $(u_i(x^*) - u_i(x_{-i}^*, y_i)) = \sum_{i=1}^N [u_i(x^*) - u_i(x_{-i}^*, y_i)] = \psi(x^*, y) \leq \sup_{y \in A(x^*)} \psi(x^*, y) \leq 0$.

Therefore $(u_i(x^*) - u_i(x_{-i}^*, y_i)) \leq 0$ for all $y_i \in A_i(x_{-i}^*)$. Hence x^* is an equilibrium point of the constrained game $\mathcal{G} = (X_i; A_i; u_i)_{i=1}^N$. □

Theorem 5.124 generalizes the corresponding result of Aubin and Ekeland (1984, pp. 350, 351), in the sense that the feasible correspondence A_i is upper semi-continuous instead of being continuous.

Note that the results of this section have been written and presented after a thorough revision of some results of an article in Tarafdar and Yuan (1993).

Chapter 6

Best Approximation and Fixed Point Theorems for Set-Valued Mappings in Topological Vector Spaces

In the first course of functional analysis we encounter the following question: Let M be a closed vector subspace of a Hilbert space H. Given an element $x \in H$, does there exist an element $Px \in M$ such that

$$\|x - Px\| = d(x, M) = \inf\{\|x - m\| : m \in M\}?$$

In other words, does there exist a best approximation to x among all the elements in M - best in the sense that $\|x - Px\| \leq \|x - z\|$ for all $z \in M$? The answer to this question provides the Projetion Theorem in Hilbert space. For a nice treatment of the subject we refer to Goldberg (1966, pp. 34–38).

More generally, let M be a nonempty closed subset of a real or complex normed space $(E, \|.\|)$. Given $u \in E$, the set (may be empty) of all $x \in M$ such that

$$\|u - x\| = \inf\{\|u - m\| : m \in M\} = d(u, M)$$

is denoted by $P_M(u)$ and each member of $P_M(u)$ is called a best M-approximant to u.

Let $f : M \to E$ be a mapping, where M and E are as above. We are interested in a point $x \in M$ which best approximates $f(x)$, i.e.

$$\|x - f(x)\| = d(f(x), M) = \inf\{\|f(x) - m\| : m \in M\}. \tag{6.1}$$

The metric projection $P_M : M \to 2^M$ is defined by $P_M(x) = \{y \in M : \|x - y\| = d(x, M)\}, x \in M$. $P_M(x)$, if nonempty, is closed and bounded and is convex if M is convex. P_M is called the metric projection on M. M is called Proximinal if $P_M(x) \neq \emptyset$ for each $x \in M$. If for each point $x \in M$, $P_M(x)$ contains at most one point, then M is called a Chebyshev set.

Now it is easy to see that the point $x \in M$ is a solution of (6.1) if and only if x is a fixed point of the setvalued mapping P_M. For fixed point theorems and best approximations, especially in Hilbert spaces, we refer to Singh et al. (1997, pp. 75–120), Park (1995), Park (1991), and Park (1987).

6.1 Single-Valued Case

Definition 6.1 Let M be a closed convex subset of a normed space $(E, \|.\|)$. Then a mapping $g : M \to E$ is called almost affine if for $x_1, x_2 \in M$ and $y \in E, 0 < \lambda < 1$, we have

$$\|g(z) - y\| \leq \lambda \|g(x_1) - y\| + (1 - \lambda) \|g(x_2) - y\|$$

where $z = \lambda x_1 + (1 - \lambda)x_2$.

Then g is called almost quasi-convex if

$$\|g(z) - y\| \leq \max(\|g(x_1) - y\|, \|g(x_2) - y\|) \qquad \text{for all } y \in M.$$

We can easily check that an almost affine map is always almost quaxi-convex.

The following theorem was proved in Carbone (1991) where Allen's Theorem Allen (1977) (see Tarafdar (1986) for more general results) and the technique of Kapoor (1973) and Sehgal, Singh, and Smithson (1987) were used. We give a new proof by using our fixed point Theorem 4.33 (Tarafdar (1987)).

Theorem 6.1 *Let M be a nonempty convex subset of a normed space $(E, \|.\|)$. Let $g : M \to M$ be a continuous, almost quasi-convex and onto mapping and $f : M \to E$ be a continuous mapping. Furthermore, let M_0 be a nonempty compact convex subset of M such that the set*

$$D = \{y \in M : \|f(y) - g(y)\| \leq \|f(y) - g(x)\| \text{ for all } x \in M_0\}$$

is empty or compact. Then there exists a point $u \in M$ such that

$$\|g(u) - f(u)\| = d(f(u), M).$$

Proof. Suppose the theorem is false.Then it follows that for each $y \in M$, the set $F(y) = \{x \in M : \|f(y) - g(y)\| > \|f(y) - g(x)\|\}$ is nonempty. Thus $F : M \to 2^M$ is setvalued with nonempty value. Moreover, for each $y \in M$, $F(y)$ is convex. To see this, let $u = \sum_{i=1}^n \lambda_i x_i, x_i \in F(y), i = 1, 2, \ldots, n, 0 \leq \lambda_i \leq 1$ and $\sum_{i=1}^n \lambda_i = 1$. Then we have $\|g(y) - f(y)\| > \|g(x_i) - f(y)\|, i = 1, 2, \ldots, n$. Hence by the almost quasi-convexity of g, we have

$$\begin{aligned} \|g(u) - f(y)\| &= \left\| g(\sum_{i=1}^n \lambda_i x_i) - f(y) \right\| \\ &\leq \max(\|g(x_1) - f(y)\|, \ldots, \|g(x_n) - f(y)\|) \\ &< \|g(y) - f(y)\|. \end{aligned}$$

This implies that $u \in F(y)$. Hence $F(y)$ is convex for each $y \in M$. Now for each

$x \in M$,

$$\begin{aligned} F^{-1}(x) &= \{y \in M : x \in F(y)\} \\ &= \{y \in M : \|f(y) - g(y)\| > \|f(y) - g(x)\|\} \\ &= \{y \in M : \|f(y) - g(y)\| \le \|f(y) - g(x)\|\}^c, \end{aligned}$$

which is a relatively open set for each $x \in M$ as f and g are continuous. Set $O_x = F^{-1}(x)$ for each $x \in M$. Then obviously $\bigcup_{x \in M} O_x = M$. (To see this, let $y \in M$. Then since $F(y) \neq \emptyset$, let $x \in F(y)$ which implies that $y \in F^{-1}(x) = O_x$.) Now we can easily check that

$$\bigcap_{x \in M_0} 0_x{}^c = \bigcap_{x \in M_0} F^{-1}(x)^c = D.$$

Thus all the conditions of a Theorem of Chapter 4 are fulfilled in this case. Hence there is a fixed point y_0 of F, i.e. $y_0 \in F(y_0)$, i.e. $\|f(y_0) - g(y_0)\| > \|f(y_0) - g(y_0)\|$, which is absurd. Thus the theorem must be true. □

The following result is also proved in Carbone (1991) by using a lemma of Fan (1969) and following the line of Kapoor (1973). In here it follows directly from above Theorem 6.1.

Corollary 6.1.1 *Let M be a nonempty compact convex subset of a normed space $(E, \|.\|)$. If $g : M \to M$ is a continuous, almost quasi-convex and onto mapping and $f : M \to E$ is a contnuous mapping, then there is a point $u \in M$ such that*

$$\|f(u) - g(u)\| = d(f(u), M).$$

Proof. We take $M = M_0$. For each $x \in M$, the set

$$F(x) = \{y \in M : \|g(y) - f(y)\| \le \|g(x) - f(y)\|$$

is a closed subset of M due to the contnuity of g and f. Hence $D = \bigcap_{x \in M} F(x)$ being a closed subset of a compact set M is compact, if nonempty. Hence the result follows from Theorem 6.1. □

The following result is proved in Prolla (1982–1983).

Corollary 6.1.2 *Let M be a nonempty compact and convex subset of a normed space $(E, \|.\|)$. If $g : M \to M$ is a continuous almost affine and onto mapping and $f : M \to E$ is a continuous mapping, then there is a point $u \in M$ such that*

$$\|g(u) - f(u)\| = d(f(u), M).$$

Proof. Noting that an almost affine mapping is always almost quasi-convex, the corollary follows from Corollary 6.1.1. □

Our next result is a variant of a result of Hayashi (1985) and is proved in Lin (1989) by using Fan's lemma (see Fan (1961)). Here we give another variant of proof by using our Fixed Point Theorem 4.33.

We need the following definition.

Definition 6.2 Let E and F be two Hausdorff (real) topological vector spaces. Let $\to$ and $\rightharpoonup$ denote respectively the convergence in given and weak topology (on E or F). A mapping $f : E \to F$ is called strongly continuous, if $f(x_\nu) \to f(x)$ whenever $x_\nu \rightharpoonup x$; and f is called weakly continuous, if $f(x_\nu) \rightharpoonup f(x)$ whenever $x_\nu \rightharpoonup x$, $\{x_\nu\}$ being a net in E.

Theorem 6.2 *Let E and F be locally convex Hausdorf topological vector spaces, $M \subset E$ be a nonempty weakly compact convex subset. Let $f, g : M \to F$ be two mappings such that f is strongly continuous and g is weakly continuous on M. Let g satisfy the following condition:*

(*a*) *$g^{-1}([y,z])$ is convex for all $y, z \in g(M)$, where $[y,z] = \{\lambda y + (1-\lambda)z : 0 \le \lambda \le 1\}$.*

Then there exists a point $u_0 \in M$ with $f(u_0) = g(u_0)$, or there exist a point $u_0 \in M$ and a continuous seminorm p on F such that for all $y \in g(X)$, $0 < p(g(u_0) - f(u_0)) \le p(y - f(u_0))$.

Proof. We repeat the argument of Lin (1989) and replace Fan's lemma by our Fixed Point Theorem 4.33. Let us assume that the second alternative does not hold. Then for each point $u \in M$, any continuous seminorm p on F with $p(g(u) - f(u)) > 0$, there exists a point $x \in M$ such that $p(g(x) - f(u)) < p(g(u) - f(u))$. Under this assumption we will prove that there is a coincidence point u_0 of f and g, i.e. $f(u_0) = g(u_0)$.

Let $\{p_\alpha : \alpha \in I\}$ be the set of all continuous seminorms on F. For each $\alpha \in I$, we define $F_\alpha = \{x \in M : p_\alpha(g(x) - f(x)) = 0\}$.

By using weak continuity of g and strong continuity of f we can easily prove that for each $\alpha \in I$, F_α is weakly closed in M. Indeed, if x_ν be a net in F_α such that $x_\nu \rightharpoonup x$, then $g(x_\nu) \rightharpoonup g(x)$ and $f(x_\nu) \to f(x)$ and thus $g(x_\nu) - f(x_\nu) \rightharpoonup g(x) - f(x)$.

We note that for any positive real number r, the set $\{y \in F : p_\alpha(y) \le r\}$ is convex, closed and hence weakly closed (see Rudin (1973, p. 64). Thus p_α is weakly lower semicontinuous and hence

$$p_\alpha(g(x) - f(x)) \le \liminf p_\alpha(g(x_\nu) - f(x_\nu)) = 0.$$

This implies that $x \in F_\alpha$ and F_α is weakly closed.

Evidently a point $u \in M$ is a coincidence point of g and f if and only if $u \in \bigcap_{\alpha \in I} F_\alpha$. Since M is weakly compact, it suffices to prove that $\bigcap_{i=1}^n F_{\alpha_i} \neq \emptyset$ for every finite subset $\{\alpha_1, \ldots, \alpha_n\}$ of I. To this end, let $\{\alpha_1, \alpha_2, \ldots, \alpha_n\}$ be a subset of I.

Then we can define $\beta \in I$ with $p_\beta = p_{\alpha_1} + p_{\alpha_2} + \cdots + p_{\alpha_n}$. Let

$$A = \{(x,u) \in M \times M : p_\beta (g(x) - f(u)) \geq p_\beta (g(u) - f(u))\}.$$

Clearly $(x,x) \in A$ for each $x \in M$. For each $x \in M$, the set

$$U = \{u \in M : p_\beta (g(x) - f(u)) \geq p_\beta (g(u) - f(u))\}$$

is weakly closed. To see this let $\{u_\nu\}$ be net in U such that $u_\nu \rightharpoonup u \in M$. Then we have from the weak continuity of g and strong continuity of f that $g(u_\nu) \rightharpoonup g(u)$ and $f(u_\nu) \to f(u)$. Thus

$$g(u_\nu) - f(u_\nu) \rightharpoonup g(u) - f(u)$$

and

$$g(x) - f(u_\nu) \to g(x) - f(u).$$

Now since p_β is weakly lower semicontinuous and $u_\nu \in U$, we obtain

$$\begin{aligned} p_\beta (g(u) - f(u)) &\leq \liminf p_\beta (g(u_\nu) - f(u_\nu)) \\ &\leq \liminf p_\beta (g(x) - f(u_\nu)) \\ &= p_\beta (g(x) - f(u)). \end{aligned}$$

Thus $u \in U$ and U is weakly closed.

We now prove that there is a point $u_0 \in M$ such that $M \times \{u_0\} \subseteq A$. Suppose that this is not true. This will imply that for each $u \in M$, the set

$$\begin{aligned} T(u) &= \{x \in M : (x,u) \notin A\} \\ &= \{x \in M : p_\beta (g(x) - f(u)) < p_\beta (g(u) - f(u))\} \end{aligned}$$

is nonempty. Now we prove that $T(u)$ is convex for each $u \in M$. Let $x_1, x_2 \in T(u)$, we will prove that for all λ with $0 \leq \lambda \leq 1, \lambda x_1 + (1-\lambda)x_2 \in T(u)$. Since $x_1, x_2 \in T(u)$, we have

$$p_\beta (g(x_i) - f(u)) < p_\beta (g(u) - f(u)), i = 1, 2.$$

Let $g(x_1) = y, g(x_2) = z$. By condition (a), $g^{-1}([y,z])$ is convex. Hence it follows that $\lambda x_1 + (1-\lambda)x_2 \in g^{-1}([y,z])$, i.e. $g(\lambda x_1 + (1-\lambda)x_2) \in [y,z]$. This implies that there is a number γ with $0 \leq \gamma \leq 1$ such that $g(\lambda x_1 + (1-\lambda)x_2) = \gamma y + (1-\gamma)z = \gamma g(x_1) + (1-\gamma) g(x_2)$. Thus

$$\begin{aligned} p_\beta (g(\lambda x_1 + (1-\lambda)x_2) - f(u)) &= p_\beta (\gamma g(x_1) + (1-\gamma)g(x_2) - f(u)) \\ &\leq \gamma p_\beta (g(x_1) - f(u)) + (1-\gamma)p_\beta (g(x_2) - f(u)) \\ &< \gamma p_\beta (g(u) - f(u)) + (1-\gamma)p_\beta (g(u) - f(u)) \\ &= p_\beta (g(u) - f(u)). \end{aligned}$$

This shows that $\lambda x_1 + (1-\lambda)x_2 \in T(u)$. Thus $T(u)$ is convex for each $u \in M$. Next we prove that for each $x \in M$, $T^{-1}(x) = \{u \in M : x \in T(u)\} = \{u \in M : p_\beta(g(x) - f(u)) < p_\beta(g(u) - f(u))\} = [u \in M : p_\beta(g(x) - f(u)) \geq p_\beta(g(u) - f(u))]^c = U^c$ relatively weakly open. Evidently $\bigcup_{x \in M} T^{-1}(x) = M$ as $T(u) \neq \emptyset$ for each $u \in M$. Thus all the conditions of Theorem 4.33 there is a fixed point $x_0 \in M$ such that $x_0 \in T(x_0)$, i.e. $p_\beta(g(x_0) - f(x_0)) < p_\beta(g(x_0) - f(x_0))$ which is impossible. Thus there must exist a point $u_0 \in M$ such that $M \times \{u_0\} \subseteq A$, i.e.

$$p_\beta(g(x) - f(u_0)) \geq p_\beta(g(u_0) - f(u_0)) \qquad \text{for all } x \in M.$$

Hence from our initial assumption it follows that $p_\beta(g(u_0) - f(u_0)) = 0$. This implies that $p_{\alpha_i}(g(u_0) - f(u_0)) = 0$, for $i = 1, 2, \ldots, n$, and $u_0 \in \bigcap_{i=1}^n F_{\alpha_i} \neq \emptyset$. This completes the proof. □

Corollary 6.2.1 *Let E be a locally convex Hausdorff topological vector space and M be a nonempty weakly compact convex subset of E. Let $f : M \to E$ be strongly continuous mapping. Then either there is a point $x_0 \in M$ such that $f(x_0) = x_0$, or there exist a point $u_0 \in M$ and a continuous seminorm p on E such that for all $x \in M$,*

$$0 < p(u_0 - f(u_0)) \leq p(x - f(u_0)).$$

Proof. We take $E = F$ and $g = I$, the identity mapping on E, and apply Theorem 6.2. □

The following celebrated result is due to Fan (1969).

Corollary 6.2.2 *Let M be a nonempty compact convex subset of a locally convex Hausdorff topological vector space E. Let $f : M \to E$ be a continuous mapping. Then either f has a fixed point in M, or there exist a point $u_0 \in M$ and a continuous seminorm p on E such that*

$$0 < p(u_0 - f(u_0)) = \inf\{p(x - f(u_0)) : x \in M\}.$$

Proof. The corollary follows directly from Corollary 6.2.1, if we replace the weak topology by the given topology of E. □

Remark 6.1 The results of this section will be particular cases of more general results which will be proved in the next section. We believe that for some readers the above presentations will be helpful.

6.2 Set-Valued Case

Since the appearance of the result of Fan (1969), (see Corollary 6.2.2), there have appeared many generalizations and applications of this result, e.g. see Carbone (1991), Carbone (1992), Ding and Tan (1992b), Fan (1984), Hayashi (1985), Hayashi (1987),

Kapoor (1973), Komiya (1986), Kong-Ding (1992), Park (1988), Park (1989), Prolla (1982–1983), Reich (1972), Reich (1978), Reich (1979), Sehgal-Singh (1985), Sehgal-Singh (1988) and Sehgal et al. (1987).

The result of Fan, i.e. Corollary 6.2.2 can be referred to as 'best approximation and fixed point result'. In the same way the results on two space settings of this section can be referred to as 'best approximation and coincidence point results'. The results of this section are mainly from Ding and Tarafdar (1995) and partly from Ding and Tan (1992b).

Throughout this section we will adapt to the following notations of Ding and Tarafdar (1995). As before, let X be a nonempty set, 2^X denote the family of all subsets of X and $\mathcal{F}(X)$ denote the family of all non-empty finite subsets of X. If X is a topological space with topology T, we shall use (X,T) and $2^{(X,T)}$ to denote the sets X and 2^X respectively with emphasis on the fact that X is equipped with the topology T. If A is a subset of a topological space (X,T), we shall denote by $\mathrm{int}_{(X,T)}$ and $\partial_{(X,T)}$ the interior and boundary of A in (X,T) respectively. Let (X,T) and (Y,S) be two topological spaces; for a setvalued mapping $F:(X,T)\to 2^{(Y,S)}$ we will tacitly assume in this that $F(x)\neq\emptyset$ for each $x\in X$. As before, a setvalued mapping $F:(X,T)\to 2^{(Y,S)}$ is said to be upper semicontinuous (respectively, lower semicontinuous) on X if for each $x_0\in X$ and for each S-open G in Y with $F(x_0)\subseteq G$ (respectively, $F(x_0)\cap G\neq\emptyset$), there exists a T-open neighbourhood U of x_0 in X such that $F(x)\subseteq G$ (respectively, $F(x)\cap G\neq\emptyset$) for all $x\in U$.

Let E be a topological vector space with topology T, will denote by $E^*=(E,T)^*$ the topological (continuous) dual of (E,T). E^* is said to separate points of E if for each $x\in E$ with $x\neq 0$, there exists an $f\in E^*$ such that $f(x)\neq 0$. We shall denote by $W=W(E,E^*)$ the weak topology of E and by $\mathcal{P}=\mathcal{P}\ (E,T)$ the family of all continuous seminorms on (E,T). If X is a nonempty subset of E, we shall denote by $\mathrm{co}(X)$ the convex hull of X and by (X,T) and (X,W) the set X equipped with the relative topology of T to X and the relative topology of W to X respectively. We shall denote by $\mathbb{R}$ the set of all real numbers and if z is a complex number, we will denote by $Re\, z$ the real part of z.

Let X be a nonempty subset of topological vector space (E,T). For each $x\in E$, the inward set and outward set of X at x, denoted by $I_X(x)$ and $O_X(x)$ respectively, are defined by

$$I_X(x)=\{x+r(y-x): y\in X \text{ and } r>0\}.$$

$$O_X(x)=\{x-r(y-x): y\in X \text{ and } r>0\}.$$

The closures of $I_X(x)$ and $O_X(x)$ in (E,T), denoted by $\mathrm{cl}\, I_x(x)$ and $\mathrm{cl}\, O_X(x)$ respectively, are called the weakly inward set and weakly outward set of X at x respectively. We shall use $Q(x)$ to denote either $\mathrm{cl}\, I_X(x)$ or $\mathrm{cl}\, O_X(x)$.

Let (E,T) and (F,S) be two topological vector spaces. Let X be a nonempty convex subset of (E,T) and let $(F,S)^*$ separate points of (F,S). Following Prolla (1982–1983) (see also Mehta-Sessa (1992)), a mapping $g: X\to F$ is said to be almost affine if for any $x, y \in X$, and for any $p \in \mathcal{P}\ (F, S)$

$$p(g(\lambda x + (1-\lambda)y) - z) \leq \lambda p(g(x) - z) + (1-\lambda)p(g(y) - z)$$

for all $z \in F$ and $\lambda \in [0,1]$. As in Carbone (1991) a mapping $g : X \to F$ is said to be almost quasi-convex if for each each $z \in F$, each $p \in \mathcal{P}\ (F, S)$ and each $r > 0$, the set $\{x \in X : p(g(x) - z) < r\}$ is convex.

Obviously, each affine mapping (i.e. $g(\lambda x + (1-\lambda)y) = \lambda g(x) + (1-\lambda)g(y)$, for all $x, y \in X$ and $\lambda \in [0,1]$) is almost affine and each almost affine mapping is almost quasi-convex, but not conversely.

6.2.1 *Some Lemmas and Relevant Results*

In this subsection we include some lemmas and relevant results (e.g. see Ding and Tan (1992b) and Ding and Tarafdar (1995).

Let (E, T) be a topological vector space for which $\mathcal{P} \neq \emptyset$. For each nonempty subset A of E and for each $p \in \mathcal{P}$, let

$$d_p(x, A) = \inf\{p(x - a) : a \in A\}$$

for each $x \in E$.

Proposition 6.1 *Let (E, T) and (F, S) be two topological vector spaces, $(F, S)^*$ separates points of (F, S) and X be a nonempty subset of (E, T). Let $g : X \to F$ satisfy the following condition: (see Hayashi (1985) and Lin (1989) or Theorem* 6.2) *$g^{-1}([u, v])$ is convex for all $u, v \in g(X)$, where $[u, v] = \{\lambda u + (1-\lambda)v : \lambda \in [0,1]\}$. Then g is almost quasi-convex.*

Proof. For any $z \in F$, $p \in \mathcal{P}(F, S)$ and $r > 0$, let $x, y \in \{x \in X : p(g(x) - z) < r\}$. Then $p(g(x) - z)) < r$ and $p(g(y) - z)) < r$. Let $u = g(x)$ and $v = g(y)$, then $x, y \in g^{-1}([u, v])$. By the assumption $g^{-1}([u, v])$ is convex. Hence we have $[x, y] \subseteq g^{-1}([u, v])$. Thus for each $\lambda \in [0,1]$ there exists $k \in [0,1]$ such that

$$g(\lambda x + (1-\lambda)y) = ku + (1-k)v = kg(x) + (1-k)g(y)$$

It follows that

$$\begin{aligned} p(g(\lambda x + (1-\lambda)y) - z) &= p(kg(x) + (1-k)g(y) - z) \\ &\leq kp(g(x) - z) + (1-k)p(g(y) - z) \\ &< kr + (1-k)r = r. \end{aligned}$$

Hence the set $\{x \in X : p(g(x) - z) < r\}$ is convex and g is almost quasi-convex. □

We will need the following result of Ding and Tan (1992b) and Aubin (1982, p. 67, Theorem 2.5.1) which we write as a lemma.

Lemma 6.1 *Let X and Y be topological spaces. Suppose that $h : X \times Y \to \mathbb{R}$ is lower semicontinuous and $f : X \to 2^Y$ is upper semicontinuous at $x_0 \in X$ such that $f(x_0)$ is compact. Then the function $x \to \inf\{h(x,y) : y \in f(x)\}$ is lower semicontinuous at x_0.*

The following is the Lemma 2.1 in Ding and Tarafdar (1995).

Lemma 6.2 *Let (E,T) and (F,S) be Hausdorff topological vector spaces, $(E,T)^*$ and $(F,S)^*$ the topological duals of (E,T) and (F,S) such that $(F,S)^*$ separates points of F. Let X be a nonempty subset of (E,T), $W(E, E^*)$ and $W(F, F^*)$ be the weak topology on E and F respectively. Suppose that $g : (X, W(E, E^*)) \to (F, (W, (F, F^*))$ is continuous, $G : (X, W(E, E^*)) \to 2^{(F,S)}$ is upper semicontinuous such that $G(x)$ is S-compact and $p \in \mathcal{P}(F,S)$. Then the function $V : (X, W(E, E^*)) \to \mathbb{R}$ defined by*

$$V(x) = d_p(g(x), G(x)) = \inf\{p(g(x) - z : z \in G(x)\}$$

is lower semicontinuous (in short, l.s.c.), i.e. $V : X \to \mathbb{R}$ is weakly l.s.c.

Proof. Define a function $h : (X, W(E, E^*)) \times (F,S) \to \mathbb{R}$ by

$$h(x,z) = p(g(x) - z), \text{ for } (x,z) \in X \times F.$$

For each $r \in \mathbb{R}$, let $A(r) = \{(x,z) \in X \times F : h(x,z) \leq r\}$. Let $\{X_\alpha, z_\alpha)\}_{\alpha\in\Delta}$ be a net in $A(r)$ and $(x,z) \in X \times F$ such that $x_\alpha \to x$ in $W(E,E^*)$-topology and $z_\alpha \to z$ in S-topology. By the continuity of g, $g(x_\alpha) \to g(x)$ in $W(F,F^*)$-topology. By the corollary of Hahn-Banach theorem (e.g. see Robertson-Robertson (1964), p. 29, Corollary 2), there exists $f^* \in (F,S)^*$ such that $f^*(g(x) - z) = p(g(x) - z)$ and $|f^*(z)| \leq p(z)$ for all $z \in F$.

Since $g(x_\alpha) - z_\alpha \to g(x) - z$ in $W(F,F^*)$-topology, we have

$$\begin{aligned} h(x,z) &= p(g(x) - z) = f^*(g(x) - z) \\ &= Ref^*(g(x) - z) \\ &= \lim_\alpha Ref^*(g(x_\alpha) - z_\alpha) \\ &\leq \liminf_\alpha |f^*(g(x_\alpha) - z_\alpha)| \\ &\leq \liminf_\alpha p(g(x_\alpha) - z_\alpha) \\ &= h(x_\alpha, z_\alpha) \leq r \end{aligned}$$

So that $(x,z) \in A(r)$. Thus $A(r)$ is closed in $(X, W(E,E^*)) \times (F,S)$ and h is l.s.c on $(X, W(E,E^*)) \times (F,S)$. Hence by Lemma 6.1 the function V is l.s.c. on $(X, W(E,E^*))$. □

The following corollary is the Lemma 4 in Ding and Tan (1992b).

Corollary 6.2.3 *Let (E,T) be a Hausdorff topological vector space whose topological dual E^* separates points of E. Let X be a nonempty subset of E, $p \in \mathcal{P}$ and $f : (X,W) \to 2^{(E,T)}$ be upper semicotinuous such that $f(x)$ is T-compact for each $x \in X$. Then the function $V : (X,W) \to \mathbb{R}$ defined by*

$$V(x) = d_p(x, f(x)) \text{ for } x \in X$$

is lower semicontnous, i.e. $V : X \to \mathbb{R}$ is weakly lower semicontinuous.

Proof. When $(E,T) = (F,S)$ and $g = I$, the identity mapping on E, Lemma 6.2 reduces to the Corollary. □

Our next lemma is the Lemma 5 in Ding and Tan (1992b).

Lemma 6.3 *Let E be a Hausdorff topological vector space whose topological dual E^* separates the points of E. Let A be a nonempty compact convex subset of E and $x \in E$. If for each $f \in E^*$, $\inf_{a\in A} |Ref(x-a)| = 0$, then $x \in A$. In particular, if $d_p(x,A) = 0$ for each continuous seminorm p on E, then $x \in A$.*

Proof. Assume that $x \notin A$. Then for each $a \in A$, there exists $f_a \in E^*$ such that $f_a(x) \neq f_a(a)$ as E^* separates the points of E. Let O_a and U_a be disjoint open convex sets containing $f_a(a)$ and $f_a(x)$ respectively. Then $f_a^{-1}(O_a)$ and $f_a^{-1}(U_a)$ are disjoint open convex sets in E containing a and x respectively. Sice A is compact, there are points $a_1, a_2, \ldots, a_n \in A$ such that $A \subseteq \bigcup_{i=1}^n f_{a_i}^{-1}(O_{a_i})$.. Let $U = \bigcap_{i=1}^n f_{a_i}^{-1}(U_{a_i})$, then U is open convex set containing x such that $U \cap A = \emptyset$. By Theorem 3.4 in Rudin (1973, p. 58), there exist $f \in E^*$ and $\gamma \in \mathbb{R}$ such that $Ref(x) < \gamma \leq Ref(a)$ for all $a \in A$. It follows that

$$\inf |Ref(x-a)| \geq \gamma - Ref(x) > 0,$$

which is a contradiction. Thus we have $x \in A$. The last assertion follows from the fact that for each $f \in E^*$, the function $p : E \to \mathbb{R}$ defined by $p(u) = |Ref(u)|$ for all $u \in E$ is a continuous seminorm on E. □

Remark 6.2 We note that E is not assumed to be locally convex in the above lemma. We also note that even when E is Hausdorff, the conclusion of Lemma 6.3 is false if E does not separate points of E; e.g., the competely metrizable vector space $\mathcal{L}^p$ with $0 < p < 1$ contains no open convex sets other than $\emptyset$ and $\mathcal{L}^p$ (see, e.g. Rudin (1973, p. 35)) and, therefore, $\mathcal{L}^p$ has no nonzero continuous linear functional and hence no nonzero continuous seminorm.

The following is Lemma 2.2 in Ding and Tarafdar (1995).

Lemma 6.4 *Let (E,T) and (F,S) be Hausdorff topological vector spaces and $(F,S)^*$ separates points of (F,S). Let X be a nonempty $W(E, E^*)$-compact subset of E, $g : (X, W(E, E^*)) \to (F, W(F, F^*))$ be continuous and $G : (X, W(E, E^*)) \to 2^{(F,S)}$ be upper semicontinuous such that for each $x \in X$, $G(x)$ is S-compact and* *convex. If for each $p \in \mathcal{P}(F,S)$, there exists $x_p \in X$ such that $d_p(g(x_p), G(x_p)) = 0$, then g and G have a coincidence point in X, i.e. there is $x_0 \in X$ such that $g(x_0) \in G(x_0)$.*

Proof. By the assumption and Lemma 6.2, for cach $p \in \mathcal{P}(F,S)$ the set $A(p) = \{x \in X : d_p(g(x), G(x)) = 0\}$ is nonempty and $W(E, E^*)$-closed. If $\{p_1, p_2, \ldots, p_n\}$ is a finite subset of $\mathcal{P}(F,S)$, then $\sum_{i=1}^n p_i \in \mathcal{P}(F,S)$ and $A\left(\sum_{i=1}^n p_i\right) \subseteq \bigcap_{i=1}^n A(p_i)$. Thus the family $\{A(p) : p \in \mathcal{P}(F,S)\}$ has the finite intersection property. By the $W(E, E^*)$-compactness of X, $\bigcap_{p \in \mathcal{P}(F,S)} A(P) \neq \emptyset$. Take any $\hat{x} \in \bigcap_{p \in \mathcal{P}(F,S)} A(p)$. Then $d_p(g(\hat{x}), G(\hat{x})) = 0$ for all $p \in \mathcal{P}(F,S)$. Since $G(\hat{x})$ is S-compact, by Lemma 6.3, $g(\hat{x}) \in G(\hat{x})$. Hence $x_0 = \hat{x}$ satisfies the requirement of the Lemma. □

Our next lemma is the minimax inequality: e.g. see Ding and Tan (1992c), Ding and Tan (1992b) and Ding and Tarafdar (1995).

Lemma 6.5 *Let X be a nonempty convex subset of a topological vector space and $\varphi : X \times X \to \mathbb{R} \cup \{-\infty, +\infty\}$ be such that*

- *(i) for each $x \in X$, $y \to \varphi(x,y)$ is lower semicontinuous on each compact subset C of X;*
- *(ii) for each $A \in \mathcal{F}(X)$ and for each $y \in \mathrm{co}(A)$, $\min_{x \in A} \varphi(x,y) \leq 0$;*
- *(iii) there exist a nonempty compact convex subset X_0 of X and a nonempty compact subset K of X such that for each $y \in X \setminus K$, there is an $x \in \mathrm{co}(X_0 \cup \{y\})$ with $\varphi(x,y) > 0$.*

Then there exists $\hat{y} \in K$ such that $\varphi(x, \hat{y}) \leq 0$ for all $x \in X$.

Proof. For each $x \in X$, let $K(x) = \{y \in K : \varphi(x,y) \leq 0.\}$ By (i), $K(x)$ is closed for each $x \in X$. We prove that family $\{K(x) : x \in X\}$ has the finite intersection property. For any fixed $\{x_1, x_2, \ldots, x_n\} \in \mathcal{F}(X)$, let

$$D = \mathrm{co}(X_0 \bigcup \{x_1, x_2, \ldots, x_n\}),$$

then D is a compact convex subset of X. Define $G : D \to 2^D$ by

$$G(x) = \{y \in D : \varphi(x,y) \leq 0\}.$$

Now by virtue of the hypothses, it is to prove that all the conditions of Fan (1961, Lemma 1) of Chapter 4 are satisfied. Hence $\bigcap_{x \in X} G(x) \neq \emptyset$; i.e. there exists $y_0 \in D$ such that $\varphi(x, y_0) \leq 0$ for all $x \in D$. By (iii), we must have $y_0 \in K$ so that $y_0 \in \bigcap_{i=1}^n K(x_i)$. This proves that $\{K(x) : x \in X\}$ has finite intersection property. Hence by virtue of compactness of K, $\bigcap_{x \in X} K(x) \neq \emptyset$. Let $\hat{y} \in \bigcap_{x \in X} K(x)$. Then $\hat{y} \in K$ and $\varphi(x, \hat{y}) \leq 0$ ofr all $x \in X$. □

The following is theorem in Hayashi (1985).

Lemma 6.6 *Let E, F be Hausdorff topological vector spaces, $X \subseteq E, Y \subseteq F$ be nonempty convex sets and Y be compact. Let $G : X \to 2^Y$ be upper semicontinuous with nonempty closed convex values and $\varphi : X \times Y \to \mathbb{R}$ be such that*

(a) *for each $x \in X$, $y \to \varphi(x, y)$ is lower semicontinuous in Y;*
(b) *for each $y \in Y$, $x \to \varphi(x, y)$ is quasi-concave in X.*

Then

$$\inf_{y \in Y} \sup_{x \in X} \varphi(x, y) \leq \sup_{u \in G(x), x \in X} \varphi(x, u).$$

Proof. For proof we refer to Yuan (1999, p. 143, Theorem 2.9.6) for more general result. □

The next two lemmas are from Ding and Tan (1992b, p. 747, Lemmas 6 and 7) and have their own interest.

Lemma 6.7 *Let (E, T) be a Hausdorff topological vector whos topological dual E^* separates points of E, X be a nonempty W-compact subset of E and $f : (X, W) \to 2^{(E,T)}$ be upper semicontinuous such that for each $x \in X$, $f(x)$ is T-compact and convex. If for each $p \in \mathcal{P}$, there exists $x_p \in X$ such that $d_p(x_p, f(x_p)) = 0$, then f has a fixed point in X.*

Proof. For each $p \in \mathcal{P}$, the set

$$A(p) = \{x \in X : d_p(x, f(x)) = 0\}$$

is nonempty as $x_p \in A(p)$ and also W-closed as $x \to d_p(x, f(x))$ is W-lower semicontinuous by Corollary 6.2.3. If $\{p_1, p_2, \ldots, p_n\}$ is a finite subset of $\mathcal{P}$, then evidently $\sum_{i=1}^n p_i \in \mathcal{P}$ and thus $A(\sum_{i=1}^n p_i) \subseteq \bigcap_{i=1}^n A(p_i)$. Hence $\{A(p) : p \in \mathcal{P}\}$ has the finite intersection property. By virtue of weak compactness of X, $\bigcap_{p \in \mathcal{P}} A(p) \neq \emptyset$. We take $u_0 \in \bigcap_{p \in \mathcal{P}} A(p)$, then $d_p(u_0, f(u_0)) = 0$ for all $p \in \mathcal{P}$. Hence by Lemma 6.3, $u_0 \in f(u_0))$ as $f(u_0)$ is T-compact and convex. □

Let X be a nonempty subset of a topological vector space (E, T). It is obvious to see that if $f : (X, W) \to 2^{(X,T)}$ is upper semicontinuous (respectively, lower semicontinuous, continuous), then $f : (X, T) \to 2^{(X,T)}$ is upper semicontinuous (respectively, lower semicontinuous, continuous). The next result of Ding and Tan (1992b) shows that the converse is also true under certain additional conditions on E and on X.

Lemma 6.8 *Let (E, T) be a Hausdorff topological vector space whose topological dual E^* separates the points of E and X be a nonempty T-compact subset of E. If $f : (X, T) \to 2(X, T)$ is upper semicontinuous (respectively, lower semicontinuous, continuous), then $f : (X, W) \to 2^{(E,T)}$ is upper semicontinuous (respectively, lower semicontinuous, continuous).*

Proof. Assume that $f : (X,T) \to 2^{(X,T)}$ is upper semicontinuous (respectively, lower semicontinuous). Let U be any T-open set in E. Then it follows that the set $A = \{x \in X : f(x) \not\subseteq U\}$ (respectively, $A = \{x \in X : f(x) \cap U = \emptyset\}$) is T-closed in X and hence A is T-compact as X is T-compact. Thus A is W-compact. Now since E^* separates points of E, it follows that W is Hausdorff. Thus A in W-closed. Thus $f : (X,W) \to 2^{(X,T)}$ is upper semicontinuous (respectively, lower semicontinuous). The proof of continuity is now obvious. □

Now we prove several approximation and fixed point theorems for continuous setvalued mappings.

The following theorem is the theorem 1 in Ding and Tan (1992b).

Theorem 6.3 *Let (E,T) be a Hausdorff topological vector space whose topological dual E^* separates points of E, X be a nonempty convex subset of E and $f : (X,W) \to 2^{(E,T)}$ be continuous on each nonempty W-compact subset C of X such that for each $x \in X$, $f(x)$ is T-compact and convex. Let X_0 be a nonempty W-compact and convex subset of X and K be a nonempty W-compact subset of X. If $p \in \mathcal{P}$ has the following property:*

for each $y \in X \setminus K$, there exists $x \in \operatorname{co}(X_0 \bigcup \{y\})$ such that $d_p(x, f(y)) < d_p(y, f(y))$,

then there exists $u \in K$ such that

$$d_p(u, f(u)) = \min\{d_p(x, f(u)) : x \in \operatorname{cl} I_X(u)\}.$$

Moreover, $u \in K \bigcap \partial X$ whenever $d_p(u, f(u)) > 0$.

Proof. We define $g : X \times X \to \mathbb{R}$ by

$$g(x,y) = d_p(y, f(y)) - d_p(x, f(y)).$$

Then we have the followings:

(a) For each fixed $x \in X$, by Theorem 2.5.2 in Aubin (1982, p. 69) the function $y \to d_p(x, f(y))$ is W-upper semicontnuous on C for each nonempty W-compact subset C of X (take $h(t,u) = p(t-u)$ for all $(t,u) \in X \times X$ to apply Theorem 2.5.2 in Aubin (1982, p. 69)). So in view of Corollary 6.2.3, $g(x,y)$ is W-lower semicontinuous function of y on C for each each each nonempty W-compact subset C of X.

(b) For each $A \in \mathcal{F}(X)$ and for each $y \in \operatorname{co}(A)$, we must have $\min_{x\in A} g(x,y) \le 0$; if this were not true, then there would exist $A = \{x_1, x_2, \ldots, x_n\} \in \mathcal{F}(X)$ and $y = \sum_{i=1}^n \lambda_i x_i \in \operatorname{co}(A)$ with $\lambda_1, \lambda_2, \ldots, \lambda_n \ge 0$ and $\sum_{i=1}^n \lambda_i = 1$ such that

$$g(x_i, y) = d_p(y, f(y)) - d_p(x_i, f(y)) > 0 \text{ for all } i = 1, \ldots, n. \tag{6.2}$$

Since $f(y)$ is T-compact, there exists $u_i \in f(y)$ for each $i = 1, \ldots, n$ such that $p(x_i - u_i) = d_p(x_i, f(y))$. Let $u = \sum_{i=1}^n \lambda_i u_i$. Cleary $u \in f(y)$ as $f(y)$ is convex.

Then it follows that

$$
\begin{aligned}
d_p(y, f(y)) &\leq p(y-u) \\
&= p\left(\sum_{i=1}^{n} \lambda_i(x_i - u_i)\right) \\
&\leq \sum_{i=1}^{n} \lambda_i p_i(x_i - u_i) \\
&= \sum_{i=1}^{n} \lambda_i d_p(x_i, f(y)) \\
&< d_p(y, f(y))
\end{aligned}
$$

the last inequality follows from (6.2). But this is impossible.

(c) By hypothesis, there exist a nonempty W-compact convex X_0 of X and a nonempty W-compact K of X such that for each $y \in X \setminus K$, there exists $x \in \operatorname{co}(X_0 \cup \{y\})$ with $g(x, y) > 0$.

Now equip E with weak topology W, then all hypotheses of Lemma 6.5 are satisfied so that there exists $u \in K$ such that $g(x, u) \leq 0$ for all $x \in X$; i.e.

$$
d_p(u, f(u)) \leq d_p(x, f(u)) \qquad \text{for all } x \in X. \tag{6.3}
$$

Now fix an arbtrary $v \in I_X(u) \setminus X$. As X is convex, there exist $z \in X$ and $r > 1$ such that $v = u + r(z - u)$. Suppose that

$$
d_p(v, f(u)) < d_p(u, f(u)). \tag{6.4}
$$

Since $f(u)$ is T-compact, there exist $z_1, z_2 \in f(u)$ such that $p(u - z_1) = d_p(u, f(u))$ and $p(v - z_2) = d_p(v, f(u))$. Let $\hat{z} = (1 - \frac{1}{r})z_1 + \frac{1}{r}z_2$, then $\hat{z} \in f(u)$ as $f(u)$ is convex. Since $z = (1 - \frac{1}{r})u + (\frac{1}{r})v \in X$, we have

$$
\begin{aligned}
d_p(z, f(u)) &\leq p(z - \hat{z}) \\
&= p\left((1 - \frac{1}{r})(u - z_1) + (\frac{1}{r})(v - z_2)\right) \\
&\leq (1 - \frac{1}{r})p(u - z_1) + (\frac{1}{r})p(v - z_2) \\
&= (1 - \frac{1}{r})d_p(u, f(u)) + (\frac{1}{r})d_p(v, f(u)) \\
&< d_p(u, f(u))
\end{aligned}
$$

by (6.4), which contradicts (6.3). Thus we must have

$$
d_p(u, f(u)) \leq d_p(x, f(u)) \qquad \text{for all } x \in I_X(u).
$$

By the continuity of p, we obtain

$$d_p(u, f(u)) \leq d_p(x, f(u)) \qquad \text{for all } x \in \operatorname{cl} I_X(u).$$

Hence $d_p(u, f(u)) = \min\{d_p(x, f(u)) : x \in \operatorname{cl} I_X(u)\}$.

Now assume that $d_p(u, f(u)) > 0$, then $f(u) \cap X = \emptyset$. Since $f(u)$ is T-compact, there exists $\overline{u} \in f(u)$ such that $p(u - \overline{u}) = d_p(u, f(u))$. Note that $\overline{u} \notin X$. If $u \in intX$, then there exists a real number λ with $0 < \lambda < 1$ such that $z = \lambda u + (1-\lambda)\overline{u} \in X$. It follows that

$$\begin{aligned} 0 < p(u - \overline{u}) &= d_p(u, f(u)) \\ &\leq d_p(z, f(u)) \leq p(z - \overline{u}) \\ &= \lambda p(u - \overline{u}) < p(u - \overline{u}) \end{aligned}$$

which is impossible. Thus $u \notin intX$. Hence $u \in K \cap \partial X$. This completes the proof. □

The following is Theorem 2 in Ding and Tan (1992b).

Corollary 6.3.1 *Let $(E, T), E^*, X, f, X_0$ and K be as in Theorem 6.3 such that for each $p \in \mathcal{P}$, the following property holds: for each $y \in X \setminus K$, there exists $x \in \operatorname{co}(X \cup \{y\})$ such that $d_p(x, f(y)) < d_p(y, f(y))$.*

Then either (a) f has a fixed point in K or (b) exist $p \in \mathcal{P}$ and $u \in K \cap \partial X$ such that $0 < d_p(u, f(u)) = \min\{d_p(x, f(u)) : x \in \operatorname{cl} I_X(u)\}$.

Proof. The conclusion of the corollary follows from Theorem 6.3 and Lemma 6.7. □

Remark 6.3 Theorem 6.3 generalizes Theorem 1 in Sehgal et al. (1987) in several ways: (1) f is setvalued; (2) continuity of f is weakened; (3) the space E need not be locally convex; (4) the noncompactness condition used here is weaker than that of Sehgal et al. (1987). In view of Lemma 6.8, Corollary 6.3.1 generalizes Fan's result (Corollary 6.2.2).

Chapter 7

Degree Theories for Set-Valued Mappings

In this section, we shall present some results on Degree Theory which will be used or will have generalizations in the next sections. With the exception of Theorem 7.7, the other results are taken from the works of Petryshyn and Fitzpatrick (1974), Mawhin (1972), Gaines and Mawhin (1977), Hetzer (1975), Dugundji (1951), Granas (1959), Nussbaum (1969, and 1971), and Vainiko and Sadovskii (1968). The proofs of these results are available from the above works and have thus been included only where it is considered necessary.

7.1 Degree Theory for Set-Valued Ultimately Compact Vector Fields

Let X denote a separated locally convex topological vector space over the real field with the additional property that for each compact subset A of X, there is a retraction of X onto the convex closure of A. This property automatically holds when X is metrizable, especially when X is a normed linear space. For any set B, let $\overline{co}B$ denote the convex closure of B and let $\bar{B}$ and ∂B denote the closure and boundary of B respectively. Let $K(B)$ and $CK(B)$ denote respectively the set of non-empty convex closed subsets of B and the set of non-empty convex compact subsets of B. If F is a set-valued mapping, then $F(B) = \bigcup_{X \in B} F(X)$.

Definition 7.1 Let X and Z be locally convex topological vector spaces over the real field.Let B be a subset of X and F a mapping defined on B taking values in the set of subsets of Z. F is said to be upper-semicontinuous at x if, given an open set V in Z with $F(x) \subset V$, there exists an open set W of X such that $x \in W$ and $F(W) \subset V$. If F is upper-semicontinuous at every point x of the domain B, F is said to be upper-semicontinuous on B or simply an upper-semicontinuous (denoted u.s.c.) mapping.

An u.s.c. mapping F defined on B is said to be a compact vector field if $(I - F)(B)$ is relatively compact, where I denotes the identity mapping.

Construction: Let $\omega \subset X$ be an open set and let $F : \bar{\Omega} \to K(X)$ be u.s.c. We define a transfinite sequence $\{K_\alpha\}$ as follows:

$$K_0 = \overline{co}F(\bar{\Omega}) \tag{7.1}$$

$$K_\alpha = \begin{cases} \overline{co}F(\bar{\Omega} \cap K_{\alpha-1}) & \text{if } \alpha \text{ is an ordinal such that } \alpha - 1 \text{ exists,} \\ \bigcap_{\beta \leq \alpha} K_\beta & \text{if } \alpha \text{ is an ordinal such that } \alpha - 1 \text{ does not exist.} \end{cases} \tag{7.2}$$

It can be easily checked that the following properties hold:

$$\textit{each } K_\alpha \textit{ is closed, convex and } K_\alpha \subset K_\beta \textit{ for all } \alpha \geq \beta; \tag{7.3}$$

$$F(K_\alpha \cap \bar{\Omega}) \subset K_\alpha \textit{ for each ordinal } \alpha. \tag{7.4}$$

Since the transfinite sequence $\{K_\alpha\}$ is non-increasing, there is an ordinal γ such that $K_\gamma = K_{\gamma+1}$ and $K_\beta = K_\gamma$ for each $\beta \geq \gamma$. We denote K_γ by $K(F, \bar{\Omega})$ or simply K when it is clearly understood which mapping is concerned. Clearly,

$$K = K_{\gamma+1} = \bar{co}F(\bar{\Omega} \cap K_\gamma) = \bar{co}F(\bar{\Omega} \cap K).$$

Thus we have,

$$K = \cap_\beta K_\beta = \bar{co}F(\bar{\Omega} \cap K). \tag{7.5}$$

Definition 7.2 A u.s.c. mapping $F : \bar{\Omega} \to K(X)$ is said to be *ultimately compact* if either $K \cap \bar{\Omega} = \emptyset$ or, if $K \cap \bar{\Omega} \neq \emptyset$, then $F(\bar{\Omega} \cap K)$ is relatively compact.

If F is an ultimately compact mapping, we shall call $(I - F)$ an *ultimately compact vector field*, where I is the identity mapping on X.

Lemma 7.1 *Let $\Omega \subset X$ be open and let $F : \bar{\Omega}K(X)$ be ultimately compact. Suppose that $0 \notin x - F(x)$ for each $x \in \partial\Omega$. Assume that $K \cap \bar{\Omega} \neq \emptyset$ and let ρ be a retraction of X onto K. Then*

$$x \in \Omega \textit{ and } x \in F(x) \Rightarrow x \in K \textit{ and} \tag{7.6}$$

$$x \in \Omega \textit{ and } x \in F(x) \Leftrightarrow x \in \overline{\rho^{-1}(\Omega)} \textit{ and } x \in F(\rho(x)). \tag{7.7}$$

Definition 7.3 Let $\Omega \subset X$ be open and let $F : \bar{\Omega} \to K(X)$ be ultimately compact with $0 \notin x - F(x)$ for each $x \in \partial\Omega$. If $K \cap \bar{\Omega}$ is empty, we define the *degree of $(I - F)$ on Ω with respect to zero*, denoted by $d(I - F, \Omega, 0)$, to be zero. If $K \cap \bar{\Omega} \neq \emptyset$, let ρ be a retraction of X onto K and define

$$d(I - F, \Omega, 0) = d_c(I - F_\rho, \rho^{-1}(\Omega), 0) \tag{7.8}$$

where the right hand term is the degree for compact set-valued vector fields defined by Ma (1972).

Remark 7.1 (a) For each $x \in \overline{\rho^{-1}(\Omega)}$, $x \in \rho^{-1}(\bar{\Omega})$ by the continuity of ρ and hence $\rho(x) \in \bar{\Omega} \cap K$. Since $F(\bar{\Omega} \cap K)$ is relatively compact, F_ρ is a compact mapping on $\overline{\rho^{-1}(\Omega)}$. (b) Again by continuity of ρ, $\partial\rho^{-1}(\Omega) = \rho^{-1}(\partial\Omega)$ and from 7.7, if X is a fixed point of F_ρ in $\overline{\rho^{-1}(\Omega)}$, $x \in \Omega \cap F(x)$ and thus $x \in K$ by 7.6. This implies that $\rho(x) = x \in \Omega$ and hence $x \notin \partial\rho^{-1}(\Omega)$ as Ω is open.

Remark 7.1 shows that the right-hand term of 7.8 is well defined. The following theorem ensures that the left-hand term of 7.8 is independent of the retraction ρ and hence the degree of $(I - F)$ on Ω with respect to zero is well defined.

Theorem 7.1 *Let $\Omega \subset X$ be open and let $F : \bar{\Omega} \to K(X)$ be ultimately compact with $0 \notin x - F(x)$ for each $x \in \partial\Omega$. Let $A \subset X$ be the convex closure of a compact set such that $K \subset A$, $A \cap \Omega \neq \emptyset$, $F(A \cap \bar{\Omega}) \subset A$ and $F(A \cap \bar{\Omega})$ is relatively compact. Let τ be retraction of X onto A. Then*

$$d_c(I - F_\tau, \tau^{-1}(\Omega), 0) = d(I - F, \Omega, 0) \tag{7.9}$$

Remark 7.2 By putting $A = K$, the above theorem shows that the degree defined in Definition 7.3 is independent of the retraction ρ and thus is well defined.

7.1.1 *Properties of the Degree of Ultimately Compact Vector Fields*

It has been shown by Petryshyn and Fitzpatrick (1974) that the degree for ultimately compact vector fields is an extension of the degree for compact vector fields defined by Ma (1972). This is stated in the following proposition. They have also shown that the usual properties of the topological degree can be extended to the case of ultimately compact vector fields. These properties are stated in the five theorems following the proposition.

Proposition 7.1 *Let $\Omega \subset X$ be open and suppose that $F : \bar{\Omega} \to K(X)$ is compact with $0 \notin x - F(x)$ for each $x \in \partial\Omega$. Then*

$$d_c(I - F, \Omega, 0) = d(I - F, \Omega, 0).$$

Theorem 7.2 (Existence of Fixed Points) *If $d(I - F, \Omega, 0) \neq 0$, there exists $x \in \Omega$ such that $x \in F(x)$ where $F : \bar{\Omega} \to K(X)$ is ultimately compact with no fixed points on the boundary of Ω.*

Theorem 7.3 (Additivity) *Let Ω be an open subset of X and $F : \bar{\Omega} \to K(X)$ be an ultimately compact mapping such that $0 \notin x - F(x)$ for each $x \in \partial\Omega$. Let Ω_1, Ω_2 be open disjoint subsets of X such that $\Omega_1 \cup \Omega_2 \subset \Omega$ and $\bar{\Omega}_1 \cup \bar{\Omega}_2 = \bar{\Omega}$. Also, suppose that $x \notin F(x)$ for each x in $\partial\Omega_1 \cup \partial\Omega_2$. Then*

$$d(I - F, \Omega, 0) = d(I - F, \Omega_1, 0) + d(I - F, \Omega_2, 0). \tag{7.10}$$

Theorem 7.4 *If Ω is an open set containing the origin, then $d(I, \Omega, 0) = 1$.*

Theorem 7.5 (Homotopy Invariance) *Let $\Omega \subset X$ be open and let $H : [0,1] \times \bar{\Omega} \to K(X)$ be u.s.c. such that $H([0,1] \times (\bar{\Omega} \cap K'))$ is relatively compact where $K' = K(H, [0,1] \times \bar{\Omega})$. If $x \notin H(t,x)$ for each $t \in [0,1]$ and each $x \in \partial\Omega$, then*

$$d(I - H_0, \Omega, 0) = d(I - H_1, \Omega, 0) \tag{7.11}$$

where $H_t = H(t, \cdot)$.

Theorem 7.6 *Let $\Omega \subset X$ be a symmetric neighbourhood of the origin and $F : \bar{\Omega} \to K(X)$ be an odd ultimately compact mapping such that $x \notin F(x)$ for each $x \in \partial\Omega$. Then $d(I - F, \Omega, 0)$ is an odd integer.*

Remark 7.3 (a) Theorem 7.3 is a slight variation of the Additivity Theorem given by Petryshyn and Fitzpatrick (1974) where it is assumed that $\Omega_1 \cup \Omega_2 = \Omega$. However, the same proof applies here with slight modifications. (b) Theorem 7.4 follows from the fact that if F is the zero mapping, it is also compact and the result follows from Proposition 7.1 and the corresponding result in the case of compact vector fields.

The following lemma is an extension of Tietze's Theorem, proved by Dugundji (1951).

Lemma 7.2 *Let X be a metric space and E a locally convex linear vector space. Let A be a closed subset of X and $f : A \to E$ be a continuous mapping on A. Then f can be extended over the whole space X such that the range of f, $f(X)$ is contained in the convex closure of $f(A)$.*

We present the following theorem of Tarafdar and Teo in (Tarafdar and Teo (1979)) which was used later in the proof of the Continuation Theorem.

Theorem 7.7 (Reduction Formula) *Let X be a metrizable, locally convex topological vector space with the additional property that, for each closed subspace E and any compact subset B of E, there exists a retraction of E onto the convex closure of B. Let $F : \bar{\Omega} \to K(X)$ be an ultimately compact mapping such that $x \notin F(x)$ for each $x \in \partial\Omega$. Let E_0 be a finite dimensional subspace of X containing the closure of $F(\bar{\Omega})$. Then*

$$d(I - F, \Omega, 0) = d(I - F|_{\bar{\Omega} \cap E_0}, \Omega \cap E_0, 0).$$

Proof. As K is a closed, convex and compact set, $K \cap E_0$ is also closed, convex and compact. Let $\rho_1 : E_0 \to K \cap E_0$ be a retraction of E_0 onto $K \cap E_0$. Now define $\rho_2 : K \cup E_0 \to K$ by

$$\rho_2(x) = \begin{cases} \rho_1(x) & \text{if } x \in E_0, \\ x & \text{if } x \in K. \end{cases}$$

As ρ_1 is a retraction of E_0 onto $K \cap E_0$, $\rho_1(x) = x$ for all x in $K \cap E_0$ and it is clear that ρ_2 is well defined in $K \cup E_0$. Also since $\partial K \cap E_0 \subset \partial(K \cap E_0)$, and

ρ_1 is continuous on E_0, ρ_2 is a continuous mapping on the closed set $K \cup E_0$. By Dugundji's Extension of Tietze's Theorem (1951), there exists an extension ρ of ρ_2, $\rho : X \to \overline{CO}\rho_2(K \cup E_0)$ such that ρ is continuous. Now as $\rho_2(K \cup E_0) = K$ which is closed and convex, ρ is a retraction of X onto K and is an extension of ρ_2 and ρ_1. Now,

$$x \in {\rho_1}^{-1}(\Omega \cap E_0) \Leftrightarrow \rho_1 x \in \Omega \text{ and } \rho_1 x \in E_0 \tag{7.12}$$

$$\Leftrightarrow x \in E_0 \text{ and } \rho x = \rho_1 x \in \Omega \tag{7.13}$$

$$\Leftrightarrow x \in E_0 \cap \rho^{-1}(\Omega). \tag{7.14}$$

Hence we have $\rho_1^{-1}(\Omega \cup E_0) = E_0 \cap \rho^{-1}(\Omega)$. By Definition 7.3,

$$d(I - F|_{\bar{\Omega}\cap E_0}, \Omega \cap E, 0) = d_c(I - F_{\rho_1}|_{(\rho_1{}^{-1}(\Omega\cap E_0))^-}, {\rho_1}^{-1}(\Omega \cap E_0), 0) \tag{7.15}$$

$$= d_c(I - F_{\rho_1}|_{(\rho^{-1}(\Omega)\cap E_0)^-}, \rho^{-1}(\Omega) \cap E_0, 0) \tag{7.16}$$

$$= d_c(I - F_{\rho}|_{(\rho^{-1}(\Omega)\cap E_0)^-}, \rho^{-1}(\Omega) \cap E_0, 0). \tag{7.17}$$

By the continuity of ρ and the hypothesis of the theorem that $(F(\bar{\Omega}))^- \subset E_0$, we have

$$(F_\rho(\rho^{-1}(\Omega))^-)^- = (F_\rho(\rho^{-1}(\bar{\Omega})))^- \tag{7.18}$$

$$\subset (F(\bar{\Omega}))^- \tag{7.19}$$

$$\subset E_0. \tag{7.20}$$

Hence, we may apply Theorem 11.1 of Ma (1972) and we have

$$d_c(I - F_\rho|_{(\rho^{-1}(\Omega)\cap E_0)^-}, \rho^{-1}(\Omega) \cap E_0, 0) = d_c(I - F_\rho, \rho^{-1}(\Omega), 0) \tag{7.21}$$

$$= d(I - F, \Omega, 0), \tag{7.22}$$

the last equality holding by Definition 7.3 as ρ is a retraction of X onto K. Hence we obtain the required result,

$$d(I - F, \Omega, 0) = d(I - F|_{E_0\cap\bar{\Omega}}, \Omega \cap E_0, 0). \qquad \square$$

7.1.2 *k-ϕ-Contractive Set Valued Mappings*

Definition 7.4 Let C be a lattice with a minimal element which we denote by zero, 0. A mapping $\phi : 2^X \to C$, where 2^X denotes the family of all subsets of X, is called a measure of non-compactness if, for any $A \subset X$, $B \subset X$, it satisfies the following properties:

$$\phi(\bar{co}A) = \phi(A), \tag{7.23}$$

$$\phi(A) = 0 \text{ if and only if } A \text{ is precompact}, \tag{7.24}$$

$$\phi(A \cup B) = \max\{\phi(A), \phi(B)\}. \tag{7.25}$$

Remark 7.4 It follows from (7.25) that $A \subset B \Rightarrow \phi(A) \leq \phi(B)$.

Definition 7.5 Let ϕ be a measure of non-compactness and we assume that the lattice C has the property that, for each $c \in C$ and $\lambda \in \mathbb{R}$ with $\lambda \geq 0$, there is defined an element $\lambda c \in C$ and $0c = 0$. A u.s.c. mapping $F : \bar{\Omega} \to CK(X)$ is called a $k - \phi$-contraction or a $k - \phi$-contractive mapping if there exists some $k \geq 0$ such that, for every subset A of $\bar{\Omega}$,

$$\phi(F(A)) \leq k\phi(A).$$

The following two propositions follow almost immediately from the definition of a $k - \phi$-contraction and will be stated without proof.

Proposition 7.2 *Let ϕ be a measure of non-compactness as given in Definition* 7.5, *with the additional property that, for any $A \subset X$, $B \subset X$,*

$$\phi(A + B) \leq \phi(A) + \phi(B) \tag{7.26}$$

If $F : \bar{\Omega} \to CK(X)$ and $G : \bar{\Omega} \to CK(X)$ are k_1- and k_2-ϕ-contractive mappings respectively, then $(F + G) : \bar{\Omega} \to CK(X)$ defined by

$$(F + G)(x) = F(x) + G(x) \text{ for each } x \in \bar{\Omega}$$

is a $(k_1 + k_2)$-ϕ-contractive mapping.

Proposition 7.3 *Let ϕ be a measure of non-compactness as in Definition* 7.5. *Let $F : \bar{\Omega} \to CK(X)$ be a k_1-ϕ-contraction and let $G : X \to X$ be a linear, continuous, single-valued mapping such that there exists $k_2 \geq 0$ with*

$$\phi(G(A)) \leq k_2\phi(A) \text{ for each } A \subset X.$$

Then $GF : \bar{\Omega} \to CK(X)$ defined by

$$GF(x) = \{G(y) : y \in F(x)\} \text{ for each } x \in \bar{\Omega}$$

is a k_1k_2-ϕ-contraction.

Remark 7.5 Linearity and continuity of G ensure that $GF(x)$ is a compact, convex subset of X for each $x \in \bar{\Omega}$.

Proposition 7.4 *Let ϕ be a measure of non-compactness as in Definition* 7.5. *If F and G are k-ϕ-contractions on $\bar{\Omega}$, then for any $\lambda \in [0, 1]$, the mapping $\lambda F + (1 - \lambda)G : \bar{\Omega} \to CK(X)$, defined by*

$$[\lambda F + (1 - \lambda)G](x) = \lambda F(x) + (1 - \lambda)G(x) \text{ for each } x \in \bar{\Omega}$$

is a k-ϕ-contraction.

Proof. For any $A \subset X$, $\lambda \in [0,1]$,

$$\phi([\lambda + (1-\lambda)G](A)) \le \phi(\bar{co}(F(A) \cup G(A))) \tag{7.27}$$
$$= \phi(F(A) \cup G(A)) \tag{7.28}$$
$$= \max\{\phi(F(A)), \phi(G(A))\} \tag{7.29}$$
$$\le k\phi(A). \tag{7.30}$$

Hence $\lambda F + (1-\lambda)G$ is also a k-ϕ-contraction. □

Theorem 7.8 *Let $\phi : 2^X \to \mathbb{R}^+ = \{t \in \mathbb{R} : t \ge 0\} \cup \{\infty\}$ be a measure of non-compactness and suppose that $F : \bar{\Omega} \to CK(X)$ is a k-ϕ-contraction with $0 \le k < 1$ and $\phi(F(\bar{\Omega})) < \infty$. If either X is quasi-complete or $\bar{\Omega}$ is complete, then F is ultimately compact.*

Proof. As $K = \bar{co}F(\bar{\Omega} \cap K)$,

$$\phi(F(\bar{\Omega} \cap K)) = \phi(K) \ge \phi(\bar{\Omega} \cap K).$$

But F is a k-ϕ-contraction for some $k \in [0,1]$. Hence,

$$\phi(\bar{\Omega} \cap K) \le \phi(F(\bar{\Omega} \cap K)) \le k\phi(\bar{\Omega} \cap K) \tag{7.31}$$

with $k \in [0,1]$.

Since $\phi(F(\bar{\Omega}))$ is finite, so are all the terms in the relation (7.31) and hence, for (7.31) to hold with $0 \le K < 1$, we must have

$$\phi(\bar{\Omega} \cap K) = \phi(F(\bar{\Omega} \cap K)) = 0.$$

Thus $\bar{\Omega} \cap K$ and $F(\bar{\Omega} \cap K)$ are precompact.

If X is quasi-complete, $F(\bar{\Omega} \cap K)$ is relatively compact. If $\bar{\Omega}$ is complete, $\bar{\Omega} \cap K$ is compact and hence $F(\bar{\Omega} \cap K)$ is relatively compact.

Hence F is an ultimately compact mapping. □

The k-ϕ-contractions as defined in Definition 7.4 are a generalizations of k-ball-contractions and k-set-contractions for multivalued mappings and are an extension of the k-ϕ-contractions for single-valued mappings. Nussbaum (1971) and Sadovskii (1972) have made contributions in these cases and more generalized multivalued k-ϕ-contractions were introduced by Petryshyn and Fitzpatrick (1974). In the following we shall recall the χ and γ measures of non-compactness and restate some of the properties of the k-ϕ-contractions for such ϕ.

Definition 7.6 Let $\{p_\alpha : \alpha \in A\}$ be a family of seminorms which define the topology on X. Given $\alpha \in A$ and $\Omega \subset X$, define

$$\chi_\alpha(\Omega) = \inf\{\epsilon > 0 : \text{ there exist } \{x_1, x_2, \cdots, x_n\} \subset X, \text{ with } \Omega \subset \bigcup_{i=1}^{n}\{y : p_\alpha(x_i - y) < \epsilon\}\},$$

$$\gamma_\alpha(\Omega) = \inf\{\epsilon > 0 : \Omega \text{ can be covered by the union of a finite number of sets, each with } p_\alpha - \text{ diameter } < \epsilon\}.$$

Let C be the set of all mappings from A into $\mathbb{R}^+$ with the usual definitions of ordering, maximum, multiplication by a real number, etc. Then C forms a lattice and the two mappings $\chi : 2^X \to C$ and $\gamma : 2^X \to C$ are defined by

$$\chi(\Omega)(\alpha) = \chi_\alpha(\Omega) \text{ and } \gamma(\Omega)(\alpha) = \gamma_\alpha(\Omega)$$

for every $\alpha \in A$ and each $\Omega \subset X$.

Remark 7.6 It can be easily verified that χ and γ are measures of non-compactness and, furthermore, they satisfy the following:

$$B \subset X \text{ is bounded if and only if } \gamma_\alpha(B) \text{ or } \chi_\alpha(B) \text{ is finite for each } \alpha \in A, \tag{7.32}$$

$$B \subset X \text{ is precompact if and only if, for each } \alpha \in A, \gamma_\alpha(B) = \chi_\alpha(B) = 0, \tag{7.33}$$

$$\chi(\lambda\Omega) = |\lambda|\chi(\Omega),\ \gamma(\lambda(\Omega)) = |\lambda|\gamma(\Omega) \text{ for any } \gamma \subset X, \lambda \in \mathbb{R}. \tag{7.34}$$

$$\text{For any } \Omega_1 \subset X, \Omega_2 \subset X, \chi(\Omega_1 + \Omega_2) \leq \chi(\Omega_1) + \chi(\Omega_2), \gamma(\Omega_1 + \Omega_2) \leq \gamma(\Omega_1) + \gamma(\Omega_2), \tag{7.35}$$

Theorem 7.9 *Let $F : \bar{\Omega} \to CK(X)$ be a k-ϕ-contraction where $0 \leq k < 1$ and $\phi = \chi$ or γ. Suppose that either X is quasi-complete or $\bar{\Omega}$ is complete and suppose that $F(\bar{\Omega})$ is bounded. Then F is ultimately compact.*

Proof. The proof follows almost identically that of Theorem 7.8. □

Suppose X is a normed linear space with norm $||\cdot||$ and the metric $d : X \times X \to \mathbb{R}^+$ is defined by $d(x, y) = ||x - y||$. If we let the norm $||\cdot||$ be the only element of A, the lattice C is isomorphic to $\mathbb{R}^+$ and χ and γ reduce to the ball- and set-measures of non-compactness respectively. Let us denote these two measures of non-compactness by χ_d and γ_d respectively.

Theorem 7.10 *(a) Let F be a k-ϕ-contraction where $\phi = \chi, \gamma, \chi_d$ or γ_d. Then for $\lambda \in \mathbb{R}$, λF is a $|\lambda|k$-ϕ-contraction.*

(b) Suppose F and G are k_1- and k_2-ϕ-contractions respectively where ϕ is χ, γ, χ_d or γ_d.

Then $(F + G)$ *is a* $(k_1 + k_2)$-ϕ-*contraction.*

Proof. These results follow immediately from (7.34) and (7.35) and Proposition 7.2. □

We note that some results of this section have been taken for presentations from Tarafdar and Teo (1979).

7.2 Coincidence Degree for Non-Linear Single-Valued Perturbations of Linear Fredholm Mappings

Let L be a linear single-valued operator between X and Z, two vector spaces over the real field, where dom$L$, the domain of L, is a subspace of X. We shall denote the kernel or null-space of L, $L^{-1}(0)$, by ker$L$, the range space of L by ImL and the quotient space $Z|\text{Im}L$, the cokernel of L, by coker$L$.

Given a vector subspace Y of a vector space E, there always exists a projection, a linear and indempotent operator, P of E onto Y and E is the direct sum of Im$P = Y$ and ker$P$. If E is a topological vector space, and P is a continuous projection, then E is the topological direct sum of ImP and ker$P$.

Definition 7.7 If X, Z, L are as above, let P and Q be continuous projections on X and Z respectively such that ImP = ker$L$ and ker$Q$ = ImL. Such a pair of projections (P, Q) will be called *exact with respect to L*.

Definition 7.8 Let L_p be the restriction of L to ker$P \cap$ dom$L$. Then L_p is an isomorphism from ker$P \cap$ dom$L$ to ImL. Let K_p : Im$L \to$ ker$P \cap$ dom$L$ be the inverse of L_p. K_p is then called the *pseudo inverse of L associated with P*.

Let $\pi : Z \to$ coker$L$ be the canonical surjection, that is $\Pi z = z + \text{Im}L$ for each $z \in Z$. Clearly, the restriction of Π to ImQ is an algebraic isomorphism. If Z is a topological vector space and coker$L$ is given the quotient topology, then Π is continuous.

The following results are almost immediate:

$$PK_p = 0, \tag{7.36}$$

$$LK_p = L_pK_p = I, \tag{7.37}$$

$$K_pL = K_pL(I - P) = K_pL_p(I - P) = I - P, \tag{7.38}$$

$$Qz = 0 <=> z \in \text{Im}L <=> \Pi z = 0, \tag{7.39}$$

where the zeros denote the null elements of the respective spaces.

Although the following two results are easy consequences of the above (e.g., see Mawhin (1972)), we are stating them with their proofs.

Proposition 7.5 *Let (P, Q) and (P', Q') be pairs of projections exact with respect to L. Then*

$$K_{p'} = (I - P')K_P \tag{7.40}$$

$$PK_{p'} + P'K_P = 0 \tag{7.41}$$

where K_p, $K_{p'}$ denote the pseudo-inverses of L associated with P and P' respectively.

Proof. From (7.37) we have the following on ImL

$$LK_p = I = LK_{p'}.$$

Hence,

$$L(K_p - K_{p'}) = 0.$$

Thus, $K_p - K_{p'}$ maps ImL into ker$L$ and we therefore have

$$K_p - K_{p'} = P'(K_p - K_{p'}) = p(K_p - K_{p'}).$$

Since $pK_p = P'K_{p'} = 0$, we obtain (7.40) from the first equality and (7.41) from the second equality. □

Proposition 7.6 *Let P, P' be projections of X onto ker$L$ and let $P'' = aP + bP'$ for some real numbers a, b. Then, P'' is a projection onto ker$L$ if and only if $a + b = 1$. If this necessary and sufficient conditions holds, the pseudo inverse of L associated with P'' is given by*

$$K_{P''} = aK_p + bK_{p'},$$

where K_p and $K_{p'}$ are the pseudo-inverses of L associated with p, p' respectively.

Proof. P'' is clearly linear and it can be checked that

$$(P'')^2 = (a + b)P''.$$

Hence P'' is indempotent if and only if $a + b = 1$.

Since ker$L$ is a subspace of X and P'' is a linear combination of P and P', the range of P'' is contained in ker$L$.

For $x \in \ker L$, $P''x = aPx + bP'x = (a + b)x = x$ and hence P'' is a projection onto ker$L$.

Thus, $(a+b) = 1$ is a necessary and sufficient condition for P'' to be a projection onto ker$L$.

Now, suppose $a + b = 1$. Then

$$\begin{aligned} K_{p''} &= (I - P'')K_p & (7.42)\\ &= (I - ap - bp')K_p & (7.43)\\ &= (I - bp')K_p & (7.44)\\ &= [(a+b)I - bp']K_p & (7.45)\\ &= aK_p + b(I - P')K_p & (7.46)\\ &= aK_p + bK_{p'} & (7.47)\end{aligned}$$

which gives the required result. □

7.2.1 *An Equivalence Theorem*

Let X and Z be vector spaces and consider the two mappings

$$L : \text{dom}L \subset X \to Z$$

$$N : \text{dom}N \subset X \to Z$$

where L is linear and dom$L$ is a subspace of X.

Theorem 7.11 *Let (P, Q) be a pair of projections exact with respect to L and let K_p and Π have the usual meanings. Suppose there exista a linear one-to-one mapping Ψ : cokerL $\to$ kerL and let the mapping M_Ψ : domN $\to X$ be defined by*

$$M_\Psi(x) = Px + \Psi\Pi Nx + K_P(I - Q)Nx. \tag{7.48}$$

Then $x \in$ domL $\cap$ domN is a solution of

$$Lx = Nx \tag{7.49}$$

if and only if it is a solution of

$$x = M_\Psi(x). \tag{7.50}$$

Proof. It is easily checked that the following equivalence relations hold:

$$\begin{aligned} Lx = Nx &\Longleftrightarrow (I - Q)Nx = Lx \text{ and } QNx = 0 & (7.51)\\ &\Longleftrightarrow K_PLx = K_P(I - Q)Nx \text{ and } \Pi Nx = 0 & (7.52)\\ &\Longleftrightarrow (I - P)x = K_P(I - Q)Nx \text{ and } \Psi\Pi Nx = 0 & (7.53)\\ &\Longleftrightarrow x - Px = \Psi\Pi Nx + K_P(I - Q)Nx & (7.54)\\ &\Longleftrightarrow x = M_\Psi(x). & (7.55)\end{aligned}$$

Hence the theorem holds. □

7.2.2 *Definition of Coincidence Degree*

Assumptions I.

(a) X and Z are normed linear spaces over the real field.

(b) L is a linear operator defined on some subspace, dom$L$, of X such that ImL is closed in Z. Also, ker$L$ and coker$L$ are of finite dimension and

$$dim \ker L = dim \operatorname{coker} L < \infty \tag{7.56}$$

where dim denotes dimension.

(c) Ω is an open, bounded subset of X and $N : \bar{\Omega} \to Z$ is a continuous mapping such that $\Pi N(\bar{\Omega})$ is bounded in Z.

(d) For a pair of projections (P, Q) exact with respect to L, the mapping $K_p(I-Q)N$ is completely continuous or compact, i.e. $K_p(I-Q)N$ is continuous and $K_p(I-Q)N(\bar{\Omega})$ is relatively compact.

(e) $0 \notin (L - N)(\text{dom}L \cap \partial\Omega)$.

Remark 7.7 (1) Assumption (b) ensures the existence of a pair of projections (P, Q), exact with respect to L

(2) Assumption (b) also ensures the existence of a continuous isomorphism ψ : coker$L \to$ ker$L$.

(3) With the quotient topology on coker$L$, coker$L$ is a normed linear space and the canonical surjection Π is continuous in this topology.

Remark 7.8 It has been shown that assumption (d) is independent of the choice of the exact pair of projections (P, Q) i.e. if (d) holds for any one pair of projections (P, Q), it holds for any other pair of projections (P', Q') exact with respect to L.

Definition 7.9 A linear operator satisfying I(b) is called a linear Fredholm mapping of index zero.

Definition 7.10 A mapping N satisfying assumptions (c) and (d) is said to be L-compact in Ω.

Assumptions II.

(a) X and Z are Banach spaces.

(b) L is a linear Fredholm mapping of index zero defined on some subspace dom$L$ of X such that ImL is a subspace of Z.

(c) Ω is an open bounded subset of X and $N : \bar{\Omega} \to Z$ is a continuous mapping such that $\Pi N(\bar{\Omega})$ is bounded.

(d) For a pair of projections (P, Q) exact with respect to L the mapping $K_P(I-Q)N$ is either (i) a k-set-contraction or (ii) a k-ball contraction for some $0 \le k < 1$.

(e) $0 \notin (L - N)(\text{dom}L \cap \partial\Omega)$.

Remark 7.9 Assumption II(D) is also independent of the choice of the exact pair of projections (P, Q).

Definition 7.11 A mapping N satisfying (c) and (d)(i) (respectively (d)(ii)) is said to be L-k-set contractive (respectively L-k-ball contractive) in $\bar{\Omega}$ for some k in $0 \leq k < 1$.

Let Ψ : coker$L \to$ ker$L$ be an orientation preserving continuous isomorphism. The mapping $M_\Psi : \bar{\Omega} \to X$ defined by

$$M_\Psi(x) = Px + \Psi\Pi Nx + K_P(I - Q)Nx$$

is compact, k-set-contractive or k-ball-contractive accordingly as N is L-compact, L-k-set-contractive or L-k-ball-contractive in $\bar{\Omega}$ respectively. Also, assumption (e) and Theorem 7.11 imply that M_Ψ has no fixed points on $\partial\Omega$. Hence, for the case of an L-compact N, the Leray-Schauder degree of $I - M_\Psi$ over Ω with respect to zero is well defined. For the other two cases, the degree of $I - M_\Psi$ has also been defined by Nussbaum [1971] and Vainikko and Sadovskii [1968].

Definition 7.12 (a) Let Assumptions I be satisfied. Then the Coincidence Degree of L and N over Ω, denoted $d[(L, N), \Omega]$, is defined to be the Leray-Schauder degree of $I - M_\Psi$ over Ω with respect to zero i.e.

$$d[(L, N), \Omega] = d(I - M_\Psi, \Omega, 0).$$

(b) Let Assumptions II be satisfied. Then the Coincidence Degree of L and N over Ω, denoted $d[(L, N), \Omega]$, is defined to be the degree of $I - M_\Psi$ over Ω with respect to zero as given by Nussbaum for the L-k-set-contractive N and as given by Vainikko and Sadovskii for the L-k-ball- contractive N.

Remark 7.10 (a) The Coincidence Degree given in Definition 7.12 is dependent only on L, N, Ω and the homotopy class of Ψ. Thus, as Ψ is chosen to be orientation preserving, the degree is dependent only on L, N, and Ω and hence is well defined.

(b) If, in addition to Assumptions I, we assume that X and Z are Banach spaces, the first case becomes a special case of the second as compact mappings are 0-set-contractions and 0-ball-contractions.

7.2.3 *Properties of the Coincidence Degree*

Unless specifically stated, Assumptions I or II will be assumed satisfied for the pair (L, N) so that $d[(L, N), \Omega]$ is defined. It has been shown that the following properties hold:

Theorem 7.12 (Existence Theorem) *If* $d[(L, N), \Omega] \neq 0$, *then there exists* $x \in$ *dom*$L \cap \Omega$ *such that* $Lx = Nx$.

Theorem 7.13 (Additivity Property) *If* Ω_1, Ω_2 *are open, disjoint subsets of* Ω *such that* $\Omega_1 \cup \Omega_2 \subset \Omega$, $\bar{\Omega}_1 \cup \bar{\Omega}_2 \subset \bar{\Omega}$ *and* $0 \notin (L - N)(\partial\Omega_1 \cup \partial\Omega_2)$, *then*

$$d[(L, N), \Omega] = d[(L, N), \Omega_1] + d[(L, N), \Omega_2].$$

Theorem 7.14 (Excision Property) *If* $\Omega_0 \subset \Omega$ *is an open set such that*

$$(L - N)^{-1}(0) \subset \Omega_0,$$

then

$$d[(L, N), \Omega] = d[(L, N), \Omega_0].$$

Theorem 7.15 (Generalized Borsuk's Theorem) *If* Ω *is symmetric about the origin and contains it, and, for every* $x \in \Omega$ *we have* $N(-x) = -Nx$*, then* $d[(L, N), \Omega$ *is odd.*

Theorem 7.16 (Homotopy Invariance) *Let Assumptions* $I(a)$ *and* (b) *be satisfied and let* Ω *be an open bounded subset of* X*. Suppose* $\tilde{N} : \bar{\Omega} \times [0, 1] \to Z$ *is L-compact in* $\bar{\Omega} \times [0, 1]$ *such that, for each* $\lambda \in [0, 1]$,

$$0 \notin [L - \tilde{N}(\cdot, \lambda)](dom\, l \cap \partial\Omega.$$

Then $d[(L, N(\cdot, \lambda)), \Omega]$ *is independent of* λ *in* $[0, 1]$.

Remark 7.11 The result holds similarly if Assumptions II(a) and (b) are satisfied and $\tilde{N}$ is a L-k-set-contraction or L-k-ball-contraction in $\Omega \times [0, 1]$.

Corollary 7.16.1 *Suppose that Assumptions I or II are satisfied. Then* $d[(L, N), \Omega]$ *depends only on* L, Ω *and the restriction of* N *to* $\partial\Omega$

Lemma 7.3 *Let* X *and* Z *be normed linear spaces with norm* $|| \cdot ||$ *and let the metric* d *be induced by the norm, i.e.* $d(x, y) = ||x - y||$. *Let Assumptions* $I(b)$, (c), (d) *and* (e) *be satisfied and we additionally assume that* $K_P(I - Q)$ *is continuous. Then there exists* $\mu > 0$ *such that*

$$\inf\{||Lx - Nx|| : x \in \partial\Omega \cap dom\, L\} \geq \mu.$$

Theorem 7.17 *Let* X, Z, L, *and* N *be as in Lemma* 7.3 *Then, for each L-compact* $N' : \bar{\Omega} \to Z$ *such that*

$$\sup\{||Nx - N'x|| : x \in \partial\Omega\} < \mu$$

where μ *is the positive number in Lemma* 7.3, *we have*

$$d[(L, N), \Omega] = d[(L, N'), \Omega].$$

Theorem 7.18 (Generalized Continuation Theorem) *Let* X, Z *and* L *satisfy Assumptions* $I(a)$ *and* (b), *and let* Ω *be an open bounded subset of* X. *Let* $N* : \bar{\Omega} \times [0, 1] \to Z$ *be L-compact in* $\bar{\Omega} \times [0, 1]$ *and let* $N = N * (\cdot, 1)$. *Suppose* $y \in ImL$ *and the following conditions hold:*

(1) $Lx \notin \lambda N * (x, \lambda) + y$ *for every* $x \in \partial\Omega \cap domL$ *and every* $\lambda \in [0, 1]$.
(2) $\Pi N * (x, 0) \neq 0$ *for evry* $x \in L^{-1}\{Y\} \cap \partial\Omega$.
(3) $d[\Pi N * (\cdot, 0)|_{L^{-1}\{Y\}}, L^{-1}\{Y\} \cap \Omega, 0] \neq 0$.

Then for each $\lambda \in [0,1]$, the equation

$$Lx = \lambda N * (x, \lambda) + y$$

has at least one solution in Ω, and the equation

$$Lx = Nx + y$$

has at least one solution in $\bar{\Omega}$.

Theorem 7.19 *Let X, Z and L satisfy Assumptions $I(a)$ and (b) and let Ω be an open bounded subset of X. Let $\tilde{N} : \bar{\Omega} \times [0,1] \to Z$ be L-compact in $\bar{\Omega} \times [0,1]$. If, for each $x \in \partial\Omega \cap dom\, L$, we have*

$$Lx \neq \tilde{N}(x, \lambda)$$

and if $d[(L, N(\cdot, 0)), \Omega] \neq 0$, then for each $\lambda \in [0,1]$, the equation $Lx \in \tilde{N}(\cdot, \lambda)$ has at least one solution in Ω.

Corollary 7.19.1 *Let Assumptions $I(a)$ and (b) be satisfied and let Ω be an open bounded set, symmetric with respect to the origin and containing it. Suppose that $\tilde{N} : \bar{\Omega} \times [0,1] \to Z$ is L-compact in $\bar{\Omega} \times [0,1]$ and that, for each $x \in \bar{\Omega}$,*

$$\tilde{N}(-x, 0) = -\tilde{N}(x, 0)$$

with $Lx \neq \tilde{N}(x, \lambda)$ for each $x \in domL \cap \partial\Omega$ and each $\lambda \in [0,1]$. Then each equation

$$Lx = \lambda N(x, \lambda) \quad \text{with} \quad \lambda \in [0,1]$$

has at least one solution in Ω.

Corollary 7.19.2 (Generalized Krasnosel'skii Theorem) *Let Assumptions $I(a)$ and (b) be satisfied, and let Ω be as in Corollary 7.19.1. Let N be a L-compact mapping in $\bar{\Omega}$ and assume that, for every $\mu \in [0,1]$ and $x \in domL \cap \partial\Omega$, we have*

$$(L - N)x \neq \mu(L - N)(-x).$$

Then the equation

$$Lx = Nx$$

has at least one solution in Ω.

Remark 7.12 Lemma 7.3, Theorems 7.17, 7.18, 7.19, Corollaries 7.19.1 and 7.19.2 can all be extended to the case when X and Z are Banach spaces and the "N" mappings are correspondingly L-k-set-contractive or L-k-ball-contractive.

Note that some results of this section were taken from Tarafdar and Teo (1979).

7.3 On the Existence of Solutions of the Equation $Lx \in Nx$ and a Coincidence Degree Theory

Let X and Z be normed linear spaces over the reals. Extensive researches have been undertaken on the study of the operator equation

$$Lx \in Nx, \tag{7.57}$$

where $L : dom\, L \subset X \to Z$ is a linear mapping and $N : \text{dom}N \subset X \to Z$ is a (possibly nonlinear) mapping. The equation (7.57) represents a wide class of problems including nonlinear ordinary, partial and functional differential equations. When L^{-1} exists, the reduced equation $x = L^{-1}Nx$ is under the scope of fixed point theory. For extensive literature for this case we refer to the survey works of Dolph and Minty (1964) and Ehrmann (1965).

When L^{-1} does not exist and X and Z are Banach spaces, the basic works on the study of the equation (7.57) are due to Cacciopoli (1946), Shimizu (1948), Cronin (1950), Bartle (1953), Vainberg and Trenogin (1962), Vainberg and Aizengendler (1968) and Nirenberg (1960). These works involve some small assumptions on N. The method for finding solutions of the equation (7.57), initiated by Cesari (1963) and Cesari (1964) and further developed by Locker (1967), Bancroft, Hale, and Sweet (1968) and Williams (1968) deals with a more general class of mappings. For application of Cesari's method to diferential equation we refer to Cesari (1969), Cesari (1971) and Hale (1969), Hale (1971).

Using an equivalence theorem which reduces the problem of existence of solutions of the equation (7.57) to that of fixed points of an auxiliary mapping and Leray-Schauder degree, Mawhin (1972) developed a degree called the coincidence degree for the pair (L, N) and applied to nonlinear differential equations (for example, see Gaines and Mawhin (1977)). In essence, Mawhin's method preserves the spirit of the works of the authors mentioned above.

In the recent past the Leray-Schauder degree theory for a single-valued compact vector field has been extended to a larger class of single-valued mappings, namely to k-set contractive vector fields by Nussbaum (1969), and Nussbaum (1971), ball condensing vector fields by Vainikko and Sadovskii (1968) and Borisovich and Sapronov (1968), ultimately compact vector fields by Sadovskiǐ (1968) (see also Sadovskiǐ (1972) and Daneš (1974)). On the other hand, Leray-Schauder degree theory has been extended to set-valued compact vector fields by Granas (1959), Cellina and Lasota (1969), Ma (1972) and to ultimately compact vector fields by Petryshyn and Fitzpatrick (1974).

The coincidence degree of Mawhin (1972) has been sharpened by Hetzer (1975a) and Hetzer (1975b) by replacing the complete continuity assumption by k-set contraction with $k < 1$ and Leray-Schauder degree by the corresponding degree of k-set contractive vector field mentioned above.

The purpose of this section is to consider the equation

$$Lx \in N(x), \tag{7.58}$$

where $L : dom\ L \subset X \to Z$ is a single-valued linear Fredholm mapping of index zero and $N : dom\ N \subset X \to CK(Z)$ is a mapping, X and Z being normed linear spaces.

Like Mawhin (1972) we have proved equivalence theorems which reduce the problem of existence of solutions of the equation (7.58) to that of fixed points of an auxiliary set-valued vector fields given by Petryshyn and Fitzpatrick (1974),and we have built up the coincidence degree theory for the pair (L, N) appearing in the equation (7.58). We have proved that this degree has all the usual properties of a degree theory. We have also extended the Rouche's theorem and the Leray-Schauder continuation principle to our context. In new section we will present the coincidence degree of Akashi (1988) by using our equivalence theorem when L is a Fredholm mapping of non-negative index.

7.3.1 *Coincidence Degree for Set-Valued $k - \phi$-Contractive Perturbations of Linear Fredholm Mappings*

In this subsection, we shall present Tarafdar and Teo's extension of the notion of coincidence degree developed by Mawhin (1972) to the case where the second mapping is set-valued. Tarafdar and Teo observed that such a degree theory will provide a method for proving the existence of solutions to the equation

$$Lx \in Nx.$$

An Equivalence Theorem of Tarafdar and Teo

Theorem 7.20 *Let X and Z be two vector spaces over the same scalar field. Let $L : domL \subset X \to Z$ be a linear mapping and $N : A \subset X \to 2^Z$ be a set-valued mapping. Further, assume that there is a linear injective (one-to-one) mapping*

$$\psi : cokerL \to kerL.$$

Then $x_0 \in domL \cap A$ is a solution of the equation

$$Lx \in Nx \tag{7.59}$$

if and only if x_0 is a fixed point of the set-valued mapping $M_\psi : A \to 2^X$ defined by

$$M_\psi x = Px + [\psi\pi + K_p(I - Q)]Nx \tag{7.60}$$

for every pair (P, Q) of exact projections with respect to L, where π and K_p have their meaning as explained in Section 7.2. In other words,

$$(L - N)^{-1}(0) = (I - M_\psi)^{-1}(0). \tag{7.61}$$

Proof. Since the images under P and ψ are contained in ker$L$ and that under K_p is in $X_{I-P} \cap \operatorname{dom} L$, it is clear that $M_\psi(A) \subset \operatorname{dom} L$. First, let us suppose that $x_0 \in A \cap \operatorname{dom} L$ with $Lx_0 \in Nx_0$. Then

$$[\psi\pi + K_p(I-Q)]Lx_0 \in [\psi\pi + K_p(I-Q)]Nx_0.$$

Hence using (7.38) and (7.39) we have

$$(I-P)x_0 \in [\psi\pi + K_p(I-Q)]Nx_0.$$

Therefore

$$x_0 \in M_\psi x_0.$$

Next, let us suppose that $x_0 \in A \cap \operatorname{dom} L$ with $x_0 \in M_\psi x_0$, that is

$$x_0 \in Px_0 + [\psi\pi + K_p(I-Q)]Nx_0. \tag{7.62}$$

Since the operator $\psi\pi + K_p(I-Q)$ is injective (see Lemma 7.6) we have

$$[\psi\pi + K_p(I-Q)]^{-1}[\psi\pi + K_p(I-Q)]Nx_0 = Nx_0. \tag{7.63}$$

Hence it follows from (7.62) and (7.63) that

$$[\psi\pi + K_p(I-Q)]^{-1}(I-P)x_0 \in Nx_0. \tag{7.64}$$

Thus

$$[\psi\pi + K_p(I-Q)]^{-1} = [(\pi/\text{Im}Q)^{-1}\psi^{-1}P + L] \tag{7.65}$$

yields $Lx_0 \in Nx_0$, where $\pi/\text{Im}Q$ denotes the restriction of π to ImQ. We now establish (7.65).

For each $z \in Z$ we have by using (7.37)

$$[(\pi/\text{Im}Q)^{-1}\psi^{-1}P + L][\psi\pi + K_p(I-Q)]z \tag{7.66}$$
$$= (\pi/\text{Im}Q)^{-1}\pi z + (I-Q)z = Qz + (I-Q)z = z. \tag{7.67}$$

Also if $x \in \text{dom}L$, then using (7.38) and (7.39) we have

$$[\psi\pi + K_p(I-Q)][(\pi/\text{Im}Q)^{-1}\psi^{-1}P + L]x = Px + (I-P)x = x. \qquad \square$$

Basic Assumptions of Tarafdar and Teo

Before we present the coincidence degree for (L, N), we shall state some assumptions of Tarafdar and Teo which they made on the mappings.

Assumptions:

(a) X is a real Banach space and Z is a real normed linear space.

(b) $L : \operatorname{dom} L \subset\to Z$ is a linear Fredholm mapping of index zero defined on a subspace $\operatorname{dom} L$ of X, that is L is linear, $\operatorname{Im} L$ is closed and

$$\dim\ \ker L = \dim\ \operatorname{coker} L < \infty,$$

where 'dim' denotes dimension.

(c) Ω is a bounded, open set in X and the set-valued mapping $N : \bar{\Omega} \to CK(Z)$ takes each x in the closure of Ω to a non-empty compact convex subset of Z.

(d) N is upper semi-continuous with $\pi N(\bar{\Omega})$ bounded in $\operatorname{coker} L$.

(e) Let (P, Q) be an exact pair of projections with respect to L and let K_p be the pseudo-inverse of L associated with P. Let ϕ be a measure of non-compactness defined on 2^X such that (i) ϕ satisfies the subadditivity condition of Proposition 7.2 and takes values in $\mathbb{R}^+ = \{t \in \mathbb{R} : t \geq 0\} \cup \{\infty\}$ or (ii) we additionally assume that Z is a Banach space and ϕ is one of χ, γ, χ_d and γ_d. We assume that with such a measure of non-compactness ϕ, $K_p(I-Q)N$ is a k-ϕ-contraction with $0 < k < 1$ and that $\phi(K_p(I-Q)N(\bar{\Omega})) < \infty$. In this case we also assume that K_p is continuous.

(f) $0 \notin (L-N)(\operatorname{dom} L \cap \partial\Omega)$ where $\partial\Omega$ denotes the boundary of Ω.

Remark 7.13 From Assumption (b), the exact pair of projections (P, Q) may be assumed continuous and will hereafter be assumed continuous. Moreover, with the quotient norm topology $\operatorname{coker} L$ is a normed space and the canonical surjection π is continuous with respect to this topology. Also, (b) is sufficient condition for the existence of a linear isomorphism $\psi : \operatorname{coker} L \to \ker L$.

Proposition 7.7 *Let Assumptions* (a) *to* (d) *hold and let* (P, Q) *and* (P', Q') *be exact pairs of continuous projections with respect to* L. *Suppose that* (P, Q) *satisfies Assumption* (e). *Then the pair* (P', Q') *also satisfies the Assumption* (e).

Proof. Writing $\pi_Q = \pi/\operatorname{Im} Q$ and $\pi_{Q'} = \pi/\operatorname{Im} Q'$ and using (7.40) we have

$$K_{p'}(I-Q')N = (I-P')K_p(I-Q')N \tag{7.68}$$

$$= (I-P')K_p(I-Q)N + (I-P')K_p(Q-Q')N \tag{7.69}$$

$$= (I-P)K_p(I-Q)N + (I-P)\tilde{K}_p({\pi_Q}^{-1} - {\pi_{Q'}}^{-1})\pi N, \tag{7.70}$$

where $\tilde{K}_p$ denotes the restriction of K_p to the finite dimensional subspace $(Q-Q')Z$. Thus $\tilde{K}_p$ is continuous. Since $\pi N(\bar{\Omega})$ is bounded in a finite dimensional subspace of X, it follows that $(I-P')\tilde{K}_p({\pi_Q}^{-1} - {\pi_{Q'}}^{-1})\pi N$ is a 0-ϕ-contraction. Hence from Proposition 7.2 and 7.3 it follows that $K_{p'}(I-Q')N$ is a k-ϕ-contraction. That $K_{p'}$ is continuous follows from (7.40) as K_p and $(I-Q')$ are continuous. Finally applying ϕ to both sides of (7.68) and using subadditivity of ϕ we can easily show that $\phi(K_{p'}(I-Q')N(\bar{\Omega})) < \infty$. □

Definition 7.13 A mapping $N : \bar{\Omega} \to CK(Z)$ satisfying (c), (d) and (e) is said to be a L-k-ϕ-contraction. (We see that this is a proper definition as Assumption (e) is independent of the choice of (P, Q).)

Tarafdar and Teo's Definition of Coincidence Degree

Proposition 7.8 *Suppose Assumptions (a) to (e) are satisfied and M_ψ is the mapping defined in Theorem 7.20 for some continuous isomorphism*

$$\psi : coker\, L \to ker\, L.$$

Then for each x in $\bar{\Omega}$, $M_\psi x$ is a compact convex subset of X and M_ψ is a k-ϕ-contraction.

Proof. Since P, Q, K_p, ψ and π are all linear and $N(x)$ is convex for each $x \in \bar{\Omega}$, it follows that $M_\psi x$ is convex for each $x \in \bar{\Omega}$. Again since P, Q, K_p and ψ are continuous and Nx is compact, $M_\psi x = Px + [\psi\pi + K_p(I - Q)]Nx$ is compact for each $x \in \bar{\Omega}$.

Now P is linear, continuous and has a finite dimensional range. Hence P is compact and is, therefore, a 0-ϕ-contraction. Also $\psi\pi N(\bar{\Omega})$ being bounded subset of a finite-dimensional subspace is relatively compact.

We now prove that $[\psi\pi + K_p(I - Q)]N$ is a k-ϕ-contraction. Let $A \subset \bar{\Omega}$. Noting that

$$[\psi\pi + K_p(I - Q)]N(A) \subset \psi\pi N(A) + K_p(I - Q)N(A)$$

we have

$$\phi([\psi\pi + K_p(I - Q)]N(A)) \tag{7.71}$$
$$\leq \phi(\psi\pi N(A) + K_p(I - Q)N(A)) \tag{7.72}$$
$$\leq \phi(\psi\pi N(A)) + \phi(K_p(I - Q)N(A)) \text{ by subadditivity of } \phi \tag{7.73}$$
$$\leq k\phi(A) \tag{7.74}$$

as $\phi(\psi\pi N(A)) = 0$, $\psi\pi N(A)$ being relatively compact. Now from Proposition 7.2 it follows that M_ψ is a k-ϕ-contraction from $\bar{\Omega}$ to $CK(X)$. □

Remark 7.14 Tarafdar and Teo noted that assumption in (e) that K_p is continuous has been used to prove that $M_\psi x$ is a compact subset for each $x \in \bar{\Omega}$. This assumption is not unrealistic. For, if in addition to the Assumption (b), $L : \operatorname{dom} L \subset X \to Z$ is a closed operator and Z is a Banach space, then K_p is continuous. To see this let $y_n \to y$, $y_n \in \operatorname{Im} L$ and $K_p y_n = x_n \to x$. Since

$$Lx_n = LK_p y_n = y_n \text{ and } x_n \in \operatorname{dom} L \cap X_{I-P},$$

we have by closedness of L that $Lx = y$ and $x \in \operatorname{dom} L$. Clearly $x \in X_{I-P}$ as X_{I-P} is closed. Hence $K_p y = K_p Lx = (I - P)x = x$ and obviously $y \in \operatorname{Im} L$ as $\operatorname{Im} L$ is closed. Thus K_p is closed. Again since $\operatorname{Im} L$ is closed, the closed graph theorem yields that K_p is continuous.

Remark 7.15 From Proposition 7.8, Tarafdar and Teo observed that if the Assumptions (a) to (f) are satisfied, M is an ultimately compact mapping (see

Theorems 7.8 and 7.9). It follows from Assumption (f) and Theorem 7.20 that $0 \notin (I - M_\psi)(\text{dom}L \cap \partial\Omega)$. Thus the degree of the ultimately compact field $(I - M_\psi)$ with respect to zero is well defined.

Definition 7.14 Let $\mathcal{L}_L$ denote the set of all continuous isomorphisms from $\text{coker}\, L$ to $\ker L$. ψ, ψ' are said to be *homotopic* in $\mathcal{L}_L$ if there exists a continuous mapping $\bar{\psi} : \text{coker}L \times [0,1] \to \ker L$ such that $\bar{\psi}(\cdot, 0) = \psi$, $\bar{\psi}(\cdot, 1) = \psi'$ and, for each $\lambda \in [0,1]$, $\bar{\psi}(\cdot, \lambda) \in \mathcal{L}_L$.

Remark 7.16 To be homotopic is an equivalence relation which partitions $\mathcal{L}_L$ into equivalence classes called homotopy classes.

The following two propositions and corollary are quoted from Gaines and Mawhin (1977):

Proposition 7.9 *ψ and ψ' are homotopic in $\mathcal{L}_L$ if and only if $\det(\psi'\psi^{-1}) > 0$.*

Corollary 7.20.1 *$\mathcal{L}_L$ is partitioned into two homotopic classes.*

Definition 7.15 $\psi : \text{coker}\, L \to \ker L$ is said to be orientation preserving if $\{\psi a_1, \psi a_2, \ldots, \psi a_n\}$ belongs to the orientation chosen in $\ker L$ where $\{a_1, a_2, \ldots, a_n\}$ is a basis for $\text{coker}\, L$ belonging to a certain chosen orientation. Otherwise, ψ is said to be orientation reversing.

Proposition 7.10 *If $\text{coker}\, L$ and $\ker L$ are oriented then ψ and ψ' are homotopic in $\mathcal{L}_L$ if and only if they are simultaneously orientation preserving or orientation reversing.*

Lemma 7.4 *Let X and Z be normed linear spaces and let Ω be a bounded open subset of X. Let $\phi : 2^X \to C$ be a measure of non-compactness as given in Assumption (e). Let $F : \bar{\Omega} \times [0,1] \to CK(X)$ be an upper semi-continuous mapping such that $\phi(F(\bar{\Omega} \times [0,1])) < \infty$ and, for some $k \in (0,1)$, we have*

$$\phi(F(A \times [0,1])) \leq k\phi(A) \text{ for every } A \subset \bar{\bar{\Omega}}.$$

Then $F((K' \cap \bar{\Omega}) \times [0,1])$ is relatively compact where

$$K' = K(F, \bar{\Omega}) \times [0,1]).$$

Proof. As $K' = K(F, \bar{\Omega} \times [0,1]) = \bar{CO}F((\bar{\Omega} \cap K') \times [0,1])$,

$$\phi(K' \cap \Omega) \leq \phi(K') = \phi(F(\bar{\Omega} \cap K') \times [0,1]) \tag{7.75}$$

$$k\phi(\bar{\Omega} \cap K'). \tag{7.76}$$

As $0 < k < 1$, and $\phi(F(\bar{\Omega} \times [0,1])) < \infty$, we have

$$\phi(K' \cap \bar{\Omega}) = \phi(F((\bar{\Omega} \cap K') \times [0,1])) = 0.$$

Hence, $(K' \cap \bar{\Omega})$ and $F((\bar{\Omega} \cap K') \times [0,1])$ are precompact and by the assumption that $\bar{\Omega}$ is complete, we conclude that $F((\bar{\Omega} \cap K') \times [0,1])$ is relatively compact. □

Theorem 7.21 *Let Assumptions (a) to (f) be satisfied. Then $d(I - M_\psi, \Omega, 0)$ as defined in Definition 7.3 depends only on L, N and the homotopy class of ψ in $\mathcal{L}_L$.*

Proof. Let (P, Q) and (P', Q') be exact pairs of projections with respect to L. Let $\psi, \psi' \in \mathcal{L}_L$ be in the same homotopy class and let $\bar{\psi} : \text{coker}L \times [0,1] \to \text{ker}L$ be the mapping in Definition 7.14. Let

$$M = P + [\psi\pi + K_p(I - Q)]N,$$

$$M' = P' + [\psi'\pi + K_{p'}(I - Q')]N.$$

For each $\lambda \in [0, 1]$, define

$$P_\lambda = (1 - \lambda)P + \lambda P',$$

$$Q_\lambda = (1 - \lambda)Q + \lambda Q'.$$

By Proposition 7.6, (P_λ, Q_λ) is an exact pair of projections with respect to L. Moreover, $P_0 = P$, $P_1 = P'$, $Q_0 = Q$, $Q_1 = Q'$, and $KP_\lambda = (1 - \lambda)K_p + \lambda K_{p'}$. Define $M^* : \bar{\Omega} \times [0,1] \to CK(X)$ by

$$M^*(x, \lambda) = P_\lambda x + [\bar{\psi}(\pi(\cdot), \lambda) + KP_\lambda(I - Q_\lambda)]Nx.$$

By Theorem 7.20 and Assumption (f),

$$x \notin M^*(x, \lambda) \text{ for every } x \in \partial\Omega, \lambda \in [0, 1].$$

Also, $M^*(\cdot, 1) = M'$ and $M^*(\cdot, 0) = M$.

We claim that $M^*((\bar{\Omega} \cap K') \times [0,1])$ is relatively compact, where

$$K' = K(M^*, \bar{\Omega} \times [0,1]).$$

Now, writing explicitly,

$$\begin{aligned} M^*(x, \lambda) &= (1 - \lambda)Px + \lambda P'x + [\bar{\psi}(\pi(\cdot), \lambda) && (7.77)\\ &\quad + \{(1 - \lambda)K_p + \lambda K_{p'}\}\{I - (1 - \lambda)Q - \lambda Q'\}]Nx && (7.78)\\ &= (1 - \lambda)Px + \lambda P'x + [\bar{\psi}(\pi(\cdot), \lambda) && (7.79)\\ &\quad + \{(1 - \lambda)K_p + \lambda(I - P')K_p\}\{I - Q + \lambda(Q - Q')\}]Nx && (7.80)\\ &= (1 - \lambda)Px + \lambda P'x + [\bar{\psi}(\pi(\cdot), \lambda) + (I - \lambda P')K_p(I - Q) && (7.81)\\ &\quad + \lambda(I - \lambda P')K_p(Q - Q')]Nx. && (7.82) \end{aligned}$$

Using the same argument as in Proposition 7.7 we can show that for each $\lambda \in [0, 1], \lambda(I - \lambda P')K_p(Q - Q')]N$ is 0-ϕ-contraction. Now by using the Assumption (e) and similar argument as in Proposition 7.8 we can show that for each $\lambda \in [0, 1]$,

$$[\bar{\psi}(\pi(\cdot), \lambda) + (I - \lambda P')K_p(I - Q) + \lambda(I - \lambda P')K_p(Q - Q')]N$$

is a k-ϕ-contraction (note that P and P' being compact maps are both 0-ϕ-contraction). Thus, it follows from Propositions 7.2 and 7.3 that for each $\lambda \in [0,1]$, $M^*(\cdot,\lambda)$ is a k-ϕ-contraction.

Now,

$$\phi(M^*(A\times[0,1])) = \phi(\bigcup_{\lambda\in[0,1]} M^*(A,\lambda)) \tag{7.83}$$

$$= \max_{\lambda\in[0,1]} \phi(M^*(A,\lambda)). \tag{7.84}$$

Since for each $\lambda \in [0,1]$, $M^*(\cdot,\lambda)$ is a k-ϕ-contraction

$$\phi(M^*(A\times[0,1])) \leq k\phi(A).$$

From the preceding lemma, $M^*((\omega \cap K')\times[0,1])$ is relatively compact. By the Homotopy Invariance Theorem given in Petryshyn and Fitzpatrick (1974),

$$d(I - M^*(\cdot,0),\Omega,0) = d(I - M^*(\cdot,1),\Omega,0)$$

or

$$d(I - M,\Omega,0) = d(I - M',\Omega,0).$$

Thus the degree of $I - M_\psi$ on Ω with respect to zero is independent of the choice of P, Q and ψ within the same homotopy class. □

Definition 7.16 Suppose that Assumptions (a) to (f) are satisfied and ψ is an orientation preserving continuous isomorphism from coker$L$ to ker$L$. Then, the coincidence degree of L and N in Ω, denoted by $d[(L,N),\Omega]$, is defined by

$$d[(L,N),\Omega] = d(I - M_\psi,\Omega,0), \tag{7.85}$$

where $M_\psi : \Omega \to CK(X)$ is defined by

$$M_\psi = P + [\psi\pi + K_p(I-Q)]N$$

and the right-hand term is the degree for the set-valued ultimately compact field $I - M_\psi$ as defined in Definition 7.3.

Remark 7.17 (a) If $X = Z$, $L = I$, then ker$L = \{0\}$ and thus coker$L = \{0\}$. This implies that Im$L = X$ and hence $P = 0$, $Q = 0$, and $K_p(I-Q) = I$ and the only isomorphism between coker$L$ and ker$L$ is the trivial one $\psi(0) = 0$. The Assumption (b) is trivially satisfied and (e) reduces to assuming that N is a k-ϕ-contraction for some k in $(0,1)$ with $\phi(N(\Omega)) < \infty$. Assumption (f) means that N has no fixed points on the boundary of Ω. As $M_\psi = N$, we have

$$d[(I,N),\Omega] = d(I - N,\Omega,0).$$

Tarafdar and Teo observed that, Assumption (e) may be replaced by the assumption that N is ultimately compact.

Tarafdar and Teo's Basic Properties of the Coincidence Degree

In this subsection, unless otherwise specified, Tarafdar and Teo assumed that Assumptions (a) to (f) were satisfied such that the Coincidence Degree was well defined.

Theorem 7.22

(a) *Existence Theorem.*
If $d[(L,N),\Omega] \neq 0$*, then* $0 \in (L-N)(domL \cap \Omega)$.

(b) *Additivity Property.*
Let Ω_1, Ω_2 *be disjoint open sets such that* $\Omega_1 \cup \Omega_2 \subset \Omega$, $\bar{\Omega}_1 \cup \bar{\Omega}_2 = \bar{\Omega}$, *and* $0 \notin (L-N)(\partial\Omega_1 \cup \partial\Omega_2)$. *Then,*

$$d[d[(L,N),\Omega] = d[(L,N),\Omega_1] + d[(L,N),\Omega_2].$$

(c) *Excision Property.*
If $\Omega_1 \subset \Omega$ *is an open set such that* $(L-N)^{-1}(0) \subset \Omega_1$ *then*

$$d[d[(L,N),\Omega] = d[(L,N),\Omega_1].$$

Proof. (a) and (b) follow from the Definition of Coincidence Degree and the corresponding properties of degree of an ultimately compact vector field given by Petryshyn and Fitzpatrick (1974). By taking $\Omega_2 = \Omega \setminus \Omega_1$, that is

$$\Omega_2 = \{x \in \Omega : x \notin \bar{\Omega}\}.$$

The result (c) follows from (a) and (b). □

Theorem 7.23 *If* Ω *is a symmetric bounded neighbourhood of the origin and* $N(-x) = -Nx$ *for all* $x \in \bar{\Omega}$, *then* $d[(L,N),\Omega]$ *is odd.*

Proof. Note that, as P, Q, K_p, ψ and π are all linear, the condition on N implies that $M_\psi(-x) = -M_\psi(x)$ for all $x \in \Omega$. Thus, by the corresponding property of degree of an ultimately compact vector field (Petryshyn and Fitzpatrick (1974)) and the definition of Coincidence Degree, $d[(L,N),\Omega]$ is odd. □

Theorem 7.24 (Homotopy Invariance) *Let Assumptions (a) and (b) be satisfied and let* Ω *be a bounded, open subset of* X. *Let* ϕ, P, Q *and* K_p *be as given in Assumption (e) and suppose* $\tilde{N} : \bar{\Omega} \times [0,1] \to CK(Z)$ *satisfy the following:*

(i) $\tilde{N}$ *is upper semi-continuous on* $\bar{\Omega} \times [0,1]$,
(ii) $\pi N(\bar{\Omega} \times [0,1])$ *is bounded,*
(iii) $\phi(K_p(I-Q)\tilde{N}(\bar{\Omega} \times [0,1])) < \infty$,
(iv) *there exists* $k \in (0,1)$ *such that, for every* $A \subset \Omega$,

$$\phi(K_p(I-Q)\tilde{N}(A \times [0,1])) < k\phi(A),$$

(*v*) *for each* $\lambda \in [0,1]$,

$$0 \notin (L - N(\cdot,\lambda))(domL \cap \partial\Omega). \tag{7.86}$$

Then, $d[(L, N(\cdot,\lambda)), \Omega]$ *is independent of* λ *in* $[0,1]$.

Proof. Let $\psi : \operatorname{coker} L \to \ker L$ be an orientation preserving continuous isomorphism. Define $M_\psi : \bar{\Omega} \times [0,1] \to CK(X)$ by

$$M_\psi(x,\lambda) = Px + [\psi\pi + K_p(I - Q)]\tilde{N}(x,\lambda).$$

Then, by Lemma 7.4 and (v), M_ψ satisfies the condition of Theorem 2.2 of Petryshyn and Fitzpatrick (1974). Hence, by the definition of Coincidence Degree,

$$d[(L, N(\cdot,0)), \Omega] = d[(L, N(\cdot,1), \Omega].$$

Now, for any $\lambda \in [0,1]$, let $\lambda' = \lambda t$ and apply the above to $\tilde{N}'(\cdot,t), t \in [0,1]$ where $\tilde{N}'(\cdot,t) = \tilde{N}(\cdot,\lambda')$. Then,

$$d[(L, \tilde{N}(\cdot,\lambda)), \Omega] = d[(L, \tilde{N}'(\cdot,1)), \Omega] \tag{7.87}$$

$$= d[(L, \tilde{N}'(\cdot,0)), \Omega] \tag{7.88}$$

$$= d[(L, \tilde{N}(\cdot,0)), \Omega]. \tag{7.89}$$

Hence, $d[(L, \tilde{N}(\cdot,\lambda)), \Omega]$ is independent of λ in [0,1]. □

Corollary 7.24.1 *Let Assumptions (a) and (b) hold and let* Ω *be an open bounded subset of* X. *Let* N *and* N' *be two* L-k-ϕ-*contractions on* $\bar{\Omega}$ *satisfying (f) such that* $Nx = N'x$ *for each* $x \in \partial\Omega$. *Then* $d[(L, N), \Omega] = d[(L, N'), \Omega]$.

Proof. Define

$$\tilde{N} : \bar{\Omega} \times [0,1] \to CK(Z)$$

by

$$\tilde{N}(x,\lambda) = (1-\lambda)Nx + \lambda N'x.$$

Then $\tilde{N}$ is clearly upper semi-continuous and satisfies all the other conditions of Theorem 7.23. Hence by Theorem 7.23,

$$d[(L, \tilde{N}), \Omega] = d[(L, \tilde{N}(\cdot,0)), \Omega] \tag{7.90}$$

$$= d[(L, \tilde{N}(\cdot,1)), \Omega] \tag{7.91}$$

and hence,

$$d[(L, N), \Omega] = d[(L, N'), \Omega].$$ □

Definition 7.17 Let X and Z be normed linear spaces with norms denoted by $\|\cdot\|$. Let x be any point of X (or Z) and let A, B be subsets of X (or Z). Then

$D^*(x, A) = \inf\{||x - a|| : a \in A\}$ is the usual distance between x and A and we define

$$d^*(A, B) = \inf\{||a - b|| : a \in A, b \in B\}$$

to be the distance between A and B. In fact, $d^*(x, A)$ is equivalent to the distance between A and the singleton $\{x\}$.

Lemma 7.5

$$d^*(x, A) + d^*(B, C) \geq d^*(x, A + C - B).$$

Proof. If $a \in A$, $b \in B$ and $c \in C$, $a + c - b \in A + C - B$ and hence, for every $a \in A$, $b \in B$, $c \in C$,

$$||x - (a + c - b)|| \geq d^*(x, A + C - B)$$

Now,

$$||x - (a + c - b)|| \leq ||x - a|| + ||b - c||$$

Hence, for every $a \in A$, $b \in B$ and $c \in C$, we have

$$d^*(x, A + C - B) \leq ||x - a|| + ||b - c||$$

and so,

$$d^*(x, A + C - B) \leq \inf\{||x - a|| : a \in A\} + \inf\{||b - c|| : b \in B, c \in C\} \quad (7.92)$$
$$= d^*(x, A) + d^*(B, C). \quad (7.93)$$

□

Lemma 7.6 *For each $x \in domL \cap \bar{\Omega}$, we have*

$$(I - M_\psi)x = [\psi\pi + K_p(I - Q)](L - N)x,$$

where $\psi\pi + K_p(I - Q)$ is an algebraic isomorphism between Z and $dom\, L$.

Proof.

$$[\psi\pi + K_p(I - Q)](L - N) = [\psi\pi + K_p(I - Q)]L - [\psi\pi + K_p(I - Q)]N \quad (7.94)$$
$$= K_p(I - Q)]L - [\psi\pi + K_p(I - Q)]N \text{ by (7.39)} \quad (7.95)$$
$$= K_pL - [\psi\pi + K_p(I - Q)]N \quad (7.96)$$
$$= I - P - [\psi\pi + K_p(I - Q)]N \text{ by (7.38)} \quad (7.97)$$
$$= I - M_\psi. \quad (7.98)$$

To show that $\psi\pi + K_p(I - Q)$ is an isomorphism, consider the equation

$$[\psi\pi + K_p(I - Q)]z = y \quad (7.99)$$

for some $y \in \text{dom}\, L$.

This is equivalent to

$$\psi\pi z = Py, \tag{7.100}$$

$$K_p(I-Q)z = (I-P)y. \tag{7.101}$$

Now, as $\ker\pi = \operatorname{Im} L = \operatorname{Im}(I-Q)$, $\psi\pi_Q$, the restriction of $\psi\pi$ to $\operatorname{Im} Q$, is an isomorphism from $\operatorname{Im} Q$ to $\ker L$ and hence (7.100) is equivalent to

$$Qz = (\psi\pi_Q)^{-1}Py \tag{7.102}$$

and since $LK_p = I$ and $LP = 0$, (7.101) is equivalent to

$$(I-Q)z = L(I-P)y = Ly \tag{7.103}$$

Hence,

$$z = Qz + (I-Q)z \tag{7.104}$$

$$= (\psi\pi_Q)^{-1}Py + Ly. \tag{7.105}$$

This shows the existence and uniqueness of the solution z of equation (7.99) for each given y in dom$L$. Hence $\psi\pi + K_p(I-Q)$ is an isomorphism from Z to dom$L$. □

Lemma 7.7 *Let Assumptions (a) to (f) be satisfied. If $M_\psi(\partial\Omega)$ is relatively compact, then there exists $\mu > 0$ such that*

$$\inf\{d^*(Lx, Nx) : x \in \partial\Omega \cap domL\} \geq \mu. \tag{7.106}$$

Proof. By Assumption (f), $d^*(Lx, Nx) > 0$ for all $x \in \partial\Omega \cap \operatorname{dom}L$. Now, suppose that for all $\mu > 0$, (7.106) does not hold. Then for each positive integer n, there exists $x_n \in \partial\Omega \cap \operatorname{dom}L$ such that

$$d^*(Lx_n, Nx_n) < \frac{1}{n}.$$

Now, $d^*(x_n, M_\psi x_n) < ||x_n - y||$ for all $y \in M_\psi x_n$.

Using the preceeding lemma and noting that $\psi\pi + K_p(I-Q)$ is a continuous linear operator from Z onto dom$L$, we have for each $z_n \in Nx_n$,

$$(\psi\pi + K_p(I-Q)(Lx_n - z_n)) = x_n - y \text{ for some } y \in M_\psi x_n.$$

Hence, for all $z_n \in Nx_n$,

$$||(\psi\pi + K_p(I-Q))(Lx_n - z_n)|| \geq d^*(x_n, M_\psi x_n.$$

If $||\psi\pi + K_p(I-Q)|| = \alpha \geq 0$,

$$d^*(x_n, M_\psi x_n \leq ||\psi\pi + K_p(I-Q)||||Lx_n - z_n|| \tag{7.107}$$

$$= \alpha||Lx_n - z_n|| \text{ for all } z_n \in Nx_n. \tag{7.108}$$

Hence,

$$d^*(x_n, M_\psi x_n \leq \alpha d^*(Lx_n, Nx_n) \tag{7.109}$$

$$< \alpha \frac{1}{n}. \tag{7.110}$$

Thus for each integer n, there exists some $u_n \in M_\psi x_n$ such that

$$||x_n - u_n|| < \frac{\alpha}{n}.$$

Now since $u_n \in M_\psi x_n \subset M_\psi(\partial\Omega)$ which is relatively compact, we can find a subsequence $\{u_{n_k}\}$ of $\{u_n\}$ such that $u_{n_k} \to u_0$ and the triangle inequality

$$||x_{n_k} - u_0|| \leq ||x_{n_k} - u_{n_k}|| + ||u_{n_k} - u_0|| \tag{7.111}$$

$$< \frac{\alpha}{n_k} + ||u_{n_k} - u_0|| \tag{7.112}$$

implies that $x_{n_k} \to u_0$ as $n_k \to \infty$. As $x_{n_k} \in \partial\Omega$ which is closed, $u_0 \in \partial\Omega$. By upper semi-continuity of M_ψ, $u_{n_k} \in M_\psi x_{n_k}$ for each n_k –implies that $u_0 \in M_\psi u_0$ which is a contradiction as $u_0 \in \partial\Omega$. Hence (7.106)holds for some $\mu > 0$. □

Remark 7.18 In Gaines and Mawhin (1977), Rouché's Theorem was extended to the context of Coincidence Degree. The following theorem is a version of Rouché's Theorem in our situation.

Theorem 7.25 *Let Assumptions (a) to (f) be satisfied and assume that $M_\psi(\partial\Omega)$ is relatively compact. Let $\mu > 0$ be such that*

$$\inf\{d^*(Lx, Nx) : x \in \partial\Omega \cap domL\} \geq \mu.$$

Then, for each L-k-ϕ-contraction $N' : \bar{\Omega} \to CK(Z)$ satisfying Assumption (f) and the following condition:

$$\sup\{d^*(Nx, N'x) : x \in \partial\Omega\} < \mu$$

we have

$$d[(L, N), \Omega] = d[(L, N'), \Omega].$$

Proof. Let $\tilde{N} : \bar{\Omega} \times [0, 1] \to CK(Z)$ be defined by

$$\tilde{N}(x, \lambda) = (1 - \lambda)Nx + \lambda N'x.$$

It can easily be verified that conditions (i) to (iv) of Theorem 7.24 are satisfied. Now,

$$d^*(Lx, \tilde{N}(x, \lambda)) = d^*(Lx, Nx - \lambda(Nx - N'x)) \tag{7.113}$$

$$\geq d^*(Lx, Nx) - \lambda d^*(Nx, N'x) \tag{7.114}$$

the last inequality following from Lemma 7.5 by putting $B = \lambda N'x$, $C = \lambda Nx$ and $A = Nx - \lambda Nx + \lambda N'x$.

Hence, for each $(x, \lambda) \in (\text{dom}L \cap \partial\Omega) \times [0, 1]$,

$$d^*(Lx, \tilde{N}(x, \lambda)) > \mu - \lambda\mu \geq 0.$$

This shows that $\tilde{N}$ satisfies the last condition of Theorem 7.24 and hence,

$$d[(L, N), \Omega] = d[(L, \tilde{N}(\cdot, 0)), \Omega] = d[(L, \tilde{N}(\cdot, 1)), \Omega] = d[(L, N'), \Omega].$$

Thus,

$$d[(L, N), \Omega] = d[(L, N'), \Omega].$$ □

A Generalized Continuation Theorem of Tarafdar and Teo and Existence Theorems

In Gaines and Mawhin (1977), the Leray-Schauder Continuation Theorem was extended to the context of Coincidence Degree. Here, Tarafdar and Teo extended it to the set-valued situation. Tarafdar and Teo also derived some existence theorems for $Lx \in Nx$.

Definition 7.18 Consider the mapping $F : X \to CK(X)$ where X is the zero-dimensional space $\{0\}$. As $CK(X)$ may only contain non-empty subsets of X, $CK(X) = \{\{0\}\}$ and hence F is the mapping $F(0) = \{0\}$. We define $d(F, \{0\}, 0) = 1$ and this degree agrees with the usual properties of the degree of an ultimately compact field F. We also set $d(F, \phi, 0) = 0$.

Definition 7.19 Let X and Z be normed linear spaces and let L be a linear Fredholm mapping of index zero. Let P, Q, K_p and ϕ be given as in Assumption (e) and let Ω be an open bounded subset of X such that Ω is complete. Let $a > 0$ and $N^* : \bar{\Omega} \times [0, a] \to CK(Z)$ be a set valued mapping. Let N^* satisfy the following conditions:

(i) N^* is upper semi-continuous on $\bar{\Omega} \times [0, a]$,
(ii) $N^*(\bar{\Omega} \times [0, a])$ is bounded,
(iii) $\phi(K_p(I - Q)N^*(\bar{\Omega} \times [0, a])) < \infty$,
(iv) there exists a positive $k < 1$ such that, for every $A \subset \bar{\Omega}$,

$$\phi(K_p(I - Q)N^*(A \times [0, a])) \leq k\phi(A).$$

Then N^* is said to be a L-k-ϕ-contraction on $\Omega \times [0, a]$.

Remark 7.19 With N^* as defined above, Tarafdar and Teo observed that for each $\lambda \in [0, a]$, $N^*(\cdot, \lambda)$ is L-k-ϕ-contraction as defined by Assumptions (c), (d) and (e). They also noted that for $a = 1$, N^* satisfies the first four conditions of the homotopy invariance theorem, Theorem 7.24.

Now, let assumptions (a) to (f) be satisfied for a pair of mappings $L : \text{dom}L \to Z$ and $N : \bar{\Omega} \to CK(Z)$ and let $N^* : \bar{\Omega} \times [0,1] \to CK(Z)$ be a L-k-ϕ-contraction on $\bar{\Omega} \times [0,1]$ such that $N^*(\cdot, 1) = N$.

Let $y \in \text{Im}L$ and consider the family of equations

$$Lx \in \lambda N^*(x, \lambda) + y. \tag{7.115}$$

An element $(x, \lambda) \in \bar{\Omega} \times [0,1]$ satisfying (7.115) is said to be a solution of (7.115). If λ is specified, any $x \in \bar{\Omega}$ satisfying the equation for that λ is also called a solution. It will be clear from the context whether a solution is an element of $\bar{\Omega}$ or $\bar{\Omega} \times [0,1]$.

Lemma 7.8 *For each $\lambda \in (0,1]$, the set of solutions of (7.115) is equal to the set of solutions of the equation*

$$Lx \in [Q + \lambda(I - Q)]N^*(x, \lambda) + y \tag{7.116}$$

and if $\lambda = 0$, every solution of (7.116) is a solution of (7.115).

Proof. If $\lambda = 0$, (7.116) reduces to

$$Lx \in QN * (x, 0) + y.$$

But $Lx = (I - Q)Lx$ which implies that

$$Lx \in (I - Q)[QN * (x, 0) + y] = \{y\}.$$

This means that $Lx = y$ or x is a solution of (7.115) for $\lambda = 0$. Let $\lambda \in (0,1]$ and let x be a solution of (7.115). Then there exists $u \in N^*(x, \lambda)$ such that

$$Lx = \lambda\mu + y.$$

Hence $u = \lambda^{-1}(Lx - y) \in \text{Im}L$. Therefore $Qu = 0$ and thus,

$$u = (I - Q)u \in (I - Q)N^*(x, \lambda).$$

Hence,

$$Lx = 0 + \lambda\mu + y = [Q + \lambda(I - Q)]u + y \in [Q + \lambda(I - Q)]N^*(x, \lambda) + y,$$

that is x is a solution of (7.116).

Conversely, let x be a solution of (7.116). Then there exists $v \in N^*(x, \lambda)$ such that

$$Lx = [Q + \lambda(I - Q)]v + y.$$

Here $0 = QLx = Qv + \lambda Q(I - Q)v + Qy = Qv$. Thus,

$$\begin{aligned} Lx &= Qv + \lambda(I - Q)v + y && (7.117)\\ &= \lambda v + y \text{ as } \; Qv = 0 && (7.118)\\ &\in \lambda N^*(x, \lambda) + y, && (7.119) \end{aligned}$$

that is x is a solution of (7.115). □

Theorem 7.26 (A Generalized Continuation Theorem of Tarafdar and Teo) *Let L and N be mappings satisfying Assumptions (a) to (f) and let N^* be a L-k-ϕ-contraction on $\bar{\Omega} \times [0,1]$ such that $N^*(\cdot, 1) = N$. Let $y \in ImL$ and we assume the following conditions hold:*

(1) $Lx \notin \lambda N^*(x, \lambda) + y$ *for every* $x \in \partial\Omega \cap domL$, $\lambda \in [0,1]$.

(2) $0 \notin \pi N^*(x, 0)$ *for every* $x \in L^{-1}\{y\} \cap \partial\Omega$.

(3) $d(g(\cdot)|_{L^{-1}\{y\}}, \Omega \cap \Omega_1 \cap L^{-1}\{y\}, 0) \neq 0$,

where the left-hand expression is the Brouwer degree for the single-valued compact field g restricted to the affine finite-dimensional space $L^{-1}\{y\}$ and g and Ω_1 are defined as follows: As kerL is a finite dimensional subspace of X, -$\psi\pi N^*(\cdot + K_p y, 0)$ defined on $((\Omega - K_p y))^- \cap kerL$ is a set-valued compact field with respect to zero (the conclusion that $0 \notin -\psi\pi N^*(x + K_p y, 0)$ for every $x \in \partial(\Omega - K_p y) \cap kerL$ follows from condition (2)). In Ma (1972), Section 5.2, it has been shown that there exists a single-valued compact field g and an open bounded set $\Omega_1 \subset kerL$ containing zero such that $g(\cdot + K_p y)$ and $-\psi\pi N^*(\cdot + K_p y, 0)$ are homotopic and

$$g(x + K_p y) = x + K_p y \text{ for all } x \text{ in } (((\Omega - K_p y))^- \setminus (\Omega_1 - K_p y)) \cap kerL.$$

Ma has also defined the degree of the set-valued compact field $-\psi\pi N^*(\cdot + K_p y, 0)$ by

$$d(-\psi\pi N^*(\cdot + K_p Y, 0)|_{kerL}, (\Omega - K_p y) \cap kerL, 0) \tag{7.120}$$
$$= d(g(\cdot + K_p y)|_{kerL}, (\Omega - K_p y) \cap (\Omega_1 - K_p y) \cap kerL, 0). \tag{7.121}$$

Then, for each $\lambda \in [0,1)$, equation (7.115) has at least one solution in Ω and for $\lambda = 1$, the equation

$$Lx \in Nx + y \tag{7.122}$$

has at least one solution in $\bar{\Omega}$.

Proof. Let $\lambda \in [0,1]$ be considered fixed. For each $x \in \bar{\Omega}$, $\mu \in [0,1]$ we define

$$\tilde{N}(x, \mu) = [Q + \lambda\mu(I - Q]N^*(x, \lambda\mu) + y.$$

Clearly $\tilde{N}$ is a L-k-ϕ-contraction in $\Omega \times [0,1]$.

Let us now consider the case where $\lambda \in [0,1]$. By condition (1) and Lemma 7.8 if $\lambda \neq 0$

$$Lx \notin \tilde{N}(x, \mu) \text{ for every } x \in \partial\Omega \cap \text{dom}L, \ \mu \in [0,1].$$

Also, if $\mu = 0$ or $\lambda = 0$

$$\tilde{N}(x, \mu) = QN * (x, 0) + y$$

and $Lx \in \tilde{N}(x,\mu)$ would imply that $Lx = y$ and $0 \in QN * (x,0)$ or $x \in L^{-1}\{y\}$ and $0 \in \pi N^*(x,0)$. Thus, by Assumption (2), $x \notin \partial\Omega$. Hence, for every $x \in \partial\Omega \cap \operatorname{dom} L$, $\mu \in [0,1]$,

$$Lx \notin \tilde{N}(x,\mu). \tag{7.123}$$

By Theorem 7.24, $d[(L,\tilde{N}(\cdot,\mu)),\Omega]$ is independent of μ in [0,1] and hence,

$$[L,\tilde{N}(\cdot,1)),\Omega] = d[L,\tilde{N}(\cdot,0)),\Omega] \tag{7.124}$$
$$= d[L,QN * (\cdot,0) + y),\Omega] \tag{7.125}$$
$$= d(I - P - [\psi\pi + K_p(I-Q)][QN * (\cdot,0) + y],\Omega,0) \tag{7.126}$$

that is

$$d[L,\tilde{N}(\cdot,1)),\Omega] = d(I - P - \psi\pi N^*(\cdot,0) - K_p y,\Omega,0). \tag{7.127}$$

Let us now consider two cases. Firstly let us assume $\ker L = 0$. Then $P = 0$, $Q = 0$, $\pi = 0$, $K_p = L^{-1}$ and hence, from (7.127), we have

$$d[L,\tilde{N}(\cdot,1)),\Omega] = d(I - L^{-1}y,\Omega,0). \tag{7.128}$$

Now, $L^{-1}\{y\} = \{L^{-1}y\}$ is a zero dimensional space and hence, for condition (3) to be satisfied, $L^{-1}\{y\} \cap \Omega \cap \Omega_1 \neq \emptyset$.

Hence, $L^{-1}y \in \Omega$ and so, as the right-hand term of (7.2, 4.2) has reduced to the degree of a single-valued mapping $I - L^{-1}y$, we have

$$d[L,\tilde{N}(\cdot,1)),\Omega] = d(I - L^{-1}y,\Omega,0) \tag{7.129}$$
$$= d(I,\Omega,L^{-1}y) \tag{7.130}$$
$$= 1. \tag{7.131}$$

From Theorem 7.22, there exists $x \in \Omega$ such that $Lx \in \tilde{N}(x,1)$, that is, for some $x \in \Omega$

$$Lx \in [Q + \lambda(I-Q)]N^*(x,\lambda) + y.$$

and by Lemma 7.8, equation (7.115) has at least one solution in Ω. Now let us consider the case where $\ker L \neq \{0\}$. By a change of variables, we have

$$d(I-P-\psi\pi N^*(\cdot,0)-K_p y,\Omega,0) = d(I-P-\psi\pi N^*(\cdot,+K_p y,0),\Omega-K_p y,0). \tag{7.132}$$

As $\ker L$ is a finite-dimensional subspace containing the range of $P + \psi\pi N^*$, one can apply Theorem 7.2 and obtain

$$d(I - P - \psi\pi N^*(\cdot + K_p y,0),\Omega - K_p y,0) \tag{7.133}$$
$$= d(I - P - \psi\pi N^*(\cdot + K_p y,0)]|_{\ker L}(\Omega - K_p y) \cap \ker L,0) \tag{7.134}$$
$$= d(-\psi\pi N^*(\cdot + K_p y,0)|_{\ker L},(\Omega - K_p y) \cap \ker L,0) \tag{7.135}$$
$$= d(g(\cdot + K_p y)|_{\ker L}(\Omega - K_p y) \cap (\Omega_1 - K_p y) \cap \ker L,0) \tag{7.136}$$

the last equality holding by definition.

By a change of variables again,

$$
\begin{aligned}
& d(g(\cdot + K_p y)|\text{ker}l, (\Omega - K_p y) \cap (\Omega_1 - K_p y) \cap \text{ker}L, 0) && (7.137)\\
&= d(g(\cdot)|_{L^{-1}y}, \Omega \cap \Omega_1 \cap L^{-1}y, 0) && (7.138)\\
&\not\equiv 0 \text{ by condition } (3) && (7.139)
\end{aligned}
$$

Hence, from (7.127), (7.132), (7.133) and (7.137)

$$d[(L, N(\cdot, 1)), \Omega] \neq 0$$

and again, we conclude from Theorems 7.22 and 7.24 that equation (7.115) has at least one solution in Ω.

Now, for $\lambda = 1$, equation (7.115) becomes

$$Lx \in Nx + y = N^*(x, 1) + y.$$

If, for every $x \in \partial\Omega \cap \text{dom}L$, (7.115) does not hold, then $Lx \notin \tilde{N}(x, \mu)$ for each $x \in \partial\Omega \cap \text{dom}L$ and each $\mu \in [0, 1]$ and the above proof can be repeated. If, however, there exists x in $\partial\Omega \cap \text{dom}L$ such that $Lx \in Nx + y$, then a solution exists in $\partial\Omega \subset \bar{\Omega}$. Hence (7.115) always has a solution in $\bar{\Omega}$.

This completes the proof of the Theorem. □

Theorem 7.27 *Let X be a Banach space, Z a normed linear space and let L be a linear Fredholm mapping of index zero from a subspace of X into Z.*

Let Ω be an open bounded subset of X and let tildeN : $\bar{\Omega} \times [0, 1] \to CK(Z)$ be a L-k-ϕ-contraction on $\bar{\Omega} \times [0, 1]$. If for each $\lambda \in [0, 1]$ and $x \in \partial\Omega \cap$ domL, we have

$$Lx \notin \tilde{N}(x, \lambda)$$

and if $d[L, \tilde{N}(\cdot, \lambda_0)), \Omega] \neq 0$ for some $\lambda_0 \in [0, 1]$, then for each $\lambda \in [0, 1]$, the equation

$$Lx \in \tilde{N}(x, \lambda) \tag{7.140}$$

has at least one solution in Ω.

Proof. By Theorem 7.24, for each $\lambda \in [0, 1]$,

$$d[(L, \tilde{N}(\cdot, \lambda)), \Omega] = d[(L, N(\cdot, \lambda_0)), \Omega] \neq 0$$

and hence by Theorem 7.22, the equation

$$Lx \in \hat{N}(x, \lambda)$$

has a solution in Ω. □

Corollary 7.27.1 (A Generalized Borsuk's Theorem) *Let X, Z and L be as in Theorem 7.27 and let Ω be a bounded open subset of X, symmetric with respect to* *the origin and containing it. Let* $\tilde{N} : \bar{\Omega} \times [0,1] \to CK(Z)$ *be a* L-k-ϕ-*contraction on* $\bar{\Omega} \times [0,1]$. *Also, suppose that* $\tilde{N}(-x,0) = -\tilde{N}(x,0)$ *for each* $x \in \bar{\Omega}$.

Then equation (7.106) *has a solution in* Ω *for each* $\lambda \in [0,1]$.

Proof. From Theorem 7.23, $d[(L, N(\cdot,0)), \Omega]$ is odd and hence different from zero. The result follows from the preceding theorem. □

Corollary 7.27.2 (A Generalized Krasnoselskii Theorem) *Let* X, Z, L *and* Ω *be as in Corollary* 7.27.1 *and let* $N : \bar{\Omega} \to CK(Z)$ *be a* L-k-ϕ-*contraction such that for each* $\lambda \in [0,1]$ *and* $x \in \partial\Omega \cap domL$, *we have*

$$[(L-N)x] \cap [\lambda(L-N)(-x)] = \emptyset. \tag{7.141}$$

The equation

$$Lx \in Nx \tag{7.142}$$

has at least one solution in Ω.

Proof. Define $N : \bar{\Omega} \times [0,1] \to CK(Z)$ by

$$\tilde{N}(x,\lambda) = (1+\lambda)^{-1}[Nx - \lambda N(-x)].$$

It can be easily verified that $\tilde{N}$ is a L-k-ϕ-contraction on $\Omega \times [0,1]$. Now, $\tilde{N}(x,0) = Nx$ and $N(x,1) = \frac{1}{2}[Nx - N(-x)]$ which is odd. We claim that $Lx \notin N(x,\lambda)$ for each $\lambda \in [0,1]$ and each $x \in \partial\Omega \cap \operatorname{dom} L$. Assuming otherwise, there exist $\lambda \in [0,1]$, $x \in \partial\Omega \cap \operatorname{dom} L$ such that

$$(1+\lambda)Lx \in Nx - \lambda N(-x);$$

that is there exist $u \in Nx$, $v \in N(-x)$ such that

$$(1+\lambda)Lx = u - \lambda v$$

or

$$Lx - u = \lambda(L(-x) - v)$$

which contradicts (7.141).

Hence the conditions of Theorem 7.27 are satisfied and thus, there is a $x \in \Omega$ such that

$$Lx \in \tilde{N}(x,0) = Nx$$

and so equation (7.142) has a solution in Ω. □

Remark 7.20 If $N : \bar{\Omega} \to Z$ is single-valued, then $d[(L,N),\Omega]$ contains the coincidence degree of Mawhin.

If $X = Z$ and N is an ultimately compact operator, then degree theorem reduces to the degree theory of Petryshyn and Fitzpatrick (1974).

Note that the results of this section have been written and presented after a thorough revision of some results of an article in Tarafdar and Teo (1979).

7.4 Coincidence Degree for Multi-Valued Mappings with Non-Negative Index

Using the equivalence theorem, Theorem 3.1, of Tarafdar and Teo (1979), Akashi (1988) has proved a new equivalence theorem and built a coincidence degree such that, even when the index of L is strictly positive, this coincidence degree is not necessarily zero. In this we dealt with this coincidence degree when L is of nonnegative index.

The following definition is due to Petryshyn and Fitzpatrick (1974).

Definition 7.20 Let ϕ be a measure of noncompactness in X and let $T : D \subset X \to \mathcal{K}(X)$ be a u.s.c. multivalued mapping.

Then T is said to be ϕ-condensing if $\phi(T(\Omega)) \not\geq \phi(\Omega)$ for all $\Omega \subset D$ such that Ω is not relatively compact. In case C is not linearly ordered, the above condition reduces to the requirement that $\phi(t(\Omega)) < \phi(\Omega)$ for each $\Omega \subset D$ which is not relatively compact.

Proposition 7.11 (Petryshyn and Fitzpatrick (1974)) *Let $D \subset X$ be closed and let $T : D \to CK(X)$ be ϕ-condensing. Then T is ultimately compact.*

Proposition 7.12 (Petryshyn and Fitzpatrick (1974)) *Let $\phi : 2^X \to \mathbf{R}^+ = \{t \in \mathbf{R}; t \geq 0\} \cup \{\infty\}$ be a measure of non-compactness and suppose that $T : D \subset X \to CK(X)$ is a k : ϕ-contraction, $0 < k < 1$, with $\phi(T(D)) \in \mathbf{R}$. Then T is ϕ-condensing if either X is quasi-complete or D is complete.*

7.4.1 *Basic Assumptions and Main Results in Akashi (1988)*

In this subsection we assume the following assumptions:

(a') X is a Banach space and Z is a real normed space.

(b') $L : D(L) \subset X \to Z$ is a linear Fredholm mapping (i.e., $R(L)$ is closed; $\dim\ker L < \infty$ and $\dim\operatorname{coker} L < \infty$) with nonnegative index of Fredholm ($\operatorname{ind} L = \dim\ker L - \dim\operatorname{coker} L \geq 0$).

(c') Ω is a bounded, open set in X and the multivalued mapping $N : \bar{\Omega} \to CK(Z)$ is u.s.c. and $N(\bar{\Omega})$ is bounded in Z. Furthermore suppose that $D(L) \cap \Omega \neq \Phi$.

(d')Let (P, Q) be an exact pair of projection with respect to L and let ϕ be a measure of non-compactness defined from 2^X into C (linearly ordered lattice) such that ϕ satisfies the sub-additivity condition ($A \subset X, B \subset X \implies \phi(A + B) \leq \phi(A) + \phi(B)$). We assume that with such a measure of non-compactness ϕ, $K_p(I - Q)N$ is ϕ-condensing and that $\phi[K_p(I - Q)N(\bar{\Omega})] < \infty$. Furthermore, we assume that K_p is bounded.

(e') $0 \notin (L - N)(D(L) \cap \partial\Omega$ where $\partial\Omega$ denotes the boundary of Ω.

Remark 7.21 Assumption (b') authorizes that we assume the continuity of the exact pair of projections (P, Q). Moreover, with the quotient norm topology, coker$L$ is a normed space and the canonical surjection π is continuous with respect to this topology. Also (b') implies that there exists a continuous linear one-to-one mapping $\Psi : \text{coker}L \to \ker L$.

Proposition 7.13 *Under the assumptions (a') to (d'), the condition (d') is independent of choice of the exact pair (P, Q) of continuous projections with respect to L.*

Proof. We follow the proof of Proposition 3.1 of Tarafdar and Teo (1979). Suppose that (P, Q) satisfies (d') and let (P', Q') be another exact pair of continuous projections with respect to L. Then by Proposition 7.5 we have

$$K_{p'}(I - Q')N = (I - P')K_p(I - Q')N \tag{7.143}$$

$$\subset (I - P')K_p(I - Q)N + (I - P')K_p(Q - Q')N \tag{7.144}$$

$$= (I - P')K_p(I - Q)N + (I - P')\tilde{K}_P(\pi_Q^{-1} - \pi_{Q'}^{-1})\pi N, \tag{7.145}$$

where K_p denotes the restriction of K_p to the finite dimensional subspace $(Q - Q')(Y)$ ($\tilde{K}_p$ is continuous); $\pi_Q = \pi/R(Q)$ and $\pi_{Q'} = \pi/r(Q')$. Since $\pi N(\bar{\Omega})$ is bounded in a finite dimensional subspace of X, it follows that

$$\phi[(I - P')\tilde{K}_P(\pi_Q^{-1} - \pi_{Q'}^{-1})\pi N(\bar{\Omega})] = 0.$$

Hence, from the subadditivity condition on ϕ, it follows that $K_{P'}(I - Q')N$ is ϕ-condensing and that $\phi[K_{P'}(I - Q')] < \infty$. □

The following proposition was used in Akashi (1988) in proving the main results.

Proposition 7.14 *If assumptions $(a'b'c'd')$ are satisfied, then for every continuous linear one-to-one mapping Ψ : coker$L$ $\to$ ker$L$ and any exact pair (P, Q) of continuous projections with respect to L, M_Ψ is a ϕ-condensing multivalued mapping.*

Proof. Since P, Q, K_P, π, Ψ are all linear and continuous and $N(x)$ is convex and compact for each $x \in \bar{\Omega}$, it follows that $M_\Psi(x)$ is convex and compact for each $x \in \bar{\Omega}$.

Now, let $A \subset \bar{\Omega}$ such that A is not relatively compact. Then,

$$M_\Psi(A) = [P + [\Psi\pi + K_P(I - Q)]N](A) \subset P(A) + \Psi\pi N(A) + K_P(I - Q)N(A)$$

and, by the subadditivity of ϕ,

$$\phi[M_\Psi(A)] \leq \phi[P(A)] + \phi[\Psi\pi N(A)] + \phi[K_P(I - Q)N(A)].$$

Now, $P(A)$ and $\Psi\pi N(A)$ are bounded subsets of finite-dimensional subspace of X and, therefore,

$$\phi[P(A)] = \phi[\Psi\pi N(A)] = 0.$$

Then,

$$\phi[M_\Psi(A)] \leq \phi[K_P(I-Q)N(A)].$$

By assumption (d') $K_P(I-Q)N$ is ϕ-condensing and, therefore, the above inequality assures that M is ϕ-condensing. □

From Proposition 7.12, we see that if assumptions $(a'b'c'd')$ are satisfied, then $M\Psi$ is an ultimately compact multivalued mapping. It follows from the assumption (e') and Theorem 7.20 that $0 \notin (I-M_\Psi)(D(L)\cap\partial\Omega)$. Thus, the topological degree of the multi-valued mapping $I-M_\Psi$ on Ω with respect to zero ($d[I-M_\Psi, \Omega, 0]$) is well defined.

Remark 7.22 If $\operatorname{ind} L = O$, $\Psi : \operatorname{coker} L \to \ker L$ is considered as an isomorphism, and if the lattice C is taken as $\mathbf{R}^+ = \{t \in \mathbf{R}; t \geq 0\} \cup \{\infty\}$ and if the multivalued mapping $K_p(I-Q)N$ is considered as a k-set-contraction with $k < l$, then the work in Akashi (1988) reduces to the work in Tarafdar and Teo (1979).

If $\operatorname{ind} L > O$, unfortunately we have the following result:

Proposition 7.15 *If* $\operatorname{ind} L > 0$ *and* $0 \notin (L-N)(D(L)\cap\partial\Omega)$ *then, for each linear one-to-one mapping* $\Psi : \operatorname{coker} L \to \ker L$*, one has* $d[I-M_\Psi, \Omega, 0] = 0$.

Proof. We follow the proof of Proposition 6.1 of Mawhin (1972) or Proposition XII.1 of Gaines and Mawhin (1977) where N is assumed to be single-valued. First note that the condition $\operatorname{ind} L > 0$ implies that there exists a linear one-to-one mapping $\Psi : \operatorname{coker} L \to \ker L$. Also, $\operatorname{ind} L > 0$ implies that $R(\Psi)$ is a proper subspace of $\ker L$. This and

$$R(I-M_\Psi) \subset R[I-P-\Psi\pi N - K_p(I-Q)N]$$

implies that $R(I-M_\Psi)$ is necessarily contained in the proper subspace of X given by

$$X' = \ker P \bigoplus R(\Psi)$$

Then, by the properties of topological degree, there exists a neighbourhood V of the origin such that

$$d[I-M_\Psi, \Omega, 0] = d[I-M_\Psi, \Omega, y]$$

for all $y \in V$. If we take y in the non-void set $V \cap C_X X'(C_X X'$ is the complement of X' in X), then y does not belong to the $R(I-M_\Psi)$ and, consequently, $d[I-M_\Psi, \Omega, 0]$. This complete the proof. □

However, this negative result can be overcome by modifying the multivalued mapping M_Ψ, related to L-N in such a way that the topological degree is no more necessarily equal to zero, as follows (cf. Proposition XII. 3 of Gaines and Mawhin (1977) in case of a single valued N):

Theorem 7.28 *Under the same notation of Theorem* 7.20 *if* $\operatorname{ind} L \geq 0$*, then:*

(*i*) *every fixed point of the multivalued mapping* $K_P N$ *is a solution of the inclusion* (7.57) *provided* L *is surjective;*

(*ii*) *if* L *is not surjective, the inclusion* (7.57) *has a solution if and only if there exists a linear one-to-one mapping* $\Psi : coker\, L \to ker\, L$ *such that the multivalued mapping*

$$\tilde{M}_\Psi = R_\Psi P + [\Psi\pi + K_P(I - Q)]N$$

has a fixed point, where $R_\Psi : ker\, L \to ker\, L$ *is a projector such that* $R(R_\Psi) = R(\Psi)$.

Proof. First suppose that L is surjective and that x is a fixed point of $K_P N$, i.e., $x \in K_P N(x)$. Thus, $L(x) \in LK_P N(x) = N(x)$, i.e., x is a solution of (7.57).

Now, suppose that L is not surjective and that x is a solution of (7.57). Then, it follows from Theorem 7.20 that

$$x \in M_\Psi(x)$$

with

$$M_\Psi = P + [\Psi\pi + K_P(I - Q)]N$$

for any linear Ψ : coker$L \to$ ker$L$ which is one-to-one. Now let V be any subspeac of ker$L$ of dimension equal to dim coker L and containing $P(x)$ (such a subspacce necessarily exists) and let R_Ψ he any projector in ker$L$ such that $R = (R_\Psi) = V$. Then necessarily

$$P(x) = R_\Psi P(x)$$

and if we take Ψ : coker$L \to$ ker$L$ linear one-to-one such that $R(\Psi) = V$ (such a linear mapping necessarily exists) then

$$x \in M_\Psi(x) = P(x) + [\Psi_\pi + K_P(I - Q)]N(x) \tag{7.146}$$

$$= R_\Psi P(x) + [\Psi_\pi + K_P(I - Q)]X(x) \tag{7.147}$$

$$= \tilde{M}_\Psi(x). \tag{7.148}$$

Conversely, if $x \in D(L) \cap \bar{\Omega}$ is a fixed point of the multivalued $\tilde{M}_\Psi = R_\Psi P + [\Psi_\pi + K_P(I - Q)]N$, i.e.,

$$x \in R_\Psi P(x) + [\Psi_\pi + K_P(I - Q)]N(x)$$

then

$$(I - P)(x) = K_p(I - Q)(z)$$
$$P(x) = R_\Psi P(x) + \Psi_\pi(z)$$

for some $z \in N(x)$. Hence,

$$L(x) = (I - Q)(z)$$

$$(I - R_\Psi)P(x) = \Psi_\pi(z) = R_\Psi \Psi_\pi(z)$$

which implies

$$L(x) = (I - Q)(z)$$
$$(I - R_\Psi)P(x) = 0, \ \ R_\Psi \Psi_\pi(z) = 0.$$

Now, $\Psi_\pi(z) = R_\Psi \Psi_\pi(z) = 0$ implies $z \in R(L)$. Hence $(I - Q)(z) = z$ and therefore, $L(x) \in N(x)$. □

Let V be a vector subspace of ker$L$ such that $\dim V = \dim \operatorname{coker} L$. Then, analogous to Proposition 7.14, Akashi (1988) obtained the following result whose proof is analogous to the proof of Proposition 7.14.

Proposition 7.16 *If assumptions $(a'b'c'd')$ are satisfied, then for every continuous linear one-to-one mapping Ψ : coker$L \to$ ker$L$ and any exact pair (P, Q) of continuous projections with respect to L, the multi-valued mapping $\tilde{M}_\Psi$ stated in Theorem 7.28 is ϕ-condensing and, for each $x \in \bar{\Omega}$, $M_\Psi(x) \in CK(X)$.*

Then, under the Assumptions $(a'b'c'd'e')$ the topological degree of the multi-valued mapping $I - \tilde{M}_\Psi$ on Ω with respect to zero is well defined. Akashi (1988) denoted it by

$$d_V[I - \tilde{M}_\Psi, \Omega, 0].$$

For each vector subspace V of ker L such that dim$V = \dim \operatorname{coker} L$, $d_V[I - \tilde{M}_\Psi, 0]$ is independent of the choice of P, Q and within the same homotopy class (here, the mappings Ψ are such that $R(\Psi) = V$).

Definition 7.21 For each vector subspace V of ker$L$ such that $\dim V = \dim \operatorname{coker} L$, let L_L^V he the set of all continuous isomorphism from coker L into V. Ψ, Ψ' are to be homotopic in L_L^V if there exists a continuous mapping $\bar{\Psi} : \operatorname{coker} L \times [0, 1] \to V$ such that $\bar{\Psi}(\cdot, 0) = \Psi$, $\bar{\Psi}(\cdot, 1) = \Psi'$ and, for each $\lambda \in [0, 1]$, $\bar{\Psi}(\cdot, \lambda) \in L_L^V$.

Remark 7.23 To be homotopic is an equivalence relation which partitions L_L^V into equivalence classes called homotopy classes.

The following two propositions and corollary are quoted from Gaines and Mawhin (1977):

Proposition 7.17 *Ψ and Ψ' are homotopic in L_L^V if and only if det$(\Psi', \Psi^{-1}) > 0$.*

Corollary 7.28.1 *L_L^V is partitioned into two homotopy classes.*

Definition 7.22 Ψ : coker $L \to V$ is said to be orientation preserving if $\{\Psi_{a_1}, \Psi_{a_2}, \ldots, \Psi_{a_n}\}$ belongs to the orientation chosen in V where $\{a_1, a_2, \ldots, a_n\}$ is a basis for coker L belonging to a certain chosen orientation. Otherwise, Ψ is said to he orientation reversing.

Proposition 7.18 *If* $\operatorname{coker} L$ *and* V *are oriented then* Ψ *and* Ψ' *are homotopic in* L_L^V *if and only if they are simultaneously orientation preserving or orientation reversing.*

Definition 7.23 Let V he as given above and suppose that assumptions $(a'b'c'd'e')$ are satisfied and Ψ is an orientation preserving continuous isomorphism from $\operatorname{coker} L$ into V. Then, the coincidence degree of L and N in Ω, denoted by $D[(L,N),\Omega]$, is defined by

$$D[(L,N),\Omega] = \begin{cases} \bigcup_{V \in \mathcal{N}_{\ker L}} d_V(I - \tilde{M}_\Psi, \Omega, 0), & \text{if } L \text{ is not surjective} \\ d(I - K_P N, \Omega, 0), & \text{if } L \text{ is surjective} \end{cases}$$

where $\mathcal{N}_{\ker L}$ is the family of all vector subspaces V of $\ker L$ such that $\dim V = \dim\ \operatorname{coker} L$.

Remark 7.24 Note that if $\operatorname{ind} L = 0$, and the projector R_Ψ is the identity mapping, then the work in Tarafdar and Teo (1979) is a particular case of the work in Akashi (1988).

7.4.2 *Akashi's Basic Properties of Coincidence Degree*

In this subsection, unless otherwise specified, Akashi (1988) assumed that assumptions $(a'b'c'd'e')$ were satisfied such that the coincidence degree was well defined.

Theorem 7.29

(*a*) (*Existence theorem*) *If*

$$D[(L,N),\Omega] \neq \{0\}$$

then (7.57) *has at least one solution in* Ω.

(*b*) (*Excision property*) *If* $\Omega_0 \subset \Omega$ *is an open set such that*

$$(L-N)^{-1}(0) \subset \Omega_0,$$

then,

$$D[(L,N),\Omega] = D[(L,N),\Omega_0].$$

(*c*) (*Addivity property*) *If* $\Omega = \Omega_1 \cup \Omega_2$ *where* Ω_1, Ω_2 *are two open sets such that* $\Omega_1 \cap \Omega_2 = \emptyset$, *then*

$$D[(L,N),\Omega] \subset D[(L,N),\Omega_1] + D[(L,N),\Omega_2].$$

Proof. This theorem follows immediately from Definition 7.23 and corresponding properties of topological degree of ultimately compact multivalued mappings (see, Petryshyn and Fitzpatrick (Petryshyn and Fitzpatrick (1974)). □

One of the most useful properties of every concept of topological degree is its invariance with respect to some type of homotopy. In the case of coincidence degree the following result was obtained in Akashi (1988):

Theorem 7.30 *If the assumptions $(a'b')$ are satisfied and if the multivalued mapping*

$$\tilde{N} : \bar{\Omega} \times [0,1] \to CK(Y)$$

is such that $(\tilde{c})$ $\tilde{N}$ *is upper semicontinuous on* $\bar{\Omega} \times [0,1]$, $(\tilde{d})$ $\pi\tilde{N}(\bar{\Omega} \times [0,1])$ *is bounded,* $(\tilde{e})$ $\phi[K_P(I-Q)\tilde{N}(\bar{\Omega} \times [0,1])] < \infty$ *and* $K_P(I-Q)\tilde{N}$ *is* ϕ*-condensing* (f) *for each* $\lambda \in [0,1]$,

$$0 \notin [L - \tilde{N}(\cdot,\lambda))(D(L) \cap \partial\Omega).$$

Then, $D[(L,\tilde{N}(\cdot,\lambda)),\Omega]$ *is independent of* λ *in* $[0,1]$.

Note that ϕ, P, Q, K_P are the same as given in assumption (d).

Proof. It is an easy consequence of Definition 7.23 and the corresponding property of topological degree of an ultimately compact vector field (Petryshyn and Fitzpatrick (Petryshyn and Fitzpatrick (1974)). □

Theorem 7.31 *If O is a symmetric bounded neighbourhood of the origin and N is odd ($N(-x) = -N(x)$ for all $x \in O$) such that $L(x) \notin N(x)$ for all $x \in \partial O \cap D(L)$, then $D[(L,N),O] \neq \{0\}$.*

Proof. Note that, how P, Q, K_P, Ψ, and R_Ψ are all linear, the condition on N implies that $\tilde{M}_\Psi$ is also odd. Thus, by the corresponding property of topological degree of an ultimately compact vector field (see Petryshyn and Fitzpatrick (1974)) and Definition 7.23, it follows that $D[(L,N),O] \neq \{0\}$. □

7.4.3 *Application to Multitivalued Boundary Value Problem for Elliptic Partial Differential Equation*

Let $G \subset \mathbf{R}^n$ be a bounded domain whose boundary ∂G is a C^∞-manifold. We will consider real-valued functions of the following type: $u : G \to \mathbf{R}$. For a multi-index $\alpha = (\alpha_1,\ldots,\alpha_n)$ and a function $u : G \to \mathbf{R}$ the symbol

$$D^\alpha u = D^{|\alpha|}u/(\partial_1^\alpha x_1 \cdots \partial_n^\alpha x_n)$$

will denote the partial derivative of u (if it exists) of the order $|\alpha| = \alpha_1 + \cdots + \alpha_n$. Let $C^m(G)$ be a space of all functions u from G into $\mathbf{R}$ which are continuous together with derivatives $D^\alpha u$, $|\alpha| \leq m$, and let

$$\tilde{C}_p^m(G) = \{u \in C^m(G) : (\Sigma_{|\infty| \leq m} \int_G |D^\alpha u(x)|^p dx)^{1/p} < \infty\}$$

for $1 < p < \infty$. In the space $\tilde{C}_p^m(G)$ we define the norm as follows:

$$||u||_{m,p} = (\Sigma_{|\infty| \leq m} \int_G |D^\alpha u(x)|^p dx)^{1/p}.$$

By $H_{m,p}(G)$ we will denote the Sobolev space which is the completion of $\tilde{C}_p^m(G)$ with respect to the norm $||\cdot||_{m,p}$. By $C_0^\infty(G)$ we will denote the space of all functions $u \in C^\infty(G) = \cap_{m=0}^\infty C^m(G)$, which have a compact support in G.

Let $u, v : G \to \mathbb{R}$ be two integrable functions. We say that the function v is the α-th weak derivative of u if, for every $f \in C_0^\infty(G)$,

$$\int_G u(x) D^\alpha f(x) dx = (-1)^{|\alpha|} \int_G v(x) f(x) dx.$$

Then we write $D^\alpha(u) = v$.

Let $L^P(G), p > 1$, he the Banach space of all measurable functions $y : G \to \mathbb{R}$, for which $\int_G |y(u)|^P du < \infty$, with the norm

$$||y||_p = (\int_G |y(u)|^P du)^{1/P}.$$

The following two facts are well known (see Nirenberg (1974)).

$$H_{m,p}(G) = \{u \in L^P(G); \tilde{D}^\alpha u \in L^P(G), |\alpha| \leq m\}. \tag{7.149}$$

Let α be such that $|\alpha| \leq m$. The mapping $\tilde{D}^\alpha : H_{m,p}(G) \to L^p(G)$ is a continuous extension of the mapping $D^\alpha : C^m(G) \to C^0(G)$. Let $C^m(\bar{G})$ be the space of all functions u from G into $\mathbb{R}$ which are uniformly continuous together with derivatives $D^\alpha(u)$ for $|\alpha| \leq m$.

In the space $C^m(\bar{G})$ we define a norm putting

$$|u|_m = \Sigma_{|x|<n} \sup_{x \in G} |D^\alpha u(x).$$

Let $C^{m+\mu}(\bar{G})$, $0 < \mu < 1$, he the Holder space with the norm

$$|u|_{m+\mu} = |u|_m + \Sigma_{|\alpha|=m} \sup \frac{|D^\alpha u(x) - D^\alpha u(y)|}{|x-y|^\mu} : x, y \in G, x \neq y.$$

We have $C^{m+\mu}(\bar{G}) \subset C^m(\bar{G})$. Note the following (see Nirenberg (1974)).

The embedding $i : C^{m+\mu}(\bar{G}) \to C^m(\bar{G})$, given by $i(u) = u$, is a completely continuous mapping. (7.150)

From the Sobolev embedding theorem (see Nirenberg (1974)) Akashi (1988) obtained the following:

Proposition 7.19 *Let $p > n$. Then, for $\mu = 1 - n/p$, the mapping $j : H_{m,p}(G) \to C^{m-1+\mu}(\bar{G})$ given as follows: $j(\tilde{u}) = u$, $u \in C^{m-1+\mu}(\bar{G})$ and $u(x) = \tilde{u}(x)$ a.e. on G, is well defined and it is a continuous mapping.*

Let $A_p : H_{m,p}(G) \to L^p(G)$ be an elliptic operator given by

$$A_p(u)(x) = \Sigma_{|\alpha|\leq m} a_\alpha(x) D^\alpha u(x),$$

where

$$a_\alpha(\cdot) \in \bigcap_{m=0}^{\infty} C^m(\bar{G}) = C^\infty(\bar{G})$$

and let $B_j : C^{m-1}(\bar{G}) \to C^0(\bar{G}); j = 1, 2, \ldots, k$ be a differential boundary operator given by

$$B_j(u)(x)\Sigma_{|\alpha|\leq m_j} b^j_\alpha(x) D^\alpha u(x),$$

where $m_j < m$, $b^j_\alpha(\cdot) \in C^\infty(\bar{G})$ for $j = l, 2, \ldots, k$ and $|\alpha| < m_j$.

For a multivalued mapping $f : \bar{G} \times \mathbb{R} \times \mathbb{R} \to 2^R$ Akashi (1988) formulated the following boundary value problem:

$$\begin{aligned} &u \in C^{m-1}(\bar{G}) \\ &A_p(u)(x) \in f(x, u(x), D^\beta u(x)) \quad \text{a.e. on } G, |\beta| < m, p > n \\ &B_j(u)(x) = 0 \quad \text{for } x \in \partial G, j = 1, 2, \ldots, m/2. \end{aligned} \tag{7.151}$$

Definition 7.24 We say that a multivalued mapping $f : \bar{G} \times \mathbb{R} \times \mathbb{R} \to 2^R$ satisfies the Caratheodory conditions if: (C_1) for each pair $(u, v) \in \mathbb{R}^2$, the mulltivalued mapping $f(\cdot, u, v)$ is measurable, i.e., for every open set $u \subset \mathbb{R}$, the set $f^{-1}(U, u, v) = \{x \in \bar{G} : f(x, u, v) \cap U \neq \emptyset\}$ is Lebesque measurable; (C_2) for each $x \in \bar{G}$, the multivalued mapping $f(x, \cdot, \cdot)$ is u.s.c.

Theorem 7.32 *Suppose that the multivalued mapping f stated in the boundary value problem* (7.151) *satisfies:*

(*i*) *the Caratheodory conditions (C_1), (C_2);*
(*ii*) *for each $(x, u, v) \in \bar{G} \times \mathbb{R} \times \mathbb{R}$, $f(x, u, v)$ is a convex set;*
(*iii*) *$|f(x, u, v)| \leq g(x)(1 + |u| + |v|)^\rho$, for each $(x, u, v) \in \bar{G} \times \mathbb{R} \times \mathbb{R}$ and some $g \in L^p(G)$ and some ρ, $\rho < 1$. Moreover, suppose that $R(A_p) = L^p(G)$ and $j(kerA_p) \subset C^\infty(G)$, where A_p is the elliptic operator stated above and that the system*

$$\begin{aligned} &A_p(u)(x) = 0 \\ &B_j(u)(x) = 0 \quad \textit{for } x \in \partial G, j = 1, 2, \ldots, m/2 \end{aligned} \tag{7.152}$$

admit only a finite number of linearly independent solutions. Then, the problem (7.151) *admits a solution.*

Proof. Let us put $X_1 = C^{m-1}(\bar{G})$; $X_2 = L^p(G)$, $p > n$ and

$$X = \{u \in C^{m-1}(\bar{G}) : H_{m,p}(G), B_j(u)_{/\partial G} = 0, j = 1, 2, \ldots, m/2\}$$

Now, let us specify the following:

$$\begin{aligned} &N : X_1 \to 2^{X_2} \\ &N(u) = \{v \in L^p(G); v(x) \in f(x, u(x), D^\beta u(x)) \quad \text{a.e. on} \quad G\} \end{aligned} \tag{7.153}$$

and

$$L : D(L) = X \to X_2; \;\; L(u) = A_p(u).$$

So, the problem (7.151) is equivalent to the equation

$$L(u) \in N(u).$$

The mapping L defined above is a linear Fredholm mapping with ind$L \geq 0$. Let P he a projection in $< X_1, |\cdot|_{m-1} >$ such that $R(P) = \ker L$. By the Banach theorem, the mapping $(L_{/X_0})^{-1} : X_2 \to < X, ||\cdot||_{m,p} >$ is continuous, where $X_1 = \ker L \bigoplus X_0$ with X_0 a closed vector subspace of $< X_1, |\cdot|_{m-1} >$. In virtue of (7.151) and Proposition 7.19, we see that the mapping K_p is a completely continuous mapping from X_2 into $< X_1, |\cdot|_{m-1} >$, where K_p is the mapping stated in the following commutative diagram:

$$\begin{array}{ccc} X_2 = L^p(G) & \xrightarrow{\quad (L_{/X_0})^{-1} \quad} & \langle X, \|\cdot\|_{m,p} \rangle \\ & \searrow & \downarrow j_{/X} \\ & & \langle X, \|\cdot\|_{m-1+\mu} \rangle \\ & K_p = i_{/X} \cdot j_{/X} \cdot (L_{/X_0})^{-1} & \downarrow i_{/X} \\ & & \langle X, \|\cdot\|_{m-1} \rangle \end{array} \tag{7.154}$$

Let $T : X_1 \to C(\bar{G}, \mathbb{R}^2)$ be a linear continous mapping given by $T(u) = (u, D^\beta u)$ for every $u \in X$, and let $S : C(\bar{G}, \mathbb{R}^2) \to 2^{X_2}$ be the multivalued mapping defined by

$$S(u) = \{Z \in L^p(G); Z(x) \in f(x, u(x), v(x)) \quad \text{a.e. on} \quad G\}.$$

Since by condition C_3, $N = S \circ T$ maps a bounded set into a bounded set, the multivalued mapping K_pN is compact since K_p is a compact linear mapping. Furthermore, by conditions (i) and (ii) we have that K_pN is u.s.c. and for each $u \in X$, $K_pN(u)$ is a convex set. The closedness of $K_pN(u)$ for each $u \in X$ follows by the upper semicontinuity and the compactness of K_pN. Now, by the surjectivity of L, the projector Q stated in the Definition 7.23 is the null operator. Then,

$$K_p(l - Q)N = K_pN.$$

Still by the surjectivity of L, coker$L = \{0\}$ and then, $\tilde{M}_\Psi = K_pN$, where $\tilde{M}_\Psi$ is the multivalued mapping stated in the Theorem 7.28. Thus, if u is a solution of

$L(u) \in N(u)$, then $u \in K_pN(u)$, and so, by condition (iii) we have

$$|u|_{m-1} \leq D(1 + |u|_{m-1})^{\rho}, \;\; \rho < 1,$$

where C is a positive constant. This implies that there exists a positive constant $\bar{C}$ such that if u is a solution of $L(u) \in N(u)$ then,

$$|u|_{m-1} \leq \bar{C}.$$

So, if we take $\delta > \bar{C}$, then, for each $u \in \partial B(O, \delta)$, we have

$$u \notin K_pN(u).$$

Let M be the multivalued mapping defined by

$$M(\lambda, u) = \lambda K_pN(u), \;\; \lambda \in [0,1], u \in X_1,$$

It is easily seen that if $u \in \partial B(0, \delta)$ then we have $u \notin M(\lambda_0, u)$ for each $\lambda_0 \in [0,1]$. So, by the homotopy property of topological degree, we have

$$d(I - K_pN, B(0,\delta), 0) = d(I, B(0,\delta), 0).$$

Now, it is well known that $d(I, B(0,\delta), 0)) = 1$ and so, by the existence property of coincidence degree, we have that the equation $L(u) \in N(u)$ admit a solution, i.e., the boundary value problem (7.151) admit a solution. □

Remark 7.25 Comparing the application above with the application 5 of Pruszko (1981), Akashi (1988) noted that while in Pruszko (1981) the mapping S was considered injective, in the case of Akashi (1988) this was exchanged by the more general condition dim ker $S < \infty$. Akashi (1988) also observed that the hypothesis (iii) of Theorem 7.32 in Akashi (1988) was more general than that of the condition (C') stated in Theorem 5.6 of Pruszko (1981).

Note that the results of this section have been written and presented after a thorough revision of some results of an article in Akashi (1988).

7.5 Applications of Equivalence Theorems with Single-Valued Mappings: An Approach to Non-Linear Elliptic Boundary Value Problems

Let D be a bounded domain in $\mathbb{R}^n$ and let L be a formally self-adjoint elliptic second order operator on $\bar{D}$ with real valued coefficients which are measurable and bounded functions on D. Assume that ker$L$ (kernel of L) is one dimensional and spanned by w. Let $f : D \times \mathbb{R} \to \mathbb{R}$ be a function such that $f(x, u(x)) \in L^2(D)$ for each $u \in L^2(D)$ and the mapping $u(x) \to f(x, u(x))$ is continuous from $L^2(D)$ into $L^2(D)$. Assume that there exist functions $h_+ \in L^2(D)$ such that

$$lim_{t \to +\infty} f(x,t) = h_+(x)$$

and that there exists a constant k such that

$$|f(x, u(x)), w)| \leq k \text{ for all } u \in L^2(D).$$

Under these conditions Landesman and Lazer (1970) have proved that the Dirichlet problem

$$(L(u))(x) = f(x, u(x)), \quad x \in D, \tag{7.155}$$

$$u(x) = 0, x \in \partial D(\text{boundary of } D) \tag{7.156}$$

has a weak solution if

$$[\int_{[w>0]} h_+(x)|w(x)|dx - \int_{[w<0]} h_-(x)|w(x)|dx] \tag{7.157}$$

$$\times [\int_{[w>0]} h_-(x)|w(x)|dx - \int_{[w<0]} h_+(x)|w(x)|dx] < 0 \tag{7.158}$$

where $[w \geq 0] = \{x \in D : w(x) \geq 0\}$. Moreover if in addition to the above conditions f, h_+ and h_- satisfy the condition

$$h_-(x) < f(x,t) < h_+(x) \quad \text{for a.e. } x \text{ in } D \text{ and } \forall t \in \mathbb{R}$$

then the condition (7.157) is also necessary for the existence of a solution of the boundary value problem (7.155). This result has been extended by Williams (1970) and also by Browder (mentioned as unpublished manuscript in Nirenberg (1970)) to the case of a higher order formally self-adjoint elliptic operator L with arbitrary finite dimensional ker$L$. In fact, various other aspects of the paper of Landesman and Lazer (1970) have been exploited in different directions by Hess (1974), Nirenberg (1970), Nirenberg (1971)), Schechter (1973), Figueiredo (1974), Gaines and Mawhin (1977), Fućik (1974b)), Fućik (1974a), Fućik and Nećas (1975), Hetzer (1975b) and many others (see Remark 7.29). Of these the particular interest is in the generalization of Landesman and Lazer's result by Figueiredo (1974) and also by Gaines and Mawhin (1977, Theorem VIII.2, p. 156). The summary of this result is as follows:

Let $D \subset \mathbf{R}^n$ be a bounded domain. With $\alpha = (\alpha_1, \alpha_2, \ldots, \alpha_n)$, $\beta = (\beta_1, \beta_2, \ldots, \beta_n)$, $\alpha_i, \beta_i \in N$ and $|\alpha| = \sum_{i=1}^n \alpha_I$, let $a_{\alpha\beta}, 0 \leq |\alpha|, |\beta| \leq m$, be real valued $L^\infty(D)$-functions. Also let $a_{\alpha\beta} = a_{\beta\alpha}$ and moreover $a_{\alpha\beta}$ with $|\alpha| = |\beta| = m$ be uniformly continuous. Assume that there exists a constant c such that $\sum_{|\alpha|=m, |\beta|=m} a_{\alpha\beta}(x)\zeta^\alpha\zeta^\beta \geq c|\zeta|^{2m}$ $\forall \zeta \in \mathbf{R}^n$ and $x \in D$. Let $f : D \times \mathbf{R} \to R$ be a function satisfying the Caratheodory condition. Assume that there are constants $\lambda > 0, \delta \in [0, 1)$ and a function $b \in L^2(D)$ such that

$$|f(x,t) \leq \lambda|t|^\delta + b(x) \quad \text{for a.e. } x \in D \text{ and } \forall t \in \mathbf{R},$$

and that there are functions $H_{+-} \in L^{2/(1-\delta)}(D)$ such that

$$\lim_{t \to +-\infty} \frac{f(x,t)}{|t|^{\delta}} = h_{+-}(x) \quad \text{for all } x \in D.$$

Under these conditions it is proved in Figueiredo (1974) and Gaines and Mawhin (1977) that there exists a $u \in H_0^m$ satisfying

$$\sum_{|\alpha| \leq m, |\beta| \leq m} \int a_{\alpha\beta}(x) D^{\alpha} u(x) D^{\beta} v(x) dx = \int_D f(x, u(x)) v(x) dx \tag{7.159}$$

for all $v \in H_0^m$, that is, u is a solution of the nonlinear elliptic boundary value problem (7.159) if for every $w \in \text{Ker}L$ with $\|w\|_{L^2(D)} = 1$,

$$\int_{[w>0]} h_+(x)|w(x)|^{1+\delta} dx - \int_{[w<0]} h_-(x)|w(x)|^{1+\delta} dx > 0. \tag{7.160}$$

Moreover, if $h_(x) < f(x,t) < h_+(x)$ holds for a.e. x in D and all $t \in \mathbf{R}$, then the condition (7.160) is necessary for the existence of a solution of (7.159). To find an analogy between conditions (7.157) and (7.160) Tarafdar (1983) observed that an analogue of condition (7.157) would be the following:

For each $w \in \text{Ker}L$ with $\|w\|_{L^2(D)} = 1$,

$$[\int_{[w>0]} h_+(x)|w(x)|^{1+\delta} dx - \int_{[w<0]} h_-(x)|w(x)|^{1+\delta} dx] \tag{7.161}$$

$$\times [\int_{[w>0]} h_-(x)|w(x)|^{1+\delta} dx - \int_{[w<0]} h_+(x)|w(x)|^{1+\delta} dx] < 0. \tag{7.162}$$

In this section Tarafdar (1983) showed that the boundary value problem (7.159) has a solution if (7.161) holds. In the sequel it will also be shown that (7.159) implies (7.161). In fact Tarafdar proved this result under a more general setting (see Theorem 7.35).

It is interesting to note that de Figueiredo (Figueiredo (1974)) has proved his result by using a perturbation argument introduced by Hess (1974); and in Gaines and Mawhin (1977) the result has been proved by using the generalized Leray-Schauder continuation theorem in the context of coincidence degree and Poincaré-Bohl Theorem, while in Tarafdar (1983) Tarafdar proved his result by using a generalized Krasnosel'skii's result which is easy to prove. Tarafdar observed that many of the applications given in Gaines and Mawhin (1977) can be obtained by Krasnosel'skii's theorem as it was done in Tarafdar (1983).

We will continue to let X and Z denote two vector spaces over the same scalar field.

An operator $P : X \to X$ is said to be an algebraic projection if P is linear and idempotent, that is, $P^2 = P$. Let $P : X \to X$ and $Q : Z - Z$ be two algebraic projections. Then the pair (P, Q) is said to be an exact pair with respect to L if the sequence $X \to^P \mathrm{dom} L \to^L Z \to^Q Z$ is exact, that is, $\mathrm{Im}\, P = \mathrm{Ker}\, L$ and $\mathrm{Im}\, L = \mathrm{Ker}\, Q$. Let $K_P = L_P^{-1}$ where L_P is the restriction of L to $\mathrm{Ker}\, P \cap \mathrm{dom}\, L$. Clearly $K_P : \mathrm{Im}\, L \to \mathrm{dom}\, L \cap \mathrm{Ker}\, P$ is a linear mapping. For each exact pair (P, Q) with respect to L we have the following:

$$PK_P = 0. \tag{7.163}$$

For each $y \in \mathrm{Im} L$,

$$LK_P(y) = L(I - P)K_P(y) = L_P(l - P)K_P(y) = y, \tag{7.164}$$

where I is the identity mapping on X.

For each $x \in \mathrm{dom}\, L$,

$$K_P L(x) = K_P L(I - P)(x) = K_P L_P(I - P)(x) = (I - P)(x). \tag{7.165}$$

$\mathrm{Coker}\, L$ denotes the quotient space $Z/\mathrm{Im}\, L$ and $\Pi : Z \to \mathrm{Coker}\, L$ the canonical surjection. Obviously

$$Q(z) = 0 <=> z \in \mathrm{Im} L <=> \pi(z) = 0. \tag{7.166}$$

We will also use the well known fact that since $\mathrm{Im}\, L = \mathrm{Ker}\, Q$, the restriction $\hat{\Pi}$ of Π to $\mathrm{Im}\, Q$ is an algebraic isomorphism. We should mention that the same symbol I will be used for the identity mapping on X as well as on Z. We believe that this will create no confusion to the reader and will be clearly understood from the context.

Equivalence Theorem

In what follows the following theorem, which is only a variant of a result due to Mawhin (see Gaines and Mawhin (1977), p. 13) and is also a particular case of equivalence Theorem 7.37, will serve as a main tool.

Theorem 7.33 *Let X and Z be two vector spaces over the same scalar field and Ω be a subset of X. Let $L : \mathrm{dom}\, L \subset X \to Z$ be a linear mapping and $N : \Omega \to Z$ be a mapping which is not necessarily linear. Further assume that there exists a mapping $\psi : \mathrm{Coker}\, L \to \ker L$ such that $\psi^{-1}(0) = \{0\}$.*

Then x is a solution of the operator equation

$$L(x) = N(x)$$

if and only if x is a fixed point of the mapping $M : \Omega \to X$ defined by

$$M(x) = P(x) + \psi\pi N(x) + K_P(I - Q)N(x), \quad x \in \Omega,$$

where (P, Q) is an exact pair of projections.

Suppose that X is a metric space and A is a bounded subset of X. Then the measure of non-compactness of A, denoted by $\alpha(A)$, is defined by $\alpha(A) = \inf\{\epsilon > 0,\ A$ can be covered with a finite number of sets of diameter less than $\epsilon\}$. A continuous mapping $f : X \to Y$ of a metric space X into a metric space Y is said to be a k-set contraction if for each bounded subset A of X, $\alpha(f(A)) \leq k\alpha(A)$ where k is a nonnegative real number.

For properties of measure of non-compactness and the degree theory of k-set contraction mapping f with $0 \leq k < 1$ we refer to Lloyd (1978). In Tarafdar (1983) it was observed that the results presented in Tarafdar (1983) would also hold if the above measure of non-compactness was replaced by ball measure (for definition see Lloyd (1978), p. 93).

Throughout the rest of this section, unless otherwise stated, X and Z will denote real Banach spaces.

Lemma 7.9 (Generalized Krasnosel'skii Theorem) *Let $L : dom L \subset X \to Z$ be a Fredholm mapping of index zero (that is L is linear and dimension $Ker L =$ dimension $coker L < \infty$), $\Omega \subset X$ a bounded open set containing the origin and symmetric with respect to the origin and $N : cl\Omega \to Z$ a mapping satisfying the following conditions ($cl\Omega$ denotes the closure of Ω):*

(i) N is continuous and $N(cl\Omega)$ is bounded;

(ii) $K_P(I-Q)N$ is a k-set contraction with $0 \leq k < 1$ where (P, Q) is an exact pair of continuous projections with respect to L, which always exists as L is a Fredholm mapping of index zero. Then if

$$(L-N)(x) \neq \mu(L-N)(-x) \tag{7.167}$$

for every $\mu \in [0,1]$ and every $x \in \partial\Omega \cap domL$ where $\partial\Omega$ denotes boundary of Ω, there exists solution $x_0 \in \Omega$ of the equation $L(x) = N(x)$.

Proof. Although the proof is similar to the one given in Gaines and Mawhin Gaines and Mawhin (1977) for compact situation we include the proof for the reason that this is basic for the results of this section and we can avoid coincidence degree.

We define $N : \mathrm{cl}\Omega \times [0,1] \to Z$ by

$$\hat{N}(x,t) = (1+t)^{-1}[N(x) - tN(-x)],\ \ x \in \mathrm{cl}\Omega, t \in [0,1].$$

We first prove that $K_P(I-Q)\hat{N}$ is a k-set contraction. Let D be a subset of $\mathrm{cl}\Omega \times [0,1]$. Then $D = C \times I$ where $C \subset \mathrm{cl}\Omega$ and $I \subset [0,1]$. We set $A = K_P(I-Q)N(C)$ and $B = K_P(I-Q)(-N(-C))$. Obviously then

$$K_P(I-Q)N(D) = K_P(I-Q)\hat{N}(C \times I) \subset \bigcup_{t\in I}(1+t)^{-1}(A+tB) \tag{7.168}$$

$$\subset \bigcup_{t\in[0,1]} (1+t)^{-1}((A \cup B) + t(A \cup B)) = \mathrm{co}(A \cup B) \tag{7.169}$$

where co$E$ denotes the convex hull of E. Hence using the well known properties of measure of non-compactness and the k-compactness of $K_P(I-Q)N$, we obtain

$$\begin{aligned}
\alpha(K_P(I-Q)\hat{N}(D)) &\le \alpha(\mathrm{co}(A\cup B)) &(7.170)\\
&= \alpha(A\cup B) &(7.171)\\
&= \max(\alpha(A),\alpha(B)) &(7.172)\\
&\le k\alpha(C) &(7.173)\\
&= k\alpha(C\times I) &(7.174)\\
&= k\alpha(D). &(7.175)
\end{aligned}$$

Let us now define the mapping $M : \mathrm{cl}\Omega\times[0,1]\to X$ by

$$\hat{M}(x,t) = P(x) + [\psi\Pi + K_P(I-Q)]\hat{N}(x,t),\ \ x\in \mathrm{cl}\Omega, t\in[0,1],$$

where ψ is a linear homeomorphism of coker$L$ onto Ker$L$ and $\Pi : Z\to$ coker$L$ is the natural surjection. Since P and $\psi\Pi\hat{N}$ are 0-set contractions, it follows from what we have proved above that $\hat{M}$ is a k-set contraction. Also $0\neq(I-\hat{M}(\cdot,t))\partial\Omega$, for otherwise by virtue of the Equivalence Theorem 7.33 we will have $x\in\partial\Omega$ and $t\in[0,1]$ such that $L(x)=\hat{N}(x,t)$, that is, $(L-N)(x)=t(L-N)(-x)$, contradicting (7.167). Hence by homotopy invariance theorem of degree theory for k-set contraction mappings, $d(I-\hat{M}_t,\Omega,0)$, the degree of $I-\hat{M}_t$ on Ω over 0, is independent of t where $\hat{M}_t(x)=\hat{M}(x,t)$. Thus $d(I-\hat{M}_0,\Omega,0)=d(I-\hat{M}_1,\Omega,0)\neq 0$ as $M_1(\cdot)=M(\cdot,1)$ is an odd mapping. Hence $d(I-\hat{M}_0,\Omega,0)$ being nonzero, there is $x_0\in\Omega$ such that $x=\hat{M}_0(x)=\hat{M}(x,0)=P(x)+[\psi\Pi+K_P(I-Q)N(x)$. By Theorem 7.33 $L(x_0)=N(x_0)$. □

Remark 7.26 The above lemma has also been obtained by Hetzer (1975b) in a similar form in the context of coincidence of degree.

Tarafdar assumed in this subsection that, unless otherwise stated, the kernel of the linear operator L is nonzero.

Theorem 7.34 *Let $L : DomL\subset X\to Z$ be a Fredholm mapping of index zero and $F : X\to Z$ be a continuous mapping (possibly nonlinear) which maps a bounded set into a bounded set. Let (P,Q) be an exact pair of continuous projections with respect to L. Assume the following:*

(*i*) *$K_P(I-Q)F$ is a k-set contraction on each closed bounded subset of X; with $0\le k<1$.*

(*ii*) *There exists $\delta\in[0,1)$, $\lambda\ge 0$ and $\nu\ge 0$ such that for each $x\in X$, $\|K_P(I-Q)F(x)\|\le\lambda\|x\|^\delta+\nu$.*

(*iii*) *For each $R>0$, there exists $t_0>0$ such that for all $t\ge t_0$, $\forall v\in KerP\cap B(R)$ and $\forall w\in KerL\cap\partial B(1)$:*

$$Q(F(tw+t^\delta v)-\mu F(-tw-t^\delta v))\neq 0 \quad \text{for all} \quad \mu\in[0,1]$$

where $\partial B(x)=\{x\in X:\|x\|=s\}$ and $B(s)=\{x\in X:\|x\|\le s\}$.

Then for each $z \in ImL$, *the equation* $L(x) = F(x) + z$ *has at least one solution.*

Proof. The proof is similar to the one of Theorem VII.4 in Gaines and Mawhin (1977), p. 141. For $z \in \operatorname{Im} L$, we define $N : X \to Z$ by

$$N(x) = F(x) + z, \ \ x \in X.$$

Let us assume that for some $x \in X$ and $\mu \in [0,1]$ we have $(L - N)(x) = \mu(L - N)(-x)$, that is,

$$L(x) - F(x) - z = \mu(-L(x) - F(-x) - z)$$

or

$$(1+\mu)L(x) = F(x) - \mu F(-x) + (1-\mu)z.$$

Now considering the direct sum decomposition $Z = \operatorname{Im}L \oplus Q(Z)$ we obtain from above

$$(1+\mu)L(x) = (I - Q)(F(x) - \mu F(-x)) + (1-\mu)z \tag{7.176}$$

and

$$Q(F(x) - \mu F(-x)) = 0. \tag{7.177}$$

From (7.176) and (7.165) we have

$$(1+\mu)(I - P)(x) = (1+\mu)K_P L(x) \tag{7.178}$$

$$= K_P(I - Q)(F(x) - \mu F(-x)) + (1-\mu)K_P z. \tag{7.179}$$

Using this and (ii) we get

$$(1+\mu)\|(I-P)(x)\| \leq \lambda\|x\|^{\delta} + \mu\lambda\| - x\|^{\delta} + (1-\mu)\|K_P z\| + (1+\mu)\nu$$

which yields

$$\|(I-P)(x)\| \leq \lambda\|x\|^{\delta} + \|K_P z\| + \nu. \tag{7.180}$$

Let $u = P(x)$ and $v = (1 - P)(x)$ and $r = \|K_P z\| + \nu$. Then (7.180) reduces to

$$\|v\| \leq \lambda\|u + v\|^{\delta} + r. \tag{7.181}$$

Let us assume that $\|u\| \neq 0$. Then from (7.181) we obtain

$$\frac{\|v\|}{\|u\|^{\delta}} \leq \lambda(1 + \frac{\|v\|}{\|u\|})^{\delta} + \frac{r}{\|u\|^{\delta}} \tag{7.182}$$

$$\leq \lambda + \frac{\lambda\delta}{\|u\|^{1-\delta}}\frac{\|v\|}{\|u\|^{\delta}} + \frac{r}{\|u\|^{\delta}}, \tag{7.183}$$

where the last inequality is obtained by applying the Mean Value Theorem to the real valued function $f(t) = (1+t)^\delta$. Let $t_1 > 0$ be the number such that whenever $\|y\| \geq t_1$, $y \in X$, we have

$$\frac{\lambda\delta}{\|y\|^{1-\delta}} \leq \frac{1}{2}.$$

Thus if $\|u\| \geq t_1$, (7.182) implies

$$\frac{\|v\|}{\|u\|^\delta} \leq 2(\lambda + rt_1{}^{-\delta} = R, \quad \text{say.}$$

Let $V = \{y \in \text{Ker}P : \|y\| \leq R\}$. Now by (iii) there exists $t_0 > 0$ such that for all $t \geq t_0$, $w \in \text{Ker}L \cap \partial B(1)$ and $y \in V$,

$$QF(tw + t^\delta y) - \mu QF(-tw - t^\delta y) \neq 0. \tag{7.184}$$

Therefore if $\|u\| \geq \max(t_1, t_0)$, we have $\|v\|/\|u\|^\delta \in V$ and $w = u/\|u\| \in \text{Ker}L \cap \partial B(1)$, and by (7.184),

$$QF(x) - \mu QF(-x) = QF(u+v) - \mu QF(-u-v) \tag{7.185}$$

$$= QF(\|u\|\frac{u}{\|u\|} + \|u\|^\delta \frac{\|v\|}{\|u\|^\delta}) - \mu QF(-\|u\|\frac{u}{\|u\|} - \|u\|^\delta \frac{\|v\|}{\|u\|^\delta}) \tag{7.186}$$

$$\neq 0 \tag{7.187}$$

which contradicts (7.177). Hence we can conclude that we have always $\|u\| < \max(t_0 t_1) = t$. Now (7.181) implies that

$$\|v\| \leq \lambda(t + \|v\|)^\delta + r. \tag{7.188}$$

Let $\bar{t}$ be the unique positive solution of the equation $\alpha - \lambda(t+\alpha)^\delta - r = 0$. Then (7.188) implies that $\|v\| \leq \bar{t}$. Thus we have obtained a priori bound of x, namely

$$\|x\| \leq t + \bar{t}.$$

Let $\bar{R} > (t + \bar{t})$ be any positive real number and $\Omega = \{x \in X : \|x\| < R\}$. Now an application of Lemma 7.9 to the triple (L, N, Ω) proves the theorem. □

Corollary 7.34.1 *Let H be a Hilbert space and $L : domL \subset H \to H$ be a closed linear mapping with dense domain and closed range with the property that $KerL = KerL^*$ or equivalently $Im\,L = (Ker\,L)^\perp$ and dim $Ker\,L < \infty$ (note that a closed self-adjoint operator L on H with dim $Ker\,L < \infty$ satisfies all these conditions). Let $F : H \to H$ be a continuous mapping which maps a bounded set into a bounded set. Noting that L is a Fredholm mapping of index zero assume that (I) and (ii) of Theorem 7.34 hold. Further assume that*

(a) for each $R > 0$, there exists $t_0 > 0$ such that for all $t \geq t_0$, $w \in KerL \cap \partial B(1)$, $v \in KerP \cap B(R)$ and $\mu \in [0,1]$:

$$(F(tw + t^\delta v) - \mu F(-tw - t^\delta v), w) \neq 0,$$

where $(\cdot,\cdot)$ denotes the inner product on H.

Then the equation $L(x) = F(x) + Z$ has a solution for every $z \in ImL$.

Proof. Clearly $H = \operatorname{Im} L \oplus \operatorname{Ker} L$ and $\operatorname{Ker} P = \operatorname{Im} L$ where P is the orthogonal projection of H onto $\operatorname{Ker} L$. Thus we can take $Q = P$. Let $\dim \operatorname{Ker} L = n$ and $w_1, w_2, \dots w_n$ be an orthonormal basis of $\operatorname{Ker} L$. Thus for each $x \in H$,

$$P(x) = Q(x) = \sum_{i=1}^{n}(x, w_i)w_i.$$

We need only to verify the condition (iii) of Theorem 7.34. Let $R > 0$ be given. Then there exists $t_0 > 0$ satisfying the condition (a) above. We claim that t_0 satisfies (iii) of Theorem 7.34.

Let $t \geq t_0$, $w \in \operatorname{Ker} L \in \partial B(1)$, $v \in \operatorname{Ker} P \cap B(R)$ and $\mu \in [0,1]$. Let $w = \sum_{i=1}^{n} \alpha_i w_i$. Then

$$(Q(F(tw + t^\delta v) - \mu F(-tw - t^\delta v), w) \tag{7.189}$$

$$= (\sum_{i=1}^{n}(F(tw + t^\delta v) - \mu F(-tw - t^\delta v), w_i)w_i, \sum_{i=1}^{n} \alpha_i w_I) \tag{7.190}$$

$$= (F(tw + t^\delta v) - \mu F(-tw - t^\delta v), w) \neq 0 \quad \text{by (a)}. \tag{7.191}$$

This implies that $Q(F(tw + t^\delta v) - \mu F(-tw - t^\delta v)) \neq 0$. Hence all the conditions of Theorem 7.34 are fulfilled. □

Remark 7.27 If the condition (a) of the above corollary is replaced by (a)′ for each $R > 0$ there exists $t_0 > 0$ such that for all $t > t_0$, $w \in \operatorname{Ker} L \cap \partial B(1)$ and $v \in \operatorname{Ker} P \cap B(R)$:

$$(F(tw + tv), w) > 0,$$

then the conclusion of the corollary still holds. This is because (a)′ implies (a). For clearly $(F(-tw - t^\delta v, w) = -F(-tw - t^\delta v, -w)) < 0$.

Theorem 7.35 *Let D be a bounded domain in $\mathbf{R}^n$ and $H = L^2(D)$. Let $L : domL \subset H \to H$ be a Fredholm mapping of index zero and be as in Corollary 7.34.1, (P, Q) being an exact pair of continuous projections with respect to L. Let $f : D \times \mathbf{R} \to \mathbf{R}$ be a function satisfying Carathéodory conditions, that is, for each fixed $u \in \mathbf{R}$, the function $x \to f(x,u)$ is measurable in D and for each $x \to \omega$ (a.e.) the function $u \to f(x,u)$ is continuous. Assume that there are constants $\lambda > 0$, $\delta \in [0,1)$ and function $b \in L^2(D)$ such that*

$$|f(x,u)| \leq \lambda |u|^\delta + b(x) \quad \textit{for a.e.} \quad x \in D \quad \textit{and} \quad \forall u \in \mathbf{R}. \tag{7.192}$$

Under these conditions it is well known that the mapping defined by $N(u)(x) = f(x, u(x))$, $u \in H$ maps H into H, is continuous, and maps a bounded set into a bounded set; in fact we have $\|N(u)\| \leq \lambda \|u\|^\delta + \|b\|$ where $\|\cdot\|$ denotes the norm in H, that is, L^2-norm.

Further assume that

(*i*) $K_P(I-Q)N$ *is a k-set contraction on each bounded subset of* H *with* $0 \le k < 1$;

(*ii*) *there are functions* h_+ , h_-, $\in L^{2/(1-\delta)}(D)$ *such that*

$$\lim_{u\to+-\infty} \frac{f(x,u)}{|u|^\delta} = h_{+-}(x);$$

(*iii*) *for each* $v \in KerL \cap \partial B(1)$ *and all* $\mu \in [0,1]$

$$\int_{[v>0]} h_+\|v\|^{1+\delta} - \int_{[v<0]} h_-\|v\|^{1+\delta} \neq \mu[\int_{[v>0]} h_-\|v\|^{1+\delta} - \int_{[v<0]} h_+\|v\|^{1+\delta}]$$

where $[v >< 0] = \{x \in D; v(x) >< 0\}$.

Then the operator equation $L(u) = N(u)$ *has a solution.*

Proof. By virtue of Corollary 7.34.1 it will suffice to prove that for every $R > 0$, there exists $t_0 > 0$ such that for all $t \ge t_0$, $w \in \text{Ker}L \cap \partial B(1)$, $v \in \text{Ker}P \cap B(R)$ and $\mu \in [0,1]$:

$$(N(tw + t^\delta v), w) \neq \mu(N(-tw - t^\delta v), w)$$

where $(\cdot,\cdot)$ denotes the inner product in $L^2(D)$. We prove by contradiction. Suppose that the above is false. Then there exists $R > 0)$ there exists $\{\mu_n\}$, $\mu_n \in [0,1]$, there exists $\{t_n\}$, $t_n > 0$ and $t_n \to \infty)$, there exists $\{v_n\}$, $v_n \in \text{Ker}P \cap B(R))$ and there exists $\{w_n\}$, $w_n \in \text{Ker}L \cap \partial B(1))$: $(N(t_nw_n + t^\delta v_n), w_n) = \mu_n(N(-t_nw_n - t_n^\delta v_n), w_n)$, that is,

$$\int_D [f(x, t_nw_n(x) + t_n^\delta v_n(x))w_n(x) - \mu_n f(x, -t_n^\delta w_n(x) - t_n^\delta v_n(x))w_n(x)]dx = 0. \tag{7.193}$$

Since dim Ker$L < \infty$ and $\mu_n \in [0,1]$, we may assume (going to subsequence if necessary), $\mu_n \to \mu \in [0,1]$, $w_n \to w$ in $\|\cdot\|$, $w_n + t_n^{\delta-1}v_n \to w$ in $\|\cdot\|$ and $w_n(x) + t_n^{\delta-1}v_n(x) \to w(x)$ a.e. in D. Thus for almost all $x \in [w > 0]$ (resp. $[w < 0]$) there exists a positive integer $n_0(x)$ such that for all $n \ge n_0(x)$

$$w_n(x) + t_n^{\delta-1}v_n(x) > \frac{w(x)}{2} (resp. < \frac{w(x)}{2})$$

Hence for a.e. in $[w > 0]$ (resp. $[w < 0]$), as $n \to \infty$

$$\begin{cases} t_nw_n(x) + t_n^\delta v_n(x) \to +\infty & \text{(resp. } -\infty\text{), and} \\ -t_nw_n(x) - t_n^\delta v_n(x) \to -\infty & \text{(resp. } +\infty\text{).} \end{cases} \tag{7.194}$$

Let us first consider the integral

$$\int_D t_n^{-\delta} f(x, -t_n w_n(x) - t_n^{\delta} v_n(x)) w_n(x) dx \tag{7.195}$$

$$= \int_{[w>0]} t_n^{-\delta} f(\cdot,\cdot) w(x) dx + \int_{[w<0]} t_n^{-\delta} f(\cdot,\cdot) w(x) dx \tag{7.196}$$

$$+ \int_D t_n^{-\delta} f(\cdot,\cdot)(w_n(x) - w(x)) dx. \tag{7.197}$$

The last integral of the right hand side of (2.12) tends to zero as $n \to \infty$. For

$$|\int_D t_n^{-\delta} f(\cdot,\cdot)(w_n(x) - w(x)) dx| \leq (\int_D (t_n^{-2\delta} f^2(\cdot,\cdot) dx)^{1/2} \|w_n - w\| \tag{7.198}$$

$$\leq [\lambda \|w_n + t_n^{\delta-1} v_n\|^{\delta} + t_n^{-\delta} \|b\|] \|w_n - w\|. \tag{7.199}$$

The first factor of (7.198) is derived from the growth condition (7.192). Again the sequence $\{t_n^{-\delta} f(x, -t_n w_n(x) - t_n^{\delta} v_n(x))\}$ is $\|\cdot\|$ bounded in $L^2(D)$ (being dominated by the first factor of the right hand side of (7.198) and due to the fact that $w_n + t_n^{\delta-1} v_n \to w$ in $\|\cdot\|$).

Similarly

$$\int_D t_n^{-\delta} f(x, t_n w_n(x) + t_n^{\delta} v_n(x)) dx \tag{7.200}$$

$$= \int_{[w>0]} t_n^{-\delta} f(\cdot,\cdot)(w(x) dx + \int_{[w<0]} t_n^{-\delta} f(\cdot,\cdot)(w(x) dx \tag{7.201}$$

$$+ \int_D t_n^{-\delta} f(\cdot,\cdot)(w_n(x) - w(x)) dx \tag{7.202}$$

and we can show in the same way as in the previous case that the last integral of the right hand side of (7.200) tends to zero as $n \to \infty$ and the sequence $\{t_n^{-\delta} f(x, t_n w_n(x) + t_n^{\delta} v_n(x))\}$ is $\|\cdot\|$ bounded in $L^2(D)$.

Now using (7.194)

$$\lim_{n\to\infty} \frac{f(x, t_n w_n(x) + t_n^{\delta} v_n(x))}{t_n^{\delta}} \tag{7.203}$$

$$= \lim_{n\to\infty} \frac{f(x, t_n w_n(x) + t_n^{\delta} v_n(x))}{|t_n w_n(x) + t_n^{\delta} v_n(x)|^{\delta}} |w_n(x) + t_n^{\delta-1} v_n(x)|^{\delta} \tag{7.204}$$

$$= h_{+-}(x) |w(x)|^{\delta} \quad \text{a.e. in} \quad [w >< 0] \tag{7.205}$$

and

$$\lim_{n\to\infty} \frac{f(x, -t_n w_n(x) - t_n^\delta v_n(x))}{t_n^\delta} \tag{7.206}$$

$$= \lim_{n\to\infty} \frac{f(x, -t_n w_n(x) - t_n^\delta v_n(x))}{|-t_n w_n(x) - t_n^\delta v_n(x)|^\delta} |-w_n(x) - t_n^{\delta-1} v_n(x)|^\delta \tag{7.207}$$

$$= h_{-+}(x)|w(x)|^\delta \quad \text{a.e. in } [w >< 0]. \tag{7.208}$$

Hence

$$\lim_n t_n^{-\delta}[f(x, t_n w_n(x) + t_n^\delta v_n(x)) - \mu_n f(x, -t_n w_n(x) - t_n^\delta v_n(x))] \tag{7.209}$$

$$= (h_{+-}(x) - \mu h_{-+}(x))|w(x)|^\delta \quad \text{a.e. in } [w >< 0]. \tag{7.210}$$

Now as we already know that the sequence $\{t_n^{-\delta}[f(x, t_n w_n(x) + t_n^\delta v_n(x)) - \mu_n f(x, -t_n w_n(x) - t_n^\delta v_n(x))]\}$ is $\|\cdot\|$ bounded in $L^2(D)$ and hence in $L^2([w >< 0])$, it converges weakly in $L^2([w >< 0])$ (going to a subsequence if necessary) to its pointwise limit given in (7.209). Hence as $n \to \infty$,

$$\int_{[w><0]} t_n^{-\delta}[f(x, t_n^\delta w_n(x) + t_n^\delta v_n(x)) - \mu_n f(x, -t_n w_n(x) - t_n^\delta v_n(x))]w(x)dx \tag{7.211}$$

$$\to \int_{[w><0]} +-(h_{+-}(x) - \mu h_{-+}(x))|w(x)|^{\delta+1}dx. \tag{7.212}$$

Now adding (7.200) and $-\mu_n$ times (7.198) and letting $n \to \infty$ and noting (7.193) and (7.211) we obtain

$$\int_{[w>0]} h_+(x))|w(x)|^{\delta+1}dx - \mu \int_{[w>0]} h_-(x))|w(x)|^{\delta+1} \tag{7.213}$$

$$- \int_{[w<0]} h_-(x))|w(x)|^{\delta+1} + \mu \int_{[w>0]} h_+(x))|w(x)|^{\delta+1}dx = 0 \tag{7.214}$$

which contradicts (iii), and the theorem is proved.

Corollary 7.35.1 *In Theorem 7.35 if the condition (iii) is replaced by either of the conditions*

(0) *for each $v \in KerL \cap \partial B(1)$*

$$[\int_{[v>0]h_+|v|}^{1+\delta} - \int_{[v<0]h_-|v|}^{1+\delta}][\int_{[v>0]h_-|v|}^{1+\delta} - \int_{[v<0]h_+|v|}^{1+\delta}] < 0,$$

(00) *for each $v \in KerL \cap \partial B(1)$ either*

$$0 \le [\int_{[v>0]h_-|v|}^{1+\delta} - \int_{[v<0]h_+|v|}^{1+\delta}] < [\int_{[v>0]h_+|v|}^{1+\delta} - \int_{[v<0]h_-|v|}^{1+\delta}],$$

or

$$[\int_{[v>0]h_+|v|}^{1+\delta} - \int_{[v<0]h_-|v|}^{1+\delta}] < [\int_{[v>0]h_-|v|}^{1+\delta} - \int_{[v<0]h_+|v|}^{1+\delta}] \leq 0,$$

(000) *for each* $v \in KerL \cap \partial B(1)$

$$\int_{[v>0]h_+|v|}^{1+\delta} - \int_{[v<0]h_-|v|}^{1+\delta} > 0,$$

then the equation $L(u) = N(u)$ *has a solution.*

Proof. Obviously each of (0) and (00) implies the condition (iii). If (000) holds, then proceeding exactly as in Theorem 7.35 we can show that for each $R > 0$, there exists $t_0 > 0$ such that for all $t \geq t_0$, $w \in \mathrm{Ker}L \cap \partial B(1)$, $v \in \mathrm{Ker}P \cap B(R)$: $(N(tw + t^\delta v), w) > 0$. This in turn implies that for all $\mu \in [0, 1]$

$$(N(+w + t^\delta v), w) \neq \mu(N(-tw - t^\delta v), w).$$

For $(N(-tw - t^\delta v), w) = -(N(-tw - t^\delta v), -w) < 0$. □

Corollary 7.35.2 *Let L and f satisfy all the conditions of Theorem* 7.35 *with* $\delta = 0$. *In addition assume that for a.e. in D and such* $u \in R$ *either*

$$j_-(x) \leq f(x, u) < h_+(x), \tag{7.215}$$

$$h_-(x) < f(x, u) \leq h_+(x). \tag{7.216}$$

Then the condition (iii) of Theorem 7.35 *is also necessary for the existence of a solution of the equation* $L(u) = N(u)$.

Proof. Let u be a solution of $L(u) = N(u)$. Let $0 \neq v \in \mathrm{Ker}L$. Then $\int_D F(x, u(x))v(x)dx = (N(u), v) = (L(u), v) = 0$, as $L(u) \in \mathrm{Ker}P$ and $v \in \mathrm{Ker}L = \mathrm{Im}P$ and P is orthogonal projection. Thus $\int_[v > 0] f(x, u(x))|v(x)|dx - \int_[v < 0] f(x, u(x))|v(x)|dx = 0$. We assume that (7.215) holds. Then

$$\int_{[v>0]} h_+(x)|v(x)|dx > \int_{[v>0]} f(x, u(x))|v(x)|dx \tag{7.217}$$

$$= \int_{[v<0]} f(x, u(x))|v(x)|dx \geq \int_{[v<0]} h_-(x)|v(x)|dx. \tag{7.218}$$

Thus $\int_{[v>0]} h_+|v| - \int_{[v<0]} h_-|v| > 0$. Also $\int_{[v>0]} h_-(x)|v(x)|dx \leq \int_{[v>0]} f(x, u(x)) \times |v(x)|dx = \int_{[v<0]} f(x, u(x))|v(x)|dx < \int_{[v<0]} h_+(x)v(x)dx$, that is, $\int_{[v>0]} h_-v - \int_{[v<0]} h_+v < 0$. Therefore condition (iii) holds for each $\mu \in [0, 1]$. Similarly we can show that (iii) holds under (7.216). □

In this context Tarafdar (1983) considered the Theorem VII.1 in Gaines and Mawhin (1977):

Theorem 7.36 *Let $L : domL \subset X \to Z$ be a Fredholm mapping of index zero and $F : X \to Z$ a continuous mapping which maps a bounded set into a bounded set. Let $P, Q)$ be an exact pair of continuous projections with respect to L. Assume the following:*

(*i*) *there exist real numbers $\lambda \geq 0$, $r \geq 0$ such that*

$$\|K_P(I-Q)F(x)\| \leq \lambda\|x\| + r \quad \textit{for all} \quad x \in X;$$

(*ii*) *there exist real numbers $\lambda \geq 0$ and $s \geq 0$ such that each possible solution x of the system of equations*

$$Q[F(x) - \mu F(-x)] = 0, \quad \mu \in [0,1],$$

satisfies the relation

$$\|P(x)\| \leq \alpha\|(I-P)(x)\| + s;$$

(*iii*) $\lambda(1+\alpha) < 1$; *and*

(*iv*) *$K_P(I-Q)F$ is a k-set contraction with $0 \leq k < 1$. Then for each $z \in ImL$, the equation $L(x) = F(x) + z$ has at least one solution.*

Proof. We define $N : X \to Z$ by

$$N(x) = F(x) + z, \quad x \in X.$$

We assume that for some $x \in X_0$ and some $\mu \in [0,1]$

$$(L-N)(x) = \mu(L-N)(-x).$$

Then using (7.165), conditions (i) to (iii) we can easily show that

$$\|x\| \leq \frac{(1+\alpha)r + \alpha\|K_P z\| + s}{1 - \lambda(1+\alpha)} = t, \quad \text{say.}$$

Let R be any number $> t$. Then the theorem is proved by applying Lemma 7.9 to (L, N, Ω) where $\Omega = \{x \in X : \|x\| < R\}$. □

Corollary 7.36.1 *Let $X = B(S, \mathbf{R^n})$, the space of bounded mappings of S into $\mathbf{R^n}$ with a norm satisfying*

$$\|x\| \geq \sup_{s \in S} |x(s)|, \quad x \in X.$$

Let L and F be as in Theorem 7.36. Assume the following:

(*i*)′ *there exists $\beta > 0$ such that for each $u \in KerL$ and each $s \in S$,*

$$\|u\| \leq \beta|u(s)|;$$

(*ii*)′ *there exists $r_1 > 0$ such that for each $x \in domL$ satisfying $|x(s)| \geq r_1$ for all $s \in S$,*

$$Q[N(x) - \mu N(-x)] \neq 0 \quad \textit{for all} \quad \mu \in [0,1].$$

(iii)' $\|K_P(I-Q)F(x)\| \le \lambda\|x\| + r$ *where* $0 \le \lambda < (1+\beta)^{-1}$ *and* $r \ge 0$; *and*

(iv)' $K_P(I-Q)F$ *is a k-set contraction on each closed bounded set with* $0 \le k < 1$.

Then for each $z \in ImL$, *the equation* $L(x) = F(x) + z$ *has a solution.*

Proof. By similar argument as in the proof of Theorem VII.2 in Gaines and Mawhin (1977) we can show that any solution x of the system of equations

$$Q[N(x) - \mu N(-x)] = 0 \quad \text{for all} \quad \mu \in [0,1]$$

satsfies the relation $\|P(x)\| \le \beta\|(I-P)(x)\| + \beta r$. Thus with $\alpha = \beta$ and $s = \beta r_1$ the condition (ii) of Theorem 7.36 is satisfied. By (iii)$'$ we have $\lambda(1+\alpha) < (1+\beta)/(1+\beta) = 1$ and the condition (iii) of Theorem 7.36 is satisfied. Thus the corollary follows from Theorem 7.36. □

Corollary 7.36.2 *Let* X *be as in Corollary* 7.36.1 *with additional condition that* $\|x\| = \sup_{s\in S}|x(s)|$ *when* x *is a constant mapping of* S *into* $\mathbf{R}^n$. *Let* $L : domL \subset X \to Z$ *be a linear mapping such that* ImL *is closed and of co-dimension* n *and* $KerL = \{x \in X : x$ *is a constant function*$\}$. *Clearly* L *is a Fredholm mapping of index zero. Let* F *be as in Corollary* 7.36.1. *Assume that conditions (ii)$'$ and (iv)$'$ hold and (iii)$'$ holds with* $\lambda < 1/2$. *Then for each* $z \in ImL$, *The equation* $L(x) = F(x) + z$ *has a solution.*

Proof. The condition (I$'$) of Corollary 7.36.1 holds with $\beta = 1$. Hence the corollary follows from Corollary 7.36.1. □

7.5.1 *Tarafdar's Application to Elliptic Boundary Value Problems*

In this section Ker L of the linear mapping L will be assumed to be non-zero.

Application 1. Let D be a bounded domain in $\mathbf{R}^n$. With $\alpha = (\alpha_1, \alpha_2, \ldots, \alpha_n)$, $\beta = (\beta_1, \beta_2, \ldots, \beta_n)$, $\alpha_i\beta_i \in N$ and $|\alpha| = \sum_{i=1}^n \alpha_i$, let $a_{\alpha\beta}$, $0 \le |\alpha|, |\beta| \le m$ be real valued $L^\infty(D)$-functions. Let $a_{\alpha\beta} = a_{\beta\alpha}$ and moreover with $|\alpha| = |\beta| = m$, $a_{\alpha\beta}$ be uniformly continuous. Assume that there exists a constant $c > 0$ such that

$$\sum_{|\alpha|\le m, |\beta|\le m} a_{\alpha\beta}(x)\xi^\alpha\xi^\beta \ge c|\xi|^{2m}, \quad \text{for all} \quad \xi \in \mathbf{R}^n \quad \text{and} \quad x \in D.$$

Let $H_0^m(D)$ be the completion of the space $C_0^\infty(D)$ under the norm

$$\|\phi\|_m = [\sum_{|\alpha|\le m} \int_D |D^\alpha\phi|^2]^{1/2}.$$

We define the bilinear form (in $C_0^\infty(\Omega)$)

$$a(u,v) = \sum_{|\alpha|=m, |\beta|=m} \int_D a_{\alpha\beta}(x) D^\alpha u(x) D^\beta v(x) dx.$$

Let $f : D \times \mathbf{R} \to \mathbf{R}$ be a function as considered in Theorem 7.35, that is, f satisfies the Carathéodory condition and the growth condition (7.192). We are interested in the existence of $u \in H_0^m(D)$ such that

$$a(u,v) = \int_D f(x, u(x))v(x)dx \tag{7.219}$$

for all $v \in H_0^m(D)$.

We define a linear mapping $L : \mathrm{dom}L \subset L^2(D) \to L^2(D)$ as follows:

$$\mathrm{dom}L = \{u \in H_0^m : v \to a(u,v) \ \text{ is continuous in } \ H_0^m \ \text{ with } \ L^2\text{-norm}\}.$$

Then using the fact that $H_0^m(D)$ is dense in $L^2(\Omega)$ and the representation theorem for functionals on Hilbert space we have, for each $u \in \mathrm{dom}L$, a unique $L(u) \in L^2(D)$ such that for all $v \in H_0^m(D)$, $a(u,v) = (Lu, v)$ where $(\cdot,\cdot)$ denotes the inner product in $L^2(D)$. $u \to Lu$ is a linear mapping of $L^2(D)$ into $L^2(D)$. Now clearly the existence of u satisfying (7.219) is equivalent to the existence of u satisfying the operator equation $L(u) = N(u)$.

Theorem 7.37 *Assume that the assumptions in* (7.219) *above hold and condition* (*ii*) *of Theorem* 7.35 *holds. Then the boundary value problem* (7.219) *has a solution if either the condition* (*iii*) *of Theorem* 7.35 *holds or any one of the conditions* (0) *to* (000) *of Corollary* 7.35.1 *holds.*

Proof. It is known from the L^2 theory of elliptic boundary value problems that L is a Fredholm mapping of index zero and K_P is compact. Hence the theorem follows from Theorem 7.35 and Corollary 7.35.1 respectively. □

Remark 7.28 In the case of $\delta = 0$ the necessary condition of Theorem 7.37 is exactly the same as in Corollary 7.35.2.

Remark 7.29 In Tarafdar (1983) it was observed that Theorem 7.37 was first proved by Landesman and Lazer (1970) with second order linear part, dim Ker$L = 1$ and $\delta = 0$. Williams (1970) generalized the result with Ker$L$ of arbitrary finite dimension and the higher order linear part. Theorem 7.37 was proved by Figueiredo (1974) with condition (000) of Corollary 7.35.1 by using a perturbation argument of Hess (1974), and was given in the above form with the same condition (000) by Gaines and Mawhin (1977) by using an extended form of the Leray-Schauder continuation theorem in terms of coincidence degree and Poincaré-Bohl Theorem.

Application 2. In Tarafdar (1983) the analogue of Theorem VIII.I in Gaines and Mawhin (1977) was considered as an application to Tarafdar's Corollary 7.36.2. Let D be a bounded domain in $\mathbf{R}^n$ and $a_{i,j} : D \to \mathbf{R}$ $(i, j = 1, 2, \ldots, n)$ be measurable and bounded functions. Assume that there exist constants m, M with $0 < m < M$ such that for all $x \in D$ and $\xi \in \mathbf{R}^n$,

$$m|\xi|^2 \le \sum_{i,j=1}^n a_{i,j}(x)\xi_i\xi_j \le M|\xi|^2$$

where $|\xi|$ is the Euclidean norm in $\mathbf{R}^n$ and that $f : \bar{D} \times \mathbf{R} \to \mathbf{R}$ is a continuous mapping, where $\bar{D} = \mathrm{cl}D$. Let $H^1 = H^{1,2}(D)$ be the completion of $C^1(D)$, the space of C^1-real functions in D under the Sobolev norm

$$\|u\|_{1,2} = \|u\|_{L^2(D)} + \sum_{i=1}^{n} \|D_i u\|_{L^2(D)}$$

where $D_i u = \partial u / \partial x_i$. Here we are interested to find the existence of $u \in H^1$ satisfying n

$$a(u,v) = \sum_{i,j=1}^{n} a_{i,j}(x) D_i u(x) D_j v(x) dx \tag{7.220}$$

$$= \int_D f(x, u(x)) v(x) dx \tag{7.221}$$

for all $v \in H^1$.

We now define a linear mapping $\tilde{L} : L^2(D) \to L^2(D)$ as follows:

$$\mathrm{Dom}\tilde{L} = \{u \in H^1 : v \to a(u,v) \quad \text{is continuous in} \quad H^1 \quad \text{in} \quad L^2\text{-norm.}\}$$

Since H^1 is dense in $L^2(D)$, we have by representation theorem for functionals that for each $u \in \mathrm{dom}L$ there is a unique $\tilde{L}(u) \in L^2(D)$ such that

$$a(u,v) = (\tilde{L}(u)v) \quad \text{for} \quad v \in H^1$$

where $(\cdot,\cdot)$ is as before the inner product in $L^2(D)$. Then clearly $u \to \tilde{L}(u)$ is a linear mapping of $L^2(D)$ into $L^2(D)$. Thus for $h \in L^2(D)$ the equation

$$a(u,v) = (h,v) \quad \text{for all} \quad v \in H^1 \tag{7.222}$$

is equivalent to $(\tilde{L}(u), v) = (h, v)$ for all $v \in H^1$ and hence to

$$L(u) = h \tag{7.223}$$

as H^1 is dense in $L^2(D)$.

It is well known from the classical result of L^2-theory of linear elliptic boundary value problems that under the assumptions made above (7.222) and hence (7.223) is solvable if and only if h satisfies the relation $\int_D h = 0$. In other words $h \in \mathrm{Im}\tilde{L}(u)$ if and only if $\int_D h = 0$. Thus $\tilde{L}h_1 = \tilde{L}h_2$ implies $h_1 - h_2$ is constant. In particular then $u \in \mathrm{Ker}\tilde{L}$ if and only if u is a constant function. Thus if we define the projection $\tilde{P} : L^2(D) \to L^2(D)$ by $\tilde{P}u = (\text{meas} \quad D)^{-1} \int_D u$, then $\mathrm{Ker}\tilde{L} = \mathrm{Im}\tilde{P}$ and $\mathrm{Im}\tilde{L} = \mathrm{Ker}\tilde{P}$. Now assuming sufficient regularity assumptions on D and using the regularization theory for (7.222) it can be shown (see for details Gaines and Mawhin (1977), p. 152) that if L is the restriction of $\tilde{L}$ to $L^{-1}(C_0(\bar{D}))$ and P is the restriction of $\tilde{P}$ to $(C_0(\bar{D}))$, then $L : C_0(\bar{D}) \to C_0(\bar{D})$ is a Fredholm mapping of index zero, (P, P) is an exact pair of continuous projections with respect to L, and there exists a constant $k > 0$ and a $\alpha \in (0, 1)$ such that for each $v \in \text{Ker}P$

$$\|K_P v\|_{C^{0,\alpha}(D)} \leq k\|v\|_{C^0(\bar{D})} \tag{7.224}$$

where

$$\|u\|_{C^{0,\alpha}(D)} = \sup_{x \in D} |u(x)| + \sup_{x,y \in D, x \neq y} \frac{|u(x) - u(y)|}{\|x - y\|^\alpha}$$

and $C^0(\bar{D})$ is the space of real continuous functions on $\bar{D}$. (7.224) implies that K_P is compact.

We now define $F : C^0(\bar{D}) \to C^0(\bar{D})$ by

$$(F(u))(x) = f(x, u(x)), \quad x \in \bar{D}.$$

We also assume that there exist $\beta \geq 0$ and $s \geq 0$ such that for all $x \in \bar{D}$ and $u \in \mathbf{R}$,

$$|f(x, u) \leq \beta|u| + s. \tag{7.225}$$

Clearly F maps a bounded set into a bounded set and is continuous.

Theorem 7.38 *Let $D \subset \mathbf{R}^n$ and $a_{i,j} : D \to \mathbf{R}$ $(i, j = 1, 2, \ldots, n)$ satisfy all assumptions made above. Let $f : \bar{D} \times \mathbf{R} \to \mathbf{R}$ be continuous satisfying* (7.225). *Further assume that*

(a) *$\beta < \frac{1}{4}k$ where k is as obtained in* (7.224);
(b) *there exists $R > 0$ such that for each $u \in C^0(\bar{D})$ satisfying $|u(x)| \geq \mathbf{R}$ for all $x \in \bar{D}$*

$$\int_D [f(x, u(x)) - \mu f(x, -u(x))] \neq 0 \quad \textit{for all} \quad \mu \in [0, 1].$$

Then the problem (7.220) *has a solution.*

Proof. Clearly (b) implies condition (ii)$'$ of Corollary 7.36.1. Since K_P is compact, $K_P(I - P)N$ is 0-set contraction and hence (iv)$'$ of Corollary 7.36.1 holds. Now from (7.225)

$$\|Fu\|_{c^0(\bar{D})} \leq \beta\|u\|_{c^0(\bar{D})} + s$$

and hence using (7.224)

$$\|K_P(I - P)F(u)\|_{c^0(\bar{D})} \leq 2k(\beta\|u\|_{c^0(\bar{D})} + s).$$

Thus from this and (a) it follows that condition (iii) of Corollary 7.36.1 holds with $\lambda < 1/2$. Hence proof of the theorem is complete. □

Corollary 7.38.1 *Let $D \subset \mathbf{R}^n$, $a_{i,j}(i, j = 1, 2, \ldots, n)$ and f be as in Theorem* 7.38 *and let* (a) *of Theorem* 7.38 *hold. Assume that there exists* (b$'$) *$R \geq 0$ such* *that for each* $u \in C^0(\bar{D})$ *satisfying* $|u(x)| \geq R$ *for all* $x \in \bar{D}$,

$$[\int_D f(x, u(x))dx][\int_D f(x, -u(x))dx] < 0.$$

Then the problem (7.220) *has a solution.*

Proof. (b′) implies (b) of the Theorem 7.38. □

Remark 7.30 In this section the application of the Lemma 7.9 in the case when $K_P N$ is non-compact but is of k-set contraction with $0 < k < 1$ has not been considered.

Note that the results of this section have been written and presented after a thorough revision of some results of an article in Tarafdar (1983).

7.6 Further Results in Coincidence Degree Theory

We shall start with the following result in Mawhin (1972) which is the basic for defining coincidence degree:

Proposition 7.20 *Let* X *and* Z *be as above,* $L : dom\, L \subset X \to Z$ *a linear mapping and* $N : \Omega \subset X \to Z$ *be a mapping (not necessarily a linear). Suppose that there exists a mapping* $\psi : coker\, L \to ker\, L$ *such that* $\psi^{-1}(0) = 0$.*Then* x *is a solution of the operator equation* $L(x) = N(x)$ *if and only if* x *is a fixed point of the mapping* $N : \Omega \to X$ *defined by*

$$M(x) = P(x) + \psi\pi N(x) + K_P(I - Q)N(x), \tag{7.226}$$

$x \in \Omega$ *where* (P, Q) *is an exact pair of algebraic projections with respect to* L.

Proof. This is just a variant of Proposition III.2 in (Gaines and Mawhin (1977), pp. 13–14), and the same proof applies. □

Let us now assume that X and Z are normed linear spaces over the real field. $L : \mathrm{dom}\, L \subset X \to Z$ is a Fredholm mapping of index > 0; i.e. $\mathrm{Im}\, L$ is closed, $\mathrm{dimker}\, L = m < \infty$, $\mathrm{dimcoker}\, L = p < \infty$ and $m - p > 0$, and $N : \Omega \subset X \to Z$ is a mapping where Ω is a bounded open set. Then clearly there exists an exact pair (P, Q) of continuous projections and a mapping $\psi : \mathrm{coker}L \to \mathrm{ker}L$ with $\psi_{-1}(0) = 0$. Further we assume that (a) πN is continuous and $N(Cl\Omega)$ is bounded, where $Cl\Omega$ denotes closure of Ω; and (b) $K_P(I-Q)N$ is completely continuous (i.e. $K_P(I-Q)N$ is continuous and $K_P(I-Q)N(Cl\Omega)$ is relatively compact (here I is the identify on Z).

With these assumptions it is easy to see that the mapping $M : Cl\Omega \to X$ defined by

$$M(x) = P(x) + \psi\pi N(x) + K_P(I - Q)N(x), \ \ x \in \mathrm{Cl}\Omega$$

is completely continuous.

Now if $0 \notin (L-N)(\partial\Omega \cap \text{dom}L)$ where $\partial\Omega$ denotes the boundary of Ω, then by Proposition 7.20, $0 \notin (I-M)(\partial\Omega)$ and hence Leray-Schauder degree $d(I-M,\Omega,0)$ of $I-M$ on Ω over 0 is well-defined where I is the identity in X. Gaines and Mawhin (1977) (also see Mawhin (1972)) has shown that $d(I-M,\Omega,0)$ is independent of the exact pair (P,Q) of continuous projections and depends only on L, N, Ω and ψ. $d(I-M,\Omega,0)$ is called the coincidence degree of the pair (L,N) and is denoted by $d[(L,N),\Omega]$. $d[(L,N)<\Omega]$ has all the important properties of Leray-Schauder degree. It is shown in (Mawhin (1972), p. 626), also in (Gaines and Mawhin (1977), p. 229), that if $\text{Ind}\, l = \text{index of} L > 0$, then $d((L,N),\Omega) = 0$ and this fact is then used to prove the existence theorem of the operator equation $L(x) == N(x)$ on the boundary $\partial\Omega$ when N is an odd mapping and an indication for possible application to elliptic boundary value problem is given (see chapter XII of Gaines and Mawhin (1977)).

In Tarfadar (1979), Tarafdar first mentioned an important fact that if $\text{Im}N \subset \text{Im}L$ and $\text{Ker}L \neq 0$, then $d[(L.N),\Omega] = 0$ even when $\text{Ind}L = 0$. So far as coincidence degree theory is concerned this is a negative result but as Tarafdar mentioned before it can be used to prove existence theorem on the boundary for odd mappings. In this section we shall present this result from Tarfadar (1979). In fact we shall give the proof of a more general result in Tarfadar (1979), and the above result will be a particular case of the latter.

Next, we shall state and prove the following proposition which is a special case of the Proposition 7.20:

Proposition 7.21 *Let X, Z, Ω, N be as in Proposition* 7.20. *Further assume that $Im N \subset L$, then x is a solution of the operator equation $L(x) = N(x)$ if anf only if x is a fixed point of the mapping $M : Cl\Omega \to X$ defined by*

$$M(x) = P(x) + K_p N(x), \quad x \in Cl\Omega, \tag{7.227}$$

where P is any algebraic projection $X \to ker L$. (Q has no effect because of (7.166).

Proof. First, let $X = M(x) = P(x) + K_P N(x)$. Then $L(X) = N(x)$ because of (7.164). Conversely, let $L(x) = N(x)$. Then $K_P N \cdot (x) = K_P L(x) = (I-P)(x)$ by (7.165). Thus $x = P(x) + K_P N(x)$. □

Let us now consider the following conditions:

(1) X and Z are Banach spaces and Ω is an open bounded subset of X.
(2) $L : (dom)L \subset X \to Z$ is a linear mapping such that $\ker L \neq 0$ and is complemented, i.e. there exists a continuous projection $P : X \to \ker L$.
(3) $N : \text{Cl}\Omega \to Z$ is a mapping with $\text{Im}\, N \subset \text{Im}\, L$.
(4) $K_P N$ is completely continuous.

Now under these conditions if $\ker L$ is finite dimensional, then clearly the mapping M defined by (7.227) is completely continuous where P appearing in the definition of M is the $\ll P \gg$ of condition (2).

If dim$ker = \infty$, then let X_n be a finite dimensional sub-space of $\ker L$. It is well known that there is a continuous projection $P_n : \ker L \to X_n$. It is clear that $P_nP : X \to X_n$ is a continuous projection. We define a mapping $M_n : Cl\Omega \to X$ by

$$M_n(x) = P_nP(x) + K_PN(x), \quad x \in Cl\Omega. \tag{7.228}$$

Since P_nP is continuous and has finite dimensional range and K_PN is completely continuous, it follows that M_n is completely continuous. IF dimker $< \infty$, we will take $X_n = X$ and $P_n = P$, i.e. we will regard $M_n = M$. Thus if $0 \notin (I - M_n)(\partial\Omega)$, then under the conditions (1)to (4), $d(I - M_n, \Omega, 0)$, the Leray-Schauder degree of $I - M_n$ on Ω over 0 is well-defined.

Lemma 7.10 *If x is a fixed point of M_n, then x is a fixed point of M.*

Proof. Let x be a fixed point of M_n. Then $x = M_n(x) = P_nP(x) + K_pN(x)$. Hence $P(x) = P_nP(x)$ by (7.163) (more precisely as $K_PN(x) \in \ker P$) and by the fact that $P_nP(x) \in \operatorname{Im} P$. Hence $x = P(x) + K_PN(x) = M(x)$. □

Theorem 7.39 *Assume that* (1), (2), (3) *and* (4) *hold and* $0 \notin (I - M_n)(\partial\Omega)$, *then* $d(I - M_n, \Omega,)) = 0$.

Proof. Since M_n is completely continuous and $\partial\Omega$ is closed, $(I - M_n)(\partial\Omega)$ is closed (e.g. see Nagumo (1951), Theorem 1, p. 499).[3] Again since $0 \notin (I - M_n)(\partial\Omega)$, there exists a convex neighborhood G of 0 such that $G \cap (I - M_n)(\partial\Omega) = \emptyset$. Hence

$$d(I - M_n, \Omega, y) = d(I - M_n, \Omega, 0) \quad \text{for all} \quad y \in G$$

(again see Nagumo (1951), Corollary 2, p. 505). Now since

$$I - M_n = I - P_nP - K_PN \quad \text{and} \quad \operatorname{Im}(I - P_nP) = X_1$$

where X_1 us defined by $X = X_1 \oplus X_n$ and $X_1 \cap X_n = 0$, it follows that $\operatorname{Im}(I - M_n) \subset X_1$. Again since $O \in X_n$, there exists $z \in G \cap X_n$ with $z \neq 0$. Thus

$$(I - M_n, \Omega, 0) = (I - M_n.\Omega, z) = 0 \quad \text{as} \quad z \notin \operatorname{Im}(I - M_n).$$

This completes the proof. □

Corollary 7.39.1 *Let $L : domL \subset X \to Z$ be a Fredholm mapping of index zero with $kerL \neq 0$ and assume that* (1), (3) *and* (4) *hold and* $0 \notin (L - N)(\partial\Omega \cap domL)$, *then the coincidence index* $d[(L, N), \Omega] = 0$.

Proof. We take $M = M_n$. Since $0 \notin (L - N)(\partial\Omega \cap \operatorname{dom}L)$, by Proposition 7.21 $0 \notin (I - M)(\partial\Omega)$. Now applying Theorem 7.39 and definition of $d[(L, N), \Omega]$ we obtain the corollary. □

[3]The object of citing Nagumo (1951) instead of the corresponding references in normed linear space is to indicate that the results can be extended in separated locally convex topological vector spaces in the line of Mawhin (1972).

Now we present the main result in Tarfadar (1979):

Corollary 7.39.2 *Assume that* (1), (2), (3) *and* (4) *hold. Further assume that* Ω *be an open bounded symmetric neighborhood of* 0 *and* N *is odd on* $\partial\Omega$ *i.e.*$N(-x) = -N(x)$ *for all* $x \in \partial\Omega$. *Then there is a solution* $x \in \partial\Omega$ *of the operator equation* $L(x) = N(x)$.

Proof. If $0 \in (I - M_n)(\partial\Omega)$. Then by Lemma 7.10 and Proposition 7.20 theorem will follow. Thus it would suffice to prove that $0 \in (I - M_n)(\partial\Omega)$. We suppose on the contrary that $0 \notin (I - M_n)(\partial\Omega)$. Then $d(I = M_n, \Omega, 0)$ is well-defined. Since N is odd on $\partial\Omega$ it follows that M_n is odd on $\partial\Omega$. Hence by Bosuk's theorem on Leray-Schauder degree $d(I - M_n, \Omega, 0) = 1(\text{mod}\, 2)$ which contradicts Theorem 7.39. This proves our corollary. □

The above results can also be used to prove some mapping theorems. We shall give a particular case in Tarfadar (1979):

Theorem 7.40 *Let* L *be a Fredholm mapping of index zero with* $kerL \notin 0$ *and* Ω *an open bounded neigbourhood of zero. Assume that* (1), (3) *and* (4) *hold and that* $N(0) = 0$. *If* $L - N$ *is one-to-one* (*on* $dom\, L \cap Cl\Omega$), *then* $Im\, N \not\subset Im\, L$.

Proof. If possible suppose that $\text{Im}\, N \subset \text{Im}\, L$. We can find a real number $\varepsilon > 0$ such that $\text{Cl}B_\varepsilon(0) \subset \text{Cl}\Omega$ where $B_\varepsilon(0) = \{x \in X : ||x|| < \varepsilon\}$ and $D(I_M, B_\varepsilon, 0)$ is well-defined and $= \pm 1$(see Gaines and Mawhin (1977), Theorem X.1, p. 190). Now since by hypothesis $N(\text{Cl}\Omega) \subset \text{Im}\, l$ we have $N(CL\, B_\varepsilon(0)) \subset \text{Im}\, L$. Hence considering N as a mapping restrictrd to $ClB_\varepsilon(0)$, it follows from Corollary 7.39.1, $d(I - M, B_\varepsilon \cdot (0), 0) = 0$ which is a contradiction. This proves the theorem. □

Remark 7.31 In Tarfadar (1979), Tarafdar mentioned that the result obtained in this section is not completely independent from the known result that, if $\text{ind}\, L > 0$, $d[(L < N), \Omega] = 0$. In fact, when $\text{Im}N \subset \text{Im}\, L$, one can consider the couple (L, N) as a couple of mapping between the vector spaces X and $\text{Im}\, L$ instead of between the vector spaces X and Z. It follows at once from the definition of coincidence degree that this degree is the same whatever the couple of spaces is chosen, and with the first choice, one has that the index of L, relative to the spaces X and $\text{Im}\, L$ is necessarily strictly positive as far as $\ker L \neq (0)$, so that the coincidence degree is zero by the known result.

Note that the results of this section have been written and presented after a thorough revision of some results of an article in Tarfadar (1979).

7.7 Tarafdar and Thompson's Theory of Bifurcation for the Solutions of Equations Involving Set-Valued Mappings

In this section we shall present some results from Tarafdar and Thompson (1985) on the theory of bifurcation for the solutions of equations involving set-valued mappings.

In the Leray-Schauder degree theory the study of the index of an isolated fixed point of a compact mapping is a basic tool for the calculation of the degree. A fundamental and well-known result of Leray and Schrauder (1934), commonly known as the Leray-Schauder principle, is that the topological index at zero of an invertible linear perturbation of the identity $I - A$ in a Banach space can be expressed in terms of multiplicity of the characteristic values of A lying in the open interval $(0, 1)$. Krasnosel'skii (Krasnosel'skii (1963)) employed this result as a basic tool to develop his bifurcation theory for equations of the form:

$$x - \mu Ax - R(x, \mu) = 0 \tag{7.229}$$

in a real Banach space X, where $A : X \to X$ is a linear compact mapping, $R : \bar{\Omega} \times R \to X$ is compact mapping, $R(x, \mu)$ is of $o(\|x\|)$ uniformly in μ in compact intervals and Ω is a bounded open neighborhood of zero in X. Rabinowitz (1971) studied the global character of the solution set of such equations and applied his results in many directions. Because of the wide scope of applicability of the bifurcation theorem of Krasnosel'skii and Rabinowitz a tremendous research interest in this field has been evidenced currently (see Crandall and Rabinowitz (1971)) and the literature cited there).

Our aim of this section is to present the results from Tarafdar and Thompson (1985) which generalize the theorems of Krasnowel'skii and Rabinowitz to the set-valued situation, i.e., to equations of the form:

$$x \in \mu Ax + R(x, \mu) \tag{7.230}$$

where A is as above, $B : \bar{\Omega} \times R \to CK(X)$ is a compact set valued mapping and B is of $o(\|x\|)$ uniformly in μ in compact intervals (see definitions below), Ω being as above and $CK(X)$ being the set of all compact convex subsets of X. From the result in Tarafdar and Thompson (1985) for equations of the form (7.230) Tarafdar and Thompson deduced the same result for equations of the form:

$$Lx \in \mu Ax + R(x, \mu) \tag{7.231}$$

where $L : X \to Z$ is a linear mapping, $A : X \to Z$ a compact linear mapping, $B : \bar{\Omega} \times R \to CK(Z)$ a compact mapping and of $o(\|x\|)$ uniformly in μ in compact intervals. The result for the equation when R is single valued was first obtained by Laloux and Mawhin (1977) (see also Gaines and Mawhin (1977)). However, even when R is single valued the result in Tarafdar and Thompson (1985) is more general than that of Laloux and Mawhin (1977) (see Remark 7.36).

As an application of the results in Tarafdar and Thompson (1985), Tarafdar and Thompson included a two point boundary value problem for a generalized ordinary differential equation in Tarafdar and Thompson (1985). Tarafdar and Thompson observed that their results in Tarafdar and Thompson (1985) would have application to problems in control theory, mathematical economics and related problems.

In all the results, from now on, X will denote a real Banach space and $CK(X)$ will denote the set of all compact convex subsets of X. A set valued mapping $N : \Omega \in X \to CK(X)$ is said to be compact if N is upper semicontinuous, and maps bounded subsets of Ω into relatively compact subsets of X. The degree theory which will be used here is due to Ma (1972) although all the results obtained for compact mappings will equally hold for more general mappings known as ultimately compact mappings by applying the degree theory for such mappings due to Petryshyn and Fitzpatrick (1974).

Index. Let Ω be an open neighborhood of $a \in X$ and $N : \bar{\Omega} \to CK(X)$ be a compact mapping. We assume that $a \in N(a)$ and there exists $\varepsilon_0 > 0$ such that $B_{\varepsilon_0}(a) \subset \Omega$ and $(I - N)^{-1}(0) \cap B_{\varepsilon_0}(a) = \{a\}$; i.e., a is an isolated zero of $I - N$, where

$$B_{\varepsilon_0}(a) = \{x \in X : \|x - a\| < \varepsilon_o\}$$

and I is Identity mapping on X. This implies that for every

$$\varepsilon \in (0, \varepsilon_0),), 0 \notin (I - N)(\delta B_{\varepsilon},(a))$$

where $\delta B_\varepsilon(a)$ denotes the boundary of $B_\varepsilon(a)$. Thus the degree $d(I - N, B_\varepsilon(a), 0)$ is defined and is independent of $\varepsilon \in (0, \varepsilon_0)$ by the excision property of the degree.

Definition 7.25 Under the above assumptions the index of the fixed point a of N is the integer $d(I - N, B_\varepsilon(a), 0)$ for any $\varepsilon \in (0, \varepsilon_0)$ and will be denoted by $i(I - N, a)$.

Remark 7.32 If N is single valued, then the index of a fixed point a of N as defined above coincides with the Leray-Schauder index.

Proposition 7.22 *Let $N : \bar{\Omega} \to CK(X)$ be a compact-napping such that $0 \notin (I - N)(\delta\Omega$ and $(I - N)^{-1}(0)$ is a finite set, say, $\{a_1, a_2, \dots, a_n\}$; i.e., the set of fixed points of N is finite.*

Then

$$d(I - N, \Omega, 0) = \sum_{j=1}^{n} I(I - N, a_j).$$

Proof. We can find positive numbers $c_j, j = 1, 2, \dots, n$ such that

$$\overline{B_{\varepsilon_j}(a_j)} \cap \overline{B_{\varepsilon_k}(a_k)} = \phi \,\text{for any}\, j, k \in \{l, 2, \dots, n\}$$

with $j \neq k$.

By the additivity and excision properties of degree we have

$$d(I - N, \Omega, 0) = d(I - N, \cup_{j=1}^{n} B_{\varepsilon j}(a_j), 0) = \sum_{j=1}^{n} i(I - N, a_j). \qquad \square$$

Definition 7.26 Let $\Omega \subset X$ be a bounded open neighborhood of the origin and $B : \bar{\Omega} \to CK(X)$ be an upper semicontinuous set valued mapping. B is said to be of order $o(\|x\|)$ if

$$\|x\|^{-1} \sup \|y\| : y \in B(x) \to 0 \,\text{as}\, \|x\| \to 0.$$

Remark 7.33 It is clear from the definition that $B(O) = 0$ and the definition of $o(\|x\|)$ coincides with the usual definition of $o(\|x\|)$ when B is single valued.

Theorem 7.41 *Let $N = A + B$ where $A : X \to X$ is a linear compact mapping with $ker(I - A) = \{0\}$ and $B : \bar{\Omega} \to CK(X)$ is a compact mapping of order $o(\|x\|)$ where $\Omega \subset X$ is a bounded open neighborhood of the origin. Then 0 is an isolated zero of $(I - N)$ and*

$$i(I - N, O) = i(I - A, O)$$

where $i(I - A, O)$ is the usual Leray-Schauder index of the fixed point 0 of A.

Proof. Since B is of order $o(\|x\|)$ and A is linear, $N(0) = 0$. We can easily verify that for each $x \in \bar{\Omega}$,

$$(I - N)(x) = (I - A)[I - (I - A)^{-1}B](x). \tag{7.232}$$

That $(I - A)^{-1}$ exists follows from Riesz theory.

Now from the fact that B is of order $o(\|x\|)$ it follows that there exists $\varepsilon_o > 0$ such that $\bar{B}_{\varepsilon_o} \subset \Omega$ 2 and for each $x \in \overline{B}_{\varepsilon_o}(O)$,

$$\sup\{\|y\| : y \in (I - A)^{-1}B(x)\} \leq \frac{1}{2}\|x\|.$$

Hence for each $\varepsilon \in (0, \varepsilon_o]$ and every $(x, \lambda) \in B_\varepsilon(0) \times [0, 1]$ we have

$$\inf\{\|x - y\| : y \in \lambda(I - A)^{-1}B(x)\} \geq \inf\{\|x\| - \|y\| : y \in \lambda(I - A)^{-1}B(x)\} \tag{7.233}$$

$$\geq \|x\| - \sup\{\|y\| : y \in \lambda(I - A)^{-1}B(x)\} \tag{7.234}$$

$$\geq \frac{1}{2}\|x\|. \tag{7.235}$$

(7.232) and (7.233) together imply that the only fixed point of N in $B_{\varepsilon_0}(0)$ is 0; i.e., $(I - N)^{-1}(0) \cap \bar{B}_{\varepsilon_0}(0) = \{0\}$.

By the homotopy invariance property

$$d(I - \lambda(I - A)^{-1}B, B_\varepsilon(0), 0) = d(I, B - \varepsilon(0), 0) = 1 \tag{7.236}$$

$$\text{for every } \varepsilon \in (0, \varepsilon_0] \text{ and } \lambda \in [0, 1]. \tag{7.237}$$

Now from (7.232), (7.236) and the product theorem of degree we have

$$d(I - N, B_\varepsilon(0), 0) = d(I - A, B_\varepsilon(0), 0)$$

for each $\varepsilon \in (0, \varepsilon_0]$; i.e.

$$i(I - N, 0) = i(I - A, 0). \qquad \square$$

7.7.1 *Characteristic Value and Multiplicity*

Let $A : X \to X$ be a linear compact mapping and $r(A) = \{\mu : \mu^{-1}$ is an eigenvalue of $A\}$. Each $\lambda \in r(A)$ is called a characteristic value of A. The multiplicity of $\mu \in r(A)$ is the integer

$$\beta(\mu) = \dim \ker[I - \mu A]^n(\mu)$$

where $n(\mu)$ is the smallest non-negative integer n such that

$$\ker[I - \mu A]^{n+1} = \ker[I - \mu A]^{n+1}.$$

Since A is compact, $\beta(\mu)$ is finite. A real number μ is said to be a regular value of A if $(I - \mu A)^{-1}$ exists and is continuous.

If $\mu \in R$ is not a characteristic value of A, $x = 0$ is an isolated zero of $I - \mu A$ and $i_{LS}(I - \mu A, O)$, the Leray-Schauder index of $I - \mu A$ at zero will be simply denoted by $i(\mu)$.

The following lemma gives us the well-known Leray-Schauder principle:

Lemma 7.11 *If $A : X \to X$ is a compact linear mapping and μ_1, μ_2 with $\mu_1 < \mu_2$ are not characteristic values of A, then $i(\mu_1) = (-1)^\beta I(\mu_2) = (-1)^\beta I(\mu_2)$ where β is the sum of multiplicities of the characteristic values of A lying in the interval $[\mu_1, \mu_2]$.*

7.7.2 *Tarafdar and Thompson's Results on the Theory of Bifurcation*

Throughout the rest of this section we will assume Ω to be an open bounded neighborhood of the origin in X.

Definition 7.27 Let $N : \bar{\Omega} \to CK(X)$ be a set valued mapping satisfying

(*o*) N is upper semicontinuous and compact on bounded subsets of $\bar{\Omega} \times R$;

(*oo*) for each $\mu \in R$,

$$0 \in N(0, \mu).$$

Thus for each $\mu \in R$, $x = 0$ is a solution of the equation

$$x \in N(x, \mu). \tag{7.238}$$

A point $(0, \mu_0)$ will be said to be a *bifurcation point* for the solution of the equation (7.238), or simply a bifurcation point of N if every neighborhood of $(0, \mu_0)$

contains at least one solution (x, μ) of the equation (7.238) with $x \neq 0$. By abuse of notation we will sometimes refer to μ_0 as the bifurcation point.

Lemma 7.12 *Let N be as above satisfying (o) and (oo). If the interval $[\mu_1, \mu_2]$ contains no bifurcation point of N, then there exists a $\delta > 0$ such that, for each $\mu \in [\mu_1, \mu_2]$ and $x \in \bar{\Omega} \cap B_\delta(0)$,*

$$x \in N(x, \mu) \Rightarrow x = 0.$$

Proof. Let $S = \{(x, \mu) \in \bar{\Omega} \times [\mu_l, \mu_2] : x \in N(x, \mu)\}$. From the compactness of N, It follows that S is a compact subset of $X \times R$. We suppose, if possible, that the lemma is not true. Then for each positive integer n, there exist $\tilde{\mu}_n \in [\mu_1 \mu_2]$, and $x_n \in \bar{\Omega} \cap B_{1/n}(0)$ such that

$$x_n \in N(x_n, \tilde{\mu}_n) \text{ and } \ x_n \neq 0.$$

Now $(x_n, \tilde{\mu}_n) \in S$ which is compact. Hence $(x_n, \tilde{\mu}_n)$ has a convergent subsequence converging to $(x_0, \tilde{\mu}_n)$. Clearly $x_0 = 0$ and $(x_0, \tilde{\mu}_0)$ is a bifurcation point of N in $[\mu_1, \mu_2]$ which contradicts the hypothesis. Hence the lemma is proved. □

Theorem 7.42 *Let N be as above satisfying (o) and (oo) and let $\mu_1, \mu_2 \in R$ with $\mu_1 < \mu_2$. Further suppose that $i(\mu_j) = I[I - M(\cdot, \mu_j), 0], j = 1, 2$ are defined and $i(\mu_1) \neq i(\mu_2)$. Then there exists $\mu_0 \in [\mu_1, \mu_2]$ such that $(0, \mu_0)$ is a bifurcation point of N.*

Proof. As $i(\mu_j), j = 1, 2$ are defined, there exist $\delta_j > 0, j = 1, 2$ such that $[I - N(\cdot, \mu_j))_{-1}(0)] \cap B_{\delta_j}(O) = \{0\}$.

Suppose that the conclusion of the theorem is false. Then by the above lemma there exists $\delta_3 > 0$ such that for each $\mu \in [\mu_l, \mu_2]$ and $x \in B_{\delta_3}(0) \cap \bar{\Omega}$,

$$x \in N(x, \mu) \Rightarrow x = 0.$$

We set $\delta_0 = \min(\delta_1, \delta_2, \delta_3)$. Then for each $\delta \subset (0, \delta_0]$, each $\lambda \in [0, 1]$, $x \in B_\delta(0)$,

$$x \in N(x, \lambda\mu_1 + (1 - \lambda)\mu_2) \Rightarrow x = 0.$$

Hence it follows that for each $\delta \in (0, \delta_0], \lambda[0, 1]$ and $x \in \partial B_\delta(0)$,

$$x \notin N(x, \lambda_1 + (1 - \lambda)\mu_2).$$

Therefore by Homotopy Invariance Theorem

$$i(\mu_1) = d[I - N(\cdot, \mu_1)B_\delta(0), 0] \tag{7.239}$$
$$= d[I - N(\cdot, \mu_2)B_\delta(0), 0] = i(\mu_2) \tag{7.240}$$

which contradicts our hypothesis. Hence the theorem is proved. □

Theorem 7.43 *Let $N : \bar{\Omega} \times R \to CK(X)$ be such that*

$$N(x, \mu) = \mu Ax + B(x, \mu)$$

where $A : X \to X$ is a linear compact (single valued) mapping and $B : \bar{\Omega} \times R \to X$ is upper semi-continuous and compact on bounded subsets of $\bar{\Omega} \times R$ with $B(x, \mu) = o(\|x\|)$ uniformly in μ on compact intervals. Then for each bifurcation point $(0, \mu_0)$ of N, μ_0 is a characteristic value of A.

Proof. Again we prove this theorem by contradiction. Suppose that $(0, \mu_0)$ is a bifurcation point of N and μ_o, is not a characteristic value. Then $A_0 = (I - \mu_0 A)^{-1}$ exists and is continuous.

Let $F : \bar{\Omega} \times R \to X$ and $G : \bar{\Omega} \times R \to CK(X)$ be defined as follows:

$$F(x, \mu) = (\mu - \mu_0) A_0 A x$$

and

$$G(x, \mu) = A_0 B(x, \mu).$$

Clearly F and G are compact on bounded subsets of $\bar{\Omega} \times \mathbb{R}$. Also from the assumption that $B(x, v) = 0(\|x\|)$ and the continuity of A_0 it follows that $A_0 B(x, \mu) = o(\|x\|)$ uniformly in μ in compact intervals.

Let $\delta > 0$ be such that $\delta \|A_0 A\| \leq \frac{1}{3}$ and $\rho > 0$ be such that whenever $(x, \mu) \in B_\rho(O) \times [\mu_0 - \delta, \mu_0 + \delta]$ we have

$$\|x\|^{-1} \sup\{\|y\| : y \in A_0 B(x, \mu)\} \leq \frac{1}{3}.$$

Then for each $x \in B_\rho(0) \cap \bar{\Omega} \backslash \{0\}$ and $\mu \in [\mu_0 - \delta, \mu_0 + \delta]$

$$\begin{aligned}
&\inf\{\|A_0 x - y\| : y \in A_0 N(x, \mu)\} && (7.241)\\
&= \inf\{\|A_0[I - \mu_0 A)x + \mu_0 A x] - \mu A_0 A x - v\| : v \in A_0 B(x, \mu)\} && (7.242)\\
&= \inf\{\|x + (\mu_0 - \mu) A_0 A x - v\| : v \in A_0 B(x, \mu)\} && (7.243)\\
&\geq \|x\| - |\mu - \mu_0| \|A_0 A\| \|x\| A \sup\{\|v\| : v \in A_0 B(x, \mu)\} && (7.244)\\
&\geq [1 - \delta \|A_0 A\| - \frac{1}{3}] \|x\|. && (7.245)
\end{aligned}$$

Hence there exist $\rho > 0$ and $\delta > 0$ such that for each

$$(x, \mu) \in \{B_\rho(0)\} \cap \{\bar{\Omega}\} \backslash \{0\}\} \times [\mu_0 - \delta, \mu_0 + \delta], x \notin N(x, \mu).$$

But this implies that μ is not a bifurcation point, which is a contradiction. Thus the theorem is proved. □

Theorem 7.44 *Let $N : \bar{\Omega} \times R \to CK(X)$ be given by $N(x, \mu) = \mu A x + B(x, \mu)$ where A and B are as in Theorem 7.43. If μ_0 is a characteristic value of A of odd multiplicity β_0, then $(0, \mu_0)$ is a bifurcation point of N.*

Proof. Since A is compact, μ_0 is an isolated characteristic value of A. Hence there exists $\varepsilon > 0$ such that μ_0, is the only characteristic value of A in $[\mu_0 - \varepsilon, \mu_0 + \varepsilon]$.

Thus from Lemma 7.11 we have

$$I(\mu_0 - \varepsilon) = i[I - (\mu_0 - \varepsilon)A_0] = (-1)^{\beta_0} i[I - (\mu_0 + \varepsilon)A, 0] \tag{7.246}$$
$$= (-1)^{\beta_0} i[I - (\mu_0 + \varepsilon). \tag{7.247}$$

Hence from the above equality and Theorem 7.41 we have

$$i[I - N(\cdot, \mu_0 - \varepsilon), 0] \neq (-1)^{\beta_0} i[I - N(\cdot, \mu_0 + \varepsilon), 0].$$

Now since β_0 is odd, we have

$$I[I - N(\cdot, \mu_0 - \varepsilon), 0] \neq (-1)^{\beta_0} i[I - N(\cdot, \mu_0 + \varepsilon), 0].$$

Hence by Theorem 7.42 there exists a bifurcation point $(0, \bar{\mu})$ of N with $\bar{\mu} \in [\mu_0 - \varepsilon, \mu_0 + \varepsilon]$. But since μ_0 is the only characteristic value in $[\mu_0 - \varepsilon, \mu_0 + \varepsilon]$, Theorem 7.43 implies that $\bar{\mu} = \mu_0$. Thus we have proved that $(0, \mu_0)$ is a bifurcation point of N. □

We note that all Theorems proved above hold with the same proof if we replace Ω by X.

Let E denote the space $X \times R$ or $\bar{\Omega} \times R$ where Ω is a bounded open neighborhood of the origin in X.

The following global version in the single valued case is due to Rabinowitz (1971).

Theorem 7.45 *Let $N : E \to CK(X)$ be such that $N(x, \mu) = \mu Ax + B(x, \mu)$ where A and B are as in Theorem 7.43 with $\bar{\Omega} \times R$ being replaced by E. Let μ_0 be a characteristic value of A of odd multiplicity. Let S denote the closure of all nontrivial solutions of (7.238). Then S contains a component C (i.e., a maximal closed and connected subset) which contains $(0, \mu_0)$ and either is unbounded or contains $(0, \tilde{\mu})$ where $\tilde{\mu}$ is a characteristic value of A and $\tilde{\mu} \neq \mu_0$.*

The proof of this theorem is exactly similar to that given by Rabinowitz (1971)(also see Crandall and Rabinowitz (1971) for the single valued case.

For the sake of completeness, the proof of the above theorem will be given after the following two lemmas. Tarafdar and Thompson (1985) used these two lemmas in the proof of the above theorem.

Lemma 7.13 (Whyburn (1958)) *Let K be a compact metric space and A and B two disjoint closed subsets of K. Then either there is a subcontinuum (i.e., a closed and connected subset) of K meeting both A and B, or there exist disjoint compact subsets $K_A \supset A$ and $K_B \supset B$ such that $K = K_A \cup K_B$.*

Lemma 7.14 *Under the hypothesis of Theorem 7.45 assume that there exists no sub-continuum C of $S \cup \{(0, \mu_0)\}$ such that either (i) C is unbounded, or (ii) C contains $(0, \bar{\mu})$ with $\mu_0 \neq \bar{\mu} \in r(L)$. Then there exists a bounded open set θ in E such that*

$$(0, \mu_0) \in \theta, \;\; \partial\theta \cap S = \emptyset,$$

and

$$\theta \cap (\{0\} \times \mathbf{R}) = \{0\} \times (\mu_0 - \delta, \mu_0 + \delta),$$

for some positive δ.

Proof. The proof is similar to that of Lemma 7.12. Let C_{μ_0} be the component of $(0, \mu_0)$ in $S \cup \{(0, \mu_0)\}$. Then C_{μ_0} is bounded by (i). Hence by the upper semi-continuity and compactness of N, C_{μ_0} is compact. Let U_δ be a δ-neighborhood of C_{μ_0}. Now from the fact that for $\lambda \notin r(L)$, by Theorem 7.41, $(0, \lambda)$ is an isolated solution of (7.238) and from (ii) it follows that for $0 < \delta < \epsilon_0$ sufficiently small U_δ contains no solution $(0, \lambda)$ with $|\lambda - \mu_0| > \delta$. Now since S is locally compact, it follows that $K = \bar{U}_| de \cap S$ is compact in the relative topology induced from E. Also $C_{\mu_0} \cap \partial U_| de = \emptyset$. Let $K_1 = C_{\mu_0}$ and $K_2 = (\partial U_| de) \cap S$. There is no sub-continuum meeting K_1 and K_2 for otherwise C_{μ_0} will not be a component. Hence by Lemma 7.13 there exist compact subsets $A \subset K_1$ and $B \subset K_2$ such that $K = A \cup B$. Let θ be an ε-neighborhood of A where ε Is less than the distance between A and B. We can easily see that θ fulfills the demand of the lemma. □

Proof of Theorem 7.45. Let C_{μ_0} be the component of $(0, \mu_0)$ in $S \cup \{0, \mu_0)\}$. Assume that the theorem is false. Then by Lemma 7.14 there is a bounded open set θ such that

$$C_{\mu_0} \subset \theta, \partial\theta \cap S = \emptyset$$

and

$$\theta \cap (\{0\} \times \mathbf{R}) \subset \{0\} \times (\mu_0 - \delta, \mu_0 + \delta)$$

for some $\delta > 0$. Let $\theta_\lambda = \{x \in X : (x, \lambda) \in \theta\}$. Now for λ satisfying $0 < |\lambda - \mu_0| \leq \delta$, $(0, \lambda)$ is an isolated solution of (7.238) (by construction of θ and Theorem 7.41). Hence there exists $\rho(\lambda) > 0$ such that $(0, \lambda)$ is the only solution of (7.238) in $\bar{B}_{\rho(\lambda)} \times \{\lambda\}$. For $\lambda > \mu_0 + \delta$ we choose $\rho(\lambda) = \rho(\mu_0 + \delta)$ and for $\lambda < \mu_0 - \delta$ we choose $\rho(\mu_0 - \delta)$. By choosing $\rho(\mu_0 - \delta)$ sufficiently small, we can assume that $\bar{B}_{\rho(\lambda)} \cap \bar{\theta}_\lambda = \emptyset$ for λ satisfying $|\lambda - \mu_0| \geq \delta$. Thus for $\lambda \neq \mu_0$ there is no solution of (7.238) in $\partial[\theta_\lambda - \bar{B}_{\rho(\lambda)}] \times \{\lambda\}$ and therefore $d(I - N(\cdot, \lambda), \theta_\lambda - \bar{B}_{\rho(\lambda)}, 0)$ is well defined for each $\lambda \neq \mu_0$. By homotopy invariance $d(I - N(\cdot, \lambda), \theta_\lambda - \bar{B}_{\rho(\lambda)}, 0) =$ constant for all $\lambda > \mu_0$. However (7.238) has no solution in $\theta_\lambda - \bar{B}_{\rho(\lambda)}$. Hence for $\lambda > \mu_0$ we have

$$(A) \quad d(I - N(\cdot, \lambda), \theta_\lambda - \bar{B}_{\rho(\lambda)}, 0) = 0.$$

Similarly (A) holds for $\lambda \geq \mu_0$. Moreover, by homotopy invariance we have

$$(B) \quad d(I - N(\cdot, \lambda), \theta_\lambda, 0) = \text{constant},$$

for $|\lambda - \mu_o| < \delta$. Let

$$\mu_0 - \delta < \underline{\mu} < \mu_0 < \bar{\mu} < \mu_0 + \delta.$$

Then from

$$\theta_{\bar{\mu}} = B_{\rho(\bar{\mu})} \cup (\theta_{\bar{\mu}} - B_{\rho(\bar{\mu})}),$$

and by additivity of degree we have

$$d(I - N(\cdot, \bar{\mu}), \theta_{\bar{\mu}}, 0) = d(I - N(\cdot, \bar{\mu}), B_{\rho(\bar{\mu})}, 0) + d(I - N(\cdot, \bar{\mu}), \theta_{\bar{\mu}} - \bar{B}_{\rho(\bar{\mu})}, 0) \tag{7.248}$$

$$= d(I - N(\cdot, \bar{\mu}), B_{\rho(\bar{\mu})}, 0) \quad \text{by} \quad (A). \tag{7.249}$$

Similarly, we can show that

$$d(I - N(\cdot, \underline{\mu}), \theta_{\underline{\mu}}, 0) = d(I - N(\cdot, \underline{\mu}), B_{\rho(\underline{\mu})}, 0).$$

Hence by using (B) we obtain

$$(C) \quad d(I - N(\cdot, \underline{\mu}), B_{\rho(\mu)}, 0) = d(I - N(\cdot, \bar{\mu}), B_{\rho(\bar{\mu})}, 0).$$

We now define the homotopy

$$\hat{N}(x, \lambda, t) = \lambda Ax + tB(x, \lambda), \quad 0 \le t \le 1.$$

As $B(x, \lambda)$ Is of $o(\|x\|)$, we can choose $\rho(\bar{\mu})$ sufficiently small to obtain

$$0 \notin (I - \hat{N})(x, \bar{\mu}, t)$$

for any

$$(x, t) \in \partial B_{\rho(\bar{\mu})} \times [0, 1].$$

Thus by homotopy invariance

$$d(I - N(\cdot, \bar{\mu}), B_{\rho(\bar{\mu})}, 0) = d(I - \hat{N}(\cdot, \bar{\mu}, t), B_{\rho(\bar{\mu})}, 0) \tag{7.250}$$

$$= d(I - \hat{N}(\cdot, \bar{\mu}, 0), B_{\rho(\bar{\mu})}, 0) \tag{7.251}$$

$$= d(I - \bar{\mu}A, B_{\rho(\bar{\mu})}, 0). \tag{7.252}$$

By repeating the same argument for $\underline{\mu}$ and using (C) we obtain

$$d(I - \underline{\mu}A, B_{\rho(\underline{\mu})}, 0) = d(I - \bar{\mu}A, B_{\rho(\bar{\mu})}, 0).$$

i.e.,

$$i(I - \underline{\mu}A, 0) = i(I - \bar{\mu}A, 0),$$

which is a contradiction in view of the fact that $\mu_0$is a characteristic value of odd multiplicity and in view of Lemma 7.11. Thus the theorem is proved. □

Remark 7.34 Tarafdar and Thompson observed that the results similar to Theorems 1.16, 1.25, 1.27, 1.40 in Rabinowitz (1971) can also be proved in the set valued case. Tarafdar and Thompson omitted these as their proofs are only repetition of the arguments given in Rabinowitz (1971).

Let X and Z be real Banach spaces. Let us now consider the equation

$$Lx \in \lambda Ax + B(x, \lambda) = N(x, \lambda) \tag{7.253}$$

where $L : X \to Z$ is a continuous linear mapping $A : X \to Z$ is a compact linear mapping and $B : \bar{\Omega} \times \mathbf{R} \to CK(Z)$ is a set valued mapping which is upper semi-continuous and is compact on bounded subset of $\bar{\Omega} \times \mathbf{R}$. We also assume that $0 \in B(0, \lambda)$ for all $\lambda \in \mathbf{R}$.

A point $(0, \mu_0)$ is said to be a bifurcation point of the equation (7.253) if every neighborhood of $(0, \mu_0)$ contains at least one solution (x, μ) of (7.253) with $x \neq 0$.

Theorem 7.46 *Let L, A and B be as above. Let $B(x, \lambda)$ be $o(\|x\|)$ uniformly in λ in compact intervals; that is,*

$$\|x\|^{-1} \sup\{\|y\| : y \in B(x, \lambda)\} \to 0 \quad as \quad \|x\| \to 0,$$

uniformly in λ in compact intervals. Assume that there exists $\bar{\mu} \in \mathbf{R}$ such that $(L - \bar{\mu}A)^{-1}$ exists and is continuous. Let $(\mu_0 - \bar{\mu})$ be a characteristic value of the compact operator

$$A_0 = (L - \bar{\mu}A)^{-1}A,$$

of odd multiplicity β_0. Then μ_0 is a bifurcation point of the equation (7.253).

Proof. Let us consider the equation

$$x \in \lambda A_0 x + (L - \bar{\mu}A)^{-1}B(x, \lambda + \bar{\mu}) = \bar{N}(x, \lambda). \tag{7.254}$$

Now (x, μ) is a solution of (7.253) if and only if $(x, \mu - \bar{\mu})$ is a solution of (7.254). Indeed,

$$\begin{aligned} Lx \in N(x,\mu) = \mu Ax + B(x,\mu) &\leftrightarrow && (7.255)\\ = Lx - \bar{\mu}Ax \in (\mu - \bar{\mu})Ax + B(x, \mu - \bar{\mu} + \bar{\mu}) &\leftrightarrow && (7.256)\\ x \in (\mu - \bar{\mu})(L - \bar{\mu}A)^{-1}Ax + (L - \bar{\mu}A)^{-1}B(x, \mu - \bar{\mu} + \bar{\mu}) &\leftrightarrow && (7.257)\\ x \in (\mu - \bar{\mu})A_0 x + (L - \bar{\mu}A)^{-1}B(x, \mu - \bar{\mu} + \bar{\mu}) = \bar{N}(x, \mu - \bar{\mu}). & && (7.258) \end{aligned}$$

Noting that $(L - \bar{\mu}A)^{-1}B(x, \lambda + \bar{\mu})$ is of $o(\|x\|)$ uniformly in λ in compact intervals, by applying Theorem 7.44 we conclude that $\mu_0 - \bar{\mu}$ is a bifurcation point of (7.254). Hence it follows from (7.255)that μ_0 is a bifurcation point of (7.253). □

Remark 7.35 Tarafdar and Thompson observes that the corresponding global version of Theorem 7.44 also holds.

Remark 7.36 Tarafdar and Thompson observed that in Theorem 7.44 instead of assuming A and $B(x, \lambda)$ to be compact it would suffice to assume $A_0 = (L - \bar{\mu}A)^{-1}A$ and $(L - \bar{\mu}A)^{-1}B$ to be compact. By doing this, Tarafdar and Thompson found that even when B is single valued their Theorem would be more general than that proved in Laloux and Mawhin (1977) in the sense that Tarafdar and Thompson were not assuming L to be a Fredholm mapping of index zero. It was noted that all the conditions in Theorem 7.44 had also been assumed implicitly by Laloux and Mawhin.

7.7.3 *Tarafdar and Thompson's Application on the Theory of Bifurcation*

Let $CK(X)$ be as defined before. Let

$$F : [0, \pi] \times \mathbf{R} \times \mathbf{R}^n \times \mathbf{R}^n \to CK(\mathbf{R}^n),$$

be upper semi-continuous and $q \in L^1[0, \pi]$ (abbreviated as L^1) be such that

$$\sup\{\|u\| : u \in F(t, \lambda, y, z), \quad \text{for some} \quad \lambda \in \mathbf{R}, y, z \in \mathbf{R}^n\} \leq q(t) \tag{7.259}$$

$\forall t \in [0, \pi]$.

Tarafdar and Thompson considered the two point boundary value problem

$$\ddot{y}(t) \in \lambda y(t) + F(t, \lambda, y(t), \dot{y}(t)), \quad \text{a.e.} \quad t \in [0, \pi] \tag{7.260}$$

$$y(0) = 0 = y(\pi) \tag{7.261}$$

where a solution $y : [0, \pi] \to \mathbf{R}^n$ satisfies $\dot{y}$ is absolutely continuous on $[0, \pi]$ and y satisfies (7.260) and (7.261).

Tarafdar and Thompson assumed that for $t \in [0, \pi]$, $\lambda \in \mathbf{R}$, $y, z \in \mathbf{R}^n$

$$\frac{\sup\{\|u\| : u \in F(t, \lambda, y, z)\}}{\|y\| + \|z\|} \to 0 \tag{7.262}$$

uniformly for $t \in [0, \pi]$ and λ bounded, as $\|y\| + \|z\| \to 0$. For $y, z \in (C[0, \pi])^n$ (abbreviated as $(C)^n$) let

$$B(y, z, \lambda) = \{f | f : [0, \pi] \to \mathbf{R}^n \quad \text{is measurable,} \quad f(t) \in F(t, \lambda, y(t), z(t))\};$$

when $z = \frac{dy}{dt}$ we abbreviate this to $B(y, \lambda)$. For $\lambda \in R$, and $y, z \in (C)^n$, let

$$H(t) = F(t, \lambda, y(t), z(t))$$

then

$$H : [0, \pi] \to CK(\mathbf{R}^n)$$

is upper semicontinuous as a composition of an upper semi-continuous function with a continuous function and hence

$$\{(t, u(t)) \in [0, \pi] \times \mathbf{R}^n : u(t) \in H(t) \forall t \in [0, \pi]\},$$

is closed and hence measurable. Thus the set $B(y,z,\lambda)$ is non empty by the results stated in Rockafeller (1976). Also B is convex valued as F is convex valued. Since $f \in B(y,z,\lambda)$, f is measurable and

$$\|f(t)\| \leq q(t) \forall t \in [0,\pi]$$

so $f \in (L^1)^n$.

Let

$$G : [0,\pi] \times [0,\pi] \to \mathbf{R}$$

be given by

$$G(x,t) = \begin{cases} \frac{(\pi - x)t}{\pi}, & 0 \leq t \leq x \leq \pi \\ \frac{x(\pi - t)}{\pi}, & 0 \leq x \leq t \leq \pi \end{cases}$$

and let $K : (L^1)^n \to X$ be given by

$$K\phi(x) = \int_0^\pi G(x,t)\phi(t)dt$$

where

$$\phi(t) = (\phi^1(t), \ldots, \phi^n(t)), \;\; \phi^i \in L^1, \;\; 1 \leq i \leq n,$$

and

$$X = (C^1[0,\pi])^n.$$

For $S \subseteq (L^1)^n$ let

$$KS = \{K\phi : \phi \in S\}.$$

For $y \in X$ let

$$N(y,\lambda) = \lambda K y + KB(y,\lambda).$$

Thus

$$N(y,\lambda) \in CK(X)$$

and finding a solution to problems (7.260), (7.261) is equivalent to finding a solution to

$$y \in N(y,\lambda) = \lambda K y + KB(y,\lambda).$$

Now

$$KB(y,\lambda) \in CK(X) \quad \text{and} \quad KB : X \times \mathbf{R} \to CK(X)$$

is upper semicontinuous. This can be seen as follows. As B is convex valued and K is linear, KB is convex valued. Since $f \in B(y,\lambda)$ implies $f \in (L^1)^n$ and

$$\|f(t)\| \leq q(t) \;\; \forall \;\; t \in [0, \pi], \|(\ddot{kf})(t)\| = \|f(t)\| \leq q(t),$$

a.e. for $t \in [0, \pi]$ so $KB(y, \lambda)$ is equicontinuous in X. If $f_i \in B(y, \lambda)$ and $Kf_i \to g$ in X then

$$\|f_I(t)\| \leq q(t)$$

so $f_i \to f$ weakly in $(L^1)^n$. Thus there exists $h_i \in (L^1)^n$, h_i are convex combinations of the f_j such that $h_i \to f \in (L^1)^n$. Thus $h_i \in B(y, \lambda)$ and $h_{j_k} \to f$, a.e. for $t \in [0, \pi]$ to some subsequence h_{i_k}. Thus as F is upper semicontinuous $f \in B(y, \lambda)$ and

$$g = Kf \in KB(y, \lambda)$$

is closed and hence compact. A similar argument using the upper semi-continuity of F with respect to (λ, y, z) shows $KB : X \times R - CK(X)$ is upper semi-continuous. Now $K : X \to X$ is completely continuous with eigenvalues

$$\frac{1}{\lambda} = \frac{1}{n^2},$$

$n = 1, 2, 3, \cdots$ all of multiplicity one.

From (7.262) we see that for $\lambda \in \mathbf{R}$ and $y \in X$

$$\frac{\sup\{\|u\| : u \in KB(y, \lambda)\}}{\|y\|} \to 0$$

as $y \to 0$ in X; here $\|u\|$ and $\|y\|$ are the X norms of u and y respectively. Thus the conditions of Theorem 7.43 are satisfied and the points $(0, n^2)$ are bifurcation points and the problem (7.260), (7.261) has nontrivial solutions (y, λ) near $(0, n^2)$.

Remark 7.37 Tarafdar and Thompson observed that the upper semicontinuity of F can be relaxed to the following:

(i) $F(t, \cdot, \cdot, \cdot)$ is upper semicontinuous for almost every $t \in [0, \pi]$
(ii) $F(\cdot, \lambda, y, z)$ is measurable for all $(\lambda, y, z) \in \mathbf{R} \times \mathbf{R}^n \times \mathbf{R}^n$
(iii) F is closed convex valued

and

(iv) for each $(\lambda, y, z) \in \mathbf{R} \times \mathbf{R}^n \times \mathbf{R}^n$

there exists a measurable function $f : [0, \pi] \to \mathbf{R}^n$ such that

$$f(t) \in F(t, \lambda, y, z)$$

for all $t \in [0, \pi]$ and there exists a fixed $q \in L^1$ such that

$$\|f(t)\| \leq q(t)$$

for almost every $t \in [0, \pi]$.

7.8 Tarafdar and Thompson's Results on the Solvability of Non-Linear and Non-Compact Operator Equations

This section will be devoted to the presentations of some results in Tarafdar and Thompson (1987) on the solvabiility of non-linear and non-compact operator equations.

The notion of p-epi mappings was introduced by Furi, Martelli, and Vignoli (1980) as follows:

If E and F are normed linear spaces, $\Omega \subset E$ is a bounded open set and $p \in F$ then a continuous mapping $f : \bar{\Omega} \to F$ with $f(x) \neq p$ for any $x \in \partial\Omega$ is called p-epi if for each compact mapping $h : \bar{\Omega} \to F$ with $h \equiv 0$ on $\partial\Omega$, the equation $f(x) = h(x) + p$ has a solution in Ω.

The authors showed in Furi et al. (1980) that the p-epi mappings have 'existence', 'boundary dependence', 'normalization', 'localization' and 'homotopy' properties similar to those of topological degree theory. The theory of p-epi mappings is based on elementary tools such as the Schauder fixed point theorem, Urysohn's lemma, etc.

Unlike the case of degree theory p-epi mappings may act between different spaces. These theories are normally used to establish the existence of solutions of nonlinear problems. However in applications such as to differential and functional differential equations the problems occur frequently as nonlinear mappings acting between different spaces. Thus the theory of p-epi mappings is directly applicable to such problems while to apply degree theory it is necessary to re-formulate the problems as non-linear self mappings acting on some space.

In this section we shall present Tarafdar and Tahompson's concept of a (p, k)-epi mapping by allowing the mapping h to be a k-set contraction rather than just a compact mapping and requiring E and F to be Banach spaces. Thus in Tarafdar and Thompson (1987), the authors dealt with a class of mappings smaller than that of p-epi mappings. However, by restricting f to be a (p, k)-epi mapping Tarafdar and Tahompson solved the equation $f(x) = p + h(x)$ for more general mappings h. The authors observed that this was necessary in some applications.

We shall also present applications of Tarafdar and Tahompson's theory of (p, k)-epi mappings to a number of problems in this section.

7.8.1 *Measure of Noncompactness and Set Contraction*

In this section we shall present the well-known concepts of measure of noncompactness and k-set contractions and some of their properties. Although most of these concepts and properties were discussed in Chapter 7, we shall repeat them with a different approach to increase the understandings of the readers. We shall also present some results from Tarafdar and Thompson (1987) which the authors proved in their paper and which they had needed for the subsequent development of their theory. Although some of the results in this section are not new, we include their proofs for the completeness of presentation of the work in Tarafdar and Thompson (1987).

Definition 7.28 Let X be a metric space and $A \subset X$ a bounded subset. Set $\Delta(A) = \{\varepsilon > 0 : A$ can be covered by a finite number of sets of diameter less than $\varepsilon\}$. Then $\alpha(A) = \inf\Delta(A)$ is defined to be the measure of noncompactness of A.

This notion of measure of noncompactness was introduced by Kuratowski (1930).

Another notion of measure of noncompactness was introduced by Gokhberg, Goldstein, and Markus (1957) in the following manner:

Let $\sigma(A) = \{\varepsilon > 0 : A$ can be covered by a finite number of balls of radius $\varepsilon\}$. Then $\beta(A) = \inf\sigma(A)$ is called the (ball) measure of non-compactness of A.

Properties of the measure of noncompactness:

Let A and B be bounded subsets of a metric space X. Then

(1) $\alpha(A) = 0$ if and only if A is relatively compact;

(2) $A \subseteq B$ implies $\alpha(A) \leq \alpha(B)$;

(3) $\alpha(A) = \alpha(\bar{A})$, where $\bar{A}$ denotes the closure of A;

(4) $\alpha(A \cup B) = \max(\alpha(A), \alpha(B))$.

Furthermore, if X is a normed space, then

(5) $\alpha(\text{Co}A) = \alpha(A)$ where CoA denotes the convex hull of A, and

(6) $\alpha(A + B) \leq \alpha(A) + \alpha(B)$.

For proof of these we refer to Lloyd (1978) or Martin (1976).

Definition 7.29 A continuous mapping $f : X \to Y$ of a metric space X into a metric space Y is said to be a k-setcontraction if for each bounded subset A of X, $\alpha(f(A)) \leq k\alpha(A)$, where $k \geq 0$, and is said to be a condensing mapping if for each noncompact bounded subset A of X, $\alpha(f(A)) < \alpha(A)$.

In the sequel we denote by $\delta(A)$ the diameter of a bounded subset A of a metric space X.

Lemma 7.15 *Let Ω be a nonempty bounded subset of a metric space X and let $A \subseteq [0,1] \times \Omega$. Let $\pi(A) = x \in \Omega : (t,x) \in A$ for some $t \in [0,1]$. Then $\alpha(\pi(A)) = \alpha(A)$.*

Proof. We first show that $\alpha(\pi(A)) \leq \alpha(A)$. Let $\varepsilon > 0$. Then there exists a finite number of subsets $D_1, D_2, \ldots, D_n$, of A with $\delta(D_i) \leq \alpha(A) + \varepsilon$ for $I = 1, 2, \ldots, n$ such that $A \subseteq \cup_{i=1}^n D_i$. Clearly $\pi(A) \subseteq \cup_{i=1}^n \pi(D_i)$ and $\delta(\pi(D_i)) \leq \delta(D_i)) \leq \alpha(A) + \varepsilon$. Thus $\alpha(\pi(A)) \leq \alpha(A) + \varepsilon$. Since $\varepsilon > 0$ is arbitrary, $\alpha(\pi(A)) \leq \alpha(A)$. Now $A \subset [0,1] \times |pi(A)$ so $\alpha(A) \leq \alpha([0,1] \times \pi(A))$. Thus it suffices to show that $\alpha([0,1] \times \pi(A)) \leq \alpha(A)$. Now given $\varepsilon > 0$ there exist a finite number of subsets D_i, of $\pi(A)$ with $\delta(D_i) \leq (\pi(A)) + \varepsilon/2$ for $i = 1, \ldots, m$ such that $\pi(A) \subseteq \cup_{i=1}^m D_i$. Let $T_i = [(i-1)/l, i/l]$ for $i = 1, \ldots, l$ where $1/l < \varepsilon/2$. Thus $[0,1] \times \pi(A) \subseteq$

$\cup_{1\leq i\leq m, 1\leq j\leq l} T_j \times D_i$ and $\delta(T_j \times D_i) \leq \alpha(\pi(A)) + \varepsilon$. Thus $\alpha([0,1] \times \pi(A)) \leq \alpha(\pi(A)) + \varepsilon$. Since ε was arbitrary, the result follows. □

Theorem 7.47 *Let X and Y be metric spaces and Ω a bounded subset of X. Let $H : [0,1] \times \Omega \to Y$ be a k-set contraction and $\phi : \Omega \to [0,1]$ be a continuous mapping. The mapping $h : \Omega \to Y$ defined by $h(x) = H(\phi(x), x)$, $x \in \Omega$ is a k-set contraction. (Throughout this section we use the metric $\rho(\cdot,\cdot)$ on $[0,1] \times X$ defined by $\rho((t,x),(\bar{t},\bar{x})) = \max(|t, -\bar{t}|, d(x,\bar{x}))$ where d is the metric on X.)*

Proof. Let $S \subset \Omega$, $\alpha(h(S)) \leq \alpha(H(\phi(S) \times S) = k\alpha(S)$ by Lemma 7.15. As h is continuous, the proof is complete. □

In proving the next theorem, the following two lemmas in Tarafdar and Thompson (1987) will be required.

Lemma 7.16 *Let A be a non-empty bounded subset of a metric space (X, ρ) and for $\varepsilon > 0$, let $B(A,\varepsilon) = \cap_{x\in A} B(x,\varepsilon)$ where $B(x,\varepsilon) = \{y \in A : \rho(x,y) < \varepsilon\}$. Then $\delta(B(A,\varepsilon)) \leq \delta(A) + 2\varepsilon$. As before, δ stands for diameter.*

Proof. Let $u, v \in B(A,\varepsilon)$. Then $u \in B(x,\varepsilon)$ and $v \in B(y,\varepsilon$ for some $x, y \in A$. Hence $\rho(u,v) \leq \rho(u,x) + \rho(x,y) + \rho(y,v) < \delta(A) + 2\varepsilon$. □

Lemma 7.17 *Let A be as in Lemma 7.16. Then $\alpha(B(A,\varepsilon)) \leq \alpha(A) + 2\varepsilon$.*

Proof. This is immediate from Lemma 7.16 and definition of α. □

For the rest of this section, unless otherwise stated, E and F will denote real Banach spaces and Ω an open bounded subset of E. We will also denote by $\Sigma_k(\bar{\Omega})$ the set of all k-set contractions $f : \bar{\Omega} \to F$.

Theorem 7.48 *Let $h : [0,1] \to \Sigma_k(\bar{\Omega})$ be a continuous mapping where $\Sigma_k(\bar{\omega})$ is equipped with the topology of uniform convergence. Let $H : [0,1] \times \bar{\Omega} \to F$ be defined by $H(t,x) = h(t)(x)$. Then H is a k-set contraction.*

Proof. Let $S \subseteq [0,1] \to \times\bar{\Omega}$.Then clearly $H(S) \subseteq H([0,1] \times \pi(S))$.Given $\varepsilon > 0$, there exists, by uniform continuity of h, points $t_1, t_2, \ldots, t_{m+1}$, with $0 = t_1 < t_2 < \cdots < t_m < T_{m+1} = 1$ such that $\|h(t)(x) - h(\tau)(x)\| < \varepsilon$ whenever $t, \tau \in [t_I, t_{I+1}], 1 \leq i \leq m, x \in \bar{\Omega}$.Thus for $(t,x) \in [0,1] \times \pi(S)$, there is $t_i, 1 \leq I \leq m$ such that $\|h(t_i)(x) - h(t)(x)\| < \varepsilon$. Hence it follows that

$$H([0,1] \times \pi(S)) \subseteq \bigcup_{i=1}^{m} B(h(t_I)(\pi(S)), \varepsilon). \tag{7.263}$$

However

$$\alpha(h(t_i)(\pi(S))) \leq k\alpha(|pi(S)) \;\; i = 1, 2, \ldots, m \leq k\alpha(S),$$

by Lemma 7.15.

Hence using (7.263), Lemma 7.17 and properties of measure of noncompactness we obtain $\alpha(H(S) \leq \alpha(H([0,1] \times \pi(S))) \leq k\alpha(S) + 2\varepsilon$. But since $\varepsilon > 0$ is arbitrary, the theorem is proved. □

Theorem 7.49 *Let E and F be normed spaces and $\Omega \subset E$ a bounded open set. Let $h : \bar{\Omega} \to F$ be a k-set contraction such that $h(x) = 0$ for all $x \in \partial\Omega$, where $\partial\Omega$ denotes the boundary of Ω. Let $\bar{h} : E \to F$ be defined by*

$$\bar{h}(x) = \begin{cases} h(x) & \text{for } x \in \bar{\Omega}, \\ 0 & \text{for } x \notin \bar{\Omega}. \end{cases}$$

Then $\bar{h}$ is a k-set contraction.

Proof. Let $S \subset E$ be a bounded set. If $S \cap \Omega = \phi$ then $\bar{h}(S) = 0$. So $\alpha(\bar{h}(S)) = 0 \leq k\alpha(S)$. Now let $S \cap \Omega \neq \phi$. Then $\alpha(\bar{h}(S \cap \Omega)) = \alpha(h(S \cap \Omega)) \leq k\alpha(S \cap \Omega) \leq k\alpha(S)$ (by property 2). Also $\bar{h}(S) \subseteq \Omega) \cup 0$. Hence $\alpha(\bar{h}(S)) \leq \alpha(\bar{h}(S \cap \Omega)) \leq k\alpha(S)$. Thus the result follows as $\bar{h}$ is continuous. □

Theorem 7.50 (1) (Fixed Point Theorem of Darbo) *If $D \subset E$ is a closed bounded convex set and $f : D \to D$ is a k-set contraction with $k \in [0,1]$, then f has a fixed point.*

(2) *If D is as above and $f : D \to D$ is a condensing mapping, then f has a fixed point.*

Proof. See Martin (1976), pp. 125–127. □

In obtaining the results of this section Tarafdar and Thompson (1987) also used the following notion of a k-proper mapping. We recall that a continuous mapping $f : X \to Y$ of a topological space X into a topological space Y is called *proper* if for every compact subset K of Y, $f^{-1}(K)$ is compact. Following this, given $k \geq 0$, we define a mapping $f : \bar{\Omega} \to F$ to be k-proper if f is continuous and $\alpha(f^{-1}(S)) \leq k\alpha(S)$, for each bounded set $S \subsetneq F$. By virtue of property 1 of the measure of noncompactness if f is a k-proper mapping for some $k > 0$, then f is proper.

Theorem 7.51 *Let $f ; \bar{\Omega} \to F$ be k_1-proper and $h : \bar{\Omega} \to F$ a k-set contraction. Then for each bounded set $S \subseteq F$,*

$$\alpha[(f-h)^{-1}(S)] \leq k_1\alpha(S) + k_1 k\alpha[(f-h)^{-1}(S)].$$

Proof. Let set $(f-h)^{-1}(S) = P$. Let $x \in P$. Then there exists $y \in S$ such that $y = f(x) - h(x)$, that is, $f(x) = y + h(x)$, that is, $x \in f^{-1}(y + h(x))$. Thus it follows that $P \subseteq f^{-1}(S + h(P))$. Hence $\alpha(P) \leq \alpha[f^{-1}(S + h(P))] \leq k_1\alpha(S + h(P)) \leq k_1[\alpha(S) + \alpha(h(P))] \leq k_1\alpha(S) + K_1 k\alpha(P)$. □

Corollary 7.51.1 *Let $f : \bar{\Omega} \to F$ be k_1-proper and $h : \bar{\Omega} \to F$ be compact (that is, h is continuous and maps each bounded set onto a relatively compact set). Then $(f-h) : \bar{\Omega} \to F$ is k_1-proper.*

Proof. Since $k = 0$, the corollary follows from Theorem 7.51. □

The next theorem of Tarafdar and Thompson (1987) will have an application in the Subsection 7.8.3 on $(p-k)$-epi mappings on the whole space. This theorem also furnishes an example of a k-proper mapping for some k.

Let $L : E \to F$ be a bounded linear Fredholm operator of index zero. Then we can write $E = M \oplus N$ and $F = U \oplus V$ where $N = N(L) =$ the kernel of L and $U = R(L) =$ the range of L, dim $N =$ dim $V < \infty$. Let $P : E \to N$ and $Q : F \to V$ be continuous projections. Also let $\tilde{L} : M \to U$ denote the restriction of L to the closed subspace M, and $\phi : N \to V$ be an isomorphism. Then it can easily be seen that $L + \phi P$ is an isomorphism.

Theorem 7.52 *Let $L : E \to F$ be a bounded linear Fredholm mapping of index zero. Then there exists a linear compact mapping K such that $L + K$ is an isomorphism and is a $k = \|\tilde{L}^{-1}\|$-proper mapping (in the sense that $\alpha((L+K)^{-1}(S)) \leq k\alpha(S)$ for each bounded set $S \subseteq F$).*

Proof. Now $K = \phi P$. Let $S \subseteq F$ be bounded and let $y \in S$. Thus $y = u + v$, where $u \in U$ and $v \in V$. Set $x = (L+K)^{-1}y = m + n$, where $m \in M$ and $n \in N$. Thus $(L+K))(m+n) = u + v$ so $Lm + Kn = u + v$. Hence $v = Kn$ and $u = Lm$ so that $(L+K)^{-1}(S) \subseteq \tilde{L}^{-1}()I-Q)(S)) + \phi^{-1}(Q(S))$.Thus $\alpha((L+K)^{-1}(S)) \leq \alpha(\tilde{(}L^{-1}((I-Q)(S))) + \alpha(\phi^{-1}(Q(S)))$. Now $\overline{\phi^{-1}(Q(S))}$ is compact so $\alpha(\phi^{-1}(Q(S))) = 0$ and $\alpha((L+K)^{-1}(S)) \leq \alpha(\tilde{L}^{-1}((I-Q)(S))) \leq \|\tilde{L}^{-1}\|\alpha((I-Q)(S))$. It suffices to show that $\alpha((I-Q)(S)) \leq \alpha(S)$. Now $(I-Q)(S) \subseteq S - Q(S)$ so $\alpha((I-Q)(S)) \leq \alpha(S) + \alpha(Q(S))$ and again since $\bar{Q}(S)$ is compact $\alpha(Q(S)) = 0$, and the result follows. □

Remark 7.38 From the proof of the above result we have $\alpha(S) = \alpha((I-Q)(S))$, since $S \subseteq (I-Q)(S) + Q(S)$ so $\alpha(S) \leq \alpha((I-Q)(S)) + \alpha(Q(S)) = \alpha((I-Q)(S))$.

7.8.2 *Epi Mappings*

As we have already indicated, throughout the rest of this section, unless otherwise stated E and F will denote real Banach spaces and Ω an open bounded subset of E.

Definition 7.30 A continuous mapping $f : \bar{\Omega} \to F$ is said to be 0-admissible (p-admissible) if $f(x) \neq 0 (f(x) \neq p \in F)$ for $x \in \partial\Omega$.

Before stating the next definition of Tarafdar and Thompson (1987) we recall Furi et al. (1980) that a 0-admissible mapping $f : \bar{\Omega} \to F$ is called 0-epi if for each compact mapping $h : \bar{\Omega} \to F$ with $h(x) \equiv 0$ on $\bar{\Omega}$ the equation $f(x) = h(x)$ has a solution in Ω. A p-admissible mapping $f : \bar{\Omega} \to F$ is called p-epi if the mapping $f - p$ defined by $(f-p)(x) = f(x) - p, x \in \bar{\Omega}$ is 0-epi.

Definition 7.31 A 0-admissible mapping $f : \bar{\Omega} \to F$ is said to be $(0,k)$-epi if for each k-set contraction $h : \bar{\Omega} \to F$ with $h(x) \equiv 0$ on $\partial\Omega$ the equation $f(x) = h(x)$

has a solution in Ω. Similarly a p-admissible mapping $f : \bar{\Omega} \to F$ is said to be (p,k)-epi if the mapping $f - p$ defined by $(f-p)(x) = f(x) - p, x \in \bar{\Omega}$, is $(0,k)$-epi.

Remark 7.39 Since every compact mapping is a k-set contraction it follows that every $(0,k)$-mapping is 0-epi and every (p,k)-epi mapping is p-epi. In fact, the class of $(p,0)$-epi mappings is strictly larger than that of (p,k)-epi mappings.

Tarafdar and Thompson (1987) gave an example (see below) of a 0-epi mapping which is not $(0,k)$-epi for some k. However, the importance of studying (p,k)-epi mappings lies in the fact that a (p,k)-epi mapping is in a sense more solvable at the point p than a p-epi mapping. Tarafdar and Thompson (1987) made it clear by introducing the following terminology.

Measure of Unsolvability:

Let $f : \bar{\Omega} \to F$ be a p-admissible mapping and $A(f,p) = \{k \geq 0 :$ there exists a k-set contraction $g : \bar{\Omega} \to F$ with $g \equiv 0$ on $\partial\Omega$ such that the equation $f(x) - p = g(x)$ has no solution in $\Omega\}$. We define

$$\gamma(f,p) = j \begin{cases} \inf A(f,p) & \text{if } A(f,p) \neq \emptyset, \\ \infty & \text{if } A(f,p) = \emptyset. \end{cases}$$

We call $\gamma(f,p)$ the measure of unsolvability of f at p. Thus $\gamma(f,p) : A(\bar{\Omega},p) \to [0,\infty]$ is a well defined mapping, where $A(\bar{\Omega},p)$ is the set of all p-admissible mapping $f : \bar{\Omega} \to F$. It then follows from the above definition that for a $(p,0)$-epi mapping $f : \bar{\Omega}, \gamma(f,p) \geq 0$ while $\gamma(f,p) \geq k$ for a (p,k)-epi mapping $f : \bar{\Omega} \to F$.

We recall that a mapping of the form $I - h$ is called a compact vector field where $h : \bar{\Omega} \to E$ is a compact mapping and I is the identity on E. (See Granas (1962).)

Lemma 7.18 *Let $(I-h) : \bar{\Omega} \to E$ be a compact vector field and assume that $h \equiv 0$ on $\partial\Omega$. Then $\gamma(f,p) \geq 1$ if and only if $p \in \Omega$, where $f = I - h$.*

Proof. First note that $\gamma(f,p)$ is defined only when f is p-admissible, that is when $f(x) = x - h(x) \neq p$ for all $x \in \partial\Omega$. This implies that $p \notin \partial\Omega$ as $h \equiv$ on $\partial\Omega$. Thus $p \in E \backslash \partial\Omega$.

Now suppose that $p \in \Omega$ and let $g : \bar{\Omega} \to E$ be a k-set contraction with $0 \leq k < 1$ and $g \equiv 0$ on $\partial\Omega$. Setting $l(x) = p + h(x) + g(x), x \in \bar{\Omega}$ we see that l is a k-set contraction. We define $\bar{l} : E \to E$ by

$$\bar{l}(x) = \begin{cases} l(x) & \text{if } x \in \bar{\Omega}, \\ p & \text{if } x \notin \bar{\Omega}. \end{cases}$$

By Theorem 7.49 $\bar{l}$ is a k-set contraction and therefore $\bar{l}(\bar{\Omega})$ is bounded. Let $M = \sup(\|\bar{l}(x)\| : x\|\bar{\Omega}\}/$ and $B = \{x \in E : \|x\| \leq M\}$. Then the restriction of $\bar{h}$ to the closed convex ball B is a selfmapping and remains a k-set contraction. Hence by Theorem 7.50 there exists a point $x_0 \in B$ such that $\bar{l}(x_0) = x_0$. Now $x_0 \in \Omega$, for otherwise $\bar{l}(x_0) = p = x_0$ which will contradict the fact that $p \in \Omega$. Thus $x_0 \in \Omega$

and $x_0 = l(x_0 = p + h(x_0 + g(x_0$, that is $f(x_0) - p = g(x_0)$ as required. Next suppose that $\gamma(f, p) \geq l$. Let $g = -h$. Then g is a 0-set contraction and therefore $f(x_0) - p - g(x_0) = 0$ for some $x_0 \in \Omega$. Hence $p = x_0 \in \Omega$, as required. □

Example 7.1 More generally, let $h : \bar{\Omega} \to E$ be a k-set contraction such that $k \in [0.1]$ and $p \notin (I - h)(\partial\Omega)$. We also assume Nussbaum (1971) degree, $d(f, \Omega, p) \neq 0$ where $f = I - h$. We choose $\varepsilon > 0$ sufficiently small that $k + \varepsilon < 1$ and consider a $[1 - (k + \varepsilon)]$-set contraction $g : \bar{\Omega} \to E$ such that $g \equiv 0$ on $\partial\Omega$. Then clearly $h + g$ is a $(1 - \varepsilon)$-set contraction and $h + g \equiv h$ on $\partial\Omega$. Hence by the boundary dependence property of the Nussbaum degree $d(I - h, \Omega, p) = d(I - (h + g), \Omega, p) \neq 0$. Hence by the existence property of the degree there exists a solution $x_0 \in \Omega$ of the equation $f(x) - p = g(x)$, that is $\gamma(f, p) \geq 1 - k$.

Example 7.2 Now we give an example where $d(f, \Omega, 0) = 0$ but f is $(0, k)$-epi, for all k. Let $E = \mathbf{R}$, the real line, $\Omega = (-2, 0) \cup (0, 2)$ and $f : \bar{\Omega} \to E$ be defined by $f(x) = x^2 - 1$. Clearly $d(f, \Omega, 0) = 0$ but $\gamma(f, 0) = \infty$. To see that $\gamma(f, 0) = \infty$ let $g : \bar{\Omega} \to E$ be continuous and let $g \equiv 0$ on $\partial\Omega$. Now we note that $d(f|_{[0,2]}, (0, 2), 0) = 1$. Since $f - g = f$ on $\partial((0, 2))$ we have $d(f - g|_{[0,2]}, (0, 2), 0) = 1$. The conclusion follows from the definition of γ.

We now return to (p, k)-epi mappings and present some of their basic properties in Tarafdar and Thompson (1987).

Existence Property:

If $f : \bar{\Omega} \to F$ is a (p, k)-epi mapping, then the equation $f(x) = p$ has a solution in Ω.

Proof. Using $h(x) \equiv 0$ on $\bar{\Omega}$ in the definition of (p, k)-epi, we obtain a solution of the equation $f(x) = p$ in Ω. □

Normalization Property:

The inclusion mapping $i : \bar{\Omega} \to E$ is (p, k)-epi for $k \in [0, 1]$ if and only if $p \in \Omega$, that is, $\gamma(i, p) \geq 1$ if and only if $p \in \Omega$.

This is a special case of Lemma 7.18 with $h \equiv 0$ in $\bar{\Omega}$.

Localization Property:

If $f : \bar{\Omega} \to F$ is $(0, k)$-epi and $f^{-1}(0)$ is contained in an open set $\Omega_1 \subset \Omega$, then f restricted to Ω_1 is also $(0, k)$-epi.

Proof. Let $h : \Omega_1 \to F$ be a k-set contraction such that $h \equiv 0$ on $\partial\Omega_1$. Define $\bar{h} : E \to F$ by

$$\bar{h}(x) = \begin{cases} h(x) & \text{if } x \in \bar{\Omega_1}, \\ 0 & \text{if } x \notin \bar{\Omega_1}. \end{cases}$$

Then $\bar{h}$ is, by Theorem 7.49, a k-set contraction and h_1, the restriction of $\bar{h}$ to $\bar{\Omega}$, is, therefore, a k-set contraction and obviously $h_1 \equiv 0$ on $\partial\Omega$. Thus the equation

$f(x) = h_1(x)$ has a solution $x)0 \equiv \Omega$. Now $f^{-1}(0) \subseteq \Omega_1$ implies that $x_0 \in \Omega_1$. Therefore $f(x) = h(x)$ has a solution $x_0 \in \omega_1$, as required. □

Homotopy Property:

Let $f : \bar{\Omega} \to F$ be $(0,k)$-epi and $h : [0,1] \times \bar{\Omega} \to F$ be an α-set contraction with $0 \leq \alpha \leq k < 1$ such that $h(0,x) = 0$ for all $x \in \bar{\Omega}$. Further let $f(x) + h(t,x) \neq 0$ for all $x \in \partial\Omega$ and for all $t \in [0,1]$. Then $f(\cdot) + h(l,\cdot) : \bar{\Omega} \to F$ is $(0, k - \alpha)$-epi.

Proof. Let $g : \bar{\Omega} \to F$ be a $(k - \alpha)$-set contraction such that $g \equiv 0$ on $\partial\Omega$. The set $S = \{x \in \bar{\Omega} : f(x) + h(t,x) = g(x) \quad \text{for some} \quad t \in [0,1]\}$ is a closed set since f, g, h are continuous and $[0,1]$ is compact. Note that g being a $(k - \alpha)$-set contraction is a k-set contraction. Hence there exists $x_0 \in \Omega$ such that $f(x_0) = g(x_0)$. Thus $f(x_0) + h(O, x_0) = g(x_0)$ and S is nonempty. Moreover S and $\partial\Omega$ are disjoint. Hence by Urysohn's Lemma there exists a continuous function $\phi : \bar{\Omega} \to [0,1]$ such that $\phi \equiv 1$ on S and $\phi \equiv 0$ on $\partial\Omega$. We now consider the function $\bar{h} : \bar{\Omega} \to F$ defined by $\bar{h}(x) = g(x) - h(\phi(x), x), x \in \bar{\Omega}$. In view of Theorem 7.47 $\bar{h}$ is a k-set contraction. Also $\bar{h} \equiv 0$ on $\partial\Omega$. Hence the equation $f(x) = \bar{h}(x) = g(x) - h(\phi(x), x)$ has a solution $x_0 \in \Omega$. This implies $x_0 \in S$. Hence $f(x) + h(l,x) = g(x)$ has a solution $x_0 \in \Omega$. □

Thus $\gamma(\cdot, p)$ has the essential features of a degree theory.

Boundary Dependence Property:

Let $f \cdot \bar{\Omega} \to F$ be $(0,k)$-epi and $g : \bar{\omega} \to F$ be an α-set contraction with $0 \leq \alpha \leq k < 1$ and $g \equiv 0$ on $\partial\Omega$. Then $(f + g) : \bar{\Omega} \to F$ is $(0, k - \alpha$-epi.

Proof. Let h be a $(k - \alpha)$-set contraction and $h \equiv 0$ on $\partial\Omega$. Then $(h - g) : \bar{\Omega} \to F$ is a k-set contraction and $(h - g) \equiv 0$ on $\partial\Omega$. Hence $f(x) = (h - g)(x)$, that is, $f(x) + g(x) = h(x)$ has a solution in Ω. □

We shall now consider the k-proper mapping which Tarafdar and Thompson (1987) introduced in Subsection 7.8.1. Since every k-proper mapping $f : \bar{\Omega} \to F$ is proper, it follows that if $f : \bar{\Omega} \to F$ is k-proper, then $f\bar{\Omega}$ is closed and hence $f(\bar{\Omega}) = \overline{f(\Omega)}$, by continuity of f.

Theorem 7.53 *If $f : \bar{\Omega} \to F$ is $(0,k)$-epi with $k \in [0,1]$ and proper, then f maps Ω onto a neighborhood of the origin. More generally, if U is the connected component of $F \backslash f(\partial\Omega)$ containing the origin, then $U \subset f(\Omega)$.*

Proof. Since f is proper, $f(\partial\Omega)$ is closed. Thus U is an open set and is path connected. Assume that $p \equiv U$ and $\phi : [0,1] \to U$ is a continuous mapping with $\phi(0) = 0$ and $\phi(1) = p$. Then taking the 0-set contraction $h(x,t) = -\phi(t)$ in the homotopy property given in Tarafdar and Thompson (1987), we see that $f(\cdot) - \phi(1) : \bar{\Omega} \to F$ is $(0,k)$-epi. Thus $f(x) - \phi(1) = 0$ has a solution in Ω, that is, $p = \phi(1) \in f(\Omega)$. □

In the following theorem by a $(p, 1^-)$-epi we mean a p-admissible mapping $f : \bar{\Omega} \to F$ such that for each condensing mapping $f : \bar{\Omega} \to F$ with $g \equiv 0 \,\text{on}\, \partial\Omega$, the equation $f(x) - p = g(x)$ has a solution in Ω.

Theorem 7.54 *Let $f : \bar{\Omega} \to F$ be continuous, injective and 1-proper. Then $f(\Omega)$ is open if and only if f is $(p, 1^-)$-epi for any $p \in f(\Omega)$.*

Proof. Let $p \in f(\Omega)$. Since f is proper, it follows from Theorem 7.47 that $f - p$ being also $(0, 0)$-epi maps Ω onto a neighbourhood of 0. Thus $f(\Omega)$ is open.

Now let $f(\Omega)$ be open. It suffices to prove that if $0 \in f(\Omega)$ then f is $(0, 1^-)$-epi. So we assume $0 \in f(\Omega)$. Now since $f(\Omega)$ is open, f is injective and 1-proper and hence proper, f^{-1} is continuous. Also since $\overline{f(\Omega)} = f(\bar{\Omega})$, we have $\partial(f(\Omega)) = f(\partial\Omega)$. Let $h : \bar{\Omega} \to F$ be condensing such that $h \equiv 0$ on $\partial\Omega$.

We define $g : F \to F$ by

$$g(y) = \begin{cases} h(f^{-1}(y)), & \text{if } y \in \overline{f(\Omega)}, \\ 0 & \text{otherwise} \end{cases}$$

Since $\partial(f(\Omega)) = f(\partial\Omega), g \equiv 0\text{on}\, \partial(f(\Omega))$. Thus it follows that g is continuous. Now let S be a bounded subset of F. Then $\alpha[g(S)] = \alpha[h(f^{-1}\{S \cap f(\bar{\Omega})\}] \leq \alpha[f^{-1}\{S \cap f(\bar{\Omega})\}] \leq \alpha(S \cap f(\bar{\Omega})) \leq \alpha(S)$. Also since $g(F) = g(\overline{f(\Omega)}) \cup \{0\} = g(f(\bar{\Omega})) \cup \{0\} = h(\bar{\Omega}) \cup \{0\}$ and h is condensing, $g(F)$ is bounded. Let $M = \{\|y\| : y \in g(F)\}$. Let $B = \{x \in F : \|x\| \leq M\}$. Then g restricted to B maps B into itself and is, by what it is proved above, a condensing mapping. Hence by Theorem 7.50 there is a point $y_0 \in B$ such that $g(y_0) = y_0$. Since $0 \in f(\Omega)$, it follows that $y_0 \in f(\Omega)$. Hence $x_0 = f^{-1}(y_0) \in \Omega$ is a solution of the equation $f(x) = h(x)$. □

Theorem 7.55 *Let $f : \bar{\Omega} \to F$ be continuous, injective and k_1-proper. Then $f(\Omega)$ is open if and only if f is (p, k)-epi for each $p \in f(\Omega)$ and each nonnegative k satisfying the condition $k_1 k < 1$.*

Proof. The argument for the 'if part' is the same as in the proof of Theorem 7.54. For the 'only if part' let $h : \bar{\Omega} \to FD$ be a k-set contraction. The rest of the argument will be the same as in the proof of Theorem 7.54 except that we need to show that the mapping g as constructed there is a $k_1 k$-set contraction. Indeed $\alpha[g(S)] = \alpha h f^{-1}\{S \cap f(\bar{\Omega})\})] \leq k\alpha[f^{-1}\{S \cap f(\bar{\Omega})\}] \leq k k_1 \alpha\{S \cap f(\bar{\Omega})\} \leq k k_1 \alpha(S)$.
□

Corollary 7.55.1 *Let $f\bar{\Omega} \to F$ be continuous, injective and k_1-proper and $f(\Omega)$ be open. Then f is (p, k)-epi for each nonnegative k satisfying $k_1 k < 1$ if and only if $p \in f(\Omega)$.*

Proof. If f is (p, k)-epi, then $p \in f(\Omega)$ by the existence property (regardless of whether $f(\Omega)$ is open or not). If $f(\Omega)$ is open, the result follows from the Theorem 7.55. □

Theorem 7.56 *Let $f : \bar{\Omega} \to F$ be continuous, injective and proper open. Then $f(\Omega)$ is open if and only if f is $(p, 0)$-epi for every $p \in f(\Omega)$.*

Proof. Again the argument is the same as in the proof of the Theorem 7.54, except that we use the fact that $g(F) = h(\bar{\Omega}) \cup \{0\}$ is relatively compact mapping. The fixed point of $g|_B : B \to B$ is guaranteed by the Schauder fixed point theorem. □

The above theorem and following corollary have been obtained by Furi et al. (1980).

Corollary 7.56.1 *Let $f : \bar{\Omega} \to F$ be continuous, injective and proper and $f(\Omega)$ open. Then f is $(p, 0)$-epi if and only if $p \in f(\Omega)$.*

Proof. The proof is similar to that of Corollary 7.55.1, except that we use Theorem 7.56 in place of Theorem 7.55. □

Remark 7.40 For each $k \geq 0$, we construct a mapping f_k such that $\gamma(f_k, 0) = k$. We consider the Hilbert space $H = \{x = \sum \alpha_i e_i : -\infty < i < \infty, \sum \alpha_i^2 < \infty\}$ where $\{e_i\}$ is the orthonormal basis, that is, $\langle e_i, e_j\rangle = 0$ if $i \neq j$ and $= 1$ if $i = j$. The norm of H is defined by $\|x\| = |\sum \alpha_i^2|^{1/2}$, where $x = \sum \alpha_i e_i$.

Let $\Omega = B(0,1) = \{x \in H : \|x\| < 1\}$. Let $f : \bar{\Omega} \to H$ be defined by $f(x)(1/2d)x, x \in \bar{\Omega}$ where $d > 0$ is a constant. Clearly $f : \bar{\Omega} \to H$ is continuous, injective and $2d$-proper and $f(\Omega)$ is open as f is a homomorphism onto $f(\bar{\Omega})(= \overline{f(\Omega)})$. Hence, by Corollary 7.62.1, f is (p, k)-epi for each nonnegative $k < 1/2d$ and each $p \in f(\Omega)$, that is $\gamma(f, p) \geq 1/2d$ for each $p \in f(\Omega)$. We now construct a $\bar{k}$-set contraction $h : \bar{\Omega} \to H$ for some positive number $\bar{k}$, with $h \equiv 0$ on $\partial\Omega$ such that $f(x) = h(x)$ has no solution in Ω, that is, $\gamma(f, 0) \leq \bar{k}$.

Let $u : \bar{\Omega} \to H$ be defined by $u(x) = \sum \alpha_i e_i \in \bar{\Omega}$ and $\phi : \bar{\Omega} \to H$ by

$$\phi(x) = (1 - \|x\|)u(x) + \|x\|x + \frac{1}{3}((1 - \|x\|)e_0, \quad x \in \bar{\Omega}$$

Finally we defined $h : \bar{\Omega} \to H$ by $h(x) = (x - \phi(x))/d, x \in \bar{\Omega}$. It is easy to verify that $h \equiv 0$ on $\partial\Omega$. We next verify that h is a $\bar{k}$-set contraction for some positive number $\bar{k}$. On simplifying $h(x) = \frac{1}{d}(1 - \|x\|)(x - u(x) - \frac{1}{3}e_0).)$. Hence for $x, y \in \bar{\Omega}$,

$$h(x) - h(y) = \frac{1}{d}(1 - \|x\|)(x - u(x) - \frac{1}{3}e_0) - (1 - \|y\|)(y - u(y) - \frac{1}{3}e_0)] \quad (7.264)$$

$$= \frac{1}{d}[(1 - \|x\|)\{(x - u(x)) - (y - u(y))\} - \{(y - u(y) - \frac{1}{3}e_0)\} \quad (7.265)$$

$$\times (\|x\| - \|y\|)] \quad (7.266)$$

$$= \frac{1}{d}[1 - \|x\|)(v(x) - v(y)) - (v(y) - \frac{1}{3}e_0)(\|x\| - \|y\|)] \quad (7.267)$$

where $v = I - u : H \to H$ is a bounded linear operator on H.

Hence

$$\|h(x) - h(y)\| \leq \frac{1}{d}[|1 - \|x\|| \|v\| \|x\| - y\| + (\|v\| \|y\| + \frac{1}{3})(\|x\| - \|y\|)] \quad (7.268)$$

$$\leq \frac{1}{d}[\|v\| + \|v\| + \frac{1}{3}]\|x - y\| \quad (7.269)$$

Thus h is $\bar{k}$-set contraction where $\bar{k} = (2\|v\| + \frac{1}{3})\frac{1}{d}$. Next we show that $f(x) = h(x)$ has no solution in $\bar{\Omega}$. Suppose that $f(x_0) = h(x_0)$ for some $x_0 \in \bar{\Omega}$.

Case 1. Let $x_0 \in \partial\Omega$ an $f(x_0) = h(x_0)$, then $x_0/2d = 0$ by definition of f and h, which is a contradiction. Hence $\|x_0\| < 1$.

Case 2. Let $\|x_0\| = \frac{1}{2}$. Then $x_0 = 2\phi(x_0)$, that is, $x_0 = u(x_0) + x_0 + \frac{1}{3}e_0$, that is, $\sum \alpha_{0+1} + e_i = 0$, where $x_0 = \sum \alpha_i e_i$. This implies that $\alpha_1 = -\frac{1}{3}$, and $\alpha_j = 0$ for $j \neq 1$. Thus $\|x_0\| = \frac{1}{3}$ which is a contradiction. Hence $\|x_0\| = t < 1$ and $t \neq \frac{1}{2}$.

Case 3. Let $\|x\| = t < 1$ and $t \neq \frac{1}{2}$. As $\phi > (0) \neq 0$ and $x_0 \neq 2\phi(x_0)$, we have $x_0 = 2(1-t)u(x_0) + 2tx_0 + \frac{2}{3}(1-t)e_0$, that is, $\sum \alpha_i e_i = 2(1-t)\sum \alpha_i e_{i+1} + 2t\sum \alpha_i e_i + \frac{2}{3}(1-t)e_0$ where $x_0 = \sum \alpha_i e_i$. Hence $(1-2t)(1-t)^{-1}\sum \alpha e_i = 2\sum \alpha_i e_{i+1} + \frac{2}{3}e_0$. Thus $(1-\delta)\sum \alpha_i e_{i+1} + \frac{2}{3}e_0$, where $\delta = t/(1-t) \neq 1$ as $t \neq \frac{1}{2}$. Hence $(1-\delta)\alpha_{i+1} = 2\alpha_i + \frac{2}{3}\delta_{0,i+1}$ where

$$\delta_{i,j} = \begin{cases} 0 & \text{if } i \neq j, \\ 1 & \text{if } i = j, \end{cases}$$

and $1 - \delta \neq 0$. Hence $|\alpha_j| \to \infty$ or $|\alpha_j| \to \infty$ as $j \to -\infty$ according as $2/(1-\delta) > 1$ or $2/(1-\delta) < 1$. But $\sum \alpha_i^2 < \infty$ which is a contradiction. Thus we conclude that $h(x_0) \neq f(x_0)$ for any $x_0 \in \bar{\Omega}$. Hence f is not $(0, \bar{k})$-epi and hence $\gamma(f, 0) \neq \bar{k}$. Thus there exists r such that $(2d)^{-1} \leq r \leq \bar{k}$ and f is $(0, s)$-epi for each $s < r$ but not $(0, s)$-epi for $s > r$. Setting $f_k = kf/r$, the result follows.

We recall that a set $Q \subset F$ is said to be star-shaped with respect to the origin if $ty \in Q$ whenever $y \in Q$ and $t \in [0, 1]$.

Theorem 7.57 *Let $f : \bar{\Omega} \to F$ be $(0, k)$-epi with $k \in [0, 1]$ and Q a star-shaped subset of F with respect to the origin such that $Q \cap f(\partial\Omega) = \emptyset$. Then for each k_l set contraction $h : \bar{\Omega} \to F$ with $h(\partial\Omega \subset Q)$ and $0 \leq k_1 \leq k, 1$, the equation $f(x) = h(x)$ has solution in Ω. In particular $Q \subset f(\Omega)$.*

Proof. It is clear that for all $x \in \partial\Omega$ and for all $t \in [0, 1], th(x) \in Q$ and $f(x) \notin Q$. Hence $f(x) \neq th(x)$ for all $x \in \partial\Omega$ and for all $t \in [0, 1]$. We define the mapping $H : [0, 1] \times \bar{\Omega} \to F$ by $H(t, x) = -th(x)$, for all $(t, x) \in [0, 1] \times \bar{\Omega}$. Clearly for each fixed $t \in [0, 1]$, the mapping $H_t : \bar{\Omega} \to F$ defined by $H_t(x) = H(t, x) = -th(x)$ is a k_1-set contraction (as $\alpha(H, (S)) = \alpha(th(S)) = t\alpha(h(S)) \leq tk_1\alpha(S) \leq k_1\alpha(S)$, for every set $S \subset \bar{\Omega}$). Hence by Theorem 7.48, $H : [0, 1] \times \bar{\Omega} \to F$ is a k_1-set contraction. Also $H(O, x) = 0$ for all $x \in \bar{\Omega}$. Thus by the homotopy property $f(\cdot) + H(1, \cdot) = f(\cdot) - h(\cdot)$ is $(0, k - k_1)$-epi. Hence $f(x) = h(x)$ has a solution in Ω.

Now let $p \in Q$. The constant mapping $h(x) = p$ for all $x \in \bar{\Omega}$ is a 0-set contraction. Hence by the above $f(x) = p$ for some $x \in \Omega$. Thus $Q \subset f(\Omega)$. □

Theorem 7.58 *Let $f : \bar{\Omega} \to F$ be continuous, 0-admissible and k_1-proper and let $f_n : f_n : \bar{\Omega} \to F$ be a sequence of $(0-k)$-epi mappings with $k_1 k < 1$ such that $f_n \to f$ uniformly. Then f is $(0,k)$-epi.*

Proof. Let $h : \bar{\Omega} \to F$ be a k-set contraction such that $h \equiv 0$ on $\partial\Omega$. For each $n = 1, 2, \ldots$, there exists $x_n \in \Omega$ such that $f_n(x_n) = h(x_n)$. Let $y_n = f(x_n) - f_n(x_n) = f(x_n) - h(x_n)$. Then $y_n \to 0$ as $n \to \infty$. Let set $S = \{y_n : n = 1, 2, \cdots\}$. Then by Theorem 7.51,

$$\alpha[(f-h)^{-1}(S)] \leq k_1 k_\alpha (f-h)^{-1}(S)]$$

as $\alpha(S) = 0$. Now as $k_1 k < 1$, this implies that $(f-h)^{-1}(S)$ is relatively compact. Hence $x_n : n = 1, 2, \cdots$ being a subset of $(f-h)^{-1}(S)$ must have a limit point x_0. Clearly $f(x_0) = h(x_0)$. □

Theorem 7.59 (Perturbation) *Let $f : \bar{\Omega} \to F$ be $(0,k)$-epi and k_1-proper and let $h : [0,1] \times \bar{\Omega} \to F$ be a k_2-set contraction such that $0 \leq k_2 \leq k < 1$ and $h(0,x) = 0$ for all $x \in \bar{\Omega}$. If $k_1 k_2 < 1$, then there exists $\varepsilon > 0$ such that $f(\cdot) - h(t,\cdot)$ is $(0, k-k_2)$-epi for all t satisfying $|t| < \varepsilon$.*

Proof. In view of the homotopy property, it will suffice to show that there exists $\varepsilon > 0$ such that $f(x) - h(t,x) \neq 0$ for all $x \in \partial\Omega$ and for all $t \in (-\varepsilon, \varepsilon)$. We assume that there is no such $\varepsilon > 0$. Then we can show that there exists a sequence $\{(t_n, x_n)\}$ with $t_n \to 0$, $x_n \in \partial\Omega$ and $f(x_n) = ht_n, x_n) = y_n$, for all n. Now using the fact that f is k_1-proper and h is a k_2-set contraction,

$$\alpha[x_n : n = 1, 2, \ldots] \leq \alpha[f^{-1}(y_n) : n = 1, 2, \ldots] \leq k_1 \alpha[y_n : n = 1, 2, \ldots] \quad (7.270)$$

$$= k_1 \alpha[h(t_n, x_n) : n = 1, 2, \ldots] \leq k_1 k_2 \alpha[(t_m, x_n) : n = 1, \ldots, \ldots] \quad (7.271)$$

$$k_1 k_2 \alpha[\{t_n\} \times \{x_m\} : m, n = 1, 2, \ldots] = k_1 k_2 \alpha[x_m : m = 1, 2, \ldots], \quad (7.272)$$

by Lemma 7.15. Since $k_1 k_2 < 1$, $\alpha[x_n : n = 1, 2, \ldots] = 0$. Hence the sequence $\{x_n\}$ has a limit point x_0. It follows that $f(x_0) = h(0, x_0) = 0$ which contradicts the fact that f is 0-admissible. □

If $f : \bar{\Omega} \to E$ is a k-set contraction mapping such that $k \in [0,1)$, I is the identity mapping on E, and $p \notin (I-f)(\partial\Omega)$, then the degree of $I - F$ at p, $d(I-f, \Omega, p)$ is well defined (see Lloyd (1978), p. 95). The following theorem is similar to the example following the definition of measure of unsolvability.

Theorem 7.60 *Let $I - f : \bar{\Omega} \to E$ be p-admissible and let $f : \bar{\Omega} \to E$ be a k-set contraction with $k \in [0,1)$. If $d(I-f, \Omega, p) \neq 0$, then for every k_1 with $0 \leq k \leq k_1 < 1$, $I - f$ is $(p, k_1 - k)$-epi.*

Proof. Let $h : \bar{\Omega} \to E$ be a $(k_1 - k)$-set contraction such that $h \equiv 0$ on $\partial\Omega$. $(I - (f + h))$ is a k_1-set contraction and clearly $f + h \equiv f$ on $\partial\Omega$. Hence by the boundary dependence property of the degree, $d(I-(f+h), \Omega, p) = d(I-f, \Omega, p) \neq 0$. Hence there exists $x_0 \in \Omega$ such that $x_0-(f+h)(x_0) = p$. Thus $x_0-f(x_0) = p+h(x_0)$, as required. □

Theorem 7.61 *Let $L : E \to F$ be bounded, linear and subjective with $\dim Ker L < \infty$. Let $g : \bar{\Omega} \to \mathbb{R}^n$ be continuous with bounded range. Let $g(x) \neq 0$ for all $x \in \partial\Omega \cap Ker L$ and $J : \mathbb{R}^n \to E$ be a linear mapping with $Im J = Ker L$. If the Brouwer topological degree $d(gJ, J^{-1}(\Omega), 0) \neq 0$, then $\gamma(M, 0) \geq \|\tilde{L}^{-1}\|$ where $E = Ker L \oplus N$, $\tilde{L}$ is the restriction of L to N, and $M : \bar{\Omega} \to F \times \mathbb{R}^n$ is defined by $M(x) = (Lx, g(x))$.*

Proof. The proof is simliar to that of Theorem 1.7 in (Furi et al. (1980)). Let $S = \tilde{L}^{-1}$. We define $A : F \times \mathbf{R^n} \to \mathbf{E}$ by $A(y, z) = Sy + Jz$. Evidently A is an isomorphism. We now consider the mapping $f = MA : A^{-1}(\Omega) \to \times\mathbf{R^n}$. Then as S is a right inverse of L and $\operatorname{Im} J = \operatorname{Ker} L$, $f(y, z) = (L(Sy + Jz), g(Sy + Jz)) = (y, g(Sy + Jz)) = (y, z) - (0, z - g(Sy + Jz)) = (I - h)(y, z)$ where $h : A^{-1}(\Omega) \to \{0\} \times \mathbf{R^n}$ is the mapping defined by $h(y, z) = (0, z - g(Sy + Jz))$. By using the properties of Leray-Schauder degree

$$d(f, A^{-1}(\Omega), 0) = d(f|_{\{0\}\times\mathbf{R}^n}, A^{-1}(\Omega) \cap (\{0\} \times \mathbf{R}^n), 0) \tag{7.273}$$

$$= d(gJ, J^{-1}(\Omega), 0) \neq 0. \tag{7.274}$$

Hence by Example 7.1, $\gamma(f, 0) \geq 1$. Now $M = fA^{-1}$ and A^{-1} is $\|\tilde{L}^{-1}\|$ proper and hence $\gamma(M, 0) \geq \gamma(f, 0)\|\tilde{L}^{-1}\|$. □

The following is an example in Tarafdar and Thompson (1987):

Example 7.3 A natural question which arises from Theorem 7.61 is the following. In Theorem 7.61 is it legitimate to replace the assumption that the Brouwer degree $d(gJ, J^{-1}(\Omega), 0) \neq 0$ merely by the assumption that $gJ : \overline{J^{-1}(\Omega)} \to F$ is 0-epi? This is not legitimate as the following example in Tarafdar and Thompson (1987) shows.

Let $\Omega = \{(x, y) \in \mathbf{R}^2 : 0 < x2 + y2 < 4\} \subseteq E = \mathbf{R}^2$ and let $J : \mathbf{R} \to E$ be given by $J(x) = (x, 0)$ so that $J^{-1}(\Omega) = (-2, 0) \cup (0, 2)$. Let $g : \bar{\Omega} \to F = \mathbf{R}$ be given by $g(x, y) = x^2 + y^2 - 1$. Thus $gJ : [-2, 2] \to \mathbf{R}$ is given by $gJ(x) = X^2 - 1$ so gJ is 0-epi on $\overline{(-2, 0) \cup (0, 2)} = \overline{J^{-1}(\Omega)}$ from Example 7.2. Let $L : E \to F$ be given by $L(x, y) = y$. Thus $\operatorname{Im} J = \operatorname{Ker} L$. Let $M(X, y) = (y, x^2 + y^2 - 1) = (L(x, y), g(x, y))$. We show that $M : \bar{\omega} \to F \times \mathbf{R}$ is not 0-epi. Define $h : \bar{\Omega} \to F \times \mathbf{R}$ by $h(x, y) = (2(x^2 + y^2)(x^2 + y^2 - 2), 0)$. Thus $h(x, y) = (0, 0)$ for all $(x, y) \in \partial\Omega$, h is continuous on $\bar{\Omega}$, and it suffices to show that $M(x, y) \neq h(x, y)$ for all $(x, y) \in \Omega$. Suppose there is a solution $(x, y) \in \Omega$, then $g(x, y) = 0$ so $x^2 + y^2 = 1$. Thus $h(x, y) = -2 \neq L(x, y) = y$ since $|y| \leq 1$.

7.8.3 *Tarafdar and Thompson's (p,k)-Epi Mappings on the Whole Space*

As before, E and F are real Banach spaces. Let $f : \Omega \to F$ be a continuous mapping of E into F. For $p \in F, f$ is said to be p-admissible if $f^{-1}(p)$ is a bounded subset of E. f is said to be (p,k)-epi if f is (p,k)-epi on the closure of any bounded open set $\Omega \subset f^{-1}(p))$, that is, $f|_{\bar{\Omega}}$, the restriction of f to $\bar{\Omega}$, is (p,k)-epi for each bounded open subset Ω containing $f^{-1}(p)$. We note that by virtue of the localization property in Tarafdar and Thompson (1987) for $(0,k)$-epi mappings it suffices to consider a sufficiently large open ball centered at the origin and containing $f^{-1}(0)$.

Homotopy Property:
If $f : E \to F$ is $(0,k^1)$-epi, $h : [0,1] \times E \to F$ is a k_2-set contraction such that $h(0,x) = 0$, for all $x \in E, 0 \leq k_2 \leq k_1 < 1$ and the set $S = \{x \in E : f(x) + h(x,t) = 0 \text{ for some } t \in (0,1]\}$ is bounded, then $f(\cdot) + h(l,\cdot)$ is $(0, k_1 - k_2)$-epi.

Proof. It is an easy consequence of the homotopy property for mappings on bounded sets in Tarafdar and Thompson (1987). □

Theorem 7.62 *Let $f : E \to F$ be $(0,k)$-epi with $k \in [0,1)$ and $Q \subset F$ be starshaped with respect to the origin. If $f^{-1}(Q)$ is bounded and $h : E \to F$ is a k_1-set contraction such that $0 \leq k_1 \leq k < 1$ and $Im\, h \subset Q$, then the equation $f(x) = h(x)$ has a solution. In particular $Im\, f \supset Q$.*

Proof. Let Ω be a ball centered at the origin such that $\Omega \subset f^{-1}(Q)$. We consider the mapping $f|_{\bar{\Omega}} : \bar{\Omega} \to F$. Then using a similar argument to that in Theorem 7.57 and the above homotopy property, the result follows. □

Corollary 7.62.1 *If $L : E \to F$ is a linear bounded mapping, then L is $(0, t/\|L^{-1}\|)$-epi for each $t \in [0,1]$ if and only if L is an isomorphism.*

Proof. Let L be an isomorphism. Then L continuous, injective and $\|L^{-1}\|$ -proper. Hence L is $(0, t/\|L^{-1}\|)$-epi for each $t \in [0,1)$, by Corollary 7.55.1. Conversely, if L is $(0, t/\|L^{-1}\|)$-epi, for all $t \in [0,1)$ then $L^{-1}(0)$ is bounded by the admissibility of L. Thus L is one-to-one. To prove that L is onto, let $p \in F$ and $Q = \{tp : 0 \leq t \leq 1\}$. Then $L^{-1}(Q)$ is clearly bounded. Hence by Theorem 7.62, $L(x) = p$ has a solution (taking h to be the constant mapping $h(x) = p$ for all $x \in E$). □

Corollary 7.62.2 *Let $f : E \to F$ be $(0,k)$-epi for some $k \in [0,1)$ and $\|f(x)\| \to \infty$ as $\|x\| \to \infty$. If $h : E \to F$ is a k_1-set contraction such that $0 \leq k_1 \leq k < 1$ and $h(E)$ is bounded, then the equation $f(x) = h(x)$ has a solution. In particular f is onto.*

Proof. The condition $\|f(x)\| \to \infty$ as $\|x\| \to \infty$ implies that $f^{-1}(S))$ is bounded for every bounded subset S of F. Since Im$h = h(E)$ is bounded, we can take a ball Q centered at the origin of F sufficiently large that $Q \supset$ Imh. Hence the Corollary follows from Theorem 7.62. □

Remark 7.41 If f and h are as in Corollary 7.62.2, then $f+h$ is $(0, k-k_1)$-epi and, therefore, f is (p,k)-epi for an $p \in F$.

Theorem 7.63 *Let $f : E \to F$ be 0-admissible and k_1-proper on bounded closed sets and let $\{f_n : f_n : E \to F\}$ be a sequence of $(0,k)$-epi mappings such that $f_n \to f$ uniformly on closed bounded subsets of E. Further assume that the sets $f_n^{-1}(0)$ are uniformly bounded and $kk_1 < 1$. Then f is $(0,k)$-epi.*

Proof. This is a direct consequence of Theorem 7.58 which is a theorem in Tarafdar and Thompson (1987). □

7.8.4 *Tarafdar and Thompson's Applications of (p, k)-Epi Mappings in Differential Equations*

Example 7.4 In this example Tarafdar and Thompson (1987) found a solution $x \in C^{-1}[0,1]$ of the following problem:

$$A \begin{cases} \dot{x}(t) = h(t, x(t), x(\phi(t)), \dot{x}(\phi(t))), \\ x(0) = d \end{cases}$$

where $\phi \in C[0,1]$ satisfies $0 \leq \phi(t) \leq t$ for all $t \in [0,1]$ and $h : [0,1] \times \mathbf{R}^3 \to \mathbf{R}$ is continuous and satisfies

$$|h(t,x,y,z)| \leq a + b|x\| + c|y| + r|z| \tag{7.275}$$

and

$$|h(t,x,y,z) - h(t,x,y,u) \leq r|z-y| \tag{7.276}$$

for all $t \in [0,1]$ and $x, y, z, u \in \mathbf{R}$, a, b, c and r being nonnegative with $r < 1$.

Furi, Marteffi and Vignoli (Furi et al. (1980)) considered the special case of this problem where $\dot{x}(t) = \mu(t)\dot{x}(\phi(t)) + h(t, x(t), x(\phi(t)))$ where $\mu : [0,1] \to \mathbf{R}$ is continuous and satisfies $\|\mu(t)\| \leq t < 1$ and $h : [0,1] \times \mathbf{R}^2 \to \mathbf{R}$ is continuous and satisfies

$$|h(t,X,y)| \leq a + b|x| + c|y|$$

where a, b, c, r and ϕ are as above. Their method cannot be extended to treat the more general problem above. To this end we present the following results in Tarafdar and Thompson (1987).

Lemma 7.19 *Let a, b, c and r be nonnegative numbers with $r < 1$. Let $x, y \in C^1[0,1]$ and $\phi \in C[0,1]$ with $0 \leq \phi(t)$, for all $t \in [0,1]$. Further assume that for all $t \in [0,1]$ we have*

$$|\dot{x}(t)\| \leq a + b|x(t)| + c|x(\phi(t))| + r|\dot{x}(\phi(t))\| \tag{7.277}$$

$$\dot{y}(t) > a + by(t) + cy(\phi(t)) + r\dot{y}(\phi(t)) \tag{7.278}$$

and

$$|x(0)| < y(0) \tag{7.279}$$

Then $|x(t)| < y(t)$, *for all* $t \in [0,1]$.

Proof. Since $\phi(0) = 0$, we obtain by using (7.277), (7.278) and (7.279)

$$\dot{y}(0) > [a + (b+c)y(0)]/(1-r) \tag{7.280}$$
$$\geq [a + (b+c)|x(0)|]/(1-r) \tag{7.281}$$
$$\geq |\dot{x}(0)|. \tag{7.282}$$

Let $\tau = \sup\{t \in [0,1] : |x(s)| < y(s) \text{ and } |\dot{x}(s)| < \dot{y}(s), 0 \leq< t\}$. By virtue of (7.279), (7.280) and continuity it follows that $0 < \tau < 1$.

Clearly

$$|x(s)| < y(s) \,\text{and}\, |\dot{x}(s)| < \dot{y}(s) \,\text{for}\, s \in [0,\tau). \tag{7.283}$$

Suppose that $|x(,\tau)| = y(,\tau)$. Since $(\dot{s}) \geq |\dot{x}(s)| \geq 0$ for $s \in [0,\tau]$ and $0 \leq \phi(s) \leq s$, then $|x(\phi(s))| \leq y(\phi(s))$ and $\dot{y}(\phi(s)) \geq |\dot{x}(\phi(s))|$ for $s \in [0,\tau]$. Thus from (7.277) and (7.278)we obtain

$$|\dot{x}(s)| < \dot{y}(s) \quad \text{for} \quad s \in [0,\tau). \tag{7.284}$$

This, together with (7.279), implies that $y(\tau) > |x(\tau)|$ which is a contradiction.

Suppose that $y(\tau) > |x(\tau)|$. Now since $|x(\phi(\tau)| < y(\phi(\tau))$ and $\dot{y}(\phi(\tau)) \geq |\dot{x}(\phi(\tau))|$, it follows from (7.277) and (7.278) that $|\dot{x}(\tau) < \dot{y}(\tau)$ which is again a contradiction. Thus the lemma is proved. □

Corollary 7.63.1 *Let* a, b, c *and* r *be nonnegative numbers with* $r < 1$. *Let* $x, y \in C^1[0,1]$ *and* $\phi \in C[0,1]$ *with* $0 \leq \phi(t)$, *for all* $t \in [0,1]$. *Assume that* x *satisfies* (7.277) *and* $\dot{y}$ *is non-decreasing. Further assume that*

$$\dot{y}(t) > a + by(t) + cy(t) + r\dot{y}(t), \text{ for all } t \in [0,1] \tag{7.285}$$

and $|x(0)| < y(0)$. *Then* $y(t) > x(t)$, *for all* $t \in [0,1]$.

Proof. Clearly $\dot{y}(0) > 0$ and so $\dot{y}(t) > 0$, for all $t \in [0,1]$. Thus $y(\phi(t)) \leq y(t)$. Again as $\dot{y}$ is non-decreasing, $\dot{y}(t) \geq \dot{y}(\phi(t))$. Hence (7.278) is satisfied and the result follows from the Lemma 7.19. □

Corollary 7.63.2 *Let* a, b, c, r, x, y *and* ϕ *be as in Lemma 7.19 and assume that* (7.277) *to* (7.279) *hold. Then* $\dot{y}(t) > \dot{x}(t)|$ *for all* $t \in [0,1]$.

Proof. This is implicit in the proof of Lemma 7.19. □

Theorem 7.64 *Let* $h : [0,1] \times \mathbf{R}^3 \to \mathbf{R}$ *be continuous and satisfy* (7.275) *and* (7.276). *Then the mapping* $H : C^1[0,1] \to C[0,1]$ *defined by*

$$H(Y)(t) = h(t, y(t), y(\phi(t)), \dot{y}(\phi(t)))$$

is an r*-set contraction.*

Proof. Let $S \subset C^1[0,1]$ be a bounded set with $\alpha(S) = \lambda$. We show that $\alpha(H(S)) \leq r\lambda$ where α is the measure of non-compactness in the relevant spaces. Denote by S' the set $\{y' : y \in S\}$ and identify S and S' as subsets of C. Let $G : C \times C \to C$ be defined by $G(u,v)(t) = h(t, u(t), u(\phi(t)), v(\phi(t)))$, for $t \in [0,1]$. As $H(S) \subset G(S,S')$ it suffices to show that $\alpha(G(S,S')) \leq r\alpha(S') \leq r\lambda$. Now the set $T = \{z : z = y(t)$ or $z = y'(t)$ for some $y \in S$ and $t \in [0,1]\}$ is bounded. So given $\varepsilon > 0$, by the uniform continuity of h on $[0,1] \times T^3$, there is $\eta > 0$ such that $|h(t,x,y,z) - h(t,p,q,z)| < \varepsilon$, whenever $|x-p| + |y-q| < 2\eta, t \in [0,1]$, and $x,y,z,p,q \in T$. As S is a bounded equicontinuous subset of C there is an η-net $u_1, \ldots, u_m$ for S in C. Thus $G(S,S') \subseteq U_{i-1}^m B(G(u,,S'),\varepsilon)$ and by Lemma 7.16 and property (iv) of the measure of non-compactness

$$\alpha(G(S,S')) \leq \max\{\alpha(G(u_i,S')) + 2\varepsilon : 1 \leq i \leq n\}$$

Thus it suffices to show that $\alpha(G(u_i,S')) \leq r\alpha(S')$ for $1 \leq i \leq m$. Let V be an arbitrary bounded subset of C. We show that $\delta(G(u_i,V))$, where δ denotes diameter in C. For $v_1, v_2 \in V$,

$$|G(u_i,v_1)(t) - G(u_i,v_2)(t) \leq r|v_l(\phi(t)) - v_2(\phi(t))|$$

so $\|G(u_i,v_i) - G(u_i,V_2)\| \leq r\|v_1 - v_2\|$ and the result follows. □

Problem A:

We now turn our attention to the following problem of Tarafdar and Thompson (1987):

$$A \begin{cases} \dot{x}(t) = h(t, x(t), x(\phi(t)), \dot{x}(\phi(t))), \\ x(0) = d \end{cases}$$

where h satisfies (7.275) and (7.276) and $\phi \in C[0,1]$ with $0 \leq \phi(t) \leq t$ for $t \in [0,1]$.

Let $M : C^1[0,1] - C[0,1] \times \mathbf{R}$ be defined by

$$M(x)(t) = (\dot{x}(t) - h(t, x(t), x(\phi(t)), \dot{x}(\phi(t))), x(0))$$

Then M is $(0,k)$-epi for each $k \in [0, 1-r)$. In particular the problem has a solution for each $d \in \mathbf{R}$.

Proof. We first note that $L : C^1[0,1] \to C[0,1] \times \mathbf{R}$ given by $(Lx)(t) = (\dot{x}(t), x(0))$ is an isomorphism with $\|L_1\| = 1$; here we use the norm in $C[0,1] \times \mathbf{R}$ defined by $\|(x,r)\| = \|x\| + |r|$ for $(x,r) \in C[0,1] \times \mathbf{R}$ where $\|x\|$ is the usual sup norm. Let $K : C^1[0,1] \to C[0,1] \times \mathbf{R}$ be defined by $K(x)(t) = (h(t, x(t), x(\phi(t)), \dot{x})\phi(t))), d)$. We claim that $S = \{x \in C^1[0,1] : Lx = \lambda K(x)$ for some $\lambda \in [0, I]\}$ is bounded. To this end let $Lx = \lambda K(x)$ where $\lambda \in [0,1]$ and $x(0) = \lambda d$. By (7.275)

$$|\dot{x}(t)| \leq at + b|x(t)| + c|x(\phi(t))| + r|\dot{x}(\phi(t))|$$

Let $y(t) = a + (b+c)y(t) + r\dot{y}(t) + 1$, that is, $\dot{y}(t) = [(1+a) + (b+c)y(t)]/(1-r)$ and $y(0) = 1 + |d|$. By solving this equation we can easily see that $\dot{y}(t)$ is non-decreasing.

Since $y(0) = 1 + |d| > |x(0)| = \lambda|d|$, by Corollaries 7.63.1 and 7.63.2 it follows that $\|x\| \leq \|y\|$ and hence S is bounded.

From Theorem 7.64 it easily follows that K is an r-set contraction and hence $H : [0,1] \times C^1[0,1] \to C[0,1] \times \mathbf{R}$ defined by $H(\lambda, x) = -\lambda K(x)$ is also an r-set contraction. By the Corollary 7.62.1 L is $(0, 1-\varepsilon)$-epi for each $\varepsilon \in [0,1)$. Hence by the homotopy property in Tarafdar and Thompson (1987) $L - K(\cdot) = M(\cdot)$ is $(0, 1-\varepsilon-r)$-epi. □

Remark 7.42 It is not difficult to see that the above result can be strengthened by replacing (7.275) by

$$|h(t, X, Y, z)| \leq \psi(|x|) + \psi(|y|) + r|z|$$

where $r \in [0,1)$ is a constant and $\psi : [0,\infty) \to (0,\infty)$ is a continuous strictly increasing function satisfying $f^{\infty}(\psi(s))^{-1}ds = \infty$. All that is required is the corresponding modification of Lemma 7.19 and its corollaries and the observation that

$$\dot{y}(t) = (2\psi(|y(t)|) + 1)/(1-r), \;\; y(0) = 1$$

has a solution on $[0,1]$ with $\dot{y}(t)$ non-decreasing.

Example 7.5 Let Ω be a bounded $C^{1,1}$ domain in $\mathbf{R}^n$, let $L^P(\Omega)$ be the spaces of Lebesgue measurable functions with the usual norms, for $1 \leq p \leq \infty$, and let $W_0^{p,q}(\Omega)$ and $W^{p,q}(\Omega)$ be the usual Sobolev spaces for $p, q = 1, 2, \cdots$. Let $a_i j \in C(\bar{\Omega}), b_i, c \in L^{\infty}(\Omega)$ for $i, j = 1, 2, \ldots, n$, and let

$$a_{ij}(x)\xi_i\xi_j \geq \lambda_0|\xi|^2$$

for all $\xi \in \mathbf{R^n}, x \in \Omega$ and, some constant $\lambda > 0$ and $c \leq 0$. Let $h : \Omega \times \mathbf{R} \times \mathbf{R}^n \times \mathbf{R}^{n^2}$ satisfy

$$|h(x, u, p, s)| \leq k\{|u| + |p| + |s|\} + g(x)$$

for some constant $k > 0$ and $g \in L^2(\Omega)$ and

$$|h(x, u, p, s) - h(x, u, p, t)| \leq k|s - t|,$$

for all $(x, u, p) \in \Omega \times \mathbf{R} \times \mathbf{R}^n$ and for all $s, t \in \mathbf{R}^{n^2}$. Moreover let h satisfy the Carathéodory conditions: that is, let $h(x, \cdot, \cdot, \cdot) : \mathbf{R} \times \mathbf{R^n} \times \mathbf{R}^{n^2}$ be continuous a.e. $x \in \Omega$ and let $h(\cdot, u, p, s) : \Omega \to \mathbf{R}$ be measurable for all $(u, p, s) \in \mathbf{R} \times \mathbf{R^n} \times \mathbf{R}^{n^2}$.

Let $B = W_0^{1,2}(\Omega) \cap W^{2,2}(\Omega)$ with the $W^{2,2}$ norm and define $L : B \to L^2(\Omega)$ by $Lu = a_{ij}D_{ij}u + b_iD_iu + cu$. We look for a solution $u \in B$ of the Dirichlet problem

$$B\{Lu = \varepsilon h(x, u, Du, D^2u) \;\text{ in}\, \Omega$$

where $e > 0$ is sufficiently small.

Define the Nemitsky operator $M : L^2(\Omega) \times (L^2(\Omega))^n \times (L^2(\Omega))^{n^2} \to L^2(\Omega)$ by $M(u, p, s)(x) = h(x, u(x), p(x), s(x))$. From the growth on h, M ic continuous (see Vainbreg (1964), p. 162). By the Kondrachov compactness theorem (see Gilbarg and

Trudinger (1983), p. 167) the mapping $K : B \to L^2(\Omega) \times (L^2(\Omega))^n$, given by $K(u) = (u, Du)$ is completely continuous. We show that the mapping $H : B \to L^2(\Omega)$ given by $H(u) = M(K(u), D^2u)$ is a $2k$-set contraction. As M, K and D^2 are continuous, H is continuous. We show that for any bounded set $S \subseteq B, H(S) \subseteq H_1(S) + H_2(S)$ where $\overline{H_2(S)}$ is compact and $\alpha(H_1(S)) \leq 2k\alpha(S)$; here δ and α represent the diameter and the measure of noncompactness, respectively, in the appropriate spaces and for $G : B \to L^2(\Omega), G(S) = \{G(u) : u \in S\}$. Fix $w \in S$ and define $H_1, H_2 : B \to L^2(\Omega)$ by $H_2(u) = M(K(u), D^2w)$ and $H_1(u) = H(u) - H_2(u)$ for all $u \in B$. As $|D^2w| \in L^2(\Omega)$ it follows that $M(\cdot, \cdot, D^2w) : L^2(\Omega) \times (L^2(\omega))^n \to L^2(\Omega)$ is continuous and hence H_2 is completely continuous. We show that $\delta(H_1(S)) \leq 2k\delta(S)$ and then $\alpha(H_1(S)) \leq 2k\alpha(S)$ as S is an arbitrary bounded set. By the Lipschitz condition on $h, |h(x, u, Du, D^2u) - h(x, u, Du, D^2w)| \leq k|D^2u - D^2w|$ so that $\|H_1(u)\| \leq k\|u\|$ and hence $\delta(H_1(S)) \leq 2k\delta(S)$. Now $\alpha(H(S)) \leq \alpha(H_1(S)) \leq \alpha(H_2(S)) + \alpha(H_2(S)) \leq \alpha(H_1(S)) \leq 2k\alpha(S)$ as required.

By Theorem 9.15 and Corollary 9.17 of Gilbarg and Trudinger (1983), $L : B \to L^2(\Omega)$ is an isomorphism. Choose $\varepsilon_0 > 0$ such that $2k\varepsilon_0 l^{-1} < 1$ where $l^{-1} = \|L^{-1}\|$. Let $S = \{(\lambda, u) \in [-1, 1] \times B : Lu = \lambda\varepsilon_0 H(u)\}$. We show that S is bounded. Let $(\lambda, u) \in S$ then $\|\lambda\varepsilon_0 H(u)\| \leq \varepsilon\|g\| + k\varepsilon\|(u)\|$ thus $\|u\| \leq l^{-1}\varepsilon_0\|g\| + k\varepsilon_0\, l^{-1}\|u\|$ and $\|u\| \leq 2l^{-1}\varepsilon_0\|g\|$. As εH is a $2\varepsilon_0 k$-set contraction for $|\varepsilon| \leq 2\varepsilon_0 k l^{-1} < 1$, and L is an isomorphism with $\|L^{-1}\| = l^{-1}$, then $L - \varepsilon H$ is $(0, \bar{k})$-epi for each $\bar{k} \in [0, l - 2\varepsilon_0 k)$, by Corollary 7.62.1 and the homotopy property of Subsection 7.8.3. Now Problem B is equivalent to finding a solution $u \in B$ of $Lu = \varepsilon H(u)$ so Problem B has a solution for each ε with $|\varepsilon| \leq \varepsilon_0$.

Problem C of the next example is well known. There are several different existence proofs in the literature. We shall give an existence proof from Tarafdar and Thompson (1987) which is based on the previous results of this section.

Example 7.6 Let X be a Banach space with norm denoted by $|\cdot|$. Let $C([0, 1], X)$ be the space of continuous functions $y : [0, 1] \to X$ with norm defined by $\|y\| = \sup\{e^{-lt}|y(t)| : 0 \leq t \leq 1\}$, where $l > 1$ is a constant given below. Let $C^1([0, 1], X)$ be the space of continuously differentiable functions y in $C([0, 1], X)$ with norm $|||y||| = \max\{\|y\|, \frac{1}{l}\|\acute{y}\|\}$. Clearly a bounded subset $S \in C^1$ is a bounded equicontinuous subset of C. Let $h : [0, 1] \times X \to X$ be uniformly continuous and assume there is a constant $k > 0$ such that $|h(t, y)| \leq k$, for all $(t, y) \in [0, 1] \times X$, and $\alpha(h(t, S)), \leq k\alpha(S)$,for all bounded subsets $S \subseteq X$.Then there is a solution $y \in C^1$ of the problem

$$C\begin{cases} y' = h(t, y), \text{ for all } t \in [0, 1], \\ y(0) = y_0. \end{cases}$$

It suffices to consider the case $y_0 = 0$. Let $C_0^1([0, 1], X) = \{y : y \in C^1, y(O) = 0\}$. Clearly C_0^1 is a closed subset of C^1.

Define $L : C_0^1 \to C$ by $(Ly)(t) = y'(t)$ and $H : C_0^1 \to C$ by $H(y)(t) = h(t, y(t))$.

If $S \subseteq C_0^1$ is a bounded set then for an arbitrary bounded subset $D \subseteq C$ and $t \in [0,1]$ let $D(t) = \{y(t) : y \in D\}$. If $S \subset C_0^1$ is a bounded set then $H(S) = \{H(y) : y \in S\}$ is a bounded equicontinuous subset of C so by variants of Lemma 2.3.1 and Theorem 2.3.1 of Martin (1976)

$$\begin{aligned} \alpha(H())) &= \sup\{e^{-lt}\alpha(H(S)(t) : 0 \le t \le 1\} && (7.286)\\ &\le \sup\{e^{-lt}k\alpha(S(t)) : 0 \le t \le 1\} && (7.287)\\ &= k\alpha(S) && (7.288)\end{aligned}$$

Clearly H is continuous and it is not difficult to show that L is an isomorphism with $\|L^{-1}\| \le l^{-1}$. Choose $l > k$. Now the problem C is equivalent to solving $Ly = H(y)$. Let $(\lambda, y) \in S = \{(\lambda, y) \in [0,1] \times C_0^1 : Ly = \lambda H(y)\}$, then for all $t \in [0,1]$, $y'(t) = \lambda h(t, y(t))$ and $y(0) = 0$. Thus $|\dot{y}t)| \le k$ and hence $|y(t)| \le k$, for all $t \in [0,1]$. Hence $|||y||| \le k$ and S is bounded.

The result now follows by an argument similar to that in Example 7.5. In particular $L - H$ is $(0, \bar{k})$-epi for $\bar{k} \in [0, \|L^{-1}\|^{-1} - k)$ and hence problem C has a solution.

Example 7.7 Let $\phi \in C^2(\mathbf{R})$ satisfy $\phi'(x) < 0$ and $\phi''(x) > 0$, for all $x \in \mathbf{R}$ and set $\Omega = \{(x,y) \in \mathbf{R}^2 : y > \phi(x)\}$. Let $h : \bar{\Omega} \times \mathbf{R}^3 \to \mathbf{R}$ be continuous. Let $\bar{y} = \phi(x)$ and $\bar{x} = \phi^{-1}(y)$ and for $u : \bar{\Omega} \to \mathbf{R}$ let

$$\|u\| - \sup\{c^{-l[x-\bar{x}+y-\bar{y}]}|u(x,y)| : (x,y) \in \bar{\Omega}\}$$

where $l > 1$ is chosen later. Let $C(\bar{\Omega}) = \{u : \bar{\Omega} \to \mathbf{R}, u \text{ is continuous}, \|u\| < \infty\}$.

Consider the Cauchy problem

$$D\begin{cases} u_{xy}(x,y) = h(x,y,u,u_x,u_y) & \text{in } \Omega,\\ u = u_x - u_y = 0 & \text{on } \partial\Omega,\end{cases}$$

where a solution u is a function satisfying $u, u_x, u_y, u_{xy} = u_{yx} \in C(\bar{\Omega})$ and D pointwise.

We assume that h satisfies

$$|h(x,y,u,p,q) - h(x,y,u,s,t) \le k\{|p-s| + |q-t|\},$$

and

$$|h(x,y,u,p,q)| \le k\{|u| + |p| + |q|\} + g(x,y)$$

for all $(x,y) \in \Omega$ and for all $p,q,r,s \in \mathbf{R}$, where $g \in C(\bar{\Omega})$. We show that problem D has a solution. By translating axes we may assume that $(0,0) \in \partial\Omega$. Let $\Omega_n = \{(x,y) \in \Omega : |x|, |y| < n\}$, for all $n = 1, 2, \ldots$, and let $C(\bar{\Omega})$ and $\|u\|$ for $u \in C(\bar{\Omega}_n)$ have the obvious interpretations. We find a solution u_n of problem

$$D_n \begin{cases} u_{xy}(x,y) = h(x,y,u,u_x,u_y) & \text{in } \Omega_n, \\ u = u_x = u_y = 0 & \text{in } \partial\Omega \cap \bar{\Omega}_n, \end{cases}$$

and show problem D has a solution by a standard diagonalization argument. Let

$$B_n = \{u : u, u_x, u_y, \quad \text{and} \quad u_{xy} = u_{yx} \in C(\bar{\Omega}_n), u = u_x = u_y = 0 \quad \text{on} \quad \partial\Omega \cap \bar{\Omega}_n\}$$

and define a norm on B_n by $|||u||| = \|u\| + \|u_x\| + \|u_y\| + l_{-1}\|u_{xy}\|$. Thus B_n is a Banach space. Let $L, N : B_n \to C(\bar{\Omega}_n)$ be given by $(Lu)(x,y) = u_{xy}(x,y)$ and $H(u)(x,y) = h(x,y,u,u_x,u_y)$, for all $u \in B_n$ and for all $(x,y) \in \bar{\Omega}_n$. It is not difficult to show that problem D_n is equivalent to solving $Lu = H(u)$. Now L is an isomorphism with $\|L^{-1}\| \leq l^{-1}$. Choose $l > 2k$. Now H is a k-set contraction. This follows by a similar argument to that in Example 7.4 since a bounded subset of B_n has compact closure in $C(\bar{\Omega}_n)$. Let $S_n = \{(\lambda, u) \in [0,1] \times B_n : Lu = \lambda H(u)\}$ then S_n is bounded independently of n. To see this $(\lambda, u) \in S_n$ then $\|\lambda H(u)\| \leq k|||u||| + \|g\|$. Since $\|L{-}1\| \leq (2k)_{-1}$ then $|||u||| \leq k_{-1}\|g\| \leq C$, a constant as $g \in C(\bar{\Omega})$. By an argument similar to that in Example 7.5, there is a solution $u_n \in B_n$ of $Lu = H(u)$. Now let $T_n = \{u \in B_n : Lu = H(u)\} = (L - H)^{-1}(\{0\})$. By Theorem 7.51, $\alpha(T_n) = 0$ since $k_1 k = \|L^{-1}\|k < 1$. Clearly $u_m|_{\bar{\Omega}_n} \in T_n$, for all $m \geq n$. Choose a subsequence of u_m denoted by $u_{m,1}$, after re-labeling, such that $u_{m,1}|_{\bar{\Omega}_1}$ converges in B_1. Proceeding inductively we may choose a subsequence of $u_{m,n}$ denoted by $u_{m,n+1}$, after re-labeling, such that $u_{m,n+1}|_{\bar{\Omega}_n}$ converges in B_{n+1}. Thus $u = \lim_{n\to\infty} u_{n,n}$ with the limit appropriately interpreted is a solution of Problem D, as required.

Remark 7.43 Problem D is similar to one studied by Hartman and Wintner (1952); however they establish existence in a bounded domain by producing a convergent sequence of approximate solutions. Thus, the existence result of Tarafdar and Thompson (1987) in a bounded domain is cleaner. It would be interesting to know if the existence of a solution in the unbounded domain could be established directly without finding solutions in an expanding sequence of bounded domains. In the case h is Lipschitz with respect to u as well as u_x and u_y the existence of a solution on all of Ω follows directly from the contraction mapping principle, as is well known.

Remark 7.44 In Fitzpatrick and Petryshyn (1979), a problem of the following form was considered:

$$Lu = g(x, u, \nabla u, D^2 u) + h(x, u, \nabla u, Lu), \quad \text{for all} \quad x \in \Omega.$$

In the above problem L is a linear second order elliptic partial differential operator from $W^{2,2}(\Omega) \cap W^{1,2}(\Omega)$ to $L^2(\Omega)$. Although there is some formal similarity with the Problem B of Example 7.5, the above problem is somewhat different from, and the results in Fitzpatrick and Petryshyn (1979) do not apply to, the Problem B of Example 7.5 of this section.

Chapter 8

Nonexpansive Types of Mappings and Fixed Point Theorems in Locally Convex Topological Vector Spaces

8.1 Nonexpansive Types of Mappings in Locally Convex Topological Vector Spaces

In this chapter we consider some results concerning nonexpansive types of mapping on locally convex topological vector spaces. Unless otherwise stated throughout this chapter (E, τ) will denote a real locally convex Hausdorff topological vector space. In here we will not consider the results explicitly on nonexpansive mappings on Banach spaces, i.e the results in Browder (1965a), Browder (1966), Kirk (1965), Opial (1967), Belluce and Kirk (1967), Bruck (1973), Dozo (1973), Kirk and Yanez (1988), Tan and Hong-Kun (1991) and in many other places(e.g. see Zeidler (1985) and Kirk (1981) and references thereof).

8.1.1 *Nonexpansive Mappings*

Some results concerning fixed theorems for nonexpansive mappings on locally convex topological vector spaces have been obtained in Taylor (1972) and also in Tarafdar (1974). In Kakutani (1938), Markov (1936), and Day (1961) fixed point theorems for commutative family of linear contnuous of self mappings on a compact convex subset of a topological vector space have been investigated. In DeMarr (1963), Belluce and Kirk (1966), Hong (1968) and in many other places the fixed point theorems for a commutative family of nonexpansive mappings on a Banach space have been considered. These theorems have been proved in locally convex topological vector spaces in Tarafdar (1975) which will be the subject matter of the next two subsections. The main tool here will be the Minkowski functional of a balanced, convex (i.e. absolutely convex) bounded subset of (E, τ).

A family $[p_\alpha : \alpha \in I]$ of seminorms defined on (E, τ) is said to be an associatd family of seminorms for τ if the family $[\rho U : \rho > 0]$, where $U = \bigcap_{i=1}^{n} U_{\alpha_i}$ and $U_{\alpha_i} = \{x \in E : p_{\alpha_i}(x) < 1\}$, forms a base of neighbourhoods of 0 for τ, I being an index set. The set U is also given by $U = \{x \in E : p(x) < 1\}$ where p is the seminorm $\max[p_{\alpha_1}, p_{\alpha_2}, \ldots, p_{\alpha_n}]$.

A family $[p_\alpha : \alpha \in I]$ of seminorms on E is said to be an augmented associated family for τ if $[p_\alpha : \alpha \in I]$ is an associated family for τ and has the further property that, given $\beta, \gamma \in I$, the seminorm $\max[p_\beta, p_\gamma] \in [p_\alpha : \alpha \in I]$. We shall denote an associated family and augmented family for τ by $A(\tau)$ and $A^*(\tau)$ respectively. It is wellknown (e.g. see Köthe (1969, p. 203)) that given a locally convex topological vector space (E, τ) there always exists a family $[p_\alpha : \alpha \in I]$ of seminorms defined on E such that $[p_\alpha : \alpha \in I] = A^*(\tau)$. Conversely each family $[p_\alpha : \alpha \in J]$ of seminorms defined on E with property that for each $x \in E$ with $x \neq 0$ there is atleast one $\alpha \in J$ such that $p_\alpha(x) \neq 0$ always determines a unique locally convex topology τ on E such that $A(\tau) = [p_\alpha : \alpha \in J]$ and $A(\tau)$ can be extended to $A^*(\tau)$ by adjoining to $A(\tau)$ all seminorms of the form $\max[p_{\alpha_1}, p_{\alpha_2}, \ldots, p_{\alpha_n}]$ for each finite subset $[\alpha_1, \alpha_2, \ldots, \alpha_n]$ of the index set J.

Definition 8.1 A mapping $f : M \to M$ of a subset M of E into itself is said to be $A(\tau)$ (respectively, $A^*(\tau)$)-nonexpansive on M if, for all $x, y \in M$,

$$p_\alpha(f(x) - f(y)) \leq p_\alpha(x - y) \quad \text{for each } p_\alpha \in A(\tau)(\text{respectively } \in A^*(\tau)).$$

It is trivial to see that f is $A^*(\tau)$-nonexpansive then f is $A(\tau)$-nonexpansive. It is also true (see Tarafdar (1974) or the begining of Chapter 2) that if f is $A(\tau)$-nonexpasive, then f is also $A^*(\tau)$-nonexpansive. Hence, instead of saying f is $A(\tau)$- or $A^*(\tau)$-nonexpansive, we say simply say that f is nonexpansive in either case.

In what follows the following construction will be crucial. Let M be a τ-bounded set in (E, τ) and let $A*(\tau) = [p_\alpha : \alpha \in I]$. Let us consider the family $\{U_\alpha : \alpha \in I\}$ where $U_\alpha = \{x \in E : p_\alpha(x) \leq 1\}$. Then the family $\{U_\alpha : \alpha \in I\}$ is a base of closed absolutely convex neighbourhoods of 0.

Since M is τ-bounded, we can select a number $\lambda_\alpha > 0$ for each $\alpha \in I$ such that $M \subseteq \lambda_\alpha U_\alpha$. Then clearly $B = \bigcap\limits_{\alpha \in I} \lambda_\alpha U_\alpha$ is τ-bounded, τ-closed and absolutely convex and $M \subseteq B$. The linear span of B in E is equal to $E_B = \bigcup_{n=1}^{\infty} nB$ and B is an abslutely convex α-body (i.e. has an algebraic interior point). The Minkowski functional of B is a norm $\|.\|_B$ on E_B. Thus E_B is a normed linear space with norm $\|.\|_B$ and closed unit ball B. The norm topology on E_B is finer than the topology on E_B induced by τ (for details see Köthe (1969, p. 252) or Horváth (1966, pp. 207–208)). Now since p_α is the Minkowski functional of U_α and $\|.\|_B$ is the Minkowski functional of B and $B \subseteq \lambda_\alpha U_\alpha$, we can easily see that for each $x \in E_B$, $p_\alpha(x) \leq \lambda_\alpha \|x\|_B$.

Thus for each $\alpha \in I$, we have

$$p_\alpha\left(\frac{x}{\lambda_\alpha}\right) \leq \|x\|_B \,. \tag{8.1}$$

We now prove that

$$\sup_{\alpha \in I} p_\alpha\left(\frac{x}{\lambda_\alpha}\right) = \|x\|_B \qquad \text{for each } x \in E_B. \tag{8.2}$$

Let $x \in E_B$. We assume that $\sup_{\alpha \in I} p_\alpha\left(\frac{x}{\lambda_\alpha}\right) < \|x\|_B$ and deduce a contradiction. Let $\sup_{\alpha \in I} p_\alpha\left(\frac{x}{\lambda_\alpha}\right) = \lambda$. Then we have $p_\alpha\left(\frac{x}{\lambda_\alpha}\right) \leq \lambda < \|x\|_B$ for each $\alpha \in I$. Now $p_\alpha\left(\frac{x}{\lambda_\alpha}\right) \leq \lambda$ implies that $\frac{x}{\lambda} \in \lambda_\alpha U_\alpha$ for each $\alpha \in I$, i.e. $\frac{x}{\lambda} \in B$. But $\|x\|_B > \lambda$ implies that $x \notin \lambda B$. Thus we have contradiction.

We are now in a position to prove the following theorem.

Theorem 8.1 *Let (E, τ) be a locally convex Hausdorf topological vector space and $A^*(\tau) = [p_\alpha : \alpha \in I]$. If f is a nonexpansive mapping on a τ-bounded subset $M \subseteq E$, then f is also nonexpansive on M with respect to the norm $\|.\|_B$ where $\|.\|_B$ has the meaning as explained above.*

Proof. Let $x, y \in M$. Then since f is nonexpansive on M,

$$p_\alpha(f(x) - f(y)) \leq p_\alpha(x - y) \qquad \text{for each } \alpha \in I.$$

Hence $\sup_{\alpha \in I} p_\alpha\left(\frac{f(x)-f(y)}{\lambda_\alpha}\right) \leq \sup_{\alpha \in I} p_\alpha\left(\frac{x-y}{\lambda_\alpha}\right)$. Thus $\|f(x) - f(y)\|_B \leq \|x - y\|_B$ from (8.2). □

We recall that the following definition was made in Chapter 2.

Definition 8.2 A subset X of E is called starshaped if there exists a point $p \in X$ such that for each $x \in X$ and real t with $0 < t < 1$, $tx + (1-t)p \in X$. p is called a star centre of X. Each convex subset of E is thus starshaped.

Remark 8.1 The following result, with M assume to be complete, is known (see Tarafdar (1974) or Chapter 2 here and Taylor (1972)). Here we have relaxed the completeness by sequential completeness.

Theorem 8.2 *Let (E, τ) and $A^*(\tau)$ be as in Theorem 8.1. Let M bo a nonempty, starshaped, τ-bounded, and τ-sequetially complete subset of E, and f be a nonexpansive mapping on M. Then 0 lies in $\|.\|_B$ - $\mathrm{cl}(I - f)M$ and hence in $\tau - \mathrm{cl}(I - f)M$ where I is the identity map on M, $\mathrm{cl}\, A$ stands for the closure of subset A of E and $\|.\|_B$ has the meaning as explained earlier.*

Proof. We have already mentioned that E_B is a normed space with norm $\|.\|_B$ and with B as the unit ball. Since the norm topology on E_B has a base of neighbourhoods of 0 consisting of τ-closed sets, namely the scalar multiples of B and M is τ-sequentially complete, we know that M is a $\|.\|_B$-sequentially complete subset of E_B (apply 18, 4.4 of Köthe (1969) to the topology on E_B induced by τ and the $\|.\|_B$-topology on E_B). Let p be the star centre of M. For each $t, 0 < t < 1$, we define

$$f_t(x) = tf(x) + (1-t)p, x \in M.$$

Then clearly f_t maps M into itself. Moreover,

$$\|f_t(x) - f_t(y)\|_B = \|t(f(x) - f(y))\|_B \leq t\,\|x - y\|_B \qquad \text{for all } x, y \in M$$

as f is nonexpansive on M with respect to the norm $\|.\|_B$ by Theorem 8.1. Thus f_t is a contraction on M with respect to the norm $\|.\|_B$. Now since M is $\|.\|_B$-complete, by Banach contraction mapping principle, f_t has a unique fixed point x_t, say, in M. By definition of f_t, we have

$$(I-f)(x_t) = x_t - \frac{1}{t}(f_t(x_t) - (1-t)p) = \left(1 - \frac{1}{t}\right)(x_t - p).$$

Hence

$$\|(I-f)(x_t)\|_B \leq |1 - \frac{1}{t}|(\|x_t\|_B + \|p\|_B) \leq 2|1 - \frac{1}{t}| \to 0$$

as $t \to 1$, because x_t and p are in the unit ball of E_B. Thus $0 \in \|.\| - \text{cl}(I-f)M \subseteq \tau - \text{cl}(I-f)M$. The last inclusion follows from the fact that the $\|.\|_B$-topology is finer than the topology induced on E_B by τ. □

Remark 8.2 This theorem incudes Theorem 2.2 in Taylor (1972) (also see Lemma 3.1 in Tarafdar (1974)). Also we note that here we have obtained a stronger result under a weaker hypothesis.

Corollary 8.2.1 *Let (E, τ) and $A^*(\tau)$ be as in Theorem 8.1. Let f be nonexpansive on a nonempty τ-sequentially complete, τ-bounded and starshaped subset M of E, and let $(I-f)$ map τ-bounded and τ-sequentially closed subsets of M into τ-sequentially closed subsets of M. Then f has a fixed point in M.*

A point $p \in \tau$-cl M is a τ-sequential limit point of M if there is exists a sequence $\{p_n\}$, $p_n \in M$, such that $p_n \to p$ in the τ-topology. M is called τ-sequentially closed if each τ-sequential limit point of M belongs to M.

Proof. Since M is τ-sequentially complete and E is Huasdorf, it follows that M is τ-sequentially closed. (Let $p_n \to p$ in the τ-topology and $p_n \in M$. The $\{p_n\}$ is a τ-Cauchy sequence and,therefore, $p \in M$.) Hence, by hypothesis $(I-f)M$ is τ-sequentially closed. By Theorem 8.2, $0 \in \|.\|_B - \text{cl}(I-f)M$. But

$$\|.\|_B - \text{cl}(I-f)M \subseteq \tau\text{–sequential–}\,\text{cl}(I-f)M$$

because it follows that each point in $\|.\|_B - \text{cl}(I-f)M$ is a τ-sequential limit point of $(I-f)M$ as $\|.\|_B$-topology is finer than the τ-topology. Hence $0 \in (I-f)M$. This completes the proof. □

Corollary 8.2.2 *Let (E, τ) and $A^*(\tau)$ be as in 8.1. Let f be nonexpansive on a nonempty, τ-sequentially compact and starshaped subset M of E. Then f has a fixed point in M.*

Proof. M being τ-sequentially compact is τ-bounded and τ-sequentially complete. Hence by Theorem 8.2 and by the reason given in Corollary 8.2.1, $0 \in \|.\|_B - \text{cl}(I-f)M \subseteq \tau$-sequential-cl$(I-f)M$. Thus there exists a sequence $\{y_n\}$, $y_n \in (I-f)M$, such that $y_n \to 0$ in the τ-topology. Now since f is nonexpansive on M, it follows that f is p_α-continuous for each $\alpha \in I$. Hence f is τ-continuous and, therefore, $(I - f)$ is τ-continuous. Then it follows that $(I - f)M$ is τ-sequentially compact as M is so. Now it is easy to see that $0 \in (I - f)M$. This completes the proof. □

Before we prove the main result of this section we need to prove two lemmas. The following result, which we write as a lemma, was proved in Göhde (1964/65, Theorem 4) in a normed space. We extend this to a locally convex topological vector space and also weaken the covexity hypothesis to the starshaped convexity.

Lemma 8.1 *Let (E, τ) and $A^*(\tau)$ be as in Theorem 8.1. Let f be nonexpansive on a nonempty, τ-closed, τ-bounded, and starshaped subset M of E. Further assume that there exists a τ-compact subset L of M such that, for each $x \in M$,*

$$\tau - \mathrm{cl}\{f^n(x) : n = 1, 2, \ldots\} \cap L \neq \emptyset.$$

Then there exists atleast one fixed point of f in L.

Proof. Let p be the star centre of M. For each t with $0 < t < 1$, we define

$$f_t(x) = tf(x) + (1 - t)p, \; x \in M.$$

Then, in exactly the same way as in the proof of Theorem 8.2 we can show that f_t is a contraction on M with respect to the norm $\|.\|_B$ where $\|.\|_B$ has the same meaning as before, i.e. as in (8.2).

For any $x \in M$, $\{f^n(x)\}$ is a $\|.\|_B$-Cauchy sequence and there are points in M which are displaced by f_t with respect to $|.\|_B$ by an arbtrary small amount. Let

$$\|f_t(x_t) - x_t\|_B \leq (1 - t).$$

Thus we have

$$\begin{aligned}
\|f(x_t) - x_t\|_B &= \|f(x_t) - \{tf(x_t) + (1 - t)p\} + f_t(x_t) - x_t\|_B \\
&\leq \|f(x_t) - tf(x_t)\|_B + (1 - t)\, \|p\|_B + \|f_t(x_t) - x_t\|_B \\
&\leq (1 - t)(\|f(x_t)\|_B + \|p\|_B + 1) \\
&\leq 3(1 - t)
\end{aligned}$$

as $f(x_t)$ and p are in the unit ball of E_B. Thus there are points in M which are displaced by f (with respect to $\|.\|_B$) by an arbitrary small amount.

By the above inequality and the $\|.\|_B$-nonexpanion of f on M (due to Theorem 8.1) we have that, for each positive integer n,

$$\left\|f^{n+1}(x_t) - f^n(x_t)\right\|_B \leq 3(1 - t).$$

Thus from (8.1), we have

$$p_\alpha\left(\frac{f^{n+1}(x_t) - f^n(x_t)}{\lambda_\alpha}\right) \leq 3(1 - t) \quad \text{for each } \alpha \in I. \tag{8.3}$$

Now, by our hypothesis, there is a $y_t \in L$ such that y_t is a τ-limit point of $\{f^n(x_t)\}$. Clearly y_t is also p_α-limit point of $\{f^n(x_t)\}$ for each $\alpha \in I$. Now let $\alpha \in I$ be arbitrary. Then, since y_t is a p_α-limit point of $\{f^n(x_t)\}$, for any $\varepsilon > 0$, there is a positive integer n such that

$$p_\alpha\left(f^n(x_t) - y_t\right) < \varepsilon. \tag{8.4}$$

Let ε be arbitrarily chosen. Then since $\lambda_\alpha > 0$, there is a positive integer m such that

$$p_\alpha\left(f^m(x_t) - y_t\right) < \lambda_\alpha\, \varepsilon. \tag{8.5}$$

From (8.3), (8.5) and the p_α-nonexpansion of f, we have

$$\begin{aligned} p_\alpha\left(\frac{f(y_t) - y_t}{\lambda_\alpha}\right) &\leq p_\alpha\left(\frac{f(y_t) - f^{m+1}(x_t)}{\lambda_\alpha}\right) + p_\alpha\left(\frac{f^{m+1}(x_t) - f^m(x_t)}{\lambda_\alpha}\right) \\ &\quad + p_\alpha\left(\frac{f^m(x_t) - y_t}{\lambda_\alpha}\right) < \varepsilon + 3(1-t) + \varepsilon. \end{aligned}$$

Since ε is arbitrary, we must hae

$$p_\alpha\left(\frac{f(y_t) - y_t}{\lambda_\alpha}\right) \leq 3(1-t). \tag{8.6}$$

Now we consider a sequence $\{t_i\}$ of real numbers such that $0 < t_i < 1$ for each i and $\lim_{i\to\infty} t_i = 1$.

As L is τ-compact, the sequence $\{y_{t_i}\}$ has a τ-cluster point y in L. Clearly y is also a p_α-cluster point of $\{y_{t_i}\}$ and hence we can select a subsequence $\{y_{t_{n_i}}\}$ of $\{y_{t_i}\}$ such that $y_{t_{n_i}} \to y$ as the p_α-topology satisfies the first axiom of countability. In view of (8.6) we have

$$\lim_{i\to\infty} p_\alpha\left(f\left(y_{t_{n_i}}\right) - y_{t_{n_i}}\right) \leq \lim_{i\to\infty} 3\lambda_\alpha\left(1 - t_{n_i}\right) = 0.$$

Again since f is p_α-nonexpansive on M, it follows that f is p_α-continuous on M. Hence $f\left(y_{t_{n_i}}\right) \to f(y)$ in the p_α-topology; i.e. $\lim_{i\to\infty} p_\alpha\left(f\left(y_{t_{n_i}}\right) - f(y)\right) = 0$. We now have

$$p_\alpha(f(y) - y) \leq p_\alpha\left(f(y) - f\left(y_{t_{n_i}}\right)\right) + p_\alpha\left(f\left(y_{t_{n_i}}\right) - y_{t_{n_i}}\right) + p_\alpha\left(y_{t_{n_i}} - y\right),$$

where $i = 1, 2, \ldots$. Taking the limit as $i \to \infty$, we have

$$p_\alpha(f(y) - y) = 0.$$

Since α is arbitrary, $p_\alpha(f(y) - y) = 0$ for each $\alpha \in I$. Again, since E is Hausdorff, $f(y) = y$. This completes the proof. □

Remark 8.3 In proving the above lemma, if we start at the outset with an arbitrary $\alpha \in I$, then it is true that f_t and $\{f^n(x)\}$ are respectively a contraction and a Cauchy sequence with respect to the seminorm p_α. But then x_t will depend on α and hence the technique of Göhde (1964/65) applied α-wise does not work. Thus it seems that the use of $\|.\|_B$, as made in the proof, is appropriate.

The next lemma was proved in DeMarr (1963) in a Banach space.

Lemma 8.2 *Let (E,τ) and $A^*(\tau)$ be as in Theorem 8.1. Let M be a nonempty, τ-compact subset of E and K the convex hull of M. If, for any $\beta \in I$, the p_β-diameter $\delta(M,\beta)$ of M is greater than 0, then there exists an element $u \in K$ such that*

$$\sup\{p_\beta(x-u) : x \in M\} < \delta(M,\beta).$$

Proof. The proof in DeMarr (1963) with slight adjustment will do. Since M is τ-compact, M is p_α-compact for each $\alpha \in I$. Thus there exist points x_0, x_1 in M such that $p_\beta(x_0 - x_1) = \delta(M,\beta)$. Let $M_\beta \subseteq M$ be the maximal so that $\{x_0, x_1\} \subseteq M_\beta$ and $p_\beta(x-y) = \delta(M,\beta)$ for all distinct $x, y \in M_\beta$. M_β is cleary nonempty. Also, since M is p_β-compact, it follows that M_β is finite. Let $M_\beta = \{x_0, x_1, \ldots, x_n\}$. We define

$$u = \sum_{k=0}^{n} \frac{1}{n+1} x_k \in K.$$

Since M is p_β-compact, there exists a point $y_0 \in M$ such that $p_\beta(y_0 - u) = \sup\{p_\beta(x-u) : x \in M\}$. Again, since $p_\beta(y_0 - u) \leq \delta(M,\beta)$ for all $k = 0, 1, \ldots, n$, we have

$$p_\beta(y_0 - u) \leq \sum_{k=0}^{n} \frac{1}{n+1} p_\beta(y_0 - x_k) \leq \delta(M,\beta).$$

Now $p_\beta(y_0 - u) = \delta(M,\beta)$ would imply that $p_\beta(y_0 - x_k) = \delta(M,\beta) > 0$ for all $k = 0, 1, \ldots, n$. But this would then imply, by definition of M_β, that $y_0 \in M_\beta$; i.e. $y_0 = x_k$ for $k = 0, 1, \ldots, n$, which would contradict that $p_\beta(y_0 - x_k) - \delta(M,\beta) > 0$ for all $k = 1, 2, \ldots, n$. Hence $p_\beta(y_0 - u) < \delta(M,\beta)$. This completes the proof. □

We now state and prove the main theorem of this section.

Theorem 8.3 *Let (E,τ) be a quasi-complete locally convex Hausdorff topological vector space and $A^*(\tau) = [p_\alpha : \alpha \in I]$. Let X be a nonempty, τ-bounded, τ-closed and convex subset of E and M be a τ-compact subset of X. let F be a nonempty commutative family of nonexpansive self mappings on X having the property that for some $f_1 \in F$ and for each $x \in X$,*

$$\tau - \mathrm{cl}\{{f_1}^n(x) : n = 1, 2, \ldots\} \cap M \neq \emptyset.$$

Then the family F has a common fixed point in M.

Proof. We take X as M in the definition of E_B. The proof proceeds in the general line of argument of Theorem 1 of Belluce and Kirk (1966). Let K be a nonempty,

τ-closed, and convex subset of X such that $f(K) \subseteq K$ for each $f \in F$. Let $x \in X$. Then $\{f_1{}^n(x)\} \subseteq K$. By hypothesis we have $K \cap M \supseteq \tau - \mathrm{cl}\{f_1{}^n(x)\} \cap M \neq \emptyset$. Applying Zorn's Lemma we obtain a subset a subset X^* of X which is minimal with respect to being nonempty, τ-closed, convex and being mapped into itself by each $f \in F$. We set $M^* = X^* \cap M$. $M^* \neq \emptyset$ by the above inclusion relation. By our Lemma 8.1 it follows that f_1 has a nonempty τ-closed fixed-point H in M^*. Now let $x \in H$ and $y = f(x)$ for any $f \in F$. Then by using the commutativity of F, we obtain, $f_1(y) = f_1[f(x)] = f[f_1(x)] = f(x) = y$. Hence $y \in H$ and $f(H) \subseteq H$ for each $f \in F$. Therefore, we are able to find a subset H^* of H which is minimal with respect to being nonempty, τ-closed, and mapped into itself by each $f \in F$. Let $g \in F$. Then g, being nonexpansive on X, is p_α-continuous for each $\alpha \in I$ and hence τ-continuous on X. Therefore $g(H^*)$ is τ-closed as H^* is τ-compact. Now for each $f \in F$, $f(g(H^*)) = g(f(H^*)) \subseteq g(H^*)$. Hence by the minimality of H^* implies that $g(H^*) = H^*$. Hence H^* is mapped onto itself by each $f \in F$.
Let W be the convex τ-closure of H^*. Then W is τ-compact, as H^* is so and E is quasi-complete. We now prove that $\delta(W, \alpha) = 0$ for each $\alpha \in I$ where as before $\delta(W, \alpha)$ is the p_α-diameter of W. we assume that $\delta(W, \beta) > 0$ for some $\beta \in I$ and deduce a contradiction. Then, by applying our Lemma 8.2 to the compact set W, there is a point $x \in W$ such that

$$\sup\{p_\beta(x - z) : z \in W\} = r < \delta(W, \beta).$$

As in Belluce and Kirk (1966) we set

$$C_1{}^\beta = \{w \in W : p_\beta(w - z) \leq r \text{ for all } z \in H^*\}$$

and

$$C_2{}^\beta = \{w \in X^* : p_\beta(w - z) \leq r \text{ for all } z \in H^*\}.$$

Then $C_1{}^\beta = C_2{}^\beta \cap W$. Since $f(H^*) = H^*$ for each $f \in F$, by using p_β-nonexpansion of each $f \in F$, we can show that $f\left(C_2{}^\beta\right) \subseteq C_2{}^\beta$. Clearly $C_2{}^\beta$ is nonemty and convex. Also $C_2{}^\beta$ is τ-closed. (For let y be a τ-limit point of $C_2{}^\beta$. Then since X^* is τ-closed, $y \in X^*$. Also y being a τ-limit point of $C_2{}^\beta$ is a p_β-limit point of $C_2{}^\beta$. Let $\varepsilon > 0$ be arbitrarily given. Then exists a $w \in C_2{}^\beta$ such that $p_\beta(y - w) < \varepsilon$. Now for any $z \in H^*$, $p_\beta(y - z) \leq p_\beta(y - w) + p_\beta(w - z) < \varepsilon + r$. Since ε is arbitrary, $p_\beta(y - z) \leq r$. Hence $y \in C_2{}^\beta$.) Hence $C_2{}^\beta = X^*$ by the minimality of X^*. Thus we obtain $C_1{}^\beta = W$. Let W' be the convex p_β-closure of H^*. Then we have $\delta(W, \beta) \leq \delta(W', \beta) = \delta(H^*, \beta)$ as $W \subseteq W'$, each τ-limit point of a set being also a p_β-limit point of the set. Hence there must be points u and v in H^* such that $p_\beta(u - v) > r$. But since $H^* \subseteq W = C_1{}^\beta$, $p_\beta(u - v) \leq r$. Thus we obtain a contradiction. Hence $\delta(W, \alpha) = 0$ for all $\alpha \in I$. Since E is Hausdorff, this implies that H^* consists of a single point which must be a fixed point of each $f \in F$. This completes the proof. □

The next theorem was proved in Hong (1968) in Banach space.

Theorem 8.4 *Let (E, τ) and $A^*(\tau)$ be as in Theorem 8.3. Let X be a nonempty τ-closed convex subset of E. Let F be a commutative family of nonexpansive mappings on X and M a τ-compact subset of X such that ther exist an $f_1 \in F$ and $x_0 \in X$ satisfying the following properties:*

(i) $\{f_1{}^n(x_0)\}$ *is τ-bounded;*
(ii) $\tau - \mathrm{cl\,co}\{f_1{}^n(x)\} \cap M \neq \emptyset$ *for every $x \in X$,*

where co$A$ stands for convex hull of A. Then the family F has a common fixed point in M.

Proof. Let $M_1 = \{f_1{}^n(x_0)\}$. Then we take this M_1 as M in the definition E_B. Let M_1 be $\|.\|_B$-bounded by the number $d \leq 2$. Let B_n denote the $\|.\|_B$-closed ball of centre $f_1{}^n$ and radius d. We now define: $D_k = \cap_n^\infty(B_n \cap K)$ and $D = \tau - \mathrm{cl}(\cup_0^\infty D_k)$.

Then we can show that D is convex, τ-closed and τ-bounded as $\cup_0{}^\infty$ is $\|.\|_B$-bounded. Moreover, D is mapped into itself by the mapping f_1. Now by applying Theorem 8.3 to the family $F = \{f_1\}$, we obtain a fixed point of f_1 in M. By condition (ii) we have that each fixed point of f_1 is contained in M. Hence The set H_1 of all fixed point of f_1 is a nonempty, τ-closed, and τ- compact of M. Also by the the commutativity of F, we have as in Theorem 8.3 that $f(H_1) \subseteq H_1$ for each $f \in F$. Furthermore, by τ-compactness of H_1 and by Zorn's lemma, there is a nonempty set H^* which is minimal with respect to being nonempty, τ-compact subset of H_1 and mapped into itself by eacy $f \in F$. Now for every $f, g \in F$, we have

$$g[f(H^*)] = f[g(H^*) \subseteq f(H^*);$$

Thus $f(H^*)$ is a nonempty τ-compact subset of H_1 and is mapped into itself by each $g \in F$. Thus by the minimality of H^*, $f(H^*) = H^*$ for eacf $f \in F$. Now we repeat the same argument as in Theorem 8.3 and establish the existence of a common fixed of F. □

8.2 Set-Valued Mappings of Nonexpansive Type

The authors of Browder (1965a), Göhde (1966) and Kirk (1965), have independently proved that a nonexpansive self mapping on a weakly compact convex subset of a Banach space with normal strcture has a fixed point. In this section we include the results proved in Husain and Tarafdar (1980), i.e. we define the concept of normal structure of a bounded convex subset of a locally convex tological vector space and also the notion of a set-valued mapping of nonexpansive type on such a space. We then prove a fixed point theorem for such mappings which include the above fixed point theorem of Browder (1965a), Göhde (1966) and Kirk (1965). We also consider another definition of nonexpansive set-valued mapping and fixed point theorem for such a mapping on a closed bounded interval.

Throughout this section as before (E,τ) will denote a locally convex Hausdorff topological space where the topology τ is generated by the family $[p_\alpha : \alpha \in I]$ of seminorms on E, I being the index set.

8.2.1 *Normal Structure and Fixed Point Theorems*

The concept of normal structure of a bounded convex set in a Banach space was first introduced in Brodskii and Milman (1948). We have introduced below the same concept for a bounded convex subset of E.

A point x of a bounded subset K of E is said to be a p_α-diametral point of K if $\delta(K,\alpha) = \sup\{p_\alpha(x-y) : y \in K\}$, as before $\delta(K,\alpha)$ is the p_α-diameter of K, i.e. $\delta(K,\alpha) = \sup\{p_\alpha(x-y) : x, y \in K\}$. A point $y \in K$ which is not a p_α-diametral point of K is called a p_α-nondiametral point of K.

Definition 8.3 A bounded convex subset K of E is said to have normal structure if every convex subset B of K containing more than one point has at least one p_α-nondiametral point of B for each $\alpha \in I$ satisfying $\delta(B,\alpha) > 0$.

Example 8.1 Let K be a convex subset of E such that K is p_α-compact for each $\alpha \in I$. Then K has normal structure.

Proof. For, suppose K does not have normal stucture. Then there are a convex subset B of K conaining more than one point and an $\alpha \in I$ with $\delta(B,\alpha) > 0$ such that B does not contain any p_α-nondiametral point. Let $x_1 \in B$. Then we can find $x_2 \in B$ such that $p_\alpha(x_1 - x_2) = \delta(B,\alpha)$. Since B is convex, $\frac{x_1+x_2}{2} \in B$. We can find $x_3 \in B$ such that $p_\alpha(x_3 - \frac{x_1+x_2}{2}) = \delta(B,\alpha)$. Continuing this process we obtain a sequence $\{x_n\}$ of points in B such that $p_\alpha\left(x_{n+1} - \frac{x_1+x_2+\cdots+x_n}{n}\right) = \delta(B,\alpha)$. Since

$$\delta(B,\alpha) = p_\alpha\left(x_{n+1} - \frac{x_1+x_2+\cdots+x_n}{n}\right) \le \frac{1}{n}\sum_{k=1}^{n} p_\alpha(x_{n+1}-x_k) \le \delta(B,\alpha),$$

it follows that $p_\alpha(x_{n+1} - x_k) = \delta(B,\alpha)$ for $k = 1, 2, \ldots, n$. This implies that $\{x_n\}$ has no p_α-Cauchy subsequence contradicting the assumption that K is p_α-compact. □

Example 8.2 Let K be a τ-compact convex subset of E. Then K has normal structure.

Since K is τ-compact, it is p_α-compact for each $\alpha \in I$. Hence K has normal structure as shown in Example 1.

For any bounded subset K of E and $x \in K$, let

$$\gamma_x(K,\alpha) = \sup\{p_\alpha(x-y) : y \in K\},$$

$$\gamma(K,\alpha) = \inf\{\gamma_x(K,\alpha) : x \in K\},$$

and

$$C(K,\alpha) = \{x \in K : \gamma_x(K,\alpha) = \gamma(K,\alpha)\}.$$

Lemma 8.3 *Let K be a nonempty weakly compact convex subset of E. Then $C(K,\alpha)$ is a nonempty closed convex subset of K for each $\alpha \in I$.*

Proof. For each positive integer n and $x \in K$, let $K_n(x,\alpha) = \{y \in K : p_\alpha(x-y) \le \gamma(K,\alpha) + \frac{1}{n}\}$. Obviously $K_n(x,\alpha)$ is nonempty, convex and p_α-closed. Let $C_n(\alpha) = \bigcap_{x\in K} K_n(x,\alpha)$. Clearly $C_n(\alpha)$ is convex and p_α-close and hence τ-closed. $C_n(\alpha)$ is also nonempty. Indeed, ther is a a $z \in K$ such that $\gamma_z(K,\alpha) \le \gamma(K,\alpha) + \frac{1}{n}$, i.e. $p_\alpha(z-x) \le \gamma(K,\alpha) + \frac{1}{n}$ for all $x \in K$. Hence $z \in C_n(\alpha)$. Now since $\{C_n(\alpha) : n = 1,2,\ldots,\}$ is a decreasing sequence of τ-closed (hence weakly closed because each $C_n(\alpha)$ is convex), convex subsets of the weakly compact set K, it follows that $\cap_n C_n(\alpha)$ is nonempty, τ-closed and convex. We complete the proof by noting that $C(K,\alpha) = \cap_n C_n(\alpha)$. □

Lemma 8.4 *Let K be as in Lemma 8.3. In addition assume that K has normal structure. Then $\delta(C(K,\alpha),\alpha) < \delta(K,\alpha)$ whenever $\delta(K,\alpha) > 0$.*

Proof. Since K has normal structure, there is a point $x \in K$ such that $\gamma_x(K,\alpha) < \delta(K,\alpha)$. If $u,v \in C(K,\alpha)$, then $p_\alpha(u-v) \le \gamma_u(K,\alpha) = \gamma(K,\alpha)$. Hence $\delta(C(K,\alpha),\alpha) \le \gamma_x(K,\alpha) < \delta(K,\alpha)$. □

Definition 8.4 Let K be a nonempty subset of E. A set-valued (or single valued) mapping $f : K \to 2^K$ with $f(x) \neq \emptyset$ for each $x \in K$ said to be nonexpansive type on K if f satisfies either of the following conditions:

(*a*) for each $\alpha \in I$, there are nonnegative real numbers $a_1(\alpha), a_2(\alpha), a_3(\alpha)$ with $a_1(\alpha) + a_2(\alpha) + a_3(\alpha) \le 1$ such that for all $x, y \in K$,

$$p_\alpha(u-v) \le a_1(\alpha)p_\alpha(x-y) + a_2(\alpha)p_\alpha(x-v) + a_3(\alpha)p_\alpha(y-u)$$

whenever $u \in f(x)$ and $v \in f(y)$;

(*b*) given $x \in K$ and $u \in f(x)$, for each $v \in f(y), y \in K$ and each $\alpha \in I$, there exists $v'(\alpha) \in f(y)$ such that $p_\alpha(u-v) \le p_\alpha(x - v'(\alpha))$;

(*c*) given $x \in K$ and real number $\varepsilon > 0$, there exists for each $\alpha \in I$ a real number $\delta(\alpha) \ge \varepsilon$ such that $p_\alpha(u-v) \le \varepsilon$ whenever $u \in f(x), v \in f(y), y \in K$ and $p_\alpha(x-y) \le \delta(\alpha)$.

Theorem 8.5 *Let K be a nonempty weakly compact convex subset of E. Assume that K has normal structure. Then for each multivalued mapping f of nonexpansive type on K, there is a point $x \in K$ such that $f(x) = \{x\}$, where $\{x\}$ denotes the set consisting of the single point x.*

Proof. By using weak compactness of K and Zorn's lemma we can find a minimal nonempty τ-closed convex subset F of K such that $f(F) \subseteq F$ (e.g see Browder

(1965a)). We assert that F is a set consisting of a single point, by showing that $\delta(F,\alpha)=0$ for each $\alpha \in I$. If possible, let $\delta(F,\alpha)>0$ for some $\alpha \in I$. Since F is weakly compact, by Lemma 8.3 $C(F,\alpha)$ is nonempty, τ-closed convex subset of F. We now prove that $f(C(F,\alpha)) \subseteq C(F,\alpha)$. To this end let $u \in f(F(C,\alpha))$ [here we note $f(A) = \bigcup_{x\in A} f(x)$ for any subset A of K]. Then there is a point $x \in C(F,\alpha)$ such that $u = f(x)$. Let $S=\{y \in F : p_\alpha(u-y) \le \gamma(F,\alpha)\}$. Clearly S is nonempty, convex and p_α-closed and hence τ-closed. Also $f(S) \subseteq S$. Let $y \in f(S)$. Then there is a $z \in S \subseteq F$ such that $y = f(z)$. If f satisfies (a), then

$$\begin{aligned} p_\alpha(u-y) &\le a_1(\alpha)p_\alpha(x-z)+a_2(\alpha)p_\alpha(x-y)+a_3(\alpha)p_\alpha(z-u)\\ &\le a_1(\alpha)\gamma(F,\alpha)+a_2(\alpha)\gamma(F,\alpha)+a_3(\alpha)\gamma(F,\alpha) \le \gamma(F,\alpha), \end{aligned}$$

because $z \in S \subseteq F$, $x \in C(F,\alpha)$ and $u, y \in F$. Hence $y \in S$ in this case. If f satsfies (b), then there is a $v'(\alpha) \in f(z) \subseteq F$ such that $p_\alpha(u-y) \le p_\alpha(x-v'(\alpha)) \le \gamma(F,\alpha)$ since $x \in C(F,\alpha)$ and $v'(\alpha) \in F$. Thus $y \in S$ in this case, too. Finally if f satisfies (c), there is a $\delta(\alpha) \ge \gamma(F,\alpha)$ such that $p_\alpha(u-y) \le \gamma(F,\alpha)$ whenever $u \in f(x), y \in f(z)$ and $p_\alpha(x-z) \le \delta(\alpha)$. Now since for all $z \in F$, $p_\alpha(x-z) \le \gamma(F,\alpha) \le \delta(\alpha)$, we have $p_\alpha(u-y) \le \gamma(F,\alpha)$. Thus $y \in S$ in this case also. Hence, we conclude that $f(S) \subseteq S$. Thus $S = F$ by the minimality of F.

Now $\gamma_u(F,\alpha) = \sup\{p_\alpha(u-y) : y \in F\} = \sup\{p_\alpha(u-y) : y \in S\} \le \gamma(F,\alpha)$. Noting that $u \in F$, we obtain that $\gamma_u(F,\alpha) = \gamma(F,\alpha)$, i.e. $u \in C(F,\alpha)$. Hence $f(C, F < \alpha)) \subseteq C(F,\alpha)$ as was to be shown. But then, $F = C(F,\alpha)$ by the minimality of F. Since $\delta(F,\alpha) > 0$, it is impossible in view of our Lemma 8.4. Thus we have shown that $\delta(F,\alpha)=0$ for all $\alpha \in I$. Since E is Hausdorff, F is a set consisting of a single point $\{x\}$, say. Whence $f(x)=\{x\}$. □.

Remark 8.4

(*i*) If $(E, \|.\|)$ is a Banach space and f is single valued and satisfies (a) with $a_2(\|.\|) = a_3(\|.\|) = 0$, then the above theorem reduces to the result of Browder (1965a), Göhde (1966) and Kirk (1965).

(*ii*) If $(E, \|.\|)$ is a Banach space, f is single valued and $C(K, |.\|)$ is a single point, then the above theorem reduces to a theorem of Wong (1974).

The following corollary shows how a set-valued mapping of nonexpansive type can arise and an application of Theorem 8.5 gives rise to a common fixed point theorem.

Corollary 8.5.1 *Let E and K be as in Theorem* 8.5. *Let $\{f_\gamma : \gamma \in J\}$ be a family of single valued mappings on K (i.e. f_γ is a mapping of K into itself for each $\gamma \in J$) satisfying either of the following conditions:*

(*i*) *for each $\alpha \in I$, there are nonnegative numbers $a_1(\alpha), a_2(\alpha), a_3(\alpha)$ with $a_1(\alpha)+a_2(\alpha)+a_3(\alpha) \le 1$ such that for all $x, y \in K$ and all $\gamma, \delta \in J$,*

$$p_\alpha(f_\gamma(x)-f_\delta(y)) \le a_1(\alpha)p_\alpha(x-y)+a_2(\alpha)p_\alpha(x-f_\delta(y))+a_3(\alpha)p_\alpha(y-f_\gamma(x));$$

(*ii*) *given* $x \in K, \alpha \in I, \gamma \in J$, *there exists for each pair* $(y \in K, \delta \in J)$ *a* $\delta' \in J$ *such that*

$$p_\alpha(f_\gamma(x) - f_\delta(y)) \leq p_\alpha(x - f_{\delta'}(y)).$$

Then the family $\{f_\gamma : \gamma \in J\}$ *has a common fixed point.*

Proof. We define the set-valued mapping $f : K \to 2^K$ by $f(x) = \{f_\gamma(x) : \gamma \in J\} = \bigcup_{\gamma \in J} f_\gamma(x), x \in K$. We can easily verify that if $\{f_\gamma : \gamma \in J\}$ satisfies (*i*), then f is nonexpansive type of (*a*) in Definition 8.4 and if $\{f_\gamma : \gamma \in J\}$ satsfies (*ii*), then f is nonexpansive type of (*b*) in Definition 8.4. Hence by Theorem 8.5 in either case there is a point $x \in K$ such that $f(x) = \{x\}$. This implies that x is a common fixed point of $\{f_\gamma : \gamma \in J\}$.

8.2.2 *Another Definition of Nonexpansive Set-Valued Mapping and Corresponding Results on Fixed Point Theorems*

In this subsection we give another definition of nonexpansive set-valued mapping which, in a single valued case, coincides with usual definition of nonexpansive mapping.

Definition 8.5 Let C be a subset of a metric space (X, ρ). A set-valued mapping $f : C \to 2^C$ with $f(x) \neq \emptyset$ for each $x \in C$ is said to be nonexpansive if given $x \in C$ and $u \in f(x)$, there is a $v_y \in f(y)$ for each $y \in C$, such that $\rho(u, v_y) < \rho(x, y)$.

Remark 8.5 This can obviously extended in locally convex spaces in terms of seminorms.

Example 8.3 Let $\{f_\alpha : \alpha \in I\}$ be a family of single valued nonexpansive self mappings on a subset C of a metric space (X, ρ) [i.e. for each $\alpha \in I, f_\alpha : C \to C$ and $\rho(f_\alpha(x), f_\alpha(y)) \leq \rho(x, y)$ for all $x, y \in C$]. Then the set-valued mapping $f : C \to 2^C$ defined by $f(x) = \{f_\alpha(x) : \alpha \in I\}(= \bigcup_{\alpha \in I} f_\alpha(x)), x \in C$ is clearly nonexpansive in the sense of our Definition 8.5.

We do not as yet know if a fixed point theorem similar to our Theorem 8.5 or the theorem of Browder (1965a), Göhde (1966) and Kirk (1965) can be proved in general for such a nonexpansive mapping on a weakly compact subset (with normal structure) of a Banach space. However, we prove the following fixed point theorem on the subsets of real line $\mathbb{R}$.

Theorem 8.6 *Let* C *be a closed, convex and bounded subset (i.e. a closed and bounded interval) of the real real line* $\mathbb{R}$. *Let* f *be a nonexpansive (in the sense of Definition* 8.5) *set-valued mapping on* C *with closed and convex vales in* C *(i.e.* $f(x)$ *is closed and convex for each* $x \in C$). *Then thetre is a point* $x_0 \in C$ *such that* $x_0 \in f(x_0)$.

Proof. Since C is compact, by using Zorn's lemma we can find a minimal nonempty colsed bounded convex set $K \subseteq C$ such that $f(K) \subseteq K$, where as before $f(K) = \bigcup_{x \in K} f(x)$, (e.g. see Browder (1965a). Noting that K is a closed bounded interval, say $[a, b]$, let z be the midpoint (centre) and r the radius of K, i.e. $r = |z - a| = |z - b|$. Let $N = N(f(z), r) = \{y \in K : |y - x| \leq r\}$ for some $x \in f(z)\}$. Since $f(z)$ and K are convex, it follows that N is convex. N is also closed. Indeed, if $y_n \in N, n = 1, 2, \ldots$ and $y_n \to y$, we can find $x_n \in f(z), n = 1, 2, \ldots$ such that $|y_n - x_n| \leq r$. Since $f(z)$ is compact, we select a subsequence $\{x_{n_i}\}$ of $\{x_n\}$ such that $x_{n_i} \to x \in f(z)$. Now from the triangle inequality

$$|y - x| \leq |y - y_{n_i}| + |y_{n_i} - x_{n_i}| + |x_{n_i} - x|,$$

it follows that $|y - x| \leq r$. Thus $x \in N$ and N is ,therefore, closed. We show that $f(N) \subseteq N$. If $y \in f(N)$, there is a $w \in N \subseteq K$ such that $y' \in f(w)$. Now since f is nonexpansive, there is a $u \in f(z)$ such that $|u - y| \leq |z - w| \leq r$, z being the centre of K and w being in K. Hence $y \in N$. Thus by the minimality of K, we have $K = N = [a, b]$. Hence $z \in f(z)$. For if $z \notin f(z)$, then since $f(z)$ is a closed and convex of $K = N = [a.b]$, it follows that either (i) $f(z) \subseteq [a, z)$, or $f(z) \subseteq (z, b]$. Clearly in case (i) $b \notin N$ and in case (ii) $a \notin N$. Thus in either case $N \neq K$ which is a contradiction.

8.3 Fixed Point Theorems for Condensing Set-Valued Mappings on Locally Convex Topological Vector Spaces

Using the concept of condensing mapping the authors in Sadovskiǐ (1967) and Lifšic and Sadovskiǐ (1968) have obtained respectively the generalizations of Schauder fixed point theorem Schauder (1930) and Tychonoff fixed point theorem Tychonoff (1935). In Daneš (1968) the generalization of Kautani's fixed point theorem Kakutani (1941) by using the concept of set-valued condensing mapping has been obtained. In Reinermann (1971) the condensing mapping defined in terms of of a measure of noncompactness (nonprecompactness) of bounded sets to obtain generalizations has also been used. Using the set-valued condensing mapping defined in terms of a measure of precompactness, Himmelberg, Porter and van Vleck Himmelberg and Vleck (1969) proved a fixed point theorem which includes the fixed point theorem of Sadovskiǐ (1967), Tychonoff (1935), Glicksberg (1952), Fan (1952) and a part of theorem of Browder (1959).

The aim of this section as is noted in Tarafdar and Výborný (1975) is to obtain a fixed point theorem which will contain the above fixed point theorems of Daneš (1968), Fan (1952), Glicksberg (1952), Himmelberg and Vleck (1969), Kakutani (1941), Sadovskiǐ (1967), Reinermann (1971), Sadovskii (1968), Schauder (1930) and Tychonoff (1935).

In the next subsection we introduced general definition of a measure of noncompactness of bounded sets in a locally convex topological vector space. In the subsequent subsections we have given various definitions of condensing set-valued mappings and have unified them in single definition and then we have proved our main fixed point theorem of this section and also obtained corollaries and theorem which are similar but more general than the corresponding results of Himmelberg and Vleck (1969).

Let $F : X \to 2^Y$ be set-valued mapping with $F(x) \neq \emptyset$ for each $x \in X$ (this will be assumed throughout this section). For $\Omega \subseteq X$, a set-valued mapping $G : \Omega \to 2^Y$ having the property that $G(x) \subseteq F(x)$ for each $x \in \Omega$ is called a subset-valued mapping of F. For $\Omega \subseteq X, F(\Omega) = \bigcup_{x\in\Omega} F(x)$.

Given two topological spaces X and Y, a set-valued mapping $F : X \to 2^Y$ is as before called upper semicontinuous if for each closed subset A of Y, $F^{-1}(A)$ is closed. F has closed graph if Y is regular and F is upper semicontinuous and closed values (see Kuratowski (1966, p. 175). A set-valued mapping $F : X \to 2^Y$ is lower semicontnuous if for each open set A of $Y, F^{-1}(A)$ is open (for more details see Chapter 2).

As before, in this section (E, τ) will denote a locally convex Hausdorff topological vector space and $[p_\alpha : \alpha \in I]$ will denote the family of seminorms which generates the topology τ. Also, any topological concept, such as closedness, pre-copactness, compactness, boundedness, and so on, will be understood as 'with respect to topology τ. In all other cases, i.e., when a topological concept is not meant with respect to τ, the corresponding topology will precede the concept; for example, p_α-precompact to mean that certain subset is precompact with respect to p_α-topology.

8.3.1 *Measure of Precompactness and Non-Precompactness*

We denote by C the class of all τ-bounded subsets of (E, τ).

Definition 8.6 $\mu = [\mu_\alpha : \alpha \in I]$ will be said to define a measure of precompactness on C, where for each $\alpha \in I$, μ_α is a set (interval) valued mapping of C into $\mathbb{R}^+$, the set of nonnegative real numbers, having the properties:

(*i*) $\mu_\alpha(\Omega) = [a, \infty)$, or $(a, \infty), a \geq 0$ for each $\Omega \in C$;
(*ii*) $\Omega_1 \subseteq \Omega_2 \in C$ implies that $\mu_\alpha(\Omega_1) \supseteq \mu_\alpha(\Omega_2)$ for each $\alpha \in I$;
(*iii*) $\mu_\alpha(\Omega) = \mu_\alpha(\mathrm{co}\,\Omega)$ for each $\Omega \in C$ where as before co stands for the convex hull of Ω;
(*iv*) $\mu_\alpha(\Omega_1 \cup \Omega_2) = \mu_\alpha(\Omega_1) \cap \mu_\alpha(\Omega_2)$ for $\Omega_1, \Omega_2 \in C$;
(*v*) $\mu_\alpha = \mathbb{R}^+$ if Ω is precompact and Ω precompact if $\mu_\alpha \supseteq (0, \infty)$, for each $\alpha \in I$.

For $\Omega \in C$, $\hat{\mu}(\Omega) = [\hat{\mu_\alpha}(\Omega) : \alpha \in I]$, where $\hat{\mu_\alpha}(\Omega) = \inf \mu_\alpha(\Omega)$, may then be regarded as a measure of nonprecompactness of Ω. Thus all the entries in the parenthesis of $\hat{\mu}(\Omega)$ are zeros if and only if Ω is precompact.

Example 8.4 (Kuratowski) For each $\Omega \in C$, we define $\lambda(\Omega) = [\lambda_\alpha(\Omega) : \alpha \in I]$, where

$$\lambda_\alpha(\Omega) = \{\varepsilon > 0 : \Omega \text{ can be covered by a finite number of sets of } p_\alpha\text{-diameter } \leq \varepsilon\}.$$

Then λ is indeed a measure of precompactness on C.

Thus taking $\hat{\lambda_\alpha}(\Omega) = \inf \lambda_\alpha(\Omega)$, we see that

(i) $\lambda_\alpha(\Omega) = [\hat{\lambda_\alpha}(\Omega), \infty)$ or $(\hat{\lambda_\alpha}(\Omega), \infty)$, (ii), (iv), and (v) follow easily. For proof of (iii) we refer to Akhmerovtal (1992) or Darbo (1955) (the proof given there for normed space applies also for seminormed space).

Example 8.5 Let $U_\alpha = \{x \in E : p_\alpha(x) \leq 1\}$. For each $\Omega \in C$, we define $\gamma(\Omega) = [\gamma_\alpha(\Omega) : \alpha \in I]$, where

$$\gamma_\alpha(\Omega) = \{\varepsilon > 0 : \text{ there exists a } p_\alpha\text{-precompact subset } S \text{ with } S + \varepsilon\, U_\alpha \supseteq \Omega\}.$$

(i) As before we take $a = \inf \gamma_\alpha(\Omega) = \hat{\gamma_\alpha(\Omega)}$. The proof of (ii) and (iv) of Definition 8.6 is trivial. We now prove (iii) and (v).

(iii) In view of (ii) it suffices to prove that $\gamma_\alpha(\Omega) \subseteq \gamma_\alpha(\operatorname{co}\Omega)$. Let $t \in \gamma_\alpha(\Omega)$. Then there exists a p_α-precompact subset S such that $S + tU_\alpha \supseteq \Omega$. Since $\operatorname{co} S + tU_\alpha \supseteq \Omega$ and $\operatorname{co} S + tU_\alpha$ is convex, $\operatorname{co} S + tU_\alpha \supseteq \operatorname{co}\Omega$. Noting that $\operatorname{co} S$ is p_α-precompact, we conclude that $t \in \gamma_\alpha(\operatorname{co}\Omega)$.

(v) Let Ω be τ-precompact. Then Ω is p_α-precompact for each $\alpha \in I$. Since $\Omega + tU_\alpha \supseteq \Omega$ for all $t \geq 0$ and $\alpha \in I$, we have $\mu_\alpha(\Omega)(= \gamma_\alpha(\Omega)) = \mathbb{R}^+$ for all $\alpha \in I$.

Next, let $\alpha \in I$ be arbitrary and $\gamma_\alpha(\Omega) \supseteq (0, \infty)$. Let $r > 0$ be any number. Since $\frac{r}{2} \in \gamma_\alpha(\Omega)$, there exists a p_α-precompact set S such that $S + \frac{r}{2}U_\alpha \supseteq \Omega$. Since S is p_α-precompact, there exists a finite set F such that $F + \frac{r}{2}U_\alpha \supseteq S$. Now $F + rU_\alpha \supseteq S + \frac{r}{2}U_\alpha \supseteq \Omega$. Thus Ω is p_α-precompact. Since α is arbitrary, Ω is τ-precompact.

Example 8.6 Let $(E, \tau), [p - \alpha : \alpha \in I]$ and C be as before. For $\Omega \in C$, we define $\nu(\Omega) = [\nu_\alpha(\Omega) : \alpha \in I]$, where

$$\nu_\alpha(\Omega) = \{\varepsilon > 0, \text{ there exists a precompact set } S \text{ such that } S + \varepsilon U_\alpha \supseteq \Omega\}.$$

The proof that ν is a measure of nonprecompactness on C is similar to that of Example 8.5. We note that for each $\Omega \in C$, $\nu_\alpha(\Omega) \subseteq \gamma_\alpha(\Omega)$ for each $\alpha \in I$.

8.3.2 *Condensing Mappings*

The authors of Himmelberg and Vleck (1969) have defined a measure of precompactness for any subset of (E, τ) in the following way.

Let $\mathcal{B}$ be base of convex neighbourhoods of 0. Then for $\Omega \subseteq E$, $Q(\Omega)$, the measure of τ-precompactness of Ω, is defined to be the collection of all $B \subseteq \mathcal{B}$

such that $S + B \supseteq \Omega$ for some τ-precompact subset S of E. With this notion of precompactness they have introduced a definition of condensing mapping.

Let X be a nonempty subset of (E, τ). Let $[p_\alpha : \alpha \in I]$ and C be as before. Let $F : X \to 2^X$ be a set-valued mapping

Definition 8.7 F is condensing with respect to Q if for each τ-bounded but not τ-precompact set $\Omega \subseteq X$ with $F(\Omega) \subseteq \Omega$ we have $Q(F(\Omega)) \not\supseteq Q(\Omega)$.

Definition 8.8 F is condensing with respect to μ if for each bounded but not τ-precompact set $\Omega \subseteq X$ with $F(\Omega) \subseteq \Omega$, there exists $\alpha \in I$ such that $\hat{\mu_\alpha}(F(\Omega)) < \hat{\mu_\alpha}(\Omega)$, where $\mu = [\mu_\alpha : \alpha \in I]$ is a measure of τ-precompactness on C.

Definition 8.9 F is condensing with respect to μ for each bounded but not precompact set $\Omega \subseteq X$ with $F(\Omega) \subseteq \Omega$, there exists $\alpha \in I$ such that $\mu_\alpha(F(\Omega)) \not\supseteq \mu_\alpha(\Omega)$.

Definition 8.10 F is condensing if for each $\Omega \subseteq X$ with $F(\Omega) \subseteq \Omega$,

(*a*) the condition that $\Omega \setminus \tau - \mathrm{cl}\,\mathrm{co}\,F(\Omega)$ is τ-compact implies the τ-compactness of $\tau - \mathrm{cl}\,\Omega$; or

(*b*) the condition that $\Omega \setminus \mathrm{co}\,F(\Omega)$ is empty or single point implies the τ-compactness of $\tau - cl\Omega$.

Definition 8.11 F is condensing if for each $\Omega \subseteq X$ with $F(\Omega) \subseteq \Omega$, the condition that $\Omega \setminus \mathrm{co}\,F(\Omega)$ is empty or single point implies that Ω is τ-precompact.

Definition 8.7 is used first in Himmelberg and Vleck (1969). For single valued mapping, Definition 8.8 has been used in Reinermann (1971) and in Stallbohm (1973) with $\mu = \lambda$, and Definition 8.10 is due to the authors of Lifšic and Sadovskiǐ (1968). Definition 8.11 is a slight variant of the one given in Daneš (1968).

(*A*) It is easy to see that Definition 8.8 implies Definition 8.9 for each measure μ.

(*B*) Definition 8.7 implies Definition 8.9 for suitable measure μ. Let Definition 8.7 hold. We index the base $\mathcal{B}$ by $\mathcal{B} = [B_\alpha : \alpha \subset I]$. Let p_α be the Minskowski functional on B_α. Let $U_\alpha = \{x \in E : p_\alpha(x) \leq 1\}$. Clearly $B_\alpha = U_\alpha$. We now consider the measure ν as defined in Example 8.6. We now show that Definition 8.9 holds with respect to this measure ν. Let Ω be any τ-bounded but not τ-precompact subset of X with $F(\Omega) \subseteq \Omega$. Then we have $Q(F(\Omega)) \not\supseteq Q(\Omega)$; that is, there exists a $B_\alpha \in \mathcal{B}$ such that $B_\alpha \in Q(F(\Omega))$ but $B_\alpha \notin Q(\Omega)$. Hence it follows that $1 \in \nu_\alpha(F(\Omega))$ but $1 \notin \nu_\alpha(\Omega)$. Also since $F(\Omega) \subseteq \Omega$, it follows from (*ii*) of Definition 8.6 that $\nu_\alpha(F(\Omega)) \not\supseteq \nu_\alpha(\Omega)$.

(*C*) Definition 8.9 with measure μ implies Definition 8.11 if F has bounded range. Let Definition 8.9 hold with a measure μ. Let $\Omega \subseteq X, F(\Omega) \subseteq \Omega$, and $\Omega \setminus \mathrm{co}\,F(\Omega) = Z$ where $Z = \emptyset$ or a single point. Obviously $\mu_\alpha(Z) = \mathbb{R}^+$ for each $\alpha \in I$.

Since $\Omega \subseteq Z \cup \mathrm{co}\,F(\Omega)$, it follows that Ω is bounded and we have for each $\alpha \in I$, $\mu_\alpha(\Omega) \supseteq \mu_\alpha(Z \cup F(\Omega))$ by (*ii*) of Definition 8.6 equal to $\mu_\alpha(Z) \cap \mu_\alpha(F(\Omega))$

by (*iv*) and (*iii*) of Definition 8.6. Again since $\Omega \supseteq Z \cup F(\Omega)$, we have for each $\alpha \in I, \mu_\alpha(\Omega) \subseteq \mu_\alpha(Z) \cap \mu_\alpha(F(\Omega))$ by (*ii*) and (*iv*) of Definition 8.6. Thus for each $\alpha \in I$, we have $\mu_\alpha(\Omega) = \mu_\alpha(Z) \cap \mu_\alpha(F(\Omega))$. From this and the fact that $\mu_\alpha(Z) = \mathbb{R}^+$ for each $\alpha \in I$, it follows that $\mu_\alpha(\Omega) = \mu_\alpha(F(\Omega))$ for each $\alpha \in I$, which in view of Definition 8.9 implies that Ω is τ-precompact.

(*D*) Obviously Definition 8.10 implies Definition 8.11.

8.3.3 *Fixed Point Theorems*

For the proof of the following elementary result see, e.g., the book of Berge (1963, Theorem 8, p. 113).

Lemma 8.5 *Let X be a compact topological space and $F : X \to 2^X$ be a upper contnuous set-valued mapping (with $F(x) \neq \emptyset$ for each $x \in X$ which has been assumed at the begining of the section), then there exists a nonempty compact subset K of X such that $F(K) = K$.*

The following corollary was given as a lemma in Himmelberg and Vleck (1969).

Corollary 8.6.1 *Let X be topological space. Let $F : X \to 2^X$ be a set-valued mapping with closed graph. If there exists a nonempty subset A of X such that $F(A) \subseteq A$ and $\mathrm{cl}\, A$ is compact, then there exists a nonempty, closed and compact subset K of X such that $K \subseteq F(K)$.*

Proof. Let $G =$Graph $F \bigcap (\mathrm{cl}\, A \times \mathrm{cl}\, A)$. Then G is closed and compact as F is upper semicontinuous and A is compact. Let G_0 be a set-valued mapping such that Graph $G_0 = G$. Clearly $G_0^{-1}(\mathrm{cl}\, A)$ is a closed subset of $\mathrm{cl}\, A$ containing A. Thus domain $G_0 = G_0^{-1}(\mathrm{cl}\, A) = \mathrm{cl}\, A$. Hence $G_0 : \mathrm{cl}\, A \to 2^{\mathrm{cl}\, A}$ is a upper semicontinuous mapping from the compact set $\mathrm{cl}\, A$. Consequently there is by Lemma 8.5 a nonempty, closed and compact subset K such that $G_0(K) = K$. This implies $F(K) \supseteq K$. □

Theorem 8.7 *Let X be a nonempty complete convex subset of a locally convex Hausdorff topological vector space E. Let $F : X \to 2^X$ be a condensing set-valued mapping in the sense of Definition 8.11 with convex values and closed graph. Then F has a fixed point.*

Proof. Unlike Sadovskiǐ (1967), Lifšic and Sadovskiǐ (1968) and Himmelberg and Vleck (1969), we will not use ordinals. Let $x \in X$. Set $A = \{x\} \cup \{\bigcup_{n=1}^{\infty} F^n(x)\}$. Then clearly $F(A) \subseteq A$ and $A \setminus \mathrm{co}\, F(A) \subseteq \{x\}$ or empty set. Since F is condensing, A is precompact. Also $\mathrm{cl}\, A \subseteq X$ and $\mathrm{cl}\, A$ is compact as X is complete. Hence by Corollary 8.6.1 there exists a nonempty compact subset K of X such that $F(K) \supseteq K$.

Let $S = \{Y \subseteq X : K \subseteq Y, F(Y) \subseteq Y$ and Y is convex$\}$. S is nonempty as $X \in S$. S is a partially ordered set with respect to the relation $\leq$ where $Y_1 \leq Y_2$ if and only if $Y_1 \supseteq Y_2$ with $Y_1, Y_2 \in S$.

We first prove that every chain in S has an upper bound in S. Let T be a chain in S. Then $Z = \bigcap_{Y \in T} Y$ is an upper bound. Clearly

$$Z \subseteq X, \quad K \subseteq Z, \quad F(Z) \subseteq Z,$$

and Z is convex. Hence $Z \in S$. Thus by Zorn's Lemma there is a maximal element $Z_0 \in S$.

We next prove that for each $Y \in S, \operatorname{co} F(Y) \in S$

(*a*) $\operatorname{co} F(Y) \subseteq X$ as $F(Y) \subseteq Y \subseteq X$ and X is convex.

(*b*) $K \subseteq \operatorname{co} F(Y)$.
Since $K \subseteq Y$ and $K \subseteq F(K)$, we have $K \subseteq F(K) \subseteq F(Y)$. Hence $K \subseteq \operatorname{co} F(Y)$.

(*c*) $F(\operatorname{co} F(Y)) \subseteq \operatorname{co} F(Y)$.
Since $F(Y) \subseteq Y$ and Y is convex, $\operatorname{co} F(Y) \subseteq Y$. Hence $F(\operatorname{co} F(Y)) \subseteq F(Y) \subseteq \operatorname{co} F(Y)$.

(*d*) $\operatorname{co} F(Y)$ is convex.

Now since for each $Y \in S, F(Y) \subseteq Y$ and Y is convex, we have $\operatorname{co} F(Y) \subseteq Y$. Thus $Y \leq \operatorname{co} F(Y)$ for each $Y \in S$. In particular $Z_0 \leq \operatorname{co} F(Z_0)$. But since Z_0 is a maximal element in S, it follows that $Z_0 = \operatorname{co} F(Z_0)$; i.e. $Z_0 \setminus F(Z_0) = \emptyset$. Hence by the condensing property of F, Z_0 is precompact. Therefore, $\operatorname{cl} Z_0 \subseteq X$ and $\operatorname{cl} Z_0$ is compact. The rest of the argument is as given in Himmelberg and Vleck (1969). Let $G =$ graph $F \cap (\operatorname{cl} Z_0 \times \operatorname{cl} Z_0)$. Then G is closed and compact subset of $X \times X$. Let G_0 be the set-valued mapping such that Graph $G_0 = G$. Since $G_0^{-1}(\operatorname{cl} Z_0)$ is a closed subset of $\operatorname{cl} Z_0$ containing Z_0, the domain $G_0 = G_0^{-1}(\operatorname{cl} Z_0) = \operatorname{cl} Z_0$. Hence $G_0 : \operatorname{cl} Z_0 \to 2^{\operatorname{cl} Z_0}$ is a set-valued mapping with convex values and compact graph (note $F = G_0$ on $\operatorname{cl} Z_0$). G_0 is also upper semicontinuous. Hence by fixed point theorem of Glicksberg (1952) and Fan (1952) (also chapter 2), G_0 has a fixed point in $\operatorname{cl} Z_0$. This fixed point is also a fixed point of F. □

Remark 8.6 The same remark as given in Himmelberg and Vleck (1969, p. 637) applies in the present situation; i.e. the theorem remains true for non-Hausdorff (non-separated) E if further assumption that X is closed is assumed. For details see Himmelberg and Vleck (1969) as quoted above.

Remark 8.7 If F is assumed to be condensing with respect to Definition 8.10, then the above theorem remains true with the completeness conition on X replaced by the condition that X is closed. The same proof applies, because in this case $\operatorname{cl} A$ and $\operatorname{cl} Z_0$ appeared in the proof would be compact directly due to the condensing of F. By remark 8.6 we can then remove the the Hausdorff condition on E as the condition that X is closed is already assumed. The resulting version of the theorem will include a fixed point theorem of Lifšic and Sadovskiĭ (1968).

Corollary 8.7.1 *Let X be a nonempty comlete convex subset of a Hausdorff locally convex topological vector space E. Let $F : X \to 2^X$ be a set-valued mapping* *with convex values, closed graph and bounded range. If F is condensing in the sense of Definition* 8.9, *then F has a fixed point.*

Proof. This follows from Theorem 8.7 and (C) following the definitions. □

Remark 8.8 In view of (B) it follows that the fixed point theorem in Himmelberg and Vleck (1969, Th. 1,p. 637) is a special case of Corollary 8.7.1.

The following theorem includes the corresponding theorem in Himmelberg and Vleck (1969).

Theorem 8.8 *Let X be a nonempty complete convex subset of a locally convex topological vector space E. Let $F : X \to 2^X$ be a lower semicontinuous set-valued mapping with closed convex values. Then F has a fixed point if either of the following conditions holds:*

- *(a) X is compact and metrizable;*
- *(b) the subspace uniformity on X is metrizable and F is condensing in the sense of Definition* 8.11.

Proof. (a) By a wellknown selection theorem of Michael (1966, Theorem 1.2), there is a continuous selection $f : X \to X$ such that $f(x) \in F(x)$ for all $x \in X$. Hence by Tychonoff fixed point theorem f has a fixed point. This fixed point is also a fixed point of F.

(b) We proceed as in the proof of Theorem 8.7 until the set Z_0 with $\operatorname{co} F(Z_0) = Z_0$ is obtained. By Corollary 2a, p. 176 in Kuratowski (1966), $F(\operatorname{cl} Z_0) \subseteq \operatorname{cl} F(Z_0) \subseteq \operatorname{cl} Z_0$. Then we apply case (a) to the mapping $F : \operatorname{cl} Z_0 \to 2^{\operatorname{cl} Z_0}$. (If required see Himmelberg and Vleck (1969, p. 641).) □

Bibliography

Adams, R. A. (1975). *Sobolev Spaces.* New York, San Francisco and London: Academic Press.

Akashi, W. Y. (1988). Equivalence theorems and coincidence degree for multivalued mappings. *Osaka J. Math.*, **25**, 33–47.

Aliprantis, C., & Brown, D. (1983). Equilibria in markets with Riesz space of commodities. *Journal of Mathematical Economics*, **11**, 189–207.

Aliprantis, C., & Brown, D. (1993). Equilibria in markets with Riesz space of commodities. *Journal of Mathematical Economics*, **11**, 189–207.

Allen, G. (1977). Variational inequalities, complementarity problems and duality theorems. *J. Math. Anal. Appl.*, **58**, 1–10.

Arrow, K., & Debreu, G. (1954). Existence of equilibrium for a competitive economy. *Econometrica*, **22**, 265–290.

Aubin, J. (1979). *Applied Functional Analysis.* New York: Wiley-Interscience.

Aubin, J. (1982). *Mathematical Methods of Game and Economic Theory* (Vol. 7, Revised ed.). Amsterdam: North-Holland.

Aubin, J. P., & Ekeland, I. (1984). *Applied Nonlinear Analysis.* New York: John Wiley and Sons.

Aubin, J. P., & Siegel, J. (1980). Fixed points and stationary points of dissipative multivalued maps. *Proc. Amer. Math. Soc.*, **78**, 391–398.

Bae, J. S., Kim, W. K., & Tan, K.-K. (1993). Another generalization of Ky Fan's minimax inequality and its applications. *Bull. Inst. Math. Acad. Sinica*, **21**, 229–244.

Baiocchi, C., & Capelo, A. (1984). *Variational and Quasivariational Inequalities.* New York: Wiley.

Bajo, I., & Liz, E. (1996). Periodic boundary value problem for first order differential equations with impulses at variable times. *J. Math. Anal. Appl.*, **204**, 65–73.

Bakker, J. W. (1980). *Mathematcal Theory of Program Correctness.* Reading, MA: Prentice-Hall International.

Bakker, J. W. de, & Zucker, J. I. (1983). Processes and the denotational semantics of concurency. *Foundations of Computer Science IV, Distributed Systems:*

Part 2, Semantics and Logic, Centre Tracts, **159**, 45–100.

Banach, S. (1922). Sur les opérations dans les ensembles abstraits et leur applications aux équations intégrales (thése présentée en juin 1920 a l' université de Léopol [= lwów] pour obtenir le grade de docteur en philosophie). *Fundamenta Math.*, **3**.

Banaś, J., & Goebel, K. (1980). Measure of noncompactness in Banach space, *Lecture Notes in Pure and Appl. Math.*, **80**. New York and Basel: Marcel Dekker.

Bancroft, S., Hale, J. K., & Sweet, D. (1968). Alternative problems for nonlinear functional equations. *J. Differential Equations*, **4**, 40–56.

Baranga, A. (1991). The contraction principle as a particular case of Kleene's fixed point theorem. *Discrete Math.*, **98**, 75–79.

Barbuti, U., & Guerra, S. (1971). On an extension of a theorem due to J. B. Diaz and F. T. Metcalf. *Rend. Accad. Naz. Lincei*, **51**, 29–31.

Bardaro, C., & Ceppitelli, R. (1988). Some further generalizations of Knaster-Kuratowski-Mazurkiewicz theorem and minimax inequalities. *J. Math. Anal. Appl.*, **132**, 484–490.

Barnsley, M. F. (1988). *Fractals Everywhere.* Academic Press.

Bartle, R. G. (1953). Singular points of functional equations. *Trans. Amer. Math. Soc.*, **75**, 366–384.

Belluce, L. P., & Kirk, W. A. (1966). Fixed point theorems for families of contraction mappings. *Pacific Journal of Mathematics*, **18**(2), 213–217.

Belluce, L. P., & Kirk, W. A. (1967). Nonexpansive mappings and fixed points in Banach spaces. *Illinois J. Math.*, **11**(3), 474–479.

Ben-El-Mechaiekh, Deguire, P., & Granas, A. (1982). Une alternative nonlineaire en analyse convex et applications. *C. R. Acad. Sci., Paris, Ser. 1 Math.*, *Ser. 1*, 257–259.

Bensousson, A., & Lions, J. L. (1973). Nouvelle formulation des problèmes de contrôle impulsionnel et applications. *C. R. Acad. Sci.*, **29**, 1189–1192.

Berge, C. (1963). *Topological Spaces.* Edinburgh/London: Oliver & Boyd.

Bewley, T. F. (1972). Existence of equilibria in economics with infinitely many commodities. *Journal of Economic Theory*, **4**, 514–540.

Bielecki, A. (1956). Une remarque sur l'application de la méthode de Banach-Cacciopoli-Tikhonov dans la théorie de l'équation $s = f(x, y, z, p, q)$. *Bull. Acad. Polon. Sci.*, **3**(4), 265–268.

Birchoff, G. (1967). *Lattice Theory* (Vol. 25). New York: Amer. Math. Soc. Publ.

Bohnenblust, H. F., & Karlin, S. (1950). On a theorem of ville. In H. W. Kuhn & A. W. Tucker (Eds.), *Contributions to the Theory of Games* (pp. 155–160). Princeton: Princeton University Press.

Border, K. C. (1985). *Fixed Point Theorems with Applications to Economics and Game Theory.* Cambridge, London, New York, New Rochelle, Melbourne, Sydney: Cambridge University Press.

Borglin, A., & Keiding, H. (1976). Existence of equilibrium actions of equilibrium, "A note on the 'new' existence theorems". *J. Math. Economics*, **3**, 313–316.

Borisovich, G. Y., & Sapronov, I. Y. (1968). A contribution to the topological theory of condensing operators. *Soviet Math. Dokl*, **9**, 1304–1308.

Boyd, D., & Wong, J. S. W. (1969). On nonlinear contractions. *Proc. Amer. Math. Soc.*, **20**, 458–464.

Brézis, J., Coron, J.-M., & Nirenberg, L. (1980). Free vibrations for a nonlinear wave equation and a theorem of P. Rabinowitz. *Commun. Pure Appl. Math.*, 667–689.

Brodskii, M. S., & Milman, D. P. (1948). On the centre of a convex set. *Dokl. Akad. Nauk. SSSR (N.S.)*, **59**, 837–840.

Brondsted, A. (1964). Conjugate convex functions in topological vector spaces. *Mat.-Fys. Medd. Dansk. Vid. Selsk*, **34**(2).

Brondsted, A. (1974). On a lemma of bishop and phelps. *Pacific J. Math.*, **55**, 335–341.

Browder, F. E. (1959). On a generalization of the Schauder fixed point theorem. *Duke Math. J.*, **26**, 291–303.

Browder, F. E. (1963). The solvability of non-linear functional equations. *Duke Math. J.*, **30**, 557–566.

Browder, F. E. (1965a). Nonexpansive nonlinear operators in a Banach space. *Proc. Nat. Acad. Sci. USA*, **54**, 1041–1044.

Browder, F. E. (1965b). Nonliear monotone operators and convex sets in Banach spaces. *Bull. Amer. Math. Soc.*, **71**, 780–785.

Browder, F. E. (1967). A new generalization of the Schauder fixed point theorem. *Math. Annalen*, **174**, 285–290.

Browder, F. E. (1968). The fixed point theory of multivalued mappings in topological vector spaces. *Mathematische Annalen*, **177**, 283–301.

Browder, F. E. (1970). Existence theorems for nonlinear partial differential equations. *Proceedings of Symposia, Amer. Math. Soc.*, Providence, R.I., **16**, 1–60.

Browder, F. E. (1976). Nonlinear operators and nonlinear equations of evolution in Banach spaces. *Proceedings of Symposia in Pure Math.*, **18**.

Browder, F. E., & Petryshyn, W. V. (1966). The solution by iteration of nonlinear fuctional equations in Banach spaces. *Bull. Amer. Math. Soc.*, **72**, 571–575.

Bruck, R. E. (1973). Properties of fixed point sets of nonexpansive mappings in Banach spaces. *Trans. Amer. Math. Soc.*, **179**, 251–262.

Burton, T. A. (1996). Integral equations, implicit functions and fixed points. *Proc. Amer. Math. Soc.*, **124**(8), 2383–2390.

Cacciopoli, R. (1946). Sulle corespondenze funzionali inverse diramate: teoria generale e applicazioni ad alcune equazioni funzionali non lineari e al problema di plateau. *Atti. Acad. Naz. Lincei, Rend. Cl. Sci. Fis. Mat. Natur.* 416-421, **24**, 258–268.

Carbone, A. (1991). A note on a theorem of prolla. *Indian J. Pure Appl. Math.*, **23**(4), 257–260.

Caristi, J., & Kirk, W. (1975). (Geometric fixed point theory and inwardness conditions,) *Proc. Conf. on Geometry of Metric and Linear Spaces,* Michigan, 1974, Lecture Notes in Mathematics 490. Berlin: Springer-Verlag.

Castains, C. (1969). Le théoreme de Dunford-Pettis généralisé de Montpellier. *Secrétariate des Mathematiques, Publication No.*, **43**.

Cellina, A., & Lasota, A. (1969). A new approach to the definition of topological degree for multivalued mappings. *Atti Acad. Naz. Lincei Rend. Cl. Sci. Fis. Mat. Natur.*, **47**(8), 434–440.

Cesari, L. (1963). Functional analysis and periodic solutions of nonlinear differential equations. *Contributions to Differential Equations*, **1**, 149–187.

Cesari, L. (1964). Functional analysis and Galerkin's method. *Michigan Math. J.*, **II**, 385–418.

Cesari, L. (1965). Existence in the large of periodic solutions of hyperbolic partial differential equations. *Archs Ration. Mech. Analysis*, **20**, 170–190.

Cesari, L. (1969). Functional analysis and differential equations. *Siam Studies in Applied Mathematics*, **5**, 143–155.

Cesari, L. (1971). Functional analysis and boundary value problems, in: *Analytic theory in differential equations*, pp. 178–194 (Vol. 183). Berlin: Springer-Verlag.

Cesari, L. (1976). *Functional analysis, nonlinear differential equations and the alternative method, pp. 1–97 — Nonlinear functional analysis and differential equations.* New York: Dekker.

Chan, D., & Pang, J. S. (1982). The generalized quasi-variational inequality problem. *Math. Oper. Res.*, **7**, 211–222.

Chang, S. Y. (1990). On the Nash equilibrium. *Soochow J. Math.*, **16**, 241–248.

Chowdhury, M. S. R. (2000). The surjectivity of upper hemi-continuous and pseudo-monotone type II operators in reflexive Banach spaces. *J. Bangladesh Math. Soc.*, **20**, 45–53.

Chowdhury, M. S. R., & Tan, K.-K. (1996). Generalization of Ky Fan's minimax inequality with applications to generalized variational inequalities for pseudo-monotone operators and fixed point theorems. *J. Math. Anal. and Appl.*, **204**, 910–929.

Chowdhury, M. S. R., & Tan, K.-K. (1997a). Generalized quasi-variational inequalities for upper semi-continuous operators on non-compact sets. *Nonlinear Analysis, Theory, Methods & Applications, Proceedings of the Second World Congress of Nonlinear Analysts*, **30:8**, 5389–5394.

Chowdhury, M. S. R., & Tan, K.-K. (1997b). Generalized variational inequalities for quasi-monotone operators and applications. *Bulletin of the Polish Academy of Sciences*, **45**(1), 25–54.

Chowdhury, M. S. R., & Tan, K.-K. (1998). Applications of pseudo-monotone operators with some kind of upper semi-continuity in generalized quasi-variational inequalities on non-compact sets. *Proceedings of the American Mathematical Society*, **126:10**, 2957–2968.

Chowdhury, M. S. R., & Tan, K.-K. (1999). Generalized quasi-variational inequalities for lower and upper hemi-continuous operators on non-compact sets. *Mathematical Inequalities & Applications*, **2:1**, 121–134.

Chu, S. C., & Diaz, J. B. (1964/65). A fixed point theorem for "in the large" application of the contraction principle. *Atti. Accad. Sci., Torino C1. Sci. Fis. Mat. Natur.*, **99**, 351–363.

Chu, S. C., & Diaz, J. B. (1965). Remarks on a generalization of Banach's principle of contraction mapping. *J. Math. Anal. Appl.*, **11**, 440–446.

Contraction principle in pseudo-uniform spaces. (1980). In F. Fadell & G. Fournier (Eds.), *Fixed point theory*, Proceedings, Sherbrooke, Quebec (pp. 300–308). Berlin: Springer-Verlag.

Cottle, R. W. (1966). Nonlinear programs with positively bounded Jacobians. *SIAM J. Appl. Math.*, **14**, 147–157.

Cottle, R. W., & Dantzig, G. B. (1968). Complementary Pivot theory of mathematical programming. *Linear Algebra and Appl.*, **1**, 103–125.

Cottle, R. W., & Yao, J. C. (1992). Pseudo-monotone complementarity problems in Hilbert spaces. *J. of Optimization Theory and Applications*, **75**, 281–295.

Crandall, M. G. (1972). A generalisation of Peano's existence theorem and flow invariance. *Proc. Amer. Math. Soc.*, **36**(1), 151–155.

Crandall, M. G., & Rabinowitz, P. H. (1971). Mathematical theory of bifurcation. In C. Bardos & D. Besis (Eds.), *Bifurcation Phenomena in Mathematical Physics and related Topics* (pp. 3–46).

Cronin, J. (1950). Branch points of solutions of equations in Banach spaces. *Trans. Amer. Math. Soc.*, **69**, 208–231.

Cubiotti, P. (1992). Finite dimensional quasi-variational inequalities associated with discontinuous functions. *Journal of Optimization Theory and Applications*, **72**, 577–582.

Cubiotti, P. (1993). An existence theorem for generalized quasi-variational inequalities. *Set-Valued Analysis*, **1**, 81–87.

Daneš, J. (1968). Some fixed point theorems. *Comment. Math. Univ. Carolinae*, **9**, 223–235.

Daneš, J. (1974). *On densifying and related mappings and their application in nonlinear functional analysis, in theory of nonlinear operators, pp. 15–56 — Proceedings of a Summer School.* Berlin: Akademic-Verlag.

Darbo, G. (1955). Punti uniti in trasformazioni a codominio non compatto. *Rend. Sem. Mat. Univ. Pavoda*, **24**, 84–92.

Day, M. M. (1961). Fixed-point theorems for compact convex sets. *Illinois J. Math.*, **5**, 585–590.

Debreu, G. (1952). A social equilibrium existence theorem. *Proc. Nat. Acad. Sci., USA*, **38**, 121–126.

Debreu, G. (1959). *Theory of Value.* New Haven and London: Yale Univ. Press.

Defigueiredo, D. G. (1967). *Topics in nonlinear functional analysis, Lecture Series No. 48.* College Park, Md.: University of Maryland Press.

Deimling, K. (1977). *Ordinary Differential Equations in Banach Spaces* (Vol. 596). Berlin: Springer-Verlag.

Deimling, K. (1992). *Multivalued differential equations.* Berlin: Walter de Gruyter & Co.

DeMarr, R. (1963). Common fixed points for commuting contraction mappings. *Pacific J. Math.*, **13**, 1139–1141.

Diedudonné, J. (1969). *Foundations of Modern Analysis.* New York: Academic Press.

Ding, W. K., X.P., & Tan, K. (1990). A selection theorem and its applications. *Bull. Austral. Math. Soc.*

Ding, X. (1991). Browder-Hartman-Stampacchia type variational inequalities for multi-valued quasi-monotone operator. *Journal of Sichuan Normal University*, **14**, 1–8.

Ding, X., Kim, W., & Tan, K. (1992). Equilibria of non-compact generalized games with l^*-majorized preference correspondences. *J. Math. Anal. Appl.*, **164:2**, 508–517.

Ding, X., & Tarafdar, E. (1994). Generalized nonlinear variational inequalities with non-monotone set-valued mappings. *Applied Mathematics Letters*, **7**, 5–11.

Ding, X., & Tarafdar, E. (1995). Existence and uniqueness of solutions for a general non-linear variational inequality. *Applied Mathematics Letters*, **8**, 31–36.

Ding, X., & Tarafdar, E. (1996). Monotone generalized variational inequalities and generalized complementarity problems. *J. of Optim. Theory Appl.*, **88**, 107–122.

Ding, X., & Tarafdar, E. (2000). Generalized variational-like inequalities with pseudo-monotone set-valued mappings. *Arch. Math.*, **74**, 302–313.

Ding, X. P., & Tan, K. K. (1990). Generalized variational inequalities and generalized quasi-variational inequalities. *J. Math. Anal. Appl.*, **148**, 497–508.

Ding, X. P., & Tan, K. K. (1992a). Fixed point theorems and equilibria of non-compact generalized games. In K. Tan (Ed.), *Fixed point theory and applications* (pp. 80–96). Singapore, London: World Scientific.

Ding, X.-P., & Tan, K.-K. (1992b). A minimax inequality with applications to existence of equilibrium point and fixed point theorems. *Colloq. Math.*, *LXIII*, 233–247.

Ding, X. P., & Tan, K. K. (1992c). A minimax inequality with applications to existence of equilibrium point and fixed point theorems. *Colloquium Math.*, **63**, 233–247.

Ding, X. P., & Tan, K.-K. (1993). On the equilibria of non-compact generalized games. *J. Math. Anal. Appl.*, **177**, 226–238.

Ding, X. P., & Tarafdar, E. (1994). Fixed point theorem and existence of equilibrium points of noncompact abstract economics. *Nonlinear World*, **1**, 319–340.

Dolph, C. L., & Minty, G. J. (1964). 'On nonlinear integral equations of the Hammerstein type' in *Integral Equations*, pp. 99–154. Madison, Wisconsin: Madison University Press.

Dozo, E. L. (1973). Multivalued nonexpansive mappings and Opial's condition. *Proc. Amer. Math. Soc.*, **38**(2), 286–292.

Dugundji, J. (1966). *Topology*. Boston: Allyn and Bacon, Inc.

Dugundji, J., & Granas, A. (1978). KKM maps and variational inequalities. *Ann. Scuola Norm Sup. Pisa cl. Sci.*, **5**, 679–682.

Dugundji, J., & Granas, A. (1982). *Fixed point theory*, volume 1. Warszawa.

Dunford, N., & Schwartz, J. T. (1958). *Linear operators*, Part 1. New York: Interscience.

Edelstein, M. (1961). An extension of Banach's contraction mapping principle. *Proc. Amer. Math. Soc.*, **12**, 7–10.

Edelstein, M. (1962). On fixed and periodic points under contractive mappings. *J. London Math. Soc.*, **37**, 74–79.

Edelstein, M. (1966). A remark on a theorem of M. A. Krasonsel'skii. *Amer. Math. Monthly*, **73**, 509–510.

Edgar, G. A. (1990). *Measure, topology and fractal geometry.* New York: Spriger-Verlag.

Ehrmann, H. (1965). *Existanzsatze für die lösungen gewisser nichtlinearer randwertaufgaben.* Berlin: S. Angew. Math. Mech. 45, 22–39: Abh. Deutsch Akad. Wiss. Berlin Kl. Math. Phys. Tech. 157–167.

Ekland, I. (1972). Sur les problèmes variationnels. *C.R. Acad. Sci., Paris*, **275**, 1057–1059.

Engelking, R. (1977). *General Topology.* Warszawa: Polish Scientific Publishers.

Existence of maximal elements and equilibria in topological vector spaces. (1984). *J. Econom.*, **13**, 305.

Fan, K. (1952). Fixed points and minimax theorems in locally convex topological linear spaces. *Proc. Nat. Acad. Sci. USA*, **38**, 131–136.

Fan, K. (1961). A generalization of Tychonoff's fixed point theorem. *Math. Ann.*, **142**, 305–310.

Fan, K. (1966). Application of a theorem concerning sets with convex sections. *Math. Annalen*, **163**, 189–203.

Fan, K. (1969). Extensions of two fixed point theorems of F. E. Browder. *Math. Z.*, **112**, 234–240.

Fan, K. (1972). A minimax inequality and applications. In O. Shisha (Ed.), *Inequalities III, Proceedings of the Third Symposium on Inequalities* (pp. 103–113). San Diego: Academic Press.

Fan, K. (1984). Some properties of convex sets related to fixed point theorems. *Math. Ann.*, **266**, 519–537.

Fang, S. C., & Peterson, E. L. (1982). Generalized variational inequalities. *Journal of Optimization Theory and Applications*, **38**, 363–383.

Figueiredo, D. G. de. (1974). Some remarks on the Dirichlet problem elliptic equations. *Univ. de Brasilia Trabalho de Mathematica*, **57**.

Fitzpatrick, P. M., & Petryshyn, W. V. (1979). Some applications of a-proper mappings. *Nonlinear Analysis, Theory Methods & Aplications*, **3**, 525–532.

Flam, S. D. (1979). Abstract economy and games. *Soochow J. Math.*, **5**, 155–162.

Florenzano, M. (1983). On the existence of equilibrium in economics with infinite dimensional commodity spaces. *J. Math. Econom.*, **12**, 189–203.

Franklin, J. (1980). *Methods of Mathematical Economics.* New York: Springer-Verlag.

Franklin, J. (1983). Mathematical methods of economics. *Bull. Amer. Math. Soc.*, 229–244.

Fućik, M. K., S., & Nećas, J. (1975). Ranges of nonlinear asymptotically linear operators. *J. Differential Equations*, **17**, 375–394.

Fućik, S. (1974a). Further remarks on a theorem by E. M. Landesman and A. C. Lazer. *Comment. Math. Univ. Carolinae*, **15**, 259–271.

Fućik, S. (1974b). Nonlinear equations with noninvertible linear part. *Czechoslovack Math. J.*, **24**, 467–495.

Furi, M., Martelli, M., & Vignoli, A. (1980). On the solvability of nonlinear operator equations in normed spaces. *Ann. Mat. Pura Appl.*, **124**, 321–343.

Gaines, R. E., & Mawhin, J. L. (1977). Coincidence degree and nonlinear differential equations. Lecture Notes in Math., edited by A. Dold and B. Eckmann, Springer-Verlag, **568**.

Gale, D., & Mas-Colell, A. (1975). An equilibrium existence theorem for general model without ordered preferences. *J. Math. Econom.*, **2**, 9–15.

Gale, D., & Mas-Colell, A. (1978). On the role of complete transitive preferences in equilibrium theory. In G. Schwödiauer (Ed.), *Equilibrium and disequilibrim in economic theory* (pp. 7–14). Dordrecht: Reider.

Gale, D., & Mas-Colell, A. (1979). Corrections to an equilibrium existence theorem for a general model without ordered preferences. *J. Math. Economics*, **6**, 297–298.

Gale, D., & Mas-Collel, A. (1975). An equilibrium existence for general model without ordered preferences. *J. Math. Econom.*, **2**, 9–15.

Gamlelin, T. W., & Greene, R. E. (1983). *Introduction to Topology.* New York: Saunders College Publishing.

Gelfand, I., & Silov, G. E. (1967). *Generalzed Functions, Vol. 3: Theory of Differential Equations.* New York: Academic Press.

Gilbarg, D., & Trudinger, N. S. (1977). *Elliptic Partial Differential Equations of Second Order.* New York: Springer-Verlag.

Gilbarg, D., & Trudinger, N. S. (1983). *Elliptic Partial Differential Equations of Second Order.* New York: Springer-Verlag.

Glicksberg, I. L. (1952). A further generalization of the kakutani fixed point theorem with applications to Nash equilibrium points. *Proc. Amer. Math. Soc.*, **3**, 170–174.

Göhde, D. (1964/65). Über fixpunkte bei stetigen selbstabbildungen mit kompakten iterierten. *Math. Nachr.*, **28**, 45–55.

Göhde, D. (1966). Zum prinzip der kontraktiven abbildung. *Math. Nachr.*, **30**, 251–258.

Gokhberg, I., Goldstein, L., & Markus, A. (1957). Investigation of some properties of bounded linear operators in connection with q-norms. *Uchen. Zap. Kishinev. Gos. Univ.*, **29**, 29–36.

Goldberg, S. (1966). *Unbounded linear operators, theory and applications.* New York, St. Louis, San Francisco, Toronto, London, Sydney: McGraw-Hill Book Company.

Granas, A. (1959). Sur la notion du degré topologique pour une certaine classe de transformations muntivalentes dans les espaces de Banach. *Bull. Acad. Polon. Sci. Sér. Sci. Math. Astronom. Phys.*, **7**, 191–194.

Granas, A. (1962). The theory of compact vector fields and some applications to the theory of functional spaces. *Rozprawy Matematyezne, Warszawa*, **30**.

Granas, A. (1990). Méthodes topoloiques en anayse convexe. *Séminaire de Mathématiques Supérieures, Les Presses de I'Université Montreal, Canada*, **110**.

Guo, D., & Lakshmikantham, V. (1988). *Nonlinear Problems in Abstract Cones.* San Diego: Academic Press.

H., B., & L., N. (1978). Characterizations of the ranges of some nonlinear operators and applications to boundary value problems. *Annali Scu. Norm. Sup. Pisa*, **5**, 225–326.

Hale, J. (1967). Periodic solutions of a class of hyperbolic equations containing a small parameter. *Archs Ration. Mech. Analysis*, **23**, 380–398.

Hale, J. K. (1969). *Ordinary Differential Equations.* New York: Wiley-Interscience.

Hale, J. K. (1971). *Aplications of alternative problems: Lecture Notes 71-1.* Providence, R.I.: Brown University.

Halpern, B. (1970). Fixed point theorems for set-valued maps in finite dimensional spaces. *Math. Ann.*, **189**, 87–98.

Hanes, E. G., & Arbib, M. A. (1986). *Algebraic approaches to program semantics.* Berlin: Springer.

Hardy, G., & Rogers, T. (1973). A generalization of fixed point theorem of Reich. *Canad. Math. Bull.*, **16**, 201–206.

Harker, P. T., & Pang, J. S. (1990). Finite dimensional variational inequalities and nonlinear complementarity problems: a survey of theory, algorithms and applications. *Math. Programming*, **48**, 161–220.

Hartman, P. (1972). On invariant sets and on a theorem of Ważewski. *Proc. Amer. Math. Soc.*, **32**(2), 511–520.

Hartman, P., & Stampacchia, C. (1966). On some nonlinear elliptic differential functional equations. *Acta Math.*, **115**, 271–310.

Hartman, P., & Wintner, A. (1952). On hyperbolic partial differential equations. *Amer. J. Math.*, **74**, 834–864.

Hayashi, S. (1985). Self-similar sets as Tarski's fixed points. *Publ. Res. Inst. Math. Sci.*, **21**, 1059–1066.

H. Brézis, L. N., & Stampacchia, G. (1972). A remark on Ky Fan's minimax principle. *Boll. Un. Mat. Ital.* (4), **6**, 293–300.

Hess, P. (1974). On a theorem by Landesman and Lazer. *Indiana Univ. Math. J.*, **23**, 827–830.

Hetzer, G. (1975a). Some applications of the coincidence degree for set-contractions to functional differential equations of neutral type. *Comm. Math. Univ. Carolinae*, **16**, 121–138.

Hetzer, G. (1975b). Some remarks on ϕ_+-operators and on the coincidence degree for a Fredholm equation with non-compact nonlinear perturbations. *Ann. Soc. Sci. Bruxelles Ser.*, **89**, 497–508.

Himmelberg, C. J. (1975). Measurable relations. *Fund. Math.*, **87**, 53–72.

Himmelberg, J. R. P., C. J., & Vleck, F. S. vav. (1969). Fixed point theorems for condensing multifunctions. *Proc. Amer. Math. Soc.*, **23**, 635–641.

Hong, D.-X. (1968). A note on fixed point theorems for a family of nonexpansive mappings. *Proc. Amer. Math. Soc.*, **19**, 1223–1224.

Horvath, C. D. (1987). Some results in a multivalued mappings and inequalities without convexity. In *Nonlinear and Convex Analysis; Lecture Notes in Pure and Applied Math. Series Vol. 107.* Springer-Verlag.

Horváth, J. (1966). *Topological vector spaces and distributions, volume 1.* Reading, Massachusetts, London. Ontario: Addison-Wesley.

Hristova, S. G., & Bainov, D. D. (1987). Existence of periodic solutions of nonlinear systems of dfferential equations with impulse effect. *J. Math. Anal. Appl.*, **125**, 192–202.

Husain, T., & Tarafdar, E. (1980). Fixed point theorems for multivalued mappings of nonexpansive type. *Yokohama Mathematical Journal*, **28**, 1–6.

Husain, T., & Tarafdar, E. (1994). Simultaneous variational inequalities minimization problems and related results. *Math. Japonica*, **39**(2), 221–231.

Husain, T., & Tarafdar, E. (1996). Corringendum to: Simultaneous variational inequalities minimization problems and related results. *Math. Japonica*, **43**(1), 175.

Hutchinson, J. E. (1981). Fractals and self-similarity. *Indiana Univ. Math. J.*, **30**, 713-747.

Itoh, S. (1977). A random fixed fixed point theorem for a multivalued contraction mapping. *Pacific J. Math.*, **60**, 85–90.

Jachymski, J., Gajek, L., & Pokarowski, P. (2000). The Tarski-Kantorovitch principle and the theory of iterated funtion systems. *Bull. Austral. Math. Soc.*, **61**(2), 247–261.

J. B. Diaz, J. B., & Metcalf, F. T. (1969). On the set of subsequential limit point of successive approximations. *Trans. Amer. Math. Soc.*, **135**, 459–485.

Jou, C. R., & Yao, J. C. (1993). Extension of generalized multi-valued variational inequalities. *Appl. Math. Lett.*, **6**(3), 21–25.

J. Parida, M. S., & Kumar, A. (1989). A variational-like inequality problem. *Bulletin of the Australian Mathematical Society*, **39**, 225–231.

Justman, M. (1978). Iterative process with 'nucleolar' restrictions. *Internat. J. Game Theory*, **6**, 189–212.

Kakutani, S. (1938). Two fixed-point theorems concerning bicompact convex sets. *Proc. Imp. Acad. Tokyo*, **14**, 242–245.

Kakutani, S. (1941). A generalization of Brouwer's fixed point theorem. *Duke Math. J.*, **8**, 457–459.

Kantorovitch, L. (1939). The method of successive approximations for functional equations. *Acta Math.*, **71**, 63–97.

Kapoor, O. P. (1973). Two applications of an intersection lemma. *J. Math. Anal. Appl.*, **41**, 226–233.

Karamardian, S. (1971). Generalized complementarity problem. *Journal of Optimization Theory and Applications*, **8**, 161–167.

Karamardian, S. (1972). The complementarity problem. *Math. Programming*, **2**, 107–129.

Karamardian, S., & Schaible, S. (1990). Seven kinds of monotone mops. *Journal of Optimization Theory and Applications*, **66**, 37–46.

Karamolegos, A., & Kravvaritis, D. (1992). Nonlinear random operator equations and inequalities in Banach spaces. *Internat. J. Math. & Math. Sci.*, **15**, 111–118.

Kaul, S., Lakshmikantham, V., & Leela, S. (1994). Extremal solutions, comparison principle and stability criteria for impulsive differential equations with variable times. *Nonlinear Anal. T. M. A.*, **22**, 1263–1270.

Kaul, S. K. (1995). The periodic boundary value problem for impulsive differential equations with variable times. *Nonlinear Times and Digest*, **2**, 107–116.

Kelley, J. L. (1955). *General Topology.* Princeton: Van Nostrand.

Kelley, J. L., & Namioka, I. (1963). *Linear Topological Spaces.* Princeton: Van Nostrand.

Khan, M. A., & Papageorgiou. (1987). On Cournot Nash equilibria in generalized quantative games with an atomless measure spaces of agents. *Proc. Amer. Math. Soc.*, **100**, 505–510.

Khan, M. A., & Yannelis, N. C. (1991). *Equilibrium theory of infinite dimensional spaces.* Berlin: Springer-Verlag.

Kim, T., Prikry, K., & Yannelis, N. C. (1989). Equilibrium in abstract economics with a measure space and with an infinite dimensional strategy space. *J. Approx. Theory*, **56**, 256–266.

Kim, T., & Richter, M. K. (1986). Nontransitive-nontotal consumer theory. *J. Econ. Theory*, **28**, 324–363.

Kim, W. (1988). Remark on a generalized quasi-variational inequality. *Proc. Amer. Math. Soc.*, **103**, 667–668.

Kim, W. K., & Tan, K. K. (1992). A variational inequality in noncompact sets and its applications. *Bull. Austral. Math. Soc.*, **46**(1), 139–148.

Kinderlehrer, D., & Stampacchia, G. (1980). *An Introduction to Variational Inequalities and Their Applications.* New York: Academic Press.

Kirk, W. A. (1965). A fixed point theorem for mappings which do not increase distances. *Amer. Math. Monthly*, **72**, 1004–1006.

Kirk, W. A. (1981). Fixed point theory for nonexpansive mappings. In E. Faddel & G. Fournier (Eds.), *Lecture Notes in Math., 886* (pp. 484–505). Berlin: Springer-Verlag.

Kirk, W. A., & Yanez, C. M. (1988). Nonexpansive and locally nonexpansive mappings in product spaces. *Nonlinear Analysis, T.M.A.*, **12**(7), 719–725.

Klein, K., & Thompson, A. C. (1984). *Theory of Correspondence: Including Applications to Mathematical Economics.* New York: Wiley.

Knaster, B. (1928). Un théorme sur les fonctions d'ensembles. *Ann. Polon. Math.*, **6**, 133–134.

Knaster, B., Kuratowski, K., & Mazurkiewicz, S. (1929). Ein beweis des fixpunktpatzes für n-dimensionale simplexe. *Fund. Math.*, **14**, 132–137.

Kneser, H. (1952). Sur une théoréme fondamental de la th eorie des jeux. *C. R. Acad. Sci. Paris*, **234**, 2418–2420.

Kothe, G. (1960). *Topologische lineare raume I.* Berlin: Springer-Verlag.

Köthe, G. (1969). *Topological vector spaces, I*, translated by D. J. H. Garling. Die Grundlehren der Mathematischen Wissenschaften, 159. Berlin, Heidelberg, New York: Springer-Verlag.

Krasnoselskii, M. A. (1963). *Topological Methods in the Theory of Nonlinear Integral Equations.* Pergamon, UK: Pergamon Press.

Krasonsel'skii, M. A. (1955). Two remarks on the method of successive approximations. *Uspehi Mat. Nauk*, **10**, 123-127.

Kuczma, M. (1968). *Functional equations in a single variable.* Warszawa: Polish Scientific Publishers.

Kuratowski, K. (1930). Sur les espaces comptets. *Fund. Math.*, **15**, 301–319.

Kuratowski, K. (1966). *Topology*, Vol. 1. New edition, revised and augmented. Translated by J. Jaworowski. New York, London, Państwowe Wydawnictwo Naukowe, Warsaw: Academic Press.

Kuratowski, K., & Ryll-Nardzewski, C. (1965). A general theorem on selectors. *Bull. Acad. Polon. Sci. Ser. Sci. Math. Astronom. Phys.*, **13**, 397–403.

Lakshmikantham, V., Bainov, D. D., & Simeonov, P. S. (1989). *Theory of Impulsive*

Differential Equations. Singapore: World Scientific.

Lakshmikantham, V., Leela, S., & Kaul, S. (1994). Comparison principle for impulsive differential equations with variable times and stability theory. *Nonlinear Anal. T. M. A.*, **22**, 499–503.

Laloux, B. A., & Mawhin, J. (1977). Multiplicity, Leray-Schauder formula and bifurcation. *J. Differential Equations*, **24**, 309–322.

Landesman, E. M., & Lazer, A. C. (1970). Nonlinear perturbations of linear elliptic boundary value problems at resonance. *J. Math. Mech.*, **19**, 609–623.

Lasota, A., & Myjak, J. (1996a). Semifractals. *Bull. Pol. Acad. Sci. Math.*, **44**, 5–21.

Lasota, A., & Myjak, J. (1996b). Semifractals. *Bull. Pol. Acad. Sci. Math.*, **44**, 5–21.

Lassonde, M., & Schenkel, C. (1992). KKM principle, fixed points and Nash equilibria. *J. Math. Anal. Appl.*, **164**, 542-548.

Leader, S. (1982). Uniformly contractive fixed points in compact metric spaces. **86**, 153–158.

Lee, C.-M. (1977). A development of contraction mapping principles in Hausdorff uniform spaces. *Transacions of Amer. Math. Soc.*, **226**, 147–159.

Leray, J., & Schrauder, J. (1934). Topologie et équations functionelles. *Ann. Ecole Norm. Sup.*, **51**, 45–78.

Lifšic, E. A., & Sadovskiĭ, B. N. (1968). A fixed-point theorem for generalized condensing operators. *Dokl. Akad. Nauk SSSR, Englih Translation: Soviet Math. Dokl.*, **183, 9**, 278-279, 1370–1372.

Lin, T.-C. (1989). Some variants of a generalization of a theorem of Ky Fan. *Bull. Polish Acad. Sci. Math.*, **37**, 629–635.

Lloyd, N. G. (1978). *Degree Theory.* Cambridge: Cambridge University Press.

Locker, J. (1967). An existence analysis for nonlinear equations in Hilbert spaces. *Trans. Amer. Math. Soc.*, **128**, 403–413.

Lovicarova, H. (1972). Periodic solutions of a weakly nonlinear wave equations in one dimension. *Czech. Math. J.*, **19**, 324–342.

Ma, T. W. (1972). Topological degree for set-valued compact vector fields in locally convex spaces. *Dissertationes Math.*, **92**, 1–43.

Mancino, O., & Stampacchia, G. (1972). Convex programming and variational inequalities. *J. Opt. Theory Appl.*, **9**, 3–23.

Markov, A. (1936). Quelques théorèmes sur les ensembles abĺiens. *C. R. Acad. Sci. URSS (NS)*, **1**, 311–313.

Martin, R. H. (1976). *Non-linear operators and differential equations in Banach spaces.* New York: John Wiley and Sons.

Maschler, M., & Peleg, B. (1976). Stable sets and stable points of set-valued dynamic systems with applications to game theory. *SIAM J. Control Optim.*, **14**, 985–995.

Mas-Colell, A. (1974). An equilibrium existence theorem without complete or transitive preferences. *J. Math. Econom.*, **1**, 237–246.

Mawhin, J. L. (1972). Equivalence theorems for nonlinear operator-equations and coincidence degree theory for some mappings in locally convex topological vector spaces. *J. Differential Equations*, **12**, 610–636.

Mehta, G., & Tarafdar, E. (1987). Infinite dimensional Gale-Debreu and a fixed point theorem of Tarafdar. *J. Econom. Theory*, **41**, 333–339.

Michael, E. (1966). A selection theorem. *Proc. Amer. Math. Soc.*, **17**, 1404–1406.

Minty, G. J. (1962). Monotone (nonlinear) operators in Hilbert space. *Duke Math. J.*, **29**, 341–346.

Minty, G. J. (1963). *Proc. Nat. Acad. Sci., USA*, **50**, 1038-1041.

Moreau, J. J. (1966). Principles extrémaux pour le probléme de la cavitation. *J. Mécanique*, **5**, 439–470.

Mosco, U. (1976). Implicit variational problems and quasi variational inequalities. In J. P. Gossez, E. J. L. Dozo, J. Mawhin, & L. Waelbroeck (Eds.), *Nonlinear operators and caculus of variations* (Vol. 543, pp. 83–156). Berlin, Heidelberg, New York: Springer-Verlag.

Nadler, J. S. B. (1969). Multi-valued contraction mappings. *Pacific J. Math.*, **30**, 475–488.

Nagumo, M. (1951). Degree of mapping in convex linear topological spaces. *Amer. J. Math.*, **73**, 497–511.

Nakamo, M. (1976). Periodic solutions of linear and nonlinear wave equations. *Archs Ration. Mech. Analysis*, **62**, 87–98.

Nash, J. (1950). Equilibrium points in n-person games. *Proc. Nat. Acad. Sci., USA*, **36**, 48–49.

Neumann, J. von. (1928a). Die axiomatisierung. *Math. Zeitschrift*, **27**, 669–752.

Neumann, J. von. (1928b). Zur theorie der gesscllschaftsspiek. *Math. Ann.*, **100**, 295–320.

Neumann, J. von. (1937). Uber ein okonomsiches gleichungssytem und eine. *Verallgemeinerung des Brouwerschen Fixpunktratzes, Erg. Math. Kolloqu.*, **8**.

Neumann, J. von, & Morgenstern, O. (1944). *Theory of Games and Economic Behavior.* Princeton, New Jersey: Princeton University Press.

Neumann, J. von, & Morgenstern, O. (1947). *Theory of Games and Economic Behavior, 2nd ed.* Princeton, New Jersey: Princeton University Press.

Nikaido, H. (1968). *Convex Structures and Economic Theory.* New York: Academic Press.

Nirenberg, L. (1960). *Functional Analysis.*

Nirenberg, L. (1970). An application of generalized degree to a class of nonlinear problems. *Troisieme Colloq. Analyse Functionelle*, 57–74.

Nirenberg, L. (1971). Generalized degree and nonlinear problems. *Contributions to Nonlinear Analysis*, edited by E. Zarantonello, Academic Press, pp. 1–9.

Nirenberg, L. (1974). *Topics in Nonlinear Analysis.* New York University: Courant Institute of Mathematical Sciences.

Nonlinear equations and inequalities in Banach spaces. (1979). *J. Math. Anal. Appl.*, **67**, 205–214.

Noor, M. (1988). General variational inequalities. *Appl. Math. Lett.*, **1**, 119–122.

Noor, M. (1992). An iterative algorithm for nonlinear variational inequalities. *Appl. Math. Lett.*, **5**, 11–14.

Noor, M. (1995). Nonconvex functions and variational inequalities. *J. Optim. Theory Appl.*, **87**, 615-630.

Nussbaum, R. D. (1969). *The fixed point index and fixed point theorem for k-set-contractions*, Doctoral dissertation. Chicago, Illinois: University of Chicago.

Nussbaum, R. D. (1971). The fixed point index for condensing maps. *Ann. Mat. Pura Appl.*, **89**, 217–258.

Opial, Z. (1967). Weak convergence of the sequence of successive approximations for nonexpansive mappings. *Bull. Amer. Math. Soc.*, **73**, 595–597.

Ortega, J. M., & Rheinboldt, W. C. (1970). *Iterative solution of nonlinear equations in several variables.* New York and London: Academic Press.

Pareto, V. (1909, 1971). *Mannuel d'economie politique*, translation by A. S. Schwier. New York: Girard and Briere, Augustus M. Kelley Publishers.

Park, S. (1987). On generalizations of Ky Fan's theorems on best approximation. *Numer. Funct. Anal. Optimiz.*, **9**, 619–628.

Park, S. (1991). *Best approximations, inward sets, and fixed points. Progress in approximation theory.* New York: Academic Press.

Park, S. (1995). Extensions of best approximation and coincidence theorems.

Pascali, D., & Shurlan, S. (1978). *Nonlinear mappings of monotone types.* The Netherlands: Sijthoff and Noordhoff International Publishers.

Petryshyn, W. V., & Fitzpatrick, P. M. (1974). A degree theory, fixed points theorems, and mappings theorems for multi-valued non-compact mappings. *Trans. Amer. Math. Soc.*, **194**, 1–25.

Picard, E. (1927). *Lecons sur quelques types simples d'equations aux dèrivèes partielles avec des applications a la physique mathèmatique.* Paris: Gauthier, Villars.

Prolla, J. a. B. (1982–1983). Fixed-point theorems for set-valued mappings and existence of best approximants. *Numer. Funct. Anal. and Optimiz.*, **5**(4), 449–455.

Pruszko, T. (1981). Topological degree methods in multi-valued boundary value problems. *Nonlinear Anal.*, **9**, 959–973.

Rabinowitz, P. H. (1967). Periodic solutions of nonlinear hyperbolic partial differential equations. *Commun. Pure Appl. Math.*, **20**, 145–205.

Rabinowitz, P. H. (1971). Some global results for nonlinear eigenvalue problems. *J. Funct. Anal.*, **7**, 487–513.

Reinermann, J. (1971). Fixpunktsätze vom Krasnoselski-typ. *Math. Z.*, **119**, 339–344.

Riesz, F., & Sz-Nagy, B. (1955). *Functional analysis.* New York: Frederick Ungar.

Rockafeller, R. (1970). *Convex Analysis.* Princeton: Princeton University Press.

Rockafeller, R. T. (1976). Integral functionals, normal integrands and measurable selections. In *Nonlinear operators and the calculus of variations, Lecture Notes in Mathematics no. 543* (pp. 157–207). Springer-Verlag.

Royden, H. L. (1970). *Real Analysis*, 2nd ed. London: Macmillan.

Rudin, W. (1973). *Functional Analysis.* New York, San Francisco, London, Sydney, Toronto: McGraw-Hill Book Company.

Rzepecki, B. (1986). On the existence of solutions to the Darboux problem for the hyperbolic differential equations in Banach spaces. *Rend. Sem. Mat. Univ. Pavoda*, **76**, 201–206.

Sadovskiĭ, B. N. (1967). A fixed-point principle. *Functional Anal. Appl.*, **1**, 151–153.

Sadovskiĭ, B. N. (1968). Measures of noncompactness and condensing operators. *Problemy Mat. Anal. Sloz. Sistem*, **2**, 89–119.

Sadovskiĭ, B. N. (1972). Ultimately compact and condensing operators. *Russian Math. Surveys*, **27**, 85–156.

Saigal, R. (1976). Extension of the generalized complementarity problem. *Mathematics of Operations Research*, **1**, 160–166.

Schaefer, H. H. (1966). *Topological Vector Spaces.* New York: Macmillan.

Schaefer, H. H. (1971). *Topological Vector Spaces.* New York: Springer-Verlag.

Schauder, J. (1930). Der fixpunktsatz in funktionalräumen. *Studia Math.*, **2**, 171–180.

Schechter, M. (1973). A nonlinear elliptic boundary value problem. *Ann. Scuola Norm. Sup. Pisa*, **27(3)**, 707–716.

Schubert, H. (1964). *Topologie.* Teubner, Stuttgart: Eine Einfuhrung.

Schwartz, L. (1957). Théorie des distributions a valeurs vectorielles. 1. *Annls Inst. Fourier Univ. Crenoble*, **7**, 1–141.

Sehgal, V. M., Singh, S. P., & Smithson, R. E. (1987). Points and some fixed point theorems for weakly compact sets. *J. Math. Anal. Appl.*, **128**, 108–111.

Shafer, W., & Sonnenschein, H. (1975). Equilibrium in abstract economies without ordered preferences. *J. Math. Econom.*, **2**, 345–348.

shendge, G., & Joshi, S. (1982). Some applications of lattice fixed point theorem. *unpublished*, 1–6.

Shih, M.-H., & Tan, K.-K. (1984). A further generalization of Ky Fan's minimax inequality and its applications. *Studia Math.*, *LXXVIII*, 279–287.

Shih, M. H., & Tan, K. K. (1985). Generalized quasi-variational inequalities in locally convex topological vector spaces. *J. Math. Anal. Appl.*, **108**, 333–343.

Shih, M.-H., & Tan, K.-K. (1986). Minimax inequalities and applications. *Contemp. Math.*, **54**, 45–63.

Shih, M.-H., & Tan, K.-K. (1987). Covering theorems of convex sets related to fixed point theorems. *Nonlinear and Convex Analysis*, eds. B. L. Lin and S. Simons, Marcel Dekker, 235–244.

Shih, M. H., & Tan, K. K. (1988a). Browder-Hartman-Stampacchia variational inequalities for multi-valued monotone operators. *J. Math. Anal. Appl.*, **134**, 431–440.

Shih, M.-H., & Tan, K.-K. (1988b). Browder-Hartman-Stampacchia variational inequalities for multi-valued monotone operators. *J. Math. Anal. Appl.*, **134**, 431–440.

Shih, M.-H., & Tan, K.-K. (1988c). A minimax inequality and Browder-Hartman-Stampacchia variational inequalities for multi-valued monotone operators. In *Proceedings of the Fourth FRANCO-SEAMS Joint Conference.* Thailand: Chiang Mai.

Shih, M.-H., & Tan, K.-K. (1989). Generalized bi-quasi-variational inequalities. *Journal of Mathematical Analysis and Applications*, **143**, 66–85.

Shimizu, T. (1948). Analytic operations and analytic operational equations. *Math. Japan.*, **1**, 36–40.

Siddiqi, Q. H. A., A. H., & Kazmi, K. R. (1994). Nonlinear variational inequalities. *Indian J. Pure Appl. Math*, **25**(9), 969–973.

Siddiqi, S. C., & Ansari, G. H. (1989). An iterative method for generalized variational inequalities. *Mathematica Japonica*, **34**, 475–481.

Simmons, G. (1963). *Introduction to Topology and Modern Analysis.* New York: McGra-Hill Book Company.

Singh, S., Watson, B., & Srivastava, P. (1997). *Fixed point theory and best approximation: The KKM-Map Principle.* Dordrecht, Boston, London: Kluwer Academic Publshers.

Singh, S. P., Tarafdar, E., & Watson, B. (1996). Variational inequalities for a pair of pseudo-monotone functions. *Far East J. Math. Sci., Special Volume, Bhattacharya Memorial Volume*, 31–52.

Singh, S. P., Tarafdar, E., & Watson, B. (1997). Variational inequalities and applications. *Indian Journal of Pure and Applied Mathematics*, **28**, 1083–1089.

Smiley, M. (1982). Existence theorem for nonlinear hyperbolic boundary value problems at resonance. *Nonlinear Analysis TMA*, **6**, 1055–1073.

Smiley, M. (1985). Existence theorems for linear hyperbolic boundary value problems at resonance. *Annali Mat. Pura Appl.*, **139**, 45–63.

Soto-Andrade, J., & Varela, F. J. (1984). Self-reference and fixed points: a discussion and an extension of lawvere's theorem. *Acta Appl. Math.*, **2**, 1–19.

Stallbohm, V. (1973). Fixpunkte nichtexpansiver abbildungen, fixpunkte kondensierender abbildungen, Fredholm'sche sätze linearer kondensierender abbildungen. *Rheinisch-Westfälische Technische Hochschule Aachen, Dr. Nat. Dissertation.*

Stoy, J. E. (1977). *Denotational Semantics: The Scott-Strachey Approach to Programming Language Theory.* MA: M. I. T., Cambridge.

Takahashi, W. (1976). Nonlinear variational inequalities and fixed point theorems. *J. Math. Soc. Japan*, **28**, 168–181.

Tan, K.-K. (1983). Comparison theorems on minimax inequalities, variational inequalities, and fixed point theorems. *J. London Math. Soc.*, **28**, 555–562.

Tan, K. K. (1991). *Lecture Notes on Topics in Topology and Tunctional Analysis*, unpublished, 1985, 1991 and 1994. Halifax, Nova Scotia: Dalhousie University.

Tan, K.-K. (1994). *Lecture Notes on Topics in Topology and Functional Analysis.* unpublished.

Tan, K. K., & Hong-Kun, X. (1991). On fixed point theorems of nonexpansive mappings in product spaces. *Proc. Amer. Math. Soc.*, **113**(4), 983–989.

Tan, K. K., & Yuan, X. (1993). A minimax inequality with applications to existence of equlibrium points. *Bull. Austral. Math. Soc.*, **47**, 483–503.

Tan, K. K., & Yuan, X. (1994). Variational inequalities on reflexive Banach spaces and applications. *J. Natural Geometry*, **5**, 43–58.

Tarafdar, E. (1974). An approach to fixed point theorems on uniform spaces. *Transactions Amer. Math. Soc.*, **191**, 209–225.

Tarafdar, E. (1975). Some fixed-point theorems on locally convex linear topological spaces. *Bull. Austral. Math. Soc.*, **13**(3), 241–254.

Tarafdar, E. (1976). On a fixed point theorem on locally convex linear topological spaces. *Monatshefte fur Mathematik*, **82**, 341–344.

Tarafdar, E. (1977). On nonlinear variational inequalities. *Proc. Amer. Math. Soc.*, **67**, 95–98.

Tarafdar, E. (1980). On the existence of solution of the equation $l(n) = n(x)$ and a generalized coincidence degree theory 1. *Commentat. Math. Univ. Carol.*, **21**, 805–823.

Tarafdar, E. (1983). An approach to nonlinear elliptic boundary value problems. *J. Austral. Math. Soc. (Series A)*, **34**, 316–335.

Tarafdar, E. (1986). Variational problems via a fixed point theorem. *Indian Journal of Mathematics*, **28**, 229–240.

Tarafdar, E. (1987). A fixed point theorem equivalent to Fan Knaster-Kuratowski-Mazurkiewic's theorem. *J. Math. Anal. Appl.*, 475–479.

Tarafdar, E. (1989). A theorem concerning sets with convex sections. *Indian Journal of Mathematics*, **31**, 225–228.

Tarafdar, E. (1990a). Nonlinear variational inequality with application to the boundary value problem for quasilinear operator in generalized divergence form. *Funkcialaj Ekvacioj*, **33**(3), 441–453.

Tarafdar, E. (1990b). An extension of fan's fixed point theorem and equilibrium point of of abstract economy. *Comment. Math. Univ. Carolinae*, **31**, 723–730.

Tarafdar, E. (1991). A fixed point theorem and equilibrium point of an abstract economy. *J. Math. Econom.*, **20**, 211–218.

Tarafdar, E. (1992). A fixed point theorem on h-spaces and equilibrium points of abstract economies. *J. Austral. Math. Soc. (Ser. A)*, **53**, 252–250.

Tarafdar, E. (1995a). Pareto optimum and equilibrium points of private ownership economies — a simpler approach without fixed point theorems. *Arab Journal of Mathematical Science*, **1**, 65–74.

Tarafdar, E. (1995b). A random version of Arrow-Debreu economy. *Nonlinear World*, **2**, 87–96.

Tarafdar, E. (1996a). Applications of Pareto optimality of a mapping to mathematical economies. In V. Lakshmikantham (Ed.), *Proceedings of the First World Congress of Nonlinear Analysts, Tampa, FL – 1992, Vol. I–IV* (pp. 2431–2439). Berlin, New York: Walter de Gruyter Publishers.

Tarafdar, E. (1996b). Pareto solution of cone inequality and Pareto optimality of a mapping. In V. Lakshmikantham (Ed.), *Proceedings of the First World Congress of Nonlinear Analysts, Tampa, FL – 1992, Vol. I–IV* (pp. 2511–2519). Berlin, New York: Walter de Gruyter Publishers.

Tarafdar, E. (1998). Pareto optimum. *Gurukula Kangri Vijnana Patrika Aryabhata*, **1**, 119–130.

Tarafdar, E., & Husain, T. (1998). An implicit function theorem for a set of mappings and its application to nonlinear hyperbolic boundary value problems at resonance. *Nonlinear Analysis TMA*, **12**, 997–1016.

Tarafdar, E., & Mehta, G. (1986). Non-linear variational inequalities and the existence of equilibrium in economics with a Riesz space of commodities. *Commentationes Mathematicae Universitatis Carolinae*, **27:2**, 259–266.

Tarafdar, E., & Teo, S. K. (1979). On the existence of solutions of the equation $lx \in nx$ and a coincidence degree theory. *J. Austral. Math. Soc. (Series A)*, **28**, 139–173.

Tarafdar, E., & Thompson, H. (1985). Bifurcation for the solutions of equations involving set-valued mappings. *Internat. J. Math. & Math. Sc.*, **8:1**, 37–48.

Tarafdar, E., & Thompson, H. (1987). On the solvabiility of non-linear non-compact operator equations. *J. Austral. Math. Soc. (Series A)*, **43**, 103–126.

Tarafdar, E., & Výborný, R. (1975). Fixed point theorems for condensing multivalued mappings on a locally convex topological space. *Bull. Austral. Math. Soc.*, **12**, 161–170.

Tarafdar, E., & Výborný, R. (1976). A generalized (multivalued) contraction mapping principle. *Research Report of Pure Mathematics* **54**, *Department of Mathematics, The University of Queensland, Brisbane, Australia*, 1–4.

Tarafdar, E., & Watson, P. J. (1999). Periodic solutions of impulsive differential equations. *Dynamics of Continuous, Discrete and Impulsive Systems*, **6**, 301–306.

Tarafdar, E., & Yuan, X. Z. (1993). Equilibria of non-compact generalized games. *Bull. Polish Academy Sciences Math.*, **41:3**, 229–239.

Tarafdar, E., & Yuan, X. Z. (1994). Non-compact generalized quasi-variational inequalities in locally convex topological vector spaces. *Nonlinear World*, **1**, 273–283.

Tarafdar, E., & Yuan, X.-Z. (1997a). Generalized variational inequalities and its applications. *Nonlinear Analysis, Theory, Methods & Applications*, **30:7**, 4171–4181.

Tarafdar, E., & Yuan, X.-Z. (1997b). Set-valued dynamical system and its applications to Pareto optima. *Acta Applicandae Mathematicae*, **46**, 93–106.

Tarafdar, E., & Yuan, X.-Z. (1997c). A remark on generalized variational inequalities in locally convex topological vector spaces. *Appl. Math. Lett.*, **10:5**, 85–90.

Tarfadar, E. (1979). A remark coincidence degree theory. *Bollettino U.M.I.*, **16-A:5**, 582–587.

Tarski, A. (1955). A lattice theoretical fixed point theorem and its applications. *Pacific J. Math.*, **5**, 285–309.

Taylor, W. W. (1972). Fixed-point theorems for nonexpansive mappings in linear topological spaces. *J. Math. Anal. Appl.*, **40**, 164–173.

Thron, W. J. (1966). *Topological structures.* New York: Holt, Rinehart.

Tian, G. (1992). On the existence of equilibria in generalized games. *Intern. J. Game Theory*, **22**, 247–254.

Toussaint, S. (1984). On the existence of equilibria in economies with infinitely many commodities and without ordered preferences. *J. Econ. Theory*, **33**, 98–115.

Tricomi, F. (1916). Un teorema sulla convergenza delle successoni formate delle successive iterate di una funzione di una variabile reale. *Giorn. Mat. Battaglini*, **54**, 1–9.

Tricomi, F. G. (1957). *Integral equations.* New York: Interscience.

Tulcea, I. C. (1986). On the equilibrium of generalized games, paper no. 696. The Centre for Mathematical Studies in Economics and Management Science.

Tulcea, I. C. (1988). On the approximation of upper semi-continous correspondences and the equilibrium of generalized games. *J. Math. Anal. Appl.*, **136**, 267–289.

Tychonoff, A. (1935). Ein fixpunktsatz. *Math. Ann.*, **111**, 767–776.

Vainberg, M. M., & Aizengendler, P. G. (1968). The theory of methods of investigation of branch points of solutions. In Plenum (Ed.), *Progress in Mathematics, Vol. II,* (pp. 1–72). New York: Plenum.

Vainberg, M. M., & Trenogin, V. A. (1962). The methods of lyapunov and schmidt in the theory of nonlinear equations and their further development. *Russian Math. Surveys*, **17**, 1–60.

Vainbreg, M. M. (1964). *Variational methods for the study of nonlinear operators.* London: Holden-Day Inc.

Vainikko, G. M., & Sadovskii, B. N. (1968). On the degree of (ball) condensing vector fields. *Problemy Mat. Anal. Slov. Sistem*, **2**, 84–88.

Vejvoda, O. (1964). Periodic solutions of a linear and weakly nonlinear wave equation in one dimension 1. *Czech. Math. J.*, **14**, 341–382.

Wagner, D. H. (1977). Survey of measurable selection theorems. *SIAM J. Control and Optimization*, **15**, 859–903.

Walter, W. (1976). A note on contraction. *SIAM Review*, **18**, 107–111.

Watson, P. J. (1998). *Topological Methods in Nonlinear Analysis*, Ph.D. thesis. St.

Lucia, Brisbane: University of Queensland.

Wilansky, A. (1970). *Topology for Analysis.* The College Division, London: Ginn A Xerox Company.

Williams, R. F. (1971). Composition of contractions. *Bol. Soc. Brasil. Mat.*, **2**, 55–59.

Williams, S. A. (1968). A connection between the Cesari and Leray-Schauder methods. *Michigan Math. J.*, **15**, 441–448.

Williams, S. A. (1970). A sharp sufficient condition for solution of nonlinear elliptic boundary value problem. *J. Differential Equation*, **8**, 580–586.

Williamson, R., & Janos, L. (1987). Constructing metrics with the Heine-Borel property. *Proc. Amer. Math. Soc.*, **100**, 567–573.

Wójtowicz, D. (1997). On the implicit Darboux problem in Banach spaces. *Bull. Austral. Math. Soc.*, **56**, 149–156.

Wong, C. S. (1974). Fixed point theorems for generalized non-expansive mappings. *J. Austral. Math. Soc.*, **18**, 265–276.

Yannelis, N., & Prabhakar, N. (1983). Existence of maximal elements and equilibria in topological vector spaces. *J. Econom.*, **12**, 233–245.

Yannelis, N. C. (1985). Maximal elements over non-compact subsets of linear topological spaces. *Economics Letters*, **17**, 133–136.

Yannelis, N. C. (1987). Equilibria in non-cooperation models of competition. *J. Econom. Theory*, **41**, 96–111.

Yen, C. (1981). A minimax inequality and its applications to variational inequalities. *Pacific J. Math.*, **97**, 477–481.

Yu, S., Lai, T., & Yao, J. (1966). Generalized nonlinear variational inequalities. *Computers Mathematics and Applications*, **32**(7), 21–27.

Yu, S., & Yao, J. (1996). On the generalized nonlinear variational inequalities. *J. Math. Anal. Appl.*, **198**, 178–193.

Yuan, G. X. Z. (1998). The study of minimax inequalities and applications to economics and variational inequalities. *Mem. Amer. Math. Soc.*, **132**(625), 1–140.

Yuan, G. X.-Z. (1999). *KKM Theory and Applications in Nonlinear Analysis.* New York: Marcel Dekker, Inc.

Zeidler, E. (1985). *Nonlinear Functional Analysis and its Applications I: Fixed-point Theorems*, translated by Peter R. Wadsack. New York Berlin Heidelberg London Paris: Sringer-Verlag.

Zhou, J., & Chen, G. (1988). Diagonal convexity conditions for problems in convex analysis and quasivariational inequalities. *J. Math. Anal. Appl.*, **132**, 213–225.

Index